Geometric Formulas

Rectangle
Area: $A = \ell w$
Perimeter: $P = 2\ell + 2w$

Square
Area: $A = s^2$
Perimeter: $P = 4s$

Parallelogram
Area: $A = bh$

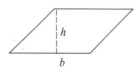

Trapezoid
Area: $A = \frac{1}{2}h(a + b)$

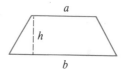

Triangle
Area: $A = \frac{1}{2}bh$

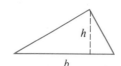

or

Area: $A = \sqrt{s(s - a)(s - b)(s - c)}$,
where $s = \frac{1}{2}(a + b + c)$
Angle sum: $A + B + C = 180°$

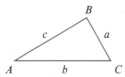

Right Triangle
Pythagorean theorem: $a^2 + b^2 = c^2$

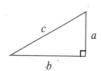

Circle
Area: $A = \pi r^2$
Circumference: $C = \pi d = 2\pi r$

Second Edition

PRECALCULUS

Second Edition

PRECALCULUS

DENNIS T. CHRISTY
Nassau Community College

WCB **Wm. C. Brown Publishers**
Dubuque, Iowa • Melbourne, Australia • Oxford, England

Book Team

Editor   *Dwala A. Canon*
Production Editor   *Eugenia M. Collins*
Designer   *K. Wayne Harms*
Art Editor   *Joseph P. O'Connell*

Wm. C. Brown Publishers
A Division of Wm. C. Brown Communications, Inc.

Vice President and General Manager   *George Bergquist*
National Sales Manager   *Vincent R. Di Blasi*
Assistant Vice President, Editor-in-Chief   *Edward G. Jaffe*
Marketing Manager   *Elizabeth Robbins*
Advertising Manager   *Amy Schmitz*
Managing Editor, Production   *Colleen A. Yonda*
Manager of Visuals and Design   *Faye M. Schilling*

Design Manager   *Jac Tilton*
Art Manager   *Janice Roerig*
Publishing Services Manager   *Karen J. Slaght*
Permissions/Records Manager   *Connie Allendorf*

Wm. C. Brown Communications, Inc.

Chairman Emeritus   *Wm. C. Brown*
Chairman and Chief Executive Officer   *Mark C. Falb*
President and Chief Operating Officer   *G. Franklin Lewis*
Corporate Vice President, Operations   *Beverly Kolz*
Corporate Vice President, President of WCB Manufacturing   *Roger Meyer*

A study guide for this textbook is available in your college bookstore. Its title is *Precalculus, Student Study Guide and Solutions Manual*. It has been written to help you review and study the course material. Ask the bookstore manager to order a copy for you if it is not in stock. For a more detailed description of the study guide, refer to the preface in this book.

About the Cover

The imperial angelfish on the cover of this text has been modified by the artist. The symbols $\div, \pi, <, >, =, \sqrt{}$ are included in the markings of the fish.

Cover illustration by *Todd Buck*
Chapter openers by Diphrent Strokes, Inc.

Copyedited by *Carol I. Beal*

Copyright © 1989, 1993 by Wm. C. Brown Communications, Inc. All rights reserved

Library of Congress Catalog Card Number: 92–53380

ISBN 0-697-12345-6

Printed in the United States of America by Wm. C. Brown Communications, Inc., 2460 Kerper Boulevard, Dubuque, IA 52001

10   9   8   7   6   5   4   3   2   1

To Ellen, Michael, and Ilene

Contents

Preface   xi

Calculator Use: To the Student   xvii

Chapter 1   Fundamentals of Algebra   1

1.1 Real Numbers and Algebraic Expressions   1
1.2 Products of Algebraic Expressions and Factoring   16
1.3 Algebraic Fractions   28
1.4 Rational Exponents and Radicals   37
1.5 Complex Numbers   49
1.6 Linear Equations and Quadratic Equations   55
1.7 Other Types of Equations   70
1.8 Inequalities   78

Chapter Overview   90
Chapter Review Exercises   93
Chapter 1 Test   95

Chapter 2   Functions and Graphs   97

2.1 Functions   97
2.2 Cartesian Coordinates and Graphs   108
2.3 Graphing Techniques   120
2.4 Operations with Functions   129
2.5 Inverse Functions   136
2.6 Variation   143

Chapter Overview   148
Chapter Review Exercises   149
Chapter 2 Test   152

Chapter 3   Polynomial and Rational Functions   153

3.1 Linear Functions   154
3.2 Quadratic Functions   165
3.3 Synthetic Division and the Remainder Theorem   175
3.4 Zeros and Graphs of Polynomial Functions   181

3.5 Additional Theorems About Zeros   192
3.6 Rational Functions   200

Chapter Overview   208
Chapter Review Exercises   209
Chapter 3 Test   210

Chapter 4 Exponential and Logarithmic Functions   211

4.1 Exponential Functions   211
4.2 Logarithmic Functions   220
4.3 Properties of Logarithms   227
4.4 Exponential and Logarithmic Equations   230
4.5 More Applications and the Number *e*   235

Chapter Overview   244
Chapter Review Exercises   246
Chapter 4 Test   247

Chapter 5 Trigonometric Functions   248

5.1 Radian and Degree Measure   249
5.2 Trigonometric Functions of Real Numbers   255
5.3 Trigonometric Ratios   267
5.4 Evaluating Trigonometric Functions of Angles   275
5.5 Graphs of Sine and Cosine Functions   286
5.6 Graphs of Other Trigonometric Functions   295
5.7 Inverse Trigonometric Functions   303
5.8 Right Triangle Applications and Harmonic Motion   311

Chapter Overview   320
Chapter Review Exercises   321
Chapter 5 Test   323

Chapter 6 Trigonometric Identities and Equations   324

6.1 Fundamental Identities   325
6.2 Sum and Difference Formulas   331
6.3 Multiple-Angle Formulas   338
6.4 Product and Sum Formulas; Reduction Formula   346
6.5 Trigonometric Equations   351

Chapter Overview   361
Chapter Review Exercises   362
Chapter 6 Test   363

Chapter 7   Further Applications of Trigonometry   365

7.1 Law of Sines   365
7.2 Law of Cosines and Area of Triangles   375
7.3 Introduction to Vectors   383
7.4 Analytic Approach to Vectors   392
7.5 Trigonometric Form of Complex Numbers   400
7.6 De Moivre's Theorem and nth Roots   406
7.7 Polar Coordinates   411

Chapter Overview   419
Chapter Review Exercises   421
Chapter 7 Test   423

Chapter 8   Analytic Geometry: Conic Sections   424

8.1 Introduction to Analytic Geometry   424
8.2 The Circle   430
8.3 The Ellipse   433
8.4 The Hyperbola   439
8.5 The Parabola   444
8.6 Classifying Conic Sections   450

Chapter Overview   453
Chapter Review Exercises   455
Chapter 8 Test   455

Chapter 9   Systems of Equations and Inequalities   457

9.1 Systems of Linear Equations in Two Variables   457
9.2 Triangular Form and Matrices   466
9.3 Determinants and Cramer's Rule   475
9.4 Solving Systems by Matrix Algebra   484
9.5 Partial Fractions   496
9.6 Nonlinear Systems of Equations   502
9.7 Systems of Linear Inequalities and Linear Programming   506

Chapter Overview   515
Chapter Review Exercises   516
Chapter 9 Test   518

Chapter 10   Discrete Algebra and Probability   519

10.1 Sequences   519
10.2 Series   525
10.3 Infinite Geometric Series   533

10.4 Mathematical Induction   537
10.5 Binomial Theorem   541
10.6 Counting Techniques   550
10.7 Probability   557

Chapter Overview   563
Chapter Review Exercises   565
Chapter 10 Test   567

Appendix   568

A.1 Scientific Notation   568
A.2 Approximate Numbers   570
A.3 Logarithmic Computations and Tables   576
A.4 Graphs on Logarithmic Paper   581

Tables   584

Table 1 Squares, Square Roots, and Prime Factors   585
Table 2 Common Logarithms   586
Table 3 Natural Logarithms (Base *e*)   588
Table 4 Exponential Functions   590
Table 5 Trigonometric Functions of Real Numbers   591
Table 6 Trigonometric Functions of Angles   594

Answers to Odd-Numbered Problems   599

Index   651

Preface

Audience

This book is intended for students who need a concrete approach to mathematics. The presentation assumes that the student has completed two years of high school algebra or a college course in intermediate algebra. However, a detailed review of all necessary ideas is given early in the text so that students whose basic skills need improvement have a wealth of helpful material.

Approach

Problem-Solving Approach

My experience is that students who take precalculus learn best by "doing." Examples and exercises are crucial since it is usually in these areas that the students' main interactions with the material take place. The problem-solving approach contains brief, precisely formulated paragraphs, followed by many detailed examples. In most cases the discussion proceeds from the specific to the general, and each section has many exercises for both in-class practice and homework. The problem sets are carefully graded and contain an unusual number of routine manipulative problems. (There is nothing more frustrating than being stuck on the beginning exercises!) So that each student may be challenged fairly, difficult problems have been included. There is also a wide variety of questions and discussions that show the student the usefulness of mathematics. Many reviewers felt that the readability of the text and the quality and abundance of the exercises and examples were outstanding features.

Functions Theme

The organization takes a functional approach to precalculus in that after developing some basic algebra, the function concept plays the unifying role in the study of polynomial, rational, exponential, logarithmic, and trigonometric functions.

Algebra Review

The first chapter, which reviews basic algebra, is particularly detailed to help ensure proficiency in basic skills.

Trigonometry Development

The organization takes a functional approach to trigonometry in that Chapter 5 focuses on functions and graphing, and the trigonometric functions are defined first through a discussion of the unit circle.

Graphics

A major component of a problem-solving approach with intuitive concept developments is a strong emphasis on graphics. Students must be given, and must be able to draw for themselves, vivid images of the relations they are analyzing and the problems they are solving. This text consistently develops graphing techniques and contains many vivid diagrams in the concept developments and in the exercise sets.

Calculus Preparation

To prepare students for calculus, there is an emphasis on explaining how to deal with some of the expressions encountered in a calculus course. For instance, Chapter 1 shows how to simplify algebraic expressions that will be found in calculus problems. Chapter 2 expands the discussion of functional notation to include the difference quotient, Chapter 5 shows problems dealing with inverse trigonometric functions that are important for integration by trigonometric substitution, and Chapter 10 considers the sense in which certain infinite geometric series converge to a sum.

Calculator Use

The text encourages the use of calculators and discusses how they can be used effectively. It is assumed in the discussion that students have scientific calculators that use the algebraic operating system (AOS). Calculator illustrations show primarily the keystrokes required on a Texas Instruments TI-30 SLR +. Tables are included and discussed in the text in case students choose not to use a calculator.

Features

- Problem-solving approach with intuitive concept developments
- Over 5,000 exercises and 500 examples
- Problem sets of graduated difficulty with enough routine problems to give students experience, confidence, and skill with basic procedures
- Extensive and varied applications problems
- Student-oriented and mathematically sound explanations
- Boxes with margin labels for important definitions and rules
- Second color to highlight important ideas
- Avoidance of awkward page and line breaks
- Chapter introductions that include an interest-getting problem that is solved as an example in the text
- Unique chapter overviews that highlight key concepts to review at the end of each chapter
- Abundant chapter review exercises and a sample chapter test
- Anticipates the needs of students who continue to calculus and other higher-level math courses
- Emphasis on functions and graphs
- Unit circle introduction to trigonometry
- Instructions on calculator use
- Important formulas and tables on endsheets
- Complete instructional package

Changes to New Edition

Many significant improvements have been made for this edition. Briefly, the major ones are as follows:

1. The coverage of trigonometry is expanded to three chapters, and Chapter 6 devotes an entire chapter to trigonometric identities and equations. Further development has also taken place in the areas of graphing, inverse trigonometric functions, vectors, and complex numbers. In keeping with the functions theme of the text, the trigonometric functions are now introduced with a unit circle approach.
2. The test point method for solving inequalities, the Gauss-Jordan method for solving systems of equations, and the *ac* method of factoring trinomials are additions to the text. The coverage of composition of functions and factoring and graphing polynomials has been significantly improved, and the discussion of slope is now included in the section on linear functions.
3. The review of algebra has been consolidated in a single chapter.
4. "Think About It" exercises, "Remember This" review exercises, and sample chapter tests are new features of the text.

Pedagogy

Chapter Introductions

In the spirit of problem solving, each chapter opens with a problem that should quickly involve students and teachers in a discussion of an important chapter concept. Some of the problems are applications, some are puzzles, and one is a proof. None of the problems require a lot of sophisticated mathematics and it is hoped that students will try to solve the problem either initially or after covering the relevant section in the text. These problems are later worked out as examples in each chapter.

Chapter Overviews
and Review Exercises

At the end of each chapter there is a detailed list of key concepts to review that is organized in a section-by-section format so the material does not seem overwhelming. There is also a collection of review exercises that not only review the basic ideas but also expose students to slightly different question wording and formats, including multiple-choice questions. From these exercises it should be easy for instructors to choose questions for review assignments or for sample tests at an appropriate level of difficulty. For convenience, sample chapter tests are also included.

"Think About It" Exercises

Each section exercise set is followed immediately by a set of "Think About It" exercises. Although some of these problems are challenging, this section is not intended as a set of "mind bogglers." Instead, the goal is to help develop critical-thinking skills by asking students to create their own examples, express concepts in their own words, extend ideas covered in the section, and analyze topics slightly out of the mainstream (for example, the golden rectangle).

Systematic Review

Students benefit greatly from a systematic review of previously learned concepts. At the end of each section exercise set there is a short set of "Remember This" exercises that review previous concepts, with particular emphasis on prerequisite skills that will be needed in the next section.

Instructional Package

- *Student Study Guide and Solutions Manual*
- *Instructor's Manual* (includes transparency masters)
- *Instructor's Solutions Manual*
- WCB TestPak 3.0
- *Test Item File*
- *Graphing Calculator Supplements*
- Computer Software
- Videotapes

Student Study Guide and Solutions Manual

Students for whom the textbook is not enough need more than just a lot of solved problems in a solutions manual. Primarily they need help focusing on the key objectives and concepts in the course. To provide some of this help, an accompanying study guide and solutions manual is available that covers the following key aspects for each section in the text.

1. Specific objectives for the section.
2. A list of important terms.
3. A summary of the key rules and formulas.
4. Detailed solutions to selected even-numbered exercises with at least one example from each exercise group. **Exercise numbers for these problems are printed in color in the text for easy identification.**
5. Margin exercises matched to the solved problems so students can check their progress.

The *Student Study Guide and Solutions Manual* also contains solutions for every other odd-numbered exercise and sample test questions (with answers) for each chapter in the text.

Instructor's Manual

The *Instructor's Manual* contains four tests for each chapter, three final exams, and the answers to the even-numbered problems in the section exercise sets. Transparency masters of the chapter introductory problems and important theorems and definitions are also included in this manual.

Instructor's Solutions Manual

This manual contains solutions to every problem in the text. These solutions are intended for the use of the instructor only and are basic outlines of possible problem solutions.

WCB TestPak 3.0

WCB TestPak 3.0 was developed expressly for WCB math texts. It is a free computerized testing service with two convenient options. First, you may use your own Apple IIe, IIc, Macintosh, or IBM PC (both 5¼″ and 3½″

disks) to produce your tests by using items available in the test bank or by editing these items, deleting them, or adding your own. Questions may be chosen by number or at random. Secondly, you may use the call-in service offered by the publisher. Contact your local WCB sales representative for details.

Test Item File   The printed *Test Item File* in an 8½″ × 11″ format contains all of the questions on the WCB TestPak. It will serve as a ready-reference if you use your own computer to generate tests. The *Test Item File* contains problems from the *Instructor's Manual* along with additional items that have been incorporated into the TestPak bank.

Graphing Calculator   *Calculator View of Precalculus for the TI-81*
Supplements   *Calculator View of Precalculus for the Casio fx 7700G*
These manuals are intended to serve as a bridge between the text and the owner's manual. This will make it easier for the student to integrate the graphing calculator into the precalculus course.

Computer Software   See your WCB representative for further details.

Videotapes   See your WCB representative for further details.

Acknowledgments

I wish to thank the many users and reviewers of my texts who have suggested improvements. At this point it is hard to separate my original ideas from the many valuable observations they made, and I am indebted to all of them. For help with this and related projects I am especially grateful to Dr. John R. Garlow, Tarrant County Junior College, who wrote the *Instructor's Solutions Manual;* Dr. John Paulling, Nicholls State University, who wrote the graphing calculator supplements; Dr. Shelba J. Mormon, North Lake College, who checked exercise solutions for accuracy; Carol Beal, who skillfully copyedited the manuscript; Earl McPeek, Dwala Canon, Linda Meehan, Eugenia M. Collins, K. Wayne Harms, Joseph P. O'Connell, and Kevin Campbell, Wm. C. Brown Publishers; and Deborah Levine, Bob Rosenfeld, and Gene Zirkel, Nassau Community College. My wife, Margaret, once again typed, proofread, and "understood," with the last contribution being irreplaceable. So, to my family, my colleagues at Nassau Community College, the staff at Wm. C. Brown Publishers, and the many users and reviewers, thank you.

Dennis T. Christy

Reviewers

Susan A. Barker
Atlantic Community College

Scott Carter
University of South Alabama

Franklin Cheek
University of Wisconsin–
Platteville

Harry Fainsilber
Dawson College

Daniel E. Flath
University of South Alabama

Pat Gilbert
Diablo Valley College

Marion Glasby
Anne Arundel Community
College

Gerald Hahm
College of St. Thomas

Thomas Hanson
Northwestern State University of
Louisiana

Rhonda Hatcher
Texas Christian University

Gail Koplin
Ocean County College

Annamarie Langlie
North Hennepin Community
College

Stanley Lukawecki
Clemson University

Jerry McDonald
Loyola University of Chicago

B. K. Michael
University of Pittsburgh

Paul O'Heron
Broome Community College

Wing Park
College of Lake County

Marilyn Peacock
Tidewater Community College

Jack Pease
West Valley Community College

Jan Rizzuti
University of Pittsburgh at
Bradford

Charles Searcy
New Mexico Highlands

Marsha Self
El Paso Community College

Richard Semmler
Northern Virginia Community
College

Marvin Shubert
Hagerstown Junior College

Deborah Vrooman
USC–Coastal Carolina College

Richard Watkins
Tidewater Community College

Earl Zwick
Indiana State University

Calculator Use /
To the Student

A scientific hand-held calculator is now standard equipment for precalculus mathematics and beyond. These ten-dollar wonders provide you with the benefits of electronic computation that is fast, accurate, and easy to learn. Most important, efficient calculator use helps you focus on important mathematical ideas. To understand and apply mathematical concepts is our fundamental aim, and calculators are marvelous aids in attaining this goal. Tables are included at the back of the book in case you choose not to use a calculator. But, since calculators are inexpensive, easy to use, and a significant learning aid, we recommend you obtain one.

A scientific calculator (the type you need) contains at least the following special features: algebraic keys x^2, $\sqrt{x}$, $1/x$, y^x or x^y, $\sqrt[x]{y}$; trigonometric keys sin, cos, tan, $\sin^{-1}$, $\cos^{-1}$, $\tan^{-1}$, degree and radian angular modes; logarithmic and exponential keys log, ln, 10^x, e^x; parentheses keys (,); a scientific notation key EE or EXP; and one memory that can store and recall.

In this book we also assume a scientific calculator using the algebraic operating system (AOS). Texas Instruments, Sharp, and Casio produce scientific calculators using this system. With AOS you can key in the problem exactly as it appears, and the calculator is programmed to use the order of operations discussed in Section 1.1. For example, since multiplication is done before addition, 2 $\boxed{+}$ 3 $\boxed{\times}$ 4 $\boxed{=}$ 14. If your calculator displays 20 when you key in this sequence, it is operating on left-to-right logic. You must then be careful to key in the problem so the correct order of operations is followed. Calculator illustrations in this text show primarily the keystrokes required on a Texas Instruments TI-30-SLR+. In any case, you should read the owner's manual that comes with your calculator to familiarize yourself with its specific keys and limitations.

One other introductory note—a calculator *computes*, that's all. You do the important part—you *think*. You analyze the problem, decide on the significant relationships, and determine if the solution makes sense in the real world. It's nice not to get bogged down in certain calculations and tables, but critical thinking has always been the main goal.

1 Fundamentals of Algebra

A race car driver must average at least 150 mi/hour for two laps around a track to qualify for the finals. The driver averages 180 mi/hour on the first lap, but mechanical trouble reduces the average speed on the second lap to only 120 mi/hour. Does the driver qualify for the finals? (See Example 10 of Section 1.3. *Hint:* average speed = total distance/total time.)

A common student lament goes something like, "I understand the new concepts, but the algebra is killing me!" In this chapter we hope to remedy this problem by reviewing *in detail* some basic rules in algebra about real numbers, exponents, factoring, fractions, radicals, complex numbers, equations, and inequalities. Success here will go a long way toward success in this course and in higher mathematics.

In this text we take a problem-solving approach which emphasizes that one learns mathematics by *doing* mathematics, while *thinking* mathematically. That is, you need to actively work through the problems (with pencil and paper), while *focusing on the definitions, relationships, and procedures* that link together all steps in the solution. In this spirit of problem solving we open each chapter with a problem. Some are applications, some are puzzles, and one is a proof. Taken together, they illustrate the varied nature of problem solving. Since none of them require a lot of sophisticated mathematics, we hope you will take a stab at an answer either initially or after covering the relevant section in the text.

1.1 Real Numbers and Algebraic Expressions

Mathematics is a basic tool in analyzing concepts in every field of human endeavor. In fact, the primary reason you have studied this subject for at least a decade is that mathematics is the most powerful instrument available in the search to understand the world and to control it. Mathematics is essential for full comprehension of technological and scientific advances, economic policies and business decisions, and the complexities of social and psychological issues. At the heart of this mathematics is algebra. Calculus, statistics, and computer science are but a few of the areas in which a knowledge of algebraic concepts and manipulations is necessary.

Algebra is a generalization of arithmetic. In arithmetic we work with specific numbers, such as 5. In algebra we study numerical relations in a more general way by using symbols, such as x, that may be replaced by a number from some collection of numbers. Since the symbols represent numbers, they behave according to the same rules that numbers must follow. Consequently, instead of studying specific numbers, we study symbolic representations of numbers and try to define the laws that govern them.

We begin our study of algebra by giving specific names to various sets* of numbers. The collection of the counting numbers, zero, and the negatives of the counting numbers is called the **integers.** Thus, the set of integers may be written as

$$\{ \ldots, -3, -2, -1, 0, 1, 2, 3, \ldots \}.$$

The set of fractions with an integer in the top of the fraction (numerator) and a nonzero integer in the bottom of the fraction (denominator) is called the **rational numbers.** Symbolically, a rational number is a number that may be written in the form a/b, where a and b are integers, with b not equal to ($\neq$) zero. The numbers $\sqrt{2}/3$ and $2/\pi$ are fractions but they are not rational numbers because they cannot be written as the quotient of two integers. All integers are rational numbers because we can think of each integer as having a 1 in its denominator. (Example: $4 = 4/1$.)

Our definition for rational numbers specified that the denominator cannot be zero. To see why, you need to know that

$$\frac{8}{2} = 4 \text{ is equivalent to saying that } 8 = 4 \cdot 2 \text{ and}$$

$$\frac{55}{11} = 5 \text{ is equivalent to saying that } 55 = 5 \cdot 11.$$

If $8/0 = a$, where a is some rational number, this would mean that $8 = a \cdot 0$. But $a \cdot 0 = 0$ for any rational number. There is no rational number a such that $a \cdot 0 = 8$. Thus, we say that $8/0$ is *undefined.*

Now consider $0/0 = a$. This is equivalent to $0 = a \cdot 0$. But $a \cdot 0 = 0$ for *any* rational number. Thus, not just one number a will solve the equation—any a will. Since $0/0$ does not name a particular number, it is also undefined. Consequently, division by zero is undefined in every case, so the denominator in a rational number cannot be zero.

To define our next set of numbers, we now consider the decimal representation of numbers. We may convert rational numbers to decimals by long division. Consider the following examples of repeating decimals. A bar is placed above the portion of the decimal that repeats.

---

*A **set** is simply a collection of objects, and we may describe a set by listing the objects or members of the collection within braces.

$$\begin{array}{r} 1.\overline{142857} \\ 7\overline{)8.000000} \\ \underline{7} \\ 1\,0 \\ \underline{7} \\ 30 \\ \underline{28} \\ 20 \\ \underline{14} \\ 60 \\ \underline{56} \\ 40 \\ \underline{35} \\ 50 \\ \underline{49} \\ 1 \end{array}$$

Figure 1.1

$$\frac{3}{4} = \begin{array}{c} 0.7500\ldots \\ \text{or} \\ 0.75\overline{0} \end{array} \qquad \frac{2}{3} = \begin{array}{c} 0.6666\ldots \\ \text{or} \\ 0.\overline{6} \end{array} \qquad \frac{8}{7} = 1.\overline{142857}$$

The decimals repeat because at some point we must perform the same division and start a cycle. For example, when converting $\frac{8}{7}$, the only possible remainders are 0, 1, 2, 3, 4, 5, and 6. In performing the division, as shown in Figure 1.1, we had remainders of 1, 3, 2, 6, 4, and 5. In the next step we must obtain one of these remainders a second time and start a cycle, or obtain 0 as the remainder, which results in repeating zeros. Thus, if a/b is a rational number, it can be written as a repeating decimal.

It is also true that any repeating decimal may be converted to a ratio between two integers, as shown in Example 1.

EXAMPLE 1   Express the repeating decimal $0.\overline{17}$ as the ratio of two integers.

Solution   First, let $x = 0.1717.\ldots$. Multiplying both sides of this equation by 100 moves the decimal two places to the right, so we obtain

$$\begin{array}{r} 100x = 17.1717\ldots \\ x = 0.1717\ldots \end{array}$$

now subtracting yields $\qquad 99x = 17 \qquad\qquad$ or $x = \dfrac{17}{99}$.

Thus, the repeating decimal $0.\overline{17}$ is equivalent to the fraction $\frac{17}{99}$. ∎

In Example 1 we multiplied by 100 because the decimal repeated after every two digits. If the decimal repeats after one digit, we multiply by 10; if it repeats every three digits, we multiply by 1,000; and so on. In summary, we have illustrated that we may define a rational number either as the quotient of two integers or as a repeating decimal.

There are decimals that do not repeat, and the set of these numbers is called the **irrational numbers.**

EXAMPLE 2

a. The numbers $\sqrt{2}$, $\sqrt{3}$, $\sqrt{5}$, $\sqrt{6}$, and $\sqrt{7}$ are irrational because they have nonrepeating decimal forms. A proof that $\sqrt{2}$ cannot be written as the quotient of two integers (and equivalently as a repeating decimal) is considered in "Think About It" Question 5.

b. The number $\sqrt{4}$ is not irrational because $\sqrt{4} = 2$, which is a rational number. (*Note:* The symbol $\sqrt{}$ denotes the nonnegative square root of a number. Thus, $\sqrt{4} \neq -2$. We discuss this concept in detail in Section 1.4.)

c. The number π, which represents the ratio between the circumference and the diameter of a circle, is a nonrepeating decimal (irrational number). The fraction $\frac{22}{7}$ is only an approximation for π ($\frac{355}{113}$ is a much better one). ∎

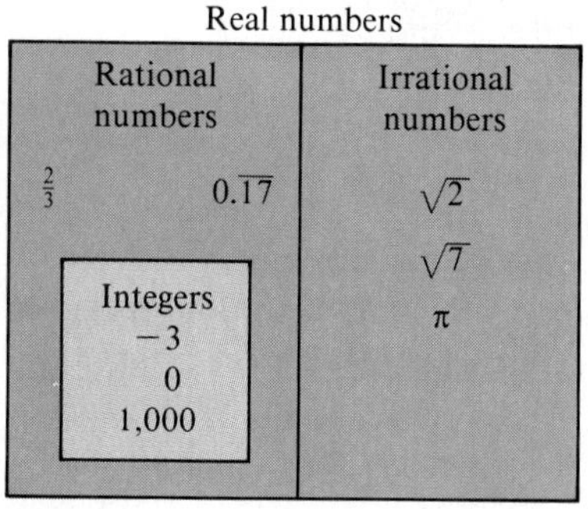

Real numbers

Rational numbers	Irrational numbers
$\frac{2}{3}$ $0.\overline{17}$	$\sqrt{2}$
	$\sqrt{7}$
Integers -3 0 $1,000$	π

Figure 1.2

Since an irrational number is a nonrepeating decimal and a rational number is a repeating decimal, there is no number that is both rational and irrational. The set of numbers that are either repeating decimals (rational numbers) or nonrepeating decimals (irrational numbers) constitutes the **real numbers.** Real numbers are used extensively in this text, and unless it is stated otherwise, you may assume that the symbols in algebra (such as x) may be replaced by any real number. Consequently, the rules that govern real numbers determine our methods of computation in algebra. A graphical illustration of the relationships among the various sets of numbers is given in Figure 1.2. Note that all of the sets we have discussed are subsets* of the set of real numbers.

We now list the most important properties of the real numbers with respect to addition and multiplication. These properties are the basis for the justification of many algebraic manipulations.

Properties of the Real Numbers

Let a, b, and c be real numbers.

	Addition	**Multiplication**
Closure Properties	$a + b$ is a unique real number.	ab is a unique real number.
Commutative Properties	$a + b = b + a$	$ab = ba$
Associative Properties	$(a + b) + c$ $= a + (b + c)$	$(ab)c = a(bc)$
Identity Properties	There exists a unique real number 0 such that $a + 0 = 0 + a = a.$	There exists a unique real number 1 such that $a \cdot 1 = 1 \cdot a = a.$
Inverse Properties	For every real number a, there is a unique real number, denoted by $-a$, such that $a + (-a)$ $= (-a) + a = 0.$	For every real number a except zero, there is a unique real number, denoted by $1/a$, such that $a \cdot \dfrac{1}{a} = \dfrac{1}{a} \cdot a = 1.$
Distributive Properties	$a(b + c) = ab + ac$ $(a + b)c = ac + bc$	

*Set A is a **subset** of set B if every element of A is an element of B.

EXAMPLE 3   Name the property of real numbers illustrated in each statement.

a. $(2 + 3) + 4 = 2 + (3 + 4)$ b. $5 + (-5) = 0$

c. $2(3 + 4) = 2 \cdot 3 + 2 \cdot 4$ d. $4 \cdot 7 = 7 \cdot 4$

Solution

a. This statement illustrates the associative property of addition. This property indicates that we obtain the same result if we change the grouping of the numbers in an addition problem.

b. This statement illustrates the addition inverse property. The number $-a$ is called the **negative** or **additive inverse** of a. Note that there is a difference between a negative number and the negative of a number. Although -5 is a negative number, the negative of -5 is 5.

c. This statement illustrates the distributive property (or more technically, the distributive property of multiplication over addition).

d. This statement illustrates the commutative property of multiplication. This property indicates that the order in which we write numbers in a multiplication problem does not affect their product.   ∎

Real Number Line

The real numbers may be interpreted geometrically by considering a straight line. Every point on the line can be made to correspond to a real number, and every real number can be made to correspond to a point. The first point that we designate is zero. It is the dividing point between positive and negative real numbers. Any number to the right of zero is called positive, and any number to the left of zero is called negative. Figure 1.3 is the result of assigning a few positive and negative real numbers to points on the line.

Figure 1.3

Order

The **trichotomy property** states that if a and b are real numbers, then either a is less than b, a is greater than b, or a equals b. By definition, a is **less than** b, written $a < b$, if and only if $b - a$ is positive; while a is **greater than** b, written $a > b$, if and only if $a - b$ is positive. We also define $a \leq b$ (read "a is **less than or equal to** b") to mean $a < b$ or $a = b$. Alternatively, we can write $b \geq a$ and say that b is **greater than or equal to** a. Relations of "less than" and "greater than" can be seen very easily on the number line

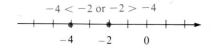

Figure 1.4

as shown in Figure 1.4. The point representing the larger number will be to the right of the point representing the smaller number. (*Note:* The definitions use the phrase "if and only if." This phrase is commonly used in mathematical definitions and theorems and the statement "*p* **if and only if** *q*" means "if *p* then *q*, and conversely, if *q* then *p*.")

Absolute Value

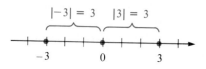

Figure 1.5

Positive and negative numbers in mathematics designate direction or indicate whether a result is above or below some reference point. However, sometimes the numerical size of a number is more important than its sign, and in such cases, the concept of absolute value is useful. The **absolute value** of a real number a, denoted $|a|$, is the distance between a and 0 on the number line. For example, $|3| = 3$ and $|-3| = 3$, as shown in Figure 1.5. Note that if $a \geq 0$, then $|a| = a$; but if $a < 0$, then $|a| = -a$ since $-a$ is positive in this case. Thus, algebraically we can define the absolute value of a real number a in tabular form as follows:

$$|a| = \begin{cases} a \text{ if } a \geq 0 \\ -a \text{ if } a < 0. \end{cases}$$

Arithmetic Operations

Arithmetic computations with real numbers may be done as follows.

Addition   To add two real numbers with the *same sign*, add their absolute values and attach the common sign. To add two real numbers with *different signs*, subtract the smaller absolute value from the larger and attach the sign of the number with the larger absolute value.

Subtraction   To subtract b from a, add to a the negative of b. In symbols, if a and b are any real numbers, $a - b = a + (-b)$.

Multiplication   To multiply two real numbers with the *same sign*, multiply their absolute values and make the sign of the product positive. To multiply two real numbers with *different signs*, multiply their absolute values and make the sign of the product negative. The product of any real number and 0 is 0.

Division   To divide a by b, where $b \neq 0$, multiply a by the reciprocal of b. In symbols, if a and b are any real numbers, with $b \neq 0$, then $a \div b = a(1/b)$. Division by 0 is undefined.

EXAMPLE 4

a. $-2 + (-4) = -6$ b. $2 + (-4) = -2$

c. $-5 - 2 = -5 + (-2) = -7$ d. $(-4)(-4) = 16$

e. $(-4)(-4)(-4) = 16(-4) = -64$

f. $(-9) \div (-3) = (-9)(-\frac{1}{3}) = 3$

g. $6 \div (-\frac{12}{5}) = 6(-\frac{5}{12}) = -\frac{5}{2}$ ■

If two or more numbers are multiplied together, each number is a **factor** of the product. In Example 4e the factor -4 is used three times. An alternative way of writing $(-4)(-4)(-4)$ is $(-4)^3$. The number $(-4)^3$, or -64, is called the third **power** of -4. In general, by a^n, where n is a positive integer, we mean to use a as a factor n times.

$$a^n = \underbrace{a \cdot a \cdot a \cdots a}_{n \text{ factors}}$$

In the expression a^n, n is called the **exponent.**

EXAMPLE 5   By definition, 2^5 means to use 2 as a factor five times. Thus, $2^5 = 2 \cdot 2 \cdot 2 \cdot 2 \cdot 2 = 32$. The number 2^5, or 32, is the fifth power of 2. The exponent in 2^5 is 5.   ■

Order of Operations

If an expression involves more than one operation, the following rule has been established to avoid confusion.

1. Perform all operations within parentheses first.
2. Evaluate powers of a number.
3. Multiply or divide working from left to right.
4. Add or subtract working from left to right.

Follow this order of operations carefully. For example, $2 + 3 \cdot 4 = 14$, not 20.

EXAMPLE 6   Evaluate $1 + 7(2 - 5)^2$.

Solution   By following the order of operations, we have

$$1 + 7(2 - 5)^2 = 1 + 7(-3)^2 = 1 + 7(9) = 1 + 63 = 64. \qquad ■$$

In algebra, two types of symbols are used to represent numbers: variables and constants. A **variable** is a symbol that may be replaced by different numbers in a particular problem. Generally, letters near the end of the alphabet, such as x, y, and z, are variables. A **constant** is a symbol that represents the same number throughout a particular problem. Numbers, such as 2, $-\sqrt{5}$, and π, never change value and are called **absolute constants.** If we do not know the fixed number until we are given specific information about the problem, the symbol is called an **arbitrary constant.** Generally k or letters near the beginning of the alphabet, such as a, b, and c, are arbitrary constants.

EXAMPLE 7   The sales tax T on a purchase is related to the price p of the item by the formula $T = kp$, where k is the sales tax rate. T and p are variables that may take on different values. The sales tax rate is fixed for any particular location so that k is a constant. For example, in New York City k is fixed at $8\frac{1}{4}$ percent. In another location k may be fixed at a different percent.   ■

An expression that combines constants and variables using the operations of addition, subtraction, multiplication, division, or extracting roots is called an **algebraic expression.** For example,

$$2x^3 - 5x + 3 \quad \text{and} \quad \frac{-b + \sqrt{b^2 - 4ac}}{2a} \quad \text{and} \quad \tfrac{1}{2}gt^2$$

are algebraic expressions.

If we are given numerical values for the symbols, we can evaluate the expression by substituting the given values and performing the indicated operations. Such substitutions are based on the **substitution property** that states that if a and b are real numbers and $a = b$, then either may replace the other without affecting the truth or falsity of the statement.

EXAMPLE 8   Find the numerical value of $-x^2 + 5yz^2$ if $x = -2$, $y = 3$, and $z = -4$.

Solution $-x^2 + 5yz^2$, when $x = -2$, $y = 3$, and $z = -4$, equals

$$-(-2)^2 + 5(3)(-4)^2 = -(4) + 5(3)(16)$$
$$= -4 + 240$$
$$= 236.$$

Pay close attention to how the expression $-x^2$ was evaluated in this example. Because powers precede multiplications in the order of operations, $-x^2$ means you must first square x and then take the negative of your result.   ■

Calculator Computation

A scientific calculator is ideal for evaluating expressions, but you need to understand the algebraic rules that have been programmed into your calculator. A calculator using the algebraic operating system (AOS) calculates in the following order (which applies whether or not a calculator is used).

1. Keys that operate on the single number in the display are done immediately. Such keys are called function keys and include square $\boxed{x^2}$; square root $\boxed{\sqrt{}}$; reciprocal $\boxed{1/x}$; sign change $\boxed{+/-}$; trigonometric $\boxed{\sin}$, $\boxed{\sin^{-1}}$, and so on; logarithmic $\boxed{\log}$, $\boxed{\ln}$; and exponential $\boxed{10^x}$, $\boxed{e^x}$.
2. Powers $\boxed{y^x}$ or $\boxed{x^y}$ and roots $\boxed{\sqrt[x]{y}}$ or $\boxed{y^{1/x}}$ follow the functions keys.
3. Multiplication $\boxed{\times}$ and division $\boxed{\div}$ have the next priority.
4. Addition $\boxed{+}$ and subtraction $\boxed{-}$ come last.

The equals key $\boxed{=}$ completes all operations. Parenthesis keys $\boxed{(}$, $\boxed{)}$ can be used to change these built-in priorities when required. Pressing close parenthesis $\boxed{)}$ automatically completes the operations since the previous open parenthesis $\boxed{(}$. Consider carefully the *logic* involved in the following examples, and keep in mind that in many cases other keystroke sequences are possible.

EXAMPLE 9   Evaluate $2x^3 + 4x^2 - x + 1$ if

a. $x = 5$
b. $x = -5$

Solution

a. The key $\boxed{y^x}$ or $\boxed{x^y}$ computes powers (here 5^3) with the $\boxed{x^2}$ key being the most direct way to square x. If $x = 5$, we can key in the problem exactly as it appears.

$$2\boxed{\times}5\boxed{y^x}3\boxed{+}4\boxed{\times}5\boxed{x^2}\boxed{-}5\boxed{+}1\boxed{=}\ \ 346$$

Instead of keying in a number several times, it can be useful to enter x just once and use the memory keys.

$$5\boxed{\text{STO}}2\boxed{\times}\boxed{\text{RCL}}\boxed{y^x}3\boxed{+}4\boxed{\times}\boxed{\text{RCL}}\boxed{x^2}\boxed{-}\boxed{\text{RCL}}\boxed{+}1\boxed{=}\ \ 346$$

b. If $x = -5$, we may have a problem computing $(-5)^3$ because some calculators restrict usage of the power key $\boxed{y^x}$ to a positive base. That is because the program for the power key often involves logarithms (see Chapter 4) which apply only to positive numbers. If your calculator works as follows,

$$5\boxed{+/-}\boxed{y^x}3\boxed{=}\text{error}$$

then you cannot use the power key with a negative base. You can evaluate $(-5)^3$ by computing 5^3 and then changing the sign to negative because there are an odd number of negative factors. If the exponent is even, the answer is positive, so there is no need to change the answer from the power key. Thus, we evaluate the given expression as follows:

$$2\boxed{\times}5\boxed{y^x}3\boxed{=}\boxed{+/-}\boxed{+}4\boxed{\times}5\boxed{+/-}\boxed{x^2}\boxed{-}5\boxed{+/-}\boxed{+}1\boxed{=}\ \ -144.$$

If your calculator accepts a negative base in conjunction with the $\boxed{y^x}$ key, then you may key in the numbers directly as in part **a.**   ∎

Combining Algebraic Expressions

Those parts of an algebraic expression separated by plus $(+)$ signs are called **terms** of the expression.

EXAMPLE 10

a. $x + \frac{2}{5}x$ is an algebraic expression with two terms, x and $\frac{2}{5}x$.

b. $-4x^3 + x^2 - 3x + 7$ may be written as $-4x^3 + x^2 + (-3x) + 7$ and is an algebraic expression with four terms, $-4x^3$, x^2, $-3x$, and 7.

c. 6 is an algebraic expression with one term.   ∎

If a term is the product of some constants and variables, the constant factor is called the **(numerical) coefficient** of the term. For example, the coefficient of the term $2x$ is 2, and the coefficient of ax^2 is a. Every term has a coefficient. If the term is x, the coefficient is 1 since $x = 1 \cdot x$. Similarly, if the term is $-x$, the coefficient is -1 since $-x = -1 \cdot x$. If two terms have identical variable factors (such as $3x^2y$ and $-2x^2y$), they are called **similar terms.** The distributive property indicates that we combine similar terms by combining their coefficients.

EXAMPLE 11

a. $3x + 7x = (3 + 7)x = 10x$
b. $yz^2 - 10yz^2 = (1 - 10)yz^2 = -9yz^2$
c. $2p - 3p + 4p = (2 - 3 + 4)p = 3p$
d. $7x + 2y - 3x = (7 - 3)x + 2y = 4x + 2y$
e. $-3x^2y + xy^2 - 7xy$ There are no similar terms, so we cannot simplify the expression.   ■

To combine algebraic expressions, it is sometimes necessary to remove the parentheses or brackets that group certain terms together. Parentheses are removed by applying the distributive property; that is, we multiply each term inside the parentheses by the factor in front of the parentheses. If the grouping is preceded by a minus sign, the factor is -1, so the sign of each term inside the parentheses must be changed. If the grouping is preceded by a plus sign, the factor is 1, so the sign of each term inside the parentheses remains the same. If there is more than one symbol of grouping, it is usually better to remove the innermost symbol of grouping first.

EXAMPLE 12   Remove the symbols of grouping and combine similar terms.

$$7 - 2[4x - (1 - 3x)]$$

Solution

$\quad 7 - 2[4x - (1 - 3x)]$
$\quad = 7 - 2[4x - 1 + 3x]$ The sign of each term in the parentheses is changed.
$\quad = 7 - 2[7x - 1]$
$\quad = 7 - 14x + 2$ Each term within the brackets is multiplied by -2.
$\quad = 9 - 14x$
$\quad (Note: 9 - 14x \neq -5x.)$ ■

Integer Exponents

Recall that a positive integer exponent is a shortcut way of expressing a repeating factor. That is,

a^n, where n is a positive integer, is shorthand for
$$\underbrace{a \cdot a \cdot a \cdots a.}_{n \text{ factors}}$$

From this definition it is easy to obtain the following laws of exponents that we will illustrate in specific terms in Example 13.

Laws of Exponents

Let a and b be any real numbers and m and n be any positive integers.

1. $a^m \cdot a^n = a^{m+n}$ 　　　　　　**2.** $(a^m)^n = a^{mn}$

3. $(ab)^n = a^n b^n$ 　　　　　　**4.** $\left(\dfrac{a}{b}\right)^n = \dfrac{a^n}{b^n}$ 　$(b \neq 0)$

5. $\dfrac{a^m}{a^n} = \begin{cases} a^{m-n} \text{ if } m > n \\ \dfrac{1}{a^{n-m}} \text{ if } n > m \end{cases}$ 　　$(a \neq 0)$

EXAMPLE 13　Simplify by the laws of exponents and check your result.

a. $2^2 \cdot 2^3$ 　　　**b.** $(2^3)^2$ 　　　**c.** $(2 \cdot 3)^2$ 　　　**d.** $2^5/2^2$ 　　　**e.** $2^2/2^5$

Solution 　　　　　　　　　　　　　　　　　　*Check*

a. $2^2 \cdot 2^3 = 2^{2+3} = 2^5 = 32$ 　　　　$2^2 \cdot 2^3 = 4 \cdot 8 = 32$

b. $(2^3)^2 = 2^{3 \cdot 2} = 2^6 = 64$ 　　　　　$(2^3)^2 = 8^2 = 64$

c. $(2 \cdot 3)^2 = 2^2 \cdot 3^2 = 4 \cdot 9 = 36$ 　　$(2 \cdot 3)^2 = 6^2 = 36$

d. $\dfrac{2^5}{2^2} = 2^{5-2} = 2^3 = 8$ 　　　　　$\dfrac{2^5}{2^2} = \dfrac{32}{4} = 8$

e. $\dfrac{2^2}{2^5} = \dfrac{1}{2^{5-2}} = \dfrac{1}{2^3} = \dfrac{1}{8}$ 　　　$\dfrac{2^2}{2^5} = \dfrac{4}{32} = \dfrac{1}{8}$ ■

We now wish to extend our definition of exponents to zero and negative integers. Note that it is meaningless to use x as a factor either zero times or a negative number of times, so we must define these exponents in a different manner. However, our guideline in these new definitions is to retain the laws of exponents developed for positive integers.

We start by considering the first law of exponents.

$$a^m \cdot a^n = a^{m+n}$$

If $n = 0$, we have

$$a^m \cdot a^0 = a^{m+0} = a^m.$$

When we multiply a^m by a^0, our result is a^m. Thus, a^0 must equal 1, and we make the following definition.

Zero Exponent

If a is a nonzero real number, then

$$a^0 = 1.$$

To obtain a definition for exponents that are negative integers, we will again consider the first law of exponents.

$$a^m \cdot a^n = a^{m+n}$$

If $m = 5$ and $n = -5$, we have

$$a^5 \cdot a^{-5} = a^{5 + (-5)} = a^0 = 1.$$

When we multiply a^5 by a^{-5}, the result is 1. Thus, a^{-5} is the reciprocal of a^5, or $a^{-5} = 1/a^5$. In general, our previous laws of exponents may be extended by making the following definition.

Negative Exponent

If a is a nonzero real number and n is a positive integer, then

$$a^{-n} = \frac{1}{a^n}.$$

EXAMPLE 14

a. $3^0 = 1$

b. If $x \neq 0$, $2x^0 = 2(1) = 2$

c. $2^{-3} = \dfrac{1}{2^3} = \dfrac{1}{8}$

d. $2^{-5} \cdot 2^3 = 2^{(-5) + 3} = 2^{-2} = \dfrac{1}{2^2}$ or $\dfrac{1}{4}$

Alternative Method

$$2^{-5} \cdot 2^3 = \frac{1}{2^5} \cdot 2^3 = \frac{2^3}{2^5} = \frac{1}{2^{5-3}} = \frac{1}{2^2} \text{ or } \frac{1}{4}$$ ∎

Note from Example 14d that there are different ways of simplifying expressions with negative exponents. An important principle to keep in mind is that any *factor* of the numerator may be made a factor of the denominator (and vice versa) by changing the sign of the exponent. For example:

$$\frac{3}{5x^{-2}} = \frac{3x^2}{5}, \quad \frac{2^{-3}}{5^{-4}} = \frac{5^4}{2^3}, \quad \frac{3^{-2}x^2}{y+1} = \frac{x^2}{3^2(y+1)},$$

$$\text{and } \left(\frac{2}{3}\right)^{-1} = \frac{2^{-1}}{3^{-1}} = \frac{3}{2}.$$

It is important to remember that this principle applies only to factors.

EXAMPLE 15 Simplify each expression. Assume variables as exponents denote integers.

a. $(6x^5y^3)(-2xy^7)$

b. $(2x^a)(7x^{2a})$

c. $\dfrac{18a^3b^2c^2}{12ab^4c}$

d. $\dfrac{2x^{r+3}}{-4x^r}$

e. $-2(x+2)^{-2}(x+2)$

f. $(3x^{-2})^{-3}$

g. $(a^2)^{1-x} \cdot a^{2x-3}$

h. $(3^{-1}x^2y^{-3}) \div (2x^{-5}y)$

Solution

a. $(6x^5y^3)(-2xy^7) = 6(-2)x^{5+1}y^{3+7} = -12x^6y^{10}$

b. $(2x^a)(7x^{2a}) = 2 \cdot 7x^{a+2a} = 14x^{3a}$

c. $\dfrac{18a^3b^2c^2}{12ab^4c} = \dfrac{18a^{3-1}c^{2-1}}{12b^{4-2}} = \dfrac{3a^2c}{2b^2}$

d. $\dfrac{2x^{r+3}}{-4x^r} = \dfrac{2x^{(r+3)-r}}{-4} = -\dfrac{1}{2}x^3$

e. $-2(x+2)^{-2}(x+2) = -2(x+2)^{-2+1} = -2(x+2)^{-1}$
$$= \dfrac{-2}{x+2}$$

f. $(3x^{-2})^{-3} = 3^{-3} \cdot x^6 = \dfrac{x^6}{3^3} = \dfrac{x^6}{27}$

g. $(a^2)^{1-x} \cdot a^{2x-3} = a^{2-2x} \cdot a^{2x-3} = a^{(2-2x)+(2x-3)} = a^{-1} = \dfrac{1}{a}$

h. $\dfrac{3^{-1}x^2y^{-3}}{2x^{-5}y} = \dfrac{x^{2-(-5)}}{3 \cdot 2y^{1-(-3)}} = \dfrac{x^7}{6y^4}$ ∎

Integer exponents are generally evaluated on a calculator with the power function key $\boxed{y^x}$ in the usual way. For example, 2^{-3} is computed as

$$2\;\boxed{y^x}\;3\;\boxed{+/-}\;\boxed{=}\quad 0.125.$$

Since negative exponents involve reciprocals, an alternative method is

$$2\;\boxed{y^x}\;3\;\boxed{=}\;\boxed{1/x}\quad 0.125.$$

An exponent of -1 is best computed with the reciprocal key. For instance, 5^{-1} is simply

$$5\;\boxed{1/x}\quad 0.2.$$

As discussed in Example 9, if your calculator does not permit usage of the power key $\boxed{y^x}$ with a negative base, you will need to use a positive base and then insert the correct sign. Thus, $(-2)^{-3}$ is computed as above with you inserting the negative sign in the answer -0.125 because 3 is an odd exponent.

EXERCISES 1.1

In Exercises 1–8 express each rational number as a repeating decimal.

1. $\frac{4}{5}$
2. $\frac{5}{4}$
3. $\frac{5}{11}$
4. $\frac{2}{9}$
5. $\frac{37}{6}$
6. $\frac{26}{11}$
7. $\frac{10}{7}$
8. $\frac{100}{99}$

In Exercises 9–18 express each repeating decimal as the ratio of two integers.

9. $0.\overline{2}$
10. $0.\overline{07}$
11. $0.\overline{321}$
12. $0.6\overline{332}$
13. $0.3\overline{0}$
14. $1.7\overline{0}$
15. $5.\overline{9}$
16. $4.8\overline{1}$
17. $2.14\overline{3}$
18. $2.1\overline{43}$

In Exercises 19–30 select all correct classifications from the following categories: real number, irrational number, rational number, integer, or none of these.

19. 0

20. $\frac{0}{3}$

21. $\frac{3}{0}$

22. $\frac{3}{4}$

23. -9

24. $\sqrt{9}$

25. $-\sqrt{9}$

26. $\sqrt{-9}$

27. $\sqrt{7}$

28. π

29. $0.\overline{01}$

30. $0.101001 \ldots$

In Exercises 31–50 name the property illustrated in the statement.

31. $2 + 7 = 7 + 2$

32. $11 + 0 = 11$

33. $4(5 \cdot 11) = (4 \cdot 5)11$

34. $6(4 + 3) = 6 \cdot 4 + 6 \cdot 3$

35. $\sqrt{2} \cdot 1 = \sqrt{2}$

36. $-7.3 + 7.3 = 0$

37. $(2.5)3 = 3(2.5)$

38. $17(\frac{1}{17}) = 1$

39. $\pi \cdot 3 + \pi \cdot 8 = \pi(3 + 8)$

40. $(5 + 3) + (2 + 1) = [(5 + 3) + 2] + 1$

41. $z(xy) = (zx)y$

42. $(xy)z = z(xy)$

43. $x + 0 = 0 + x$

44. $(-z) + z = 0$

45. $y(z + x) = (z + x)y$

46. $y \cdot 1 = 1 \cdot y$

47. $x(1/x) = 1$ if $x \neq 0$

48. $ax + ay = a(x + y)$

49. $y + (x + z) = (x + z) + y$

50. $(x + z) + y = x + (z + y)$

In Exercises 51–60 evaluate each expression by performing the indicated operations.

51. $(-2) \div (6 - 5)$

52. $(5 - 3)(5 + 2)$

53. $-4 + 2(3 - 8)$

54. $(-4 + 2)(3 - 8)$

55. $(-3)^3$

56. $(-2)^4$

57. $2 - 8(3 - 7)^2$

58. $(-5 + 2)^2 + (-2 - 6)^2$

59. $-5(11 - 6) - [7 - (11 - 19)]$

60. $7 - 3[2(13 - 5) - (5 - 13)]$

In Exercises 61–70 evaluate each expression after setting $x = -2$, $y = 3$, and $z = -4$.

61. $x + y + z$

62. $4x - z$

63. $(x - y)^2 \div (y - x)^2$

64. $2x^3 + 3x^2 - x + 1$

65. $x - 2(3y - 4z)$

66. $(x - 2)(3y - 4z)$

67. $(2y - x)x - z$

68. $(2y - x)(x - z)$

69. $(x + y + z)(x + y - z)$

70. $x + y + z(x + y - z)$

In Exercises 71–80 simplify each expression.

71. $x - 2(x + y) + 3(x - y)$

72. $k + (m - k) - (m - 2k)$

73. $-2(a - 2b) - 7(2a - b)$

74. $2a - (7a + 3) + (4a - 6)$

75. $(b^3 - 2b^2 + 3b + 4) - (b^2 + 2b - 1)$

76. $(4x^3 + 7x^2y^2 - 2y^3) + (-7x^3 + 2x^2y^2 + 2y^3)$

77. $2[a + 5(a + 2)] - 6$

78. $10 - 4[3x - (1 - x)]$

79. $3a - [3(a - b) - 2(a + b)]$

80. $-[x - (x + y) - (x - y) - (y - x)]$

In Exercises 81–90 evaluate each expression and check your result with a calculator.

81. $3^2 \cdot 3^3$

82. $(3^2)^3$

83. 7^0

84. 3^{-2}

85. $(4/3)^{-1}$

86. $5^{-4} \cdot 5$

87. $\dfrac{2}{2^4}$

88. $\dfrac{2}{2^{-2}}$

89. $\dfrac{6^{-1} \cdot 2^{-4}}{3^{-2} \cdot 2^{-3}}$

90. $\dfrac{2^{-3} \cdot 7^0}{4^{-1}}$

In Exercises 91–110 use the laws of exponents to perform the operations and write the result in the simplest form that contains only positive exponents.

91. $x^4 \cdot x^3$

92. $(x^3)^4$

93. $(-2p)^2$

94. $(-5y^5z)^3$

95. $c^9 \div c^3$

96. $9x^2 \div x^3$

97. $(x + y) \div (x + y)^4$

98. $(-3x^2)(-3x)^2$

99. $x^5 \cdot x^{-3}$

100. $t^{-1} \div t$

101. $\left(\dfrac{y}{5}\right)^{-2}$

102. $(1 - n)^{-1}(1 - n)$

103. $(4x^0)(-2x^{-3})$

104. $[3(x + h)^{-1}]^2$

105. $(2a^{-1}/x)^{-2}$

106. $2a^{-1}/a$

107. $\dfrac{(-x^3)^2(4yz)}{(-2x^2)(2y^2z^3)^3}$

108. $\dfrac{3^{-1}x^2y^{-3}}{9x^{-2}y^{-3}}$

109. $\dfrac{(xyz)^{-1}}{x^{-2}yz^{-3}}$

110. $\dfrac{(2ax^2)^{-2}(a^3x^{-1})^2}{2(ax)^{-1}(ax^5)}$

In Exercises 111–130 use the laws of exponents to simplify each expression. Variables as exponents denote integers.

111. $2^x \cdot 2^y$

112. $5^a \cdot 5$

113. $2/2^n$

114. $5^{2x}/5^x$

115. $(5^x)^{2x}$

116. $3^a \cdot 3^{-a}$

117. $y^{2a} \cdot y^{5a}$

118. $x^{b+2} \cdot x^{2b}$

119. $(a - b)^x(a - b)^y$

120. $[(b - a)^x]^y$

121. $a^x \cdot a^{-x}$

122. $2x^{1-a} \cdot x^{a-1}$

123. $\dfrac{y^x}{y^{x+1}}$

124. $\dfrac{x^{2n+2}}{x^2}$

125. $(x^a)^{p-1} \cdot x^{a-1}$

126. $(b^x)^{1-a} \div b^x$

127. $\dfrac{x^{2a}y^{a+1}}{x^a y^{a+2}}$

128. $\dfrac{x^a + b^y b}{x^a - b^y b^{-1}}$

129. $\dfrac{(1 - x)^{2a}(1 - x)^2}{(1 - x)^a}$

130. $\dfrac{(y + 2)^{1+m}}{(y + 2)^m(y + 2)}$

In Exercises 131–136 determine if the given calculator sequence produces the correct answer to the problem. If not, do two things. First, write the arithmetic expression being computed by the given sequence. Then change the sequence so it computes the given problem.

131. $\dfrac{5 + 4}{3}$; $5\boxed{+}4\boxed{\div}3\boxed{=}$

132. $\dfrac{6}{7} - 3$; $6\boxed{\div}7\boxed{-}3\boxed{=}$

133. $2 \div \dfrac{15}{9}$; $2\boxed{\div}15\boxed{\div}9\boxed{=}$

134. $\dfrac{2}{9(15)}$; $2\boxed{\div}9\boxed{\times}15\boxed{=}$

135. $2 \cdot 3^5$; $2\boxed{\times}3\boxed{y^x}5\boxed{=}$

136. $2^{3(5)}$; $2\boxed{y^x}3\boxed{\times}5\boxed{=}$

In Exercises 137–140 answer true or false. If false, give a specific counterexample.

137. All rational numbers are integers.
138. All rational numbers are real numbers.
139. All real numbers are irrational numbers.
140. All integers are irrational numbers.

141. The **equal-monthly-payment formula** is used by banks when they finance many common loans. The problem is to find the equal monthly installment, E, that will pay off a loan of P dollars with interest in n months. The formula used by the bank is

$$E = \frac{Pr}{1 - (1 + r)^{-n}}$$

where r is the monthly interest rate. Suppose that you obtain a \$10,000 car loan for 24 months with 12 percent annual interest charged on the unpaid balance. Use a calculator to determine the amount of the equal monthly installment that will pay off this debt. [*Note:* If the annual interest rate is 12 percent, the monthly interest rate (r) is 12 percent/12, or 1 percent.]

142. Find, by two methods, the area of the rectangle and explain how the rectangle can be used to illustrate geometrically the distributive property.

a	Area 1	Area 2
	b	c

THINK ABOUT IT

1. **a.** A common student error is to assume that $-a$ must represent a negative number. Explain why this assumption is incorrect.
   b. Explain the difference in meaning between $-a^2$ and $(-a)^2$.
   c. Why is a^{-n} the multiplicative inverse of a^n if $a \neq 0$?

2. The Greek mathematician Archimedes (250 B.C.) is usually considered one of the greatest mathematicians of all time. Archimedes considered his most important achievement to be the discovery that whenever a sphere is circumscribed by a cylinder, the ratio of their volumes is 3:2. He even asked that this figure and ratio be engraved on his tombstone. If the volume of a cylinder is $\pi r^2 h$ and the volume of a sphere is $\frac{4}{3}\pi r^3$, verify Archimedes' discovery.

3. Use a dictionary to find meanings for the words *commute, associate,* and *distribute* that are in agreement with the concepts expressed by the commutative, associative, and distributive properties stated in this section.

4. What is the 1993rd digit to the right of the decimal point in the decimal expansion of $\frac{22}{7}$?

5. A prime number is a positive integer other than 1 that is exactly divisible only by itself and 1. The *fundamental theorem of arithmetic* states that every integer greater than 1 can be written as the product of prime factors in exactly one way if we disregard the order of the factors. For example, the prime factorization of 12 is $2 \cdot 2 \cdot 3$. Explain how the assumption that $\sqrt{2}$ may be written in the form a/b, where a and b are positive integers, contradicts the fundamental theorem of arithmetic.

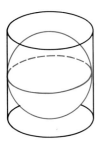

REMEMBER THIS

1. What is the greatest common factor of 18 and 12?
2. Which numbers in the set $\{81, 50, 2^4, 2^6, 2^9\}$ are perfect squares?
3. Which numbers in the set $\{-2, \frac{1}{3}, \sqrt{7}, 0, 6\}$ are integers?
4. Find a pair of integers whose product is -12 and whose sum is -1.
5. Find (a) the sum and (b) the product of $2x^3y^3$ and $5x^3y^3$.

6. Simplify $(x^3 - 5x^2 + 25x) + (3x^2 - 15x + 75)$.
7. True or false: All irrational numbers are real numbers.
8. Express $a - b$ as a sum.
9. Name the property of real numbers illustrated by $(a + b)(a - b) = (a - b)(a + b)$.
10. $2a + 2b + ka + kb$ may be written as $(2a + 2b) + (ka + kb)$ or as $(2a + ka) + (2b + kb)$. Name the properties of real numbers that allow us to reorder and regroup this expression in any form that is useful in a problem.

1.2 Products of Algebraic Expressions and Factoring

To multiply algebraic expressions, we use the laws of exponents and the distributive properties.

$$a(b + c) = ab + ac \quad \text{and} \quad (a + b)c = ac + bc$$

Example 1 shows the basic procedure.

EXAMPLE 1 Multiply.

a. $3x(x^2 - 2x + 1)$ **b.** $(2c^2 + t)ct$

Solution

a. $3x(x^2 - 2x + 1) = (3x)(x^2) + (3x)(-2x) + (3x)(1)$
$$= 3x^3 - 6x^2 + 3x$$

b. $(2c^2 + t)ct = (2c^2)(ct) + (t)(ct)$
$$= 2c^3t + ct^2$$ ∎

Since division is defined in terms of multiplication, a similar procedure is used to simplify division problems in which the divisor is a single term. For example, we may divide $a + b$ by c as follows:

$$(a + b) \div c = (a + b) \cdot \frac{1}{c} = \frac{a}{c} + \frac{b}{c}.$$

This result shows that

$$\frac{a + b}{c} = \frac{a}{c} + \frac{b}{c}.$$

EXAMPLE 2 Divide $\dfrac{2xh + h^2 + 2h}{h}$.

Solution $\dfrac{2xh + h^2 + 2h}{h} = \dfrac{2xh}{h} + \dfrac{h^2}{h} + \dfrac{2h}{h}$
$$= 2x + h + 2$$ ∎

If both factors in the multiplication contain more than one term, the distributive property must be used more than once. For example, no matter what expression is inside the parentheses,

$$(\rule{1cm}{2mm})(x + 2) \qquad \text{means} \qquad (\rule{1cm}{2mm})x + (\rule{1cm}{2mm})2.$$

Thus,

$$(x + 3)(x + 2) \qquad \text{means} \qquad (x + 3)x + (x + 3)2.$$

Using the distributive property the second time, we get

$$(x + 3)x + (x + 3)2 = x^2 + 3x + 2x + 6 = x^2 + 5x + 6.$$

Therefore,

$$(x + 3)(x + 2) = x^2 + 5x + 6.$$

EXAMPLE 3 Multiply.

a. $(2t + 3)(5t - 1)$ **b.** $(x^n - 2)(x^n - 1)$

Solution

a. $(2t + 3)(5t - 1) = (2t + 3)(5t) + (2t + 3)(-1)$
$$= 10t^2 + 15t - 2t - 3$$
$$= 10t^2 + 13t - 3$$

b. $(x^n - 2)(x^n - 1) = (x^n - 2)(x^n) + (x^n - 2)(-1)$
$$= x^{2n} - 2x^n - x^n + 2$$
$$= x^{2n} - 3x^n + 2$$ ∎

Notice that this method of multiplication is equivalent to multiplying each term of the first factor by each term of the second factor and then combining similar terms. This observation leads to the arrangement in Example 4, which is good for multiplying longer expressions since similar terms are placed under each other.

EXAMPLE 4 Multiply $(c^2 - 5c + 25)(c + 3)$.

Solution

$$
\begin{array}{r}
c^2 - 5c + 25 \\
c + 3 \\
\hline
c^3 - 5c^2 + 25c \\
3c^2 - 15c + 75 \\
\hline
c^3 - 2c^2 + 10c + 75
\end{array}
$$

add

This line equals $c(c^2 - 5c + 25)$.
This line equals $3(c^2 - 5c + 25)$.

∎

When multiplying expressions that contain two terms, a mental shortcut is often used, as shown in Example 5. This method is called the FOIL method, and the letters F, O, I, and L denote the products of the first, outer, inner, and last terms, respectively.

EXAMPLE 5   Multiply using the FOIL method: $(x^3 + y^3)^2$.

Solution $(x^3 + y^3)^2 = (x^3 + y^3)(x^3 + y^3)$. Then

$$\begin{array}{cccc} F + & O & + I & + L \\ x^6 + & x^3y^3 & + x^3y^3 & + y^6 \\ = x^6 + & 2x^3y^3 & + & y^6 \end{array}$$

[*Note:* $(x^3 + y^3)^2 \neq x^6 + y^6$, for this result leaves out the middle term, $2x^3y^3$.]   ∎

We now consider the process of factoring, which reverses the process of multiplication, as shown below.

If we change a product into a sum, we are multiplying.

$$(x + 5)(x + 1) \xrightarrow{\text{multiplying}} x^2 + 6x + 5$$

If we change a sum into a product, we are factoring.

$$(x + 5)(x + 1) \xleftarrow{\text{factoring}} x^2 + 6x + 5$$

In general, the procedures for factoring are not as straightforward as those for multiplication, and we limit our discussion of factoring to algebraic expressions called polynomials. Examples of polynomials are

$$64x^6 - y^6, \quad 9x^2 + 24x + 16, \quad \text{and} \quad 12ct^2 + 9c^2t,$$

and a **polynomial** is defined as an algebraic expression which may be written as a finite sum of terms that contain only nonnegative integer exponents on the variables. Thus, algebraic expressions like

$$4x^{-2}, \quad \sqrt{x} - 1, \quad \text{and} \quad 5x^{1/3} + 5$$

are not polynomials. A polynomial with just one term is called a **monomial;** one that contains exactly two terms is a **binomial;** and one with exactly three terms is a **trinomial.** The **degree of a monomial** is the sum of the exponents on the variables in the term. The **degree of a polynomial** is the same as the degree of its highest monomial term. Thus, the degree of $5x^4y^2$ is 6, the degree of $2xy^2$ is 3, and the degree of $5x^4y^2 + 2xy^2$ is 6, since $5x^4y^2$ is the highest-degree term.

There are several techniques that enable us to factor certain polynomials, and you need to consider carefully each of the factoring methods that follow.

Common Factors

The first method that should always be employed in factoring a polynomial is to attempt to find a factor that is common to each of the terms. For example:

Polynomial	Common Factor
$9s + 6t$	3
$15x^2 + (-5x)$ (or $15x^2 - 5x$)	$5x$
$2a^2b + 8ab^2$	$2ab$

In each case we attempt to pick the greatest common factor that will divide into each term of the polynomial. Therefore, although 5 is a common factor of $15x^2 - 5x$, a preferable common factor is $5x$ since $5x$ is the largest factor that divides both terms. After we determine the greatest common factor or **GCF,** we express each term in the polynomial as a product with the GCF as one factor. Then, factor out the GCF using the distributive property. You will notice that the directions in factoring problems use the phrase "factor completely." This expression directs us to continue factoring until the polynomial contains no factors of two or more terms that can be factored again. The restrictions we place on the form of the factors will determine if a polynomial is factorable, so it is important to note that, unless otherwise specified, we are interested only in polynomial factors with integer coefficients.

EXAMPLE 6   Factor completely.

a. $12ct^2 + 9c^2t$ **b.** $x^{n+2} + x^n$

c. $s(c + 2) - t(c + 2)$

Solution

a. The greatest common factor of $12ct^2$ and $9c^2t$ is $3ct$.

$$12ct^2 + 9c^2t = 3ct \cdot 4t + 3ct \cdot 3c$$
$$= 3ct(4t + 3c) \qquad \text{Distributive property}$$

b. By noting that $x^{n+2} = x^n \cdot x^2$, determine that the GCF is x^n and then factor as follows.

$$x^{n+2} + x^n = x^n \cdot x^2 + x^n \cdot 1$$
$$= x^n(x^2 + 1) \qquad \text{Distributive property}$$

c. The greatest common factor is the binomial $c + 2$. Factoring out the GCF using the distributive property gives

$$s(c + 2) - t(c + 2) = (c + 2)(s - t).$$ ■

Difference of Squares

A second method of factoring comes from reversing the following product.

$$(x + y)(x - y) \overset{\text{multiplying}}{\underset{\text{factoring}}{\longleftrightarrow}} x^2 - y^2$$

Note that the factors on the left differ only in the operation between x and y, while the result of the multiplication is two perfect squares with a minus sign between them. (Remember that x^2, x^4, x^6, x^{even}, and 1, 4, 9, 16, 25, and so on, are perfect squares.) Thus, in factoring two perfect squares with a minus sign between them, we obtain two factors that consist of the sum and the difference of the square roots of each of the squared terms.

EXAMPLE 7   Factor completely.

a. $x^2 - 81$ b. $y^4 - x^4$

Solution

a. $x^2 = x \cdot x$
   $81 = 9 \cdot 9$
   Therefore, $x^2 - 81 = (x + 9)(x - 9)$.
b. $y^4 = y^2 \cdot y^2$
   $x^4 = x^2 \cdot x^2$
   Therefore, $y^4 - x^4 = (y^2 + x^2)(y^2 - x^2)$. But

$$y^2 - x^2 = (y + x)(y - x).$$

Therefore, $y^4 - x^4 = (y^2 + x^2)(y + x)(y - x)$.   ■

Sums or Differences of Cubes

To factor sums or differences of cubes, we rely heavily on these factoring models.

$$a^3 + b^3 = (a + b)(a^2 - ab + b^2)$$
$$a^3 - b^3 = (a - b)(a^2 + ab + b^2)$$

The idea is to identify appropriate replacements for a and b in the expression to be factored, and then substitute in these formulas.

EXAMPLE 8   Factor completely.

a. $8x^3 + 27$ b. $64y^6 - x^6$

Solution

a. Here we use the formula for the sum of two cubes. Since $8x^3 = (2x)^3$ and $27 = (3)^3$, we replace a with $2x$ and b with 3. The result is

$$\begin{aligned} 8x^3 + 27 &= (2x)^3 + 3^3 \\ &= (2x + 3)[(2x)^2 - (2x)(3) + (3)^2] \\ &= (2x + 3)(4x^2 - 6x + 9). \end{aligned}$$

b. First, factor $64y^6 - x^6$ as the difference of squares.

$$64y^6 - x^6 = (8y^3)^2 - (x^3)^2$$
$$= (8y^3 + x^3)(8y^3 - x^3)$$

Now replace a by $2y$ and b with x in the formulas for both the sum and difference of cubes.

$$64y^6 - x^6 = (2y + x)[(2y)^2 - 2yx + x^2](2y - x)$$
$$[(2y)^2 + 2yx + x^2]$$
$$= (2y + x)(2y - x)(4y^2 - 2yx + x^2)$$
$$(4y^2 + 2yx + x^2)$$ ∎

Factoring Trinomials (Leading Coefficient 1)

In some cases trinomials of the form $ax^2 + bx + c$ can be factored (with integer coefficients) by reversing the FOIL multiplication process shown earlier. Example 9 discusses this factoring procedure in the simplest case when the leading coefficient a equals 1.

EXAMPLE 9   Factor completely $x^2 - 5x - 6$.

Solution
The first term, x^2, is the result of multiplying the first terms in the FOIL method. Thus,

$$x^2 - 5x - 6 = (x + ?)(x + ?).$$

The last term, -6, is the result of multiplying the last terms in the FOIL method. Since -6 equals $(-6)(1)$, $(6)(-1)$, $(-3)(2)$, and $(3)(-2)$, we have four possibilities. We want the pair whose sum is -5, which is the coefficient of the middle term. The combination of -6 and 1 satisfies this condition and produces the middle term of $-5x$.

$$(x - 6)(x + 1)$$
$$x + (-6x) = -5x$$

Thus, $x^2 - 5x - 6 = (x - 6)(x + 1)$. We summarize our method in this example with the factoring formula

$$x^2 + (a + b)x + ab = (x + a)(x + b)$$

which you can use as a model for factoring such expressions. ∎

Factoring Trinomials (General Case)

When the coefficient of the squared term is not 1, the trinomial is harder to factor. The reason for this difficulty can be seen in the following product.

$$(ax + b)(cx + d) = (ac)x^2 + (ad + bc)x + bd$$

In factoring these more complicated trinomials, if you keep in mind that the middle term is the sum of the inside product and the outside product, then a little trial and error will eventually produce the answer. Although there are often many combinations to consider, experience and practice will help you *mentally* eliminate many of the possibilities.

EXAMPLE 10   Factor completely $9y^2 + 12y + 4$.

Solution

The first term, $9y^2$, is the result of multiplying the first terms in the FOIL method.

$$9y^2 + 12y + 4 \stackrel{?}{=} \begin{array}{c} (9y + ?)(y + ?) \\ \text{or} \\ (3y + ?)(3y + ?) \end{array}$$

The last term, 4, is the result of multiplying the last terms in the FOIL method. Since 4 equals $(4)(1)$ and $(2)(2)$, we have two possibilities. We eliminate negative factors of 4 since the middle term is positive.

$$9y^2 + 12y + 4 \stackrel{?}{=} \begin{array}{c} (9y + 1)(y + 4) \\ (9y + 4)(y + 1) \\ (9y + 2)(y + 2) \\ (3y + 4)(3y + 1) \\ (3y + 2)(3y + 2) \end{array}$$

The middle term, $12y$, is the sum of the inner and the outer terms.

$$(3y + 2)(3y + 2)$$
$$6y + 6y = 12y$$

Thus, $9y^2 + 12y + 4 = (3y + 2)(3y + 2)$ or $(3y + 2)^2$.

Note: Very often when the first term and the last term of the trinomial to be factored are perfect squares, the factored form is also a perfect square. Try this possibility first. The factoring model in such cases (called perfect square trinomials) is

$$x^2 + 2xy + y^2 = (x + y)^2.$$

■

It should be noted that many trinomials cannot be factored with integer coefficients, and if you suspect that you are trying to find a factorization that might not exist, then apply the following test.

Factoring Test for Trinomials

A trinomial of the form $ax^2 + bx + c$, where a, b, and c are integers, is factorable into binomial factors with integer coefficients if and only if $b^2 - 4ac$ is a perfect square.

The expression $b^2 - 4ac$ is called the **discriminant,** and when we derive the discriminant in Section 1.6, the rationale for this test will be easy to understand.

EXAMPLE 11   Calculate $b^2 - 4ac$ and determine if $x^2 - 10x - 30$ can be factored into binomial factors with integer coefficients.

Solution   In this trinomial $a = 1$, $b = -10$, and $c = -30$. Therefore,

$$b^2 - 4ac = (-10)^2 - 4(1)(-30)$$
$$= 220.$$

Since 220 is not a perfect square, $x^2 - 10x - 30$ cannot be factored into binomial factors with integer coefficients.   ■

A polynomial is said to be **irreducible** over the set of integers if it cannot be written as the product of two polynomials of positive degree with integer coefficients. Thus, we say $x^2 - 10x - 30$ is irreducible over the set of *integers.* Note, however, that $x^2 - 10x - 30$ is not irreducible over the set of *real numbers* since we can show that

$$x^2 - 10x - 30 = [x - (5 + \sqrt{55})][x - (5 - \sqrt{55})]$$

by using techniques that we will develop. Similarly,

$$x^2 - 7 = (x + \sqrt{7})(x - \sqrt{7})$$

if we allow real number coefficients in the factors. Thus, whether a polynomial is irreducible depends on the number system from which the coefficients may be selected.

Factoring by Grouping

Factoring by grouping is not really a different way of factoring for it relies on the methods we have already discussed. However, before those methods can be applied, it is sometimes necessary to rewrite or group together terms in the expression to be factored. For example,

$$3xy + 2y + 3xz + 2z$$

can be factored by rewriting the expression as

$$(3xy + 2y) + (3xz + 2z).$$

If we factor out the common factor in each group, we have

$$y(3x + 2) + z(3x + 2).$$

Now $3x + 2$ is a common factor and the final result is

$$(3x + 2)(y + z).$$

You might check that in this case different groupings of the original expression achieve the same result.

EXAMPLE 12   Factor completely $x^3 - 5x^2 - 4x + 20$.

Solution   If we factor x^2 from the first two terms and -4 from the last two terms, we have

$$x^3 - 5x^2 - 4x + 20 = (x^3 - 5x^2) + (-4x + 20)$$
$$= x^2(x - 5) - 4(x - 5).$$

Then $x - 5$ is a common binomial factor, so

$$x^2(x - 5) - 4(x - 5) = (x - 5)(x^2 - 4)$$
$$= (x - 5)(x + 2)(x - 2).$$

Thus, $x^3 - 5x^2 - 4x + 20 = (x - 5)(x + 2)(x - 2)$. Note in the solution that we factor out -4 instead of 4 from the grouping on the right to reach the goal of obtaining a common binomial factor.   ∎

EXAMPLE 13   Factor completely $x^2 + y^2 - z^2 - 2xy$.

Solution   We need to recognize that if we group together $x^2 - 2xy + y^2$, it factors into the perfect square $(x - y)^2$. We then have $(x - y)^2 - z^2$, which factors as the difference of two squares. Thus,

$$x^2 + y^2 - z^2 - 2xy = (x^2 - 2xy + y^2) - z^2$$
$$= (x - y)^2 - z^2$$
$$= (x - y + z)(x - y - z).$$ ∎

Factoring by grouping is a key step in the **ac method** for factoring the trinomial $ax^2 + bx + c$. Example 14 explains in detail the steps for this method, which is especially useful if our previous method of reversing FOIL results in many possible factorizations to consider.

EXAMPLE 14   Factor completely $12x^2 - x - 20$.

Solution   Follow the steps below.

Step 1   **Find two integers whose product is ac and whose sum is b.** In the trinomial $a = 12$, $b = -1$, and $c = -20$. Thus, $ac = 12(-20) = -240$, and we look for two integers whose product is -240 and whose sum is -1. With a little trial and error we find that

$$-16(15) = -240 \quad \text{and} \quad -16 + 15 = -1,$$

so the required integers are -16 and 15.

Step 2   **Replace b by the sum of the two integers from step 1 and then distribute x.**

$$12x^2 - x - 20 = 12x^2 + (-16 + 15)x - 20$$
$$= 12x^2 - 16x + 15x - 20$$

Step 3   **Factor by grouping.**

$$(12x^2 - 16x) + (15x - 20) = 4x(3x - 4) + 5(3x - 4)$$
$$= (3x - 4)(4x + 5)$$

Thus, $12x^2 - x - 20 = (3x - 4)(4x + 5)$.   ∎

Summary

The basic formulas we discussed in this section are listed in the box for reference purposes. Some people find it helpful to stick very closely to these forms, while others have a feel for the basic patterns involved and find that the formulas only complicate the issue. You decide which group you belong to. As a final note, remember you can always check your factoring by multiplying out your answer.

Factoring and Product Models

1. Common factor:
$$ax + ay = a(x + y)$$

2. Difference of squares:
$$x^2 - y^2 = (x + y)(x - y)$$

3. Sum of cubes:
$$a^3 + b^3 = (a + b)(a^2 - ab + b^2)$$

4. Difference of cubes:
$$a^3 - b^3 = (a - b)(a^2 + ab + b^2)$$

5. Trinomial (leading coefficient 1):
$$x^2 + (a + b)x + ab = (x + a)(x + b)$$

6. Perfect square trinomial:
$$x^2 + 2xy + y^2 = (x + y)^2$$

7. General trinomial:
$$(ac)x^2 + (ad + bc)x + bd = (ax + b)(cx + d)$$

EXERCISES 1.2

In Exercises 1–30 perform the multiplication or division and combine similar terms.

1. $2(x - y)$
2. $-6(z + y)$
3. $-5x(x^3 - x^2 - x)$
4. $y(3y^2 + 5y - 6)$
5. $-2xyz(4x - y + 7z)$
6. $4x^2yz^3(3x^3 - 2yz + 5xz^4)$
7. $(p^2 + q^2)p^2q^2$
8. $(a^2 - b^2)ab$
9. $(24n^2x + 18nx^2) \div 6nx$
10. $(6dt^2 - 12d^2t) \div 2dt$
11. $\dfrac{2hx + h^2}{h}$
12. $\dfrac{6xh + 3h^2}{h}$
13. $2x^4\left(\dfrac{3}{x} + \dfrac{4}{x^2} - \dfrac{5}{x^3}\right)$
14. $-12x^2\left(3x^2 - \dfrac{1}{3} + \dfrac{3}{4x} - \dfrac{1}{x^2}\right)$

15. $(a + 3)(a + 4)$
16. $(z - 2)(z - 6)$
17. $(3x - 4)(2x - 1)$
18. $(5y - 4x)(3y + x)$
19. $(k - 2)^2$
20. $(2x + 3y)^2$
21. $(y - 4)(y^2 + 5y - 1)$
22. $(2a - 1)(a^3 + a + 1)$
23. $(x - y)(x^2 + xy + y^2)$
24. $(2y + 3z)(y^2 - 2yz + z^2)$
25. $(2y - 1)(3y + 2)(1 - y)$
26. $(x + 1)^3$
27. $[(x - y) - 1]^2$
28. $[3 + (4a - b)]^2$
29. $(x^2 - 2xy + y^2)(x^2 + 2xy + y^2)$
30. $(x^2 - x - 1)(x^2 + x + 1)$

In Exercises 31–38 factor out the greatest common factor.

31. $yx + yz$ **32.** $ab - ac$

33. $9a^2x^2 + 3ax$ **34.** $22x^3y^3 - 2x^2y^2$

35. $b(a - c) + d(a - c)$

36. $2x(y + z) - 5y(y + z)$

37. $3x(x + 5) + 2(x + 5)$ **38.** $(x + 3)^2 - (x + 3)$

In Exercises 39–44 factor each difference of squares.

39. $36x^2 - 1$ **40.** $100 - 81y^2$

41. $25p^2 - 49q^2$ **42.** $9y^2 - x^2$

43. $36r^6 - k^4$ **44.** $25x^{12} - 36a^{10}$

In Exercises 45–50 factor each sum or difference of cubes.

45. $y^3 + 27$ **46.** $x^3 + 64$

47. $1 - x^3$ **48.** $8x^3 - 27y^3$

49. $27t^3 - 125$ **50.** $a^3b^3 + c^3$

In Exercises 51–62 factor each trinomial.

51. $a^2 + 5a + 4$ **52.** $y^2 + 7y + 10$

53. $x^2 + x - 12$ **54.** $p^2 - 3p - 18$

55. $7z^2 - 13z + 6$ **56.** $5r^2 - 4r - 12$

57. $x^2 + 5xy + 4y^2$ **58.** $2a^2 - 7ad + 6d^2$

59. $16y^2 - 24y + 9$ **60.** $9x^2 + 24x + 16$

61. $9a^2 - 12ab + 4b^2$ **62.** $16x^2y^2 + 8xy + 1$

In Exercises 63–68 factor by grouping.

63. $5x + 5y + ax + ay$ **64.** $2a + 2b + ka + kb$

65. $3mn + 6m + n + 2$ **66.** $4xy - 6x - 6y + 9$

67. $x^5 + x^3 + x^2 + 1$ **68.** $x^3 - 3x^2 - x + 3$

In Exercises 69–90 factor each expression completely.

69. $x - xy^2$ **70.** $n^3 - n$

71. $3k^2 - 6k - 24$ **72.** $5y^2 - 15y - 50$

73. $6 + x - x^2$ **74.** $9 - 8y - y^2$

75. $3x^3 - 12x^2 + 9x$ **76.** $2x^3y - 18x^2y + 40xy$

77. $4x^4 - 144x^2$ **78.** $2x^5 - 162x$

79. $2x^5 + 54x^2$ **80.** $x^4 - 27x$

81. $27(x - 1)^2 - 48x^2$ **82.** $2(x - 2)^3 + 2$

83. $x^6 - 1$ **84.** $y^6 - x^6$

85. $x^8 - y^8$ **86.** $1 - x^{12}$

87. $x^2 + 4x + 4 - y^2$ **88.** $a^2 + x^2 - 2ax - 1$

89. $x^2 - y^2 - z^2 - 2yz$ **90.** $4x^2 + y^2 + 4xy - 1$

In Exercises 91–94 determine whether each trinomial can be factored into binomial factors with integer coefficients by calculating $b^2 - 4ac$.

91. $6x^2 + 7x + 2$ **92.** $12s^2 - s + 1$

93. $12y^2 - 11y - 18$ **94.** $18x^2 + 19x - 12$

In Exercises 95–100 factor each trinomial using the *ac* method.

95. $6x^2 + 7x - 5$ **96.** $6y^2 + 5y - 6$

97. $9b^2 - 25b - 6$ **98.** $12x^2 + 19x - 18$

99. $12x^2 - 29xy + 10y^2$ **100.** $18t^2 - 3kt - 10k^2$

In Exercises 101–110 find each product. Assume variables as exponents denote integers.

101. $(x^n + 5)(x^n + 2)$ **102.** $(3x^n - 2)(x^n - 1)$

103. $(z^a + 3)^2$ **104.** $(t^b - 7)^2$

105. $(x^a - y^b)(x^a + y^b)$ **106.** $(x^a + y^b)^2$

107. $(a^{bx} + a^{-bx})^2$

108. $(a^{bx} + a^{-bx})(a^{bx} - a^{-bx})$

109. $(x^n + y^n)(x^{2n} - x^ny^n + y^{2n})$

110. $(x^n - y^k)(x^{2n} + x^ny^k + y^{2k})$

In Exercises 111–120 factor each expression completely. Assume variables as exponents denote integers.

111. $x^{n+1} + x^n$ **112.** $y^{3n} - y^{2n}$

113. $y^{2n} - 16$ **114.** $x^{4n} - 49$

115. $1 - t^{3m}$ **116.** $x^{3n} + y^{3n}$

117. $x^{m+2} - x^m$ **118.** $y^{4m} - a^{4m}$

119. $x^{2n+1} + x^{n+1} - 20x$ **120.** $x^{2m+2} + x^{m+2} - 30x^2$

121. Show that $\dfrac{(x + h)^2 + 1 - (x^2 + 1)}{h}$

simplifies to $2x + h$.

122. Show that $\dfrac{(x + h)^3 - x^3}{h}$

simplifies to $3x^2 + 3xh + h^2$.

123. Simplify $\dfrac{[1 - (-3 + h)^2] - [1 - (-3)^2]}{h}$.

124. Simplify $\dfrac{(2 + h)^3 - 2^3}{h}$.

125. Simplify $\dfrac{[(x + h) + 1]^2 - (x + 1)^2}{h}$.

126. Simplify

$\dfrac{[128(t + h) - 16(t + h)^2] - [128t - 16t^2]}{h}$.

127. Verify through multiplication the factoring model $a^3 + b^3 = (a + b)(a^2 - ab + b^2)$.

128. Show that for any three consecutive integers the square of the middle one is always one more than the product of the smallest and the largest.

129. An odd number is a positive integer that can be written in the form $2k + 1$, where k is an integer. Show that the product of two odd numbers is odd.

THINK ABOUT IT

1. A common student error is to expand $(a + b)^2$ as $a^2 + b^2$. Explain how the square shown below with side length $a + b$ may be used to illustrate geometrically the correct expansion of $(a + b)^2$.

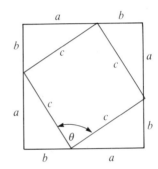

2. If the product of $kx - 3$ and $kx + 1$ equals $k^2x^2 - 6x - 3$, then find k.

3. **a.** Multiply $(x - 1)(x + 1)$.
    b. Multiply $(x - 1)(x^2 + x + 1)$.
    c. Multiply $(x - 1)(x^3 + x^2 + x + 1)$.
    d. Look at your results from parts a–c. Without multiplying, simplify the product $(x - 1)(x^7 + x^6 + x^5 + x^4 + x^3 + x^2 + x + 1)$.
    e. What is a factoring model for $x^n - 1$?

4. Factor $x^6 - 64$ completely by the following methods.
    Method 1   First, factor as a difference of squares and then use the formulas for the sum and the difference of cubes.
    Method 2   First, factor as a difference of cubes. Then use the difference of squares formula on *both* factors and match the result from the first method. Applying the difference of squares formula to the trinomial factor requires some ingenuity.

5. The Pythagorean relation is one of the most famous ideas in mathematics. Several hundred different proofs of this theorem have been recorded, and the National Council of Teachers of Mathematics published a book,

The Pythagorean Proposition (1968), that presents 370 demonstrations of this statement. Early proofs, which were geometric in nature, gradually gave way to analytical proofs that used algebra to verify geometric ideas. Two well-known proofs of this variety follow.

a. Consider the square with side length $a + b$. Calculate the area of the square in two ways and establish the Pythagorean relation $c^2 = a^2 + b^2$. (*Note:* It must be shown that $\theta = 90°$.)

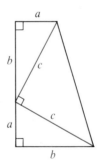

b. The trapezoid in the diagram was used by James Garfield (the twentieth president of the United States) to prove the Pythagorean theorem. Once again the idea is to find the area in two ways and establish $c^2 = a^2 + b^2$. Try it.

REMEMBER THIS

1. Divide $-\frac{7}{4} \div \frac{21}{16}$.
2. Add $\frac{2}{7} + \frac{3}{5}$.
3. Find the least common denominator of $\frac{17}{18}$ and $\frac{5}{24}$.
4. Express $\frac{17}{11}$ as a repeating decimal.
5. Evaluate $\dfrac{x - y}{y - x}$ if $x = -2$ and $y = 7$.

6. Evaluate $\dfrac{x^2 - 3x}{x^2 - 2x}$ if $x = 2$.
7. Evaluate $(-3)^{-2} \cdot (-3)^{-1}$.
8. Evaluate $\dfrac{3^{-2} \cdot 4^0}{6^{-1}}$.
9. Simplify $x^{b+3} \cdot x^{3b}$.
10. Factor $a^2 + ab$, $a^2 - b^2$, and $ab - b^2$.

1.3 Algebraic Fractions

Algebraic fractions are the quotients of algebraic expressions, and the same principles that govern a fraction in arithmetic also apply when the numerator and the denominator contain algebraic expressions. Two fraction principles of particular importance follow.

Fraction Principles

Let a, b, c, d, and k be real numbers with b, d, and $k \neq 0$.

Equality of fractions $\dfrac{a}{b} = \dfrac{c}{d}$ if and only if $ad = bc$

Fundamental principle $\dfrac{ak}{bk} = \dfrac{a}{b}$

The fundamental principle can be established from the criterion for the equality of two fractions since $(ak)b$ is equal to $(bk)a$. As its name suggests, the fundamental principle is applied often, and two common uses of this principle are shown below.

Simplifying fractions $\dfrac{6}{8} = \dfrac{3 \cdot 2}{4 \cdot 2} = \dfrac{3}{4}$

Fundamental principle

Adding fractions $\dfrac{1}{8} + \dfrac{3}{4} = \dfrac{1}{8} + \dfrac{3 \cdot 2}{4 \cdot 2} = \dfrac{1}{8} + \dfrac{6}{8} = \dfrac{7}{8}$

When simplifying fractions by the fundamental principle, it is important to recognize that we may divide out only nonzero factors of the numerator and the denominator. Thus, we have a general procedure for expressing a fraction in simplest form.

To Simplify Fractions

1. Factor completely the numerator and the denominator of the fraction.
2. Divide out nonzero factors that are common to the numerator and the denominator according to the fundamental principle.

Example 1 illustrates this procedure.

EXAMPLE 1 Express in simplest form.

a. $\dfrac{x^2 - 3x - 4}{x^2 - 2x - 8}$

b. $\dfrac{(x^2 + 2) - (a^2 + 2)}{x - a}$

Solution   Factor and use the fundamental principle.

a. $\dfrac{x^2 - 3x - 4}{x^2 - 2x - 8} = \dfrac{(x + 1)(x - 4)}{(x + 2)(x - 4)} = \dfrac{x + 1}{x + 2}, \qquad x \ne 4$

b. $\dfrac{(x^2 + 2) - (a^2 + 2)}{x - a} = \dfrac{x^2 - a^2}{x - a} = \dfrac{(x + a)(x - a)}{x - a}$

$\qquad\qquad\qquad\qquad\qquad = x + a, \qquad x \ne a$ ∎

In Example 1 the restrictions $x \ne 4$ and $x \ne a$ are necessary because we may divide out only nonzero factors since division by zero is undefined. With the understanding that such restrictions always apply, we will not continue to list them from this point on.

When working with fractions, it is important to keep in mind what happens in addition and subtraction if we change the order of the terms. In addition $a + b = b + a$. However, in subtraction $a - b = -(b - a)$. Thus,

$$\frac{a + b}{b + a} = 1 \qquad \text{while} \qquad \frac{a - b}{b - a} = \frac{-(b - a)}{b - a} = -1.$$

Although $a - b$ does not equal $b - a$, the expressions differ only by a factor of -1 and can be simplified accordingly.

EXAMPLE 2   Express in simplest form $\dfrac{x^2 - y^2}{y - x}$.

Solution   Since $y - x = -(x - y)$ we proceed as follows:

$$\frac{x^2 - y^2}{y - x} = \frac{(x + y)(x - y)}{(-1)(x - y)}$$

$$= \frac{x + y}{-1} \text{ or } -x - y.$$ ∎

Multiplication

In arithmetic we know that the product of two or more fractions is the product of their numerators divided by the product of their denominators. In symbols, this principle is

$$\frac{a}{b} \cdot \frac{c}{d} = \frac{ac}{bd} \qquad b, d \ne 0.$$

Similarly, we will use this procedure for multiplying algebraic fractions. To express products in simplest form, factoring and the fundamental principle are usually required.

EXAMPLE 3   Simplify $\dfrac{5x^2 + 15x}{4 - x^2} \cdot \dfrac{12 - 6x}{(3 + x)^2}$.

Solution   We factor, multiply, and then simplify as shown.

$$\frac{5x^2 + 15x}{4 - x^2} \cdot \frac{12 - 6x}{(3 + x)^2} = \frac{5x(x + 3)}{(2 + x)(2 - x)} \cdot \frac{6(2 - x)}{(3 + x)^2}$$

$$= \frac{5 \cdot 6x(x + 3)(2 - x)}{(2 + x)(2 - x)(3 + x)(3 + x)}$$

$$= \frac{30x}{(2 + x)(3 + x)} \qquad \blacksquare$$

Division

To divide two fractions, we must recall that division is defined in terms of multiplication. That is, to divide a by b, we multiply a by the reciprocal of b. Note the reciprocal of a fraction can be found by inverting the fraction. For example, the reciprocal of $\frac{2}{7}$ is $\frac{7}{2}$. Thus, to divide two fractions, we invert the fraction by which we are dividing to find its reciprocal and then we multiply.

$$\frac{a}{b} \div \frac{c}{d} = \frac{a}{b} \cdot \frac{d}{c} \qquad b, d, \frac{c}{d} \neq 0$$

EXAMPLE 4   Simplify $\dfrac{n + 1}{2^{n + 1}} \div \dfrac{n}{2^n}$.

Solution   We convert to multiplication and simplify.

$$\frac{n + 1}{2^{n + 1}} \div \frac{n}{2^n} = \frac{n + 1}{2^{n + 1}} \cdot \frac{2^n}{n} = \frac{1}{2^{(n + 1) - n}} \cdot \frac{n + 1}{n} = \frac{n + 1}{2n} \qquad \blacksquare$$

Addition and Subtraction

In arithmetic we know that the sum (or difference) of two or more fractions that have the same denominator is given by the sum (or difference) of the numerators divided by the common denominator. In symbols, the principle is

$$\frac{a}{b} \pm \frac{c}{b} = \frac{a \pm c}{b} \qquad b \neq 0$$

We add or subtract algebraic fractions in a similar way, as shown in Example 5.

EXAMPLE 5   Simplify $\dfrac{2x + 1}{x - 3} - \dfrac{x - 2}{x - 3}$.

Solution   The fractions have the same denominator, so

$$\frac{2x + 1}{x - 3} - \frac{x - 2}{x - 3} = \frac{(2x + 1) - (x - 2)}{x - 3}$$

$$= \frac{2x + 1 - x + 2}{x - 3}$$

$$= \frac{x + 3}{x - 3}.$$

Note: Be careful in this and other subtraction problems to remove correctly the parentheses in an expression like $- (x - 2)$.   ■

When fractions have different denominators, we can change them into equivalent fractions with the same denominator and then add them. In such cases computation is simpler if we use the smallest possible common denominator, called the **least common denominator** or **LCD.** One way of finding the LCD follows.

To Find the LCD

> 1. Factor *completely* each denominator.
> 2. The LCD is the product of all the different factors with each factor raised to the highest power to which it appears in any one factorization.

The procedure is illustrated in Example 6.

EXAMPLE 6   Find the LCD of $\dfrac{2}{x^2 - 1}$ and $\dfrac{1}{x^2 + 2x + 1}$.

Solution   First, factor each denominator completely.

$$x^2 - 1 = (x + 1)(x - 1)$$
$$x^2 + 2x + 1 = (x + 1)(x + 1) = (x + 1)^2$$

The LCD will contain the factors $x + 1$ and $x - 1$. The highest power of $x + 1$ is 2, and the highest power of $x - 1$ is 1. Thus,

$$\text{LCD} = (x + 1)^2(x - 1).$$ ■

We can now combine several principles of fractions and formulate a general procedure for adding or subtracting two or more fractions.

To Add or Subtract Fractions

1. Completely factor each denominator and find the LCD.
2. For each fraction, obtain an equivalent fraction by applying the fundamental principle and multiplying the numerator and the denominator of the fraction by the factors of the LCD that are not contained in the denominator of that fraction.
3. Add or subtract the numerators and divide this result by the common denominator.

EXAMPLE 7   Simplify $\dfrac{2a + 3}{a^2 + a} - \dfrac{a}{a^2 - 1}$.

Solution   The denominators $a^2 + a$ and $a^2 - 1$ factor as $a(a + 1)$ and $(a + 1)(a - 1)$, respectively. Since the highest power of each different factor is 1, the LCD is $a(a + 1)(a - 1)$. Then,

$$\frac{2a + 3}{a^2 + a} - \frac{a}{a^2 - 1} = \frac{2a + 3}{a(a + 1)} - \frac{a}{(a + 1)(a - 1)}$$

$$= \frac{(2a + 3)(a - 1)}{a(a + 1)(a - 1)} - \frac{a(a)}{(a + 1)(a - 1)(a)}$$

$$= \frac{2a^2 + a - 3 - a^2}{a(a + 1)(a - 1)}$$

$$= \frac{a^2 + a - 3}{a(a + 1)(a - 1)}.$$

■

We often need to add or subtract two fractions for which the LCD is simply the product of the denominators. This special situation is best handled by the following formula.

$$\frac{a}{b} \pm \frac{c}{d} = \frac{ad \pm bc}{bd} \qquad b, d \neq 0$$

We can verify this property from our previous methods since

$$\frac{a}{b} \pm \frac{c}{d} = \frac{a \cdot d}{b \cdot d} \pm \frac{c \cdot b}{d \cdot b} = \frac{ad \pm bc}{bd}.$$

EXAMPLE 8   Simplify $\dfrac{x + h + 1}{x + h} - \dfrac{x + 1}{x}$.

Solution   We apply the above formula and then simplify.

$$\frac{x + h + 1}{x + h} - \frac{x + 1}{x} = \frac{(x + h + 1)x - (x + h)(x + 1)}{(x + h)x}$$

$$= \frac{x^2 + hx + x - x^2 - x - hx - h}{(x + h)x}$$

$$= \frac{-h}{(x + h)x}$$

■

Complex Fractions

A **complex fraction** is a fraction in which the numerator or the denominator, or both, involve fractions. A procedure for simplifying complex fractions follows.

To Simplify Complex Fractions

1. Find the LCD of all the fractions that appear in the numerator and the denominator of the complex fraction.
2. Multiply the numerator and the denominator of the complex fraction by this LCD and simplify your results.

EXAMPLE 9   Simplify.

a. $\dfrac{x^{-1} + y^{-1}}{(x/y) - (y/x)}$
   b. $\left(\dfrac{1}{x + h} - \dfrac{1}{x} \right) \div h$

Solution

a. $x^{-1} = 1/x$, $y^{-1} = 1/y$, and the LCD is xy.

$$\frac{\dfrac{1}{x} + \dfrac{1}{y}}{\dfrac{x}{y} - \dfrac{y}{x}} = \frac{xy\left(\dfrac{1}{x} + \dfrac{1}{y} \right)}{xy\left(\dfrac{x}{y} - \dfrac{y}{x} \right)}$$

$$= \frac{y + x}{x^2 - y^2}$$

$$= \frac{y + x}{(x + y)(x - y)}$$

$$= \frac{1}{x - y}$$

b. The LCD is $x(x + h)$.

$$\frac{\dfrac{1}{x + h} - \dfrac{1}{x}}{h} = \frac{x(x + h)\left(\dfrac{1}{x + h} - \dfrac{1}{x} \right)}{x(x + h) \cdot h}$$

$$= \frac{x - (x + h)}{x(x + h)h}$$

$$= \frac{-h}{x(x + h)h}$$

$$= \frac{-1}{x(x + h)}$$

∎

EXAMPLE 10 Solve the problem in the chapter introduction on page 1.

Solution Let d represent the distance around the track. Since $d = rt$, the times required for the first and second laps are denoted by $d/180$ and $d/120$, respectively. Then,

$$\text{average speed} = \frac{\text{total distance}}{\text{total time}} = \frac{2d}{\dfrac{d}{180} + \dfrac{d}{120}}.$$

Multiplying each term in the numerator and the denominator by 360 gives

$$\frac{360(2d)}{360\left(\dfrac{d}{180} + \dfrac{d}{120}\right)} = \frac{720d}{2d + 3d} = \frac{720d}{5d} = 144.$$

Thus, the average speed is 144 mi/hour and the driver does not qualify for the finals. Note that the average speed is closer to 120 mi/hour because more *time* was needed for the second lap. ∎

EXERCISES 1.3

In Exercises 1–20 express each fraction in simplest form.

1. $\dfrac{4x - 4}{11x - 11}$

2. $\dfrac{y^2 + 2y}{y^2 + 3y}$

3. $\dfrac{4b}{12b^2 + 4b}$

4. $\dfrac{20z^2 - 5z}{10z}$

5. $\dfrac{y - x}{x - y}$

6. $\dfrac{(x - 4)^2}{4 - x}$

7. $\dfrac{x - a}{ax(a - x)}$

8. $\dfrac{-x + a}{x - a}$

9. $\dfrac{1 - x}{x^2 - 1}$

10. $\dfrac{i - 1}{-(1^2 - i^2)}$

11. $\dfrac{a^2 - 3a - 4}{a^2 - 8a + 16}$

12. $\dfrac{(x^2 + 1) - (a^2 + 1)}{x - a}$

13. $\dfrac{49 - 14y + y^2}{49 - y^2}$

14. $\dfrac{2m^2 - 5m + 2}{4m - 2}$

15. $\dfrac{6x^3 - 3x^2 - 30x}{2x^2 - 4x - 16}$

16. $\dfrac{z^3 - z^2 - 6z}{z^3 + z^2 - 12z}$

17. $\dfrac{x^3 - a^3}{x - a}$

18. $\dfrac{x^4 - a^4}{x - a}$

19. $\dfrac{x^3 - (x - 1) \cdot 3x^2}{x^6}$

20. $\dfrac{x^4(-2) - (3 - 2x)4x^3}{x^8}$

In Exercises 21–38 express each product or quotient in simplest form.

21. $\dfrac{2x}{5y} \cdot \dfrac{5y^2}{8x^3}$

22. $\dfrac{14z}{10y} \cdot \dfrac{15y}{35z}$

23. $\dfrac{5a}{a^2 - 9} \cdot \dfrac{6a - 18}{15a^2}$

24. $\dfrac{n^2 - n - 12}{n^2} \cdot \dfrac{n}{n - 4}$

25. $\dfrac{(y - 1)^2}{(y + 2)^2} \cdot \dfrac{y^2 - 4}{y^2 - 1}$

26. $(x + 4)^2 \cdot \dfrac{x}{16 - x^2}$

27. $\dfrac{x^2 + 3x - 4}{x^2 - 3x - 4} \cdot \dfrac{x^2 - 5x + 4}{x^2 + 5x + 4}$

28. $\dfrac{2a^2 - 7a + 6}{a^2 - a - 6} \cdot \dfrac{5a - 15}{a^2 - 4}$

29. $\dfrac{xy}{z} \div \dfrac{x^2y}{z}$

30. $\dfrac{zy^2}{2x} \div \dfrac{yz^2}{x}$

31. $\dfrac{n^2 + n - 2}{n - 3} \div (n + 2)$

32. $\dfrac{4x^2 - 9}{4x - 9} \div (2x - 3)$

33. $\dfrac{x^4 - 1}{7x + 7} \div \dfrac{x^3 + x}{x^2 - 1}$

34. $\dfrac{3b^2 + 6b - 24}{b^2 - 7b + 10} \div \dfrac{3b^2 + 4b}{b^3 - 5b^2}$

35. $\dfrac{1 - a}{x^2 + x} \div \dfrac{a^2 - 1}{x^2 - 1}$

36. $\dfrac{x^2 - 7x + 10}{5x - 25} \div \dfrac{4 - 2x}{25 - x^2}$

37. $\dfrac{x^3 - y^3}{(x - y)^3} \div \dfrac{x^2 + xy + y^2}{x^2 - 2xy + y^2}$

38. $\dfrac{2x^2 - 2ax + 2a^2}{(ax)^3} \div \dfrac{a^3 + x^3}{ax^3}$

In Exercises 39–62 combine each expression into a single fraction in simplest form.

39. $\dfrac{4a}{xy} + \dfrac{2a}{xy} - \dfrac{3a}{xy}$

40. $\dfrac{1}{a^2x^2} - \dfrac{5}{a^2x^2} + \dfrac{11}{a^2x^2}$

41. $\dfrac{3x + 4}{x + 2} - \dfrac{2x + 5}{x + 2}$

42. $\dfrac{4y + 1}{2y - 6} - \dfrac{2y + 7}{2y - 6}$

43. $\dfrac{2n - 3}{n - 6} - \dfrac{n + 1}{6 - n}$

44. $\dfrac{2x - 5}{4 - 3x} + \dfrac{2 - x}{3x - 4}$

45. $\dfrac{4a}{5} + a$

46. $z + \dfrac{1}{z}$

47. $\dfrac{1}{3x} + \dfrac{1}{4x}$

48. $\dfrac{5}{8y} - \dfrac{2}{12y}$

49. $\dfrac{2}{k^3} - \dfrac{3}{k^2} + \dfrac{5}{k}$

50. $\dfrac{16}{a^2b} + \dfrac{1}{ab} - \dfrac{6}{ab^2}$

51. $\dfrac{2x + 3}{5x} - \dfrac{2x - 1}{10x} + \dfrac{4}{x}$

52. $\dfrac{7}{n^2} - \dfrac{5n - 2}{n} + 6$

53. $\dfrac{2 - y}{9y + 6} + \dfrac{y - 2}{6y + 4}$

54. $\dfrac{2x + 3}{2x^3 - 4x^2} - \dfrac{1}{x - 2}$

55. $\dfrac{3a}{a - 3} - \dfrac{3}{a}$

56. $\dfrac{d}{b + d} - \dfrac{b}{b - d}$

57. $\dfrac{3}{2x - 3} - \dfrac{2}{9 - 4x^2}$

58. $3k + 1 + \dfrac{5}{k + 2}$

59. $\dfrac{n}{n - 1} - \dfrac{1}{n} + 1$

60. $\dfrac{4x}{x^2 + 2x - 3} - \dfrac{5x}{x^2 + 5x + 6}$

61. $\left(\dfrac{1}{x} + \dfrac{1}{y}\right) \div \left(\dfrac{x}{y} - \dfrac{y}{x}\right)$

62. $\left(4 + \dfrac{1}{a - 1}\right) \div \left(\dfrac{2}{a - 1} + 3\right)$

In Exercises 63–72 change the complex fraction to a fraction in simplest form.

63. $\dfrac{3 + \frac{1}{4}}{4 - \frac{2}{5}}$

64. $\dfrac{\frac{5}{18} + \frac{11}{12}}{\frac{7}{6} - \frac{2}{9}}$

65. $\dfrac{1 + (1/x)}{1 - (1/x)}$

66. $\dfrac{x + (1/a)}{(2/a) - x}$

67. $\dfrac{n + 1 + (2/n)}{n - 1 - (2/n)}$

68. $\dfrac{z + 4 - (5/z)}{z + 1 - (2/z)}$

69. $\left(\dfrac{y^2}{x^2} - 1\right) \div \left(\dfrac{y}{x} - 1\right)$

70. $\left(k - \dfrac{1}{x^2}\right) \div \left(2 - \dfrac{k}{x}\right)$

71. $\dfrac{1 + \dfrac{3}{x + 1}}{\dfrac{4}{x^2 - 1}}$

72. $\dfrac{\dfrac{b}{b - a} - \dfrac{a}{b + a}}{\dfrac{b^2 + a^2}{b^2 - a^2}}$

In Exercises 73–102 perform the indicated operations and/or simplify.

73. $\dfrac{x + y - z}{y^2 - x^2 + xz - yz}$

74. $\dfrac{a^3 - 8}{a^3 - 2a^2 + 4a - 8}$

75. $\dfrac{2(2 + h) - 3(2 + h)}{2h(2 + h)}$

76. $\dfrac{(2 + h)^2 - 4}{(2 + h) - 2}$

77. $\dfrac{(x^2 - x) - (a^2 - a)}{x - a}$

78. $\dfrac{(x^2 + 2x) - (a^2 + 2a)}{x - a}$

79. $\dfrac{(x + 1)^2 - x \cdot 2(x + 1)}{(x + 1)^4}$

80. $\dfrac{(x^2 - 1)^2(-2) - (-2x)(x^2 - 1)4x}{(x^2 - 1)^4}$

81. $\dfrac{x - y}{4a} \div \left(\dfrac{y - x}{a} \div \dfrac{x - y}{a^2}\right)$

82. $\left(\dfrac{x - y}{4a} \div \dfrac{y - x}{a}\right) \div \dfrac{x - y}{a^2}$

83. $(x - y)\left(2 - \dfrac{2x - y}{x - y}\right)$

84. $\left(\dfrac{x^2}{1 - x^2} + \dfrac{2x}{1 + x}\right) \div \dfrac{3}{x^2 - 1}$

85. $\dfrac{5}{y + 1} - \dfrac{1}{y^2 - 1} + \dfrac{2}{y - 1}$

86. $\dfrac{2a}{a^2 - 1} - \dfrac{a + 1}{a - 1} + 7$

87. $\dfrac{y}{y - x} - \dfrac{x}{x + y} + \dfrac{x^2 + y^2}{x^2 - y^2}$

88. $\dfrac{a}{a^2 + ab} - \dfrac{b}{a^2 - b^2} + \dfrac{a + b}{ab - b^2}$

89. $\dfrac{1}{h}\left(\dfrac{1}{x + h} - \dfrac{1}{x}\right)$

90. $\dfrac{1}{h}\left(\dfrac{2}{(x + h)^2} - \dfrac{2}{x^2}\right)$

91. $\left[\dfrac{x + h}{x + h + 1} - \dfrac{x}{x + 1}\right] \div h$

92. $\left[\dfrac{3}{(x + h)^2} - \dfrac{3}{x^2}\right] \div h$

93. $\dfrac{\dfrac{(x + h)^2 + 1}{x + h} - \dfrac{x^2 + 1}{x}}{h}$

94. $\dfrac{\dfrac{x + h - 4}{x + h + 1} - \dfrac{x - 4}{x + 1}}{h}$

95. $\dfrac{3P - \dfrac{(3 + \pi)3P}{6 + \pi}}{6}$

96. $\dfrac{4P - \dfrac{2P}{2 + \pi} - \dfrac{4P}{2 + \pi}}{2}$

97. $\dfrac{n + 1}{3^{n + 1}} \div \dfrac{n}{3^n}$

98. $\left(\dfrac{n + 2}{n + 1} \cdot \dfrac{1}{4^n}\right) \div \left(\dfrac{n + 1}{n} \cdot \dfrac{1}{4^{n - 1}}\right)$

99. $\dfrac{1}{x} - \dfrac{1}{a^x - 1}$ **100.** $\dfrac{2}{x^a} + \dfrac{3}{y^b}$

101. $\dfrac{1}{x^n} + \dfrac{2}{x^{n + 1}}$ **102.** $\dfrac{a}{x^{n - 1}} + \dfrac{b}{x^n}$

In Exercises 103–116 perform the indicated operations and write the result in the simplest form that contains only positive exponents.

103. $-(4 - x^2)^{-1}(8 - 4x)$

104. $2(x - y)^{-3} \div 3(2y - 2x)^{-2}$

105. $\dfrac{5x - 5}{2}\left(\dfrac{x - 1}{4}\right)^{-1}$

106. $\dfrac{3x + 6}{x^2 + 4x} \div \left(\dfrac{x^2 + 3x - 4}{x^2 - 4}\right)^{-1}$

107. $2 + 3y^{-1} - 4y^{-2}$

108. $(x - 4)^{-2} - 5(x - 4)^{-1}$

109. $4(3x + 1)^{-1} - (4x - 1)(3x + 1)^{-2}$

110. $(2x^{-3} - 3x^{-2} + x^{-1})(x^{-2} - x^0)$

111. $\dfrac{x^{-1} + y^{-1}}{(xy)^{-1}}$ **112.** $\dfrac{xy^{-1} + x^{-1}y}{x^{-1}y - xy^{-1}}$

113. $(x^{-1} + y^{-1})^{-1}$ **114.** $1 + (x + x^{-1})^{-1}$

115. $\left(\dfrac{a^x + a^{-x}}{2}\right)^2 - \left(\dfrac{a^x - a^{-x}}{2}\right)^2$

116. $1 - \left(\dfrac{a^x - a^{-x}}{a^x + a^{-x}}\right), \quad x > 0$

117. When resistors R_1 and R_2 are connected in parallel, their combined resistance is given by

$$\dfrac{1}{\dfrac{1}{R_1} + \dfrac{1}{R_2}}.$$

Change this complex fraction to a fraction in simplest form.

118. A lawyer drives from her home to work at an average speed of 30 mi/hour. On the way home she takes the same route and averages 20 mi/hour. What is her average speed for the round trip?

119. We can generalize the idea in Exercise 118. If we let s_g represent the speed going and s_h represent the speed coming back, then the following formula can be used to find the average speed for the round trip.

$$s_{\text{avg}} = \dfrac{2}{\dfrac{1}{s_g} + \dfrac{1}{s_h}}$$

Change this complex fraction to a fraction in simplest form.

120. a. The harmonic mean of two numbers a and b is given by

$$\dfrac{2}{\dfrac{1}{a} + \dfrac{1}{b}}.$$

Simplify this expression.

b. Verify that the average speed in Exercise 118 is the harmonic mean of 30 and 20.

c. A teacher invests $6,000 in a growth-oriented mutual fund at $10 a share. Six months later another $6,000 is invested in the same fund at $15 a share. Find the average-per-share purchase price and verify that it is the harmonic mean of 10 and 15.

THINK ABOUT IT

1. For what values of x does $\dfrac{3x + 9}{5x + 15} = \dfrac{3}{5}$?

2. A common student error is contained in the following "simplification." Find the mistake and discuss how to avoid such an error.

$$\frac{x^2}{x - 1} - \frac{2x + 1}{x - 1} = \frac{x^2 - 2x + 1}{x - 1} = \frac{(x - 1)^2}{x - 1}$$
$$= x - 1, \quad \text{if } x \neq 1.$$

3. Which statement is true for all values of the unknowns (for which it is defined)?

   a. $\dfrac{1}{n} - \dfrac{1}{n + 1} = \dfrac{1}{n(n + 1)}$

   b. $\dfrac{2x + y}{x + y} = 2$

   c. $\dfrac{1/a}{1/b} = \dfrac{1}{ab}$

   d. $\dfrac{1}{p} + \dfrac{1}{q} = \dfrac{2}{p + q}$

4. Simplify $\left(1 - \dfrac{1}{n}\right)\left(1 - \dfrac{1}{n + 1}\right)\left(1 - \dfrac{1}{n + 2}\right) \cdots$
$\left(1 - \dfrac{1}{n + 8}\right)\left(1 - \dfrac{1}{n + 9}\right).$

5. a. Show that if n is any real number, then
$$n = \left(\frac{n + 1}{2}\right)^2 - \left(\frac{n - 1}{2}\right)^2.$$

   b. Determine the fallacy in the following argument concerning prime numbers: From the result in part **a** we may conclude that every odd integer n greater than 1 is the difference of the squares of two positive integers a and b, where $a = (n + 1)/2$ and $b = (n - 1)/2$. Consequently,
$$n = a^2 - b^2 = (a + b)(a - b).$$
   Thus, n has positive integer factors $a + b$ and $a - b$, so there exists no odd positive integer greater than 1 that is a prime number.

REMEMBER THIS

1. Insert the correct order symbol ($<$, $>$, $=$).

   a. $\sqrt{9} + \sqrt{16} \underline{\quad} \sqrt{9 + 16}$ b. $\sqrt{3^2 + 4^2} \underline{\quad} 7$

2. Find the product of $a + b$ and $a - b$.

3. Simplify x^{-3}/x^{-5}.

4. Simplify $\left(\dfrac{x^2 y^{-3}}{x^{-1} y}\right)^{-2}$.

5. Simplify $-(x + a) + 3(x - a)$.

6. Expand $(x - 2)^3$.

7. Factor $x^3 - 8$.

8. Express $9.\overline{5}$ as the ratio of two integers.

9. Insert the correct order symbol ($<$, $>$, $=$):
$(\sqrt{4})^3 \underline{\quad} \sqrt{4^3}$.

10. Which number in the set $\{\sqrt{100}, \sqrt{1000}, \sqrt{-9}, -\sqrt{9}\}$ is an irrational number?

1.4 Rational Exponents and Radicals

In Section 1.1 we extended our definition of exponents to include zero and negative integers by defining them so they satisfied the laws of exponents that we developed for positive integers. We will now follow the same procedure to develop a consistent definition for rational exponents.

If the laws of exponents are to hold for rational numbers, consider the result of the following product:

$$9^{1/2} \cdot 9^{1/2} = 9^{1/2 + 1/2} = 9^1.$$

Here we have two equal factors whose product is 9. Since the square root of a number is one of its two equal factors, $9^{1/2}$ must mean the square root of 9. That is,

$$9^{1/2} = \sqrt{9}.$$

Similarly,

$$8^{1/3} \cdot 8^{1/3} \cdot 8^{1/3} = 8^{1/3 + 1/3 + 1/3} = 8.$$

The factor $8^{1/3}$ is one of three equal factors whose product is 8. Since the cube root of a number is one of its three equal factors, $8^{1/3}$ must mean the cube root of 8.

$$8^{1/3} = \sqrt[3]{8}$$

In general, our previous laws of exponents may be extended by making the following definitions.

Definitions of $a^{1/n}$ and $\sqrt[n]{a}$

For any positive integer n,

$$a^{1/n} = \sqrt[n]{a} \begin{cases} \text{for } a \geq 0, \text{ if } n \text{ is even} \\ \text{for any real number } a, \text{ if } n \text{ is odd} \end{cases}$$

where

$$\sqrt[n]{a} = b \text{ if and only if } b^n = a \begin{cases} \text{for } a \geq 0, b \geq 0, \text{ if } n \text{ is even} \\ \text{for any real number } a, \text{ if } n \text{ is odd.} \end{cases}$$

It is important to note the following ideas concerning the root of a number.

1. In the expression $\sqrt[n]{a}$, which is called a **radical,** we say $\sqrt{}$ is the **radical sign,** a is the **radicand,** and n is the **index** of the radical. The index is usually omitted from the square root radical.

2. There are two square roots for a positive number such as 4 since $(2)(2) = 4$ and $(-2)(-2) = 4$. To avoid ambiguity, we define the **principal square root** of a positive number to be its *positive* square root. Thus, the principal square root of 4 is 2 and not -2. The radical sign, $\sqrt{}$, is used to symbolize the principal square root, so $\sqrt{4} = 2$. To symbolize the negative square root of 4 (which is -2), we use $-\sqrt{4}$. In general, when n is even, $\sqrt[n]{a}$ means the nonnegative nth root of a.

3. No real number is the square root of a negative number such as -4 since the product of two equal real numbers is never negative. In general, when n is even, the nth root of a negative number does not exist in the set of real numbers.

4. The square root of any positive number that is not a perfect square is an irrational number. Consequently, if we wish to express the square root as a decimal, we can only *approximate* the number to some desired number of significant digits. In most cases we will leave these numbers in radical form.

EXAMPLE 1 Simplify.

a. $9^{1/2}$ **b.** $8^{1/3}$ **c.** $(-8)^{1/3}$ **d.** $-8^{1/3}$ **e.** $16^{-1/4}$

Solution

a. $9^{1/2} = \sqrt{9} = 3$ **b.** $8^{1/3} = \sqrt[3]{8} = 2$

c. $(-8)^{1/3} = \sqrt[3]{-8} = -2$ **d.** $-8^{1/3} = -\sqrt[3]{8} = -2$

e. $16^{-1/4} = \dfrac{1}{16^{1/4}} = \dfrac{1}{\sqrt[4]{16}} = \dfrac{1}{2}$ ■

When the numerator in the rational exponent is not 1, we can evaluate the expression by considering the following law of exponents:

$$(a^m)^n = a^{mn}.$$

For example, consider

$$9^{3/2}.$$

Rewrite the expression in either of these forms and simplify.

$$
\begin{array}{ccc}
(9^{1/2})^3 & \text{or} & (9^3)^{1/2} \\
= (\sqrt{9})^3 & \text{or} & \sqrt{9^3} \\
= (3)^3 & \text{or} & \sqrt{729} \\
= 27 & &
\end{array}
$$

Thus, $9^{3/2} = 27$. (*Note:* Although either form may be used, you should notice that if the root results in a rational value, it is easier to find the root first.) This example leads us to the following definition.

Rational Exponent

If m and n are integers with $n > 0$, and if m/n represents a reduced fraction such that $a^{1/n}$ represents a real number, then

$$a^{m/n} = (a^{1/n})^m = (a^m)^{1/n}$$

or equivalently

$$a^{m/n} = (\sqrt[n]{a})^m = \sqrt[n]{a^m}.$$

Notice that in this definition we avoid the situation where the base is negative and the exponent is a reduced fraction in which n is an even number. This is done because such a condition deals with numbers that are not real numbers and, in such a case, the laws of exponents do not necessarily hold.

EXAMPLE 2 Simplify.

a. $8^{2/3}$ **b.** $49^{3/2}$ **c.** $(-32)^{4/5}$ **d.** $125^{-2/3}$

Solution

a. $8^{2/3} = (8^{1/3})^2 = (\sqrt[3]{8})^2 = (2)^2 = 4$

b. $49^{3/2} = (49^{1/2})^3 = (\sqrt{49})^3 = (7)^3 = 343$

c. $(-32)^{4/5} = [(-32)^{1/5}]^4 = (\sqrt[5]{-32})^4 = (-2)^4 = 16$

d. $125^{-2/3} = \dfrac{1}{125^{2/3}} = \dfrac{1}{(125^{1/3})^2} = \dfrac{1}{(\sqrt[3]{125})^2} = \dfrac{1}{(5)^2} = \dfrac{1}{25}$ ■

EXAMPLE 3   Use the laws of exponents to simplify.

a. $6^{7/5} \cdot 6^{8/5}$ **b.** $(a^8)^{-3/16}$ **c.** $\sqrt{2}/\sqrt[3]{2}$ **d.** $(x^{1/a})^{a/3}$

Solution

a. $6^{7/5} \cdot 6^{8/5} = 6^{7/5 + 8/5} = 6^{15/5} = 6^3 = 216$

b. $(a^8)^{-3/16} = a^{8(-3/16)} = a^{-3/2} = 1/a^{3/2}$ or $1/\sqrt{a^3}$

c. $\sqrt{2}/\sqrt[3]{2} = 2^{1/2}/2^{1/3} = 2^{1/2 - 1/3} = 2^{1/6}$ or $\sqrt[6]{2}$

d. $(x^{1/a})^{a/3} = x^{(1/a)(a/3)} = x^{1/3}$ or $\sqrt[3]{x}$ ■

EXAMPLE 4   Simplify by performing the indicated operations and writing your results with only positive exponents. Also, express the result in radical form.

a. $\dfrac{2^{-1/2} \cdot 3x^{1/3}}{2 \cdot 3^{1/4}x}$ **b.** $\left(\dfrac{4x^{-1}y^{1/5}}{32x^5y^0}\right)^{1/3}$

Solution

a. $\dfrac{2^{-1/2} \cdot 3x^{1/3}}{2 \cdot 3^{1/4}x} = \dfrac{3^{1 - 1/4}}{2^{1 - (-1/2)}x^{1 - (1/3)}}$

$= \dfrac{3^{3/4}}{2^{3/2}x^{2/3}}$ or $\dfrac{\sqrt[4]{3^3}}{\sqrt{2^3}\,\sqrt[3]{x^2}}$

b. $\left(\dfrac{4x^{-1}y^{1/5}}{32x^5y^0}\right)^{1/3} = \left(\dfrac{y^{1/5}}{8x^{5 - (-1)} \cdot 1}\right)^{1/3}$

$= \left(\dfrac{y^{1/5}}{8x^6}\right)^{1/3}$

$= \dfrac{y^{(1/5)(1/3)}}{8^{1(1/3)}x^{6(1/3)}}$

$= \dfrac{y^{1/15}}{2x^2}$ or $\dfrac{\sqrt[15]{y}}{2x^2}$ ■

EXAMPLE 5   Simplify by performing the indicated operations and writing your results with only positive exponents. Also, express the result in radical form.

a. $\dfrac{x^{1/2} \cdot 2 - (2x - 1) \cdot \frac{1}{2}x^{-1/2}}{x}$ **b.** $\frac{1}{2}x^{-1/2} - x^{-3/2}$

Solution

a. The main problem is how to take care of the expression $\frac{1}{2}x^{-1/2}$ or $1/(2x^{1/2})$. One way to simplify this expression is to multiply it by $2x^{1/2}$, since the resulting product is 1. Thus, in this case we will multiply both the numerator and the denominator of the given fraction by $2x^{1/2}$.

$$\left(\frac{2x^{1/2}}{2x^{1/2}}\right)\left(\frac{x^{1/2}\cdot 2 - (2x-1)\cdot\frac{1}{2}x^{-1/2}}{x}\right) = \frac{4x - (2x-1)(1)}{2x^{3/2}}$$

$$= \frac{2x+1}{2x^{3/2}} \text{ or } \frac{2x+1}{2\sqrt{x^3}}$$

b. $\frac{1}{2}x^{-1/2} - x^{-3/2} = \dfrac{1}{2x^{1/2}} - \dfrac{1}{x^{3/2}}$

$$= \frac{1}{2x^{1/2}}\left(\frac{x}{x}\right) - \frac{1}{x^{3/2}}\left(\frac{2}{2}\right)$$

$$= \frac{x}{2x^{3/2}} - \frac{2}{2x^{3/2}}$$

$$= \frac{x-2}{2x^{3/2}} \text{ or } \frac{x-2}{2\sqrt{x^3}} \qquad ∎$$

Expressions with negative exponents are often simplified by factoring out from each term the smallest power of the variable along with other common numerical factors, as discussed in Section 1.2. The next example gives a simple illustration of this technique in part **a** and then provides an alternative solution to Example 5b by using this method in part **b**.

EXAMPLE 6   Simplify each expression by factoring and then writing the result with only positive exponents.

a. $10x^{-4} + 15x^{-2}$ **b.** $\frac{1}{2}x^{-1/2} - x^{-3/2}$

Solution

a. The smallest power of x is x^{-4} and the GCF of 10 and 15 is 5. Thus, we may simplify by factoring out $5x^{-4}$ as follows:

$$10x^{-4} + 15x^{-2} = 5x^{-4}\cdot 2 + 5x^{-4}\cdot 3x^2$$
$$= 5x^{-4}(2 + 3x^2)$$
$$= \frac{5(2 + 3x^2)}{x^4}.$$

b. In this case $x^{-3/2}$ is the smallest power of x, so

$$\tfrac{1}{2}x^{-1/2} - x^{-3/2} = x^{-3/2} \cdot \tfrac{1}{2}x - x^{-3/2} \cdot 1$$
$$= x^{-3/2}(\tfrac{1}{2}x - 1)$$
$$= \frac{\tfrac{1}{2}x - 1}{x^{3/2}}$$
$$= \frac{x - 2}{2x^{3/2}} \quad \text{or} \quad \frac{x - 2}{2\sqrt{x^3}}. \qquad \blacksquare$$

Although it is possible to use fractional exponents for any operation required with radicals, computations are often performed by use of radical symbolism. For this reason we now state three essential properties of radicals.

Properties of Radicals

> (a, b, $\sqrt[n]{a}$, and $\sqrt[n]{b}$ denote real numbers.)
> **1.** $(\sqrt[n]{a})^n = a$ (by definition of $\sqrt[n]{a}$)
> **2.** $\sqrt[n]{a} \cdot \sqrt[n]{b} = a^{1/n} \cdot b^{1/n} = (a \cdot b)^{1/n} = \sqrt[n]{a \cdot b}$
> **3.** $\dfrac{\sqrt[n]{a}}{\sqrt[n]{b}} = \dfrac{a^{1/n}}{b^{1/n}} = \left(\dfrac{a}{b}\right)^{1/n} = \sqrt[n]{\dfrac{a}{b}}$ ($b \neq 0$)

With the aid of these properties we can now use radicals or rational exponents, whichever is more convenient. When working with radicals, we must be able to change radicals to simplest form, which is accomplished as follows.

Simplifying Radicals

> To simplify a radical:
> **1.** Remove any factor of the radicand whose indicated root can be taken exactly.
> **2.** Reduce the index of the radical as far as possible.
> **3.** Eliminate any fractions in the radicand and all radicals from any denominators. (This procedure is called "rationalizing the denominator.")

EXAMPLE 7   Express each radical in simplest form. Assume that x and y denote positive numbers.

a. $\sqrt{8x^9y^6}$

b. $\sqrt[6]{x^4}$

c. $\sqrt[3]{\tfrac{7}{9}}$

d. $\sqrt{2x/y}$

Solution

a. Rewrite $8x^9y^6$ as the product of a perfect square and another factor and simplify. Remember that even powers are perfect squares.

$$\sqrt{8x^9y^6} = \sqrt{(4x^8y^6)(2x)} = \sqrt{4x^8y^6}\sqrt{2x} = 2x^4y^3\sqrt{2x}$$

b. Express $\sqrt[6]{x^4}$ in exponential form, reduce the rational exponent, and then convert back to radical form.

$$\sqrt[6]{x^4} = x^{4/6} = x^{2/3} = \sqrt[3]{x^2}$$

c. Rewrite $\frac{7}{9}$ as an equivalent fraction whose denominator is a perfect cube and simplify.

$$\sqrt[3]{\frac{7}{9}} = \sqrt[3]{\frac{7 \cdot 3}{9 \cdot 3}} = \sqrt[3]{\frac{21}{27}} = \frac{\sqrt[3]{21}}{\sqrt[3]{27}} = \frac{\sqrt[3]{21}}{3}$$

d. Rewrite $2x/y$ as an equivalent fraction whose denominator is a perfect square and simplify.

$$\sqrt{\frac{2x}{y}} = \sqrt{\frac{2x \cdot y}{y \cdot y}} = \sqrt{\frac{2xy}{y^2}} = \frac{\sqrt{2xy}}{\sqrt{y^2}} = \frac{\sqrt{2xy}}{y}$$ ∎

In Example 7 we assumed x and y denoted positive numbers, which meant expressions like $\sqrt{y^2}$ simplified to y. Without this assumption it is necessary to use $\sqrt{y^2} = |y|$. For example, $\sqrt{(-2)^2} \neq -2$; instead $\sqrt{(-2)^2} = |-2| = 2$. Keep in mind that while $(\sqrt[n]{a})^n = a$ for any positive integer n,

$$\sqrt[n]{a^n} = \begin{cases} a, & \text{if } n \text{ is odd} \\ |a|, & \text{if } n \text{ is even.} \end{cases}$$

In the remaining examples we will continue to assume variables represent positive numbers, and the next example shows how to operate on radicals.

EXAMPLE 8   Perform the indicated operations and simplify. Assume all variables represent positive real numbers.

a. $2\sqrt{32} + 4\sqrt{\frac{1}{2}}$ **b.** $\sqrt{50x^3y} - \sqrt{8xy^3}$

c. $4\sqrt[3]{6x} \cdot 3\sqrt[3]{3x}$ **d.** $\sqrt[3]{48} \div \sqrt[3]{2}$

e. $(3\sqrt{2} + \sqrt{3})(3\sqrt{2} - \sqrt{3})$

Solution   Use the properties of radicals developed in this section together with the methods for operating on polynomials.

a. $2\sqrt{32} = 2\sqrt{16 \cdot 2} = 2\sqrt{16}\sqrt{2} = 2 \cdot 4\sqrt{2} = 8\sqrt{2}$

$$4\sqrt{\frac{1}{2}} = 4\sqrt{\frac{1 \cdot 2}{2 \cdot 2}} = 4\sqrt{\frac{2}{4}} = \frac{4\sqrt{2}}{\sqrt{4}} = \frac{4\sqrt{2}}{2} = 2\sqrt{2}$$

Then, $2\sqrt{32} - 4\sqrt{\frac{1}{2}} = 8\sqrt{2} - 2\sqrt{2} = (8 - 2)\sqrt{2} = 6\sqrt{2}$.

b. $\sqrt{50x^3y} = \sqrt{(25x^2)(2xy)} = \sqrt{25x^2}\sqrt{2xy} = 5x\sqrt{2xy}$

$\sqrt{8xy^3} = \sqrt{(4y^2)(2xy)} = \sqrt{4y^2}\sqrt{2xy} = 2y\sqrt{2xy}$

Then, $\sqrt{50x^3y} - \sqrt{8xy^3} = 5x\sqrt{2xy} - 2y\sqrt{2xy} = (5x - 2y)\sqrt{2xy}$.

c. $4\sqrt[3]{6x} \cdot 3\sqrt[3]{3x} = 4 \cdot 3\sqrt[3]{6x \cdot 3x} = 12\sqrt[3]{18x^2}$

d. $\dfrac{\sqrt[3]{48}}{\sqrt[3]{2}} = \sqrt[3]{\dfrac{48}{2}} = \sqrt[3]{24} = \sqrt[3]{8} \cdot \sqrt[3]{3} = 2\sqrt[3]{3}$

e. $(3\sqrt{2} + \sqrt{3})(3\sqrt{2} - \sqrt{3}) = 9 \cdot 2 - 3\sqrt{6} + 3\sqrt{6} - 3 = 15$ ■

We have discussed how to simplify a radical so the radicand contains no fractions. A similar procedure may be used to ensure that no radicals appear in a denominator. For example, $1/\sqrt{2}$ can be converted to $\sqrt{2}/2$ as follows.

$$\frac{1}{\sqrt{2}} = \frac{1}{\sqrt{2}} \cdot \frac{\sqrt{2}}{\sqrt{2}} = \frac{\sqrt{2}}{2}$$

One instance when the latter form is more useful occurs with the addition of radicals, as shown below.

$$3\sqrt{2} + \frac{1}{\sqrt{2}} = 3\sqrt{2} + \frac{\sqrt{2}}{2} = \frac{7}{2}\sqrt{2}$$

Thus, we adopt the condition that a simplified radical cannot have radicals in the denominator. When we change the denominator from a radical (irrational number) to a rational number, the procedure is called rationalizing the denominator.

EXAMPLE 9   Rationalize the denominator.

a. $\dfrac{2}{\sqrt{5}}$ **b.** $\dfrac{7}{4\sqrt{3}}$ **c.** $\dfrac{9}{2\sqrt{18}}$ **d.** $\dfrac{9}{\sqrt{3x}}$

Solution

a. $\dfrac{2}{\sqrt{5}} = \dfrac{2\sqrt{5}}{\sqrt{5}\sqrt{5}} = \dfrac{2\sqrt{5}}{5}$

b. $\dfrac{7}{4\sqrt{3}} = \dfrac{7\sqrt{3}}{4\sqrt{3}\sqrt{3}} = \dfrac{7\sqrt{3}}{4 \cdot 3} = \dfrac{7\sqrt{3}}{12}$

c. $\dfrac{9}{2\sqrt{18}} = \dfrac{9\sqrt{2}}{2\sqrt{18}\sqrt{2}} = \dfrac{9\sqrt{2}}{2\sqrt{36}} = \dfrac{9\sqrt{2}}{2 \cdot 6} = \dfrac{3\sqrt{2}}{4}$

d. $\dfrac{9}{\sqrt{3x}} = \dfrac{9\sqrt{3x}}{\sqrt{3x}\sqrt{3x}} = \dfrac{9\sqrt{3x}}{3x} = \dfrac{3\sqrt{3x}}{x}$ ■

If the denominator contains square roots and is the sum of two terms, we rationalize the denominator by multiplying the numerator and the denominator of the fraction by the difference of the same two terms, and vice versa. This procedure is effective since if a and b are nonnegative rational numbers, then

$$(\sqrt{a} + \sqrt{b})(\sqrt{a} - \sqrt{b}) = a - \sqrt{ab} + \sqrt{ab} - b = a - b,$$

which is a rational number. The expressions $\sqrt{a} + \sqrt{b}$ and $\sqrt{a} - \sqrt{b}$ are called **conjugates** of each other.

EXAMPLE 10   Rationalize the denominator: $\dfrac{2 + \sqrt{3}}{4\sqrt{3} - \sqrt{2}}$.

Solution   The conjugate of $4\sqrt{3} - \sqrt{2}$ is $4\sqrt{3} + \sqrt{2}$. Then,

$$\frac{2 + \sqrt{3}}{4\sqrt{3} - \sqrt{2}} = \frac{(2 + \sqrt{3})(4\sqrt{3} + \sqrt{2})}{(4\sqrt{3} - \sqrt{2})(4\sqrt{3} + \sqrt{2})}$$

$$= \frac{8\sqrt{3} + 2\sqrt{2} + 4 \cdot 3 + \sqrt{6}}{16 \cdot 3 + 4\sqrt{6} + (-4\sqrt{6}) + (-2)}$$

$$= \frac{8\sqrt{3} + 2\sqrt{2} + 12 + \sqrt{6}}{46}.$$ ∎

In some cases, instead of rationalizing the denominator, it is useful to **rationalize the numerator.** The idea is the same and the following example illustrates the procedure.

EXAMPLE 11   Rationalize the numerator: $\dfrac{\sqrt{x} - \sqrt{a}}{x - a}$.

Solution   The conjugate of $\sqrt{x} - \sqrt{a}$ is $\sqrt{x} + \sqrt{a}$. Then,

$$\frac{\sqrt{x} - \sqrt{a}}{x - a} = \frac{(\sqrt{x} - \sqrt{a})(\sqrt{x} + \sqrt{a})}{(x - a)(\sqrt{x} + \sqrt{a})}$$

$$= \frac{x - a}{(x - a)(\sqrt{x} + \sqrt{a})} = \frac{1}{\sqrt{x} + \sqrt{a}}.$$ ∎

Now that we have shown how to perform the basic operations with radicals, we consider a problem that requires several of the ideas discussed. It is essential that you understand this type of problem if you intend to continue to higher mathematics.

EXAMPLE 12   Write as a single fraction in simplest form

$$x \cdot \frac{2x}{2\sqrt{x^2 + 5}} + \sqrt{x^2 + 5}.$$

Solution

$$x \cdot \frac{2x}{2\sqrt{x^2 + 5}} + \sqrt{x^2 + 5} = \frac{x^2}{\sqrt{x^2 + 5}} + \sqrt{x^2 + 5}\left(\frac{\sqrt{x^2 + 5}}{\sqrt{x^2 + 5}}\right)$$

$$= \frac{x^2 + (x^2 + 5)}{\sqrt{x^2 + 5}}$$

$$= \frac{2x^2 + 5}{\sqrt{x^2 + 5}} \quad \text{or} \quad \frac{(2x^2 + 5)\sqrt{x^2 + 5}}{x^2 + 5}$$ ∎

EXERCISES 1.4

In Exercises 1–10 calculate the expression.

1. $8^{2/3} - 2^{-1} + 4^0$

2. $(\frac{1}{9})^{-2} + 2(3)^0 - 27^{2/3}$

3. $4(16)^{-1/2} + (\frac{1}{64})^{-2/3} - 16^0$

4. $(3^2 + 4^2)^{-1/2}$

5. $2^{1/2} \cdot 2^{3/2}$ **6.** $(-4)^{7/5} \cdot (-4)^{8/5}$

7. $3^{1/2} \cdot 27^{1/2}$ **8.** $2^{1/3} \cdot 4^{1/3}$

9. $\dfrac{5^{1/2}}{5^{3/2}}$ **10.** $\dfrac{6^{1/5} \cdot 6^{3/5}}{6^{-1/5}}$

In Exercises 11–50 simplify each expression by performing the indicated operations and writing your result with only positive exponents. Also, show your result in radical form. (*Note:* When the denominator in a fractional exponent is even, assume that the base is a positive number.)

11. $x^{4/3} \cdot x^{-1}$ **12.** $x^{1/2} \cdot x^{2/3}$ **13.** $\dfrac{x^{-1/3}}{x^{5/6}}$

14. $\dfrac{x^{-1/4}}{x^{1/2}}$ **15.** $(8x^3y^6)^{1/3}$ **16.** $(3^{1/2}x^{3/4})^{-2}$

17. $(4a^{-2}b^6)^{-1/2}$ **18.** $\left(\dfrac{27x^{-3}}{64y^{-3}}\right)^{-1/3}$

19. $\left(\dfrac{49x^{-4}y^6}{25x^{-2}y^{-10}}\right)^{-3/2}$ **20.** $\left(\dfrac{x^{1/4}y^{-2/3}z^0}{x^{3/4}y^{-2/3}z^{1/2}}\right)^{-3/2}$

21. $\dfrac{3x^{1/2}y^{-2/3}}{5y^{-1}} \cdot \dfrac{20y^{1/4}}{27x^{-2}}$

22. $\dfrac{7^{-1}x^{1/4}}{3^{-2}y^{-1}} \div \dfrac{2(3x^{-1/2})^2}{(49y)^0}$ **23.** $\sqrt[3]{2} \cdot \sqrt[4]{2}$

24. $\sqrt[3]{3} \cdot \sqrt{3}$ **25.** $\dfrac{16}{\sqrt[3]{16}}$

26. $\dfrac{\sqrt[4]{2}}{\sqrt[8]{16}}$ **27.** $\sqrt{3\sqrt{3}}$

28. $\sqrt[4]{\sqrt[3]{5}}$ **29.** $\sqrt[m]{\sqrt[n]{7}}$

30. $\sqrt[m]{3} \cdot \sqrt[n]{3}$ **31.** $\sqrt[3]{2} \cdot \sqrt[n]{3}$

32. $\sqrt{2} \div \sqrt[4]{5}$ **33.** $(x^{a/2}y^{a/5})^{1/a}$

34. $(a^{3/n})^{n/6}(a^{n/2})^{4/n}$ **35.** $k^{1/x}k^{-y/x}$

36. $(x^{1/q})^{1-q}$ **37.** $(x^{1/q})^p - {}^1x^{(1/q)-1}$

38. $(a^{x^2 + 2x} \cdot a)^{1/(x+1)}$ **39.** $(x^{1/2} - x^{-1/2})^2$

40. $(x^{2/3} + x^{-2/3})^2$ **41.** $(x^{1/2} + y^{1/2})^2$

42. $(a^{1/2} - b^{3/2})^2$ **43.** $x^{-1/2} - \frac{1}{2}x^{-3/2}$

44. $\frac{2}{3}x^{-1/3} + \frac{1}{3}x^{-4/3}$ **45.** $\dfrac{x^{1/3} + x^{-2/3}}{x^{1/3} - x^{-2/3}}$

46. $\dfrac{x^{1/2}}{2x^{-3/2} + 3x^{-1/2}}$

47. $(x + 2)(-\frac{1}{2}x^{-3/2}) + x^{-1/2} \cdot 1$

48. $x^{1/2} \cdot [-(x + 2)^{-2}] + (x + 2)^{-1} \cdot \frac{1}{2}x^{-1/2}$

49. $\dfrac{x^{1/2} \cdot 1 - (x + 2) \cdot \frac{1}{2}x^{-1/2}}{x}$

50. $\dfrac{1 + \frac{1}{2}(1 + x^2)^{-1/2}(2x)}{x + (1 + x^2)^{1/2}}$

In Exercises 51–62 express each radical in simplest form (assume that x and y represent positive real numbers).

51. $\sqrt{12x^7y^4}$ **52.** $\sqrt{54x^{11}y^5}$ **53.** $\sqrt[4]{9x^6y^7}$

54. $\sqrt[5]{16x^3y^5}$ **55.** $\sqrt[4]{25}$ **56.** $\sqrt[9]{8}$

57. $\sqrt[6]{81y^8}$ **58.** $\sqrt[4]{4x^{10}}$ **59.** $\sqrt{9/x}$

60. $\sqrt{5/(3x^2y)}$ **61.** $\sqrt[3]{x/y}$ **62.** $\sqrt[4]{2x/y^5}$

In Exercises 63–80 express each sum or difference in simplest form (assume that x and y represent positive real numbers).

63. $3\sqrt{72} - 5\sqrt{50}$ **64.** $3\sqrt{28} - 10\sqrt{63}$

65. $2\sqrt{50} - 6\sqrt{\frac{1}{2}}$ **66.** $4\sqrt{\frac{1}{5}} + 2\sqrt{125}$

67. $6\sqrt{\frac{1}{6}} - \sqrt{\frac{2}{3}}$ **68.** $5\sqrt{\frac{1}{3}} - 3\sqrt{\frac{1}{12}}$

69. $\sqrt[3]{\frac{1}{9}} + 2\sqrt[3]{-375}$ **70.** $3\sqrt[4]{32} - 5\sqrt[4]{2}$

71. $\sqrt{8x} + \sqrt{72x}$ **72.** $9\sqrt{3y} - 4\sqrt{12y}$

73. $\sqrt{20xy^2} - 2\sqrt{45x^3}$ **74.** $\sqrt{x^3y/2} + \frac{1}{2}\sqrt{32xy^7}$

75. $\sqrt{50x^3y} + x\sqrt{xy/2} - 2\sqrt{9x^3y/2}$

76. $2\sqrt{x/y} - \sqrt{y/x} + 3\sqrt{1/(xy)}$

77. $2\sqrt{2x/y} - 4\sqrt{y/(2x^3)} + 5\sqrt{\frac{1}{8}x^3y}$

78. $\sqrt[3]{54x^4y} - \sqrt[3]{xy^4/4}$ **79.** $\sqrt[4]{x/y} - \sqrt[4]{xy^3}$

80. $\sqrt{4 + 4x} + \sqrt{16 + 16x}$

In Exercises 81–92 express each product or quotient in simplest form (assume that x and y represent positive real numbers).

81. $(\sqrt{15} + \sqrt{60}) \div \sqrt{3}$ **82.** $-3\sqrt{3}(4 - 2\sqrt{27})$

83. $(1 + \sqrt{2})(1 - \sqrt{2})$ **84.** $(1 + \sqrt{2})^2$

85. $\sqrt{15x} \cdot \sqrt{6x}$ **86.** $\sqrt{5x^3y} \cdot \sqrt{10y}$

87. $\sqrt[3]{4x^2y} \cdot \sqrt[3]{6x^2y^3}$ **88.** $4\sqrt[4]{27x^3} \cdot 5\sqrt[4]{3x^5}$

89. $\sqrt{18} \div \sqrt{3x}$ **90.** $-2\sqrt{54x^3y} \div 4\sqrt{3xy}$

91. $\sqrt[4]{7x^2} \div \sqrt[4]{2x}$ **92.** $\sqrt[3]{-3y} \div \sqrt[3]{48y}$

In Exercises 93–102 rationalize the denominator.

93. $\dfrac{2}{\sqrt{3}}$ **94.** $\dfrac{7}{2\sqrt{18}}$ **95.** $\dfrac{1 - \sqrt{3}}{\sqrt{3}}$

96. $\dfrac{\sqrt{3}}{1 - \sqrt{3}}$ **97.** $\dfrac{2 - \sqrt{2}}{2 + \sqrt{2}}$ **98.** $\dfrac{\sqrt{3} + \sqrt{5}}{2\sqrt{3} - 7\sqrt{5}}$

99. $\dfrac{\sqrt{x}}{\sqrt{x} + \sqrt{y}}$ **100.** $\dfrac{2\sqrt{y}}{x - 2\sqrt{y}}$ **101.** $\dfrac{x - 1}{\sqrt{x^2 - 1}}$

102. $\dfrac{x}{\sqrt{2x^2 + x}}$

In Exercises 103–108 rationalize the numerator.

103. $\dfrac{\sqrt{x} - 2}{x - 4}$ **104.** $\dfrac{\sqrt{x} - \sqrt{9}}{x - 9}$

105. $\dfrac{\sqrt{x + h} - \sqrt{x}}{h}$ **106.** $\dfrac{\sqrt{1 + h} - 1}{h}$

107. $\dfrac{\sqrt{2x + 2h + 1} - \sqrt{2x + 1}}{h}$

108. $\dfrac{\sqrt{x + h - 1} - \sqrt{x - 1}}{h}$

In Exercises 109–118 simplify each expression by factoring out the smallest power of the variable from each term and writing your result with only positive exponents.

109. $x^{-5} + x^{-2}$ **110.** $x^{-8} + 3x^{-4}$
111. $12x^{-1} + 8x^{-5}$ **112.** $5x^{-2} - 10x^{-1}$
113. $3x^{1/3} - 15x^{-2/3}$ **114.** $4x^{-1/3} + 12x^{-4/3}$
115. $2x + \frac{1}{2}x^{-1/2}$ **116.** $x^2 - \frac{1}{3}x^{-2/3}$
117. $x^2(x - 5)^{1/2} - x(x - 5)^{-1/2}$
118. $-x^4(1 - x^2)^{-1/2} + 3x^2(1 - x^2)^{1/2}$

In Exercises 119–130 write each expression as a single fraction in simplest form (assume that all variables represent positive real numbers).

119. $600\sqrt{5} + \dfrac{360,000}{120\sqrt{5}}$ **120.** $900\sqrt{2} + \dfrac{540,000}{300\sqrt{2}}$

121. $\dfrac{\sqrt{3/x}(1 + 3/x)}{\sqrt{3}/(2x^{3/2})}$

122. $-\dfrac{1}{2}\sqrt{\dfrac{x}{y}} \cdot \dfrac{x\sqrt{y/x} - y}{x^2}$

123. $1 - \dfrac{1}{2\sqrt{x + 1}}$ **124.** $\dfrac{u}{2\sqrt{u + 1}} + \sqrt{u + 1}$

125. $a \cdot \dfrac{a}{\sqrt{a^2 + b^2}} + b \cdot \dfrac{b}{\sqrt{a^2 + b^2}}$

126. $\dfrac{2x}{2\sqrt{x^2 - 1}} - \dfrac{1}{x\sqrt{x^2 - 1}}$

127. $\pi\left(h \cdot \dfrac{-2h}{2\sqrt{4R^2 - h^2}} + \sqrt{4R^2 - h^2}\right)$

128. $\sqrt{R^2 - x^2} + (R + x)\dfrac{-x}{\sqrt{R^2 - x^2}}$

129. $\dfrac{x \cdot \dfrac{2x}{2\sqrt{x^2 + 1}} - \sqrt{x^2 + 1}}{x^2}$

130. $\dfrac{\sqrt{1 - x^2} - x \cdot \dfrac{-x}{\sqrt{1 - x^2}}}{1 - x^2}$

131. Evaluate $x^2 - 4x - 3$ if $x = 2 + \sqrt{7}$.
132. Evaluate $x^2 - 4x - 3$ if $x = 2 - \sqrt{7}$.
133. Evaluate $t^2 - 10t - 2$ if $t = 5 - 2\sqrt{7}$.
134. Evaluate $t^2 - 10t - 2$ if $t = 5 + 2\sqrt{7}$.

In Exercises 135 and 136 simplify each expression after setting

$$x_1 = \dfrac{-b + \sqrt{b^2 - 4ac}}{2a} \quad \text{and} \quad x_2 = \dfrac{-b - \sqrt{b^2 - 4ac}}{2a}.$$

135. $x_1 + x_2$ **136.** $x_1 \cdot x_2$

A useful definition for proving statements about absolute value is $|a| = \sqrt{a^2}$. Use this definition in Exercises 137 and 138.

137. Show that $|ab| = |a| \cdot |b|$.

138. Show that $\left|\dfrac{a}{b}\right| = \dfrac{|a|}{|b|}$.

THINK ABOUT IT

1. State a property that may be used to simplify $\sqrt[m]{\sqrt[n]{a}}$. Then apply this property to simplify $\sqrt[5]{\sqrt[4]{2}}$.
2. State a property that may be used to simplify $\sqrt[m]{a}\,\sqrt[n]{a}$. Then apply this property to simplify $\sqrt[5]{3} \cdot \sqrt[4]{3}$.
3. Determine the fallacy in the following argument:
   $(a^m)^{1/n} = a^{m/n}$ implies $[(-5)^2]^{1/2} = (-5)^1 = -5$,
   but $[(-5)^2]^{1/2} = (25)^{1/2} = \sqrt{25} = 5$.
   Therefore $-5 = 5$.

4. Use the product model
   $(a - b)(a^2 + ab + b^2) = a^3 - b^3$
   to rationalize the numerator: $\dfrac{\sqrt[3]{x + h} - \sqrt[3]{x}}{h}$.

5. When drawing a rectangle, what should be the ratio of the length to the width to achieve the most satisfying visual effect? The answer is obviously subjective, but there is a leading candidate for the title of most pleasing. This rectangle is called the **golden rectangle,**

and the ratio of the length to the width in this figure is called the **golden ratio.** Consider the buildings sketched within rectangles whose side ratios are indicated. Which figure appeals the most to you?

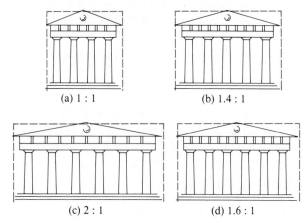

(a) 1 : 1   (b) 1.4 : 1

(c) 2 : 1   (d) 1.6 : 1

If you selected figure (d) you are in good company, for this figure has the same side ratios as the Parthenon in Athens. This fact and other examples of the golden rectangle in art and architecture are nicely illustrated in the Time-Life book by David Bergamini entitled *Mathematics* (1963, 1970). The ratio 1.6:1 (or simply 1.6) is actually an approximation of the golden ratio. The exact number can be determined by

considering the simple geometric method used to construct a golden rectangle, which is shown below. We start with square *ABCD* and extend this square to form golden rectangle *ABEF* by drawing arc *CF*, which is centered at the midpoint *M* of *AD*.

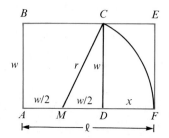

a. Show in rectangle *ABEF* that $\ell/w = (1 + \sqrt{5})/2$, which is the exact value of the golden ratio. [*Hint:* Note that $\ell = (w/2) + r$.]

b. Show that $w/x = (1 + \sqrt{5})/2$, which means that rectangle *DCEF* is also a golden rectangle.

c. Subtracting 1 from the golden ratio yields its reciprocal. Verify this.

d. In some literature the golden ratio is defined as w/ℓ instead of ℓ/w. Using this definition, what is the exact value of the golden ratio in simplest form?

REMEMBER THIS

1. Which numbers in the set $\{\sqrt{9}, \sqrt{-9}, -\sqrt{9}, \sqrt{12}\}$ are rational numbers? Which numbers in this set are real numbers?

2. True or false: Every real number is either a rational number or an irrational number.

3. Evaluate y^2 if $y = \sqrt{83}$.

4. Subtract $\dfrac{2x}{x^2 - 1} - \dfrac{x + 1}{x - 1}$.

5. Simplify $\dfrac{a^{-1}}{x^{-1} - a^{-1}}$.

6. Subtract $\sqrt{18} - \sqrt{50}$.

7. Rationalize the denominator $\dfrac{5}{2 - \sqrt{3}}$.

8. Expand $(a + b)^2$.

9. Multiply $(a + by)(a - by)$.

10. Express a^6 as the product of two factors with a^4 as one factor.

1.5 Complex Numbers

The real number system is the basis for most of the work both in this course and in calculus. However, we occasionally have use for an extension of the real numbers which is helpful in many applications of mathematics. This new kind of number is called a *complex number.* As we will see, complex numbers are made up of two components and can, therefore, convey more information about a quantity. For example, a complex number is often used in physics to describe a force, since such a numerical representation indicates not only the magnitude but also the direction of the force. In electronics complex numbers are used extensively to designate voltage, current, and other electrical quantities, for this representation indicates both the strength and time (or phase) relationships of the quantities.

Mathematically we extend the number system a step beyond the real numbers by considering numbers such as $\sqrt{-1}$. Such numbers are not real numbers, for the product of two equal real factors never results in a negative number. We thus introduce a different type of number, called an *imaginary number.* The basic unit in imaginary numbers is $\sqrt{-1}$ and it is designated by i. Thus, by definition we have

$$i = \sqrt{-1} \quad \text{and} \quad i^2 = -1.$$

When working with the square roots of negative numbers, the first step is to express these imaginary numbers as the product of a real number and i. For example:

$$\sqrt{-4} = i\sqrt{4} \quad\;\; = 2i$$
$$\sqrt{-12} = i\sqrt{4}\sqrt{3} = 2i\sqrt{3}.$$

Using real numbers and imaginary numbers, we may extend the number system to include a number like $3 + 2i$, and we define this new type of number as follows.

Definition of Complex Number

> A number of the form $a + bi$, where a and b are real numbers and $i = \sqrt{-1}$, is called a complex number.

The number a is called the **real part** of $a + bi$, and b is called the **imaginary part** of $a + bi$. Note that both the real part and the imaginary part of a complex number are real numbers. Two complex numbers are **equal** if and only if their real parts are equal and their imaginary parts are equal. That is,

$$a + bi = c + di \text{ if and only if } a = c \text{ and } b = d.$$

Since a and/or b may be zero, the complex numbers include the real numbers and the imaginary numbers. Figure 1.6 illustrates the relationship among the various sets of numbers.

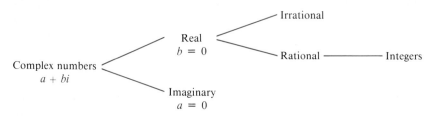

Figure 1.6

In many ways computations with complex numbers are similar to computations with real numbers that contain square roots. The important difference is that we must always express the imaginary part in terms of i, remembering that $i^2 = -1$. The following examples illustrate how to perform the various operations with complex numbers.

EXAMPLE 1 Express in terms of i and combine.

    a. $\sqrt{-100} + \sqrt{-64}$ **b.** $3\sqrt{-3} - 2\sqrt{-48}$

Solution

  a. $\sqrt{-100} + \sqrt{-64} = i \cdot \sqrt{100} + i \cdot \sqrt{64}$
$$= 10i + 8i$$
$$= 18i$$

  b. $3\sqrt{-3} - 2\sqrt{-48} = 3i \cdot \sqrt{3} - 2i \cdot \sqrt{16} \cdot \sqrt{3}$
$$= 3i\sqrt{3} - 8i\sqrt{3}$$
$$= -5i\sqrt{3}$$

$\blacksquare$

EXAMPLE 2 Perform the indicated operation and simplify $\sqrt{-2} \cdot \sqrt{-32}$.

Solution

$$\sqrt{-2} \cdot \sqrt{-32} = (i \cdot \sqrt{2})(i \cdot \sqrt{16} \cdot \sqrt{2})$$
$$= (i\sqrt{2})(4i\sqrt{2}) = 4 \cdot i^2 \cdot 2$$
$$= 4(-1)2 \quad (\text{since } i^2 = -1)$$
$$= -8$$

Note that the property of radicals $\sqrt{x}\sqrt{y} = \sqrt{xy}$ does not hold in this case. That is, it is *not* true that

$$\sqrt{-2}\sqrt{-32} = \sqrt{(-2)(-32)} = \sqrt{64} = 8.$$

The correct result is -8. This is why we must be careful when we work with complex numbers. Always remember to express the complex numbers in terms of i before performing computations.

$\blacksquare$

EXAMPLE 3   Perform the indicated operation and simplify $(2 - i\sqrt{3})^2$.

Solution

$$
\begin{aligned}
(2 - i\sqrt{3})^2 &= (2 - i\sqrt{3})(2 - i\sqrt{3}) \\
&= 4 - 2i\sqrt{3} - 2i\sqrt{3} + i^2 \cdot 3 \\
&= 4 - 4i\sqrt{3} + (-1)3 \\
&= 1 - 4i\sqrt{3}
\end{aligned}
$$ ∎

EXAMPLE 4   Evaluate $x^2 - 8x + 25$ if $x = 4 - 3i$.

Solution   Substitute $4 - 3i$ for x and simplify.

$$
\begin{aligned}
x^2 - 8x + 25 &= (4 - 3i)^2 - 8(4 - 3i) + 25 \\
&= 16 - 24i + 9i^2 - 32 + 24i + 25 \\
&= 9i^2 + 9 \\
&= 0
\end{aligned}
$$ ∎

Conjugates and Division

To understand the procedure for dividing complex numbers, we first must define the conjugate of a complex number and then consider some basic properties of conjugates. The complex conjugate, or simply **conjugate,** of $a + bi$ is $a - bi$. It is standard notation to denote the conjugate of a number by placing a bar above the number. For example,

$$\overline{3 + 2i} = 3 - 2i \quad \text{and} \quad \overline{-4 - 0i} = -4 + 0i.$$

We now state some of the basic properties of conjugates that will be needed either here or in Section 3.4.

Properties of Conjugates

> If z and w are complex numbers, then
> 1. $z \cdot \bar{z}$ is a real number.
> 2. $\bar{z} = z$ if and only if z is a real number.
> 3. $\overline{z + w} = \bar{z} + \bar{w}$.
> 4. $\overline{z \cdot w} = \bar{z} \cdot \bar{w}$.
> 5. $\overline{z^n} = (\bar{z})^n$ for any positive integer n.

Illustrations of these properties with specific complex numbers are given in the exercises, which also outline proofs for properties 2, 3, and 4. (The proof of property 5 requires the use of mathematical induction, which is discussed in the last chapter of the text.) Property 1 is our main interest at this time because it provides the rationale for the procedure for dividing two complex numbers. In words, property 1 states that the product of a complex number and its conjugate is always a real number. To prove this property, we first let $z = a + bi$. Then

$$
\begin{aligned}
z \cdot \bar{z} &= (a + bi)(a - bi) \\
&= a^2 - abi + abi - b^2i^2 \\
&= a^2 + b^2.
\end{aligned}
$$

Since a and b are real numbers, $a^2 + b^2$ is a real number, which proves the property. Because $z \cdot \bar{z}$ is a real number, the quotient w/z can be written in the form $a + bi$ by multiplying both the numerator and the denominator by $\bar{z}$. Example 5 shows this procedure.

EXAMPLE 5   Write the quotient $\dfrac{4 + 3i}{1 - 2i}$ in the form $a + bi$.

Solution   As discussed, we divide two complex numbers by multiplying the numerator and the denominator by the conjugate of the denominator. Thus,

$$\frac{4 + 3i}{1 - 2i} = \frac{4 + 3i}{1 - 2i} \cdot \frac{1 + 2i}{1 + 2i}$$

$$= \frac{4 + 8i + 3i + 6i^2}{1 + 2i - 2i - 4i^2}$$

$$= \frac{-2 + 11i}{5} = -\frac{2}{5} + \frac{11}{5}i. \qquad \blacksquare$$

EXAMPLE 6   The reciprocal, or multiplicative inverse, of a complex number z is $1/z$. Write the reciprocal of $4i$ in the form $a + bi$.

Solution   The reciprocal of $4i$ is $1/4i$. Since the conjugate of $0 + 4i$ is $0 - 4i$, we have

$$\frac{1}{4i} = \frac{1}{4i} \cdot \frac{-4i}{-4i} = \frac{-4i}{-16i^2} = \frac{-4i}{16} = -\frac{1}{4}i.$$

Thus, the reciprocal of $4i$ is $-\frac{1}{4}i$, or $0 - \frac{1}{4}i$ in standard form. We can verify our result as follows:

$$(4i)(-\tfrac{1}{4}i) = -1 \cdot i^2 = -1(-1) = 1. \qquad \blacksquare$$

Powers of i

To simplify i^n, where n is a positive integer, consider the cyclic pattern contained in the following simplifications.

$$i^1 = i \qquad\qquad i^5 = i^4 \cdot i = 1 \cdot i = i$$
$$i^2 = -1 \qquad\qquad i^6 = i^4 \cdot i^2 = 1(-1) = -1$$
$$i^3 = i^2 \cdot i = (-1)i = -i \qquad\qquad i^7 = i^4 \cdot i^3 = 1(-i) = -i$$
$$i^4 = i^2 \cdot i^2 = (-1)(-1) = 1 \qquad\qquad i^8 = (i^4)^2 = 1^2 = 1$$

Continuing this pattern to higher powers of i gives the cyclic property shown in Figure 1.7. Note in particular that $i^n = 1$ if n is a multiple of 4 because we may use this observation to simplify large powers of i, as shown in Example 7.

EXAMPLE 7   Simplify each power of i.

a. i^{21} **b.** i^{95}

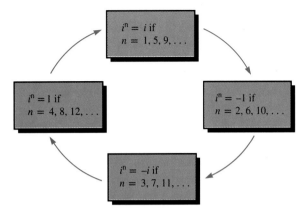

Figure 1.7   Powers of i^n

Solution

a. The largest multiple of 4 that is less than or equal to the exponent is 20. Therefore, rewrite i^{21} with i^{20} as one factor and then simplify.

$$i^{21} = i^{20} \cdot i = (i^4)^5 \cdot i = (1)^5 \cdot i = i$$

b. Because $i^{92} = (i^4)^{23} = 1$, we have

$$i^{95} = i^{92} \cdot i^3 = 1 \cdot (-i) = -i. \qquad \blacksquare$$

In summary, note that we have been operating with complex numbers according to the following definitions.

Operations on Complex Numbers

1. Two complex numbers are added by adding separately their real parts and their imaginary parts.

$$(a + bi) + (c + di) = (a + c) + (b + d)i$$

2. Two complex numbers are subtracted by subtracting separately their real parts and their imaginary parts.

$$(a + bi) - (c + di) = (a - c) + (b - d)i$$

3. Two complex numbers are multiplied as two binomials are multiplied, with i^2 being replaced by -1.

$$(a + bi)(c + di) = ac + adi + bci + bdi^2$$
$$= (ac - bd) + (ad + bc)i$$

4. Two complex numbers are divided by multiplying the numerator and the denominator by the conjugate of the denominator.

$$\frac{a + bi}{c + di} = \frac{(a + bi)(c - di)}{(c + di)(c - di)} = \frac{(ac + bd) + (bc - ad)i}{c^2 + d^2}$$

EXERCISES 1.5

In Exercises 1–20 simplify each expression in terms of i.

1. $\sqrt{-25}$ **2.** $\sqrt{-49}$

3. $2\sqrt{-16}$ **4.** $5\sqrt{-121}$

5. $\sqrt{-22}$ **6.** $-3\sqrt{-7}$

7. $\frac{1}{2}\sqrt{-12}$ **8.** $-3\sqrt{-98}$

9. $2 - \sqrt{-128}$ **10.** $-3 + \frac{1}{2}\sqrt{-162}$

11. $\sqrt{(-4)^2 - 4(1)(5)}$ **12.** $\sqrt{(12)^2 - 4(9)(8)}$

13. $\sqrt{(2)^2 - 4(-1)(-3)}$ **14.** $\sqrt{(-5)^2 - 4(2)(7)}$

15. $\dfrac{-(-1) - \sqrt{(-1)^2 - 4(2)(3)}}{2(2)}$

16. $\dfrac{-(0) + \sqrt{(0)^2 - 4(1)(2)}}{2(1)}$

17. $\dfrac{-(4) + \sqrt{(4)^2 - 4(1)(5)}}{2(1)}$

18. $\dfrac{-(-2) - \sqrt{(-2)^2 - 4(1)(5)}}{2(1)}$

19. $\dfrac{-(2) - \sqrt{(2)^2 - 4(5)(3)}}{2(5)}$

20. $\dfrac{-(-2) + \sqrt{(-2)^2 - 4(-5)(-3)}}{2(-5)}$

In Exercises 21–40 perform the indicated operations and write each result in the form $a + bi$.

21. $2\sqrt{-9} - 3\sqrt{-9}$

22. $4\sqrt{-1} - 2\sqrt{-4}$

23. $\sqrt{-8} + \sqrt{-18} - \sqrt{-50}$

24. $4\sqrt{-27} - 2\sqrt{-48} + 3\sqrt{-75}$

25. $(-1 - 2i) + (-2 + i)$

26. $(11 + 2i) - (-5 - 7i)$

27. $(3 - i) - (-2 + 5i)$

28. $(6 + 2i) + (2 - 4i)$

29. $\sqrt{-3} \cdot \sqrt{-12}$ **30.** $\sqrt{-5} \cdot \sqrt{-18}$

31. $(2i)^2$ **32.** $(-i)^2$

33. $(4 + 2i)(1 - 3i)$ **34.** $(2 - 5i)(3 - i)$

35. $(2 + 3i)(2 - 3i)$ **36.** $(6 - 5i)(6 + 5i)$

37. $(1 - 2i)^2$ **38.** $(5 + i)^2$

39. $(-3 + \frac{1}{3}i)^2$ **40.** $(-1 - \frac{1}{2}i)^2$

In Exercises 41–50 evaluate the expression for the given value of x.

41. $x^2 + 4$ for $x = -2i$ **42.** $x^2 - 4$ for $x = 2i$

43. $x^2 - 4x + 5$ for $x = 2 + i$

44. $x^2 - 4x + 5$ for $x = 2 - i$

45. $2x^2 + x + 1$ for $x = -1 + 2i$

46. $2x^2 + x + 1$ for $x = -1 - 2i$

47. $3x^2 - 2x + 1$ for $x = \dfrac{1}{3} - i\dfrac{\sqrt{2}}{3}$

48. $3x^2 - 2x + 1$ for $x = \dfrac{1}{3} + i\dfrac{\sqrt{2}}{3}$

49. $x^3 - 27$ for $x = 3i$ **50.** $x^4 - 81$ for $x = -3i$

In Exercises 51–56 find the conjugate of the given complex number.

51. $3 + 4i$ **52.** $5 - 7i$ **53.** i

54. $-4i$ **55.** -7 **56.** 7

In Exercises 57–66 write each quotient in the form $a + bi$.

57. $\dfrac{1}{1 + 2i}$ **58.** $\dfrac{-3}{2 - 3i}$ **59.** $\dfrac{-2}{5i}$

60. $\dfrac{1}{i}$ **61.** $\dfrac{i}{-1 - 4i}$ **62.** $\dfrac{-2i}{6 - 2i}$

63. $\dfrac{2 + i}{3 + i}$ **64.** $\dfrac{2 + 5i}{4 + 2i}$ **65.** $\dfrac{\sqrt{2} - i}{\sqrt{2} + i}$

66. $\dfrac{\frac{1}{2} + i}{\frac{1}{2} - i}$

In Exercises 67–70 find the reciprocal and write the answer in the form $a + bi$.

67. $-i$ **68.** $3i$ **69.** $1 + i$ **70.** $-2 - 2i$

In Exercises 71–78 simplify each power of i.

71. i^3 **72.** i^4

73. i^7 **74.** i^9

75. i^{36} **76.** i^{43}

77. i^{93} **78.** i^{100}

In Exercises 79 and 80 verify that **(a)** $\overline{z + w} = \bar{z} + \bar{w}$, **(b)** $\overline{z \cdot w} = \bar{z} \cdot \bar{w}$, and **(c)** $\overline{z^2} = (\bar{z})^2$ for the given values of z and w.

79. $z = 1 + i, w = 2 - 3i$

80. $z = 4 - i, w = 5 + 2i$

In Exercises 81–84 let $z = a + bi$ and $w = c + di$, and use the definition of conjugate to prove each of the following properties.

81. $\overline{z + w} = \bar{z} + \bar{w}$ **82.** $\overline{z \cdot w} = \bar{z} \cdot \bar{w}$

83. If z is a real number, then $\bar{z} = z$.

84. If $\bar{z} = z$, then z is a real number.

85. Use the fact that $\overline{z^2} = \overline{z} \cdot \overline{z}$ to prove that $\overline{z^2} = (\overline{z})^2$ for any complex number z.

86. Complex numbers are used extensively in electrical engineering to designate voltage (E), current (I), and resistance (R). However, since i usually symbolizes current in this discipline, complex numbers are written in the form $a + bj$ with $j = \sqrt{-1}$. In the following exercises use Ohm's law, $E = IR$, to determine the value of the missing variable.

   a. $I = 1 - 2j$ amperes, $R = 2 + 3j$ ohms

   b. $E = 2$ volts, $I = 1 + 4j$ amperes

   c. $R = 7 - 5j$ ohms, $E = 1 - j$ volts

THINK ABOUT IT

1. If $p = 5 + 2i - qi$, where p and q are real numbers, then find p and q.

2. If $x^2 + 2 = bx$ when $x = 1 + i$, then find b.

3. Evaluate $x^3 - 2x^2 + 3x - 6$ if $x = i\sqrt{3}$.

4. If m and n are positive integers and $i^m = i^n$, then describe the relation between m and n.

5. Consider the product

$$\sqrt{-1} \cdot \sqrt{-1} = \sqrt{(-1)(-1)} = \sqrt{1} = 1$$

but $\sqrt{-1} \cdot \sqrt{-1} = (\sqrt{-1})^2 = -1.$

Therefore, $1 = -1$! What is wrong?

REMEMBER THIS

1. Express $\sqrt{(-4)^2 - 4(1)(-3)}$ in simplest radical form.

2. Factor $5x^2 - 11x + 2$.

3. Factor $5x^2 - 11x$.

4. Add $\dfrac{-c}{a} + \dfrac{b^2}{4a^2}$.

5. Which numbers in the set $\{\sqrt{0}, \sqrt{-23}, -\sqrt{7}, \sqrt{1000}\}$ are irrational numbers?

6. Evaluate $\dfrac{3}{2x - 1}$ if $x = -\dfrac{2}{5}$.

7. Find the least common denominator of $\dfrac{2}{3t}$ and $\dfrac{5}{3t - 1}$.

8. For what values of x does $-(x - 5) = 5 - x$?

9. Can $12x^2 - x - 18$ be factored into binomial factors with integer coefficients?

10. True or false: A correct calculator sequence for evaluating 8^{-1} is $8 \boxed{1/x}$.

1.6   Linear Equations and Quadratic Equations

Setting up and solving equations is often an essential part of the mathematical analysis of a problem. An **equation** is a statement of equality between two expressions. To solve an equation means to find the set of all values for the variable that make the equation a true statement. This set is called the **solution set** of the equation. Equations may be always true, always false, or their truth may depend on the value substituted for the variable. If the equation is a true statement for all admissible values of the variable, as is $2x + x = 3x$, then it is called an **identity**. A **conditional equation** is an equation in which some replacements for the variable make the statement true, while others make it false. For example, the conditional equation $x + 5 = 21$ is true if $x = 16$ and false otherwise. We call 16 the **solution** or **root** of the equation. The concept of solving equations is usually associated with conditional equations.

One of the simplest equations to solve is a linear equation. By definition, a **first-degree** or **linear equation** is an equation that can be written in the form

$$ax + b = 0,$$

where a and b are real numbers with $a \neq 0$. We solve a linear equation by isolating the variable on one side of the equation as described below.

Solution of Linear Equations

Two equations are **equivalent** when they have the same solution set. We may change a linear equation to an equivalent equation of the form $x = $ number by performing any combination of the following steps:

1. Adding or subtracting the same expression to (from) both sides of the equation
2. Multiplying or dividing both sides of the equation by the same nonzero number

EXAMPLE 1   Solve the equation $6(x + 1) = 14x + 2$.

Solution

$$
\begin{aligned}
6(x + 1) &= 14x + 2 \\
6x + 6 &= 14x + 2 \\
6x + 6 - 6 &= 14x + 2 - 6 &&\text{Subtract 6 from both sides.} \\
6x &= 14x - 4 \\
6x - 14x &= 14x - 14x - 4 &&\text{Subtract } 14x \text{ from both sides.} \\
-8x &= -4 \\
\frac{-8x}{-8} &= \frac{-4}{-8} &&\text{Divide both sides by } -8. \\
x &= \frac{1}{2}
\end{aligned}
$$

Thus, $\frac{1}{2}$ is the solution of the equation and the solution set is $\{\frac{1}{2}\}$. We can check the solution by replacing x with $\frac{1}{2}$ in the original equation.

$$6(\tfrac{1}{2} + 1) \overset{?}{=} 14(\tfrac{1}{2}) + 2$$

$$6(\tfrac{3}{2}) \overset{?}{=} 7 + 2$$

$$9 \overset{\checkmark}{=} 9$$

■

EXAMPLE 2   Solve and check $\dfrac{3x}{8} - \dfrac{1}{4} = \dfrac{x + 5}{2}$.

Solution   The usual procedure for solving equations with fractions is to remove the fractions by multiplying both sides of the equation by the LCD. In this equation the LCD is 8, so

$$8\left(\frac{3x}{8} - \frac{1}{4}\right) = 8\left(\frac{x + 5}{2}\right)$$

$$3x - 2 = 4x + 20$$

$$-22 = x.$$

Check

$$\frac{3(-22)}{8} - \frac{1}{4} \overset{?}{=} \frac{-22 + 5}{2}$$

$$-\frac{17}{2} \overset{\checkmark}{=} -\frac{17}{2}$$

Thus, the solution set is $\{-22\}$.   ∎

Note that in Examples 1 and 2 each linear equation had exactly one solution. We can anticipate this result since the general form of a linear equation, $ax + b = 0$ with $a \neq 0$, may be solved as follows:

$$ax + b = 0$$

$$ax = -b \qquad \text{Add } -b \text{ to both sides.}$$

$$x = \frac{-b}{a}. \qquad \text{Divide both sides by } a \neq 0.$$

Thus, the linear equation $ax + b = 0$, $a \neq 0$, has exactly one solution, $-b/a$. The requirement that $a \neq 0$ is important, as shown in Example 3.

EXAMPLE 3   Solve $3(x + 3) = 3x$.

Solution   If we proceed in the usual way, we have

$$3(x + 3) = 3x$$

$$3x + 9 = 3x$$

$$9 = 0.$$

When we subtract $3x$ from both sides of $3x + 9 = 3x$, the result is $9 = 0$. This statement is never true and indicates the original equation has no solution. When the solution set contains no elements, we say the solution set is $\emptyset$, called the **empty set.** Note that when an equation may be written in the form $ax + b = 0$, with $a = 0$, then either the equation has no solution (if $b \neq 0$) or the equation is an identity (if $b = 0$).   ∎

The procedures for solving a linear equation are fundamental to the solutions of a wide variety of equations. For example, fractional equations involving unknown denominators often lead to linear equations, as shown in Examples 4 and 5.

EXAMPLE 4   Solve and check $\dfrac{2}{3a} = \dfrac{3}{2a - 1}$.

Solution   As in Example 2, we first remove the fractions by multiplying both sides of the equation by the LCD, which is $3a(2a - 1)$ in this equation. Note that this step requires the restrictions that $a \neq 0$ and $a \neq \frac{1}{2}$ to ensure that we are multiplying both sides of the equation by a nonzero number.

Check

$$3a(2a - 1)\frac{2}{3a} = 3a(2a - 1)\frac{3}{2a - 1}$$

$$2(2a - 1) = 9a$$

$$4a - 2 = 9a$$

$$-2 = 5a$$

$$\frac{-2}{5} = a$$

$$\frac{2}{3(-2/5)} \overset{?}{=} \frac{3}{2(-2/5) - 1}$$

$$\frac{2}{(-6/5)} \overset{?}{=} \frac{3}{(-9/5)}$$

$$\frac{5}{-3} \overset{\checkmark}{=} \frac{5}{-3}$$

The LCD is nonzero when $a = -\frac{2}{5}$, and the check of this solution is shown above. Thus, the solution set is $\left\{-\frac{2}{5}\right\}$. ∎

EXAMPLE 5 Solve and check $\dfrac{2}{x + 1} + \dfrac{3}{x - 1} = \dfrac{-4}{x^2 - 1}$.

Solution The LCD is $(x + 1)(x - 1)$ and if $x \neq -1$ and $x \neq 1$, then

$$(x + 1)(x - 1)\left(\frac{2}{x + 1} + \frac{3}{x - 1}\right) = (x + 1)(x - 1)\frac{-4}{x^2 - 1}$$

$$2(x - 1) + 3(x + 1) = -4$$

$$2x - 2 + 3x + 3 = -4$$

$$5x + 1 = -4$$

$$5x = -5$$

$$x = -1.$$

The restriction $x \neq -1$ eliminates -1 as a solution, so no number satisfies the original equation and the solution set is $\emptyset$. We verify that -1 is not a solution in the following check.

Check

$$\frac{2}{(-1) + 1} + \frac{3}{(-1) - 1} \overset{?}{=} \frac{-4}{(-1)^2 - 1}$$

$$\frac{2}{0} + \frac{3}{-2} \neq \frac{-4}{0}$$

∎

An equation that can be put in the form $ax^2 + bx + c = 0$, where a, b, and c are real numbers with $a \neq 0$, is called a **second-degree** or **quadratic equation**. For example,

$$2x^2 - 7x + 6 = 0$$

$$4t^2 = 9t - 2$$

$$(x - 1)^2 = 5$$

are quadratic equations. Depending on the particular equation, different procedures are used for solving quadratic equations, and you need to consider carefully each of the methods that follow.

The Factoring Method

A technique that can often be used to solve a second-degree equation relies on factoring. For example, let us try to find the numbers that make the equation $x^2 + 5x + 4 = 0$ a true statement.

First, factor the expression on the left-hand side of the equation.

$$(x + 4)(x + 1) = 0$$

We now have the situation where the product of two factors is zero. In multiplication zero is a special number as outlined in the following principle.

Zero Product Principle

> For any numbers a and b,
>
> $$ab = 0 \text{ if and only if } a = 0 \text{ or } b = 0.$$

By applying this principle, we know

$$(x + 4)(x + 1) = 0 \text{ if and only if } x + 4 = 0 \text{ or } x + 1 = 0.$$

Since $x + 4 = 0$ when $x = -4$ and $x + 1 = 0$ when $x = -1$, the solutions are -4 and -1, and the solution set is $\{-4, -1\}$. To catch any mistakes, it is recommended that solutions be checked in the original equation, so

$$x^2 + 5x + 4 = 0 \qquad\qquad x^2 + 5x + 4 = 0$$
$$(-4)^2 + 5(-4) + 4 \overset{?}{=} 0 \qquad\qquad (-1)^2 + 5(-1) + 4 \overset{?}{=} 0$$
$$0 \overset{\checkmark}{=} 0 \qquad\qquad\qquad 0 \overset{\checkmark}{=} 0.$$

To summarize, quadratic equations that are factorable (with integer coefficients) are usually solved as follows.

Factoring Method for Solving Quadratic Equations

> 1. If necessary, change the form of the equation so that one side is 0.
> 2. Factor the nonzero side of the equation.
> 3. Set each factor equal to zero and obtain the solution(s) by solving the resulting equations.
> 4. Check each solution by substituting it in the original equation.

EXAMPLE 6   Solve $4x^2 = x$.

Solution

$$4x^2 = x$$
$$4x^2 - x = 0$$
$$x(4x - 1) = 0$$
$$x = 0 \qquad 4x - 1 = 0$$
$$x = \tfrac{1}{4}$$

$$Check \quad 4x^2 = x \qquad 4x^2 = x$$
$$4(0)^2 \overset{?}{=} 0 \qquad 4(\tfrac{1}{4})^2 \overset{?}{=} \tfrac{1}{4}$$
$$4(0) \overset{?}{=} 0 \qquad 4(\tfrac{1}{16}) \overset{?}{=} \tfrac{1}{4}$$
$$0 \overset{\checkmark}{=} 0 \qquad \tfrac{1}{4} \overset{\checkmark}{=} \tfrac{1}{4}$$

Thus, 0 and $\frac{1}{4}$ are the solutions of the equation, and the solution set is $\{0,\frac{1}{4}\}$.   ∎

The Quadratic Equation $ax^2 + c = 0$

The easiest quadratic equation to solve is one in which there is no x term and the equation has the form $ax^2 + c = 0$ with $a \neq 0$. For example, to solve $2x^2 - 10 = 0$, we merely transform the equation so that x^2 is on one side of the equation by itself, giving $x^2 = 5$. We then conclude the solutions are $\sqrt{5}$ and $-\sqrt{5}$ by applying the following property.

Square Root Property

If n is any real number, then

$$x^2 = n \text{ implies } x = \sqrt{n} \text{ or } x = -\sqrt{n}.$$

This property is a key step in the general solution to quadratic equations, and we can prove this property as follows:

$$x^2 = n$$
$$x^2 - n = 0$$
$$(x - \sqrt{n})(x + \sqrt{n}) = 0 \qquad \text{Factoring over the set of complex numbers}$$

$$x - \sqrt{n} = 0 \quad \text{or} \quad x + \sqrt{n} = 0 \qquad \text{Zero product principle}$$
$$x = \sqrt{n} \quad \text{or} \qquad x = -\sqrt{n}.$$

We usually abbreviate these solutions as $\pm\sqrt{n}$. Note that there are two real number solutions if $n > 0$, two (conjugate) complex number solutions if $n < 0$, and one solution (namely, 0) if $n = 0$.

EXAMPLE 7   Solve each equation by using the square root property.

a. $x^2 = 27$ **b.** $5(x - 3)^2 + 20 = 0$

Solution

a. $x^2 = 27$ implies $x = \pm\sqrt{27}$. Since $\sqrt{27} = \sqrt{9}\sqrt{3} = 3\sqrt{3}$ the solution set is $\{\pm 3\sqrt{3}\}$.

b. First, transform the equation so $(x - 3)^2$ is on one side of the equation by itself.

$$5(x - 3)^2 + 20 = 0$$
$$5(x - 3)^2 = -20$$
$$(x - 3)^2 = -4$$

Now apply the square root property and simplify.

$$x - 3 = \pm\sqrt{-4}$$
$$x = 3 \pm 2i \qquad \text{since } \sqrt{-4} = 2i$$

Thus, the solution set is $\{3 \pm 2i\}$.   ∎

Completing the Square

In Example 7b we used the square root property to solve a quadratic equation of the form

$$(x + \text{constant})^2 = \text{constant}.$$

Actually, any quadratic equation may be placed in this form by a technique called completing the square. Consider an expression like

$$x^2 + 8x.$$

What constant needs to be added to make this expression a perfect square? Since

$$(x + k)^2 = x^2 + 2kx + k^2$$

we set $2k = 8$, so $k = 4$ and $k^2 = 16$. Thus, adding 16 gives

$$x^2 + 8x + 16 = (x + 4)^2.$$

In general, we complete the square for $x^2 + bx$ by adding $(b/2)^2$, which is the square of one half of the coefficient of x. Example 8 shows how completing the square can solve a quadratic equation.

EXAMPLE 8   Solve the equation $x^2 - 4x - 3 = 0$ by completing the square.

Solution   First, rearrange the equation with the x terms to the left of the equals and the constant to the right.

$$x^2 - 4x = 3$$

Now complete the square on the left. Half of -4 is -2 and $(-2)^2 = 4$. Add 4 to both sides of the equation and proceed as follows:

$$x^2 - 4x + 4 = 3 + 4$$
$$(x - 2)^2 = 7$$
$$x - 2 = \pm \sqrt{7}$$
$$x = 2 \pm \sqrt{7}.$$

The solutions are $2 + \sqrt{7}$ and $2 - \sqrt{7}$, and the solution set is abbreviated $\{2 \pm \sqrt{7}\}$.   ∎

Quadratic Formula

Completing the square is a useful technique in many situations because it often converts an expression to a standard form that is easy to analyze. In this case, the completing the square method works so well that if we apply it to the general quadratic equation $ax^2 + bx + c = 0$, we obtain a powerful formula (the quadratic formula) that solves all quadratic equations. We now derive this formula with the particular equation

$2x^2 - 5x + 1 = 0$ being solved on the left to illustrate in specific terms what is happening on the right.

Particular Equation	General Equation
$2x^2 - 5x + 1 = 0$	$ax^2 + bx + c = 0$ (with $a \neq 0$)

We shall attempt to make the left side of the equation a perfect square so that we can use the square root property. To begin, if the coefficient of x^2 is not 1, we must divide by this coefficient before proceeding as before.

$$x^2 - \frac{5}{2}x + \frac{1}{2} = 0 \qquad x^2 + \frac{b}{a}x + \frac{c}{a} = 0$$

Subtract the constant term from both sides of the equation.

$$x^2 - \frac{5}{2}x = \frac{-1}{2} \qquad x^2 + \frac{b}{a}x = \frac{-c}{a}$$

We now add a number to both sides of the equation to make the left side a perfect square. The desired number can always be found by taking half the coefficient of the x term and squaring the result.

Half of $\dfrac{-5}{2}$ is $\dfrac{-5}{4}$ and $\qquad$ Half of $\dfrac{b}{a}$ is $\dfrac{b}{2a}$ and

$\left(\dfrac{-5}{4}\right)^2$ is $\dfrac{25}{16}$. $\qquad\qquad\qquad$ $\left(\dfrac{b}{2a}\right)^2$ is $\dfrac{b^2}{4a^2}$.

$$x^2 - \frac{5}{2}x + \frac{25}{16} \qquad x^2 + \frac{b}{a}x + \frac{b^2}{4a^2}$$

$$= \frac{-1}{2} + \frac{25}{16}. \qquad = \frac{-c}{a} + \frac{b^2}{4a^2}.$$

Write the left side as a perfect square and simplify the right side.

$$\left(x - \frac{5}{4}\right)^2 = \frac{17}{16} \qquad \left(x + \frac{b}{2a}\right)^2 = \frac{b^2 - 4ac}{4a^2}$$

Use the square root property.

$$x - \frac{5}{4} = \pm\sqrt{\frac{17}{16}} \qquad x + \frac{b}{2a} = \pm\sqrt{\frac{b^2 - 4ac}{4a^2}}$$

Simplify the result.

$$x - \frac{5}{4} = \frac{\pm\sqrt{17}}{4} \qquad x + \frac{b}{2a} = \frac{\pm\sqrt{b^2 - 4ac}}{2a}$$

$$x = \frac{5}{4} + \frac{\pm\sqrt{17}}{4} \qquad x = \frac{-b}{2a} + \frac{\pm\sqrt{b^2 - 4ac}}{2a}$$

$$= \frac{5 \pm \sqrt{17}}{4} \qquad = \frac{-b \pm \sqrt{b^2 - 4ac}}{2a}$$

Thus, our two solutions x_1 and x_2 are

$$x_1 = \frac{5 + \sqrt{17}}{4} \qquad\qquad x_1 = \frac{-b + \sqrt{b^2 - 4ac}}{2a}$$

$$x_2 = \frac{5 - \sqrt{17}}{4} \qquad\qquad x_2 = \frac{-b - \sqrt{b^2 - 4ac}}{2a}$$

and we have derived the following formula.

Quadratic Formula

> If $ax^2 + bx + c = 0$ and $a \neq 0$, then
> $$x = \frac{-b \pm \sqrt{b^2 - 4ac}}{2a}.$$

Note that this formula results in two solutions.

$$x_1 = \frac{-b + \sqrt{b^2 - 4ac}}{2a} \quad \text{and} \quad x_2 = \frac{-b - \sqrt{b^2 - 4ac}}{2a}$$

We can solve *any* quadratic equation with this formula by identifying the values of a, b, and c and substituting these numbers into the formula.

EXAMPLE 9   Solve the equation $2x^2 - x + 3 = 0$ by using the quadratic formula.

Solution   In this equation $a = 2$, $b = -1$, and $c = 3$. Therefore,

$$x = \frac{-b \pm \sqrt{b^2 - 4ac}}{2a} = \frac{-(-1) \pm \sqrt{(-1)^2 - 4(2)(3)}}{2(2)}$$

$$= \frac{1 \pm \sqrt{1 - 24}}{4}$$

$$= \frac{1 \pm \sqrt{-23}}{4}$$

$$= \frac{1 \pm i\sqrt{23}}{4}.$$

$$x_1 = \frac{1 + i\sqrt{23}}{4} \qquad x_2 = \frac{1 - i\sqrt{23}}{4}$$

The solutions of this equation are (conjugate) complex numbers, and the solution set is $\left\{ \dfrac{1 \pm i\sqrt{23}}{4} \right\}$. ∎

As with linear equations, the procedures for solving quadratic equations are fundamental to the solution of a wide variety of equations. Example 10 shows that fractional equations may lead to quadratic equations, and the methods of this section will be needed to solve many of the equations in Section 1.7.

EXAMPLE 10   Solve $\dfrac{2}{x} + \dfrac{10}{x^2} = 1$.

Solution   First, we remove the fractions by multiplying both sides of the equation by the LCD, which is x^2 in this equation. Thus, if $x \neq 0$,

$$x^2\left(\frac{2}{x} + \frac{10}{x^2}\right) = x^2 \cdot 1$$
$$2x + 10 = x^2$$
$$0 = x^2 - 2x - 10.$$

Then, by the quadratic formula

$$x = \frac{-(-2) \pm \sqrt{(-2)^2 - 4(1)(-10)}}{2(1)} = \frac{2 \pm \sqrt{44}}{2}$$
$$= \frac{2 \pm 2\sqrt{11}}{2} = 1 \pm \sqrt{11}.$$

The restriction $x \neq 0$ does not affect the proposed solution and both numbers check in the original equation. Thus, the solution set is $\{1 \pm \sqrt{11}\}$.   ∎

The Discriminant

Notice from the examples that the nature of the solutions to a quadratic equation depends on the value of $b^2 - 4ac$ (called the **discriminant**), which appears under the radical in the quadratic formula. That is:

When a, b, c Are Rational and	The Solutions Are
$b^2 - 4ac < 0$	conjugate complex numbers
$b^2 - 4ac = 0$	real, rational, equal numbers
$b^2 - 4ac > 0$ and a perfect square	real, rational, unequal numbers
$b^2 - 4ac > 0$, not a perfect square	real, irrational, unequal numbers

For example, in the equation $3x^2 + 5x + 2 = 0$ we find $b^2 - 4ac = (5)^2 - 4(3)(2) = 1$. Since 1 is greater than 0 and a perfect square, we know that there are two different solutions that are rational numbers. As another example, in the equation $2x^2 - x + 3 = 0$ (Example 9), $b^2 - 4ac = (-1)^2 - 4(2)(3) = -23$. Since -23 is less than 0, our solutions are conjugate complex numbers. Thus, without solving the equation, we can determine the nature of the solutions by finding the value of $b^2 - 4ac$.

Applications

Now that we can solve some equations, we can determine the solutions for some word problems. The most difficult part is usually to describe the situation mathematically; but this is a realistic obstacle. Problems you wish to solve from your own areas of interest will probably not start out as math problems but will be stated in words. It's up to you to analyze the situation and apply the appropriate formulas. When solving word problems, it is recommended by both mathematics and reading specialists that you use the following steps.

To Solve a Word Problem

1. **Read the problem several times.** The first reading is a preview and is done quickly to obtain a general idea of the problem. The objective of the second reading is to determine exactly what you are asked to find. Write this down. Finally, read the problem carefully and note what information is given. If possible, display the given information in a sketch or chart.
2. **Let a variable represent an unknown quantity** (which is usually the quantity you are asked to find). Write down precisely what the variable represents. If there is more than one unknown, represent these unknowns in terms of the original variable when possible.
3. **Set up an equation** that expresses the relationship between the quantities in the problem.
4. **Solve the equation.**
5. **Answer the question.**
6. **Check the answer** by interpreting the solution in the context of the word problem.

EXAMPLE 11   Percentage Problem: The total cost (including tax) of a new car is $14,256. If the sales tax rate is 8 percent, how much do you pay in taxes?

Solution   Let x represent the price of the car; then

$$\text{Total cost} = \text{price of car} + \text{tax}$$
$$14{,}256 = x \qquad\qquad + 0.08x$$
$$14{,}256 = 1.08x$$
$$\frac{14{,}256}{1.08} = x$$
$$13{,}200 = x.$$

Tax $= 8$ percent of $\$13{,}200 = 0.08(\$13{,}200) = \$1{,}056$,
or tax $= \$14{,}256 - \$13{,}200 = \$1{,}056$. ∎

EXAMPLE 12   Proportion Problem: A cylindrical tank holds 480 gal when it is filled to its full height of 8 ft. When the gauge shows that it contains water at a height of 3 ft 1 in., how many gallons are in the tank?

Solution   A **ratio** is a comparison of two quantities by division; a **proportion** is a statement that two ratios are equal. If we let x represent the amount of water in the tank and set up two equal ratios that compare like measurements, we have the proportion

$$\frac{x}{480 \text{ gal}} = \frac{37 \text{ in.}}{96 \text{ in.}} . \qquad (\textit{Note: } 3 \text{ ft } 1 \text{ in.} = 37 \text{ in.})$$

To find x multiply both sides of this equation by 480 and simplify.

$$x = \left(\frac{37}{96}\right) \cdot 480$$

$$x = 185$$

Thus, there are 185 gal of water in the tank when the water is at a height of 3 ft 1 in. ■

EXAMPLE 13   Mixture Problem: A chemist has two acid solutions, one 30 percent acid and the other 70 percent acid. How much of each solution must be used to obtain 100 ml of a solution that is 41 percent acid?

Solution   Let x be the amount of the first solution in the mixture.

Solution	Percent Acid	Amount of Solution (ml)	=	Quantity of Acid (ml)
First solution	30	x		$0.3x$
Second solution	70	$100 - x$		$0.7(100 - x)$
New solution	41	100		$0.41(100)$

$$\begin{array}{ccc}
\text{Quantity of acid} & \text{Quantity of acid} & \text{Quantity of acid} \\
\text{in first solution} & + \quad \text{in second solution} & = \quad \text{in new solution}
\end{array}$$

$$\begin{aligned}
0.3x + 0.7(100 - x) &= 0.41(100) \\
0.3x + 70 - 0.7x &= 41 \\
-0.4x &= -29 \\
x &= \frac{-29}{-0.4} = 72.5
\end{aligned}$$

Thus, the chemist must mix 72.5 ml of the first solution with 27.5 ml of the second solution to obtain the desired mixture. ■

EXAMPLE 14   Projectile Problem: The height (y) of a ball thrown vertically up from the roof of a building 64 ft high with an initial velocity of 48 ft/second is given by the formula $y = 64 + 48t - 16t^2$. When will the ball strike the ground?

Solution   When the ball strikes the ground, the height (y) of the ball is 0. Therefore, we want to find the values of t for which $64 + 48t - 16t^2$ equals 0.

$$\begin{aligned}
0 &= 64 + 48t - 16t^2 \\
&= 16(4 + 3t - t^2) \\
&= 16(4 - t)(1 + t)
\end{aligned}$$

Since the first factor 16 cannot be 0, there are two ways that $0 = 16(4 - t)(1 + t)$ can be true.

$$4 - t = 0 \qquad 1 + t = 0$$
$$\text{or}$$
$$4 = t \qquad t = -1$$

The ball hits the ground 4 seconds later. We reject -1 as a solution since for this problem a negative time has no physical significance.   ■

EXERCISES 1.6

In Exercises 1–30 solve each equation and check your answer.

1. $3y - 14 = -4y + 7$ **2.** $3z - 8 = 13z - 9$

3. $8x - 10 = 3x$ **4.** $15 - 6y = 0$

5. $9 - 2y = 27 + y$ **6.** $15x + 5 = 2x + 14$

7. $4x - x = 3x$ **8.** $x - 1 = x + 3$

9. $2(x + 4) = 2(x - 1) + 3$

10. $2(x + 3) - 1 = 2x + 5$

11. $3[2x - (x - 2)] = -3(3 - 2x)$

12. $2y - (3y - 4) = 4(y + \frac{3}{4})$

13. $\dfrac{x - 2}{7} = -1$ **14.** $\dfrac{2y - 7}{9} = 5$

15. $\dfrac{x - 9}{3} = \dfrac{3x}{2} - \dfrac{11}{6}$ **16.** $\dfrac{2x - 5}{3} - \dfrac{x}{4} = \dfrac{1}{2}$

17. $\dfrac{3}{7} - \dfrac{n - 5}{14} = \dfrac{11n}{14}$ **18.** $\dfrac{5}{3a} - \dfrac{7}{4a} = \dfrac{5}{6}$

19. $\dfrac{2 + x}{6x} - 2 = \dfrac{3}{5x}$ **20.** $\dfrac{9y - 5}{7} = \dfrac{2y - 4}{3}$

21. $\dfrac{k}{k + 2} = \dfrac{2}{3}$ **22.** $\dfrac{x + 4}{x + 1} = \dfrac{x + 2}{x + 3}$

23. $\dfrac{z - 1}{z - 2} = \dfrac{z + 1}{z - 3}$ **24.** $\dfrac{1}{n - 3} + \dfrac{2}{3 - n} = \dfrac{1}{2}$

25. $\dfrac{2}{2 - x} + \dfrac{x}{x - 2} = 1$

26. $\dfrac{4}{x + 1} - \dfrac{3}{x - 1} = \dfrac{-6}{x^2 - 1}$

27. $\dfrac{2}{a^2 + a} = \dfrac{1}{a} - \dfrac{2}{a + 1}$ **28.** $\dfrac{3}{k + 2} - 2 = \dfrac{-5}{2k + 4}$

29. $\dfrac{5x}{x - 2} - \dfrac{4x}{2x - 7} = 3$ **30.** $3 - \dfrac{t - 4}{t - 3} = \dfrac{4t - 1}{2t + 3}$

In Exercises 31–38 solve each equation by using the factoring method.

31. $x^2 - 5x = 0$ **32.** $y^2 = 2y$

33. $y^2 - 2y - 8 = 0$ **34.** $z^2 - 3z - 18 = 0$

35. $3x^2 - 16x + 5 = 0$ **36.** $5y^2 - 11y + 2 = 0$

37. $3r^2 = 5r - 2$ **38.** $2x^2 = 7x + 4$

In Exercises 39–46 solve each equation by using the square root property.

39. $3x^2 + 24 = 0$ **40.** $2y^2 + 48 = 0$

41. $(b - 5)^2 = 4$ **42.** $(x + 3)^2 = 100$

43. $(x + 1)^2 = -1$ **44.** $(a - 1)^2 = -16$

45. $5(k - 7)^2 + 35 = 0$ **46.** $3(z + 4)^2 + 42 = 0$

In Exercises 47–50 solve each equation by completing the square.

47. $x^2 - 2x + 2 = 0$ **48.** $x^2 + 4x + 5 = 0$

49. $3x^2 - 4x - 2 = 0$ **50.** $5x^2 - 3x - 4 = 0$

In Exercises 51–60 solve each equation by using the quadratic formula.

51. $x^2 + 5x + 4 = 0$ **52.** $x^2 - 3x + 2 = 0$

53. $x^2 - 4 = x$ **54.** $x^2 + 2 = 2x$

55. $3x^2 - 2x = 6$ **56.** $2x^2 + x = 14$

57. $3x^2 = 2x - 1$ **58.** $4x^2 = 12x - 9$

59. $4x^2 + x = 3$ **60.** $5x^2 - 28x = 12$

In Exercises 61–80 solve each equation. In all cases, choose an efficient method for solving the particular quadratic equation involved in the problem.

61. $x^2 - x - 6 = 0$ **62.** $x^2 - 7x + 8 = 0$

63. $5r^2 + 20 = 16$ **64.** $4y^2 + 25 = 20y$

65. $3n^2 = 2n$ **66.** $4(a + 3)^2 = 9$

67. $t(t - 2) = 4$ **68.** $(x + 2)^2 = 9x$

69. $x^2 + 540 = 69x$ **70.** $76 = 100t - 16t^2$

71. $\dfrac{x^2}{2} - 5x = 0$ **72.** $\dfrac{x^2}{3} + \dfrac{x}{2} = 1$

73. $3x - 3 = \dfrac{2}{x}$ **74.** $2 - \dfrac{3}{x^2} = \dfrac{1}{x}$

75. $y + \dfrac{1}{y} = \dfrac{34}{15}$ **76.** $1 - \dfrac{900}{x^2} = 0$

77. $\dfrac{n-1}{2} = \dfrac{8}{n-1}$ **78.** $\dfrac{7t}{4} = \dfrac{t^2+6}{t}$

79. $\dfrac{x-3}{x-2} + \dfrac{2}{x^2-2x} = \dfrac{2}{x}$ **80.** $1 + \dfrac{12}{c^2-4} = \dfrac{3}{c-2}$

In Exercises 81–90 use the discriminant to determine the nature of the solutions of the equation.

81. $2x^2 - x + 2 = 0$ **82.** $-3x^2 + x + 1 = 0$
83. $x^2 + 2x + 1 = 0$ **84.** $5x^2 - 7x + 2 = 0$
85. $x^2 - 10x - 9 = 0$ **86.** $-2x^2 + 4x + 9 = 0$
87. $4x^2 = 2x + 5$ **88.** $3x^2 = -4x - 1$
89. $x^2 - 5x = -7$ **90.** $-2x^2 - x = -2$

91. The sum of the three angles in a triangle is 180°. The second angle is 20° more than the first and the third angle is twice the first. What is the measure of each of the angles in the triangle?

92. The length of a rectangle is three times its width. If the perimeter is 160 in., determine the area of the rectangle.

93. A magazine and a newsletter cost $1.10. If the magazine cost $1.00 more than the newsletter, how much did each cost?

94. If the total cost (including tax) of a new car is $13,054, how much do you pay in taxes if the sales tax rate is 7 percent?

95. A cylindrical tank holds 370 gal when it is filled to its full height of 9 ft. When the gauge shows that it contains oil at a height of 6 ft 7 in., how many gallons (to the nearest gallon) are in the tank?

96. If 72 gal of water flows through a feeder pipe in 40 minutes, how many gallons of water will flow through the pipe in 3 hours?

97. A map has a scale of 1.5 in. = 7 mi. To the nearest tenth of a mile, what distance is represented by 20 in. on the map?

98. A chemist has 5 liters of a 25 percent sulfuric acid solution. He wishes to obtain a solution that is 35 percent acid by adding a solution of 75 percent acid to his original solution. How much of the more concentrated acid must be added to achieve the desired concentration?

99. One bar of tin alloy is 35 percent pure tin and another bar is 10 percent pure tin. How many pounds of each must be used to make 95 lb of a new alloy that is 20 percent pure tin?

100. An alloy of copper and silver weighs 40 lb and is 20 percent silver. How much silver must be added to produce a metal that is 50 percent silver?

101. A 20 percent antifreeze solution is a solution that consists of 20 percent antifreeze and 80 percent water. If we have 20 qt of such a solution, how much

water should be added to obtain a solution that contains 10 percent antifreeze?

102. A race car driver must average 100 mi/hour for two laps around the track if he is to qualify for the finals. Because of mechanical trouble he is only able to average 60 mi/hour for the first lap. What speed must he average for the second lap if he is to qualify for the finals?

103. One of the classic problems from the history of mathematics is the ingenious method used by the Greek mathematician Eratosthenes to estimate the circumference of the earth in about 200 B.C. Eratosthenes obtained his estimate by noting the following information, which is illustrated here. Alexandria (where Eratosthenes was librarian) was 500 mi due north of the city of Syene. At noon on June 21 he knew from records that the sun cast no shadow at Syene, which meant that the sun was directly overhead. At the same time in Alexandria he measured from a shadow that the sun was 7.5° south from the vertical. By assuming that the sun was sufficiently far away for the light rays to be parallel to the earth, he then determined that angle AOB was 7.5°. Why? Complete the line of reasoning and obtain Eratosthenes' estimate for the circumference. You will find the result remarkably close to the modern estimate of 24,900 mi, which is obtained with the same basic procedure but with more accurate measurements.

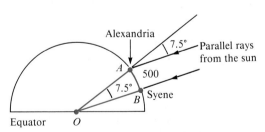

104. Consider this illustration of a piece of a broken chariot wheel found by a group of archaeologists. What is the radius of the wheel?

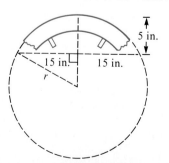

105. The age at which Diophantus (the Greek father of algebra) died is preserved in the following riddle: "Diophantus's youth lasted $\frac{1}{6}$ of his life. He grew a beard after $\frac{1}{12}$ more. After $\frac{1}{7}$ more of his life Diophantus married; five years later he had a son. The son lived exactly $\frac{1}{2}$ as long as his father, and Diophantus died just four years after his son." At what age did Diophantus die?

106. The perimeter of a rectangle is 100 ft and the area is 400 ft². Find the dimensions of the rectangle.

107. A rectangular plate is to have an area of 30 in.² and the sides of the plate are to differ in length by 4 in. To the nearest tenth of an inch, find the dimensions of the plate.

108. When a uniform border is added to a rug that is 6 ft by 4 ft, the area increases by 11 ft². Find the width of the border.

109. The height (y) of a projectile that is shot directly up from the ground with an initial velocity of 96 ft/second is given by the formula $y = 96t - 16t^2$.
  a. When will the projectile strike the ground after firing?
  b. At what time after firing is the projectile 80 ft off the ground?
  c. Physically, why are there two solutions to part b?

110. The height (y) above water of a diver t seconds after she steps off a platform 10 ft high is given by the formula $y = 10 - 16t^2$. When will the diver hit the water?

111. The height (y) of a ball that is thrown down from a roof 200 ft high with an initial velocity of 50 ft/second is given by the formula $y = 200 - 50t - 16t^2$. To the nearest tenth of a second, when will the ball hit the ground?

112. The height (y) of a projectile that is shot directly up from the ground with an initial velocity of 100 ft/second is given by the formula $y = 100t - 16t^2$. To the nearest hundredth of a second, when will the projectile initially attain a height of 50 ft?

113. From a square sheet of metal an open box is made by cutting 2-in. squares from the four corners and folding up the ends. To the nearest tenth of an inch, how large a piece of metal should be used if the box is to have a volume of 100 in.³?

114. When two resistors R_1 and R_2 are connected in series, their combined resistance is given by $R = R_1 + R_2$. If the resistors are connected in parallel, their combined resistance is given by $1/R = (1/R_1) + (1/R_2)$. Find the value of resistors R_1 and R_2 if their combined resistance is 32 ohms when connected in series and 6 ohms when connected in parallel.

115. Show that the sum of the solutions of the equation $ax^2 + bx + c = 0$ (with $a \neq 0$) is $-b/a$ and that the product of the solutions is c/a.

THINK ABOUT IT

1. a. Create a quadratic equation with solution set $\left\{\frac{2}{3}, -5\right\}$.
  b. If $4x^2 - 3x = k$, then what value for k leads to roots that are equal?

2. a. Create a word problem that may be solved by the equation
$$1{,}000 - x = 4x,$$
where x represents the unknown requested in the problem.
  b. In a word problem give two reasons why we check the answer by interpreting the solution in the context of the word problem instead of checking in the equation that has been set up to solve the problem.

3. a. Consider the following statement: "If two fractions are equal and have equal numerators, they also have equal denominators." Now consider the following equation, which we wish to solve for x.

$$\frac{x-1}{x+1} + 2 = \frac{3x+1}{x+5}$$

$$\frac{x-1}{x+1} + \frac{2(x+1)}{x+1} = \frac{3x+1}{x+5}$$

$$\frac{x-1+2(x+1)}{x+1} = \frac{3x+1}{x+5}$$

$$\frac{3x+1}{x+1} = \frac{3x+1}{x+5}$$

By the above statement we conclude that $x + 1 = x + 5$ or, upon subtracting x from both sides, that $1 = 5$. What is wrong? What value of x satisfies the equation?

  b. In the equation $ax = b$, a and b are constants. What combination of values for a and b results in an equation with exactly one solution? No solution? Infinitely many solutions?

4. In a "Think About It" exercise in Section 1.4 we discussed the golden ratio. One way to determine the exact value for this number is to answer the following question from Euclid's *Elements:* "Divide a line segment such that the ratio of the large part to the whole is equal to the ratio of the small part to the large."

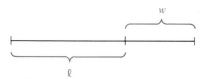

Determine the exact value of the golden ratio by setting up a proportion based on this condition and the segment shown above. Use the definition that the golden ratio equals ℓ/w and set w equal to 1 unit.

5. Consider the following expression, which continues indefinitely:

$$x = 1 + \cfrac{1}{1 + \cfrac{1}{1 + \text{etc.}}}$$

This expression is called a **continued fraction.** Continued fractions have been studied by many of the great mathematicians since the seventeenth century and are particularly useful in the analysis of numbers. For example, the continued fraction representation of π shows the following ratios to be succeedingly better approximations for this number: 22/7, 333/106, 355/113, and 103,993/33,102. The list could be continued, but for most purposes 355/113 is adequate since this ratio is accurate to seven significant digits. Determine the real number x in the continued fraction above. The solution requires a clever substitution.

REMEMBER THIS

1. Simplify $(\sqrt[3]{x^2 - 17})^3$.
2. Expand $(x + 4)^2$.
3. Expand $(2 - \sqrt{x})^2$.
4. Multiply $(2 + \sqrt{7})(2 - \sqrt{7})$.
5. Factor completely $9x^3 - 4x$.
6. Evaluate $|b - a|$ if $a = -5$ and $b = -12$.

7. Explain geometrically why $|-2| = 2$.
8. Express $\sqrt{-72} + \sqrt{-18}$ in the form $a + bi$.
9. Express $\dfrac{1}{3 - 2i}$ in the form $a + bi$.
10. List all correct classifications of the number $\sqrt[3]{-8}$ from the following categories: imaginary number, real number, rational number, irrational number, and integer.

1.7 Other Types of Equations

An equation in which the unknown appears under a radical is called a **radical equation.** Use these steps to solve such equations.

1. Isolate a radical on one side of the equation.
2. Raise both sides of the equation to the power that matches the index of the radical.
3. Solve the resulting equation and check all solutions in the *original* equation.

The check in step 3 is essential because this procedure is based on the following principle.

Principle of Powers

If P and Q are algebraic expressions, then the solution set of the equation $P = Q$ is a subset of the solution set of $P^n = Q^n$ for any positive integer n.

Thus, every solution of $P = Q$ is a solution of $P^n = Q^n$; but solutions of $P^n = Q^n$ may or may not be solutions of $P = Q$, so checking *is* necessary. Solutions of $P^n = Q^n$ that do not satisfy the original equation are called **extraneous solutions,** and Example 1 illustrates this possibility.

EXAMPLE 1 Solve $x - \sqrt{x + 1} = 1$.

Solution We must first isolate the radical on one side of the equation.

$$x - \sqrt{x + 1} = 1$$
$$x - 1 = \sqrt{x + 1}$$

Now square both sides of the equation and solve for x.

$$x^2 - 2x + 1 = x + 1$$
$$x^2 - 3x = 0$$
$$x(x - 3) = 0$$
$$x = 0 \quad \text{or} \quad x = 3$$

Check

$$0 - \sqrt{0 + 1} \overset{?}{=} 1$$
$$0 - 1 \overset{?}{=} 1$$
$$-1 \neq 1 \quad \text{Extraneous solution}$$

$$3 - \sqrt{3 + 1} \overset{?}{=} 1$$
$$3 - 2 \overset{?}{=} 1$$
$$1 \overset{\checkmark}{=} 1$$

Only 3 is a solution of the original equation, so the solution set is $\{3\}$. ∎

EXAMPLE 2 Solve the equation $\sqrt[3]{x^2 + 15} = 4$.

Solution The index of the radical is 3, so we raise both sides of the equation to the third power and then solve for x.

$$\sqrt[3]{x^2 + 15} = 4$$
$$(\sqrt[3]{x^2 + 15})^3 = 4^3$$
$$x^2 + 15 = 64$$
$$x^2 = 49$$
$$x = \pm 7$$

Check

$$\sqrt[3]{(7)^2 + 15} \overset{?}{=} 4$$
$$\sqrt[3]{64} \overset{?}{=} 4$$
$$4 \overset{\checkmark}{=} 4$$

$$\sqrt[3]{(-7)^2 + 15} \overset{?}{=} 4$$
$$\sqrt[3]{64} \overset{?}{=} 4$$
$$4 \overset{\checkmark}{=} 4$$

Since both solutions check, the solution set is $\{7, -7\}$. ∎

EXAMPLE 3 Solve the equation $\sqrt{x} + \sqrt{x + 5} = 5$.

Solution To solve the equation, we first isolate one of the radicals.

$$\sqrt{x} + \sqrt{x + 5} = 5$$
$$\sqrt{x + 5} = 5 - \sqrt{x}$$
$$(\sqrt{x + 5})^2 = (5 - \sqrt{x})^2 \qquad \text{Square both sides of the equation.}$$
$$x + 5 = 25 - 10\sqrt{x} + x$$
$$-20 = -10\sqrt{x}$$
$$2 = \sqrt{x}$$
$$(2)^2 = (\sqrt{x})^2 \qquad \text{Square both sides of the equation.}$$
$$4 = x$$

Check

$$\sqrt{4} + \sqrt{4 + 5} \stackrel{?}{=} 5$$
$$2 + 3 \stackrel{?}{=} 5$$
$$5 \stackrel{\checkmark}{=} 5$$

Thus, the solution set is $\{4\}$. Note that it is sometimes necessary to raise both sides of the equation to some power more than once. ∎

EXAMPLE 4 Solve the equation $x + \sqrt{x} - 2 = 0$.

Solution To solve the equation, we must first isolate the radical.

$$x + \sqrt{x} - 2 = 0$$
$$x - 2 = -\sqrt{x}$$
$$(x - 2)^2 = (-\sqrt{x})^2 \qquad \text{Square both sides of the equation.}$$
$$x^2 - 4x + 4 = x$$
$$x^2 - 5x + 4 = 0$$
$$(x - 4)(x - 1) = 0$$
$$x = 4 \quad \text{or} \quad x = 1$$

Check

$$4 + \sqrt{4} - 2 \stackrel{?}{=} 0 \qquad\qquad 1 + \sqrt{1} - 2 \stackrel{?}{=} 0$$
$$4 + 2 - 2 \stackrel{?}{=} 0 \qquad\qquad 1 + 1 - 2 \stackrel{?}{=} 0$$
$$4 \neq 0 \quad \text{Extraneous solution} \qquad\qquad 0 \stackrel{\checkmark}{=} 0$$

Therefore, 1 is the solution of the equation, and the solution set is $\{1\}$. ∎

In Example 4 the solution led to a quadratic equation. Although we often encounter quadratic equations in the process of solving other equations, $x + \sqrt{x} - 2 = 0$ is a special case. Equations that are not themselves quadratic but which are equivalent to equations having the form

$$at^2 + bt + c = 0 \qquad (a \neq 0)$$

are called **equations with quadratic form**. In Example 4 if we let $t = \sqrt{x}$, then $x + \sqrt{x} - 2 = 0$ becomes

$$t^2 + t - 2 = 0$$

with solution

$$(t + 2)(t - 1) = 0$$
$$t = -2 \quad \text{or} \quad t = 1.$$

Now we can resubstitute $\sqrt{x}$ for t and solve for x.

$$\sqrt{x} = -2 \quad \text{or} \quad \sqrt{x} = 1$$
$$x = 4 \quad \text{or} \quad x = 1$$

As in our solution above, only 1 is a root of the equation. The extraneous root $x = 4$ is easy to pick out here because $\sqrt{x}$ cannot be -2. To spot equations with quadratic form, look for the exponent in one term to be double the exponent in another term.

EXAMPLE 5   Solve $x^{-2} - 4x^{-1} - 3 = 0$.

Solution   First, note that

$$x^{-2} - 4x^{-1} - 3 = (x^{-1})^2 - 4(x^{-1}) - 3 = 0,$$

so if we let $t = x^{-1}$, we have

$$t^2 - 4t - 3 = 0.$$

By the quadratic formula,

$$t = \frac{4 \pm \sqrt{28}}{2} = 2 \pm \sqrt{7}.$$

Now we resubstitute x^{-1} for t and solve for x.

$$x^{-1} = 2 \pm \sqrt{7}$$
$$x = \frac{1}{2 \pm \sqrt{7}} = \frac{2 \pm \sqrt{7}}{-3}$$

Since x^{-1} may be any real number except 0, both roots check, and the solution set is $\left\{ \dfrac{2 \pm \sqrt{7}}{-3} \right\}$.   ∎

Higher-Degree Polynomial Equations

We have considered how to solve linear and quadratic equations which are first- and second-degree polynomial equations. Higher-degree polynomial equations are usually much harder to solve, and we will consider this topic in detail in Chapter 3. For now, however, it is sometimes possible to factor higher-degree polynomials into linear and/or quadratic factors with integer coefficients so that the equations can be solved by using the zero product principle. The idea is the same as with second-degree equations. First, rearrange the equation (if necessary) so that zero is on one side by itself, then factor. Now set each factor equal to zero and obtain solutions by solving the resulting equations.

EXAMPLE 6   Solve $x^5 = 10x^3 - 9x$.

Solution   First, rewrite the equation so that one side is 0.

$$x^5 = 10x^3 - 9x$$
$$x^5 - 10x^3 + 9x = 0$$

Now factor completely.

$$x(x^4 - 10x^2 + 9) = 0$$
$$x(x^2 - 9)(x^2 - 1) = 0$$
$$x(x + 3)(x - 3)(x + 1)(x - 1) = 0$$

Setting each factor equal to zero gives

$$x = 0 \quad \bigg| \quad \begin{matrix} x + 3 = 0 \\ x = -3 \end{matrix} \quad \bigg| \quad \begin{matrix} x - 3 = 0 \\ x = 3 \end{matrix} \quad \bigg| \quad \begin{matrix} x + 1 = 0 \\ x = -1 \end{matrix} \quad \bigg| \quad \begin{matrix} x - 1 = 0 \\ x = 1. \end{matrix}$$

Thus, the solution set is $\{0, -3, 3, -1, 1\}$. ∎

Absolute Value Equations

Recall from Section 1.1 that the absolute value of a real number a, denoted $|a|$, is defined by

$$|a| = \begin{cases} a \text{ if } a \geq 0 \\ -a \text{ if } a < 0 \end{cases}$$

and that $|a|$ is the distance between a and 0 on the number line. For example, $|-4| = 4$ as shown in Figure 1.8(a), while Figure 1.8(b) shows the general geometric interpretation of $|a|$.

Figure 1.8

We can extend the idea that an absolute value represents a distance on the number line. Given any two points a and b, the distance between them on the number line is $|a - b|$ [see Figure 1.9(a)]. For example, the

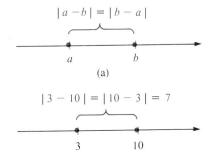

Figure 1.9

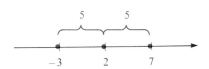

Figure 1.10

distance between 3 and 10 is $|3 - 10| = |-7| = 7$ [see Figure 1.9(b)]. Note that $|a - b| = |b - a|$, so it does not matter whether the smaller or larger number is labeled a. The next two examples show how the definition of absolute value may be used to solve equations containing absolute value.

EXAMPLE 7   Solve $|x - 2| = 5$.

Solution   By definition, $|a|$ equals either a or $-a$. Thus, $|x - 2|$ equals either $x - 2$ or $-(x - 2)$. Setting these expressions equal to 5, we have

$$x - 2 = 5 \qquad -(x - 2) = 5$$
$$\text{or}$$
$$x = 7 \qquad\qquad x = -3.$$

Thus, 7 and -3 are the solutions of the equation, and the solution set is $\{7, -3\}$. We can also arrive at the answers by geometric considerations. $|x - 2|$ can be interpreted as meaning the distance between 2 and some number x on the number line. When solving $|x - 2| = 5$, we are looking for the two numbers that are 5 units from 2 on the number line. As shown in Figure 1.10 the desired numbers are 7 and -3. ∎

EXAMPLE 8   Solve $|2x + 1| = |3 - x|$.

Solution   By definition, $|2x + 1| = \pm(2x + 1)$ and $|3 - x| = \pm(3 - x)$. Setting these expressions equal to each other produces two distinct cases:

$$2x + 1 = 3 - x \qquad \text{or} \qquad 2x + 1 = -(3 - x).$$

Finally, solving each of these equations separately gives

$$2x + 1 = 3 - x \qquad \text{or} \qquad 2x + 1 = -(3 - x)$$
$$3x = 2 \qquad\qquad\qquad 2x + 1 = -3 + x$$
$$x = \tfrac{2}{3} \qquad\qquad\qquad x = -4.$$

Thus, the solution set is $\{\tfrac{2}{3}, -4\}$. [*Note:* The given equation is also satisfied when $-(2x + 1) = -(3 - x)$ or when $-(2x + 1) = 3 - x$. However, these equations are equivalent to the two we solved and need not be considered.] ∎

Formulas

The translation of a relationship into mathematical symbols often results in a general equation or formula. For example, the formula

$$C = \tfrac{5}{9}(F - 32)$$

is often used to convert from degrees Fahrenheit to degrees Celsius. Although equations with more than one variable are very different from the conditional equations we have studied, the equation-solving techniques we learned may be used to rearrange formulas so they may be solved for any variable in the equation. For instance, since the formula $C = \frac{5}{9}(F - 32)$ is not very efficient for converting degrees Celsius to degrees Fahrenheit, we may solve for F as follows:

$$C = \tfrac{5}{9}(F - 32)$$
$$\tfrac{9}{5}C = F - 32 \qquad \text{Multiply both sides by } \tfrac{9}{5}.$$
$$\tfrac{9}{5}C + 32 = F. \qquad \text{Add 32 to both sides.}$$

Now we may select the more efficient version of the formula, depending on the given information.

EXAMPLE 9 Solve for y' (read "y prime"): $\dfrac{1}{2\sqrt{x}} + \dfrac{y'}{2\sqrt{y}} = 0$.

Solution

$$\frac{1}{2\sqrt{x}} + \frac{y'}{2\sqrt{y}} = 0$$

$$\frac{y'}{2\sqrt{y}} = \frac{-1}{2\sqrt{x}} \qquad \text{Subtract } \tfrac{1}{2\sqrt{x}} \text{ from both sides.}$$

$$y' = \frac{-2\sqrt{y}}{2\sqrt{x}} = \frac{-\sqrt{xy}}{x} \qquad \text{Multiply both sides by } 2\sqrt{y} \text{ and simplify.} \qquad\blacksquare$$

EXERCISES 1.7

In Exercises 1–36 solve each equation and check the solution.

1. $\sqrt{x + 5} = 2$

2. $\sqrt{2x + 1} = 7$

3. $\sqrt{x - 5} = -3$

4. $\sqrt{x - 2} + 3 = 0$

5. $2\sqrt{3x} + 3 = 13$

6. $3\sqrt{2x} - 4 = 5$

7. $\sqrt[4]{2x + 1} = \sqrt[4]{3x - 5}$

8. $\sqrt[4]{x^2 + x + 6} = \sqrt[4]{x^2 + 3x + 2}$

9. $\sqrt[3]{x + 1} = -2$

10. $\sqrt[3]{x^2 - 17} = 4$

11. $\sqrt{x^2 + 9} = x + 1$

12. $\sqrt{x^2 - x + 1} = x - 1$

13. $\sqrt{8 - x^2} + x = 0$

14. $\sqrt{x + 1} + 1 = x$

15. $\sqrt{x - 2} + x = 4$

16. $\sqrt{x + 4} - 4 = x$

17. $2\sqrt{x + 1} = 2 - x$

18. $2\sqrt{x - 1} = x - 1$

19. $\sqrt{2x^2 - 2} + x - 1 = 0$

20. $\sqrt{x - 4} = 5 - \sqrt{x + 1}$

21. $\sqrt{x - 8} + \sqrt{x} = 2$

22. $\sqrt{x + 5} - 9 = \sqrt{x - 4}$

23. $\sqrt{x} - \sqrt{2x + 1} = -1$

24. $\sqrt{4x + 2} + \sqrt{2x} - \sqrt{2} = 0$

25. $\sqrt{2x} + \sqrt{2x + 3} = 3$

26. $\dfrac{10}{\sqrt{x - 5}} = \sqrt{x - 5} - 3$

27. $\sqrt{x^2} = x$

28. $\sqrt{(2x - 1)^2} = 2x - 1$

29. $\dfrac{x}{\sqrt{x^2 + 16}} - 1 = 0$

30. $\dfrac{1}{4} \cdot \dfrac{2x}{2\sqrt{x^2 + 16}} - \dfrac{1}{5} = 0$

31. $\dfrac{1}{\sqrt{2}} = \dfrac{6 - a}{\sqrt{5(9 + a^2)}}$

32. $\dfrac{2a + 3}{\sqrt{5a^2 + 45}} = 1$

33. $x + 2\sqrt{x} - 3 = 0$ **34.** $x - 10\sqrt{x} + 9 = 0$

35. $x + 23 = 10\sqrt{x}$

36. $4x + 3 = 8\sqrt{x}$

In Exercises 37–64 solve each equation.

37. $x^4 - 6x^2 + 5 = 0$ **38.** $x^4 + 3 = 4x^2$

39. $4x^4 + 1 = 4x^2$ **40.** $3x^4 + 14x^2 - 5 = 0$

41. $x^4 - 4x^2 + 1 = 0$ **42.** $x^4 - 6x^2 + 4 = 0$

43. $(x + 3)^2 - 5(x + 3) + 4 = 0$
44. $(x - 1)^2 - (x - 1) = 0$
45. $4(x - 2)^2 + 9 = 12(x - 2)$
46. $3(x + 1)^2 + 13(x + 1) = 10$
47. $x^{-2} + x^{-1} = 0$ **48.** $3x^{-2} + 1 = 4x^{-1}$
49. $x^{1/2} + 2 = 3x^{1/4}$ **50.** $x^{1/3} - 1 = 2x^{-1/3}$
51. $y^3 + 9y^2 + 14y = 0$ **52.** $4a^3 - 13a^2 = -3a$
53. $3a^3 = 3a$ **54.** $16y^3 = 9y$
55. $x^2(2x - 1)(3x + 1) = 0$
56. $x^3(1 - 2x)(4x + 5) = 0$
57. $x^4 + 4 = 5x^2$
58. $x^6 = 10x^4 - 9x^2$
59. $3x^3 - 11x^2 = 3x - 11$
60. $2x^3 - x^2 - 8x + 4 = 0$
61. $x^3 - 8 = 0$ **62.** $x^3 + 1 = 0$
63. $x^6 - 1 = 0$ **64.** $x^6 = 64$

In Exercises 65–80 solve each absolute value equation.

65. $|x - 1| = 2$ **66.** $|x - 3| = 7$
67. $|\frac{1}{2}x - 9| = 8$ **68.** $|3x + 7| = 0$
69. $2|x| + 7 = 10$ **70.** $3|x - 1| + 2 = 11$
71. $|x| = x$ **72.** $|x| = -x$
73. $|x - 5| = 5 - x$ **74.** $|x - 10| = x - 10$
75. $|2x - 1| = x$ **76.** $|-x + 2| = x + 1$
77. $|x - 3| = |x + 1|$ **78.** $|1 - x| = |3x - 1|$
79. $|2x + 1| = |3 + x|$ **80.** $|6x - 4| = |2x - 7|$

In Exercises 81–102 solve each equation for the variable indicated.

81. $x + 2y = 4$ for y **82.** $7x - y = 2$ for x
83. $x = \sqrt[3]{y + 2}$ for y **84.** $t = \pi\sqrt{L/g}$ for g

85. $A = \frac{1}{2}h(b + c)$ for b **86.** $a = p(1 + rt)$ for t
87. $a = p + prt$ for p
88. $A = \pi(R^2 - r^2)$ for R^2
89. $R = \dfrac{kL}{d^2}$ for d^2 **90.** $S = \dfrac{a}{1 - r}$ for r
91. $C = \dfrac{E}{R + S}$ for R **92.** $I = \dfrac{2E}{R + 2r}$ for r
93. $Z = \dfrac{Z_1 Z_2}{Z_1 + Z_2}$ for Z_2 **94.** $\dfrac{E}{e} = \dfrac{R + r}{r}$ for r
95. $\dfrac{1}{f} = \dfrac{1}{a} + \dfrac{1}{b}$ for a **96.** $\dfrac{1}{R} = \dfrac{1}{R_1} + \dfrac{1}{R_2}$ for R_2
97. $y = \dfrac{x + 3}{x}$ for x **98.** $x = \dfrac{y - 1}{y - 2}$ for y
99. $2x + xy' + y - 2yy' = 0$ for y'
100. $8x + 3xy' + 3y - 2yy' = 0$ for y'
101. $3x^2 + 3[x^2y' + y(2x)] + 3y^2y' = 0$ for y'
102. $\dfrac{2}{3\sqrt[3]{x}} + \dfrac{2y'}{3\sqrt[3]{y}} = 0$ for y'

103. If the perimeter in this right triangle is 14 cm, find x.

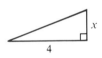

104. The perimeter in a right triangle is 36 m and the height is 3 m longer than the base. Find the base in the triangle.

THINK ABOUT IT

1. A common student error is to rewrite $\sqrt{x^2 + 4}$ as $x + 2$. Find all values of x for which
$$\sqrt{x^2 + 4} = x + 2$$
is a true statement.
2. Give five examples of equations with quadratic form.
3. Create a third-degree polynomial equation with solution set $\{4, 0, -3\}$.
4. **a.** Solve for x: $4^{1/4} + x^{1/3} = \dfrac{2}{2 - \sqrt{2}}$.

b. If $R > 0$, solve this equation for x.
$$(4x)\left(\dfrac{-2x}{2\sqrt{R^2 - x^2}}\right) + 4\sqrt{R^2 - x^2} = 0$$
5. Determine the value of p when the equation
$$\sqrt{(x - 3)^2 + y^2} = |x + 3|$$
is written in the form $y^2 = 4px$.

REMEMBER THIS

1. Insert the correct order symbol ($<, >, =$):
   $-5 \underline{\hspace{1cm}} -10$.
2. Why is $2 \geq 2$ a true statement?
3. What is the sign of a product of nonzero factors if the number of negative factors is odd?
4. For what values of x will $\sqrt{x}$ be a real number?
5. Evaluate $\dfrac{12 - y}{y - 4}$ when $y = 4$.
6. Simplify $\dfrac{x^3 + a^3}{x^2 - a^2}$.

7. Solve $\dfrac{7}{x} + \dfrac{4}{x^2} = 2$.
8. Express $\dfrac{-(-2) \pm \sqrt{(-2)^2 - 4(2)(5)}}{2(2)}$ in the form $a + bi$.
9. Use the discriminant to determine the nature of the solutions of $6x^2 - 4 = 3x$.
10. The height (y) of a projectile that is shot directly up from the ground with an initial velocity of 176 ft/second is given by $y = 176t - 16t^2$. When will the projectile strike the ground after firing?

1.8 Inequalities

The solution to a problem often requires us to use the inequality signs ($<, \leq, >, \geq$) when expressing relationships. For instance, in Example 3 we will determine the temperatures in degrees Fahrenheit at which water turns to steam by solving the inequality

$$\tfrac{5}{9}(F - 32) \geq 100.$$

As with equations, a **solution** of an inequality is a value for the variable that makes the inequality a true statement, and the set of all such solutions is called the **solution set.** To solve an inequality like $\tfrac{5}{9}(F - 32) \geq 100$, we use basically the same method that we developed to solve linear equations, but an important exception is discussed in Example 1.

EXAMPLE 1 $4 < 8$. If we multiply both sides of the inequality by -3, the result, $-12 < -24$, is a false statement. We can make this statement true by changing the "less than" to "greater than" so that the result is $-12 > -24$. In general, if we multiply an inequality by a negative number, we must change the sense of the inequality if the statement is to be true. The sense of the inequality means the direction in which the inequality symbol is pointing. ∎

With this example in mind, we now state the properties of inequalities that are the basis for solving inequalities.

Properties of Inequalities

Let a, b, and c be real numbers.

Comment

1. If $a < b$, then
$a + c < b + c$.

The sense of the inequality is preserved when the same number is added to (or subtracted from) both sides of an inequality.

2. If $a < b$ and $c > 0$, then
$ac < bc$.

The sense of the inequality is preserved when both sides of an inequality are multiplied (or divided) by the same positive number.

3. If $a < b$ and $c < 0$, then
$ac > bc$.

The sense of the inequality is reversed when both sides of an inequality are multiplied (or divided) by the same negative number.

4. If $a < b$ and $b < c$, then
$a < c$.

This property is called the **transitive property.**

Similar properties may be stated for the other inequality signs. We can prove any of these properties by using the appropriate definition from Section 1.1 and the fact that the sum and product of any two positive numbers is positive. For example, to prove that

$$\text{if } a < b \text{ and } c < 0, \text{ then } ac > bc,$$

we recall that $a < b$ if and only if $b - a$ is positive. Similarly, $c < 0$ implies $0 - c$, or $-c$, is positive. Then

$$\underbrace{(b - a)}_{\text{pos}}\underbrace{(-c)}_{\text{pos}} = \underbrace{-bc + ac}_{\text{pos}}.$$

Thus, $ac - bc$ is positive, so $bc < ac$, or alternatively $ac > bc$, which proves the property.

EXAMPLE 2   Solve the inequality $2x + 1 < 5x - 8$.

Solution   We isolate x on one side of the inequality in the following sequence of equivalent (same solution set) inequalities.

$$2x + 1 < 5x - 8$$
$$2x < 5x - 9 \qquad \text{Subtract 1 from both sides.}$$
$$-3x < -9 \qquad \text{Subtract } 5x \text{ from both sides.}$$
$$\frac{-3x}{-3} > \frac{-9}{-3} \qquad \text{Divide both sides by } -3 \text{ and change the sense of the inequality.}$$
$$x > 3$$

Figure 1.11

Thus, all real numbers greater than 3 make the original inequality a true statement, and the solution set, written in set-builder notation,* is $\{x: x > 3\}$. The graph of this solution set is the set of all points on the real number line representing numbers greater than 3 (see Figure 1.11). Note that we put a parenthesis at the point representing the number 3. This denotes that 3 is not a solution to the problem. When the endpoint is a member of the solution set, a bracket is used (as shown in Example 3). ∎

EXAMPLE 3 Water turns to steam when the temperature is at least 100 degrees Celsius. At what temperatures in degrees Fahrenheit will water turn to steam if the temperature scales are related by the formula $C = \frac{5}{9}(F - 32)$?

Solution If water turns to steam when $C \geq 100$, then by the given formula this event occurs when the Fahrenheit temperature satisfies

$$\tfrac{5}{9}(F - 32) \geq 100.$$

The easiest way to solve this inequality is to multiply both sides by $\frac{9}{5}$ to clear fractions. Doing this, we have

$$\tfrac{9}{5} \cdot \tfrac{5}{9}(F - 32) \geq \tfrac{9}{5} \cdot 100$$
$$F - 32 \geq 180$$
$$F \geq 212.$$

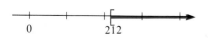

Figure 1.12 Fahrenheit temperatures for steam

Therefore, water turns to steam when the Fahrenheit temperature is at least 212 degrees. We express this result in a graph as shown in Figure 1.12. ∎

EXAMPLE 4 Solve the inequality $x > x + 1$.

Solution If we proceed in the usual way, we have

$$x > x + 1$$
$$x - x > x - x + 1 \qquad \text{Subtract } x \text{ from both sides.}$$
$$0 > 1.$$

When we subtract x from both sides, the resulting inequality is false. Since the original inequality is *equivalent* to a false statement, no replacements of x will make it true, and the solution set is $\emptyset$. ∎

Inequalities can be used to describe an interval between two numbers. For example, suppose the average (a) of your grades must be between 80 and 90 if you are to receive a grade of B in the course. We may describe this interval by using a pair of inequalities.

$$80 < a \quad \text{and} \quad a < 90$$

The inequality $80 < a$ specifies that the average must be above 80, while $a < 90$ specifies that the average must be below 90. The two inequalities

*Set-builder notation writes sets in the form $\{x: x$ has property $P\}$, which is read "the set of all elements x such that x has property P." The colon : is read "such that."

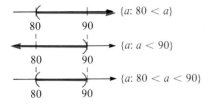

Figure 1.13

are both true only for averages between 80 and 90 (see Figure 1.13). The statement "$80 < a$ and $a < 90$" is usually expressed in compact form and written $80 < a < 90$.

Intervals such as $\{a: 80 < a < 90\}$ may be written more simply by using **interval notation.** With this notation, we merely write $(80,90)$ to specify the set of numbers between 80 and 90. More generally, if a and b are real numbers with $a < b$, then by definition,

$$(a,b) = \{x: a < x < b\}.$$

The set (a,b) is called the **open interval** from a to b, and the numbers a and b are called **endpoints** of the interval. As with graphs, interval notation uses a bracket if the endpoint is included in the set and a parenthesis if it is not. When an interval extends indefinitely, the symbols ∞, read "infinity," and $-\infty$ are used as shown in the following chart, which summarizes our methods for specifying intervals. ∞ and $-\infty$ are not real numbers but convenient symbols to help us designate intervals that are unbounded in the positive or negative direction.

Type of Interval	Interval Notation	Set Notation	Graph
Open interval	(a,b)	$\{x: a < x < b\}$	
Closed interval	$[a,b]$	$\{x: a \le x \le b\}$	
Half-open interval	$[a,b)$	$\{x: a \le x < b\}$	
	$(a,b]$	$\{x: a < x \le b\}$	
Infinite interval	$(-\infty,a)$	$\{x: x < a\}$	
	$(-\infty,a]$	$\{x: x \le a\}$	
	(a,∞)	$\{x: x > a\}$	
	$[a,\infty)$	$\{x: x \ge a\}$	
	$(-\infty,\infty)$	$\{x: x \text{ is a real number}\}$	

This chart shows that **intervals** are sets of real numbers that may be represented graphically as line segments, half lines, or the entire number line.

EXAMPLE 5   Solve $-4 < 5 - 3x < 4$ and write the solution in interval notation.

Solution   Although the given inequality is actually an abbreviated form of a pair of inequalities, $-4 < 5 - 3x$ and $5 - 3x < 4$, we can remain in compact form provided we use the properties of inequalities carefully.

$$-4 < 5 - 3x < 4$$
$$-9 < -3x < -1 \qquad \text{Subtract 5 from each member.}$$
$$3 > x > \frac{1}{3} \qquad \text{Divide each member by } -3 \text{ and change the sense of the inequalities.}$$

Figure 1.14

Figure 1.15

Thus, the solution set is the interval $(\frac{1}{3}, 3)$, as graphed in Figure 1.14. Note that we do not write $(3, \frac{1}{3})$ since a must be less than b when writing (a,b). ∎

EXAMPLE 6   Designate the set of numbers graphed in Figure 1.15 by using (a) set-builder notation and (b) interval notation.

Solution

a. We need to specify the set of numbers that are either less than -3 or greater than 1, so in set-builder notation we write

$$\{x: x < -3 \text{ or } x > 1\}.$$

We cannot write $1 < x < -3$ since this statement means $x < -3$ and $x > 1$ *simultaneously*, and no number satisfies both inequalities at the same time.

b. To use interval notation, we first note that the **union** of two sets A and B is the set of all elements belonging to A or B and is denoted by $A \cup B$. Then in interval notation the set of numbers less than -3 is written $(-\infty, -3)$, while the set of numbers greater than 1 is written $(1, \infty)$. Thus, the set of numbers belonging to either of these intervals is

$$(-\infty, -3) \cup (1, \infty).$$ ∎

Intervals on the number line are frequently specified in higher mathematics by inequalities that contain absolute value expressions. The most important such expression is of the form

$$|a| < b,$$

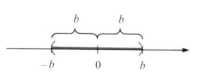

Figure 1.16 $|a| < b$ means a is between $-b$ and b.

where b is some positive number. Remember that $|a|$ gives the distance between 0 and a on the number line. Geometrically then, $|a| < b$ says a is contained in an open interval that begins and ends b units from 0. As shown in Figure 1.16 this condition means that a must be between $-b$ and b. Thus, we have the following important result, which indicates how we algebraically handle these expressions.

> If $b > 0$, then $|a| < b$ is equivalent to $-b < a < b$.

The following examples show the procedure for applying this rule.

EXAMPLE 7   Solve $|x - 4| < 3$.

Solution   Applying the above rule with $a = x - 4$ and $b = 3$, we rewrite

$$|x - 4| < 3$$

as

$$-3 < x - 4 < 3.$$

Now add 4 to each member to obtain

$$1 < x < 7.$$

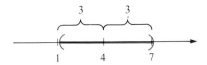

Figure 1.17 $|x - 4| < 3$ means x is between 1 and 7.

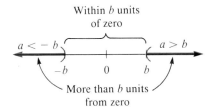

Figure 1.18

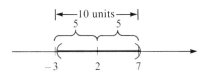

Figure 1.19 $|x - 4| > 3$ means $x > 7$ or $x < 1$.

Thus, any number between 1 and 7 satisfies the given inequality, and the solution set is $(1,7)$. We can also arrive at the answer by geometric considerations. $|x - 4| < 3$ can be interpreted as meaning that the distance between 4 and some number x on the number line must be less than 3 units. Figure 1.17 shows the interval defined by being within 3 units of the number 4 starts at 1 and ends at 7.   ∎

Another common expression for specifying intervals on the number line is of the form

$$|a| > b,$$

where b is some positive number. Once again remember that $|a|$ gives the distance between 0 and a on the number line. Geometrically then, $|a| > b$ says a is located more than b units from zero. Figure 1.18 shows this condition means that a is either to the right of b or to the left of $-b$. Thus, we have the following algebraic rule for working with these expressions.

If $b > 0$, then $|a| > b$ is equivalent to $a > b$ or $a < -b$.

Example 8 shows how this rule can be applied.

EXAMPLE 8   Solve $|x - 4| > 3$.

Solution   Applying the above rule with $a = x - 4$ and $b = 3$, we rewrite

$$|x - 4| > 3$$

as

$$x - 4 > 3   \text{ or }   x - 4 < -3$$

which simplifies to

$$x > 7   \text{ or }   x < 1$$

so the solution set is $(-\infty,1) \cup (7,\infty)$. This answer checks with our geometric interpretation since the numbers located more than 3 units from 4 on the number line are either to the left of 1 or the right of 7 (see Figure 1.19).   ∎

Quadratic inequalities are inequalities that may be expressed in the forms

$$ax^2 + bx + c > 0 \qquad\qquad ax^2 + bx + c \geq 0$$
$$ax^2 + bx + c < 0 \qquad \text{or} \qquad ax^2 + bx + c \leq 0,$$

where a, b, and c are real numbers with $a \neq 0$. A method for solving such inequalities that relies on factoring is shown in Example 9.

EXAMPLE 9   Solve $x^2 + 2x - 3 > 0$.

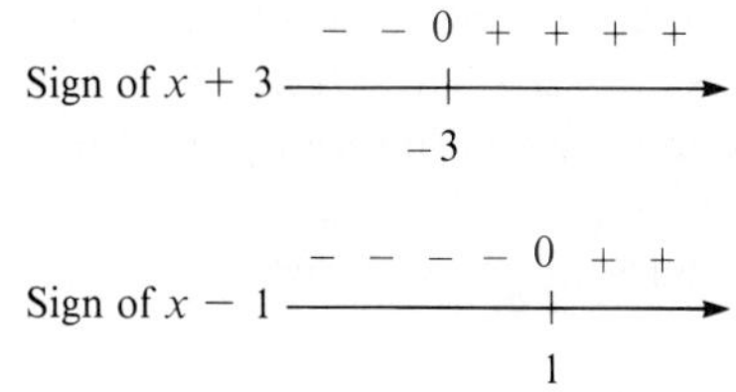

Sign of $x + 3$

$$-\ -\ 0\ +\ +\ +\ +$$
$$-3$$

Sign of $x - 1$

$$-\ -\ -\ -\ 0\ +\ +$$
$$1$$

Figure 1.20

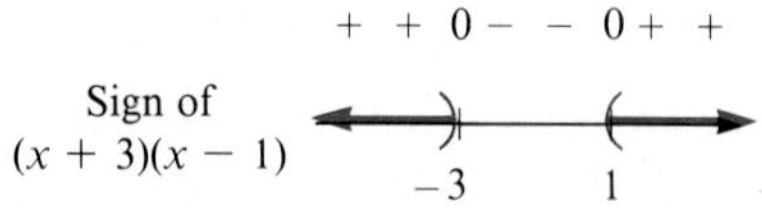

Sign of $(x + 3)(x - 1)$

$$+\ +\ 0\ -\ -\ 0+\ +$$
$$-3\quad 1$$

Figure 1.21

Solution   As with second-degree equations, we want zero by itself on one side of the inequality. In this case, the right side is already zero, so factor the left side.

$$x^2 + 2x - 3 > 0$$
$$(x + 3)(x - 1) > 0$$

We now make up the chart in Figure 1.20, which shows the signs of both factors for the various values of x. Note the importance of the numbers -3 and 1 in the figure. Real numbers that make $ax^2 + bx + c$ zero are called **critical numbers,** and the solution hinges on these numbers. The sign of $(x + 3)(x - 1)$ is now easy to determine, and this product is positive provided x is to the left of -3 or to the right of 1. Thus, $(x + 3)(x - 1)$ is greater than zero if $x < -3$ or $x > 1$, as shown in Figure 1.21. We designate the solution set in interval notation by writing

$$(-\infty, -3) \cup (1, \infty)$$

as discussed in Example 6.   ∎

Note in Example 9 that the two critical numbers separate the number line into three intervals, namely $(-\infty, -3)$, $(-3, -1)$, and $(-1, \infty)$. Throughout each interval also note that all numbers produce the same sign in $x^2 + 2x - 3$. Thus, an alternative approach for solving quadratic inequalities is to simply find the sign of a convenient number in each of the intervals determined by the critical numbers. The solution set may then be determined by comparing the resulting sign to the inequality in question, as shown in Example 10.

EXAMPLE 10   Solve $1 \geq x^2$.

Solution   First, rewrite the inequality so one side is zero. Then factor.

$$1 \geq x^2$$
$$1 - x^2 \geq 0$$
$$(1 + x)(1 - x) \geq 0$$

Because $(1 + x)(1 - x)$ equals 0 when $x = -1$ and when $x = 1$, the critical numbers are -1 and 1. These numbers separate the number line into the intervals $(-\infty, -1)$, $(-1, 1)$, and $(1, \infty)$, and Figure 1.22 shows whether a true statement results when a specific number in each of these intervals is tested.

Figure 1.22

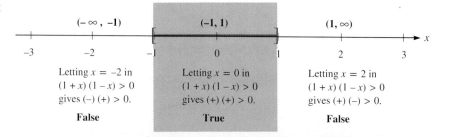

All numbers in each interval lead to the same result as the specific number tested, so $(1 + x)(1 - x) > 0$ is a true statement on the interval $(-1,1)$. Because the inequality in question is greater than or equal to zero, -1 and 1 are also solutions, and the solution set is the interval $[-1,1]$. ∎

In the next example the quadratic formula provides an efficient method for finding the critical numbers which are irrational.

EXAMPLE 11   Solve $x^2 - 2x - 1 > 0$.

Solution   First, determine the critical numbers by using the quadratic formula to find when $x^2 - 2x - 1$ equals zero.

$$x = \frac{-(-2) \pm \sqrt{(-2)^2 - 4(1)(-1)}}{2(1)}$$

$$= \frac{2 \pm \sqrt{8}}{2} = \frac{2 \pm 2\sqrt{2}}{2} = 1 \pm \sqrt{2}$$

The critical numbers are $1 + \sqrt{2}$ and $1 - \sqrt{2}$, which are approximately equal to 2.4 and -0.4, respectively. Now we may test in the intervals determined by these critical numbers, as shown in Figure 1.23.

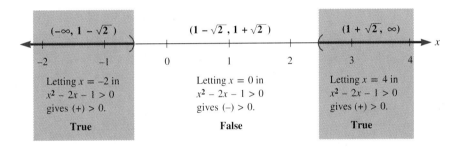

Figure 1.23

The tests show the inequality $x^2 - 2x - 1 > 0$ is true for all x such that $x < 1 - \sqrt{2}$ or $x > 1 + \sqrt{2}$. Thus the solution set is $(-\infty, 1 - \sqrt{2}) \cup (1 + \sqrt{2}, \infty)$. ∎

The methods for solving quadratic inequalities may be applied to higher-degree polynomial inequalities. This extension is based on the definition that critical numbers are real numbers that make the *polynomial* in question zero, and the principle that throughout each of the intervals determined by the critical numbers, the sign of the polynomial remains the same.

EXAMPLE 12   Solve $x(2x - 3)(x + 2) < 0$.

Solution   The inequality is already in a workable form; that is, it is factored and one side is zero. Therefore we go straight to the analysis in Figure 1.24. Critical numbers are 0, $\frac{3}{2}$, and -2·because $x(2x - 3)(x + 2)$ equals zero when $x = 0$, $x = \frac{3}{2}$, and $x = -2$.

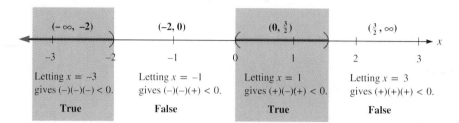

Figure 1.24

Thus, $x(2x - 3)(x + 2) < 0$ is true if $x < -2$ or $0 < x < \frac{3}{2}$, and the solution set is $(-\infty, -2) \cup (0, \frac{3}{2})$.   ■

Inequalities involving quotients of polynomials may also be solved by our current methods if we define critical numbers in such problems to be real numbers that make either the numerator zero or the denominator zero. When analyzing quotients, it is important to remember that division by zero is undefined, so critical numbers that make the denominator zero are never included in the solution set.

EXAMPLE 13   For what values of x will $\sqrt{\dfrac{3x - 1}{x + 1}}$ be a real number?

Solution   For the square root to be a real number, the expression under the radical must be greater than or equal to 0. Thus, we need to solve

$$\frac{3x - 1}{x + 1} \geq 0.$$

The numerator is zero when $x = \frac{1}{3}$, and the denominator is zero when $x = -1$. So $\frac{1}{3}$ and -1 are critical numbers, and we may test in the intervals determined by these two numbers, as shown in Figure 1.25.

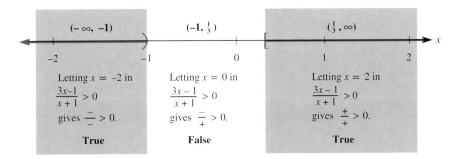

Figure 1.25

We see in Figure 1.25 that $\dfrac{3x-1}{x+1} > 0$ is true if $x < -1$ or $x > \frac{1}{3}$. The inequality in question is $\geq$, so we also determine that $\frac{1}{3}$ is included and -1 is excluded from the solution set, because the quotient is 0 when $x = \frac{1}{3}$ and undefined when $x = -1$. Thus, the solution set is $(-\infty, -1) \cup [\frac{1}{3}, \infty)$. ∎

EXAMPLE 14   Solve $\dfrac{5x+1}{x+1} \geq 2$.

Solution   First, rewrite the inequality so the right side is zero. Then simplify the resulting expression on the left side into a single fraction.

$$\frac{5x+1}{x+1} \geq 2$$

$$\frac{5x+1}{x+1} - 2 \geq 0$$

$$\frac{5x+1-2(x+1)}{x+1} \geq 0$$

$$\frac{3x-1}{x+1} \geq 0$$

This inequality is solved in Example 13, and in interval notation the solution set is $(-\infty, -1) \cup [\frac{1}{3}, \infty)$. ∎

EXERCISES 1.8

In Exercises 1–14 solve the inequality and graph each solution.

1. $2x + 9 < 0$
2. $3 - 4x > 0$
3. $13 - 2x < 7 - 3x$
4. $16x - 7 \leq 17x - 4$
5. $-2x + 5 > 5x - 7$
6. $1 - 4x \geq 2x + 3$
7. $\dfrac{x}{2} + 4 < 8 + x$
8. $24 - 13x \geq -12.5x + 26$
9. $x > x - 1$
10. $x < x - 1$
11. $3x < 4x$
12. $-x \geq 2x$
13. $2(x + 3) \geq 7 + 2x$
14. $7 - 4y \leq 4(7 - y)$

In Exercises 15–24 complete the chart and provide the alternative designations for the given set of real numbers.

	Interval Notation	Set Notation	Graph
15.	$[1,4)$		
16.	$(-\infty,1)$		
17.		$\{x: x \le 3\}$	
18.		$\{x: -1 < x \le 1\}$	
19.			$-3 \quad 0$
20.			
21.	$[0,\infty)$		
22.		$\{x: 0 < x\}$	
23.			$-2 \quad 2$
24.			$-2 \quad 2$

In Exercises 25–30 solve each inequality and write the solution set in interval notation.

25. $-2 < x - 4 < 2$
26. $0 < x + 7 < 10$
27. $-5 \le 2 - x \le 7$
28. $-0.1 \le 1 - x \le 0.1$
29. $-1 \le \dfrac{5 - 2x}{3} < \dfrac{1}{2}$
30. $-\dfrac{9}{4} < \dfrac{3x + 2}{-2} \le -\dfrac{1}{4}$

In Exercises 31–50 solve the inequality. In Exercises 31–40 use both an algebraic and a geometric interpretation.

31. $|x| < 4$
32. $|x| \le \frac{1}{2}$
33. $|x - 5| \le 3$
34. $|x - 3| < 0.01$
35. $|x - a| < 1$
36. $|x - a| < d, (d > 0)$
37. $|x| > 2$
38. $|x| > 0$
39. $|x - 5| \ge 3$
40. $|x| < -2$
41. $|3x + 5| < 8$
42. $|6x - 7| \le 10$
43. $|1 - 2x| \le 13$
44. $|3 - 4x| < 5$
45. $|2x - 1| > 3$
46. $|1 - x| + 2 \ge 4$
47. $|x| < x$
48. $x < |x|$
49. $|3x - 7| < x$
50. $|3x - 7| > x$

In Exercises 51–104 solve each inequality.

51. $(x + 1)(x - 2) > 0$
52. $(x - 3)(x + 2) < 0$
53. $y(1 - y) \ge 0$
54. $t(1 - 3t) \le 0$
55. $x^2 + x - 6 < 0$
56. $x^2 - 4x - 5 > 0$
57. $1 - x^2 \le 0$
58. $z^2 - 9 \ge 0$
59. $y^2 - 3 > 0$
60. $7 - x^2 < 0$
61. $k^2 > 5k$
62. $2x \le x^2$
63. $b^2 + b \le 2$
64. $10 - c^2 > -3c$
65. $2x^2 \ge 7x + 4$
66. $3y^2 + 2 < 5y$
67. $x^2 + 4 \ge 0$
68. $x^2 + 9 \le 0$
69. $(a - 1)^2 \le 0$
70. $(3x - 1)^2 > 0$
71. $x^2 - 4x - 3 > 0$
72. $x^2 - 2x - 5 < 0$
73. $3x^2 \le 4x + 2$
74. $5x^2 \ge 3x + 4$
75. $2x^2 - x + 3 < 0$
76. $2x^2 - x + 3 > 0$
77. $x(x - 1)(x - 2) \ge 0$
78. $b(4b + 3)(3b - 4) > 0$
79. $a(2a - 1)^2 < 0$
80. $y^2(1 - 3y) > 0$
81. $(1 - y)(2 - y)(3 - y)^2 < 0$
82. $k(4k + 3)^2(3k - 4) \ge 0$
83. $t^3 > 4t$
84. $x^4 \le 4x^2$
85. $x^4 - 5x^2 \le -4$
86. $x^3 + 8x^2 + 12x < 0$
87. $c^5 + 9c \ge 10c^3$
88. $a^6 + 9a^2 > 10a^4$
89. $2x^3 - 5x^2 < 2x - 5$
90. $3n^3 - 2n^2 - 6n + 4 > 0$
91. $\dfrac{t - 3}{t + 2} \le 0$
92. $\dfrac{x + 2}{x - 3} > 0$
93. $\dfrac{(x - 1)(x + 1)}{x} \le 0$
94. $\dfrac{a(4 - a)}{2a + 3} \ge 0$
95. $\dfrac{x^2 - 4}{3 - x} \ge 0$
96. $\dfrac{t^2 + 3t - 4}{3t + 2} \le 0$
97. $\dfrac{1}{a} < 2$
98. $\dfrac{2}{x} > 5$
99. $\dfrac{1}{4} \ge \dfrac{7}{7 - x}$
100. $\dfrac{1}{y + 2} \le \dfrac{1}{3}$
101. $\dfrac{2m}{m - 4} \le 3$
102. $\dfrac{-x}{2x + 1} \ge 1$
103. $\dfrac{3}{x - 2} \le \dfrac{3}{x + 3}$
104. $\dfrac{2}{x - 1} < \dfrac{1}{x + 1}$

Intervals appear in a natural way whenever we analyze data based on measurements. Measurements of one kind or another are essential to both scientific and nontechnical work. Weight, distance, time, volume, and temperature are but a few of the quantities for which measurements are required. However, any reading is an approximation that can only be as accurate as the measuring instruments allow. Thus, if we record a man's weight as 164 lb, we guarantee only that his exact weight is somewhere in the interval [163.5 lb, 164.5 lb). Use this concept to answer Exercises 105 and 106.

105. The circumference of the earth through the North and South Poles is about 25,000 mi. Use interval notation to write an interval that contains the earth's exact polar circumference. If we use a more accurate measuring system and record the polar circumference as 24,860 mi, which interval now contains the exact distance?

106. To the nearest inch, the height of a woman is 69 in. Use interval notation to write an interval that contains the woman's exact height. If we use a more accurate measuring device and record her height as 69.3 in., which interval now contains the woman's exact height?

107. In Lobachevskian geometry the sum of the three angles in a triangle is always less than 180°. If the second angle is 20° more than the first, and the third is twice the first, how many degrees might there be in the first angle?

108. If we categorize any temperature in adults above 98.6 degrees Fahrenheit as a "fever," then what is the temperature range of a fever in degrees Celsius? Use $F = \frac{9}{5}C + 32$.

109. Suppose a student's average grade on four tests must be in the interval [74.5,79.5) to receive a grade of C^+. If the student has grades of 81, 70, and 76 so far, find all possible grades for the last test that result in a grade of C^+.

110. What set of real numbers has the following property: The square of the number is smaller than the number itself?

111. What set of real numbers satisfies the following condition: When the number and its square are added together, the sum is between 6 and 42?

112. The length of a rectangle is 2 ft more than the width, and the area is less than 63 ft². What are the possibilities for the length in such a case?

113. The height (y) of a projectile that is shot directly up from the ground with an initial velocity of 96 ft/second is given by the formula $y = 96t - 16t^2$. During what period is the projectile more than 80 ft off the ground?

114. For what values of x will $\sqrt{4 - x^2}$ be a real number?

115. For what values of x will $\sqrt{x^2 - 4}$ be a real number?

116. For what values of x will $\sqrt{\dfrac{x + 1}{x - 2}}$ be a real number?

117. If $x^2 + 9 = kx$, then what values for k lead to roots that are not real numbers?

118. A projectile fired vertically up from the ground with an initial velocity of 128 ft/second will hit the ground 8 seconds later, and the speed of the projectile in terms of the elapsed time t equals $|128 - 32t|$. For what values of t is the speed of the projectile less than 80 ft/second?

In Exercises 119–122 let a, b, c, and d be real numbers and prove each statement by using the definition of less than as demonstrated in this section.

119. If $a < b$, then $a + c < b + c$.

120. If $a < b$ and $c > 0$, then $ac < bc$.

121. If $a < b$ and $b < c$, then $a < c$.

122. If $a < b$ and $c < d$, then $a + c < b + d$.

123. The geometric mean of a and b is given by $\sqrt{ab}$. Show that the geometric mean is always less than or equal to the arithmetic mean. That is, show $\sqrt{ab} \leq (a + b)/2$ for all $a \geq 0$, $b \geq 0$. (*Hint:* Start with $x^2 \geq 0$, which is true for all values of x, and replace x by $\sqrt{b} - \sqrt{a}$. Then work your way to the desired result.)

124. Use $|x| = \sqrt{x^2}$ to show that
$$|a + b| \leq |a| + |b|.$$
(*Hint:* $a^2 + 2ab + b^2 \leq |a|^2 + 2|a||b| + |b|^2$.)

THINK ABOUT IT

1. **a.** Create a quadratic inequality whose solution set is $[-4,6]$.
    b. Create an inequality involving a quotient that is less than or equal to 0 whose solution set is $(-\infty,-5) \cup [2,\infty)$.
2. Give examples of values for the unknowns that show the following statements are *false*.
    a. If $a < b$, then $ac < bc$.
    b. If $a < b$, then $a^2 < b^2$.
    c. If $a > 0$, then $a^2 \geq a$.
    d. If $a < 0$ and $b > 1$, then $ab < -1$.
3. Take two identical sheets of notebook paper. Form two cylinders from these rectangular ($y > x$) sheets: one with the length as the base [diagram (a)], the other with the width as the base [diagram (b)]. Show that the cylinder formed in (a) will always have the greater volume. (*Note*: $V = \pi r^2 h$.)

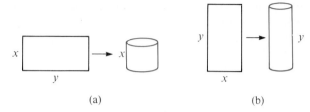

(a)   (b)

4. Use inequalities and absolute value to describe the set of all real numbers x such that x is within d units of a but not equal to a.
5. **a.** Use the property $\left|\dfrac{a}{b}\right| = \dfrac{|a|}{|b|}$ to solve
    $$\left|\frac{1}{x}\right| > 3.$$
    b. Solve $\left|\dfrac{x}{x-2}\right| \geq 3$.

REMEMBER THIS

1. Express the set of real numbers in interval notation.
2. Solve $2x^4 + 6x^2 = 7x^3$.
3. Solve $\dfrac{3x}{x+2} - 2 = \dfrac{2x-3}{2x-1}$.
4. Divide $\dfrac{9x^2 - 6x + 1}{3x - 9} \div \dfrac{9x - 3}{9 - x^2}$.
5. Simplify $\dfrac{x^{1/2}y^0z}{xyz^{-1}}$.
6. Solve for y: $x^2 + y^2 = 1$.
7. Solve for x: $\dfrac{1}{x} - \dfrac{1}{a} = \dfrac{1}{b}$.
8. Determine by checking if $5 - i$ is a solution of $x^2 + 26 = 10x$.
9. For what values of x will $\sqrt{4 - x^2}$ be a real number?
10. The perimeter in a right triangle is 30 cm and the base is 7 cm longer than the height. Find the base and the height in the triangle.

CHAPTER OVERVIEW

Section	Key Concepts to Review
1.1	• Definitions of integers, rational numbers, irrational numbers, real numbers, $a < b$, $a > b$, $a \leq b$, absolute value, factor, power, exponent, variable, constant, algebraic expression, terms, (numerical) coefficient, and similar terms
	• Relationships among the various sets of numbers
	• Statements of basic properties of real numbers
	• Methods to add, subtract, multiply, and divide real numbers
	• Order of operations
	• Methods to evaluate an algebraic expression
	• Methods to add or subtract algebraic expressions

Section	Key Concepts to Review
	• Guidelines for using a scientific calculator • Laws of exponents (m and n denote integers) **1.** $a^m \cdot a^n = a^{m+n}$ **2.** $(a^m)^n = a^{mn}$ **3.** $(ab)^n = a^n b^n$ **4.** $\left(\dfrac{a}{b}\right)^n = \dfrac{a^n}{b^n}$ $(b \neq 0)$ **5.** $\dfrac{a^m}{a^n} = a^{m-n} = \dfrac{1}{a^{n-m}}$ $(a \neq 0)$ **6.** $a^0 = 1$ $(a \neq 0)$ **7.** $a^{-n} = \dfrac{1}{a^n}$ $(a \neq 0)$
1.2	• Definitions of polynomial, monomial, binomial, trinomial, degree of a monomial, and degree of a polynomial • Methods to multiply various types of algebraic expressions • FOIL multiplication method • Methods to factor an expression that contains a factor common to each term, that is the difference of squares, that is a trinomial, or that is the sum or difference of cubes • Method to determine if $ax^2 + bx + c$ can be factored with integer coefficients • Factoring and product models
1.3	• Definitions of LCD and complex fractions • Methods to simplify, multiply, divide, add, and subtract fractions; to find the LCD; and to simplify a complex fraction • Equality principle: $\dfrac{a}{b} = \dfrac{c}{d}$ if and only if $ad = bc$ $(b, d \neq 0)$ • Fundamental principle: $\dfrac{ak}{bk} = \dfrac{a}{b}$ $(b, k \neq 0)$ • Operations principles $(b, d, c/d \neq 0)$ $\dfrac{a}{b} \cdot \dfrac{c}{d} = \dfrac{ac}{bd}$ $\qquad$ $\dfrac{a}{b} \pm \dfrac{c}{b} = \dfrac{a \pm c}{b}$ $\dfrac{a}{b} \div \dfrac{c}{d} = \dfrac{a}{b} \cdot \dfrac{d}{c}$ $\qquad$ $\dfrac{a}{b} \pm \dfrac{c}{d} = \dfrac{ad \pm bc}{bd}$
1.4	• Definitions of $a^{1/n}$, index and radicand of a radical, principal square root, nth root of a number, and conjugates • If m/n represents a reduced fraction such that $a^{1/n}$ represents a real number, then $\quad a^{m/n} = (a^{1/n})^m = (a^m)^{1/n}$, or equivalently, $a^{m/n} = (\sqrt[n]{a})^m = \sqrt[n]{a^m}$ • Methods to simplify, add, subtract, multiply, and divide radicals • Methods to rationalize the denominator or the numerator • Properties of radicals ($a, b, \sqrt[n]{a}, \sqrt[n]{b}$ denote real numbers) **1.** $(\sqrt[n]{a})^n = a$ $\qquad\qquad$ **2.** $\sqrt[n]{a} \cdot \sqrt[n]{b} = \sqrt[n]{ab}$ **3.** $\dfrac{\sqrt[n]{a}}{\sqrt[n]{b}} = \sqrt[n]{\dfrac{a}{b}}$ $(b \neq 0)$ $\qquad$ **4.** $\sqrt[n]{a^n} = \begin{cases} a, & \text{if } n \text{ is odd} \\ |a|, & \text{if } n \text{ is even} \end{cases}$

Section	Key Concepts to Review
1.5	• Definitions of imaginary number, complex number, equality for complex numbers, and the conjugate of a complex number
	• Methods to add, subtract, multiply, and divide complex numbers
	• Properties of conjugates
	• Methods to find powers of i
	• Relationships among the various sets of numbers
	• $i = \sqrt{-1}$ and $i^2 = -1$
1.6	• Definitions of equation, conditional equation, identity, equivalent equation, solution set, linear equation, proportion, and quadratic equation
	• Methods to obtain an equivalent equation
	• Methods to solve an equation containing fractions
	• Zero product principle: $a \cdot b = 0$ if and only if $a = 0$ or $b = 0$
	• Square root property: $x^2 = n$ implies $x = \sqrt{n}$ or $x = -\sqrt{n}$
	• Methods to solve a quadratic equation by the factoring method, by the square root property, by completing the square, and by using the quadratic formula
	• Quadratic formula: If $ax^2 + bx + c = 0$ and $a \neq 0$, then $$x = \frac{-b \pm \sqrt{b^2 - 4ac}}{2a}$$
	• Method to determine the nature of the solutions to a quadratic equation from the discriminant $(b^2 - 4ac)$
	• Guidelines for setting up and solving word problems
1.7	• Definitions of radical equation and equations with quadratic form
	• Principle of powers
	• Methods to solve radical equations, equations with quadratic form, higher-degree polynomial equations that factor into linear and/or quadratic factors, and absolute value equations
	• If $b \geq 0$, then $|a| = b$ is equivalent to $a = b$ or $a = -b$
	• Methods to obtain an equivalent formula
1.8	• Definitions of intervals and critical numbers
	• Properties of inequalities
	• Methods to write intervals of numbers using inequalities or interval notation
	• Methods to solve absolute value inequalities and inequalities involving polynomials and quotients of polynomials
	• If $b > 0$, then $|a| < b$ is equivalent to $-b < a < b$
	• If $b > 0$, then $|a| > b$ is equivalent to $a > b$ or $a < -b$

CHAPTER REVIEW EXERCISES

In Exercises 1–30 perform the indicated operations and/or simplify.

1. $\dfrac{(1-x)^{a+b}}{(1-x)^{a-b}}$

2. $\dfrac{2k+1}{k-1} - \dfrac{3k-4}{k-1}$

3. $\dfrac{10ax^2}{21y^4} \cdot \dfrac{7y}{5ax}$

4. $\dfrac{2}{x} + \dfrac{3}{y}$

5. $\dfrac{x^2+1}{4x+4} \cdot \dfrac{(x+1)^2}{x^2+x}$

6. $(x^n-3)(x^n-1)$

7. $\dfrac{(3x)^2}{y}\left(\dfrac{2zy}{x^2}\right)^3$

8. $\dfrac{(x+h)^4 - x^4}{h}$

9. $(2x+4)(4+x-x^2)$

10. $2^{3x}/2^x$

11. $2^0 - (4/9)^{-3/2}$

12. $\sqrt{4/3} + \sqrt{3/4}$

13. $\dfrac{6R^2}{\sqrt{3R^2}},\ R>0$

14. $\dfrac{a^{-2/3}}{\sqrt[3]{a}}$

15. $\dfrac{1-(1/s)}{1+(1/s)}$

16. $\left(\dfrac{x}{1+x} - \dfrac{1-x}{x}\right) \div \left(\dfrac{x-1}{x} - \dfrac{x}{x+1}\right)$

17. $\left[\dfrac{2}{x+h+1} - \dfrac{2}{x+1}\right] \div h$

18. $\dfrac{2t}{t^2+5t+6} + \dfrac{5t}{t^2+2t-3}$

19. $\dfrac{(x+y)^2 - 4xy}{x^2-y^2}$

20. $\dfrac{(2x^2-1)-(2a^2-1)}{x-a}$

21. $\dfrac{(x^2+x)-(a^2+a)}{x-a}$

22. $(x^{1/2}+x^{-1/2})^2$

23. $x^{1/3} + \tfrac{1}{2}x^{-1/3}$

24. $\dfrac{3}{x^{-3}} - \dfrac{2x^2}{x^{-1}}$

25. $(x-3)(\tfrac{1}{2}x^{-1/2}) + x^{1/2} \cdot 1$

26. $[(x+h)^{1/3} - x^{1/3}] \cdot$
$[(x+h)^{2/3} + x^{1/3}(x+h)^{1/3} + x^{2/3}]$

27. $\dfrac{1}{\sqrt{1-x^2}} - x \cdot \dfrac{-x}{\sqrt{1-x^2}} - \sqrt{1-x^2}$

28. $\sqrt{R^2-y^2}(-1) + (R-y) \cdot \dfrac{-2y}{2\sqrt{R^2-y^2}}$

29. $\dfrac{u \cdot \dfrac{2u}{2\sqrt{u^2+4}} - \sqrt{u^2+4}}{u^2}$

30. $\dfrac{\dfrac{1}{\sqrt{x+h}} - \dfrac{1}{\sqrt{x}}}{h}$

In Exercises 31–60 solve each equation or inequality.

31. $3(x-6) = -2(12-3x)$

32. $1-x = 1+x$

33. $1-x \le 1+x$

34. $-x+1 > 3$

35. $\dfrac{20-x}{3} = \dfrac{x}{5}$

36. $|x| - 1 = \dfrac{|x|}{2}$

37. $|x| < 5$

38. $|x| > 3$

39. $5 - 2(a-2) = 2 - 3a$

40. $|5-x| = 2$

41. $|2x-3| \ge 1$

42. $85 = \tfrac{1}{2}(17)(-7+L)$

43. $3 - 120{,}000x^{-2} = 0$

44. $2\sqrt{x+1} = x-2$

45. $(x+5)^2 + 9 = 0$

46. $\sqrt{2x+1} - 3 = 0$

47. $\dfrac{x}{3} - \dfrac{5}{3} = 2x^{-1}$

48. $\dfrac{x}{2\sqrt{25+x^2}} - \dfrac{1}{4} = 0$

49. $x(x+2) = 0$

50. $x^2 + 3x \ge 4$

51. $b^3 < 9b$

52. $(x-3)^2 - 3(x-3) - 4 = 0$

53. $2x^2 + 5x = 12$

54. $(x-2)^2 \ge 0$

55. $a^3 - 5a^2 = a - 5$

56. $x + \sqrt{x} - 12 = 0$

57. $\dfrac{x+1}{x-1} < 1$

58. $\dfrac{x}{2x-5} < 0$

59. $\dfrac{2}{x+3} \ge \dfrac{1}{x-1}$

60. $1 - \dfrac{12}{c^2-4} = \dfrac{3}{c+2}$

In Exercises 61–64 factor each of the expressions completely.

61. $t^3 - t^3x^2$

62. $x^2 - 8x + 12$

63. $x^3 + x^2 + x + 1$

64. $2x(h-k) + (h^2-k^2) - (h-k)$

65. Simplify $10x^{-6} - 15x^{-3}$ by factoring and then writing the result with only positive exponents.

66. For what values of x does $\dfrac{5x-10}{x-2} = 5$?

In Exercises 67 and 68 answer true or false. If false, give a specific counterexample.

67. All real numbers are rational numbers.

68. All integers are rational numbers.

69. Express $4.\overline{71}$ as the ratio of two integers.

70. Name the property of real numbers illustrated in each statement.

   a. $(x+y) + z = z + (x+y)$

   b. $\tfrac{1}{9} \cdot 8 + \tfrac{1}{9} \cdot 2 = \tfrac{1}{9}(8+2)$

71. The number 3.14 is a member of which of the following collections of numbers: reals, rationals, irrationals, integers?

72. Calculate $b^2 - 4ac$ for $12x^2 - 23x - 18$. Can the expression be factored into binomial factors with integer coefficients?

73. Use inequalities and interval notation to describe the sets of numbers illustrated.

a.

b.

74. Find the value of the expression $-x^2 + 2xy^3$ when $x = -1$ and $y = 3$.

75. Evaluate $x^{3/2} + 2x^0$ when $x = 4$.

76. Evaluate $x^2 - 4x + 2$ if $x = 2 - \sqrt{3}$.

77. Evaluate $x^2 + 4x - 5$ for $x = 2 - i$.

78. Perform the indicated operations and simplify.
a. $(2 + 3i^2) - (3 - 2i)$
b. $1 \div (3 + i)$
c. $(4 - i)(4 + i)$
d. $\sqrt{(-8)^2 - 4(1)(41)}$

79. Solve for y: $x = \dfrac{2y + 1}{1 - y}$.

80. Solve for y': $12x - 5xy' - 5y + 3 - y' = 0$.

81. Solve for y': $-\frac{1}{4}x^{-4/3} - \frac{1}{3}y^{-4/3}y' = 0$.

82. If $s = x - x^2$, solve for x in terms of s.

83. Rationalize the denominator: $(x - 2)/\sqrt{x^2 - 4}$.

84. Rationalize the numerator: $\sqrt{x - 2}/(x^2 - 4)$.

85. Rationalize the numerator in the answer to Exercise 30.

86. Use the definition $|a| = \sqrt{a^2}$ to show that $|3x| = 3|x|$.

87. Show that $\dfrac{(2x + 1)(2x - 1) - (x^2 - x) \cdot 2}{(2x + 1)^2}$

simplifies to $\dfrac{2x^2 + 2x - 1}{(2x + 1)^2}$.

88. Show that $\dfrac{[(x + h)^2 - (x + h)] - [x^2 - x]}{h}$

simplifies to $2x + h - 1$.

89. Simplify

$$\dfrac{(x - 2y)\left(\dfrac{y}{x - 2y}\right) - (-y)\left(x + \dfrac{2y}{x - 2y}\right)}{(x - 2y)^2}.$$

90. Simplify $n\left(\dfrac{2}{n}\right) + \frac{1}{2}n^2\left(\dfrac{2}{n}\right)^2$
$+ \frac{1}{4}(4n^3 - n)\left(\dfrac{2}{n}\right)^3 + \frac{1}{8}(2n^4 - n^2)\left(\dfrac{2}{n}\right)^4$.

91. When resistors R_1, R_2, and R_3 are connected in parallel, their combined resistance is given by

$$\dfrac{1}{\dfrac{1}{R_1} + \dfrac{1}{R_2} + \dfrac{1}{R_3}}.$$

Change this complex fraction to a fraction in simplest form.

92. A jet travels half the distance to its destination at an average speed of 400 mi/hour. If the jet averages 600 mi/hour for the second half, what is the average speed for the trip?

93. For what values of x will $\sqrt{x^2 + 2x - 3}$ be a real number?

94. The sum of a number and its reciprocal is $\frac{25}{12}$. What are the numbers?

95. One side of a right triangle is 3 ft longer than the other, and the hypotenuse is 15 ft. What is the length of the shortest side in the triangle?

96. A student needs an average of 90 or better to get an A in a course. Her first four test grades were 94, 91, 88, and 97. What possible grades on her fifth test will result in an A grade?

97. A real estate broker's commission for selling a house is 6 percent of the selling price. If you want to sell your house and receive $80,000, then what must be the selling price of the house?

98. Triangle ABC is similar to triangle EDC. If $\overline{BD} = 25$ ft, $\overline{AB} = 9$ ft, and $\overline{DE} = 11$ ft, then find $\overline{BC}$ to the nearest foot.

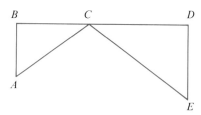

99. A diver jumps from a cliff that is 64 ft above water. Her height above water (y) after t seconds is given by the formula $y = 64 - 16t^2$. When will the diver hit the water? During what period is the diver at least 28 ft above water?

100. a. Because of a company's financial crisis, all employees' weekly salaries are cut a fixed percent, p. What percent raise, x, is now needed to return the employees' weekly salaries to their original level? [*Hint:* Solve the equation $S(1 - p)(1 + x) = S$ for x.]

   b. Use the answer to part a and determine the raise needed when the pay cut is 20 percent.

In Exercises 101–110 select the choice that answers the question or completes the statement.

101. If $x^{-2} = -36$, then x equals

    a. ± 6 **b.** $\pm 1/6$ **c.** $\pm 6i$ **d.** $\pm i/6$

102. If $\sqrt{x^2 + 3} = 2x$, then x equals

    a. both 1 and -1 **b.** only 1

    c. only -1 **d.** no solution

103. Which statement illustrates the commutative property of multiplication?

    a. $a(b + c) = (b + c)a$

    b. $a(b + c) = ab + ac$

    c. $(ab)c = a(bc)$

    d. $a \cdot 1 = a$

104. The reciprocal of $3 + (1/n)$ is

    a. $(3n + 1)/3$ **b.** $n/(3n + 1)$

    c. $(3n + 1)/n$ **d.** $3 + n$

105. $\dfrac{(R + r)^2 E^2 - RE^2 \cdot 2(R + r)}{(R + r)^4}$ simplifies to

    a. $\dfrac{rE^2(r - 2R)}{(R + r)^4}$ **b.** $\dfrac{E^2(r - 2R)}{(R + r)^3}$

    c. $\dfrac{E^2(r - R)}{(R + r)^3}$ **d.** $\dfrac{E^2(r - R)^2}{(R + r)^4}$

106. Which number is an irrational number?

    a. $-\sqrt{4}$ **b.** $\sqrt{-4}$ **c.** $\sqrt[3]{8}$ **d.** $\sqrt{8}$

107. The expression $\dfrac{\sqrt{2} + 1}{\sqrt{2} - 1}$ is equivalent to

    a. 3   **b.** -1 **c.** $5 + \sqrt{2}$ **d.** $3 + 2\sqrt{2}$

108. Pick the statement that describes the set of numbers illustrated.

    a. $|x - 2| < 1$ **b.** $|x + 2| > 1$

    c. $|x - 1| > 2$ **d.** $|x + 1| < 2$

109. The solution set to $x^2 - 6 < x$ is

    a. $(-2, 3)$ **b.** $(-\infty, -2) \cup (3, \infty)$

    c. $(-\infty, -3) \cup (2, \infty)$ **d.** $(-3, 2)$

110. If $\dfrac{1}{a} = \dfrac{1}{x} - \dfrac{1}{b}$, then x equals·

    a. $\dfrac{ab}{a - b}$ **b.** $\dfrac{a + b}{ab}$ **c.** $a + b$ **d.** $\dfrac{ab}{a + b}$

CHAPTER 1 TEST

1. List all correct classifications of the number 0 from the following categories: real number, irrational number, rational number, or integer.

2. Simplify $\dfrac{x^{a+2}y^a}{x^{a+1}y^{2a}}$.

3. Factor completely $2x^3 - 8x$.

4. Simplify $\dfrac{(x + h)^2 - 3 - (x^2 - 3)}{h}$.

5. Subtract $\dfrac{2x}{x^2 - 1} - \dfrac{x - 1}{x + 1}$.

6. Simplify $\dfrac{b^{-1}}{a^{-1} - b^{-1}}$.

7. Evaluate $27^{2/3} - 3^{-1} + 9^0$.

8. Add $x^{1/3} + x^{-1/3}$.

9. Rationalize the denominator $\dfrac{5 + \sqrt{5}}{5 - \sqrt{5}}$.

10. Express $\dfrac{2 + i}{3 - i}$ in the form $a + bi$.

In Questions 11–14 solve each equation or inequality.

11. $\dfrac{3 + x}{7x} - 1 = \dfrac{5}{4x}$ **12.** $3x^3 + 2x = -5x^2$

13. $t^2 + t < 20$ **14.** $\dfrac{2}{y} > 3$

15. Evaluate $x^2 - 8x + 24$ if $x = 4 - 3i$.

16. Solve for C_1: $\dfrac{1}{C} = \dfrac{1}{C_1} + \dfrac{1}{C_2}$.

17. For what values of x will $\sqrt{x^2 + x}$ be a real number?

18. A race car driver averages 180 mi/hour on the first lap around a track, but mechanical trouble reduces the average speed on the second lap to only 90 mi/hour. What is the average speed for the first two laps?

19. If the perimeter in the right triangle below is 12 ft, find x.

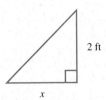

2 ft

x

20. The height (y) of a projectile that is shot directly up from the ground with an initial velocity of 80 ft/second is given by the formula $y = 80t - 16t^2$. To the nearest tenth of a second, when will the projectile initially attain a height of 44 ft?

2 Functions and Graphs

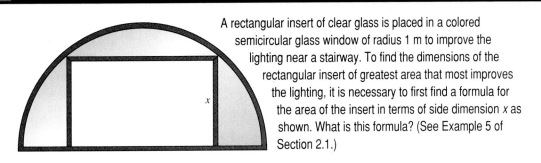

A rectangular insert of clear glass is placed in a colored semicircular glass window of radius 1 m to improve the lighting near a stairway. To find the dimensions of the rectangular insert of greatest area that most improves the lighting, it is necessary to first find a formula for the area of the insert in terms of side dimension x as shown. What is this formula? (See Example 5 of Section 2.1.)

We now introduce the idea of a function and its graph. This concept will then become the central theme for the next few chapters as we investigate the behavior of the "elementary functions" studied in higher mathematics. These important functions are called polynomial, rational, exponential, logarithmic, and trigonometric functions. This chapter starts us on the road to analyzing relationships with formulas, tables, and graphs, and when done, it will be easy to see why the function concept is so important.

2.1 Functions

One of the most important considerations in mathematics is determining the relationship between two variables. For example:

The postage required to mail a package is a function of the weight of the package.

The bill from an electric company is a function of the number of kilowatt-hours of electricity that are purchased.

The current in a circuit with a fixed voltage is a function of the resistance in the circuit.

The demand for a product is a function of the price charged for the product.

The perimeter of a square is a function of the length of the side of the square.

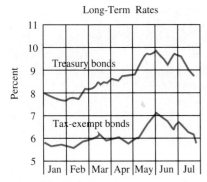

Figure 2.1

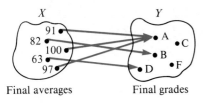

Figure 2.2 Function

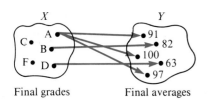

Figure 2.3 Not a function

In each of these examples we determine the relationship between the two variables by finding a rule that establishes a correspondence between values of each variable. For example, if we know the length of the side of a square, we can determine the perimeter by the formula $P = 4s$. In this case the rule is a formula or equation. Sometimes the rule is given in tabular form. For example, consider the formula table below. This rule assigns to each final average (a) a final grade for the course.

$$\text{Final grade} = \begin{cases} A & \text{if} & 90 \le a \le 100 \\ B & \text{if} & 80 \le a < 90 \\ C & \text{if} & 70 \le a < 80 \\ D & \text{if} & 60 \le a < 70 \\ F & \text{if} & 0 \le a < 60 \end{cases}$$

Since formulas and tables are not always applicable, it is sometimes best to give the rule verbally or to make a list that shows the correspondence. The relationship between students and their social security numbers is such a case. Finally, you are undoubtedly familiar with the type of graph shown in Figure 2.1, which is commonly used to specify relationships in a quick and vivid way.

Whether the rule is given by formula or table, graphically, or by a list, the rule is most useful if we obtain *exactly one* answer whenever we use it. For example, the rule in the formula table above assigns to each final average exactly one final grade. Once we compute the final average, the rule tells us exactly what grade to assign for the course. Some typical assignments are shown in Figure 2.2. Assignments are represented as arrows from the points that represent final averages to the points that represent final grades. However some correspondences do not always give us exactly one answer. For example, if we reverse the assignments in Figure 2.2, then we cannot determine a unique final average when the final grade is A, as shown in Figure 2.3.

We wish to define a function so the correspondence in Figure 2.2 is a function, while the correspondence in Figure 2.3 is not a function. We do this as follows:

Definition of a Function

> A **function** is a rule that assigns to each element x in a set X exactly one element y in a set Y. In this definition, set X is called the **domain** of the function, and the set of all elements of Y that correspond to elements in the domain is called the **range** of the function.

We clarify these definitions in Example 1.

EXAMPLE 1   The correspondence in Figure 2.2 assigns to *each* final average *exactly one* final grade, so it is a function with domain $\{63,82,91,97,100\}$ and range $\{A,B,D\}$. Note that the range may or may not be set Y since the definition of a function does not require that all elements of Y be used. Meanwhile, the correspondence in Figure 2.3 is not a function for two reasons. First, an A grade corresponds to three elements in set Y, and second, the final grades of C and F in set X are not associated with any element in set Y.   ■

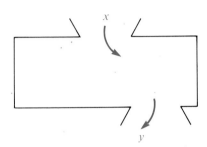

Figure 2.4

The analogy between a function and a computing machine may help to clarify the concept. Consider Figure 2.4, which shows a machine processing domain elements (x values) into range elements (y values). In goes an x value, out comes exactly one y value. With this in mind, you should have a clear image of the three features of a function: (1) the domain (the input), (2) the range (the output), and (3) the rule (the machine). When y is a function of x as just described, the value of y depends on the choice for x, so we call x the **independent variable** and y the **dependent variable.**

It is common practice to refer to a function by stating only the rule; no domain or range is specified. Thus we say, "Consider the circumference function $C = \pi d$." The domain is then assumed to be the collection of values for d that are interpretable in the problem. Since the length of the diameter of a circle must be positive in this case, the domain is the set of positive real numbers. For functions defined by algebraic equations like $y = 1/\sqrt{x-1}$, the domain is the set of all real numbers for which a real number exists in the range. Thus we exclude from the domain values for the independent variable (x) that result in an even root of a negative number or in division by zero. For $y = 1/\sqrt{x-1}$ we want $x - 1 > 0$ so that the domain is the interval $\{x: x > 1\}$, or $(1,\infty)$ in interval notation. We illustrate these ideas in the following examples.

EXAMPLE 2   Indicate whether or not the given equation determines y as a function of x.

a. $y = x^2$ **b.** $x^2 + y^2 = 1$ **c.** $x = y^3$

Solution

a. The equation $y = x^2$ determines y as a function of x, since to each real number input for x there corresponds exactly one y value.

b. Since $x^2 + y^2 = 1$ implies $y = \pm\sqrt{1 - x^2}$, it is possible for an x value to correspond to two different y values. For example, if $x = 0$, then $y = 1$ or $y = -1$. Thus, $x^2 + y^2 = 1$ does not determine y as a function of x. (However, $x^2 + y^2 = 1$ is an important equation, as shown in Section 2.2.)

c. $x = y^3$ implies $y = \sqrt[3]{x}$. Thus, the assignment of a real number to x results in exactly one output for y, so the equation determines y as a function of x.   ■

EXAMPLE 3   Determine the domain and range of each function.

a. $y = x^2 + 2$ **b.** $y = 4$ **c.** $y = \sqrt{1 - x^2}$

Solution

a. For $y = x^2 + 2$ the domain is the set of all real numbers, since the assignment of any real number to x results in a real number output for y. For all x, $x^2 \geq 0$, so y, or $x^2 + 2$, is greater than or equal to 2. Thus, the range is the interval $[2, \infty)$.

b. In the equation $y = 4$ (or $y = 4 + 0 \cdot x$) the y value is 4 regardless of the x chosen. This type of function in which the y value does not change is called a **constant function**. The domain is the set of all real numbers and the range is $\{4\}$.

c. For the output of $y = \sqrt{1 - x^2}$ to be a real number, x must satisfy $1 - x^2 \geq 0$. As shown in Example 10 of Section 1.8, $1 - x^2 \geq 0$ is true in the interval $[-1, 1]$, which is the domain. The corresponding y values vary from 0 (when $x = \pm 1$) to 1 (when $x = 0$), so the range is the interval $[0, 1]$.

(*Note:* The range of a function is usually more difficult to determine than the domain, but this topic will become easier in the next section when we show how to read the range from the graph of a function.)   ■

EXAMPLE 4   A piece of wire 46 in. long is bent into a rectangle of sides x and y units. Express y as a function of x. Express the area (A) of the rectangle as a function of x. Find the domain of the function.

Solution   First, express y as a function of x by using the perimeter formula

$$P = 2x + 2y$$
$$46 = 2x + 2y$$
$$23 = x + y$$
$$23 - x = y.$$

To express the area (A) as a function of x we now replace y by $23 - x$ in the area formula

$$A = x \cdot y$$
$$= x(23 - x).$$

Since x and $23 - x$ must both be positive, the domain is the interval $(0, 23)$.   ■

EXAMPLE 5   Solve the problem in the chapter introduction on page 97.

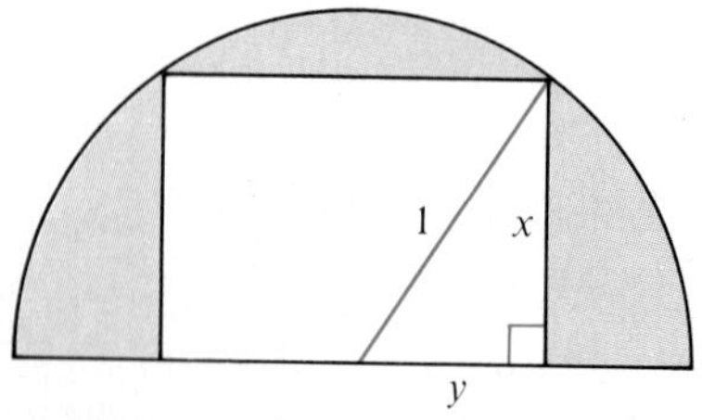

Solution   First, consider the sketch of the situation in Figure 2.5 and note that we strategically placed the radius of the semicircle so as to form a right triangle inside the rectangle. We may then use the Pythagorean theorem to write y in terms of x as follows:

$$x^2 + y^2 = 1^2$$
$$y^2 = 1 - x^2$$
$$y = \sqrt{1 - x^2}. \quad \text{Since } y \text{ is positive}$$

Figure 2.5

Then the area formula in terms of x is

$$A = 2xy$$
$$= 2x\sqrt{1 - x^2}.$$

In the context of this problem, the meaningful replacements for x are the real numbers between 0 and 1, so the domain is the interval $(0,1)$. ∎

EXAMPLE 6 Each month a salesperson earns $500 plus 7 percent commission on sales above $2,000. Find a rule that expresses the monthly earnings (e) of the salesperson in terms of the amount (a) of merchandise sold during the month.

Solution If the salesperson sells less than or equal to $2,000 worth of merchandise for the month, he earns $500. Thus,

$$e = \$500 \quad \text{if} \quad \$0 \le a \le \$2,000.$$

If the salesperson sells above $2,000, he earns $500 plus 7 percent of the amount above $2,000. Thus,

$$e = \$500 + 0.07(a - \$2,000) \quad \text{if} \quad a > \$2,000.$$

The following rule may then be used to determine the monthly earnings of the salesperson when we know the amount of merchandise sold:

$$e = \begin{cases} \$500 & \text{if} \quad \$0 \le a \le \$2,000 \\ \$500 + 0.07(a - \$2,000) & \text{if } a > \$2,000. \end{cases}$$

The domain of the function is $\{a: a \ge \$0\}$, and the range is $\{e: e \ge \$500\}$. ∎

Functions as Ordered Pairs

In mathematical notation we use **ordered pairs** to show the correspondence in a function. For example, consider the equation $y = 2x + 1$.

If x Equals	Then $y = 2x + 1$	Thus, the Ordered Pairs
2	$2(2) + 1 = 5$	$(2,5)$
1	$2(1) + 1 = 3$	$(1,3)$
0	$2(0) + 1 = 1$	$(0,1)$
-1	$2(-1) + 1 = -1$	$(-1,-1)$
-2	$2(-2) + 1 = -3$	$(-2,-3)$

In the pairs that represent the correspondence, we list the values of the independent variable first and the values of the dependent variable second. Thus, the order of the numbers in the pair is significant. The pairing $(2,5)$ indicates that when $x = 2$, $y = 5$; $(5,2)$ means when $x = 5$, $y = 2$. In the equation $y = 2x + 1$, $(2,5)$ is an ordered pair that makes the equation a true statement; so $(2,5)$ is said to be a solution of the equation; $(5,2)$ is not a solution of this equation.

EXAMPLE 7   Which of the following ordered pairs are solutions of the equation $y = 3x - 2$?

a. $(-1, -5)$ **b.** $(-5, -1)$

Solution

a. $(-1, -5)$ means that when

$$x = -1, y = -5$$
$$y = 3x - 2$$
$$(-5) \overset{?}{=} 3(-1) - 2$$
$$-5 \overset{\checkmark}{=} -5.$$

Thus, $(-1, -5)$ is a solution.

b. $(-5, -1)$ means that when

$$x = -5, y = -1$$
$$y = 3x - 2$$
$$(-1) \overset{?}{=} 3(-5) - 2$$
$$-1 \neq -17.$$

Thus, $(-5, -1)$ is not a solution. ∎

The representation of a correspondence as a set of ordered pairs gives us a different perspective of the function concept. In higher mathematics the following definition of a function is very useful.

Alternate Definition of a Function

> A **function** is a set of ordered pairs in which no two different ordered pairs have the same first component. The set of all first components (of the ordered pairs) is called the **domain** of the function. The set of all second components is called the **range** of the function.

In this section we have defined a function in terms of (1) a rule and (2) ordered pairs. Since the function concept is so important, you should consider both definitions and satisfy yourself that these definitions are equivalent.

EXAMPLE 8   State whether the set of ordered pairs is a function.

a. $\{(-1, 0), (0, 0), (1, 0)\}$ **b.** $\{(0, -1), (0, 0), (0, 1)\}$

Solution

a. This set of ordered pairs is a function because the first component in the ordered pairs is always different. The domain is $\{-1, 0, 1\}$ and the range is $\{0\}$.

b. This set of ordered pairs is not a function because the number 0 is the first component in more than one ordered pair. ∎

Functional Notation

A useful notation commonly used with functions allows us to represent more conveniently the value of the dependent variable for a particular value of the independent variable. In this notation we write an equation like $y = 2x$ as $f(x) = 2x$, where we replace the dependent variable y with the symbol $f(x)$. The symbol representing the independent variable appears in the parentheses. The term $f(x)$ is read "f of x" or "f at x" and means the value of the function (the y value) corresponding to the value of x. Similarly, $f(3)$ is read "f of 3" or "f at 3" and means the value of the function when $x = 3$. To find $f(3)$, we substitute 3 for x in the equation $f(x) = 2x$.

$$f(3) = 2 \cdot 3 = 6 \quad \text{therefore} \quad f(3) = 6$$

EXAMPLE 9   If $y = f(x) = (x + 2)/x$, find $f(1), f(2)$, and $f(0)$.

Solution

$$y_{\text{when } x = 1} = f(1) = \frac{1 + 2}{1} = \frac{3}{1} = 3$$

$$y_{\text{when } x = 2} = f(2) = \frac{2 + 2}{2} = \frac{4}{2} = 2$$

$$y_{\text{when } x = 0} = f(0) = \frac{0 + 2}{0} = \frac{2}{0} \quad \text{undefined} \qquad ■$$

EXAMPLE 10   If $f(x) = x^2$, show that $f(a + b)$ does not equal $f(a) + f(b)$ for all a and b.

Solution   To determine $f(a + b), f(a)$, and $f(b)$, replace x in the function $f(x) = x^2$ by $a + b, a$, and b, respectively.

$$f(a + b) = (a + b)^2 = a^2 + 2ab + b^2$$
$$f(a) = a^2$$
$$f(b) = b^2$$

Since $a^2 + 2ab + b^2 \neq a^2 + b^2, f(a + b) \neq f(a) + f(b)$. ■

EXAMPLE 11   If $f(x) = \begin{cases} 2 & \text{if} \quad x < 0 \\ x + 1 & \text{if} \quad 0 \leq x < 3, \end{cases}$ find

a. $f(2)$ **b.** $f(-2)$ **c.** $f(3)$

Solution

a. Since 2 is in the interval $0 \leq x < 3$, use $f(x) = x + 1$, so $f(2) = 2 + 1 = 3$.
b. Since -2 is less than 0, use $f(x) = 2$, so $f(-2) = 2$.
c. Since 3 is not in the domain of the function, $f(3)$ is undefined. ■

EXAMPLE 12   The difference quotient of a function $y = f(x)$ is defined as

$$\frac{f(x + h) - f(x)}{h}, \quad h \neq 0.$$

Computing this ratio is an important consideration when analyzing the rate of change of a function in calculus. Find the difference quotient for $f(x) = x^2 + 2x$.

Solution   If $f(x) = x^2 + 2x$, we have

$$f(x + h) = (x + h)^2 + 2(x + h) = x^2 + 2xh + h^2 + 2x + 2h.$$

Then

$$\frac{f(x + h) - f(x)}{h} = \frac{(x^2 + 2xh + h^2 + 2x + 2h) - (x^2 + 2x)}{h}$$

$$= \frac{x^2 + 2xh + h^2 + 2x + 2h - x^2 - 2x}{h}$$

$$= \frac{2xh + h^2 + 2h}{h}$$

$$= \frac{2xh}{h} + \frac{h^2}{h} + \frac{2h}{h}$$

$$= 2x + h + 2. \qquad \blacksquare$$

In conclusion, here are some important ideas to keep in mind about functional notation.

Functional Notation Concepts

1. If $y = f(x)$ and a is in the domain of f, then $f(a)$ means the value of y when $x = a$. Thus, evaluating $f(a)$ often requires nothing more than a *substitution* of the value a for x.
2. $f(a)$ is a y value; a is an x value. Hence, ordered pairs for the function defined by $y = f(x)$ all have the form $(a, f(a))$.
3. In functional notation we use the symbols f and x more out of custom than necessity, and other symbols work just as well. The notations $f(x) = 2x$, $f(t) = 2t$, $g(y) = 2y$, and $h(z) = 2z$ all define exactly the same function if x, t, y, and z may be replaced by the same numbers.

EXERCISES 2.1

In Exercises 1–4 determine whether or not the given correspondence is a function.

1.

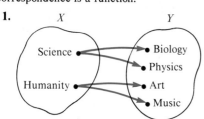

2.

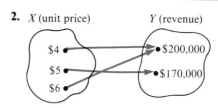

3.

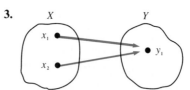

4.

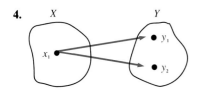

In Exercises 5–12 indicate whether or not the given equation determines y as a function of x.

5. $x + y = 3$　　　　　　**6.** $x = y^4$
7. $|y| = x$　　　　　　　**8.** $y = x^4$
9. $y = -\sqrt{1 - x^2}$　　　**10.** $x^2 + y^2 = 4$
11. $y = \pm\sqrt{x}$　　　　　**12.** $x^2y = 1$

In Exercises 13–30 find the domain and range of the given function.

13. $C = 2\pi r$ (circumference of a circle)
14. $C = 100n$ (cost of n radial tires)
15. $y = x^2 + 3$　　　　**16.** $f(x) = x^2 + 3, x < 0$
17. $g(x) = (x + 3)^2$　　**18.** $y = 3 - x^2$
19. $y = 1 - |x|$　　　　**20.** $y = |1 - x|$
21. $f(x) = \dfrac{1}{x + 1}$　　　**22.** $h(x) = \dfrac{1}{x^2 + 1}$
23. $f(x) = 0$　　　　　**24.** $y = x^{1/3}$
25. $y = \sqrt{x - 1}$　　　**26.** $y = 1/\sqrt{x + 5}$
27. $y = -\sqrt{4 - x^2}$　　**28.** $y = \sqrt{x^2 - 9}$

29. $y = \begin{cases} 1 & \text{if } x \le 0 \\ x & \text{if } 0 < x \le 4 \end{cases}$

30. $f(x) = \begin{cases} x + 1 & \text{if } -2 \le x < 1 \\ x - 1 & \text{if } 1 \le x < 5 \end{cases}$

In Exercises 31–34 determine which of the ordered pairs are solutions of the equation.

Equation	Ordered Pairs
31. $y = 3x + 1$	$(4,1), (-2,-5), (1,4), (-5,-2)$
32. $y = \|x\|$	$(-1,-1), (1,1), (-1,1), (1,-1)$
33. $y = x^2$	$(-1,-1), (1,1), (-1,1), (1,-1)$
34. $y = 1/x$	$(2,0.5), (0.5,2), (-1,-1), (0,0)$

35. Fill in the missing component in each of the following ordered pairs so they are solutions of the equation $y = -3x + 7$: (0,), (,0), (-5,), (,5).
36. Fill in the missing component in each of the following ordered pairs so they are solutions of the equation $y = \sqrt{x}$: (,4), (4,), (0,), (,$\sqrt{5}$).
37. If $(a,1)$ is a solution of the equation $y = 2^x$, then find the value of a.
38. If $(0,b)$ is a solution of the equation $y = 7x + 5$, then find the value of b.

39. If $y = |x - 1|$, then determine the range element that corresponds to a domain element of -1.
40. If $y = -x - 1$, then determine the domain element that corresponds to a range element of 6.

In Exercises 41–46 state which sets of ordered pairs are functions.

41. $\{(1,-3), (2,0), (3,1)\}$
42. $\{(-3,1), (0,2), (1,3)\}$
43. $\{(2,-1), (3,0), (4,-1)\}$
44. $\{(-1,2), (0,3), (-1,4)\}$
45. $\{(2,1), (2,2), (2,3)\}$　　**46.** $\{(1,2), (2,2), (3,2)\}$

In Exercises 47 and 48 state the domain and range of the function.

47. $\{(-2,4), (-1,1), (1,1), (2,4)\}$
48. $\{(0,5), (1,6), (2,7), (3,8)\}$
49. If $f(x) = -2x + 7$, find $f(-2)$, $f(1)$, and $f(5)$.
50. If $f(t) = t^2 + 1$, find $f(-1)$, $f(0)$, and $f(1)$.
51. If $g(x) = 2x^2 - x + 4$, find $g(3)$, $g(0)$, and $g(-1)$.
52. If $h(t) = -t^2$, find $h(5)$, $h(1)$, and $h(-5)$.
53. If $f(x) = (x + 1)/(x - 2)$, find $f(-3)$, $f(1)$, and $f(2)$.
54. If $h(y) = y/(y + 1)$, find $h(1)$, $h(0)$, and $h(-1)$.
55. If $f(x) = x - 1$, for what value of x will
　　a. $f(x) = 2$?　　**b.** $f(x) = 0$?　　**c.** $f(x) = -2$?
56. If $g(x) = 2x + 7$, for what value of x will
　　a. $g(x) = 5$?　　**b.** $g(x) = 0$?　　**c.** $g(x) = -3$?
57. If $f(x) = x + 1$ and $g(x) = x^2$, find
　　a. $f(1) + g(0)$　　　**b.** $f(-2) - g(-3)$
　　c. $4g(0) + 5f(1)$　　**d.** $2f(3) - 3g(2)$
　　e. $f(2) \cdot g(1)$　　**f.** $\dfrac{f(1)}{g(2)}$　　**g.** $[f(1)]^2$
　　h. $g[f(1)]$　　**i.** $f[g(1)]$　　**j.** $f[g(x)]$
58. If $f(x) = x + 1$, find
　　a. $f(2)$　　　　**b.** $f(3)$
　　c. $f(2 + 3)$　　**d.** Does $f(2 + 3) = f(2) + f(3)$?
　　e. $f(a)$　　　　**f.** $f(b)$
　　g. $f(a + b)$　　**h.** Does $f(a + b) = f(a) + f(b)$?
59. If $f(x) = x + 1$, find
　　a. $f(2)$　　**b.** $f(-2)$　　**c.** Does $f(-2) = -f(2)$?
60. If $g(x) = x^2$, find
　　a. $g(2)$　　**b.** $g(-2)$
　　c. Does $g(x) = g(-x)$ for all values of x?
61. If $f(x) = \begin{cases} x - 1 & \text{if } x < 0 \\ -1 & \text{if } x > 0 \end{cases}$, find
　　a. $f(-1)$　　**b.** $f(1)$　　**c.** $f(0)$
62. If $f(x) = \begin{cases} -1 & \text{if } x > 0 \\ 0 & \text{if } x = 0 \\ 1 & \text{if } x < 0 \end{cases}$, find
　　a. $f(1)$　　**b.** $f(0)$　　**c.** $f(-1)$　　**d.** $f(0.0001)$

63. If $g(x) = \begin{cases} x & \text{if} \quad x \geq 0 \\ -x & \text{if} \quad x < 0, \text{find} \end{cases}$

    a. $g(3)$ **b.** $g(0)$ **c.** $g(-3)$

64. If $h(x) = \begin{cases} x & \text{if} \quad 0 \leq x \leq 1 \\ -1 & \text{if} \quad x > 1, \text{find} \end{cases}$

    a. $h(3)$ **b.** $h(1)$ **c.** $h(\frac{1}{2})$ **d.** $h(0)$ **e.** $h(-3)$

65. If $f(x) = 2x^2 - 1$, find
    a. $f(x + h)$ **b.** $f(x + h) - f(x)$
    c. $\dfrac{f(x + h) - f(x)}{h}$

66. If $f(x) = 1 - x$, find
    a. $f(1 + h)$ **b.** $f(1 + h) - f(1)$
    c. $\dfrac{f(1 + h) - f(1)}{h}$

In Exercises 67–76 write the difference quotient for each function in simplest form (see Example 12).

67. $f(x) = x^2$ 68. $f(x) = 2x^2 + 3$
69. $f(x) = 4 - x^2$ 70. $f(x) = x^2 + 3x + 5$
71. $f(x) = 2x$ 72. $f(x) = 4x - 1$
73. $f(x) = 1$
74. $f(x) = \sqrt{x}$ (*Note:* Rationalize the numerator.)
75. $f(x) = \dfrac{1}{x}$ 76. $f(x) = x^3$

In Exercises 77–101 find a formula or table that defines the functional relationship between the two variables; in each case indicate the domain of the function.

77. Express the area (A) of a square in terms of the length of its side (s).
78. Express the area (A) of a rectangle of width 5 as a function of its length (ℓ).
79. Express the length (ℓ) of a rectangle of width 5 in terms of its area (A).
80. Express the area (A) of a circle as a function of its radius (r).
81. Express the length of the side (s) of a square in terms of its perimeter (P).
82. Express the area (A) of a square as a function of the perimeter (P) of the square.
83. Express the distance (d) that a car will travel in t hours if the car is always traveling at 40 mi/hour.
84. Express the cost (c) of y gal of gasoline if the gas cost $1.17/gal.
85. Express the earnings (e) of an electrician in terms of the number (n) of hours worked if she makes $28 per hour.

86. Express the interest (i) earned on $1,000 at 6 percent simple interest as a function of the time (t).
87. Express the earnings (e) of a real estate agent who receives a 6 percent commission as a function of the sale price (p) of a house.
88. Express the monthly cost (c) for a checking account as a function of the number (n) of checks serviced that month if the bank charges 10 cents per check plus a 75-cent maintenance charge.
89. The total cost of producing a certain product consists of paying $400 per month for rent plus $5 per unit for material. Express the company's monthly total costs (c) as a function of the number of units (x) they produce that month.
90. A company plans to purchase a machine to package its new product. If the company estimates that the machine will cost $5,000 per year plus $1 to package each unit, express the yearly cost (c) of using the machine as a function of the number of units (x) packaged during the year.
91. A reservoir contains 10,000 gal of water. If water is being pumped from the reservoir at a rate of 50 gal/minute, write a formula expressing the amount (a) of water remaining in the reservoir as a function of the number (n) of minutes the water is being pumped.
92. Express the monthly cost of renting a computer as a function of the number (n) of hours the computer is used if the company charges $200 plus $100 for every hour the computer is used during the month.
93. Express the federal income tax (t) for a single person as a function of his taxable income (i) if the person's taxable income is between $2,000 and $4,000 inclusive and the tax rate is $310 plus 19 percent of the excess over $2,000.
94. Express the monthly earnings (e) of a salesperson in terms of the cash amount (a) of merchandise sold if the salesperson earns $600 per month plus 8 percent commission on sales above $10,000.
95. Express the monthly cost (c) of an electric bill in terms of the number (n) of electrical units purchased (the unit of measure is the kilowatt-hour) if the person used no more than 48 units and the electric company has the following rate schedule:

Amount	Charge
First 12 units or less	$1.75
Next 36 units at	3.82 cents/unit

96. Express the monthly cost (c) of a gas bill in terms of the number (n) of units purchased (1 unit equals 100 ft³ of gas) if the customer used no more than 14 units and the gas company has the following rate schedule:

Amount	Charge
First 2 units or less	$2.06
Next 6 units at	40 cents/unit
Next 6 units at	26.5 cents/unit

97. A long strip of galvanized sheet metal 12 in. wide is to be formed into an open gutter by bending up the edges to form a gutter with rectangular cross section. Write the cross-sectional area of the gutter as a function of the depth (x).

98. A farmer encloses an area by connecting 2,000 ft of fencing in the shape of a rectangle of sides x and y units. Express y as a function of x. Find a formula for the enclosed area (A) as a function of x.

99. A parking lot is adjacent to a building and is to have fencing on three sides, the side on the building requiring no fencing. If 100 yd of fencing is used to form a rectangular parking lot, find a formula expressing the area (A) of the lot as a function of the length (x) of the side of the fencing that is perpendicular to the building.

100. A rectangular sheet of tin 8 in. by 15 in. is to be used to make an open top box by cutting a small square of tin from each corner and bending up the sides. Find an equation for the volume (v) of the box in terms of the length (x) of the side of the square that is to be cut from each corner (see the diagram).

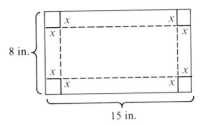

101. A rectangular billboard is to display 100 ft² of advertising material with borders that are 3 ft on the top and bottom and 5 ft on the sides. Express the total area (A) of the billboard as a function of the width (w) of the advertised material (see the diagram).

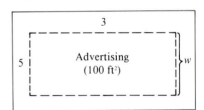

THINK ABOUT IT

1. Write five *different* functions that have the same domain and the same range as the function $\{(3,6),(4,7),(5,8)\}$.

2. **a.** Give two examples of a function for which $f(a + b)$ does not equal $f(a) + f(b)$ for all a and b.
   b. Give two examples of a function for which $f(a + b)$ does equal $f(a) + f(b)$ for all a and b.

3. Describe a realistic situation that may be analyzed by the given function.
   a. $y = 3x$, where $x > 0$.
   b. $y = 2x + 10$, where x is a nonnegative integer.

4. The combined federal, state, and local tax rates for a couple place them in the 40-percent tax bracket, so they invest significantly in a tax-free money market fund. Write a formula that expresses the equivalent taxable yield (t) for this couple's investment as a function of the tax-free yield (x). Then use this formula to find the equivalent taxable yield for this investment if the money market fund is currently paying 5.31 percent tax-free.

5. It is often useful to formulate (at least mentally) a verbal description of a function. For example, the function defined by the equation $f(x) = 2x + 5$ may be expressed in words as follows: For each real number, double it, and then add 5 to the result. Write verbal descriptions for the following functions.
   a. $f(x) = x^2 + 3$ **b.** $f(x) = (x + 3)^2$
   c. $f(x) = 1 - x^3$ **d.** $f(x) = -|x|$
   e. $f(x) = \sqrt{x + 2}$ **f.** $f(x) = \dfrac{1}{x - 4}$

REMEMBER THIS

1. Evaluate $|y_2 - y_1|$ if $y_1 = -6$ and $y_2 = -11$.
2. Evaluate $(x_2 - x_1)^2 + (y_2 - y_1)^2$ if $x_1 = 7$, $y_1 = -5$, $x_2 = -1$, and $y_2 = 5$.
3. Find the length of the hypotenuse in a right triangle whose legs measure 12 ft and 5 ft.
4. Express $\sqrt{48}$ in simplest radical form.
5. Express $\sqrt{(-2 - 4)^2 + [-3 - (-1)]^2}$ in simplest radical form.

6. If $x < 5$ then $|x - 5|$ simplifies to _____ .
7. If $f(x) = 1 - x^2$, find $f(-3)$.
8. If $f(x) = \begin{cases} x, & \text{if } x \geq 3 \\ 2, & \text{if } x < 3 \end{cases}$, find $f(3)$.
9. What is the domain of $y = \sqrt{x + 4}$?
10. What is the range of $y = x^2$?

2.2 Cartesian Coordinates and Graphs

One of the sources of information and insight about a relationship is a picture that describes the particular situation. Pictures or graphs are often used in business reports, laboratory reports, and newspapers to present data quickly and vividly. Similarly, it is useful to have a graph that describes the behavior of a particular function, for this picture helps us see the essential characteristics of the relationship.

We can pictorially represent a function by using the **Cartesian coordinate system.** This system was devised by the French mathematician and philosopher René Descartes and is formed from the intersection of two real number lines at right angles. The values for the independent variable (usually x) are represented on a horizontal number line and values for the dependent variable (usually y) on a vertical number line. These two lines are called **axes,** and they intersect at their common zero point, which is called the **origin** [see Figure 2.6(a)].

This coordinate system divides the plane into four regions called **quadrants.** The quadrant in which both x and y are positive is designated the first quadrant. The remaining quadrants are labeled in a counterclockwise direction. Figure 2.6(b) shows the name of each quadrant as well as the sign of x and y in that quadrant.

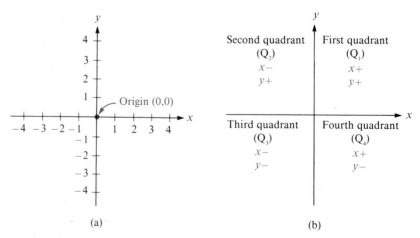

(a)   (b)

Figure 2.6

Any ordered pair can be represented as a point in this coordinate system. The first component indicates the distance of the point to the right or left of the vertical axis. The second component indicates the distance of the point above or below the horizontal axis. These components are called the **coordinates** of the point.

EXAMPLE 1   Represent the ordered pairs $(0,-3)$, $(-4,-2)$, $(\pi,-\frac{5}{2})$, and $(-\frac{4}{3},\sqrt{2})$ as points in the Cartesian coordinate system.

Solution   See Figure 2.7. ∎

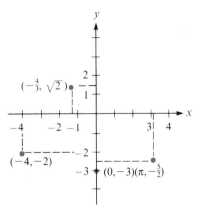

Figure 2.7

When working with the Cartesian coordinate system, it is often necessary to determine the distance between two points in the system. It is easiest to calculate the distance between two points on a horizontal or vertical line. If the points lie on the same horizontal line, we find the distance between them by taking the absolute value of the difference between their x-coordinates. If two points lie on the same vertical line, we calculate the distance between them by taking the absolute value of the difference between their y-coordinates.

EXAMPLE 2   Find the distance between the given points.

a. $(-4,3)$ and $(2,3)$ **b.** $(1,1)$ and $(1,-3)$

Solution   See Figure 2.8.

a. The points $(-4,3)$ and $(2,3)$ have the same y-coordinate, so

$$d = |x_2 - x_1| = |2 - (-4)| = |6| = 6.$$

b. The points $(1,1)$ and $(1,-3)$ have the same x-coordinate, so

$$d = |y_2 - y_1| = |-3 - 1| = |-4| = 4. \qquad ∎$$

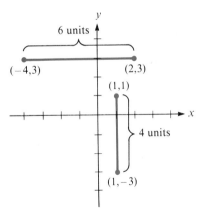

Figure 2.8

If two points do not lie on the same horizontal or vertical line, we can find the distance between them by drawing a horizontal line through one point and a vertical line through the other, as illustrated in Figure 2.9. The x-coordinate at the point where the two lines intersect will be x_2. The y-coordinate will be y_1. The distance between (x_1,y_1) and (x_2,y_1) will be $|x_2 - x_1|$ because they lie on the same horizontal line. The distance between (x_2,y_2) and (x_2,y_1) will be $|y_2 - y_1|$ because they lie on the same vertical line. The vertical and horizontal lines meet at a right angle (90°) and thus the resulting triangle is a right triangle.

The distance d is the length of the hypotenuse in the right triangle, and by the Pythagorean theorem we get

$$d^2 = |x_2 - x_1|^2 + |y_2 - y_1|^2$$
$$d = \sqrt{(x_2 - x_1)^2 + (y_2 - y_1)^2}.$$

Note that we no longer need the absolute value symbols since the square of any real number is never negative. Thus, we have the following formula.

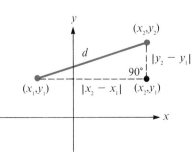

Figure 2.9

Distance Formula

The distance d between the points $P_1(x_1,y_1)$ and $P_2(x_2,y_2)$ is given by

$$d = \overline{P_1P_2} = \sqrt{(x_2 - x_1)^2 + (y_2 - y_1)^2}.$$

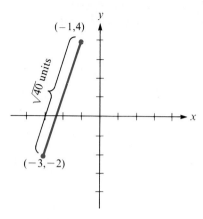

Figure 2.10

EXAMPLE 3   Find the distance between $(-3,-2)$ and $(-1,4)$.

Solution

$$\begin{aligned}
d &= \sqrt{(x_2 - x_1)^2 + (y_2 - y_1)^2} \\
&= \sqrt{[-1 - (-3)]^2 + [4 - (-2)]^2} \\
&= \sqrt{(2)^2 + (6)^2} \\
&= \sqrt{4 + 36} \\
&= \sqrt{40} \quad \text{or} \quad 2\sqrt{10} \quad \text{(See Figure 2.10.)}
\end{aligned}$$
∎

EXAMPLE 4   Show that the points $(-4,-2)$, $(-3,2)$, and $(1,1)$ are vertices of an isosceles triangle. Is this triangle a right triangle?

Solution   Label $(-4,-2)$ as point A, $(-3,2)$ as B, and $(1,1)$ as C. Then

$$\begin{aligned}
\overline{AB} &= \sqrt{[-4 - (-3)]^2 + (-2 - 2)^2} = \sqrt{17} \\
\overline{BC} &= \sqrt{(-3 - 1)^2 + (2 - 1)^2} = \sqrt{17} \\
\overline{AC} &= \sqrt{(-4 - 1)^2 + (-2 - 1)^2} = \sqrt{34}.
\end{aligned}$$

Since the lengths of two sides are equal, the triangle is isosceles. Since $(\sqrt{34})^2 = (\sqrt{17})^2 + (\sqrt{17})^2$ the Pythagorean theorem holds, and ABC is a right triangle with angle B as the right angle (see Figure 2.11). ∎

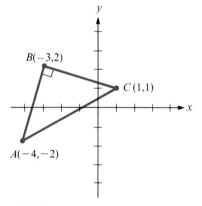

Figure 2.11

We now consider one of the most useful ideas in mathematics. With the Cartesian coordinate system we can represent any ordered pair of real numbers by a particular point in the system. This enables us to draw a geometric picture (graph) of a function.

Graph

The graph of a function is the set of all points in a coordinate system that correspond to ordered pairs in the function.

There are many techniques associated with determining the graph of a function. One technique is simply to assign values to the independent variable and obtain a list of ordered pair solutions. By plotting enough of these solutions, we can establish a trend and then complete the graph by following the established pattern. However, we cannot possibly list all the solutions of most equations because they are an infinite set of ordered pairs. Determining how many and which points to plot is a difficult decision. Therefore, as we proceed in this section and succeeding chapters, we develop the more efficient method of determining the essential characteristics of the graph from the form of the equation. For example, in

Section 3.1 we will prove that the graph of a function of the form $y = f(x)$ $= mx + b$, where m and b are real number constants, is a straight line. For now let us use this fact and illustrate many of the important ideas of graphing by considering functions that require only straight lines for their graphs.

EXAMPLE 5   Graph the function $f(x) = -2x + 4$.

Solution

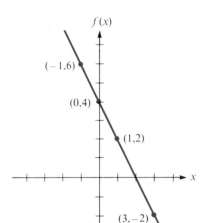

$f(x)$

$(-1,6)$

$(0,4)$

$(1,2)$

$(3,-2)$

x

Figure 2.12

If $x =$	Then $f(x) = -2x + 4$	Thus, the Ordered-Pair Solutions
3	$-2(3) + 4 = -2$	$(3,-2)$
1	$-2(1) + 4 = 2$	$(1,2)$
0	$-2(0) + 4 = 4$	$(0,4)$
-1	$-2(-1) + 4 = 6$	$(-1,6)$

The horizontal axis is used to represent the values of the independent variable and thus is called the x-axis. The vertical axis is named for the dependent variable, so it is labeled the $f(x)$- or y-axis. The function is of the form $y = f(x) = mx + b$, with $m = -2$ and $b = 4$. Thus, the graph of $f(x) = -2x + 4$ is a straight line that passes through the points in the above table (see Figure 2.12).   ■

EXAMPLE 6   Graph the function $y = 4$ (or $y = 4 + 0x$).

Solution

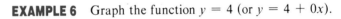

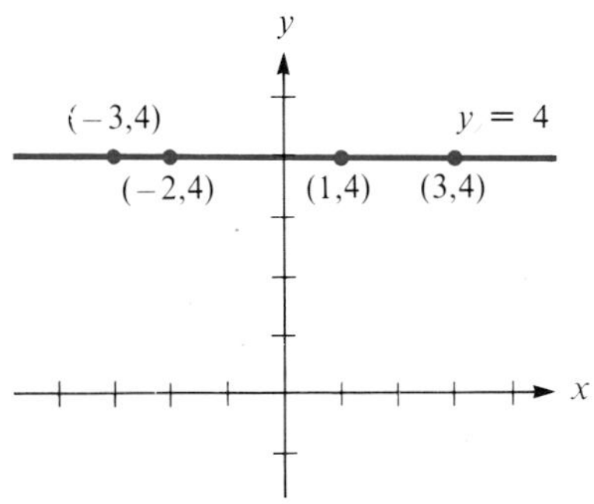

y

$(-3,4)$ $y = 4$

$(-2,4)$ $(1,4)$ $(3,4)$

x

Figure 2.13

If $x =$	Then $y = 4 + 0 \cdot x$	Thus, the Ordered-Pair Solutions
3	$4 + 0(3) = 4$	$(3,4)$
1	$4 + 0(1) = 4$	$(1,4)$
-2	$4 + 0(-2) = 4$	$(-2,4)$
-3	$4 + 0(-3) = 4$	$(-3,4)$

Note that no matter what number we substitute for x, the resulting value for y is 4 (see Figure 2.13). Since the value of the function does not change, $y = 4$ is called a **constant function**. A physical example of a constant function is the acceleration of a falling body as a function of time ($a = 32$) since the acceleration is always 32 ft/second^2, regardless of the time. In every case the graph of a constant function is a horizontal line.   ■

EXAMPLE 7   Graph the function defined as follows:

$$f(x) = \begin{cases} x & \text{if } x \geq 0 \\ -x & \text{if } x < 0. \end{cases}$$

Solution

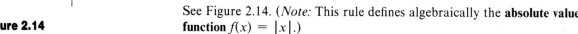

If $x =$	Then $f(x) = \begin{cases} x & \text{if } x \geq 0 \\ -x & \text{if } x < 0 \end{cases}$	Thus, the Ordered-Pair Solutions
3	since $3 \geq 0$, $f(3) = 3$	(3,3)
2	since $2 \geq 0$, $f(2) = 2$	(2,2)
1	since $1 \geq 0$, $f(1) = 1$	(1,1)
0	since $0 \geq 0$, $f(0) = 0$	(0,0)
-1	since $-1 < 0$, $f(-1) = -(-1) = 1$	$(-1,1)$
-2	since $-2 < 0$, $f(-2) = -(-2) = 2$	$(-2,2)$
-3	since $-3 < 0$, $f(-3) = -(-3) = 3$	$(-3,3)$

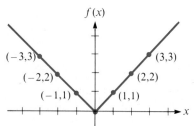

Figure 2.14

See Figure 2.14. (*Note:* This rule defines algebraically the **absolute value function** $f(x) = |x|$.) ∎

EXAMPLE 8   Graph the function defined as follows:

$$f(x) = \begin{cases} 1 - x & \text{if } x < 0 \\ -1 & \text{if } x \geq 0. \end{cases}$$

Solution   If $x \geq 0$, $f(x) = -1$, which is a constant function whose graph is a horizontal line with such ordered pairs as $(0,-1)$, $(1,-1)$, $(2,-1)$, and so on. If $x < 0$, $f(x) = 1 - x$, which graphs as a line with ordered pairs $(-1,2)$, $(-2,3)$, $(-3,4)$, and so on. The graph is given in Figure 2.15. Note that we draw a solid circle at $(0,-1)$ and an open circle at $(0,1)$ to show that $(0,-1)$ is part of the graph, while $(0,1)$ is not. ∎

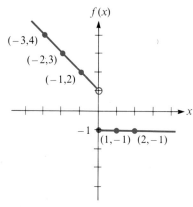

Figure 2.15

EXAMPLE 9   Graph the function $y = \dfrac{|x - 10|}{x - 10}$.

Solution   From Section 1.7 we know

$$|x - 10| = \begin{cases} x - 10 & \text{if } x - 10 \geq 0 \\ -(x - 10) & \text{if } x - 10 < 0. \end{cases}$$

Since division by 0 is undefined, $x \neq 10$. If $x > 10$, we have

$$\frac{|x - 10|}{x - 10} = \frac{x - 10}{x - 10} = 1.$$

Similarly, if $x < 10$, we have

$$\frac{|x - 10|}{x - 10} = \frac{-(x - 10)}{x - 10} = -1.$$

Thus, $y = \dfrac{|x - 10|}{x - 10}$ is equivalent to $y = \begin{cases} 1 & \text{if } x > 10 \\ -1 & \text{if } x < 10. \end{cases}$

The function is graphed in Figure 2.16. Note the open circles above and below $x = 10$, which indicate the function is not defined at this point.

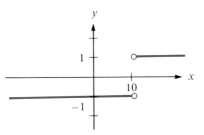

Figure 2.16

We have said that a graph helps us see the essential characteristics of a function. However, the benefits derived from such a picture are directly related to your ability to read the graph. Essential features like the domain and range of a function stand out, but you have to know what you are looking for. For example, consider Figures 2.17 and 2.18, which are the graphs from Examples 7 and 9. In each case the domain and range are specified in the figure. The domain is given by the variation of the graph in a horizontal direction, while the range is given by the variation in the vertical direction.

Here is another example illustrating how you can read important information from a graph.

EXAMPLE 10   Consider the graph of $y = f(x)$ in Figure 2.19.
a. What is the domain of the function?
b. What is the range of the function?
c. Determine $f(0)$.
d. For what values of x does $f(x) = 0$?
e. For what values of x is $f(x) < 0$?

Solution
a. The x values in the figure start at -2 and keep going to the right. Thus, the domain is the interval $[-2, \infty)$.
b. The minimum y value is -4 and the graph extends indefinitely in the positive y direction so there is no maximum value. Thus, the range is $[-4, \infty)$.
c. The notation $f(0)$ means we want the y value when $x = 0$. The ordered pair $(0,4)$ tells us $y = 4$ when $x = 0$. Thus, $f(0) = 4$.
d. Here $f(x)$ or y is 0 and we look at the ordered pairs $(-2,0)$, $(2,0)$, and $(6,0)$. Thus, $f(x) = 0$ when $x = -2$, 2, or 6.
e. The y values are less than 0 when the figure is below the x-axis. As indicated in color, $f(x) < 0$ for $2 < x < 6$.

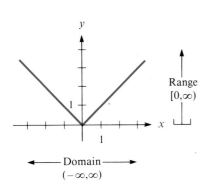

Figure 2.17

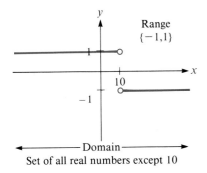

Figure 2.18

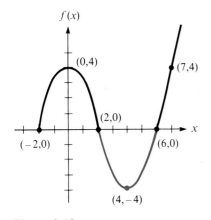

Figure 2.19

Examples 5 to 9 showed functions whose graphs require only lines. Graphs are by no means limited to lines, however, and the next two examples consider the parabola—a figure that is associated with the squaring function.

EXAMPLE 11   Graph the squaring function $y = x^2$. Indicate on the graph the domain and range of the function.

Solution

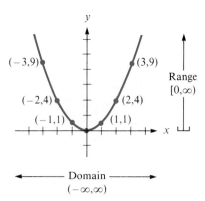

If $x =$	Then $y = x \cdot x = x^2$	Thus, the Ordered-Pair Solutions
3	$y = 3(3) = 9$	(3,9)
2	$y = 2(2) = 4$	(2,4)
1	$y = 1(1) = 1$	(1,1)
0	$y = 0(0) = 0$	(0,0)
-1	$y = -1(-1) = 1$	(−1,1)
-2	$y = -2(-2) = 4$	(−2,4)
-3	$y = -3(-3) = 9$	(−3,9)

Figure 2.20

The cuplike curve in Figure 2.20 is called a **parabola.** In every case the graph of a function of the form $y = ax^2 + bx + c$, where a, b, and c are real numbers ($a \neq 0$), is a parabola. We discuss these functions in detail in Section 3.2. As shown in the graph, the domain is the set of all real numbers and the range is the interval $[0,\infty)$.   ■

EXAMPLE 12   Graph the function $f(x) = 4 - x^2$. Indicate on the graph the domain and range of the function.

Solution

If $x =$	Then $f(x) = 4 - x^2$	Thus, the Ordered-Pair Solutions
3	$f(3) = 4 - (3)^2 = -5$	(3,−5)
2	$f(2) = 4 - (2)^2 = 0$	(2,0)
1	$f(1) = 4 - (1)^2 = 3$	(1,3)
0	$f(0) = 4 - (0)^2 = 4$	(0,4)
-1	$f(-1) = 4 - (-1)^2 = 3$	(−1,3)
-2	$f(-2) = 4 - (-2)^2 = 0$	(−2,0)
-3	$f(-3) = 4 - (-3)^2 = -5$	(−3,−5)

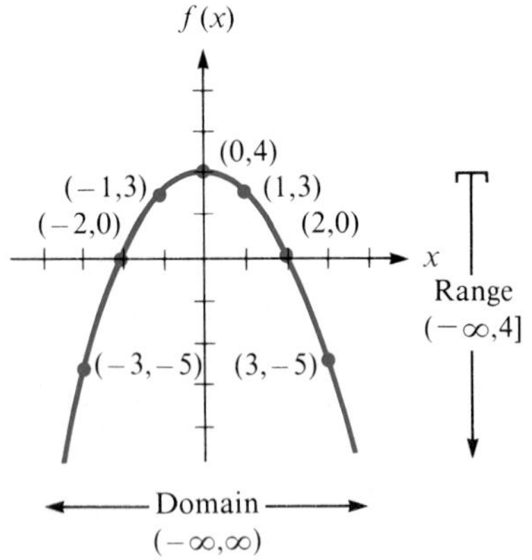

Figure 2.21

The graph of $f(x) = 4 - x^2$, shown in Figure 2.21, is a parabola. The domain is the set of all real numbers and the range is the interval $(-\infty,4]$.   ■

In a function no two ordered pairs may have the same first component. Graphically, this means that a function cannot contain two points that lie on the same vertical line. This observation leads to the following simple test for determining if a graph represents a function.

Vertical Line Test

Imagine a vertical line sweeping across the graph. If the vertical line at any position intersects the graph in more than one point, the graph is not the graph of a function.

EXAMPLE 13   Use the vertical line test to determine which graphs in Figure 2.22 represent the graph of a function.

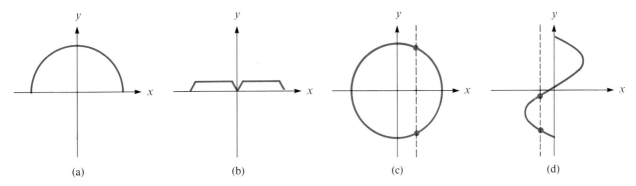

(a)   (b)   (c)   (d)

Figure 2.22

Solution   By the vertical line test graphs (a) and (b) represent functions, whereas (c) and (d) do not.   ∎

In (c) of Figure 2.22 we saw that the graph of a circle is not the graph of a function. However, circles are important graphs, and in particular, we can use the distance formula to show that the equation of a circle with center at the origin and radius r is

$$x^2 + y^2 = r^2.$$

To prove this, consider Figure 2.23 and note that (x,y) is a point on the circle if and only if the distance from (x,y) to the origin is equal to the radius r. Thus, by the distance formula, the equation for the specified circle is

$$\sqrt{(x - 0)^2 + (y - 0)^2} = r$$

or, after squaring both sides of the equation,

$$x^2 + y^2 = r^2.$$

This result allows us to graph without difficulty the type of equations given in the next example.

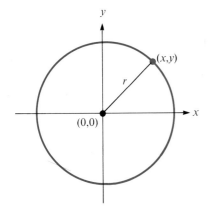

Figure 2.23

EXAMPLE 14   Graph each equation and indicate the domain and range of the functions in parts b and c.

a. $x^2 + y^2 = 4$ **b.** $y = \sqrt{4 - x^2}$ **c.** $y = -\sqrt{4 - x^2}$

Solution

a. The equation $x^2 + y^2 = 4$ fits the form $x^2 + y^2 = r^2$, so the graph is a circle centered at the origin. Also, $r^2 = 4$, so the radius is $\sqrt{4}$, or 2. By drawing a circle that satisfies these conditions, we obtain the graph in Figure 2.24(a).

b. We start with $y = \sqrt{4 - x^2}$ and square both sides of the equation to obtain $y^2 = 4 - x^2$, or $x^2 + y^2 = 4$, which is the equation from part a. However, in the original equation y is restricted to positive numbers or zero, so the graph is a semicircle as shown in Figure 2.24(b). From the graph we read that the domain is $[-2,2]$ and the range is $[0,2]$.

c. As in part b, the graph of $y = -\sqrt{4 - x^2}$ is a semicircle. But y is restricted to negative numbers or zero in this case, so we draw the semicircle in Figure 2.24(c), which shows that the domain is $[-2,2]$ and the range is $[-2,0]$.

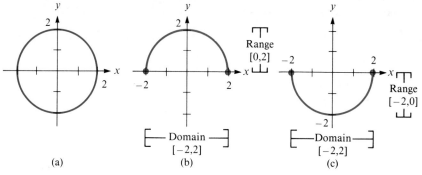

Figure 2.24

In conclusion, keep in mind the following ideas that are associated with graphing:

1. **Coordinate (or analytic) geometry bridges the gap between algebra and geometry.** The fundamental principle in this scheme is that every ordered pair that satisfies an equation corresponds to a point in its graph, and every point in the graph corresponds to an ordered pair that satisfies the equation. Thus, a picture may be drawn for each equation relating x and y, which helps us see the essential characteristics of the relationship. On the other hand, a graph can often be represented as an equation, so that the powerful methods of algebra can be used to analyze these curves. This simple, yet brilliant, idea is one of the most useful in mathematics.

2. All necessary information must be shown in the graph so you can see the relationship without consulting any supplementary material. In particular, be sure to identify the appropriate units on both axes. A graph must be able to "stand" by itself!

3. It is not necessary to scale the two axes in the same manner. In fact, different scales are frequently desirable. It is important that you be aware that different scales can have a dramatic effect on the apparent behavior of a relationship.

4. In this section we discussed the Cartesian (or rectangular) coordinate system. There are other important systems that can be used to obtain a picture of a relationship. If you continue in mathematics, you are sure to encounter some of them (for example, see Appendix A.4). No one system is ever more correct than another, just more advantageous for the particular situation being analyzed.

EXERCISES 2.2

1. Graph the following ordered pairs.
   a. (3,1)   b. (−3,4)   c. (0,2)
   d. (−3,0)   e. (−1,−2)   f. (2,−3)
   g. (−2,−3)   h. (−1,2)   i. ($\sqrt{2}$,3)
   j. (1,−π)

2. Approximate (use integers) the ordered pairs corresponding to the points shown in the graph.

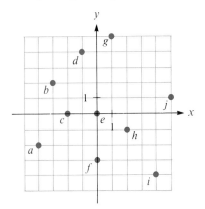

3. Represent the ordered pairs as points in the Cartesian coordinate system and find the distance between them.
   a. (−1,3), (−1,1)   b. (2,−4), (2,−1)
   c. (4,1), (0,1)   d. (−3,−2), (2,−2)
   e. (−12,5), (0,0)   f. (0,0), (−4,1)
   g. (4,−1), (−2,−3)   h. (0,2), (4,−3)
   i. (−1,−1), (−2,−2)   j. (5,2), (1,3)
   k. (−2,5), (4,−3)   l. (6,−2), (−6,3)

4. Three vertices of a square are (3,3), (3,−1), and (−1,3). Find the ordered pair corresponding to the fourth vertex. What is the perimeter of the square?

5. Three vertices of a rectangle are (4,2), (−5,2), and (−5,−4). Find the ordered pair corresponding to the fourth vertex. What is the perimeter of the rectangle?

6. By finding the lengths of *AB, BC,* and *AC,* show that the points $A(−7,−7)$, $B(−1,1)$, and $C(2,5)$ lie on a straight line.

7. Show that the points (3,5), (−3,7), and (−6,−2) are vertices of a right triangle.

8. Show that the points (−8,−2), (−11,3), and (3,8) are vertices of an isosceles triangle. Is this triangle a right triangle?

9. Find the radius of a circle with center at (0,0) and that passes through the point (2,3).

10. Find the area of a circle whose diameter extends from (−1,3) to (3,1).

In Exercises 11–34 graph the function and indicate the range of the function on the graph.

11. $y = x$ 12. $y = −3x$
13. $f(x) = x + 3$ 14. $g(x) = x − 1$
15. $y = −x + 2$ 16. $y = 5x − 5$
17. $f(x) = 1$ 18. $h(x) = π$
19. $y = x^2 + 1$ 20. $f(x) = x^2 − 4$
21. $f(x) = 1 − x^2$ 22. $y = x^2 − 2x + 1$
23. $f(x) = x^3$ 24. $y = x^3 − 2$
25. $y = |x − 1|$ 26. $g(x) = |x + 3|$
27. $y = \dfrac{|x|}{x}$ 28. $f(x) = \dfrac{|x − 2|}{x − 2}$
29. $f(x) = \sqrt[3]{x}$ 30. $y = \sqrt{x}$
31. $h(x) = 1 − \sqrt{x}$ 32. $y = \sqrt{1 − x}$
33. $y = \sqrt{1 − x^2}$ 34. $f(x) = −\sqrt{9 − x^2}$

In Exercises 35 and 36 graph each equation and indicate the domain and range of the function in part b.

35. a. $x^2 + y^2 = 1$ b. $y = −\sqrt{1 − x^2}$
36. a. $x^2 + y^2 − 25 = 0$ b. $y = \sqrt{25 − x^2}$

37. The bracket notation [x] means the greatest integer less than or equal to x. For example, [4.3] = 4, [π] = 3, [−½] = −1, and [2] = 2. If $f(x) = [x]$, which is called the **greatest integer function**, find
   a. $f(\frac{1}{2})$ b. $f(\sqrt{2})$ c. $f(−1.4)$ d. $f(−π)$
   e. the graph of this function for $−2 \le x < 3$

38. A projectile fired vertically up from the ground with an initial velocity of 128 ft/second will hit the ground 8 seconds later, and the speed of the projectile in terms of the elapsed time *t* equals $|128 − 32t|$. Graph the function $s = |128 − 32t|$.

39. The height (y) above water of a diver t seconds after she steps off a platform 100 ft high is given by the formula $y = 100 - 16t^2$. Graph this function.

40. The charge for the first 3 minutes of a Monday station-to-station call from New York to Los Angeles depends on the time of day that the call is made. The following formula table indicates the charge in terms of 24-hour time (that is, 1700 hours is equivalent to 5 P.M.):

$$\text{Charge} = \begin{cases} \$0.85 \text{ if} & 0 \text{ hours} \le t < 0800 \text{ hours} \\ \$1.45 \text{ if } 0800 \text{ hours} \le t \le 1700 \text{ hours} \\ \$0.85 \text{ if } 1700 \text{ hours} < t \le 2400 \text{ hours.} \end{cases}$$

Graph this function.

41. Because of the need for energy conservation, the local power company advertises to "save a watt." To this end, they calculate the efficiency of air conditioners by finding the ratio between the cooling ability of the machine (Btu) and the amount of watts of electricity that it requires (that is, efficiency = Btu/watt). Their rating in terms of the efficiency (e) of the air conditioner is as follows:

$$\text{Rating} = \begin{cases} 0 \text{ (flunk)} & \text{if } 0 \le e < 6 \\ 1 \text{ (pass)} & \text{if } 6 \le e < 8 \\ 2 \text{ (good)} & \text{if } 8 \le e < 10 \\ 3 \text{ (very good) if } 10 \le e. \end{cases}$$

Graph this function.

In Exercises 42–50 graph the function and indicate the domain and range of the function on the graph.

42. $f(x) = \begin{cases} 1 & \text{if } x > 2 \\ 0 & \text{if } x \le 2 \end{cases}$

43. $g(x) = \begin{cases} 1 & \text{if } x \ge 0 \\ -1 & \text{if } x < 0 \end{cases}$

44. $h(x) = \begin{cases} x & \text{if } x \ge 1 \\ 1 & \text{if } x < -1 \end{cases}$

45. $f(x) = \begin{cases} x & \text{if } -2 \le x < 0 \\ 0 & \text{if } x = 0 \\ -x & \text{if } x > 0 \end{cases}$

46. $y = \begin{cases} -x & \text{if } 0 < x < 3 \\ 2 & \text{if } x = 0 \\ x + 1 & \text{if } x < 0 \end{cases}$

47. $y = \begin{cases} x^2 & \text{if } 0 \le x \le 1 \\ x & \text{if } x > 1 \end{cases}$

48. $f(x) = \begin{cases} \sqrt{x} & \text{if } x \le 0 \\ -x^2 & \text{if } x > 0 \end{cases}$

49. $g(x) = \begin{cases} 1 & \text{if } x < 0 \\ x^2 + 1 & \text{if } x \ge 0 \end{cases}$

50. $h(x) = \begin{cases} |x| & \text{if } x < 0 \\ \sqrt{x} & \text{if } x \ge 0 \end{cases}$

In Exercises 51 and 52 find the domain and range of the functions shown.

51.

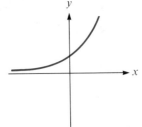

52.

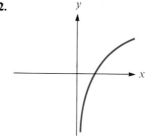

53. Consider the graph of $y = f(x)$ given here.
 a. What is the domain of the function?
 b. What is the range of the function?
 c. Determine $f(0)$ and $f(5)$.
 d. For what value(s) of x does $f(x) = 4$?
 e. Solve $f(x) = 0$.
 f. For what value(s) of x is $f(x) > 0$?
 g. For what value(s) of x is $f(x) < 0$?
 h. For what value(s) of x is $|f(x)| < 5$?

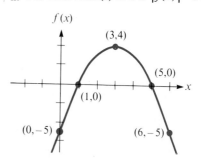

54. Consider the graph of $y = f(x)$ given on page 119.
 a. What is the domain of the function?
 b. What is the range of the function?
 c. Determine $f(a)$ and $f(c)$.
 d. Solve $f(x) = a$.
 e. For what value(s) of x does $f(x) = 0$?
 f. For what value(s) of x is $f(x) < 0$?
 g. For what value(s) of x is $f(x) > 0$?
 h. For what value(s) of x is $|f(x)| < a$?

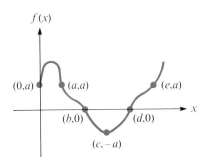

In Exercises 55–64 determine which of the graphs represent the graph of a function?

55.

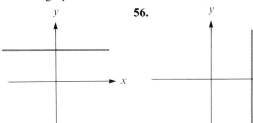

56.

57.

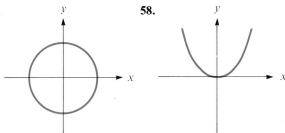

58.

59.

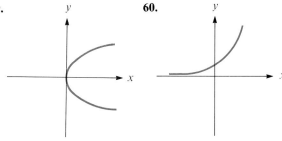

60.

61.

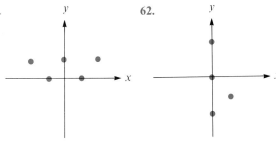

62.

63.

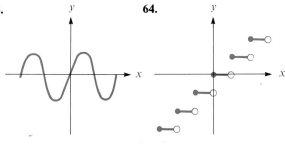

64.

THINK ABOUT IT

1. Give two examples of a function for which $f(-x) = f(x)$ for all real numbers x. In each case, graph the function and describe how the graph is related to the y-axis.

2. Graph $y = |x|$ and $y = -|x|$ on the same coordinate system. How are the graphs related? In general, what is the relationship between the graphs of $y = f(x)$ and $y = -f(x)$?

3. **a.** Graph $y = x^2 + 2$ and $y = x^2 - 2$ on the same coordinate system. How do the graphs relate to $y = x^2$? In general, what is the effect of adding a constant *after* applying the squaring function?

    b. Graph $y = (x + 2)^2$ and $y = (x - 2)^2$ on the same coordinate system. How do these graphs relate to $y = x^2$? In general, what is the effect of adding a constant *before* applying the squaring function?

4. Describe a realistic situation that would produce each graph.

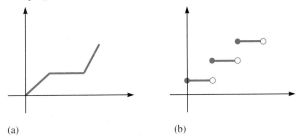

(a)   (b)

5. Find x, given that the point $(x,0)$ is equidistant from $(-4,-4)$ and $(2,3)$.

REMEMBER THIS

1. Is $|-x| = |x|$ a conditional equation or an identity?
2. Express $2 - x^2$ as a sum.
3. If $f(x) = x^5$, does $f(-2) = -f(2)$?

4. If $f(x) = 5x - 1$, find (a) $2f(3)$ and (b) $\frac{1}{2}f(3)$.

5. If $f(x) = x^2$, find (a) $f(a + 2)$ and (b) $f(a) + 2$.
6. Does the equation $x = y^5$ determine y as a function of x?

7. If $g(x) = \dfrac{1}{x + 1}$, for what value of x does $g(x) = 2$?

8. If $f(x) = \dfrac{x + 5}{x - 4}$, find the domain of f.

9. If $f = \{(6,-2),(7,-1),(8,0)\}$, find the range of f.
10. If the base in a triangle measures twice the height, then express the area of the triangle as a function of the base.

2.3 Graphing Techniques

There are many techniques associated with graphing functions. In the last section we concentrated primarily on plotting points in conjunction with graphs that required only lines or parabolas. We now mention some useful general approaches that revolve around two central themes: (1) how we can use the idea of symmetry to "cut in half" the job of graphing unfamiliar functions and (2) how we can graph variations of familiar functions by somehow adjusting a known curve. In the previous section we drew the graphs shown in Figure 2.25. From this point on, it is important that you memorize the graphs of these important types of functions.

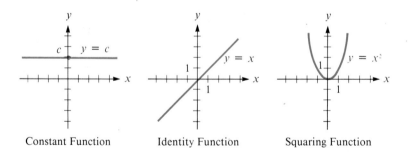

Constant Function   Identity Function   Squaring Function

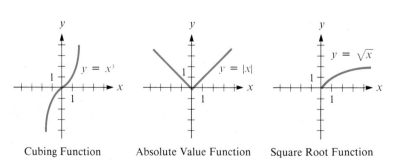

Cubing Function   Absolute Value Function   Square Root Function

Figure 2.25

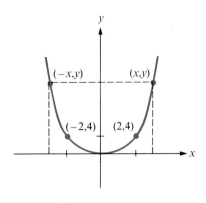

Figure 2.26

Symmetry

Consider the graph of $y = x^2$ in Figure 2.26. Notice that the y-axis divides the parabola into two segments such that if the curve were folded over on this line, its left half would coincide with its right half. In such a case we say $y = x^2$ has **symmetry about the y-axis.** Such symmetry is a nice feature for this or any other curve. It means we have to plot only the curve for $x \geq 0$ and then reflect our result over to the other side of the y-axis to complete the graph.

EXAMPLE 1   Complete the graph in Figure 2.27 for $x < 0$ if f is symmetric about the y-axis.

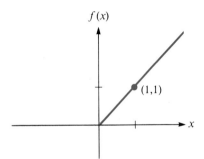

Figure 2.27

Solution   Just reflect the given segment over to quadrant 2 to obtain the graph for $x < 0$. The completed graph is shown in Figure 2.28.

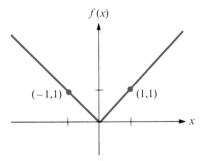

Figure 2.28   ■

Now the question is how can we anticipate y-axis symmetry by looking at an equation. Consider Figure 2.28 again and note that y-axis symmetry results when x values that are opposite in sign share the same y value. For example, in the case of $y = x^2$, values of both 2 and -2 for x lead to a y value of 4; and in general, both (x,y) and $(-x,y)$ are on the graph. This observation leads us to the following rule for recognizing y-axis symmetry.

Even Function

> A function f is called an even function when $f(-x) = f(x)$ for all x in the domain of f. The graph of an even function is symmetric about the y-axis.

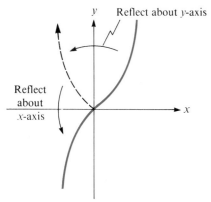

a.

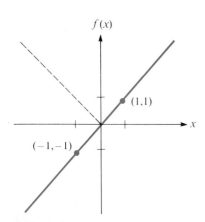

b.

Figure 2.29

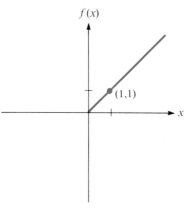

Figure 2.30

EXAMPLE 2   Show that the absolute value function $f(x) = |x|$ is an even function.

Solution

$$f(x) = |x| \text{ and } f(-x) = |-x| = |x| = f(x)$$

Since $f(-x) = f(x)$ for any x, the absolute value function is an even function. In case you did not recognize it, Example 1 is the graph of $y = |x|$, which has y-axis symmetry. ∎

Another important type of symmetry is shown in the graph of $y = x^3$ in Figure 2.29(a). In this graph every point (x,y) in the first quadrant has a matching point $(-x,-y)$ in the third quadrant equidistant from the origin. In such a case we say $y = x^3$ has **symmetry about the origin.** Origin symmetry is a little more subtle than y-axis symmetry in that we must reflect the given segment twice in our shortcut graphing approach. As shown in Figure 2.29(b), we reflect the curve for $x \geq 0$ about both the x- and y-axes (in either order) to complete the graph.

EXAMPLE 3   Complete the graph in Figure 2.30 for $x < 0$ if f is symmetric about the origin.

Solution   Reflect the given segment over to quadrant 2 and then reflect the result about the x-axis and down to quadrant 3. The completed graph is shown in Figure 2.31. ∎

Figure 2.31

We can anticipate origin symmetry because it results when x values that are opposite in sign produce y values that are opposite in sign. For example, in the case of $y = x^3$, values of 2 and -2 for x produce y values of 8 and -8, respectively; and in general, both (x,y) and $(-x,-y)$ are on the graph. Thus, we have the following rule for recognizing origin symmetry.

Odd Function

> A function f is called an odd function when $f(-x) = -f(x)$ for all x in the domain of f. The graph of an odd function is symmetric about the origin.

EXAMPLE 4   Show that the identity function $f(x) = x$ is an odd function.

Solution

$$f(x) = x \quad \text{and} \quad f(-x) = -x = -f(x)$$

Thus, $f(-x) = -f(x)$ for any x and the identity function is an odd function. Example 3 is the graph of the identity function that has origin symmetry. ■

EXAMPLE 5   Is the function $y = (x^2 - 1)/x^3$ an even function, an odd function, or neither? Figure 2.32 is the graph of this function for $x \geq 0$. Complete the graph.

Solution

$$f(x) = \frac{x^2 - 1}{x^3}$$

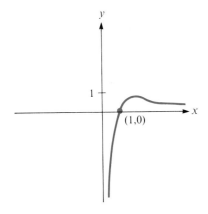

Figure 2.32

and

$$f(-x) = \frac{(-x)^2 - 1}{(-x)^3} = -\frac{x^2 - 1}{x^3} = -f(x)$$

Since $f(-x) = -f(x)$, f is an odd function that is symmetric about the origin. Reflecting the given curve about both the x- and y-axes gives the completed graph in Figure 2.33. ■

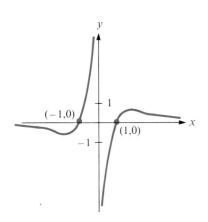

Figure 2.33

Translations

We now come to a second major consideration in graphing: How can we graph variations of a familiar function by manipulating a known curve? Consider Figure 2.34, which shows four variations of the absolute value function $f(x) = |x|$. Do you see a pattern?

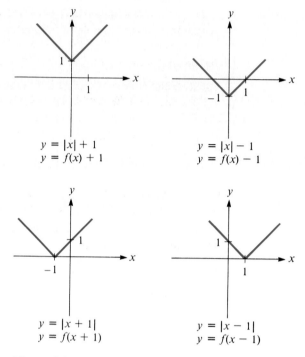

$$y = |x| + 1$$
$$y = f(x) + 1$$

$$y = |x| - 1$$
$$y = f(x) - 1$$

$$y = |x + 1|$$
$$y = f(x + 1)$$

$$y = |x - 1|$$
$$y = f(x - 1)$$

Figure 2.34

All four graphs have the basic $\vee$ shape that characterizes the absolute value function. Our job is merely to translate the $\vee$ to the right spot. Note that when we add or subtract the 1 *after* applying the absolute value rule, the effect is to move the $\vee$ up (if adding) or down (if subtracting) a distance of 1 unit. If we add or subtract the 1 *before* applying the absolute value rule, the $\vee$ moves to the left (if adding) or to the right (if subtracting) a distance of 1 unit. Be careful with the left-right shifts—they are deceiving. The graph of $y = |x - 1|$ is 1 unit to the right (not the left) of $y = |x|$ because x must be one larger in $y = |x - 1|$ than in $y = |x|$ to produce the same y value. For example, $y = 0$ when $x = 1$ in

$y = |x - 1|$, while $y = 0$ when $x = 0$ in $y = |x|$. These ideas generalize to other functions and constants and provide us with the following guidelines.

Vertical and Horizontal Shifts

Let c be a positive constant.

1. The graph of $y = f(x) + c$ is the graph of f raised c units.
2. The graph of $y = f(x) - c$ is the graph of f lowered c units.
3. The graph of $y = f(x + c)$ is the graph of f shifted c units to the left.
4. The graph of $y = f(x - c)$ is the graph of f shifted c units to the right.

EXAMPLE 6   Use the graph of $y = x^2$ to graph each of the following functions.

a. $y = x^2 + 3$ **b.** $y = (x - 2)^2$ **c.** $y = (x + 1)^2 - 2$

Solution   The graph of the squaring function $y = x^2$ is a parabola. The rules above tell us to move this basic shape as follows (see Figure 2.35):

a. The constant 3 is added *after* the squaring rule, which means we move the parabola up 3 units.
b. The constant 2 is subtracted *before* the squaring rule, which moves the parabola 2 units to the right.
c. In this case we move the parabola 1 unit to the left and 2 units down.

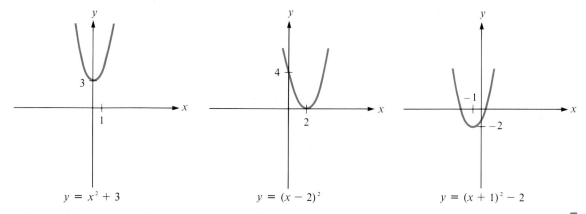

$y = x^2 + 3$ $y = (x - 2)^2$ $y = (x + 1)^2 - 2$

Figure 2.35

Reflecting and Stretching

Consider Figures 2.36 and 2.37, which illustrate the graphs of $y = |x|$, $y = -|x|$, $y = 2|x|$, and $y = \frac{1}{2}|x|$. These graphs show what happens when $|x|$ is multiplied by some constant. The most dramatic effect comes when the sign of the function is changed. As shown in Figure 2.36, switching from $|x|$ to $-|x|$ reflects the graph about the x-axis. Multiplying by a positive constant, such as 2, causes the curve to climb faster, while multiplying by a constant like $\frac{1}{2}$ flattens out the graph (see Figure 2.37). We summarize this pattern in the following rule.

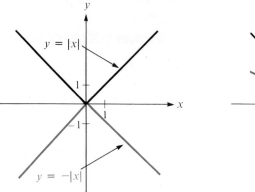

Figure 2.36   **Figure 2.37**

To Graph $y = cf(x)$

> **Reflecting**   The graph of $y = -f(x)$ is the graph of f reflected about the x-axis.
> **Stretching**   If $c > 1$, the graph of $y = cf(x)$ is the graph of f stretched by a factor of c.
> **Shrinking**   If $0 < c < 1$, the graph of $y = cf(x)$ is the graph of f flattened out by a factor of c.

EXAMPLE 7   Use the graph of $y = x^2$ to graph $y = 2 - x^2$.

Solution   To graph $y = -x^2$, we reflect over the basic $\cup$ shape about the x-axis as in Figure 2.38(a). Now $2 - x^2$ is the same as $-x^2 + 2$, so we raise the graph in Figure 2.38(a) up 2 units to get our answer [see Figure 2.38(b)].

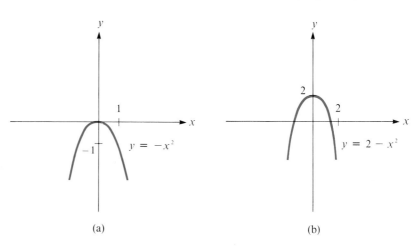

(a)   (b)

Figure 2.38   ■

EXAMPLE 8   Consider the graph of $y = f(x)$ in Figure 2.39. Use this graph to sketch $y = 2f(x)$ and $y = \frac{1}{2}f(x)$.

Solution   To graph $y = 2f(x)$, we just double each y value in f. Note in particular that the maximum y in f is 1, while the biggest y in $2f$ is 2. Similarly we graph $y = \frac{1}{2}f(x)$ by halving the y values in f. Both graphs are shown in Figure 2.40.

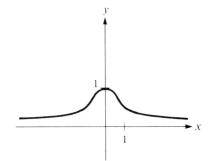

Figure 2.39

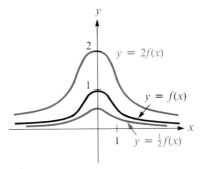

Figure 2.40   ■

EXERCISES 2.3

In Exercises 1–8 classify each of the functions in the figures shown as even, odd, or neither of these.

1.

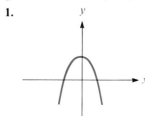

2.

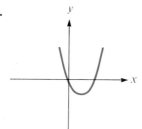

3.

4.

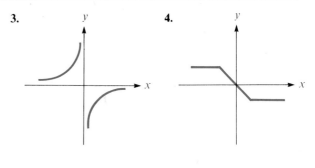

5.

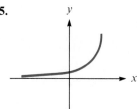

6.

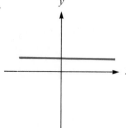

7.

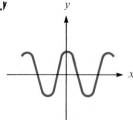

8.

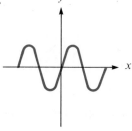

In Exercises 9–12 complete the graphs for $x < 0$ if (a) f is an even function and (b) f is an odd function.

9.

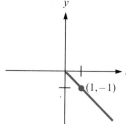

10.

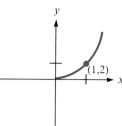

11.

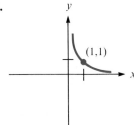

12.

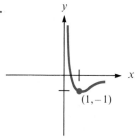

In Exercises 13–22 classify each of the functions as even, odd, or neither of these.

13. $y = -x$

14. $f(x) = x + 1$

15. $f(x) = x^2 - 4$

16. $g(x) = 2$

17. $g(x) = x^2 - x$

18. $y = x^5$

19. $y = x^6$

20. $f(x) = 2 - |x|$

21. $f(x) = 1/x$

22. $y = 1/(x^3 + 1)$

23. Is the function $y = 1/(x^2 + 1)$ an even function, an odd function, or neither? The graph of this function for $x \geq 0$ follows. Complete the graph.

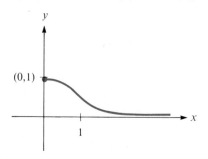

24. Is the function $y = x/(x^2 + 1)$ an even function, an odd function, or neither? The graph of this function for $x \geq 0$ follows. Complete the graph.

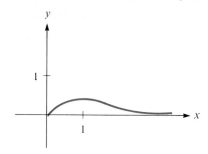

In Exercises 25–30 use the graph of $y = |x|$ to graph the given functions.

25. $y = |x| + 2$

26. $y = |x + 3|$

27. $y = -|x - 1|$

28. $y = 3|x|$

29. $y = |x - 2| - 1$

30. $y = 1 - |x + 1|$

In Exercises 31–36 use the graph of $y = x^2$ to graph the given functions.

31. $y = (x + 2)^2$

32. $y = x^2 - 2$

33. $y = 2x^2$

34. $y = \frac{1}{2}x^2$

35. $y = 1 - x^2$

36. $y = (x + 4)^2 - 3$

In Exercises 37–40 use the graph of $y = x^3$ to graph each function.

37. $y = x^3 + 1$

38. $y = (x + 1)^3$

39. $y = (x - 2)^3 - 1$

40. $y = -x^3$

In Exercises 41–44 use the graph of $f(x) = \sqrt{x}$ that follows to graph each function.

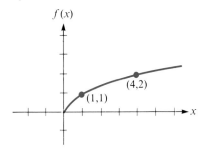

41. $f(x) = \sqrt{x-1}$ **42.** $f(x) = \sqrt{x} - 1$
43. $f(x) = 2 - \sqrt{x}$ **44.** $f(x) = 2 + \sqrt{x+2}$

In Exercises 45–50 use the graph of $y = f(x)$ that follows to graph each function.

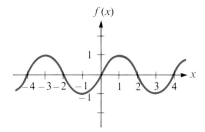

45. $y = 2f(x)$ **46.** $y = f(x) + 2$
47. $y = f(x-1)$ **48.** $y = -f(x)$
49. $y = -\frac{1}{2}f(x+4)$ **50.** $y = 3f(x-1) - 1$

THINK ABOUT IT

1. Classify the following function as even, odd, or neither of these.

$$f(x) = \begin{cases} 1 & \text{if } x > 1 \\ x & \text{if } -1 \leq x \leq 1 \\ -1 & \text{if } x < -1 \end{cases}$$

2. Give two graphical examples of an odd function with a graph that does not pass through the origin.
3. The point $(-4,3)$ lies on the graph of $f(x) = (x+2)^2 - 1$. What are the translated coordinates of this point if the graph of f is shifted so that the minimum point of the parabola is at the origin?

4. Graph $y = x^2 + 6x$ by completing the square on the right-hand side of this equation and using the methods of this section.
5. Graph $y = \sqrt{x}$ and $y = \sqrt{-x}$ on the same coordinate system. How do the graphs relate to each other? In general, how may the graph of $y = f(-x)$ be obtained from the graph of $y = f(x)$?

REMEMBER THIS

1. Find the intersection of sets A and B if $A = \{0,1,2,3\}$ and $B = \{2,3,4,5\}$.

2. Find the sum of $\dfrac{1}{x}$ and $\dfrac{x}{1+x}$.

3. Find the product of $3x^2 - 2$ and $-x + 6$.
4. Find the difference when $7x - 2$ is subtracted from $-2x + 3$.
5. Find the quotient when x is divided by $\sqrt{x}$.

6. **a.** If $f(x) = \sqrt{x}$, what is the domain of f?
   b. If $f(x) = 1/\sqrt{x}$, what is the domain of f?
7. If $g = \{(0,2),(-3,1),(-5,-1)\}$, find the domain of g.
8. Find $f(x + h)$ if $f(x) = 3x^2 - x + 5$.
9. Graph $h(x) = \begin{cases} -x, & \text{if } x < -1 \\ 3, & \text{if } x \geq 1. \end{cases}$
10. Find the circumference of a circle whose radius extends from $(2,-3)$ to $(-4,-3)$.

2.4   Operations with Functions

We may add, subtract, multiply, and divide functions to form new functions. For example, suppose that an item that costs \$5 to manufacture sells for \$7. The total cost of x items is given by the function

$$C(x) = 5x.$$

The total sales revenue from x items is given by

$$S(x) = 7x.$$

Since profit is the difference between sales and cost, we form the profit function as the difference of the two other functions.

$$P(x) = S(x) - C(x)$$
$$= 7x - 5x$$
$$= 2x$$

In general, for two functions f and g, the functions $f + g, f - g, f \cdot g, f/g$ are defined as follows:

$$(f + g)(x) = f(x) + g(x)$$
$$(f - g)(x) = f(x) - g(x)$$
$$(f \cdot g)(x) = f(x) \cdot g(x)$$
$$\left(\frac{f}{g}\right)(x) = \frac{f(x)}{g(x)} \qquad g(x) \neq 0$$

The domain of the resulting function is the intersection of the domains of f and g. The domain of the quotient function excludes any x for which $g(x) = 0$. In effect, the definitions say that for any x at which both functions are defined, we combine the y values.

EXAMPLE 1   If $f(x) = x$ and $g(x) = \sqrt{x}$, find $(f + g)(x)$, $(f - g)(x)$, $(f \cdot g)(x)$, and $(f/g)(x)$. In each case give the domain of the resulting function.

Solution   Using the definitions, we have

$$(f + g)(x) = f(x) + g(x) = x + \sqrt{x}$$
$$(f - g)(x) = f(x) - g(x) = x - \sqrt{x}$$
$$(f \cdot g)(x) = f(x) \cdot g(x) = x\sqrt{x} \quad \text{or} \quad x^{3/2}$$
$$\left(\frac{f}{g}\right)(x) = \frac{f(x)}{g(x)} = \frac{x}{\sqrt{x}} \quad \text{or} \quad \sqrt{x}.$$

The domain of f is the set of all real numbers and the domain of g is $[0,\infty)$. The domain of $f + g, f - g,$ and $f \cdot g$ is the set of numbers common to both domains, which is $[0,\infty)$. In the quotient function $x \neq 0$. Thus, the domain of f/g is the interval $(0,\infty)$.   ∎

EXAMPLE 2   If $f = \{(0,1), (1,2), (2,3), (3,4)\}$ and $g = \{(2,0), (3,1), (4,-5), (5,-3)\}$, find $f + g, f - g, f \cdot g,$ and f/g.

Solution   Both functions are defined at $x = 2$ and $x = 3$. Thus,

$$f + g = \{(2, 3 + 0), (3, 4 + 1)\} = \{(2,3), (3,5)\}$$
$$f - g = \{(2, 3 - 0), (3, 4 - 1)\} = \{(2,3), (3,3)\}$$
$$f \cdot g = \{(2, 3 \cdot 0), (3, 4 \cdot 1)\} = \{(2,0), (3,4)\}$$
$$f/g = \{(3, \tfrac{4}{1})\} = \{(3,4)\}; \text{ since } \tfrac{3}{0} \text{ is undefined, we do not list } (2, \tfrac{3}{0}). \quad ∎$$

There is another important way of combining functions that has no counterpart in arithmetic. The operation is called **composition.** Basically, composition is a substitution that causes a "chain reaction" in which two functions are applied in succession.

Consider the problem of determining the area of a square whose perimeter is 20 in. You would probably reason as follows: If the perimeter is 20 in., then the side is 5 in., so the area is 25 in.² Given a specific value for the perimeter (P), we first determine the side (s) by using the function

$$s = \frac{P}{4}.$$

The result of this step is then substituted in the function

$$A = s^2$$

to determine the area. Let us put these formulas in functional notation and use the analogy of a function as a machine. Consider Figure 2.41, in which $s = f(P) = P/4$ and $A = g(s) = s^2$. The output of the f machine feeds the input of the g machine. Thus, we first apply the f function to determine s and then apply the g function to determine A. The area is given by the function

$$A = g[f(P)].$$

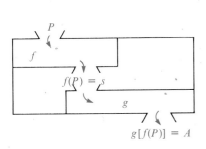

Figure 2.41

This function is said to be a *composition* of f and g. By substitution, we have

$$A = g[f(P)] = g\left(\frac{P}{4}\right) = \left(\frac{P}{4}\right)^2.$$

The symbol "∘" is also used for composition. We could write

$$A = (g \circ f)(P).$$

In general, the composite functions of functions f and g are defined as follows:

$$(g \circ f)(x) = g[f(x)]$$
$$(f \circ g)(x) = f[g(x)].$$

EXAMPLE 3   If $f(x) = x - 1$ and $g(x) = x^2$, find $(g \circ f)(x)$ and $(f \circ g)(x)$.

Solution

$$(g \circ f)(x) = g[f(x)] = g(x - 1)$$
$$= (x - 1)^2 = x^2 - 2x + 1$$
$$(f \circ g)(x) = f[g(x)] = f(x^2)$$
$$= x^2 - 1$$

Note that the order in which the functions are applied is important. That is,

$$(g \circ f)(x) \neq (f \circ g)(x). \qquad \blacksquare$$

A number x must satisfy two requirements to belong to the domain of $g \circ f$.

1. Since we first apply the f function, x must be in the domain of f.
2. The output from step 1, $f(x)$, must be in the domain of g.

Thus, the domain of $g \circ f$ is the set of all real numbers x in the domain of f for which $f(x)$ is in the domain of g. This set consists of all the values in the domain of f for which $g \circ f$ is defined.

EXAMPLE 4 If $f(x) = \sqrt{x}$ and $g(x) = x^2$, find $(g \circ f)(x)$ and $(f \circ g)(x)$. Indicate the domain of each function.

Solution

$$(g \circ f)(x) = g[f(x)] = g(\sqrt{x}) = (\sqrt{x})^2 = x$$

Domain of f: $[0,\infty)$
$(g \circ f)(x) = x$ is defined for all real numbers.

Thus, the domain of $g \circ f$ is $[0,\infty)$.

$$(f \circ g)(x) = f[g(x)] = f(x^2) = \sqrt{x^2} \quad \text{or} \quad |x|$$

Domain of g: $(-\infty,\infty)$
$(f \circ g)(x) = |x|$ is defined for all real numbers.

Thus, the domain of $f \circ g$ is the set of all the real numbers. ∎

EXAMPLE 5 If $f = \{(0,1), (1,2), (2,3), (3,4)\}$ and $g = \{(2,0), (3,1), (4,-5), (5,-3)\}$, find $f \circ g$ and $g \circ f$. Indicate the domain of each composite function.

Solution To determine $f \circ g$, we first apply the g function. As illustrated below, the range elements 0 and 1 are in the domain of f. Thus, $f \circ g = \{(2,1), (3,2)\}$ and the domain of $f \circ g$ is $\{2,3\}$.

$$
\begin{array}{ll}
2 \xrightarrow{g} & 0 \xrightarrow{f} 1 \\
3 \rightarrow & 1 \rightarrow 2 \\
4 \rightarrow & -5 \\
5 \rightarrow & -3
\end{array}
$$

To find $g \circ f$, we first apply the f function. As illustrated below, the range elements 2, 3, and 4 are in the domain of g. Thus, $g \circ f = \{(1,0), (2,1), (3,-5)\}$ and the domain of $g \circ f$ is $\{1, 2, 3\}$.

$$
\begin{array}{l}
0 \xrightarrow{f} 1 \\
1 \rightarrow 2 \xrightarrow{g} 0 \\
2 \rightarrow 3 \rightarrow 1 \\
3 \rightarrow 4 \rightarrow -5
\end{array}
$$

∎

EXAMPLE 6   An oil leak is spreading over a plane surface in the shape of a circle. The radius of this circle is increasing at a constant rate of 5 cm/ second so the radius of this spill t seconds after the start of the leak may be expressed by $r = g(t) = 5t$. If function f expresses the area of this circular spill as a function of r so that $A = f(r) = \pi r^2$, then find and interpret $(f \circ g)(t)$.

Solution   By definition $(f \circ g)(t) = f[g(t)]$, and in this problem $g(t) = 5t$. Thus,

$$(f \circ g)(t) = f[g(t)] = f(5t) = \pi(5t)^2 = 25\pi t^2.$$

To interpret this result, if function f expresses A in terms of r, and function g expresses r in terms of t, then the function $f \circ g$ expresses A in terms of t. Thus, the area A of this circular spill t seconds after the start of the leak is given by

$$A = (f \circ g)(t) = 25\pi t^2. \qquad \blacksquare$$

In calculus it is often important to recognize a function as the composition of simpler functions. Although more than one answer may be possible in such problems, the next example shows a useful way for viewing an expression that involves a power.

EXAMPLE 7   If $h(x) = (2x + 3)^3$, find simpler functions f and g so that $h(x) = (f \circ g)(x)$.

Solution   Because $(f \circ g)(x) = f[g(x)]$, the g rule is applied first and we will refer to g as the "inner" function. Look to let $g(x)$ equal an expression that is set off by a grouping symbol, particularly when such expressions may be written as the base in an exponential expression. From this perspective, we note $h(x)$ equals the third power of $2x + 3$ so if we define the inner function g by

$$g(x) = 2x + 3$$

and let f be the cubing function defined by

$$f(x) = x^3,$$

then

$$f[g(x)] = f(2x + 3) = (2x + 3)^3 = h(x).$$

Thus, $(f \circ g)(x) = h(x)$ as desired. (*Note:* It is also true that $(f \circ g)(x) = h(x)$ if $g(x) = 2x$ and $f(x) = (x + 3)^3$, and other choices are also possible. However, viewing $(2x + 3)^3$ as the third power of $2x + 3$ is the type of straightforward and useful analysis that is desired in these problems.) $\blacksquare$

EXERCISES 2.4

In Exercises 1–12 find $(f + g)(x)$, $(f - g)(x)$, $(f \cdot g)(x)$, $(f/g)(x)$, $(f \circ g)(x)$, and $(g \circ f)(x)$. Indicate the domain of each function.

1. $f(x) = 2x; g(x) = x - 1$
2. $f(x) = x - 2; g(x) = x + 2$
3. $f(x) = 4x^2 - 5; g(x) = -x + 3$
4. $f(x) = -3x + 5; g(x) = 1 + 2x - x^2$
5. $f(x) = x^2; g(x) = 1$
6. $f(x) = x; g(x) = 1/x$
7. $f(x) = 1/x; g(x) = x/(1 + x)$
8. $f(x) = \sqrt{x}; g(x) = -x^2$
9. $f(x) = x^2 - 1; g(x) = \sqrt{x + 1}$
10. $f(x) = \sqrt[3]{x - 5}; g(x) = x^3 + 5$
11. $f = \{(0,2), (1,3), (2,4), (3,5)\}$
    $g = \{(2,-1), (3,0), (4,1), (5,2)\}$
12. $f = \{(-3,5), (0,1), (2,6), (4,11)\}$
    $g = \{(0,0), (2,2), (5,5), (-3,-3)\}$
13. If $f = \{(1,3), (2,-2), (3,5)\}$ and
    $g = \{(2,-1), (3,4), (4,0)\}$, find
    a. $(f + g)(2)$ b. $(f \cdot g)(3)$
    c. $(g \cdot f)(1)$ d. $(g \circ f)(1)$
14. If $f = \{(1,5), (-2,4), (-3,0)\}$ and
    $g = \{(0,2), (-3,1), (-5,-1)\}$, find

    a. $\left(\dfrac{f}{g}\right)(-3)$ b. $\left(\dfrac{g}{f}\right)(-3)$

    c. $(f \circ g)(-3)$ d. $(g \circ f)(-3)$

In Exercises 15–22 use the graphs of f and g given below to evaluate each expression.

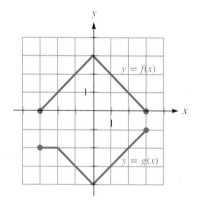

15. $(f + g)(1)$ 16. $(f - g)(2)$
17. $(f \cdot g)(-2)$ 18. $(f/g)(0)$
19. $(f \circ g)(1)$ 20. $(g \circ f)(1)$
21. $(g \circ f)(0)$ 22. $(f \circ g)(2)$
23. If $f(x) = 9x - 5$ and $g(x) = \frac{1}{9}x + \frac{5}{9}$, find
    $(f \circ g)(x)$ and $(g \circ f)(x)$.

24. If $f(x) = x^3 + 5$ and $g(x) = \sqrt[3]{x - 5}$, find $(f \circ g)(x)$ and $(g \circ f)(x)$.
25. If $g(u) = u^3$ and $u = f(x) = 5x - 4$, find $(g \circ f)(x)$.
26. If $g(u) = \sqrt{u}$ and $u = f(x) = 2x^2 + 9$, find $(g \circ f)(x)$.
27. If $f(r) = \pi r^2$ and $r = g(t) = 6t$, find $(f \circ g)(t)$.
28. If $f(r) = \frac{4}{3}\pi r^3$ and $r = g(t) = \frac{5}{2}t$, find $(f \circ g)(t)$.

In Exercises 29–38 find simpler functions f and g so that $h(x) = (f \circ g)(x)$. (See Example 7.)

29. $h(x) = (4x - 1)^3$ 30. $h(x) = (x^2 + 1)^2$

31. $h(x) = \left(\dfrac{x - 1}{x + 1}\right)^{1/2}$ 32. $h(x) = (5x - 10)^{-4}$

33. $h(x) = \sqrt[3]{2x + 1}$ 34. $h(x) = \sqrt{1 - x^2}$
35. $h(x) = 2(3 - x)^4$ 36. $h(x) = (x^3 - 8)^{1/3} + 5$
37. $h(x) = (1 - x)^3 + 6(1 - x)^2$

38. $h(x) = \dfrac{1}{3x - 5}$

39. If $h(x) = \sqrt{2x^2 + 2}$ and $f(x) = \sqrt{x}$, find $g(x)$ so that $h(x) = (f \circ g)(x)$.
40. If $h(x) = 3x^2 - 1$ and $g(x) = x^2$, find $f(x)$ so that $h(x) = (f \circ g)(x)$.
41. If $f(x) = 2x - 3$ and $g(x) = 5x + b$, find b so that $(f \circ g)(x) = (g \circ f)(x)$.
42. If $f(x) = ax + b$ and $g(x) = 5x - 2$, find conditions on a and b so that $(f \circ g)(x) = (g \circ f)(x)$.

In Exercises 43–46 find $(f \circ g \circ h)(x)$.

43. $f(x) = 2x; g(x) = x^2; h(x) = x - 2$
44. $f(x) = x - 2; g(x) = 2x; h(x) = x^2$
45. $f(x) = \sqrt{x}; g(x) = 1/x; h(x) = 1 - x$
46. $f(x) = 1/x; g(x) = 1 - x; h(x) = \sqrt{x}$
47. Show that the sum of two odd functions is an odd function.
48. Show that the product of two odd functions is an even function.
49. Show that the product of an even function and an odd function is an odd function.
50. If f and g are odd functions, is $f \circ g$ an odd function or an even function? Verify your answer.
51. a. There are 1,760 yd in 1 mi. Write a function that converts miles to yards.
    b. Write a function that converts yards to feet.
    c. Use parts a and b and composition to construct a function that converts miles to feet.
52. a. Express the diagonal of a square as a function of the side.
    b. Express the side of a square as a function of the perimeter.

c. Use parts a and b and composition to construct a function that expresses the diagonal of a square as a function of the perimeter.

53. Helium is pumped into a spherical balloon. The radius of this sphere is increasing at a constant speed of 0.5 cm/second so the radius of this balloon t seconds after the start of inflation may be expressed by $r = g(t) = \frac{1}{2}t$. If function f expresses the volume of this balloon

as a function of r so that $V = f(r) = \frac{4}{3}\pi r^3$, then find and interpret $(f \circ g)(t)$.

54. An oil leak is spreading over a plane surface in the shape of a circle. If the radius of this circle is increasing at a constant rate of 8 cm/second, then express the area A of this circular spill as a function of time t, where t represents the time in seconds from the beginning of the leak.

THINK ABOUT IT

1. If $f(x) = \sqrt{x - 7}$ and $g(x) = \sqrt{2 - x}$, consider the domains of the functions and explain why it does not make sense to form $(f + g)(x)$.

2. If $f(x) = \sqrt{x - 7}$ and $g(x) = 2 - x^2$, explain why it does not make sense to form $(f \circ g)(x)$.

3. **a.** Give two examples of functions for which $(f \circ g)(x)$ does not equal $(g \circ f)(x)$.
  b. Give two examples of functions for which $(f \circ g)(x)$ does equal $(g \circ f)(x)$.

4. If $f(x) = 4x - 9$ and $(f \circ g)(x) = x$ for all values of x, then find $g(x)$.

5. Use the graphs below and graph $y = (f + g)(x)$.

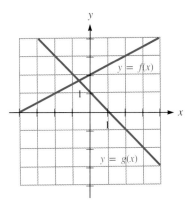

REMEMBER THIS

1. Solve for y: $x = \dfrac{4 - y}{9}$.

2. Solve for y: $x = \sqrt[3]{y - 5}$.

3. Simplify $\dfrac{1}{1/(x + 4)} - 4$.

4. Does $\sqrt{x^2} = x$ for all real numbers x? Explain.

5. Is the set of ordered pairs $\{(4,2),(0,0),(4,-2)\}$ a function?

6. Is the following graph the graph of a function?

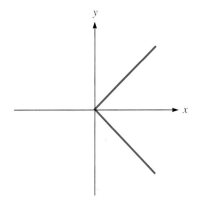

7. Graph $y = x$.

8. Graph $y = 3 - |x - 2|$.

9. If $f(x) = \sqrt{x - 6}$, find the range of f.

10. In each case is the graph of f symmetric with respect to the y-axis?
  a. $f(x) = (x - 4)^2$ **b.** $f(x) = x^2 - 4$

2.5 Inverse Functions

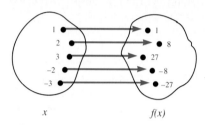

Figure 2.42

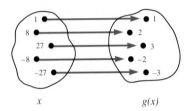

Figure 2.43

Consider Figure 2.42, which illustrates a few of the assignments in the function $f(x) = x^3$. Now consider Figure 2.43, which illustrates a few of the assignments in the function $g(x) = \sqrt[3]{x}$.

Note that the functions f and g have reverse assignments. For example, $f(2) = 8$ and $g(8) = 2$. Two functions with exactly reverse assignments are called **inverse functions** of each other. Thus, the inverse of $f(x) = x^3$ is $g(x) = \sqrt[3]{x}$, and vice versa. In the notation of ordered pairs, the reverse assignments of inverse functions means that when (a,b) is an element of a function, (b,a) is an element of the inverse function. Since the components in the ordered pairs are reversed, the domain of a function f is the range of its inverse function and the range of f is the domain of its inverse.

EXAMPLE 1   A function f consists of the ordered pairs $(7,-2)$, $(9,1)$, and $(-3,2)$.

a. Find the ordered pairs in the inverse function.
b. Compare the domain and range of the two functions.

Solution

a. Since the inverse function reverses the assignments, the inverse function (say g) consists of the ordered pairs $(-2,7)$, $(1,9)$, and $(2,-3)$.
b. The domain of f is $\{7,9,-3\}$ and the range of f is $\{-2,1,2\}$. The domain of g is $\{-2,1,2\}$ and the range of g is $\{7,9,-3\}$. Thus, the domain of f is the range of g and the range of f is the domain of g. ∎

When we reverse the assignments in a function, the result is not always a function. For example, consider the function $y = f(x) = x^2$. Since $f(2) = 4$ and $f(-2) = 4$, an inverse of f would have to assign both 2 and -2 to the number 4. The definition of a function excludes this type of assignment. Thus, there is no function that precisely reverses the assignments of the function $y = x^2$. This example suggests the following guideline for judging if the inverse of a function is a function.

One-to-One Function

A function is **one-to-one** when each x value in the domain is assigned a different y value so that no two ordered pairs have the same second component. If f is one-to-one, then the inverse, denoted f^{-1}, is a function; and if f is not one-to-one, then the inverse is not a function.

EXAMPLE 2   Consider the function $\{(4,-3), (5,-3), (6,-3)\}$. Is the inverse of this function a function?

Solution   Since -3 appears as the second component in more than one ordered pair, the function is not one-to-one. Thus, the inverse of this function is not a function. ∎

In a one-to-one function no two ordered pairs have the same second component. Graphically, this means that a one-to-one function cannot contain two points that lie on the same horizontal line. This observation leads to the following simple test for determining from a graph if a function has an inverse function.

Horizontal Line Test

> Imagine a horizontal line sweeping down the graph of a function. If the horizontal line at any position intersects the graph in more than one point, the function is not one-to-one and its inverse is not a function.

EXAMPLE 3   Which functions graphed in Figure 2.44 have an inverse function?

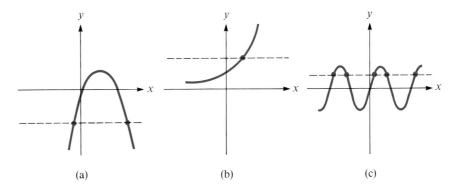

(a)   (b)   (c)

Figure 2.44

Solution

   a. This function does not have an inverse function since a horizontal line may intersect the graph at two points.
   b. This function has an inverse function because no horizontal line intersects the graph at more than one point.
   c. This function does not have an inverse function since a horizontal line may intersect the graph at many points.   ■

   Graphically, there is another useful property of inverse functions. Because the x- and y-coordinates change places, the graphs of inverse functions are related in that each one is the reflection of the other across the line $y = x$. Figure 2.45 shows this relationship between the square root and squaring functions for $x \geq 0$.

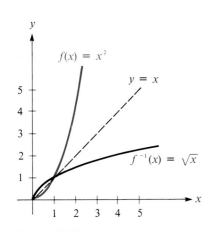

Figure 2.45

EXAMPLE 4   Consider the graph of $y = f(x)$ in Figure 2.46. Use this graph to sketch $y = f^{-1}(x)$.

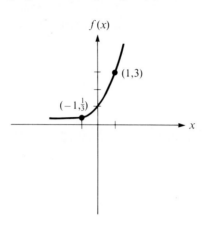

Figure 2.46

Solution   To graph $y = f^{-1}(x)$, we reflect the curve for $y = f(x)$ about the line $y = x$. Both graphs are shown in Figure 2.47.   ∎

If a function has an inverse function and is defined by an equation, then the fact that f and f^{-1} make reverse assignments means that we can find the rule for the inverse function by reversing the roles of x and y in the defining equation and solving for y. For example, the inverse of the function defined by

$$y = 2x - 5$$

is defined by

$$x = 2y - 5$$

which, when solved for y, becomes

$$y = \tfrac{1}{2}x + \tfrac{5}{2}.$$

Thus, if $f(x) = 2x - 5$, then $f^{-1}(x) = \tfrac{1}{2}x + \tfrac{5}{2}$. A complete statement of the procedure for finding an equation for f^{-1} follows:

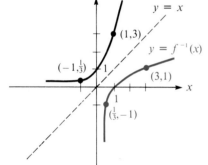

Figure 2.47

To Find an Inverse Function

> 1. Start with a one-to-one function $y = f(x)$ and interchange x and y in this equation.
> 2. Solve the resulting equation for y, and then replace y by $f^{-1}(x)$.
> 3. Define the domain of f^{-1} to be equal to the range of f.

EXAMPLE 5   If $f(x) = \dfrac{7 - x}{3}$, find

a. $f^{-1}(x)$

b. $f^{-1}(0)$

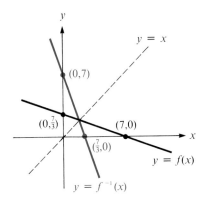

Figure 2.48

Solution

a. The function $f(x) = \dfrac{7 - x}{3}$, or $f(x) = -\dfrac{1}{3}x + \dfrac{7}{3}$, graphs as a straight line (see Figure 2.48) so the function is one-to-one and the inverse is a function. To find an equation for f^{-1}, write the equation for f as

$$y = \frac{7 - x}{3}$$

and then proceed as follows:

$$x = \frac{7 - y}{3} \qquad \text{Interchange } x \text{ and } y.$$

$$\left. \begin{array}{l} 3x = 7 - y \\ y = -3x + 7 \end{array} \right\} \text{ Solve for } y.$$

$$f^{-1}(x) = -3x + 7. \qquad \text{Replace } y \text{ by } f^{-1}(x).$$

The domain of f^{-1} is equal to the range of f, which is the set of all real numbers.

b. To find $f^{-1}(0)$, replace x by 0 in the equation for f^{-1}.

$$f^{-1}(0) = -3(0) + 7$$
$$= 7 \qquad\qquad\blacksquare$$

EXAMPLE 6   If $f(x) = \sqrt{x + 1}$, find $f^{-1}(x)$.

Solution   The graph of $f(x) = \sqrt{x + 1}$ in Figure 2.49 shows the function is one-to-one and the range is the interval $[0,\infty)$. We start with $y = f(x)$, so first write the equation as

$$y = \sqrt{x + 1}.$$

Then interchange x and y and solve for y.

$$x = \sqrt{y + 1}$$
$$x^2 = y + 1$$
$$x^2 - 1 = y$$

Now replace y by $f^{-1}(x)$, so

$$f^{-1}(x) = x^2 - 1.$$

Finally, although $y = x^2 - 1$ is defined for any real number, we need to match the domain of f^{-1} to the range of f so that the functions will be inverses of each other. Thus, we restrict x as follows:

$$f^{-1}(x) = x^2 - 1, \ x \geq 0. \qquad\qquad\blacksquare$$

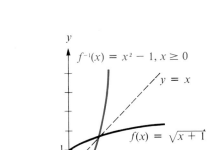

Figure 2.49

When finding an inverse function, you should be aware of two additional concepts. First, it may be easy to show that a function is one-to-one so that an inverse exists, but finding an equation for the inverse may be difficult or impossible because of the algebra involved. (For instance, try finding f^{-1} if $f(x) = x^3 + x$.)

Second, note in Example 5 that $f^{-1}(0) = 7$ and $f(7) = (7 - 7)/3$ $= 0$. In general, inverse functions offset each other, so after finding $f^{-1}(x)$ you may wish to check your answer and test that applying f and f^{-1}, one after the other, brings you back to your original number. This observation also leads to the following important definition for inverse functions, which may be used to prove that two functions are inverses of each other as shown in Example 7.

Definition of Inverse Functions

Two functions f and g are said to be inverses of each other provided

$$(f \circ g)(x) = f[g(x)] = x \quad \text{for all } x \text{ in the domain of } g$$

and

$$(g \circ f)(x) = g[f(x)] = x \quad \text{for all } x \text{ in the domain of } f.$$

EXAMPLE 7   Verify that $f(x) = 5x - 4$ and $g(x) = \frac{1}{5}x + \frac{4}{5}$ are inverses of each other.

Solution   By the above definition, we first show that $f[g(x)] = x$ for all real numbers (which is the domain of g).

$$
\begin{aligned}
f[g(x)] &= f(\tfrac{1}{5}x + \tfrac{4}{5}) \\
&= 5(\tfrac{1}{5}x + \tfrac{4}{5}) - 4 \\
&= x + 4 - 4 \\
&= x
\end{aligned}
$$

Next we verify that $g[f(x)] = x$ for all real numbers (which is the domain of f).

$$
\begin{aligned}
g[f(x)] &= g(5x - 4) \\
&= \tfrac{1}{5}(5x - 4) + \tfrac{4}{5} \\
&= x - \tfrac{4}{5} + \tfrac{4}{5} \\
&= x
\end{aligned}
$$

Thus, f and g are inverses of each other.   ■

The fact that f and f^{-1} must satisfy $f[f^{-1}(x)] = x$ provides an alternative method for finding an equation for f^{-1} that does not involve all the symbol switching of our previous method. For example, to redo Example 6 we start with $f(x) = \sqrt{x + 1}$ and find $f[f^{-1}(x)]$ by replacing x by $f^{-1}(x)$ to obtain

$$f[f^{-1}(x)] = \sqrt{f^{-1}(x) + 1}.$$

Then by definition, $f[f^{-1}(x)]$ equals x, so

$$
\begin{aligned}
x &= \sqrt{f^{-1}(x) + 1} \\
x^2 &= f^{-1}(x) + 1 \\
x^2 - 1 &= f^{-1}(x),
\end{aligned}
$$

which agrees with our previous result. Of course, the restriction $x \geq 0$ applies as discussed in Example 6. If you feel comfortable working with functional notation, then this method may be helpful.

Finally, scientific calculators use the concept of inverse functions with several important keys, particularly with respect to the logarithmic and exponential functions we are to consider. Pressing the $\boxed{\text{INV}}$ (or $\boxed{\text{2nd}}$) key gives the inverse of the next key pressed or some alternative function programmed for that key. Carefully read the owner's manual to your calculator to learn the second functions associated with some of the keys.

EXERCISES 2.5

In Exercises 1–6 find the inverse of the function. Also, find the domain and range of both functions.

1. $\{(1,7), (2,8), (3,9)\}$
2. $\{(4,-1), (8,-2), (12,-3)\}$
3. $\{(-2,10), (0,0), (5,-1)\}$
4. $\{(-7,1), (-4,11), (1,13)\}$
5. $\{(4,4), (5,5), (6,6)\}$
6. $\{(1,1), (2,\frac{1}{2}), (\frac{1}{2},2)\}$

In Exercises 7–10 determine if the inverse of the function is a function.

7. $\{(1,4), (2,4), (3,4)\}$
8. $\{(-4,4), (-5,5), (-6,6)\}$
9. $\{(0,0), (5,0.2), (0.2,5)\}$
10. $\{(-1,1), (0,0), (1,1)\}$

In Exercises 11–18 determine if the inverse of the function graphed is a function.

11.

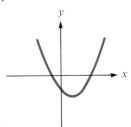

12.

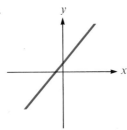

13.

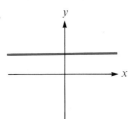

14.

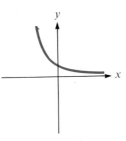

15.   16.

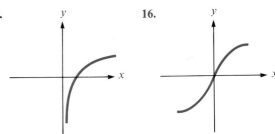

17.   18

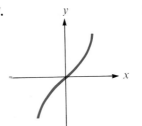
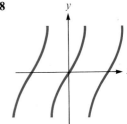

In Exercises 19 and 20 use the graph of $y = f(x)$ to graph $y = f^{-1}(x)$.

19. $f(x)$ 20. $f(x)$

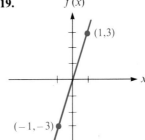

(1,3)

(-1,-3)

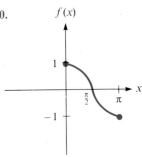

1

$\frac{\pi}{2}$ π

-1

In Exercises 21–38 find the inverse function and graph both the function and its inverse function on the same set of axes. If the given function is not one-to-one so no inverse function exists, state this.

21. $y = x - 10$

22. $g(x) = -x + 5$

23. $f(x) = \dfrac{1}{3}x - \dfrac{7}{3}$

24. $y = \dfrac{4 - x}{11}$

25. $y = x^2$

26. $y = x^3$

27. $f(x) = 2$

28. $f(x) = |x|$

29. $y = \dfrac{1}{x}$

30. $f(x) = \dfrac{1}{x + 3}$

31. $h(x) = \sqrt[3]{x - 4}$

32. $g(x) = \sqrt{x}$

33. $f(x) = x^2 - 4, x \geq 0$

34. $g(x) = 1 - x^2, x < 0$

35. $y = \sqrt{x + 3}$

36. $f(x) = 2 + \sqrt{x}$

37. $f(x) = \sqrt{1 - x^2}$

38. $h(x) = \sqrt{1 - x^2}, x \geq 0$

39. If $g(x) = 7x - 3$, find $g^{-1}(-1)$.

40. If $f(x) = \dfrac{1}{x - 1}$, find $f^{-1}(9)$.

41. Let $f(x) = \sqrt{x - 1}$. Find $f^{-1}(x)$. Find the domain and range for f and for f^{-1}.

In Exercises 42–47 verify that the given functions are inverses of each other.

42. $f(x) = x + 5, g(x) = x - 5$

43. $f(x) = 3x - 2, g(x) = \frac{1}{3}x + \frac{2}{3}$

44. $f(x) = \sqrt[3]{x}, g(x) = x^3$

45. $f(x) = \dfrac{1}{x + 4}, g(x) = \dfrac{1 - 4x}{x}$

46. $f(x) = \dfrac{1}{x + 2}, g(x) = \dfrac{1}{x} - 2$

47. $f(x) = \sqrt{x + 1}, x \geq -1,$
$g(x) = x^2 - 1, x \geq 0$

48. If $f(x) = \dfrac{1}{x}$, show that f is its own inverse. That is, show $f[f(x)] = x$.

49. If $f(x) = \dfrac{2x + 1}{3x - 2}$, show that f is its own inverse.

50. If $f(x) = x^2$ and $g(x) = \sqrt{x}$, find $f[g(x)]$ and $g[f(x)]$. For what values of x does $g[f(x)] = x$?

51. Use the definition of inverse functions to determine whether $f(x) = 3x - 1$ and $g(x) = \frac{1}{3}x + 1$ are inverses of each other.

52. The function $y = \frac{5}{9}(x - 32)$ converts degrees Fahrenheit (x) to degrees Celsius (y). Find the inverse function. What formula does the inverse function represent?

53. The function $y = x^2, x > 0$ gives the formula for the area (y) of a square in terms of the side length (x). Find the inverse function. What formula does this function represent?

54. A manager's weekly salary is cut by 20 percent during a financial crisis and then raised by 20 percent when the crisis is over. Let c and r be functions that define the cut and the raise in salary, respectively, and show r is not the inverse of c. That is, show $(r \circ c)(x) \neq x$. What is the percent gain or loss in the manager's weekly salary?

THINK ABOUT IT

1. Give two graphical examples of an odd function that is not a one-to-one function.

2. If f is an even function, then does f^{-1} exist? Explain.

3. If $f(x) = 9^x$, find $f^{-1}(3)$.

4. If $f(x) = 3x$ and $g(x) = 2x - 4$, then show that
$$(f \circ g)^{-1}(x) = (g^{-1} \circ f^{-1})(x) \text{ for all values of } x.$$

5. If function g is the inverse of function f, then the rule in g must "undo" the rule in f so that $(g \circ f)(x) = x$. Find the inverse function for each of the following functions by undoing *in words* the rule in f. For example, if $f(x) = 5x - 2$, then the rule in f is "for each real number, multiply it by 5 and then subtract 2 from the result." In words, the rule to undo f says "for each real number, add 2 to it and then divide the result by 5," so f^{-1} is defined by $f^{-1}(x) = \dfrac{x + 2}{5}$.

a. $f(x) = 2x - 7$

b. $f(x) = \dfrac{x + 5}{4}$

c. $f(x) = \dfrac{x}{4} + 5$

d. $f(x) = -x + 4$

e. $f(x) = (x + 2)^3$

f. $f(x) = \sqrt[3]{x - 6}$

REMEMBER THIS

1. Simplify $\dfrac{(2x)(3y)}{\left(\dfrac{1}{2}d\right)^2}$.

2. Solve for k: $36 = \dfrac{4k}{7^2}$.

3. Solve for y: $xy = k$.
4. Express the sales tax (T) on a purchase as a function of the price of the item if the sales tax rate is 6 percent.
5. Express the length (ℓ) of a rectangle of area 30 square units as a function of its width (w).

6. Find the domain and range of the function graphed below.

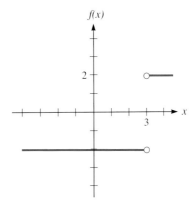

7. Is the function defined by $y = 2/x$ even, odd, or neither of these?
8. Graph $y = (x - 1)^2 + 1$.
9. If $f(x) = 2x^2 - 5$ and $g(x) = 1 - x$, find $f(5) - 2g(-3)$.
10. If $f(x) = 1/x$ and $g(x) = 3x + 8$, find $(g \circ f)(x)$.

2.6 Variation

In many scientific laws the functional relationship between variables is stated in the language of variation. The statement "y varies *directly* as x" means that there is some positive number k such that $y = kx$. The constant k is called the **variation constant.**

EXAMPLE 1   The perimeter of a square varies directly as the side. Algebraically, we write $P = ks$. In this case we know that $k = 4$. ■

EXAMPLE 2   The circumference of a circle varies directly as the diameter. Algebraically, we write $C = kd$. In this case we know that $k = \pi$. ■

EXAMPLE 3   The sales tax (T) on a purchase varies directly as the price (p) of the item. Algebraically, we write $T = kp$. The value of k depends on the sales tax rate in a given location. ■

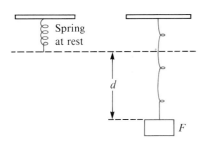

Figure 2.50

EXAMPLE 4   Hooke's law states that the distance (d) a spring is stretched varies directly as the force (F) applied to the spring. Algebraically, we write $d = kF$. The value of k depends on the spring in question (see Figure 2.50). ■

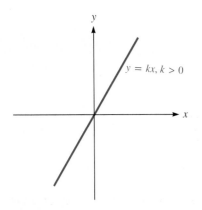

Figure 2.51

Figure 2.51 shows the graph of the variation equation $y = kx$, $k > 0$. Note that the graph is a straight line through the origin and that as x increases, y increases. We may determine the value of k if one ordered pair in the relation other than (0,0) is known. This value of k may then be used to find other corresponding values of the variables.

EXAMPLE 5   If y varies directly as x and $y = 21$ when $x = 3$, write y as a function of x. Determine the value of y when $x = 10$.

Solution   Since y varies directly as x, we have

$$y = kx.$$

To find k, replace y by 21 and x by 3.

$$21 = k \cdot 3$$
$$7 = k$$

Thus, $y = 7x$. When $x = 10$, $y = 7(10) = 70$.   ∎

EXAMPLE 6   In a spring to which Hooke's law applies, a force of 20 lb stretches the spring 6 in. How far will the spring be stretched by a force of 13 lb?

Solution   From Example 4 we have

$$d = kF.$$

To find k, replace d by 6 and F by 20.

$$6 = k \cdot 20$$
$$0.3 = k$$

Thus, $d = 0.3F$. When $F = 13$, $d = 0.3(13) = 3.9$ in.   ∎

When the product of corresponding values of the variables is a constant k, we say we have **inverse variation.** The statement "y varies inversely as x" means that there is some positive number k (the variation constant) such that

$$xy = k \text{ or } y = \frac{k}{x}.$$

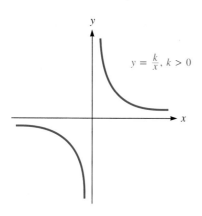

Figure 2.52 shows the graph of the variation equation $y = k/x$ for $k > 0$. The graph in this figure, which is called a **hyperbola,** consists of two disconnected curves, known as **branches.** The graph shows the variation equation $y = k/x$ is meaningless if $x = 0$ or $y = 0$, and that on each branch as x increases, y decreases.

EXAMPLE 7   If y varies inversely with x and $y = 27$ when $x = 2$, write y as a function of x. Find y when $x = 9$.

Solution   Since y varies inversely as x, we have

$$y = \frac{k}{x}.$$

To find k, replace y by 27 and x by 2.

$$27 = \frac{k}{2}$$
$$54 = k$$

Thus, $y = 54/x$. When $x = 9$, $y = \frac{54}{9} = 6$.   ■

EXAMPLE 8   The speed (S) of a gear varies inversely as the number (n) of teeth. Gear A, which has 12 teeth, makes 400 rpm. How many revolutions per minute are made by a gear with 32 teeth that is connected to gear A?

Solution   Since S varies inversely with n, we have

$$S = \frac{k}{n}.$$

To find k, replace S by 400 and n by 12.

$$400 = \frac{k}{12}$$
$$4{,}800 = k$$

Thus, $S = 4{,}800/n$. When $n = 32$, $S = 4{,}800/32 = 150$ rpm.   ■

We may extend the concept of variation to include direct and inverse variation of variables raised to specified powers and relationships that involve more than two variables.

EXAMPLE 9   If y varies directly as x^2 and inversely as z^3 and $y = 18$ when $x = 3$ and $z = 2$, determine the value of y when $x = 1$ and $z = 5$.

Solution   Since y varies directly as x^2 and inversely as z^3, we have

$$y = \frac{kx^2}{z^3}.$$

To find k, replace y by 18, x by 3, and z by 2.

$$18 = \frac{k(3)^2}{(2)^3}$$
$$18 = \frac{9k}{8}$$
$$16 = k$$

Thus, $y = 16x^2/z^3$. When $x = 1$ and $z = 5$,

$$y = \frac{16(1)^2}{(5)^3} = \frac{16}{125}.$$ ■

EXAMPLE 10   Newton's law of gravitation states that the gravitational attraction between two objects varies directly as the product of their masses and inversely as the square of the distance between their centers of mass. What will be the change in attraction between the two objects if both masses are doubled and the distance between their centers is cut in half?

Solution   If F represents the gravitational attraction, m_1 and m_2 represent the masses of the objects, and d represents the distance between the centers of mass, then algebraically, we write Newton's law as

$$F = \frac{km_1m_2}{d^2}.$$

If the masses are doubled and the distance is cut in half, we have

$$F = \frac{k(2m_1)(2m_2)}{(\frac{1}{2}d)^2}$$

$$= \frac{4km_1m_2}{\frac{1}{4}d^2}$$

$$= \frac{16km_1m_2}{d^2}.$$

Thus, the gravitational attraction becomes 16 times as great.   ∎

Finally, we point out that any variation problem may also be solved by using a proportion. For example, if y varies directly as x, we know that $y = kx$. Thus, $y/x = k$ and there is a constant ratio between any corresponding values of x and y. This means that

$$\frac{y_1}{x_1} = \frac{y_2}{x_2} \quad \text{or} \quad \frac{y_1}{y_2} = \frac{x_1}{x_2}.$$

The equation $y_1/y_2 = x_1/x_2$ is called a **proportion** and may be used to solve the problem. For this reason, the variation constant k is often called the constant of proportionality, and the expression "varies directly as" is often replaced by "is proportional to." However, the language of variation usually provides a more convenient and informative statement of a relationship.

EXERCISES 2.6

In Exercises 1–30 write an equation relating the variables and solve the problem.

1. If y varies directly as x and $y = 14$ when $x = 6$, write y as a function of x. Determine the value of y when $x = 10$.

2. If y varies directly as x and $y = 2$ when $x = 5$, write y as a function of x. Determine the value of y when $x = 11$.

3. If y varies directly as the square of x and $y = 3$ when $x = 2$, write y as a function of x. Find y when $x = 4$.

4. If y varies directly as x^3 and $y = 5$ when $x = 3$, write y as a function of x. Find y when $x = 2$.

5. If y varies inversely as x and $y = 9$ when $x = 8$, write y as a function of x. Find y when $x = 24$.

6. If y varies inversely as x and $y = 3$ when $x = 7$, write y as a function of x. Find y when $x = 21$.

7. If y varies inversely as x^3 and $y = 3$ when $x = 2$, write y as a function of x. Find y when $x = 3$.

8. If y varies inversely as the square of x and $y = 2$ when $x = 4$, write y as a function of x. Find y when $x = 8$.

9. If y varies directly as x and z and $y = 105$ when $x = 7$ and $z = 5$, find y when $x = 10$ and $z = 2$.

10. If y varies directly as x and inversely as z and $y = 10$ when $x = 4$ and $z = 3$, find y when $x = 7$ and $z = 15$.

11. If y varies directly as x and inversely as z^2 and $y = 36$ when $x = 4$ and $z = 7$, find y when $x = 9$ and $z = 9$.

12. If y varies inversely as x^2 and z^3 and $y = 0.5$ when $x = 3$ and $z = 2$, find y when $x = 2$ and $z = 3$.

13. In a spring to which Hooke's law applies (see Examples 4 and 6), a force of 15 lb stretches the spring 10 in. How far will the spring be stretched by a force of 6 lb?

14. The weight of an object on the moon varies directly with the weight of the object on earth. An object that weighs 114 lb on earth weighs 19 lb on the moon. How much will a person who weighs 174 lb on earth weigh on the moon?

15. The amount of garbage produced in a given location varies directly with the number of people living in the area. It is known that 25 tons of garbage is produced by 100 people in 1 year. If there are 8 million people in New York City, how much garbage is produced by New York City in 1 year?

16. Property tax varies directly as assessed valuation. The tax on property assessed at $12,000 is $400. What is the tax on property assessed at $40,000?

17. If the area of a rectangle remains constant, the length varies inversely as the width. The length of a rectangle is 9 in. when the width is 8 in. If the area of the rectangle remains constant, find the width when the length is 24 in.

18. The speed of a gear varies inversely as the number of teeth. Gear A with 48 teeth makes 40 rpm. How many revolutions per minute are made by a gear with 120 teeth that is connected to gear A?

19. The speed of a pulley varies inversely as the diameter of the pulley. The speed of pulley A, which has an 8-in. diameter, is 450 rpm. What is the speed of a 6-in. diameter pulley connected to pulley A?

20. The time required to complete a certain job varies inversely as the number of machines that work on the job (assuming each machine does the same amount of work). It takes five machines 55 hours to complete an order. How long will it take 11 machines to complete the same job?

21. The volume of a sphere varies directly as the cube of its radius. The volume is 36π cubic units when the radius is 3 units. What is the volume when the radius is 5 units?

22. The distance an object falls due to gravity varies directly as the square of the time of fall. If an object falls 144 ft in 3 seconds, how far did it fall the first second?

23. The weight of an object varies inversely as the square of the distance from the object to the center of the earth. At sea level (4,000 mi from the center of the earth) a man weighs 200 lb. Find his weight when he is 200 mi above the surface of the earth.

24. The intensity of light on a plane surface varies inversely as the square of the distance from the source of light. If we double the distance from the source to the plane, what happens to the intensity?

25. The exposure time for photographing an object varies inversely as the square of the lens diameter. What will happen to the exposure time if the lens diameter is cut in half?

26. The resistance of a wire to an electrical current varies directly as its length and inversely as the square of its diameter. If a wire 100 ft long with a diameter of 0.01 in. has a resistance of 10 ohms, what is the resistance of a wire of the same length and material but 0.03 in. in diameter?

27. The general gas law states that the pressure of an ideal gas varies directly as the absolute temperature and inversely as the volume. If $P = 4$ atm when $V = 10$ cm^3 and $T = 200$ K, find P when $V = 30$ cm^3 and $T = 250$ K.

28. The safe load of a beam (the amount it supports without breaking) that is supported at both ends varies directly as the width and the square of the height and inversely as the distance between supports. If the width and height are doubled and the distance between supports remains the same, what is the effect on the safe load?

29. Coulomb's law states that the magnitude of the force that acts on two charges q_1 and q_2 varies directly as the product of the magnitude of q_1 and q_2 and inversely as the square of the distance between them. If the magnitude of q_1 is doubled, the magnitude of q_2 is tripled, and the distance between the charges is cut in half, what happens to the force?

30. Newton's law of gravitation states that the gravitational attraction between two objects varies directly as the product of their masses and inversely as the square of the distance between their centers of mass. What will be the change in attraction between the two objects if both masses are cut in half and the distance between their centers is doubled?

THINK ABOUT IT

1. The area of a circle varies directly as the square of the diameter. What is the variation constant in this relation?

2. **a.** Describe two instances not discussed in this section in which one variable varies *directly* as another variable.

   b. Describe two instances not discussed in this section in which one variable varies *inversely* as another variable.

3. Solve the problem in Exercise 14 by setting up a *proportion*.

4. Solve the problem in Exercise 17 by setting up a *proportion*.

5. The intensity of light on a plane surface varies inversely as the square of the distance from the source of light. By what factor must the distance from the source of light be changed in order for the intensity of light on a plane surface to double? If you are reading a book 4 ft from an electric light and wish the intensity of light on a page to double, then how far should the page be from the light?

REMEMBER THIS

1. If $(0,b)$ is a solution of the equation $y = 7x + 5$, find the value of b.

2. Graph $f(x) = -3x + 4$.

3. Is the inverse of the function graphed below a function?

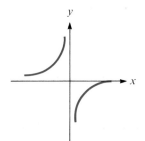

4. If $f(x) = \dfrac{x}{6} - 1$, find $f^{-1}(x)$.

5. If $h(x) = 3(5x - 2)^5$, find simpler functions f and g so that $h(x) = (f \circ g)(x)$.

6. If $f(x) = 1/x$ and $g(x) = x^2$, find $(g - f)(x)$. What is the domain of $g - f$?

7. Find the distance between $(6,-2)$ and $(3,4)$.

8. Graph $y = \begin{cases} x^2 & \text{if} \quad 0 \le x \le 2 \\ 4 & \text{if} \quad x > 2. \end{cases}$

9. If $f(x) = 6$, compute $\dfrac{f(x + h) - f(x)}{h}$ with $h \ne 0$ in simplest form.

10. Express the area (A) of a circle as a function of its circumference (C).

CHAPTER OVERVIEW

Section	Key Concepts to Review
2.1	• Rule definitions of a function and its domain and range • Ordered pair definitions of a function and its domain and range • The term $f(x)$ is read "f of x" or "f at x" and means the value of the function (the y value) corresponding to the value of x. • If a is in the domain of f, then ordered pairs for the function defined by $y = f(x)$ all have the form $(a, f(a))$. *Note: a* is an x value and $f(a)$ is a y value.
2.2	• Definition of the graph of a function • Distance formula: $d = \sqrt{(x_2 - x_1)^2 + (y_2 - y_1)^2}$ (for any two points) • Statement of the fundamental principle in coordinate (or analytic) geometry • Vertical line test

Section	Key Concepts to Review
	• Basic graphs: Let a, b, c, m, and r be real-number constants with $a \neq 0$. $\quad y = mx + b$: straight line $\qquad\qquad\quad y = ax^2 + bx + c$: parabola $\quad y = c$: horizontal line $\qquad\qquad\quad\; x^2 + y^2 = r^2$: circle (center $(0,0)$, radius r) $\quad y = |x|$: $\vee$ shape
2.3	• Definitions of an even function and an odd function • The graph of an even function is symmetric about the y-axis. • The graph of an odd function is symmetric about the origin. • Methods to graph variations of a familiar function by using vertical and horizontal shifts, reflecting, stretching, and shrinking
2.4	• Definitions of $f + g$, $f - g$, $f \cdot g$, and f/g for two functions f and g • The symbol "$\circ$" denotes the operation of composition. The composite functions are $(f \circ g)(x) = f[g(x)]$ and $(g \circ f)(x) = g[f(x)]$. • Methods to determine the domain of $f + g$, $f - g$, $f \cdot g$, f/g, $f \circ g$, and $g \circ f$
2.5	• Definitions of a one-to-one function and inverse functions • The special symbol f^{-1} is used to denote the inverse function of f. • Methods to determine if the inverse of a function is a function and how to find f^{-1}, if it exists • Horizontal line test • f and f^{-1} interchange their domain and range. • The graphs of f and f^{-1} are reflections of each other across the line $y = x$.
2.6	• The statement "y varies directly as x" means there is some positive number k (variation constant) such that $y = kx$. • The statement "y varies inversely as x" means there is some positive number k such that $xy = k$ or $y = k/x$.

CHAPTER REVIEW EXERCISES

1. Fill in the missing component in each of the following ordered pairs so they are solutions of the equation $y = -2x + 4$: $(0, \;)$, $(\;, -1)$, $(-1, \;)$, $(\;, 0)$, $(3, \;)$.
2. Give three ordered pairs in f if $f(x) = 2$.
3. Is the set of ordered pairs $\{(-1,3), (0,3), (1,3)\}$ a function?
4. Write a formula expressing the circumference of a circle as a function of its radius.
5. Is the following graph the graph of a function? If so, is the inverse also a function?

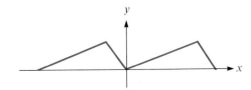

6. If $f(x) = x - 1/x$, find $f(0)$.
7. If $g(x) = 2x + 1$, for what values of x does $g(x) = 0$?
8. Three vertices of a rectangle are $(-4, -3)$, $(-4, 2)$, and $(3, 2)$. Determine the ordered pair corresponding to the fourth vertex. What is the perimeter of the rectangle?
9. If y varies directly as x and $y = 5$ when $x = 12$, write y as a function of x. Determine the value of y when $x = 20$.
10. Find the distance between $(-2, 1)$ and $(3, -1)$.

In Exercises 11–12 state the domain of the function.

11. $g(x) = 1/[x(x + 1)]$ 12. $y = \sqrt{x + 7}$

13. Write a formula that expresses the area of a square as a function of the diagonal. (*Hint:* Use the drawing and the Pythagorean theorem.)

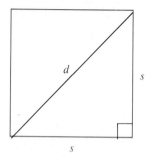

14. The speed of a pulley varies inversely as the diameter of the pulley. The speed of pulley A, which has a 7-in. diameter, is 320 rpm. What is the speed of a 5-in. diameter pulley that is connected to pulley A?

15. Are the following points the vertices of a right triangle: $(1,-3)$, $(8,-7)$, and $(5,-1)$?

In Exercises 16–21 let $f(x) = x - 1$ and $g(x) = 2x$.

16. Evaluate $3f(3) - 4g(-2)$.
17. Does $f[g(1)] = g[f(1)]$?
18. Does $f(a + b) = f(a) + f(b)$?
19. Does $g(a + b) = g(a) + g(b)$?
20. Is f an even function, an odd function, or neither?
21. Is g an even function, an odd function, or neither?

In Exercises 22–25 use the graph of $y = x^2$ to graph each function.

22. $y = (x + 1)^2$ 23. $y = x^2 + 1$
24. $y = -x^2$ 25. $y = (x - 2)^2 - 1$
26. Is the function $y = 3/x^2$ even, odd, or neither of these? The following figure gives the graph of this function for $x \geq 0$. Complete the graph.

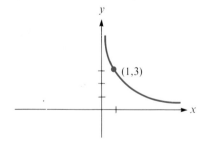

27. The intensity of light on a surface is inversely proportional to the square of the distance from the source of light. You are reading a book 6 ft from an electric light. If you move 2 ft closer, how many times greater is the intensity of light on the book?

28. A circle of radius 5 has its center at $(-1,-1)$. Is the point $(3,2)$ on this circle?
29. Express the area (A) of a circle as a function of its diameter (d).

In Exercises 30–33 consider the function
$$f(x) = \begin{cases} -x & \text{if } 0 \leq x \leq 3 \\ 3 & \text{if } x > 3. \end{cases}$$

30. Find $f(3)$.   31. Graph the function.
32. What is the domain?   33. What is the range?

In Exercises 34–37 use the following graph of $y = f(x)$ to graph each function.

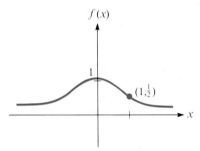

34. $y = 2f(x)$ 35. $y = -\frac{1}{2}f(x)$
36. $y = f(x - 2)$ 37. $y = f(x) - 2$

In Exercises 38–41 graph the given function.

38. $y = -3$ 39. $f(x) = 2x - 4$
40. $g(x) = 9 - x^2$ 41. $y = |x + 5|$

In Exercises 42 and 43 classify each function as even, odd, or neither of these.

42. $f(x) = |x|/x$ 43. $y = x^4 + 1$
44. If $f = \{(3,-1), (4,-2)\}$, find f^{-1}. Determine the domain and range of each function.
45. If $g(x) = -\frac{1}{2}x + \frac{5}{2}$, find
    a. $g^{-1}(x)$ b. $g^{-1}(-1)$

In Exercises 46–53 use the following figure.

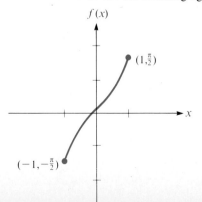

46. What is the domain?
47. What is the range?
48. Is f even, odd, or neither?
49. Find $f(0)$.
50. Solve $f(x) = 0$.
51. Solve $f(x) > 0$.
52. Solve $f(x) \le 0$.
53. Graph $y = f^{-1}(x)$.
54. Write a different function that has the same domain and range as the function $\{(1,4), (2,5)\}$.

In Exercises 55–58 compute in simplest form
$$\frac{f(x + h) - f(x)}{h} \text{ with } h \ne 0.$$

55. $f(x) = 3x^2 - 1$
56. $f(x) = -3$
57. $f(x) = 3 - x$
58. $f(x) = 1/(x + 1)$
59. Express the area of an equilateral triangle as a function of its side length (x).
60. A rectangle of sides x and y units is inscribed in a circle of radius 5 units. Express the area of the rectangle as a function of x. What is the domain of the function?
61. If $f(x) = x^2$ and $g(x) = 1/x$, find $(f \cdot g)(x)$. What is the domain of $f \cdot g$?
62. A rectangular field is completely enclosed by x feet of fencing. If the length of the rectangular region is twice the width, then express the area of the field as a function of x.
63. Express the monthly earnings (e) of a salesperson in terms of the cash amount (a) of merchandise sold if the salesperson earns \$500 per month plus 9 percent commission on sales above \$2,000.
64. If $h(x) = 3(1 - x)^2$, find simpler functions f and g so that $h(x) = (f \circ g)(x)$.

In Exercises 65–72 use the following figure.

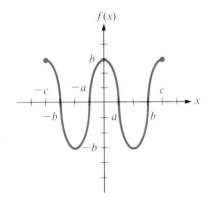

65. What is the domain?
66. What is the range?
67. Is f even, odd, or neither?
68. Find $f(b)$.
69. Solve $f(x) = b$.
70. Solve $f(x) < 0$.
71. Graph $y = -\frac{1}{2}f(x)$.
72. Graph $y = f(x - c)$.

73. Find the domain and range of the functions shown.

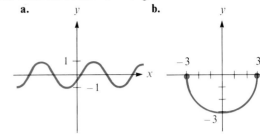

a.   b.

74. If $f(x) = x - 2$, find $f^{-1}(x)$. Find the domain and range for f and for f^{-1}.
75. Use the composition definition of inverse functions and verify that $f(x) = \frac{1}{2}x - 3$ and $g(x) = 2x + 6$ are inverses of each other.

In Exercises 76–85 select the choice that answers the question or completes the statement.

76. Which number is not in the domain of $y = \dfrac{x + 2}{x + 3}$?
  a. 2   **b.** -2 **c.** -3 **d.** 3
77. The graph of $y = f(x) - 3$ is the graph of $y = f(x)$ shifted 3 units
  a. up   **b.** down   **c.** left   **d.** right
78. If $f(x) = 5x^2$, then the difference quotient
$$\frac{f(x + h) - f(x)}{h} \text{ with } h \ne 0 \text{ in simplest form is}$$
  a. $10x + 5h$ **b.** $5h$ **c.** 1   **d.** $10xh + 5h$
79. The area of a circle whose radius extends from $(-2,5)$ to $(5,1)$ is
  a. 11π **b.** 25π **c.** 45π **d.** 65π
80. If y varies inversely with x, then when x is doubled y is
  a. decreased by 2   **b.** increased by 2
  c. multiplied by 2   **d.** divided by 2
81. If $f = \{(0,-3), (-1,-2), (2,-1)\}$ and $g = \{(-1,2), (-2,1), (-3,3)\}$, then $(g \circ f)(-1)$ equals
  a. -1 **b.** 1   **c.** 2   **d.** -2
82. The graph of an odd function is symmetric with respect to the
  a. x-axis   **b.** y-axis   **c.** origin
83. If $f(x) = 2x + b$ and $f(-2) = 3$, then b equals
  a. 7   **b.** -1 **c.** 1   **d.** -8
84. If a square field is completely enclosed by x feet of fencing, then the area of the field as a function of x equals
  a. x^2 **b.** $x^2/4$ **c.** $4x^2$ **d.** $x^2/16$
85. If $f(x) = 2x$, then which one of the following is true?
  a. $f(-a) = -f(a)$ **b.** $f(-a) = f(a)$
  c. $f(ab) = f(a) \cdot f(b)$ **d.** $f(a/b) = f(a)/f(b)$

CHAPTER 2 TEST

1. If $f(x) = \dfrac{2}{x - 9}$, find the domain of f.

2. What is the range of the function defined by $y = 6 - |x|$?

3. If $f(x) = x^2$ and $g(x) = 3x - 2$, find $3f(4) + \frac{1}{2}g(-6)$.

4. Compute $\dfrac{f(x + h) - f(x)}{h}$ with $h \neq 0$ in simplest form if $f(x) = x^2 - 5x + 3$.

5. A rectangular area of sides x and y units is enclosed using 800 ft of fencing. Find a formula for the enclosed area as a function of x.

6. Find the area of a circle whose radius extends from $(-1,2)$ to $(0,-1)$.

7. Is the following graph the graph of a function?

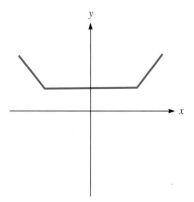

8. Is the inverse of the function graphed below a function?

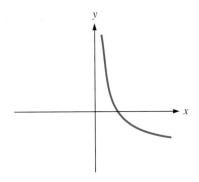

9. Find the domain and range of the function graphed below.

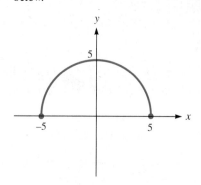

10. Graph $y = \sqrt{x - 3}$.

11. Graph $y = (x + 2)^2 - 1$.

12. Graph $y = \begin{cases} x & \text{if } x \geq 0 \\ 2 & \text{if } x < 0. \end{cases}$

13. Is the function defined by $f(x) = 3 - x^2$ even, odd, or neither of these?

14. If $f = \{(0,-3),(1,2),(-1,0)\}$ and $g = \{(0,5),(1,-4),(2,0)\}$, find $f + g$.

15. Find $(g \circ f)(x)$ if $f(x) = 3x^2 - 1$ and $g(x) = 2x + 5$.

16. If $h(x) = \sqrt{25 - x^2}$, find simpler functions f and g so that $h(x) = (f \circ g)(x)$.

17. If $f(x) = x^3 - 5$, find $f^{-1}(x)$.

18. Use the composition definition of inverse functions and verify that $f(x) = 7x - 2$ and $g(x) = \frac{1}{7}x + \frac{2}{7}$ are inverses of each other.

19. In a spring to which Hooke's law applies, a force of 25 lb stretches the spring 4 in. How far will the spring be stretched by a force of 40 lb?

20. The exposure time for photographing an object varies inversely as the square of the lens diameter. What will happen to the exposure time if the lens diameter is tripled?

3 Polynomial and Rational Functions

A real-life version of the Monopoly game for small investors is the purchase of a single family home for use as a rental property. An important benefit of this investment is that one can *depreciate* the cost of the building (but not the land) for tax benefits while the home (hopefully) *appreciates* in value. Usually the investment depreciates linearly by the straight- line method, which means the cost of the building is deducted equally over a specified number of years as defined in the tax laws. If such an investment depreciates linearly from $200,000 to $0 in 30 years, then how much is the investment worth (for tax purposes) after 7 years? (See Example 10 of Section 3.1.)

n this chapter we continue with the theme of a function and discuss many ideas associated with a large and important class of functions called polynomial functions. Basically, a polynomial function is characterized by terms of the form

$$(\text{real number})x^{\text{nonnegative integer}}$$

so that a polynomial function may contain terms like 4, $2x$, $-3x^2$, $\sqrt{5}x^3$, and so on. More technically, a **polynomial function of degree n** is a function of the form

$$y = P(x) = a_n x^n + a_{n-1}x^{n-1} + \cdots + a_1 x + a_0 \qquad (a_n \neq 0),$$

where n is a nonnegative integer and $a_n, a_{n-1}, \ldots, a_1$, and a_0 are real-number constants. For example, the function $P(x) = 5x^3 + 2x^2 - 1$ is a polynomial function of degree 3 in which $a_3 = 5$, $a_2 = 2$, $a_1 = 0$, and $a_0 = -1$. Since n must be a nonnegative integer, note that functions with terms such as $x^{1/2}$ (or $\sqrt{x}$) and x^{-2} (or $1/x^2$) are not polynomial functions.

Although $y = 1/x^2$ is not a polynomial function, it can be expressed as a quotient of two polynomials and is called a **rational function.** This chapter concludes by showing how to graph rational functions. We begin by noting that the constant function $y = P(x) = a_0$ (with $a_0 \neq 0$) is a polynomial function of degree 0. For technical reasons the constant function $P(x) = 0$ is called the zero polynomial function and is not assigned a degree. We now consider polynomial functions of degree 1.

3.1   Linear Functions

First-degree polynomial functions are called linear functions so the following definition applies.

Definition of Linear Function

A function of the form
$$f(x) = ax + b,$$
where a and b are real numbers with $a \neq 0$, is called a **linear function.**

The term *linear* is used because such functions graph as straight lines as shown later in this section. The concept of the slope of a line is fundamental to analyzing linear functions, so we begin with this topic.

Slope of a Line

The slope of a line is a measure of its steepness or inclination with respect to the horizontal axis. To find this measure, pick *any* two distinct points on the given line, calculate the change in the dependent variable, and divide it by the change in the independent variable. This ratio is called the *slope of the line*.

Definition of Slope

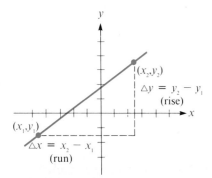

If (x_1, y_1) and (x_2, y_2) are any two distinct points on a nonvertical line, as shown in Figure 3.1, then
$$\text{slope} = m = \frac{\Delta y}{\Delta x} = \frac{y_2 - y_1}{x_2 - x_1}.$$
(*Note:* The symbol Δ is the Greek capital letter delta. The symbol is used to indicate a change.)

The slope of a line is a fundamental notion in analytic geometry as a measure of the inclination or steepness of a line. In this context we often refer to Δy as the "rise" and Δx as the "run," so
$$\text{slope} = \frac{\text{rise}}{\text{run}}.$$

Figure 3.1

With this interpretation, it is natural to use the idea of slope to analyze such concrete situations as the steepness of a ramp, the pitch of a roof, or the inclination of a ski slope.

EXAMPLE 1   Find the slope of a line containing the points (2,1) and (5,4).

Solution   If we label (2,1) as point 1 (Figure 3.2),

$$x_1 = 2 \qquad y_1 = 1 \qquad x_2 = 5 \qquad y_2 = 4$$

$$m = \frac{\Delta y}{\Delta x} = \frac{y_2 - y_1}{x_2 - x_1} = \frac{4 - 1}{5 - 2} = \frac{3}{3} = 1.$$

A slope of 1 means that as x increases 1 unit, y increases 1 unit. Notice that the slope is unaffected by the way in which we label the points. If we label (5,4) as point 1,

$$x_1 = 5 \qquad y_1 = 4 \qquad x_2 = 2 \qquad y_2 = 1$$

$$m = \frac{\Delta y}{\Delta x} = \frac{y_2 - y_1}{x_2 - x_1} = \frac{1 - 4}{2 - 5} = \frac{-3}{-3} = 1. \qquad \blacksquare$$

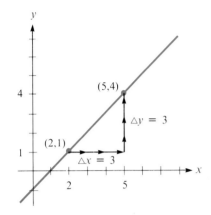

Figure 3.2

EXAMPLE 2   Find the slope of the line containing the points $(-1,-1)$ and $(3,-6)$.

Solution   (See Figure 3.3.)

$$x_1 = -1 \qquad y_1 = -1 \qquad x_2 = 3 \qquad y_2 = -6$$

$$m = \frac{\Delta y}{\Delta x} = \frac{y_2 - y_1}{x_2 - x_1} = \frac{-6 - (-1)}{3 - (-1)} = \frac{-5}{4}$$

A slope of $-\frac{5}{4}$ means that as x increases 4 units, y decreases 5 units. (*Note:* A negative slope indicates that y is decreasing as x increases. If the right end of a line segment is lower than the left end, the slope of the segment is negative. If the right end is higher, the slope of the segment [and of the whole line] is positive.) $\qquad \blacksquare$

The special cases discussed in the next example occur when the slope formula is applied to lines that are horizontal or vertical.

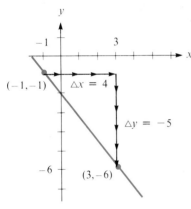

Figure 3.3

EXAMPLE 3   Find the slope of the line containing the following pairs of points.

a. $(-1,4)$ and $(5,4)$ **b.** $(3,2)$ and $(3,-4)$

Solution

a. The line through the given points is horizontal as shown in Figure 3.4. If we let $x_1 = -1$, $y_1 = 4$, $x_2 = 5$, and $y_2 = 4$, then

$$m = \frac{\Delta y}{\Delta x} = \frac{y_2 - y_1}{x_2 - x_1} = \frac{4 - 4}{5 - (-1)} = \frac{0}{6} = 0.$$

The slope of every horizontal line is zero since the numerator of the slope ratio (Δy) is always zero.

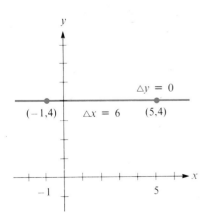

Figure 3.4

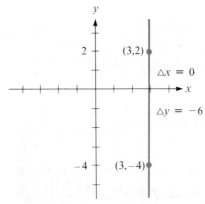

Figure 3.5

b. Figure 3.5 shows the line through the given points is vertical. The slope formula is not meaningful in this case since

$$x_1 = 3 \qquad y_1 = 2 \qquad x_2 = 3 \qquad y_2 = -4$$

$$m = \frac{\Delta y}{\Delta x} = \frac{y_2 - y_1}{x_2 - x_1} = \frac{-4 - 2}{3 - 3} = \frac{-6}{0}$$

m is undefined.

The slope of every vertical line is undefined since the denominator of the slope ratio (Δx) is always zero. ■

In summary we have considered lines with slopes that are positive, negative, zero, or undefined, and we have seen that the features shown in Figure 3.6 are associated with these cases.

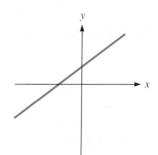

Slope: positive
Line is higher on the right.
y increases as x increases.

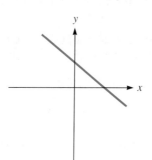

Slope: negative
Line is lower on the right.
y decreases as x increases.

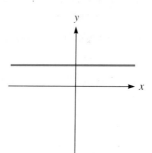

Slope: zero
Line is horizontal.
y remains constant as x increases.

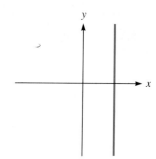

Slope: undefined
Line is vertical.
x remains constant as y increases.

Figure 3.6

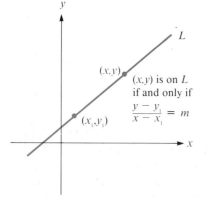

Figure 3.7

Equations of a Line

Two important formulas for an equation of a line, called the *point-slope equation* and the *slope-intercept equation,* are used often when working with linear functions, so we now develop these formulas. Consider any nonvertical line L with slope m that passes through the point (x_1, y_1) as shown in Figure 3.7. (We eliminate vertical lines since their slope is undefined.) Then, any other point (x, y) with $x \neq x_1$ is on L if and only if the slope of the line segment joining (x, y) and (x_1, y_1) is m; that is, if and only if

$$\frac{y - y_1}{x - x_1} = m.$$

If we write this equation in the form

$$y - y_1 = m(x - x_1),$$

then (x_1, y_1) also satisfies the equation, which means that the points on L match the solutions of $y - y_1 = m(x - x_1)$. We summarize this result as follows:

Point-Slope Equation

An equation of the line with slope m passing through (x_1, y_1) is

$$y - y_1 = m(x - x_1).$$

This equation is called the **point-slope** form of the equation of a line.

EXAMPLE 4   Find an equation for the line that contains the point $(1,3)$ and whose slope is 2.

Solution   We are given that $x_1 = 1$, $y_1 = 3$, and $m = 2$. Substituting these numbers in the point-slope equation, we have

$$y - 3 = 2(x - 1)$$
$$y - 3 = 2x - 2$$
$$y = 2x + 1.$$ ∎

EXAMPLE 5   Find an equation for the line that contains the points $(1,-2)$ and $(4,5)$. Write the answer in the form $y = ax + b$.

Solution   First, we find the slope.

$$m = \frac{y_2 - y_1}{x_2 - x_1} = \frac{5 - (-2)}{4 - 1} = \frac{7}{3}$$

Now use the point-slope equation with $m = \frac{7}{3}$ and either $(1,-2)$ or $(4,5)$ for (x_1, y_1). Using $x_1 = 4$ and $y_1 = 5$, we have

$$y - 5 = \tfrac{7}{3}(x - 4),$$

which is converted to the form requested as follows:

$$y - 5 = \tfrac{7}{3}x - \tfrac{28}{3}$$
$$y = \tfrac{7}{3}x - \tfrac{13}{3}.$$ ∎

There are many ways of writing the equation of a line. The point-slope equation is used extensively for finding the equation of a line, but it is not very helpful for graphing lines or interpreting linear relationships when the equation is known. To develop a form that is useful in such cases, consider Figure 3.8 and note that any nonvertical line L must cross the y-axis at some point. The x-coordinate of this point is zero and we will label the y-coordinate b. This point $(0,b)$, where the line crosses the y-axis, is called

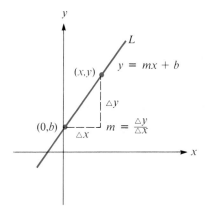

Figure 3.8

the **y-intercept.** If we apply the point-slope equation in the case when (x_1, y_1) is the y-intercept, we find that (x, y) is on L if and only if

$$y - b = m(x - 0)$$

or

$$y = mx + b.$$

To emphasize the geometric aspects of our discussion, we state this result as follows:

Slope-Intercept Equation

> The graph of the equation
>
> $$y = mx + b$$
>
> is a line with slope m and y-intercept $(0, b)$. The equation $y = mx + b$ is called the **slope-intercept** form of the equation of a line.

Note that a linear function f may be defined by an equation of the form $y = f(x) = mx + b$ with $m \neq 0$, so that it follows from the above result that the graph of a linear function is a line.

EXAMPLE 6   Find the slope and the y-intercept of the line defined by the equation $2x + 3y = 6$.

Solution   First, transform the equation to the form $y = mx + b$.

$$2x + 3y = 6$$
$$3y = -2x + 6$$
$$y = -\tfrac{2}{3}x + 2$$

Matching this equation to the form $y = mx + b$, we conclude that $m = -\tfrac{2}{3}$ and $b = 2$. Thus, the slope is $-\tfrac{2}{3}$ and the y-intercept is $(0, 2)$.   ■

EXAMPLE 7   Graph the line whose slope is $-\tfrac{2}{3}$ and whose y-intercept is $(0, 4)$. Also, find an equation for the line.

Solution   We are given the y-intercept so we know one point on the line is $(0, 4)$. To find another point, we may interpret a slope of $-\tfrac{2}{3}$ to mean that when x increases 3 units, y decreases 2 units. By starting at $(0, 4)$ and going 3 units to the right and 2 units down, we obtain a second point on the line at $(3, 2)$. Drawing a line through these two points produces the graph in Figure 3.9. By substituting $m = -\tfrac{2}{3}$ and $b = 4$ in the slope-intercept equation $y = mx + b$, we determine that

$$y = -\tfrac{2}{3}x + 4$$

is an equation for the line.   ■

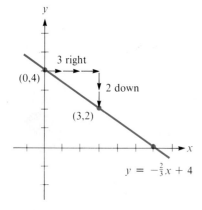

Figure 3.9

EXAMPLE 8   Find the equation that defines the linear function f if $f(1) = 6$ and $f(-1) = 2$.

Solution   If $f(1) = 6$, then when $x = 1$, $y = 6$. Similarly, if $f(-1) = 2$, then when $x = -1$, $y = 2$. Thus, $(1,6)$ and $(-1,2)$ are points on the graph. First, calculate the slope.

$$m = \frac{y_2 - y_1}{x_2 - x_1} = \frac{6 - 2}{1 - (-1)} = 2$$

Now use the point-slope form with one of the points, say $(1,6)$, as follows:

$$\begin{aligned} y - y_1 &= m(x - x_1) \\ y - 6 &= 2(x - 1) \\ y - 6 &= 2x - 2 \\ y &= 2x + 4. \end{aligned}$$

The equation that defines the function f is

$$f(x) = 2x + 4.$$ ∎

EXAMPLE 9   Find an equation that defines the linear relationship between Fahrenheit and Celsius temperatures if the freezing point of water is 32° F and 0° C and the boiling point is 212° F and 100° C.

Solution   Arbitrarily, let F be the independent variable (like x) and C be the dependent variable (like y). Then, find the slope.

$$m = \frac{\Delta C}{\Delta F} = \frac{100 - 0}{212 - 32} = \frac{100}{180} = \frac{5}{9}$$

Now use the point-slope equation with C replacing y and F replacing x. By selecting $C_1 = 0$ and $F_1 = 32$, we have

$$C - 0 = \tfrac{5}{9}(F - 32),$$

so an equation that defines the linear relationship is

$$C = \tfrac{5}{9}(F - 32).$$

(*Note:* If the roles of F and C are reversed in step 1, then $m = \Delta F/\Delta C$ and the resulting equation is $F = \tfrac{9}{5}C + 32$.)   ∎

EXAMPLE 10   Solve the problem in the chapter introduction on page 153.

Solution   If we let y represent the depreciated value of the building, then we are given that $y = 200{,}000$ when $t = 0$, and $y = 0$ when $t = 30$. From these corresponding values of y and t, we calculate the slope.

$$m = \frac{\Delta y}{\Delta t} = \frac{200{,}000 - 0}{0 - 30} = -\frac{20{,}000}{3} \qquad (\approx -\$6{,}667)$$

Since $(0, 200{,}000)$ is the y-intercept and $-\tfrac{20{,}000}{3}$ is the slope, the equation of the line is

$$y = -\tfrac{20{,}000}{3}t + 200{,}000.$$

When $t = 7$, we have

$$y = -\tfrac{20,000}{3}(7) + 200,000$$
$$= \$153,333.33.$$

Thus, the value of the building (for tax purposes) after 7 years is $153,333. Note that the depreciation benefit in each year is about $6,667.   ■

Parallel and Perpendicular Lines

We often work with parallel and perpendicular lines in analytic geometry, and the slope concept may be used to establish conditions that parallel and perpendicular lines must satisfy.

Conditions for Parallel and Perpendicular Lines

1. Two nonvertical lines are parallel if and only if their slopes are equal.
2. Two nonvertical lines are perpendicular if and only if the product of their slopes is -1.

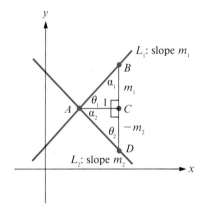

Figure 3.10

Intuitively, it is easy to comprehend that parallel lines have the same slope or inclination, and the proof of statement 1 is outlined in Exercise 80. We can establish statement 2 by using the definition of slope and elementary geometry. Consider Figure 3.10 in which L_1 is perpendicular to L_2 and two right triangles have been formed as shown. From the definition of slope, the length of BC (denoted $\overline{BC}$) is m_1, while $\overline{CD} = -m_2$ (since m_2 is negative). Then since L_1 is perpendicular to L_2, both α_1 and α_2 are complementary to θ_1, so

$$\alpha_1 = \alpha_2 \quad \text{and} \quad \theta_1 = \theta_2$$

and the triangles are similar. In such triangles, corresponding side lengths are proportional, so

$$\frac{\overline{BC}}{\overline{AC}} = \frac{\overline{AC}}{\overline{CD}} \quad \text{or} \quad \frac{m_1}{1} = \frac{1}{-m_2}.$$

It then follows that $m_1 \cdot m_2 = -1$, which establishes that if L_1 and L_2 are nonvertical perpendicular lines, then the product of their slopes is -1. By reversing each of the steps given, we can also show that if $m_1 \cdot m_2 = -1$, then L_1 and L_2 are perpendicular, thereby establishing statement 2.

EXAMPLE 11   Find an equation for the line passing through the point $(-4,1)$ that is

a. parallel to $5x - 2y = 3$ **b.** perpendicular to $5x - 2y = 3$

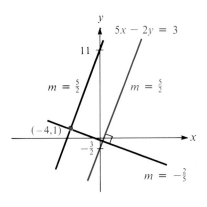

$5x - 2y = 3$

$m = \frac{5}{2}$ $m = \frac{5}{2}$

$(-4,1)$

$m = -\frac{2}{5}$

Figure 3.11

Solution   (See Figure 3.11.)

a. First, find the slope of the given line by changing the equation to the form $y = mx + b$.

$$5x - 2y = 3$$
$$-2y = -5x + 3$$
$$y = \tfrac{5}{2}x - \tfrac{3}{2}$$

By inspection, the slope of this line is $\frac{5}{2}$. Since the slopes of parallel lines are equal, we want an equation for the line through $(-4,1)$ with slope $\frac{5}{2}$. Using the point-slope equation, we have

$$y - 1 = \tfrac{5}{2}(x - (-4))$$
$$y = \tfrac{5}{2}x + 11.$$

b. As shown in part a, the slope of $5x - 2y = 3$ is $\frac{5}{2}$. It follows from statement 2 that the slope m of every line perpendicular to the given line must satisfy $\frac{5}{2}m = -1$, so $m = -\frac{2}{5}$. Then we find an equation for the line through $(-4,1)$ with slope $-\frac{2}{5}$ as follows:

$$y - y_1 = m(x - x_1)$$
$$y - 1 = -\tfrac{2}{5}(x - (-4))$$
$$y = -\tfrac{2}{5}x - \tfrac{3}{5}.$$ ∎

EXAMPLE 12   Show that if two nonvertical lines $A_1x + B_1y + C_1 = 0$ and $A_2x + B_2y + C_2 = 0$ are perpendicular, then $A_1A_2 + B_1B_2 = 0$.

Solution   First, write the given equations in the form $y = mx + b$. Since neither line is vertical, $B_1 \neq 0$ and $B_2 \neq 0$, which means

$$A_1x + B_1y + C_1 = 0 \text{ is equivalent to } y = \frac{-A_1}{B_1}x - \frac{C_1}{B_1}$$

and

$$A_2x + B_2y + C_2 = 0 \text{ is equivalent to } y = \frac{-A_2}{B_2}x - \frac{C_2}{B_2}.$$

By inspection, the slopes of the lines are $-A_1/B_1$ and $-A_2/B_2$. Then, if the lines are perpendicular, the product of their slopes is -1, so

$$\left(\frac{-A_1}{B_1}\right)\left(\frac{-A_2}{B_2}\right) = -1$$
$$A_1A_2 = -B_1B_2$$
$$A_1A_2 + B_1B_2 = 0.$$

This last equation is the result we needed to establish. (*Note:* The equation $Ax + By + C = 0$ with A and B not both zero is often called the **general form** of the equation of a line.)   ∎

EXERCISES 3.1

In Exercises 1–10 is the function a polynomial function? If so, state the degree.

1. $f(x) = x^3 + \sqrt{5}x - 3$
2. $f(x) = x^3 + 5\sqrt{x} - 3$
3. $f(x) = 4x - 3$
4. $f(x) = 7 + 4x - x^2$
5. $f(x) = x^2 + x^{1/2} - 1$
6. $f(x) = x^{-3} + 2x^2 - x + 3$
7. $f(x) = \dfrac{1}{x}$
8. $f(x) = 3^{-1}$
9. $f(x) = \pi$
10. $f(x) = x^{100} - 1$

In Exercises 11–20 find the slope of the line determined by each pair of points.

11. $(1,2)$ and $(3,4)$
12. $(3,-1)$ and $(5,-7)$
13. $(5,1)$ and $(-2,-3)$
14. $(-1,2)$ and $(0,0)$
15. $(4,-4)$ and $(2,7)$
16. $(-5,-1)$ and $(-2,-4)$
17. $(-1,-3)$ and $(2,-3)$
18. $(-2,0)$ and $(5,0)$
19. $(1,3)$ and $(1,-1)$
20. $(-3,2)$ and $(-3,-2)$

In Exercises 21–26 find an equation for the line that has the given slope m and passes through the given point.

21. $m = 5$; $(4,3)$
22. $m = -4$; $(2,-1)$
23. $m = \frac{1}{2}$; $(-2,0)$
24. $m = -\frac{3}{4}$; $(-5,2)$
25. $m = -\frac{1}{2}$; $(0,0)$
26. $m = 0$; $(1,2)$

In Exercises 27–30 find an equation for the line that contains the given points.

27. $(3,2)$ and $(4,1)$
28. $(-4,3)$ and $(-2,5)$
29. $(-3,-3)$ and $(-4,-2)$
30. $(2,-3)$ and $(3,-3)$

In Exercises 31–40 find the slope and the y-intercept of the line defined by the following equations.

31. $y = x + 7$
32. $y = -\frac{2}{3}x - 5$
33. $y = 5x$
34. $y = -x$
35. $y = -2$
36. $y = 0$
37. $3y + 2x = -2$
38. $3x + 5y = 2$
39. $6x - y = -7$
40. $2x - 7y = -4$

In Exercises 41–46 graph the line whose slope and y-intercept are given and find an equation for the line.

41. $m = \frac{1}{2}$; $(0,3)$
42. $m = 5$; $(0,1)$
43. $m = -3$; $(0,-1)$
44. $m = -\frac{3}{4}$; $(0,-2)$
45. $m = 0$; $(0,-5)$
46. $m = 0$; $(0,0)$

In Exercises 47–52 find the equation that defines the function if f is a linear function or a constant function.

47. $f(3) = 0$ and $f(0) = -2$
48. $f(0) = -1$ and $f(2) = 0$
49. $f(-1) = -3$ and $f(2) = -7$
50. $f(-6) = -1$ and $f(-2) = -5$
51. $f(-3) = 1$ and $f(0) = 1$
52. $f(4) = -6$ and $f(-1) = -6$

In Exercises 53–56 determine the given function value; f is a linear function or a constant function.

53. $f(3) = 1$ and $f(5) = -3$; find $f(4)$.
54. $f(-2) = 0$ and $f(1) = 6$; find $f(0)$.
55. $f(-4) = -1$ and $f(-1) = -3$; find $f(0)$.
56. $f(-5) = 2$ and $f(1) = 2$; find $f(5)$.

In Exercises 57–68 find an equation for the line passing through the given point that is

a. parallel to the given line
b. perpendicular to the given line

57. $y = x - 2$; $(1,2)$
58. $y = 3x + 1$; $(-2,-1)$
59. $y = -2x + 7$; $(3,0)$
60. $y = -x$; $(-4,2)$
61. $x + 7y = 2$; $(0,0)$
62. $3x - 2y = 5$; $(0,-5)$
63. $5x - 3y = 2$; $(1,-3)$
64. $4x - y = -8$; $(-1,-4)$
65. $3x + 12y = 10$; $(5,2)$
66. $x + 5y = -3$; $(-2,3)$
67. $2y - 3x + 1 = 0$; $(\frac{1}{2},-\frac{1}{3})$
68. $4x - y - 1 = 0$; $(-\frac{3}{4},-\frac{1}{2})$

69. Approximate the slope of each line in the figure.

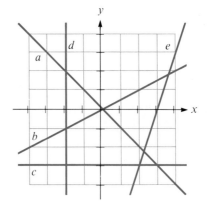

70. Approximate the slope of each line in the figure.

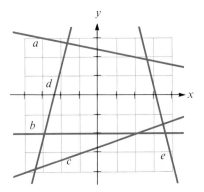

71. A moving company charges $46 to move a certain machine 10 mi and $58 to move the same machine 30 mi.
    a. Find an equation that defines this relationship if it is linear.
    b. What will it cost to move the machine 25 mi?
    c. What is the minimum charge for moving the machine?
    d. What is the rate for each mile the machine is moved?

72. The total cost of producing a certain item consists of paying rent for the building and paying a fixed amount per unit for material. The total cost is $250 if 10 units are produced and $330 if 30 units are produced.
    a. Find an equation that defines this relationship if it is linear.
    b. What will it cost to produce 100 units?
    c. How much is paid in rent?
    d. What is the cost of the material for each unit?

73. A ball thrown vertically down from a building has a velocity of -252 ft/second when $t = 1$ second and -316 ft/second when $t = 3$ seconds.
    a. Find an equation that defines this relationship if it is linear.
    b. Find the velocity of the ball when $t = 4$ seconds.
    c. What is the initial velocity of the ball?
    d. What is the significance of whether the velocity is positive or negative?

74. A spring that is 24 in. long is compressed to 20 in. by a force of 16 lb and to 15 in. by a force of 36 lb.
    a. Find an equation that defines this relationship if it is linear.
    b. What is the length of the spring if a force of 28 lb is applied?
    c. How much force is needed to compress the spring to 10 in.?

75. Show that the points $A(-4,-2)$, $B(-3,2)$, and $C(1,1)$ are the vertices of a right triangle by using the concept of slope.

76. Show that the points $A(-7,-7)$, $B(-1,1)$, and $C(2,5)$ lie on a straight line by using the concept of slope.

77. If the line through $(-2,-3)$ and $(a,4)$ is perpendicular to the graph of $y = 4 - x$, find a.

78. Write an equation for the line with y-intercept $(0,1)$ and parallel to the graph of $y = x$.

79. If two nonvertical lines $A_1x + B_1y + C_1 = 0$ and $A_2x + B_2y + C_2 = 0$ are parallel, then show that $A_1B_2 = A_2B_1$.

80. Use the illustration below and the fact that L_1 is parallel to L_2 if and only if $\theta_1 = \theta_2$ to prove the following statements.
    a. If two nonvertical lines are parallel, then their slopes are equal.
    b. If two nonvertical lines have equal slopes, then the lines are parallel.

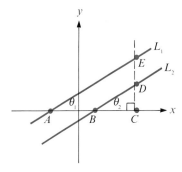

81. For the function $f(x) = mx + b$, compute $\dfrac{f(x + h) - f(x)}{h}$ with $h \neq 0$. Interpret the meaning of your result.

82. Linear functions are often used to approximate the relationship between two variables so that predictions can be made about future events or trends. The history of the record for the 1-mi run is an interesting case in point. Following are a table and graph that show the evolution of the record in this event. Many people are surprised to see that the times have not been leveling off. Instead, they have dropped in a rather predictable linear pattern.

a. The color line in the figure nicely "fits" the points and can be used to predict with some reliability what will happen in the future in this event. This line passes through the points (1900,258) and (1950,239). Write an equation for this line.

b. Use the answer to part a and predict the time for the mile run in the year 2000.

c. Use the answer to part a and predict the year in which the record will be 3:42.

Time (seconds) vs Year

Time	Runner	Year
4:36.5	Richard Webster, Britain	1865
4:29	William Chinnery, Britain	1868
4:28.8	W.C. Gibbs, Britain	1868
4:26	Walter Slade, Britain	1874
4:24.5	Walter Slade, Britain	1875
4:23.2	Walter George, Britain	1880
4:21.4	Walter George, Britain	1882
4:19.4	Walter George, Britain	1882
4:18.4	Walter George, Britain	1884
4:18.2	Fred Bacon, Scotland	1894
4:17	Fred Bacon, Scotland	1895
4:15.6	Thomas Conneff, U.S.	1895
4:15.4	John Paul Jones, U.S.	1911
4:14.6	John Paul Jones, U.S.	1913
4:12.6	Norman Taber, U.S.	1915
4:10.4	Paavo Nurmi, Finland	1923
4.09.2	Jules Ladoumegue, France	1931
4:07.6	Jack Lovelock, New Zealand	1933
4:06.8	Glen Cunningham, U.S.	1934
4:06.4	Sydney Wooderson, Britain	1937
4:06.2	Gunder Haegg, Sweden	1942
4:06.2	Arne Andersson, Sweden	1942
4:04.6	Gunder Haegg, Sweden	1942
4:02.6	Arne Andersson, Sweden	1943
4:01.6	Arne Andersson, Sweden	1944
4:01.4	Gunder Haegg, Sweden	1945
3:59.4	Roger Bannister, Britain	1954
3:58	John Landy, Australia	1954
3:57.2	Derek Ibbotson, Britain	1957
3:54.5	Herb Elliott, Australia	1958
3:54.4	Peter Snell, New Zealand	1962
3:54.1	Peter Snell, New Zealand	1964
3:53.6	Michel Jazy, France	1965
3:51.3	Jim Ryun, U.S.	1966
3:51.1	Jim Ryun, U.S.	1967
3:50	Filbert Bayi, Tanzania	1975
3:49.4	John Walker, New Zealand	1975
3:48.9	Sebastian Coe, Britain	1979
3:48.8	Steve Ovett, Britain	1980
3:48.5	Sebastian Coe, Britain	1981
3:48.4	Steve Ovett, Britain	1981
3:48.3	Sebastian Coe, Britain	1981
3:46.3	Steve Cram, Britain	1985

THINK ABOUT IT

1. Create a triangle from the intersection of three lines with positive slope such that the origin is inside the triangle. Write an equation for each of the three lines and display your solution graphically.
2. Line L_1 passes through $(1,-2)$ and $(4,3)$ and line L_2 passes through $(-5,k)$ and $(-2,1)$.
   a. Find k if L_1 is parallel to L_2.
   b. Find k if L_1 is perpendicular to L_2.
3. In each of the following cases, if a linear function is used to approximate the relationship between the given variables, state whether you would expect the slope of the line to be positive or negative.

a. An individual's height and weight
b. The number of hours that a runner practices and his time for a 1-mi race
c. A student's cumulative average and the number of hours per week that she watches television
d. The number of automobile accidents and the amount of insurance premiums that the driver has to pay
e. The percentage of nitrogen in a fertilizer and the height to which a treated plant will grow
f. The amount of alcohol consumed by an individual and the length of time in which he responds to a given stimulus

4. The following figure shows five lines whose equations are as follows:
   a. $y = ax$
   b. $y = bx + c$
   c. $y = dx + e$
   d. $y = mx + k$
   e. $y = px + n$

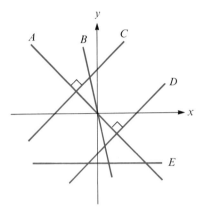

If $d = n = 0$, $b = m$, and $a < k < p$, then match each equation to the corresponding line.

5. Two points A and B are said to be *symmetric about the line L* if line segment AB is perpendicular to L and points A and B are equidistant from L (see illustration (a)). Use this definition to show in illustration (b) that the points (a,b) and (b,a) are symmetric about the line $y = x$. (*Note:* This exercise demonstrates why the graphs of f and f^{-1} are symmetric about the line $y = x$.)

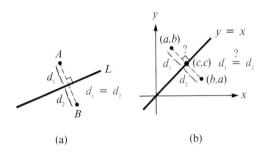

(a)   (b)

REMEMBER THIS

1. Graph $y = (x - 3)^2 - 5$ by translating the graph of $y = x^2$.

2. Graph $y = 4 - x^2$.

3. Solve $x^2 - 6x - 2 = 0$.

4. Solve $x^2 + 2x - 3 = 0$.

5. Solve $x^2 + 2x - 3 > 0$.

6. Solve $x^2 + 3 < 0$.

7. If $f(t) = 112t - 16t^2$, find $f(\frac{9}{2})$.

8. What number should be added to $x^2 - x$ to make the expression a perfect square?

9. What is the y-coordinate of all of the points on the x-axis?

10. What is the range of the function graphed below?

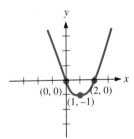

3.2   **Quadratic Functions**

We now consider polynomial functions of degree 2, which are called quadratic functions.

Definition of Quadratic Function

A function of the form

$$f(x) = ax^2 + bx + c,$$

where a, b, and c are real numbers with $a \neq 0$, is called a **quadratic function**.

To see the essential properties of a quadratic function consider the graphs of $f(x) = x^2 - 4x + 3$ (Figure 3.12) and $f(x) = 4x - 2x^2$ (Figure 3.13). We obtain these graphs by finding some ordered pairs that satisfy the relationships as shown.

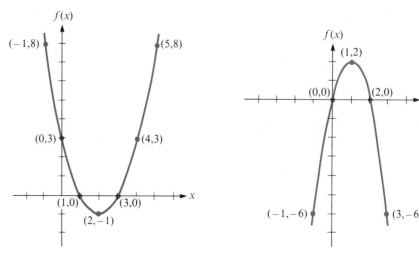

Figure 3.12   **Figure 3.13**

x	$x^2 - 4x + 3 = f(x)$	Solutions
-1	$(-1)^2 - 4(-1) + 3 = 8$	$(-1,8)$
0	$(0)^2 - 4(0) + 3 = 3$	$(0,3)$
1	$(1)^2 - 4(1) + 3 = 0$	$(1,0)$
2	$(2)^2 - 4(2) + 3 = -1$	$(2,-1)$
3	$(3)^2 - 4(3) + 3 = 0$	$(3,0)$
4	$(4)^2 - 4(4) + 3 = 3$	$(4,3)$
5	$(5)^2 - 4(5) + 3 = 8$	$(5,8)$

x	$4x - 2x^2 = f(x)$	Solutions
-1	$4(-1) - 2(-1)^2 = -6$	$(-1,-6)$
0	$4(0) - 2(0)^2 = 0$	$(0,0)$
1	$4(1) - 2(1)^2 = 2$	$(1,2)$
2	$4(2) - 2(2)^2 = 0$	$(2,0)$
3	$4(3) - 2(3)^2 = -6$	$(3,-6)$

The graph in both instances is a curve that is called a **parabola.** For $f(x) = x^2 - 4x + 3$, the parabola has a minimum turning point and opens upward like a cup. This occurs because the coefficient of the dominant term (x^2) is a positive number. In contrast, the graph of $f(x) = 4x - 2x^2$ has a turning point at the maximum y value. The parabola opens down since the coefficient of x^2 is a negative number. In general:

1. If a is positive ($a > 0$), the parabola opens upward and turns at the lowest point on the graph.
2. If a is negative ($a < 0$), the parabola opens downward and turns at the highest point on the graph.

Another essential property of this graph is that a vertical line that passes through the turning point divides the parabola into two segments such that if the curve were folded over on this line, its left half would

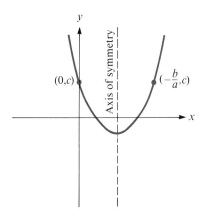

Figure 3.14

coincide with its right half (see Figure 3.14). This line is called the **axis of symmetry** for the parabola. To find the equation of the axis of symmetry, note that this line is halfway between any pair of points on the parabola that have the same y-coordinate. The simplest pair of points to work with is the pair of points on the graph of $y = ax^2 + bx + c$ whose y-coordinate is c. Thus,

$$y = ax^2 + bx + c.$$

Let $y = c$; then
$$c = ax^2 + bx + c$$
$$0 = ax^2 + bx$$
$$0 = (ax + b)x$$
$$ax + b = 0 \qquad x = 0$$
$$ax = -b$$
$$x = \frac{-b}{a}.$$

The x-coordinate halfway between these two x values is

$$x = \frac{0 + (-b/a)}{2}$$

so the equation of the axis of symmetry is

$$x = \frac{-b}{2a}.$$

Since the maximum or minimum point lies on the axis of symmetry, the x-coordinate of this turning point is $-b/2a$. We determine the y-coordinate of this point by finding the y value when $x = -b/2a$. Thus, the coordinates of the turning point of the parabola (which is also called the **vertex**) are

$$\left(\frac{-b}{2a}, f\left(\frac{-b}{2a} \right) \right).$$

The last important property to the graph of a quadratic function that we consider is the coordinates of the points where the curve crosses the axes. The point where the parabola crosses the y-axis (that is, the y-intercept) can be found by substituting 0 for x in the equation $y = ax^2 + bx + c$.

$$y = a(0)^2 + b(0) + c$$
$$= c$$

Thus, the y-intercept is always $(0,c)$.

To find the points where the parabola crosses the x-axis (that is, the x-intercepts), we substitute 0 for y in the equation $y = ax^2 + bx + c$. Thus, the x-coordinates of the x-intercepts are found by solving the equation $ax^2 + bx + c = 0$.

EXAMPLE 1   Graph the function defined by $f(x) = x^2 - 7x + 6$ and indicate

a. the coordinates of the x- and y-intercepts
b. the equation of the axis of symmetry
c. the coordinates of the maximum or minimum point
d. the range of the function

Solution

a. To find the x-intercepts, set $f(x) = 0$ and solve the resulting equation.

$$x^2 - 7x + 6 = 0$$
$$(x - 6)(x - 1) = 0$$
$$x - 6 = 0 \qquad x - 1 = 0$$
$$x = 6 \qquad x = 1$$

x-intercepts: $(1,0)$ and $(6,0)$; y-intercept: $(0,6)$ since $c = 6$

b. Axis of symmetry:

$$x = \frac{-b}{2a} = \frac{-(-7)}{2(1)} = \frac{7}{2}$$

c. Since $a > 0$, we have a minimum point. The x-coordinate of the minimum point is $\frac{7}{2}$ since the minimum point lies on the axis of symmetry. We find the y-coordinate of the minimum point by finding the y value when $x = \frac{7}{2}$.

$$f(x) = x^2 - 7x + 6$$
$$f(\tfrac{7}{2}) = (\tfrac{7}{2})^2 - 7(\tfrac{7}{2}) + 6$$
$$= \tfrac{49}{4} - \tfrac{49}{2} + 6$$
$$= \tfrac{49}{4} - \tfrac{98}{4} + \tfrac{24}{4}$$
$$= -\tfrac{25}{4}$$

Minimum point: $(\frac{7}{2}, -\frac{25}{4})$

d. As illustrated in Figure 3.15, the range is the interval $[-\frac{25}{4}, \infty)$.   ■

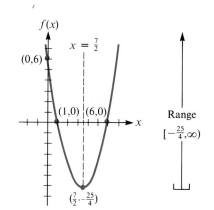

Figure 3.15

EXAMPLE 2   Graph the function defined by $f(x) = 3 - x - x^2$ and indicate

a. the coordinates of the x- and y-intercepts
b. the equation of the axis of symmetry
c. the coordinates of the maximum or minimum point
d. the range of the function

Solution

a. To find the x-intercepts, set $y = 0$ and solve the resulting equation

$$3 - x - x^2 = 0.$$

By the quadratic formula (with $a = -1$, $b = -1$, and $c = 3$), we have

$$x = \frac{-(-1) \pm \sqrt{(-1)^2 - 4(-1)(3)}}{2(-1)}$$

$$= \frac{1 \pm \sqrt{13}}{-2}.$$

x-intercepts: $\left(\dfrac{1 + \sqrt{13}}{-2}, 0\right)$ and $\left(\dfrac{1 - \sqrt{13}}{-2}, 0\right)$, which are about $(-2.3, 0)$ and $(1.3, 0)$

y-intercept: $(0,3)$ since $c = 3$

b. Axis of symmetry:

$$x = \frac{-b}{2a} = \frac{-(-1)}{2(-1)} = -\frac{1}{2}$$

c. Since $a < 0$, we have a maximum point. The x-coordinate of the maximum point is $-\frac{1}{2}$, since the maximum point lies on the axis of symmetry. We find the y-coordinate of the maximum point by finding $f(-\frac{1}{2})$, the value of the function when $x = -\frac{1}{2}$.

$$f(x) = 3 - x - x^2$$
$$f(-\tfrac{1}{2}) = 3 - (-\tfrac{1}{2}) - (-\tfrac{1}{2})^2$$
$$= 3 + \tfrac{1}{2} - \tfrac{1}{4}$$
$$= \tfrac{13}{4}$$

Maximum point: $(-\frac{1}{2}, \frac{13}{4})$

d. As illustrated in Figure 3.16, the range is the interval $(-\infty, \frac{13}{4}]$.

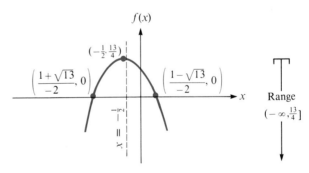

Figure 3.16

The vertex of a parabola can be found more efficiently in some cases by employing the graphing techniques discussed in Section 2.3. For instance, in Example 6c of that section we graphed $y = (x + 1)^2 - 2$ by moving the graph of $y = x^2$, 1 unit to the left and 2 units down, which placed the vertex of the parabola at $(-1,2)$. More generally, any quadratic function may be placed in the form

$$f(x) = a(x - h)^2 + k \qquad \text{(with } a \neq 0\text{)}$$

by completing the square, and in this form our graphing techniques tell us that the graph of f is the graph of $y = ax^2$ (a parabola) shifted so the vertex is the point (h,k) and the axis of symmetry is the vertical line $x = h$

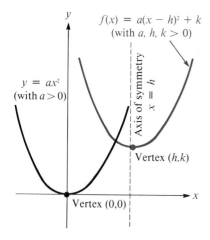

$f(x) = a(x - h)^2 + k$
(with a, h, $k > 0$)

$y = ax^2$
(with $a > 0$)

Axis of symmetry
$x = h$

Vertex (h,k)

Vertex $(0,0)$

Figure 3.17

(see Figure 3.17). Example 3 illustrates this result using the functions analyzed in Examples 1 and 2.

EXAMPLE 3   Determine the vertex and axis of symmetry of the graph of f by matching the function to the form $f(x) = a(x - h)^2 + k$.

   a. $f(x) = x^2 - 7x + 6$ **b.** $f(x) = 3 - x - x^2$

Solution

a. First, write the equation as

$$f(x) = (x^2 - 7x \quad) + 6.$$

As discussed in Section 1.6, we complete the square for $x^2 + bx$ by adding $(b/2)^2$, which is the square of one-half of the coefficient of x. Half of -7 is $-\frac{7}{2}$ and $(-\frac{7}{2})^2 = \frac{49}{4}$. Now inside the parentheses both add $\frac{49}{4}$ (so we can complete the square) and subtract $\frac{49}{4}$ (so $f(x)$ does not change), and proceed as follows:

$$\begin{aligned} f(x) &= (x^2 - 7x + \tfrac{49}{4} - \tfrac{49}{4}) + 6 \\ &= (x^2 - 7x + \tfrac{49}{4}) - \tfrac{49}{4} + 6 \\ &= (x - \tfrac{7}{2})^2 - \tfrac{25}{4}. \end{aligned}$$

Matching this result to the form $f(x) = a(x - h)^2 + k$, we determine $h = \frac{7}{2}$ and $k = -\frac{25}{4}$, so the vertex of the parabola is $(\frac{7}{2}, -\frac{25}{4})$ and the axis of symmetry is $x = \frac{7}{2}$.

b. To complete the square, the coefficient of x^2 must be 1, so first factor out -1 from the x terms and then proceed as in part a.

$$\begin{aligned} f(x) &= 3 - x - x^2 \\ &= -1(x^2 + x) + 3 \\ &= -1(x^2 + x + \tfrac{1}{4} - \tfrac{1}{4}) + 3 \\ &= -1(x^2 + x + \tfrac{1}{4}) + (-1)(-\tfrac{1}{4}) + 3 \\ &= -1(x + \tfrac{1}{2})^2 + \tfrac{13}{4} \end{aligned}$$

Once again, we match this equation to the form $f(x) = a(x - h)^2 + k$, which gives $h = -\frac{1}{2}$ and $k = \frac{13}{4}$. Thus, the vertex is $(-\frac{1}{2}, \frac{13}{4})$ and the axis of symmetry is $x = -\frac{1}{2}$. ∎

The following examples illustrate how quadratic functions can be applied in practical situations. Notice, in particular, the significance of the maximum or minimum value, which is an important topic in applied mathematics.

EXAMPLE 4   The height (y) of a projectile shot vertically up from the ground with an initial velocity of 144 ft/second is given by the formula $y = 144t - 16t^2$. Graph the function defined by this formula and indicate

   a. when the projectile will strike the ground
   b. when the projectile attains its maximum height
   c. the maximum height

Solution

a. When the projectile hits the ground, the height (y) of the projectile is 0. Thus, we want to find the values of t for which $144t - 16t^2$ equals 0.

$$144t - 16t^2 = 0$$
$$16t(9 - t) = 0$$
$$16t = 0 \qquad 9 - t = 0$$
$$t = 0 \qquad t = 9$$

The projectile will hit the ground 9 seconds later.

b. Since $a < 0$, we have a maximum point. Axis of symmetry:

$$t = \frac{-b}{2a} = \frac{-(144)}{2(-16)} = \frac{-144}{-32} = 4.5$$

The projectile reaches its highest point when $t = 4.5$ seconds.

c. We can find the maximum height by finding the value of y when $t = 4.5$ (or $t = \frac{9}{2}$) seconds.

$$y = f(t) = 144t - 16t^2$$
$$f(\tfrac{9}{2}) = 144(\tfrac{9}{2}) - 16(\tfrac{9}{2})^2$$
$$= 72 \cdot 9 - 16(\tfrac{81}{4})$$
$$= 324$$

Thus, the projectile reaches a maximum height of 324 ft, and Figure 3.18 shows the graph of this function. ∎

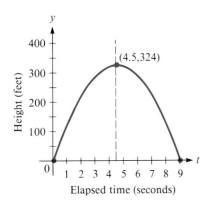

Figure 3.18

EXAMPLE 5 A businessperson wants to fence in a parking lot in which one side of the lot is bounded by a building. If a total of 300 ft of fencing is available, what are the dimensions of the largest rectangular parking lot that can be enclosed with the available fencing?

Solution First, draw a sketch of the situation (Figure 3.19). If we let x represent the length of one side of the lot, the opposite side also has length x. The length of the side opposite the building is then $300 - 2x$. We wish to maximize the area of this rectangular region, which is given by the following quadratic function.

$$\text{Area} = x(300 - 2x)$$
$$A = 300x - 2x^2$$

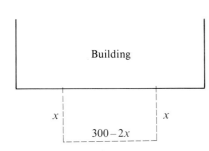

Figure 3.19

Since $a < 0$, we have a maximum point. Axis of symmetry:

$$x = \frac{-b}{2a} = \frac{-300}{2(-2)} = \frac{-300}{-4} = 75$$

Length of the side opposite building:

$$300 - 2x = 300 - 2(75) = 300 - 150 = 150$$

The area is a maximum when the dimensions are 75 ft by 150 ft. ∎

In Section 1.8 we discussed the meaning and importance of inequalities. We now extend our ability to work with these statements by considering a graphical approach to quadratic inequalities. They are expressed (with $a \neq 0$) in the forms

$$ax^2 + bx + c > 0, \qquad ax^2 + bx + c \geq 0,$$
$$ax^2 + bx + c < 0, \quad \text{or} \quad ax^2 + bx + c \leq 0.$$

For example, let us find the values of x for which $x^2 - x - 6 > 0$ is a true statement.

1. Sketch the graph of $y = x^2 - x - 6$ with emphasis on the x-intercepts and on whether the curve has a maximum or minimum turning point.
   a. To find the x-intercepts, set $y = 0$ and solve the resulting equation.

$$0 = x^2 - x - 6$$
$$0 = (x - 3)(x + 2)$$
$$x = 3 \qquad x = -2$$

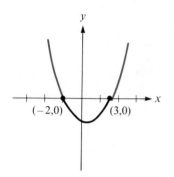

Figure 3.20

   x-intercepts: $(3,0)$ and $(-2,0)$
   b. Since $a > 0$, the graph has a minimum turning point and opens up (see Figure 3.20).
2. To solve the inequality $x^2 - x - 6 > 0$, we want only those values of x for which $x^2 - x - 6$ (or y) is positive. Observing the graph, we see that values of x that are to the right of 3 (that is, $x > 3$) or values of x that are to the left of -2 (that is, $x < -2$) make $x^2 - x - 6$ (or y) positive. Therefore, the solution set in interval notation is $(-\infty, -2) \cup (3, \infty)$.

EXAMPLE 6   For what values of x is the inequality $4 + 3x - x^2 > 0$ a true statement?

Solution

a. Sketch the graph of $y = 4 + 3x - x^2$.
   (1) To find the x-intercepts, set $y = 0$ and solve the resulting equation, $0 = 4 + 3x - x^2$.

$$4 + 3x - x^2 = 0$$
$$(4 - x)(1 + x) = 0$$
$$4 - x = 0 \qquad 1 + x = 0$$
$$x = 4 \qquad x = -1$$

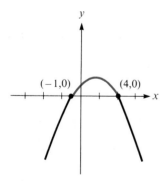

Figure 3.21

   x-intercepts: $(-1,0)$ and $(4,0)$

   (2) Since $a < 0$, the graph has a maximum turning point and opens down (see Figure 3.21).
b. To solve the inequality $4 + 3x - x^2 > 0$, we want only those values of x for which $4 + 3x - x^2$ (or y) is positive. Observing the graph, we see that values of x between -1 and 4 (that is, $-1 < x < 4$) make $4 + 3x - x^2$ (or y) positive. Thus, the solution set is the interval $(-1, 4)$.

EXAMPLE 7   Solve the inequality $2x^2 + 5x \leq 2$.

Solution

a. $2x^2 + 5x \le 2$ in the form $ax^2 + bx + c \le 0$ is $2x^2 + 5x - 2 \le 0$. Therefore, we want to sketch the graph of $y = 2x^2 + 5x - 2$.

   (1) To find the x-intercepts, set $y = 0$ and solve the resulting equation, $0 = 2x^2 + 5x - 2$.

$$x = \frac{-b \pm \sqrt{b^2 - 4ac}}{2a} = \frac{-(5) \pm \sqrt{(5)^2 - 4(2)(-2)}}{2(2)}$$

$$= \frac{-5 \pm \sqrt{25 + 16}}{4}$$

$$= \frac{-5 \pm \sqrt{41}}{4}$$

$$x_1 = \frac{-5 + \sqrt{41}}{4} \qquad x_2 = \frac{-5 - \sqrt{41}}{4}$$

x-intercepts: $\left(\dfrac{-5 + \sqrt{41}}{4}, 0\right)$ and $\left(\dfrac{-5 - \sqrt{41}}{4}, 0\right)$

   (2) Since $a > 0$, the graph has a minimum point and opens up (see Figure 3.22).

b. To solve the inequality $2x^2 + 5x - 2 \le 0$, we want only those values of x for which $2x^2 + 5x - 2$ (or y) either equals 0 or is negative. Observing the graph, we see that $2x^2 + 5x - 2$ is less than or equal to zero in the interval

$$\left[\frac{-5 - \sqrt{41}}{4}, \frac{-5 + \sqrt{41}}{4}\right]. \qquad \blacksquare$$

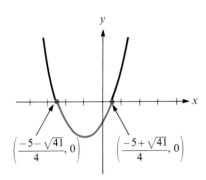

Figure 3.22

The x-intercept labels in the figure: $\left(\dfrac{-5 - \sqrt{41}}{4}, 0\right)$ and $\left(\dfrac{-5 + \sqrt{41}}{4}, 0\right)$

EXAMPLE 8   Solve the inequality $x^2 + 2 < 0$.

Solution

a. Sketch the graph of $y = x^2 + 2$.

   (1) To find the x-intercepts, set $y = 0$ and solve the resulting equation.

$$0 = x^2 + 2$$
$$-2 = x^2$$
$$\pm\sqrt{-2} = x$$

The solutions are not real numbers. Therefore, there are no x-intercepts.

   (2) Since $a > 0$, the graph has a minimum turning point and opens up (see Figure 3.23).

b. To solve the inequality $x^2 + 2 < 0$, we want only those values of x for which $x^2 + 2$ (or y) is negative. Observing the graph, we see $x^2 + 2$ (or y) is never negative. Therefore, there are no values of x for which the inequality will be a true statement, and the solution set is $\emptyset$. $\qquad \blacksquare$

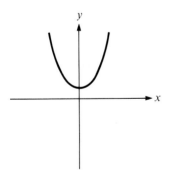

Figure 3.23

Note from the examples that only a very rough sketch of the graph, which includes the x-intercepts and the direction that the curve opens, is needed. In particular, it is not necessary to find the coordinates of the turning point of the parabola.

EXERCISES 3.2

In Exercises 1–20 graph the quadratic function defined by the equation. On the graph indicate

a. the coordinates of the x- and y-intercepts
b. the equation of the axis of symmetry
c. the coordinates of the maximum or minimum point
d. the range of the function

1. $y = x^2 - 3x + 2$ **2.** $y = x^2 + 6x + 5$
3. $f(x) = x^2 - 4$ **4.** $g(x) = 3x^2 - 3$
5. $y = x^2 - 2x$ **6.** $f(x) = 2x^2 - 3x$
7. $g(x) = 4x - x^2$ **8.** $y = 1 - x^2$
9. $y = x^2 + 2x + 1$ **10.** $y = x^2 - 4x + 4$
11. $f(x) = -x^2 + x + 2$ **12.** $f(x) = -x^2 + 3x - 4$
13. $y = x^2 - 5$ **14.** $y = 2 - x^2$
15. $y = 5 - 4x^2$ **16.** $y = 2x^2 - 3$
17. $f(x) = 2x^2 - x - 1$ **18.** $y = 3x^2 - 5x + 1$
19. $y = 3 + 6x - 2x^2$ **20.** $g(x) = -3x^2 + 5x + 2$

In Exercises 21–30 determine the vertex and the axis of symmetry of the graph of f by matching the function to the form $f(x) = a(x - h)^2 + k$.

21. $f(x) = (x + 2)^2 - 3$ **22.** $f(x) = -2(x - 1)^2 + 5$
23. $f(x) = 3x^2 - 4$ **24.** $f(x) = 3(x - 4)^2$
25. $f(x) = x^2 - 2x + 2$ **26.** $f(x) = x^2 + 6x + 5$
27. $f(x) = 3x^2 - 9x + 1$ **28.** $f(x) = 2x^2 + 4x - 5$
29. $f(x) = 2 + x - x^2$ **30.** $f(x) = 1 - 3x - 4x^2$

31. What is the domain of all the functions in Exercises 1–30?

32. The height of a ball thrown directly up from a roof 144 ft high with an initial velocity of 128 ft/second is given by the formula $y = 144 + 128t - 16t^2$. What is the maximum height attained by the ball? When does the ball hit the ground?

33. The height of a projectile shot vertically upward from the ground with an initial velocity of 96 ft/second is given by the formula $y = 96t - 16t^2$. What is the maximum height attained by the projectile? When does the projectile hit the ground?

34. The total sales revenue for a company is estimated by the formula $S = 8x - x^2$, where x is the unit selling price (in dollars) and S is the total sales revenue (1 unit equals $10,000). What should be the unit selling price if the company wishes to maximize the total revenue?

35. In a 120-volt line having a resistance of 10 ohms, the power W in watts when the current I is flowing is given by the formula $W = 120I - 10I^2$. What is the maximum power that can be delivered in this circuit?

36. Find two positive numbers whose sum is 8 and whose product is a maximum.

37. Find two positive numbers whose sum is 20 and whose product is a maximum.

38. What positive number exceeds its square by the largest amount?

39. Find two positive numbers whose sum is 20 such that the sum of their squares is a minimum.

40. The sum of the base and altitude of a triangle is 12 in. Find each dimension if the area is to be a maximum.

41. A rectangular field is adjacent to a river and is to have fencing on three sides, the side on the river requiring no fencing. If 400 yd of fencing is available, what are the dimensions of the largest rectangular section that can be enclosed with the available fencing?

42. A farmer wants to make a rectangular enclosure along the side of a barn and then divide the enclosure into two pens with a fence constructed at a right angle to the barn. If 300 yd of fencing is available, what are the dimensions of the largest section that can be enclosed with the available fencing?

43. A sheet of metal 12 in. wide and 20 ft long is to be made into a gutter by turning up the same amount of material on each edge at right angles to the base. Determine the amount of material that should be turned up to maximize the volume that the gutter can carry.

44. Find the maximum possible area of a rectangle with a perimeter of 100 ft.

45. A bus tour charges a fare of $10 per person and carries 200 people per day. The manager estimates that she will lose 10 passengers for each $1 increase in fare. Find the most profitable fare for her to charge.

46. An airline offers a charter flight at a fare of $100 per person if 100 passengers sign up. For *each* passenger above 100, the fare for *each* passenger is reduced by 50 cents. What is the maximum total revenue that the airline can obtain if the plane has 200 seats?

In Exercises 47–66 solve the quadratic inequality.

47. $x^2 - 4x - 5 > 0$ **48.** $x^2 + x - 6 < 0$
49. $x^2 < 4x + 5$ **50.** $x^2 + 5 < 6x$
51. $-x^2 + x + 2 \leq 0$ **52.** $-10 \geq 3x - x^2$
53. $x^2 \geq 4$ **54.** $x^2 - 5 \leq 0$
55. $x^2 < 3x$ **56.** $0 > 5x - 10x^2$
57. $2x^2 + x \leq 6$ **58.** $4 + x > x^2$
59. $3x^2 - 2x < 5$ **60.** $5x^2 \geq 4x - 3$
61. $x^2 + 2x > -1$ **62.** $4x^2 \leq 12x - 9$
63. $x^2 - 5x + 7 < 0$ **64.** $x^2 + 1 \geq 0$
65. $1 - x^2 \geq 0$ **66.** $-2x^2 + x > 4$

THINK ABOUT IT

1. If the parabola $y = x^2 + 6x + k$ has its vertex on the x-axis, what is the value of k?
2. Find an equation that defines the quadratic function f with the given properties.
   a. The graph of f passes through the origin and has its vertex at $(1, -4)$.
   b. The intercepts of the graph of f are $(1,0)$, $(3,0)$, and $(0,-2)$.
3. Create a quadratic inequality with the given solution set.
   a. $[-2,5]$ b. $(-\infty,0) \cup (3,\infty)$
   c. $(-\infty,\infty)$ d. $\emptyset$

4. If f is a linear function and g is a quadratic function, then is the function $f \circ g$ linear, quadratic, or neither? Also answer this question for the function $g \circ f$. In both cases, prove your answer is correct.
5. By completing the square, place the quadratic function defined by $f(x) = ax^2 + bx + c$ with $a \neq 0$ in the form $f(x) = a(x - h)^2 + k$ and show that the vertex is at

$$\left(\frac{-b}{2a}, \frac{4ac - b^2}{4a} \right).$$

REMEMBER THIS

1. Subtract $-8x^2 - 24x$ from $-8x^2 + x - 4$.
2. If $f(x) = 5x^3 - x^2 + 3x - 2$, find $f(-\frac{1}{2})$.
3. If $f(x) = (x + 1)(2x^2 - 5x + 3) - 8$, find $f(-1)$.
4. Write $(x + 1)(2x^2 - 5x + 3) - 8$ in the form $ax^3 + bx^2 + cx + d$.
5. What is the remainder when 266 is divided by 9?
6. When dividing 266 by 9 in the previous exercise, how may we use the quotient and the remainder to check that we divided correctly?

7. Find the slope of the line containing the points $(2,-3)$ and $(-6,0)$.
8. Find an equation of the line passing through the points $(2,-3)$ and $(-6,0)$.
9. Find the slope and the y-intercept of the line given by $5x - 2y = 8$.
10. Write an equation for the line with slope $-\frac{1}{3}$ that passes through $(0,-6)$.

3.3 Synthetic Division and the Remainder Theorem

To discuss polynomial functions of degree greater than 2, we must first consider the division of polynomials. We divide one polynomial by another in a manner similar to the division of two integers. Consider the following arrangement for dividing $3x^3 + x - 4$ by $x + 4$.

$$
\begin{array}{r}
3x^2 - 12x + 49 \qquad \text{(quotient)} \\
(\text{divisor}) \quad x + 4 \overline{)\, 3x^3 + 0x^2 + x - 4} \quad \text{(dividend)} \\
\underline{3x^3 + 12x^2} \\
-12x^2 + x - 4 \\
\underline{-12x^2 - 48x} \\
49x - 4 \\
\underline{49x + 196} \\
-200 \quad \text{(remainder)}
\end{array}
$$

First, we arrange the terms of the dividend and the divisor in descending powers of x. If a term is missing, write 0 as its coefficient. Then divide the first term of the dividend by the first term of the divisor to obtain the first term of the quotient. Next, we multiply the entire divisor by the first term of the quotient and subtract this product from the dividend. Use the remainder as the new dividend and repeat the above procedure until the remainder is of lower degree than the divisor.

As with the division of numbers,

$$\text{dividend} = (\text{divisor})(\text{quotient}) + \text{remainder}$$

or

$$\frac{\text{dividend}}{\text{divisor}} = \text{quotient} + \frac{\text{remainder}}{\text{divisor}}.$$

For example, the division of 7 by 3 may be expressed as

$$7 = 3 \cdot 2 + 1 \text{ or } \frac{7}{3} = 2 + \frac{1}{3}.$$

In the case of the preceding polynomial division, we may write

$$3x^3 + x - 4 = (x + 4)(3x^2 - 12x + 49) + (-200)$$

or

$$\frac{3x^3 + x - 4}{x + 4} = 3x^2 - 12x + 49 + \frac{-200}{x + 4}.$$

EXAMPLE 1   Divide $2x^4 + 6x^3 - 5x^2 - 1$ by $2x^2 - 3$. Express the answer in the form $\dfrac{\text{dividend}}{\text{divisor}} = \text{quotient} + \dfrac{\text{remainder}}{\text{divisor}}.$

Solution   The division is performed below. Note that $0x$ is inserted to help align like terms vertically and the division process stopped when the remainder $9x - 4$ was of lower degree than the divisor $2x^2 - 3$.

$$
\begin{array}{r}
x^2 + 3x - 1 \\
2x^2 - 3 \overline{)\, 2x^4 + 6x^3 - 5x^2 + 0x - 1} \\
\text{subtract}\quad 2x^4 \qquad\quad -3x^2 \\
\hline
6x^3 - 2x^2 + 0x - 1 \\
\text{subtract}\qquad 6x^3 \qquad\quad - 9x \\
\hline
-2x^2 + 9x - 1 \\
\text{subtract}\qquad -2x^2 \qquad + 3 \\
\hline
9x - 4
\end{array}
$$

$2x^4/2x^2 = x^2$, $6x^3/2x^2 = 3x$ and $-2x^2/2x^2 = -1$.

This line is $x^2(2x^2 - 3)$.

This line is $3x(2x^2 - 3)$.

This line is $-1(2x^2 - 3)$.

The answer in the form requested is

$$\frac{2x^4 + 6x^3 - 5x^2 - 1}{2x^2 - 3} = x^2 + 3x - 1 + \frac{9x - 4}{2x^2 - 3}.$$ ∎

The division of any polynomial by a polynomial of the form $x - b$ is of theoretical and practical importance. This division may be performed by a shorthand method called **synthetic division.** Consider the arrangement for dividing $2x^3 + 5x^2 - 1$ by $x - 2$.

$$
\begin{array}{r}
2x^2 + 9x + 18 \\
x - 2 \overline{\smash{\big)}\ 2x^3 + 5x^2 + 0x - 1} \\
\underline{2x^3 - 4x^2} \\
9x^2 + 0x - 1 \\
\underline{9x^2 - 18x} \\
18x - 1 \\
\underline{18x - 36} \\
35
\end{array}
$$

When the polynomials are written with terms in descending powers of x, there is no need to write all the x's. Only the coefficients are needed. Also, notice that the encircled coefficients entailed needless writing. Using only the necessary coefficients, we may abbreviate this division as follows:

$$
\begin{array}{r}
\ 2 \quad\ \ 9 \quad\ \ 18 \\
-2\ \overline{\smash{\big)}\ 2 \quad\ \ 5 \quad\ \ 0 \quad -1} \\
\ -4 \ \ -18 \ \ -36 \\
\hline
\ 9 \quad\ 18 \quad\ 35.
\end{array}
$$

If we bring down 2 as the first entry in the bottom row, all the coefficients of the quotient appear. The arrangement may then be shortened to

$$
\begin{array}{llll}
-2\rfloor & \boxed{\ 2 \qquad 5 \qquad 0 \qquad -1\ } & \leftarrow \ \text{coefficients of dividend} & \text{(row 1)} \\[2mm]
& \quad\ \ \ -4 \quad -18 \quad -36 & & \text{(row 2)} \\[2mm]
\text{coefficients of quotient} \rightarrow & \boxed{\ 2 \qquad 9 \qquad 18 \qquad 35\ } & \leftarrow \ \text{remainder} & \text{(row 3)}
\end{array}
$$

Finally, if we replace -2 by 2, which is the value of b, we may change the sign of each number in row 2 and add at each step instead of subtracting. The final arrangement for synthetic division is then as follows:

$$
\begin{array}{llllll}
2\rfloor & 2 & 5 & 0 & -1 & \text{(row 1)} \\
& & 4 & 18 & 36 & \text{(row 2)} \\
\hline
& 2 & 9 & 18 & 35 & \rightarrow \ \text{remainder} \quad \text{(row 3)} \\
& \downarrow & \downarrow & \downarrow & & \\
\text{quotient} \rightarrow & 2x^2 & + 9x & + 18. & &
\end{array}
$$

Let us summarize the procedure for synthetic division. First, write the coefficients of the terms in the dividend. Be sure the powers are in descending order and enter a 0 as the coefficient of any missing term. Write the value of b (in this case 2) to the left of the dividend. Bring down the first dividend entry (2). Multiply this number by $b(2)$ and place the result (4) under the next number (5) in row 1. Add, then multiply the resulting 9 by $b(2)$. Place the result (18) under the next number (0) in row 1. Add, then multiply the resulting 18 by $b(2)$. Place the result (36) under the next number (-1) in row 1. Add and obtain 35. The last number 35 is the remainder. The other numbers in row 3 are the coefficients of the quotient.

The degree of the polynomial in the quotient is always one less than the degree of the polynomial in the dividend.

A summary cannot do justice to the simple procedure in synthetic division. Careful consideration of Examples 2 and 3 should help clarify the process. Remember, synthetic division applies only when we divide by a polynomial of the form $x - b$.

EXAMPLE 2   Use synthetic division to divide $x^5 - 1$ by $x - 1$. Express the result in the form dividend = (divisor)(quotient) + remainder.

Solution   We use 0's as the coefficients of the missing x^4, x^3, x^2, and x terms. We divide by $x - 1$ so that $b = 1$ (not -1). The arrangement is as follows:

$$
\begin{array}{r|rrrrrr}
1 & 1 & 0 & 0 & 0 & 0 & -1 \\
  &   & 1 & 1 & 1 & 1 & 1 \\
\hline
  & 1 & 1 & 1 & 1 & 1 & 0 \quad \leftarrow \text{ remainder}
\end{array}
$$

quotient $\longrightarrow$ $x^4 + x^3 + x^2 + x + 1.$

The answer in the form requested is

$$x^5 - 1 = (x - 1)(x^4 + x^3 + x^2 + x + 1) + 0. \qquad \blacksquare$$

EXAMPLE 3   Use synthetic division to divide $4x^3 - x^2 + 2$ by $x + 3$. Express the result in the form dividend = (divisor)(quotient) + remainder.

Solution   We use 0 as the coefficient of the missing x term. We divide by $x + 3$ or $x - (-3)$ so that $b = -3$ (not 3). The arrangement is as follows:

$$
\begin{array}{r|rrrr}
-3 & 4 & -1 & 0 & 2 \\
   &   & -12 & 39 & -117 \\
\hline
   & 4 & -13 & 39 & -115 \quad \leftarrow \text{ remainder}
\end{array}
$$

quotient $\longrightarrow$ $4x^2 - 13x + 39.$

The answer in the form requested is

$$4x^3 - x^2 + 2 = (x + 3)(4x^2 - 13x + 39) - 115. \qquad \blacksquare$$

In Example 3 we found that the function

$$P(x) = 4x^3 - x^2 + 2$$

may be written as

$$P(x) = (x + 3)(4x^2 - 13x + 39) - 115.$$

When $x = -3$, the factor $x + 3 = 0$; thus,

$$P(-3) = 0 - 115$$
$$= -115.$$

The value of the function when $x = -3$ is the same as the remainder obtained when $P(x)$ is divided by $x - (-3)$. This discussion suggests the following theorem.

Remainder Theorem

If a polynomial $P(x)$ is divided by $x - b$, the remainder is $P(b)$.

We now prove this theorem. Let $Q(x)$ and r represent the quotient and remainder when $P(x)$ is divided by $x - b$. Then

$$\text{dividend} = (\text{divisor})(\text{quotient}) + \text{remainder}$$
$$P(x) = (x - b)Q(x) + r.$$

This statement is true for all values of x. If $x = b$, then

$$P(b) = (b - b)Q(b) + r$$
$$= 0 \cdot Q(b) + r.$$

Thus,

$$P(b) = r.$$

We know that $P(b)$ may be found by substituting b for x in the function. The remainder theorem provides an alternative method. That is, we find $P(b)$ by determining the remainder when $P(x)$ is divided by $x - b$. Since this remainder may be obtained by synthetic division, this approach is often simpler than direct substitution. In the next section other advantages of the remainder theorem method will be discussed.

EXAMPLE 4 If $P(x) = 3x^5 + 5x^4 + 7x^3 - 4x^2 + x - 24$, find $P(-2)$ by (a) direct substitution and (b) the remainder theorem.

Solution

a. By direct substitution, we have

$$P(-2) = 3(-2)^5 + 5(-2)^4 + 7(-2)^3 - 4(-2)^2 + (-2) - 24$$
$$= -96 + 80 - 56 - 16 - 2 - 24$$
$$= -114.$$

b. By the remainder theorem, we have

$$
\begin{array}{r|rrrrrr}
-2 & 3 & 5 & 7 & -4 & 1 & -24 \\
   &   & -6 & 2 & -18 & 44 & -90 \\
\hline
   & 3 & -1 & 9 & -22 & 45 & -114.
\end{array}
$$

Since the remainder is -114, $P(-2) = -114$. ■

EXAMPLE 5 If $P(x) = 2x^4 - 5x^3 + 11x^2 - 3x - 5$, find $P(-\tfrac{1}{2})$ by (a) direct substitution and (b) the remainder theorem.

Solution

a. By direct substitution, we have

$$P(-\tfrac{1}{2}) = 2(-\tfrac{1}{2})^4 - 5(-\tfrac{1}{2})^3 + 11(-\tfrac{1}{2})^2 - 3(-\tfrac{1}{2}) - 5$$
$$= \tfrac{1}{8} + \tfrac{5}{8} + \tfrac{11}{4} + \tfrac{3}{2} - 5$$
$$= 0.$$

b. By the remainder theorem, we have

$$
\begin{array}{r|rrrrr}
-\frac{1}{2} & 2 & -5 & 11 & -3 & -5 \\
  &   & -1 & 3 & -7 & 5 \\
\hline
  & 2 & -6 & 14 & -10 & 0.
\end{array}
$$

Since the remainder is 0, $P(-\frac{1}{2}) = 0$.   ∎

Notice in Examples 4 and 5 that the solution is obtained more easily by the remainder theorem method.

EXERCISES 3.3

In Exercises 1–6 perform the indicated divisions by using long division. Express the result in the form dividend = (divisor)(quotient) + remainder.

1. $(x^2 + 7x - 2) \div (x + 5)$
2. $(x^2 - 4) \div (x - 1)$
3. $(6x^3 - 3x^2 + 14x - 7) \div (2x - 1)$
4. $(4x^3 + 5x^2 - 10x + 4) \div (4x - 3)$
5. $(3x^4 + x - 2) \div (x^2 - 1)$
6. $(2x^3 - 3x^2 + 10x - 5) \div (x^2 + 5)$

In Exercises 7–12 perform the indicated divisions by using long division. Express the answer in the form
$$\frac{\text{dividend}}{\text{divisor}} = \text{quotient} + \frac{\text{remainder}}{\text{divisor}}.$$

7. $\dfrac{x^2 - 5}{x + 1}$
8. $\dfrac{x^2 + 3x - 4}{x - 3}$
9. $\dfrac{3x^4 - 5x^2 + 7}{x^2 + 2x + 1}$
10. $\dfrac{2x^4 - 8x^3 - 7x^2 + 1}{2x^2 - 5}$
11. $\dfrac{x^3 + 1}{x(x - 1)}$
12. $\dfrac{x^3 - 5}{(x - 1)^2}$

In Exercises 13–22 perform the indicated division by using synthetic division. Express the result in the form dividend = (divisor)(quotient) + remainder.

13. $(x^3 - 5x^2 + 2x - 3) \div (x - 1)$
14. $(x^3 + x^2 - 9x - 6) \div (x - 3)$
15. $(2x^3 + 9x^2 - x + 14) \div (x + 5)$
16. $(3x^3 - 5x^2 - 18x + 9) \div (x + 2)$
17. $(7 + 6x - 2x^2 - x^3) \div (x + 3)$
18. $(4 - x + 3x^2 - x^3) \div (x - 4)$
19. $(2x^3 + x - 5) \div (x + 1)$
20. $(3x^4 - x^2 + 7) \div (x + 3)$
21. $(x^4 - 16) \div (x - 2)$
22. $(x^3 + 27) \div (x - 3)$

In Exercises 23–30 find the given function value by (a) direct substitution and (b) the remainder theorem.

23. $P(x) = x^3 - 4x^2 + x + 6$; find $P(2)$ and $P(3)$.
24. $P(x) = 2x^3 + 5x^2 - x - 7$; find $P(4)$ and $P(-3)$.
25. $P(x) = 2x^4 - 7x^3 - x^2 + 4x + 11$; find $P(2)$ and $P(-2)$.
26. $P(x) = x^4 - 5x^3 - 4x^2 + 17x + 15$; find $P(-1)$ and $P(5)$.
27. $P(x) = 2x^4 + 5x^3 - 20x - 32$; find $P(-2)$ and $P(2)$.
28. $P(x) = 2x^4 - x^3 + 2x - 1$; find $P(\frac{1}{2})$ and $P(-\frac{1}{2})$.
29. $P(x) = 6x^4 + 2x^3 - 5x - 5$; find $P(\frac{1}{3})$ and $P(-\frac{1}{3})$.
30. $P(x) = 3x^4 + x^2 - 7x + 2$; find $P(0.1)$ and $P(-0.1)$.

THINK ABOUT IT

1. The dividend is x^2, the quotient is $x - 4$, and the remainder is 16. What is the divisor?
2. Find the remainder when $7x^{15} + 4x^8 - 5$ is divided by $x + 1$.
3. If $x^3 - x^2 + ax - 8$ is exactly divisible by $x - 4$, then find a.
4. If $x^5 - 32$ is exactly divisible by $x + a$, then find a.
5. Find the given function value by the direct substitution method and by the remainder theorem method.
   a. $P(x) = x^4 - 14x^2 + 45$; $P(-\sqrt{5})$
   b. $P(x) = x^4 - 16$; $P(2i)$
   c. $P(x) = ax^2 + bx + c$; $P(k)$
   d. $P(x) = ax^3 + bx^2 + cx + d$; $P(k)$

REMEMBER THIS

1. Is the function $f(x) = x - \sqrt{2}$ a polynomial function?
2. If the roots of $ax^2 + bx + c = 0$ are real numbers r_1 and r_2, what are the x-intercepts of the graph of $y = ax^2 + bx + c$?
3. If $P(x) = (x - b)Q(x)$, find $P(b)$.
4. What is the degree of $P(x) = 3x^4 - x^2 + 7$?
5. What is the conjugate of $-6 - 4i$?

6. Which numbers in the set $\{-\sqrt{2}, \sqrt{-2}, \sqrt[3]{-2},$ $-2 + \sqrt{3}, -2 + i\sqrt{3}\}$ are irrational numbers?
7. If $f(-5) = 0$, $f(0) = 5$, and f is a linear function, then find $f(9)$.
8. What is the slope of the line perpendicular to the graph of $3x - y = 12$?
9. If $f(x) = 4(x + 6)^2 - 5$, what is the equation of the axis of symmetry of the graph of f?
10. Find the coordinates of the vertex of the graph of $y = 3x - x^2$.

3.4 Zeros and Graphs of Polynomial Functions

A **zero** of a function f is a value of x for which $f(x) = 0$. For example, the zeros of the function

$$f(x) = x^2 - 5x + 4$$

are 1 and 4 since

$$f(1) = (1)^2 - 5(1) + 4 = 0$$

and

$$f(4) = (4)^2 - 5(4) + 4 = 0.$$

Zeros is merely a new name applied to a familiar concept. Depending on the frame of reference, consider three closely related names applied to the preceding example.

1. The **solutions** or **roots** of the **equation** $x^2 - 5x + 4 = 0$ are 1 and 4.
2. The **zeros** of the **function** $f(x) = x^2 - 5x + 4$ are 1 and 4.
3. The **x-intercepts** of the **graph** of the function $y = x^2 - 5x + 4$ are $(1,0)$ and $(4,0)$.

Finding zeros for polynomial functions of degree greater than 2 is often difficult. There is no easy formula (like the quadratic formula) that can be used to produce them, and numerical methods that require a computer are frequently used to approximate such zeros. Nevertheless, there are some theorems you should be aware of concerning zeros of polynomial functions. The first one we consider is called the factor theorem.

Factor Theorem

If b is a zero of the polynomial function $y = P(x)$, then $x - b$ is a factor of $P(x)$; and conversely, if $x - b$ is a factor of $P(x)$, then b is a zero of $y = P(x)$.

We now prove the first part of this theorem. As in the proof of the remainder theorem, let $Q(x)$ and r represent the quotient and remainder when $P(x)$ is divided by $x - b$. Then

$$P(x) = (x - b)Q(x) + r.$$

By the remainder theorem, $r = P(b)$, so we have

$$P(x) = (x - b)Q(x) + P(b).$$

If b is a zero of $y = P(x)$, then $P(b) = 0$. Thus,

$$P(x) = (x - b)Q(x) + 0$$

and $x - b$ is a factor of $P(x)$. The proof of the second part of this theorem is left as Exercise 63. Following are two examples that illustrate the usefulness of the factor theorem.

EXAMPLE 1   Write a polynomial function (in factored form) of degree 3 with zeros of 2, -4, and 3.

Solution   If 2, -4, and 3 are zeros of $y = P(x)$, then by the factor theorem, $x - 2$, $x - (-4)$, and $x - 3$ are factors of $P(x)$. Thus, a possible polynomial function is $P(x) = (x - 2)(x + 4)(x - 3)$.   ∎

EXAMPLE 2   If $P(x) = x(x + 2)(x - 5)^2$, what are the zeros of the function?

Solution   Since x, $x + 2$, and $x - 5$ are factors of $P(x)$, we conclude by the factor theorem that 0, -2, and 5 are zeros of the function. Remember that from a slightly different viewpoint we can also conclude that 0, -2, and 5 are roots or solutions of the equation $x(x + 2)(x - 5)^2 = 0$, and that the graph of $P(x) = x(x + 2)(x - 5)^2$ intercepts the x-axis at 0, -2, and 5.   ∎

Note in Example 2 that the function $P(x) = x(x + 2)(x - 5)^2$ is shorthand for $P(x) = x(x + 2)(x - 5)(x - 5)$. In this case we have four factors but only three zeros, with the zero of 5 being repeated from two different factors. It is useful to adopt a convention that indicates how many factors of $P(x)$ result in the same zero. Using this convention in this example, we say 5 is a **zero of multiplicity** 2, since there are two factors of $x - 5$. Similarly, for $P(x) = (x - 3)^4$ we say 3 is a zero of multiplicity 4. In general, the multiplicity of zero b is given by the highest power of $x - b$ that is a factor of $P(x)$.

EXAMPLE 3   If $P(x) = x^4(x - 2)(x + 7)^3$, find the zeros of the function as well as the multiplicity of each zero. What is the degree of the polynomial function?

Solution $P(x)$ contains only three different factors: x, $x - 2$, and $x + 7$. Thus, 0, 2, and -7 are the three distinct zeros of the function. However, since x appears as a factor four times, 0 is a zero of multiplicity 4.

Similarly, $(x + 7)^3$ means -7 is a zero of multiplicity 3, while $(x - 2)^1$ means 2 is a zero of multiplicity 1 (called a simple zero). The degree of the polynomial function is given by the sum of the multiplicities of the zeros. Thus, the polynomial function is of degree 8.   ■

Knowing the multiplicity of a real-number zero is useful when graphing polynomial functions. Before we see how, let us first consider the graphs of polynomial functions of the form

$$y = x^n$$

in which y equals a power of x. We have already graphed such functions for $n = 1$, 2, and 3, and the graphs of $y = x$, $y = x^2$, and $y = x^3$ are shown in Figure 3.24 for reference purposes.

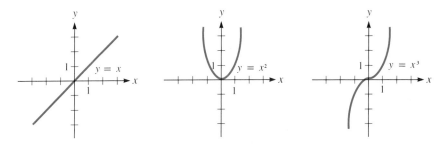

Figure 3.24

For larger values of n, the graph of $y = x^n$ is similar to the graph of $y = x^2$ if n is even, and similar to the graph of $y = x^3$ if n is odd. To see this, consider Figure 3.25. Note that when n is even the graph is symmetric with respect to the y-axis and passes through $(-1,1)$, $(0,0)$, and $(1,1)$. When n is odd the graph is symmetric with respect to the origin and passes through $(-1,-1)$, $(0,0)$, and $(1,1)$. In both cases as the exponent increases, the graph becomes flatter on the interval $[-1,1]$ and rises or falls more quickly for $|x| > 1$. To graph certain variations of $y = x^n$ we may use the graphing techniques discussed in Section 2.3 as shown in the next example.

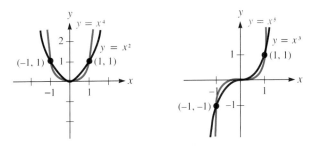

Figure 3.25

EXAMPLE 4   Use the graph of $f(x) = x^4$ to graph each function.

a. $y = -x^4$ **b.** $y = 3x^4$ **c.** $y = (x + 1)^4 - 2$

Solution

a. The graph of $y = -f(x)$ is the graph of $y = f(x)$ reflected about the x-axis. In Figure 3.26(a) the graph of $y = x^4$ has been reflected about the x-axis to obtain the graph of $y = -x^4$.

b. If $f(x) = x^4$, then $y = 3x^4 = 3f(x)$. To graph $y = 3f(x)$ we triple each y-value in f and stretch the graph of f by a factor of 3. Thus, the graph in Figure 3.26(b) is the graph of $y = 3x^4$.

c. The graphs of $y = f(x + c)$ and $y = f(x) + c$ are horizontal and vertical translations of the graph of f, respectively. To graph $y = (x + 1)^4 - 2$ we translate the graph of $y = x^4$, 1 unit to the left and 2 units down to obtain the graph in Figure 3.26(c).

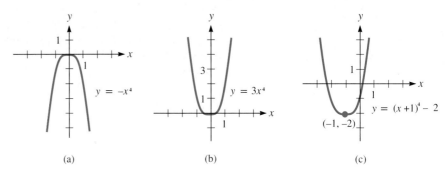

(a)   (b)   (c)

Figure 3.26

We have already indicated that the real-number zeros of $y = P(x)$ also give the x-intercepts in the graph of $y = P(x)$. In the case of $y = ax^n$, note that the only real-number zero is 0 and that when n is odd, the graph crosses the x-axis at $x = 0$, while the graph turns around and stays on the same side of the x-axis if n is even. This behavior occurs because x-values change sign as x passes through 0, so y changes sign when n is odd and y keeps the same sign when n is even. This type of analysis is applicable to all x-intercepts and indicates that the multiplicity of each real-number zero reveals whether the graph crosses the x-axis at such intercepts according to the following theorem.

Graph of $y = P(x)$ near x-intercepts

> If b is a real-number zero with multiplicity n of $y = P(x)$, then the graph of $y = P(x)$ crosses the x-axis at $x = b$ if n is odd, while the graph turns around and stays on the same side of the x-axis at $x = b$ if n is even.

To use this theorem, we need to express $P(x)$ in factored form. Then the x-intercepts (or real-number zeros) can be obtained from the factor theorem, while the behavior of the graph at an x-intercept, say $(b,0)$, can be determined from the multiplicity of zero b (or the highest power of $x - b$ that is a factor of $P(x)$). We illustrate this theorem in the next two examples. The smooth, unbroken type of curves shown in the examples are characteristic of the graphs of polynomial functions.

EXAMPLE 5   Graph $y = (x + 1)(x - 2)^2$ based on information obtained from the intercepts.

Solution   By setting $x = 0$, we determine that the y-intercept is $(0,4)$. Since $x + 1$ is a factor with an odd exponent, we conclude that $(-1,0)$ is an x-intercept at which the graph crosses the x-axis. Since $(x - 2)^2$ is a factor with an even exponent, we conclude that $(2,0)$ is an x-intercept at which the graph touches the x-axis and then turns around. We also note that as x gets very large, so does y; while as x increases in magnitude in the negative direction, y becomes very small. Figure 3.27(a) illustrates our results so far. Finally, since the function is a polynomial function, we draw a smooth, unbroken curve that satisfies the conditions in Figure 3.27(a), and we obtain the graph in Figure 3.27(b).

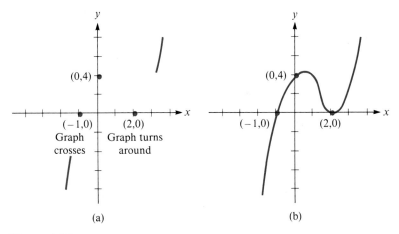

Figure 3.27   ∎

With the aid of the polynomial graphs drawn to this point, we now discuss some important properties of the graphs of polynomial functions.

Continuity   The graph of a polynomial function is a continuous curve. This property guarantees that polynomial graphs cannot have breaks as illustrated in Figure 3.28.

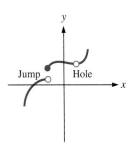

Figure 3.28

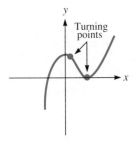

Figure 3.29

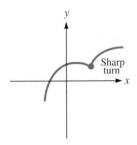

Figure 3.30

Turning Points Points at which the graph changes from rising to falling, or vice versa, are called **turning points,** and it can be shown that a polynomial function of degree n has *at most* $n - 1$ turning points. For example, Figure 3.29 shows the two turning points for the third-degree polynomial graphed in Example 5. Note that it is possible for a polynomial function of degree n to have less than $n - 1$ turning points. For instance, $y = x^3$ is a polynomial function of degree 3 with no turning points. All turns in polynomial graphs are rounded turns so a sharp turn (or corner) as illustrated in Figure 3.30 cannot occur in these graphs.

Behavior for Large $|x|$ The graph of the nth-degree polynomial function

$$y = a_n x^n + \cdots + a_1 x + a_0$$

resembles the graph of $y = a_n x^n$ when $|x|$ is large. This behavior occurs because for x-values far from the origin, the leading term $a_n x^n$ is much larger than the sum of all other terms in the polynomial. Thus, the behavior of a polynomial graph for large $|x|$ depends on whether n is even or odd, and on the sign of a_n as specified in Figure 3.31.

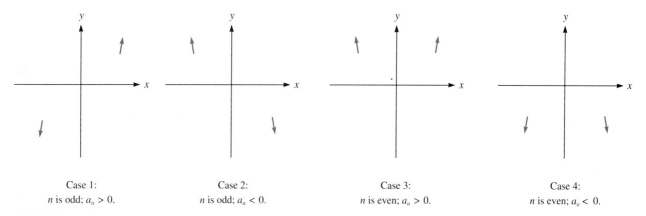

Case 1: n is odd; $a_n > 0$.

Case 2: n is odd; $a_n < 0$.

Case 3: n is even; $a_n > 0$.

Case 4: n is even; $a_n < 0$.

Figure 3.31

EXAMPLE 6 Graph $y = x^5 - 4x^3$ based on information obtained from the intercepts.

Solution By setting $x = 0$, we determine that $(0,0)$ is a y- (and x-) intercept. To find x-intercepts, we next factor the polynomial as follows:

$$y = x^5 - 4x^3 = x^3(x^2 - 4) = x^3(x + 2)(x - 2).$$

Since x^3, $x + 2$, and $x - 2$ are factors, we conclude the x-intercepts are $(0,0)$, $(-2,0)$, and $(2,0)$. All of these intercepts are derived from factors with odd exponents, so the graph crosses the x-axis at each of these points.

By using this information and observing that the graph behaves like $y = x^5$ (or case 1 in Figure 3.31) for large values of $|x|$, we draw the graph in Figure 3.32(a). Keep in mind that our current methods produce only a rough sketch of the graph. A more detailed graph of $y = x^5 - 4x^3$ that is obtained by methods from calculus is given in Figure 3.32(b). Note that this graph indicates turning points when $x = \pm 2\sqrt{15}/5$ and an unanticipated wavy pattern between the turning points. Determining such subtleties is best left to a calculus course.

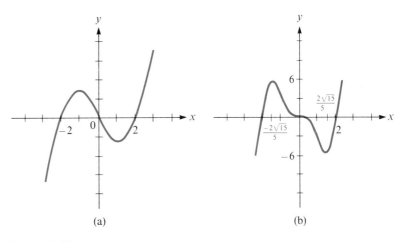

(a)   (b)

Figure 3.32   ■

We now discuss some theorems that are fundamental to the theory of finding zeros of polynomial functions. You need a good grasp of number systems to really understand these statements, and it might be helpful if you review the relationships among the various sets of numbers, which are given in Sections 1.1 and 1.5.

Fundamental Theorem of Algebra

Every polynomial function of degree $n \geq 1$ with complex number coefficients has at least one complex zero.

Although the proof of this theorem is beyond the scope of this book, we can prove an important corollary of this theorem that answers a basic question: How many zeros are there for a polynomial function of degree n?

Number of Zeros Theorem

Every polynomial function of degree $n \geq 1$ has exactly n complex zeros, where zeros of multiplicity k are counted k times.

For instance, in Example 3 we saw that $P(x) = x^4(x - 2)(x + 7)^3$ is a polynomial function of degree 8 with distinct zeros of 0, 2, and -7.

However, the multiplicities of these zeros are 4, 1, and 3, respectively, so that if we take into account the idea of multiplicity, then this eighth-degree polynomial function has exactly eight zeros. More generally, we can prove the above theorem by first noting that if $y = P(x)$ is a polynomial function of degree $n \geq 1$ with leading coefficient a_n, then the fundamental theorem of algebra guarantees that $y = P(x)$ has at least one complex zero, say c_1. By the factor theorem, since c_1 is a zero, $x - c_1$ is a factor of $P(x)$, so

$$P(x) = (x - c_1)Q_1(x),$$

where polynomial $Q_1(x)$ has degree $n - 1$ and leading coefficient a_n. If the degree of $Q_1(x)$ is at least 1, then once again the fundamental theorem of algebra guarantees that $y = Q_1(x)$ has at least one zero, say c_2, so as above

$$Q_1(x) = (x - c_2)Q_2(x),$$

where polynomial $Q_2(x)$ has degree $n - 2$ and leading coefficient a_n. Then combining results, we have

$$P(x) = (x - c_1)(x - c_2)Q_2(x).$$

This process is continued for a total of n times until $Q_n(x) = a_n$. Thus, $P(x)$ can be factored into n linear factors and written as

$$P(x) = a_n(x - c_1)(x - c_2) \cdots (x - c_n)$$

so that, by the factor theorem, $y = P(x)$ has n zeros: $c_1, c_2, \ldots, c_n$. Furthermore, no other number, say c, distinct from $c_1, c_2, \ldots, c_n$ can be a zero since

$$P(c) = a_n(c - c_1)(c - c_2) \cdots (c - c_n) \neq 0$$

because none of the factors is zero. Thus, every polynomial function of degree $n \geq 1$ has exactly n (not necessarily distinct) complex zeros.

EXAMPLE 7   If 2 is a zero of $P(x) = x^3 - 2x^2 + 3x - 6$, find the other zeros.

Solution   Since $y = P(x)$ is a polynomial function of degree 3, there are exactly three (not necessarily distinct) zeros. We are given that 2 is one of these zeros, and if 2 is a zero of $y = P(x)$, then by the factor theorem, $x - 2$ is a factor of the polynomial. We can then use synthetic division to find another factor of $P(x)$ as follows:

$$
\begin{array}{r|rrrr}
2\rfloor & 1 & -2 & 3 & -6 \\
 & & 2 & 0 & 6 \\
\hline
 & 1 & 0 & 3 & 0.
\end{array}
$$

The coefficients of another factor of $P(x)$ are given in the bottom row of this synthetic division. Thus,

$$P(x) = (x - 2)(x^2 + 3).$$

Setting the factor $x^2 + 3$ equal to 0 gives us the other two zeros of $i\sqrt{3}$ and $-i\sqrt{3}$.   ∎

Note in Example 7 that the zeros $i\sqrt{3}$ and $-i\sqrt{3}$ are conjugates of each other. We can anticipate that the two remaining zeros are conjugate pairs from our work with quadratic equations in Section 1.6, and in general, the following theorems show that when certain restrictions are placed on the coefficients of a polynomial, then zeros of polynomial functions always occur in conjugate pairs.

Conjugate-Pair Theorems

> 1. If a complex number $a + bi$ is a zero of a polynomial function of degree $n \geq 1$ with *real*-number coefficients, then its conjugate $a - bi$ is also a zero.
> 2. Let a, b, and c be rational numbers. If an irrational number of the form $a + b\sqrt{c}$ is a zero of a polynomial function of degree $n \geq 1$ with *rational*-number coefficients, then $a - b\sqrt{c}$ is also a zero.

The restrictions on the coefficients of the polynomial are crucial conditions in these theorems. For example, $P(x) = x - \sqrt{2}$ has only one zero, namely $\sqrt{2}$. Note that the conjugate $-\sqrt{2}$ is not also a zero, and that the second conjugate-pair theorem does not apply to $P(x) = x - \sqrt{2}$ because the polynomial does not have rational-number coefficients.

We can prove the first conjugate-pair theorem by using the properties of complex conjugates discussed in Section 1.5 (which you may need to review). We start with a polynomial function of degree $n \geq 1$ with real-number coefficients that we may write as

$$P(x) = a_n x^n + a_{n-1} x^{n-1} + \cdots + a_1 x + a_0.$$

Now if $z = a + bi$ is a zero of $y = P(x)$, then $P(z) = 0$, so

$$0 = a_n z^n + a_{n-1} z^{n-1} + \cdots + a_1 z + a_0.$$

We then take the conjugate of both sides of the equation (noting that the conjugate of 0 is 0) and proceed as follows:

$$0 = \overline{a_n z^n + a_{n-1} z^{n-1} + \cdots + a_1 z + a_0}$$
$$= \overline{a_n z^n} + \overline{a_{n-1} z^{n-1}} + \cdots + \overline{a_1 z} + \overline{a_0} \qquad \text{(Conjugate of a sum = Sum of conjugates)}$$
$$= \overline{a_n}\,\overline{z^n} + \overline{a_{n-1}}\,\overline{z^{n-1}} + \cdots + \overline{a_1}\,\overline{z} + \overline{a_0} \qquad \text{(Conjugate of a product = Product of conjugates)}$$
$$= a_n \overline{z^n} + a_{n-1} \overline{z^{n-1}} + \cdots + a_1 \overline{z} + a_0 \qquad \text{(Conjugate of a real number = The real number)}$$
$$= a_n (\overline{z})^n + a_{n-1} (\overline{z})^{n-1} + \cdots + a_1 \overline{z} + a_0. \qquad \text{(Conjugate of a power = Power of the conjugate)}$$

This last equation tells us that $P(\overline{z}) = 0$, so $\overline{z}$ is also a zero of $y = P(x)$, which proves the theorem.

EXAMPLE 8 If $-2i$ and $2 + \sqrt{7}$ are zeros of the polynomial function $P(x) = x^4 - 4x^3 + x^2 - 16x - 12$, find the other zeros.

Solution $P(x)$ has rational-number coefficients. Thus, both conjugate-pair theorems apply, and if $-2i$ and $2 + \sqrt{7}$ are zeros, then $2i$ and $2 - \sqrt{7}$ are also zeros. There are no other zeros since the polynomial has degree 4.   ■

EXAMPLE 9   Write a polynomial function (in factored form) of the lowest possible degree that has rational coefficients and zeros of 0, $1 - i$, and $\sqrt{2}$.

Solution   Since the polynomial has rational coefficients, we utilize both conjugate-pair theorems. Thus, in addition to the three zeros given, $1 + i$ and $-\sqrt{2}$ are also zeros. The lowest possible degree for the polynomial is then degree 5, and in factored form one possibility is

$$P(x) = x(x - \sqrt{2})(x + \sqrt{2})(x - (1 + i))(x - (1 - i)).$$ ■

In conclusion, let us discuss briefly why the fundamental theorem of algebra is so central to the theory of polynomial equations. During the historical growth of equation solving, the solution to certain types of equations continued to force mathematicians to develop new types of numbers. For example, if an equation uses just positive integers (like $x + 5 = 1$), it might be necessary to leave the positive integers for a solution. In the case of $x + 5 = 1$, an extension to -4 and the negative integers is needed. Integers were not enough for equations like $2x + 5 = 0$, so rational numbers were required. Similarly, $x^2 - 5 = 0$ brings an extension to irrational (or real) numbers, while $x^2 + 5 = 0$ can be solved only by introducing imaginary (or complex) numbers. However, the process stops here. The fundamental theorem of algebra guarantees that every polynomial equation involving complex numbers can be solved using only complex numbers. No extension of the number system is necessary. The mathematician Carl Friedrich Gauss was the first to prove this theorem. He did so in his doctoral thesis in 1799 when he was 20.

EXERCISES 3.4

In Exercises 1–4 find the zeros of each polynomial function. In each case state the degree of the polynomial function and the multiplicity of each zero.

1. $P(x) = (x + 3)^2$ 2. $P(x) = x(x - 2)^3$
3. $P(x) = (x + 4)(x + \sqrt{2})(x - \sqrt{2})$
4. $P(x) = 4x^3(x - 1)^5(x + 6)$

In Exercises 5–10 write a polynomial function (in factored form) with the given degree and zeros.

5. degree 3; zeros are 1, 2, and 3
6. degree 3; zeros are $\sqrt{2}$, $-\sqrt{2}$, and 0
7. degree 4; zeros are -4, 0, i, and $-i$
8. degree 5; zeros are 5, $2 \pm i$, and $1 \pm \sqrt{3}$
9. degree 4; 1 and 5 are both zeros of multiplicity 2
10. degree 6; -2 is a zero of multiplicity 1, 0 is a zero of multiplicity 2, and 5 is a zero of multiplicity 3

In Exercises 11–14 use the graph of $y = x^3$ to graph the given functions.

11. $y = x^3 - 1$ 12. $y = 1 - x^3$
13. $y = (x - 2)^3$ 14. $y = (x + 2)^3 - 5$

In Exercises 15–18 use the graph of $y = x^4$ to graph the given functions.

15. $y = (x + 2)^4$ 16. $y = x^4 - 4$
17. $y = 2 - (x - 1)^4$ 18. $y = -\frac{1}{2}x^4$

In Exercises 19–22 use the graph of $y = x^5$ to graph the given functions.

19. $y = -x^5$ 20. $y = 2x^5$
21. $y = (x + 1)^5 + 4$ 22. $y = 1 - (x - 1)^5$

In Exercises 23–26 use the graph of $y = x^6$ to graph the given functions.

23. $y = \frac{1}{4}x^6$ **24.** $y = -x^6$

25. $y = 2 - x^6$ **26.** $y = (x + 1)^6 + 3$

In Exercises 27–36 graph the functions based on information obtained from the intercepts.

27. $y = (x - 2)(x + 1)$ **28.** $y = (x - 2)(x + 1)^2$

29. $y = (x - 2)^2(x + 1)$ **30.** $y = (x - 2)^2(x + 1)^2$

31. $y = (x^2 - 1)(x^2 - 4)$ **32.** $y = x^4 - 3x^2 + 2$

33. $y = x^3 - 2x^2 - 3x$ **34.** $y = x^3 - 4x$

35. $y = x^2 - x^3$ **36.** $y = x^3 + x^2$

In Exercises 37–42 write a polynomial function (in factored form) with rational coefficients of the lowest possible degree with the given zeros.

37. 2 and $\sqrt{3}$ **38.** i and 0

39. 0, $2 + 3i$, and $4 - \sqrt{3}$ **40.** $-5, -\sqrt{5}$, and $5i$

41. 3, -3, and $2i$ **42.** $3i$ and $1 + \sqrt{2}$

In Exercises 43–56 one or more zeros are given for each of the following polynomial functions. Find the other zeros.

43. $P(x) = x^2 - 4x - 3; 2 + \sqrt{7}$

44. $P(x) = 3x^2 + 4x - 2; \dfrac{-2 - \sqrt{10}}{3}$

45. $P(x) = x^2 + 2; i\sqrt{2}$

46. $P(x) = 3x^2 - 2x + 1; \dfrac{1}{3} - \dfrac{i\sqrt{2}}{3}$

47. $P(x) = 2x^3 - 11x^2 + 28x - 24; \frac{3}{2}$ and $2 + 2i$

48. $P(x) = x^4 + 13x^2 - 48; \sqrt{3}$ and $-4i$

49. $P(x) = 2x^3 + 5x^2 + 4x + 1; -1$

50. $P(x) = 3x^3 - 2x^2 - 10x + 4; 2$

51. $P(x) = x^4 - 6x^3 + 7x^2 + 12x - 18$; 3 is a zero of multiplicity 2

52. $P(x) = 2x^4 - 5x^3 + 11x^2 - 3x - 5; 1$ and $-\frac{1}{2}$

53. $P(x) = x^4 - 4; \sqrt{2}$

54. $P(x) = x^4 - 14x^2 + 45; -\sqrt{5}$

55. $P(x) = x^4 - 1; i$

56. $P(x) = x^4 - 16; 2i$

57. If -1 is a zero of $y = x^3 - x^2 - 10x - 8$, find all intercepts and then graph the function.

58. Graph the function in Exercise 51 based on information obtained from the intercepts.

In Exercises 59–62 $y = P(x)$ is a polynomial function with real coefficients. Answer true or false.

59. Every polynomial function of degree 3 has at least one real zero.

60. Every polynomial function of degree 4 has at least one real zero.

61. Every polynomial function of degree $n \geq 1$ has exactly n (not necessarily different) real zeros.

62. Every polynomial function of degree $n \geq 1$ has exactly n (not necessarily different) complex zeros.

63. Show that if $x - b$ is a factor of the polynomial $P(x)$, then b is a zero of $y = P(x)$.

THINK ABOUT IT

1. For what value of k is $x - 2$ a factor of $2x^4 - 5x^3 + kx^2 - 5x + 2$?

2. Write a polynomial function (in factored form) with *real* coefficients of lowest possible degree with zeros of 0, $-\sqrt{3}$, and $3i$.

3. If $f(x) = ax^3 + bx^2 + cx + d$, where a, b, c, and d are real numbers with $a \neq 0$, then explain why the graph of f must have at least one x-intercept.

4. Graph $y = x^4 - 4x^2 + 3$ and then solve the following inequalities by analyzing this graph.
   a. $x^4 - 4x^2 + 3 \geq 0$ **b.** $x^4 + 3 < 4x^2$

5. Create a third-degree polynomial inequality with the given solution set.
   a. $(-\infty, -2] \cup [1,5]$ **b.** $(-2,3) \cup (3,\infty)$

REMEMBER THIS

1. Which numbers in the set $\{\frac{2}{3}, -\frac{1}{2}, \sqrt{2}/2, -5, 0\}$ are rational numbers?

2. Simplify $(p/q)^{n-1} \cdot q^n$.

3. **a.** Simplify $5(-x)^n$ if n is a positive even integer.
   b. Simplify $5(-x)^n$ if n is a positive odd integer.

4. If $P(x) = 5x^3 + 28x^2 + 8x - 8$, find $P(\frac{2}{5})$ by using the remainder theorem.

5. Solve $5x^2 + 30x + 20 = 0$.

6. Graph the line whose slope is $\frac{3}{2}$ and whose y-intercept is $(0, -3)$. Also, find an equation for the line.

7. Solve $x^2 > 6x - 6$.

8. Find the range of the function defined by $y = x^2 - 6x + 9$.

9. Find two positive numbers whose sum is 16 and whose product is a maximum. Show your work.

10. What is the remainder when $2x^4 - 3x^3 + 16x - 25$ is divided by $x + 4$?

3.5 Additional Theorems About Zeros

The following theorem is used to obtain a list of *possible* rational zeros of a polynomial function with integer coefficients. We emphasize the word *possible*. The zeros may all be either irrational or complex. This theorem states only that if there are rational zeros, they satisfy the requirement indicated.

Rational-Zero Theorem

> If p/q, a rational number in lowest terms, is a zero of the polynomial function with integer coefficients,
>
> $$P(x) = a_n x^n + a_{n-1} x^{n-1} + \cdots + a_1 x + a_0 \qquad (a_n \neq 0),$$
>
> then p is an integral factor of the constant term a_0 and q is an integral factor of the leading coefficient a_n.

To prove this theorem, we begin by noting that if p/q is a zero of $y = P(x)$, then

$$a_n(p/q)^n + a_{n-1}(p/q)^{n-1} + \cdots + a_1(p/q) + a_0 = 0.$$

If we multiply both sides of this equation by q^n, we have

$$a_n p^n + a_{n-1} p^{n-1} q + \cdots + a_1 p q^{n-1} + a_0 q^n = 0.$$

Next, we subtract $a_0 q^n$ from both sides of the equation yielding

$$a_n p^n + a_{n-1} p^{n-1} q + \cdots + a_1 p q^{n-1} = -a_0 q^n,$$

and finally, we factor the common factor p from each term on the left to obtain

$$p(a_n p^{n-1} + a_{n-1} p^{n-2} q + \cdots + a_1 q^{n-1}) = -a_0 q^n.$$

Since p is a factor of the left-hand side of the equation, p must also be a factor of $-a_0 q^n$. But by hypothesis, p/q is in lowest terms, so p and q have no common factor, and p is not a factor of q^n. Thus, p is a factor of a_0.

To prove that q is a factor of a_n, we rewrite our second equation as

$$q(a_{n-1} p^{n-1} + a_{n-2} p^{n-2} q + \cdots + a_0 q^{n-1}) = -a_n p^n$$

and reason as above. In practice, the rational-zero theorem is not difficult to apply.

EXAMPLE 1 List the possible rational zeros of the function

$$P(x) = 2x^3 - x^2 - 6x - 3.$$

Solution The constant term a_0 is -3. The possibilities for p are the integers that are factors of 3. Thus,

$$p = \pm 3, \pm 1.$$

The leading coefficient a_n is 2. The possibilities for q are the integers that are factors of 2. Thus,

$$q = \pm 2, \pm 1.$$

The possible rational zeros p/q are then

$$3, \tfrac{3}{2}, 1, \tfrac{1}{2}, -\tfrac{1}{2}, -1, -\tfrac{3}{2}, -3.$$ ∎

Since we are often faced with a large list of possible rational zeros, we need some help in narrowing down the possibilities. Consider the theorem that follows, which helps us chip away at the problem. Note that it is a statement about real zeros. Thus, it applies to zeros that are rational or irrational.

Descartes' Rule of Signs

> The maximum number of positive real zeros of the polynomial function $y = P(x)$ is the number of changes in sign of the coefficients in $P(x)$. The number of changes in sign of the coefficients in $P(-x)$ is the maximum number of negative real zeros. In both cases, if the number of zeros is not the maximum number, then it is less than this number by a multiple of 2.

We illustrate this theorem in the following example.

EXAMPLE 2   Find the zeros of $P(x) = 2x^3 + 5x^2 + 4x + 1$.

Solution

a. Use Descartes' rule of signs to determine the maximum number of positive and negative real zeros.

$$P(x) = 2x^3 + 5x^2 + 4x + 1$$

All the coefficients in $P(x)$ are positive. There are no positive real zeros since there are no changes in sign.

$$P(-x) = 2(-x)^3 + 5(-x)^2 + 4(-x) + 1$$
$$= -2x^3 + 5x^2 - 4x + 1$$

There are three sign changes in $P(-x)$. At most, there may be three negative real zeros. By decreasing this number by 2, we determine that one negative real zero is the only other possibility.

b. Determine the possible rational zeros. The constant term a_0 is 1. The possibilities for p are the integers that are factors of 1. Thus,

$$p = \pm 1.$$

The leading coefficient a_n is 2. The integers that are factors of 2 give the possibilities for q. Thus,

$$q = \pm 2, \pm 1.$$

The possible rational zeros p/q are then

$$1, \tfrac{1}{2}, -\tfrac{1}{2}, -1.$$

From step a we eliminate the positive possibilities, leaving

$$-\tfrac{1}{2}, -1.$$

c. Use synthetic division to test whether one of the possibilities, say -1, is a zero.

$$
\begin{array}{r|rrrr}
-1 & 2 & 5 & 4 & 1 \\
   &   & -2 & -3 & -1 \\
\hline
   & 2 & 3 & 1 & 0
\end{array}
$$

Since the remainder is zero, $P(-1) = 0$. Thus, -1 is a zero of the function.

d. The factor theorem states that if -1 is a zero, $x - (-1)$ or $x + 1$ is a factor. The coefficients of the other factor are given in the bottom row of the synthetic division. That is,

$$2x^3 + 5x^2 + 4x + 1 = (x + 1)(2x^2 + 3x + 1).$$

We now find other zeros by setting the factor $2x^2 + 3x + 1$ equal to 0. The big advantage is that after finding a zero we may lower by 1 the degree of the equation we are solving. When we reach a second-degree equation, as in this case, we may apply the quadratic formula or (sometimes) the factoring method.

$$2x^2 + 3x + 1 = 0$$
$$(2x + 1)(x + 1) = 0$$
$$2x + 1 = 0 \qquad x + 1 = 0$$
$$x = -\tfrac{1}{2} \qquad x = -1$$

Thus, the zeros are -1, -1, and $-\tfrac{1}{2}$. Remember that it is possible for a number to be counted as a zero more than once, and in this case we say -1 is a zero of multiplicity 2. Based on the x- and y-intercepts and the multiplicity of each zero, the function is graphed in Figure 3.33. ■

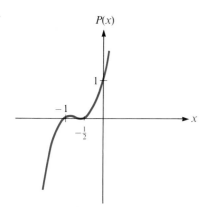

$P(x)$

Figure 3.33

For the next example we add another useful rule, which permits us to establish upper and lower bounds on the zeros of a polynomial function.

Upper and Lower Bounds Theorem

1. If we divide a polynomial $P(x)$ synthetically by $x - b$, where $b > 0$ and all the numbers in the bottom row have the same sign, then there is no zero greater than b. We say b is an **upper bound** for the zeros of $y = P(x)$.
2. If we divide a polynomial $P(x)$ synthetically by $x - c$, where $c < 0$ and the numbers in the bottom row alternate in sign, then there is no zero less than c. We say c is a **lower bound** for the zeros of $y = P(x)$.

For the purposes of these tests, zero may be denoted as $+0$ or -0.

A rationale for these tests is contained in the next example.

EXAMPLE 3   Find the zeros of $P(x) = 2x^4 - 5x^3 + 11x^2 - 3x - 5$.

Solution

a. Use Descartes' rule of signs to determine the maximum number of positive and negative real zeros.

$$P(x) = 2x^4 - 5x^3 + 11x^2 - 3x - 5$$

There are three sign changes in $P(x)$. Thus, the number of positive real zeros is three or one.

$$P(-x) = 2(-x)^4 - 5(-x)^3 + 11(-x)^2 - 3(-x) - 5$$
$$= 2x^4 + 5x^3 + 11x^2 + 3x - 5$$

There is one sign change in $P(-x)$, so Descartes' rule of signs guarantees exactly one negative *real* zero in this case.

b. Determine the possible rational zeros. The constant term a_0 is -5. The possibilities for p are the integers that are factors of -5. Thus,

$$p = \pm 5, \pm 1.$$

The leading coefficient a_n is 2. The integral factors of 2 give the possibilities for q. Thus,

$$q = \pm 2, \pm 1.$$

The possible rational zeros are then

$$5, \tfrac{5}{2}, 1, \tfrac{1}{2}, -\tfrac{1}{2}, -1, -\tfrac{5}{2}, -5.$$

c. We test negative zeros first, because step a indicated that there is, at most, one negative zero. If we find a negative zero, we then switch to the positive possibilities. Pick -1, a number in the middle of the negative choices. If it is not a zero, it may be a lower bound that eliminates other possibilities.

$$
\begin{array}{r|rrrrr}
-1 & 2 & -5 & 11 & -3 & -5 \\
   &   & -2 & 7 & -18 & 21 \\
\hline
   & 2 & -7 & 18 & -21 & 16 \\
\end{array}
$$

The remainder is 16, so -1 is not a zero. We tested a negative possibility and the numbers in the bottom row alternate in sign. Thus, there is no zero less than -1 as specified in statement 2 of the above theorem. To see why, note that if we test a negative choice, say c, greater in absolute value than -1, then the numbers in the bottom row of the synthetic division will continue to alternate in sign with respective numbers of greater absolute value (after the first entry) than our previous bottom row $(2, -7, 18, -21, 16)$. Thus $P(c) > 16$ when

$c < -1$, so -1 is a lower bound for the zeros of $y = P(x)$. This eliminates $-\frac{5}{2}$ and -5, so try $-\frac{1}{2}$.

$$
\begin{array}{r|rrrrr}
-\frac{1}{2} & 2 & -5 & 11 & -3 & -5 \\
 & & -1 & 3 & -7 & 5 \\
\hline
 & 2 & -6 & 14 & -10 & 0
\end{array}
$$

Since the remainder is zero, $P(-\frac{1}{2}) = 0$. Thus, $-\frac{1}{2}$ is a zero of the function.

d. Since $-\frac{1}{2}$ is a zero, $x - (-\frac{1}{2})$ is a factor and we write

$$
\begin{aligned}
2x^4 &- 5x^3 + 11x^2 - 3x - 5 \\
&= (x + \tfrac{1}{2})(2x^3 - 6x^2 + 14x - 10).
\end{aligned}
$$

We now try to find the values of x for which $2x^3 - 6x^2 + 14x - 10$ equals 0. We found the one negative zero, so we switch to the positive possibilities. Pick 1. If it is not a zero, the numbers in the bottom row in the synthetic division may have the same sign. Such a result would establish 1 as an upper bound for the zeros (thereby eliminating $\frac{5}{2}$ and 5), since any synthetic division by $x - b$, where $b > 1$, would continue to produce bottom-row entries that have the same sign and even larger absolute values (after the first entry) than those that result from the division involving $b = 1$.

$$
\begin{array}{r|rrrr}
1 & 2 & -6 & 14 & -10 \\
 & & 2 & -4 & 10 \\
\hline
 & 2 & -4 & 10 & 0
\end{array}
$$

Therefore, 1 is a zero of the function.

e. We now use the quadratic formula to complete the solution.

$$
2x^2 - 4x + 10 = 0
$$
$$
x^2 - 2x + 5 = 0
$$
$$
x = \frac{-(-2) \pm \sqrt{(-2)^2 - 4(1)(5)}}{2(1)} = \frac{2 \pm \sqrt{-16}}{2}
$$
$$
= \frac{2 \pm 4i}{2} = 1 \pm 2i
$$

The zeros are $-\frac{1}{2}$, 1, $1 + 2i$, and $1 - 2i$. ∎

EXAMPLE 4   The volume of a rectangular packaging carton is to be 30 ft³. If the length and width are respectively 3 ft and 1 ft greater than the height, show that only one set of dimensions is possible for the given carton. What are these dimensions? (See Figure 3.34.)

Solution   If x represents the height of the carton, then $x + 3$ and $x + 1$ represent the length and width, respectively. Since the formula for the volume is $V = \ell wh$ and the volume of the carton is 30 ft³, we have

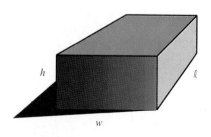

Figure 3.34

$$(x + 3)(x + 1)x = 30$$
$$x^3 + 4x^2 + 3x = 30$$
$$x^3 + 4x^2 + 3x - 30 = 0.$$

The possible positive rational zeros are then

$$30,15,10,6,5,3,2,1.$$

With the aid of our theorems and some trial and error, we determine that 2 is a solution as shown in the following synthetic division.

$$
\begin{array}{r|rrrr}
2 & 1 & 4 & 3 & -30 \\
 &   & 2 & 12 & 30 \\
\hline
 & 1 & 6 & 15 & 0 \\
\end{array}
$$

The two other roots are then determined by solving $x^2 + 6x + 15 = 0$. However, since the discriminant ($b^2 - 4ac$) is less than 0, these roots are not real numbers. Thus, the only real-number solution is 2, and the box has unique dimensions of 5 ft by 3 ft by 2 ft. ∎

Up to this point we have limited our discussion to finding rational zeros of polynomial functions of degree greater than 2. Unfortunately, irrational zeros of higher degree polynomial functions are usually very difficult to find. Complicated formulas are available for polynomial functions of degrees 3 and 4, and it can be shown that no formula exists for degree 5 and greater. In such cases, numerical methods that utilize a computer are employed to approximate irrational zeros, and this topic is usually considered in detail in a course in numerical analysis. For our purposes, we need only illustrate the following two theorems that enable us to approximate zeros of polynomial functions.

Intermediate Value Theorem for Polynomials

Let $y = P(x)$ be a polynomial function. If a and b are real numbers with $a < b$ and if d is any number between $P(a)$ and $P(b)$, inclusive, then there is at least one number c in the interval $[a,b]$ such that $P(c) = d$.

This result is easy to see geometrically. Consider Figure 3.35 and keep in mind that the graph of a polynomial function is an unbroken curve. Through any number d between $P(a)$ and $P(b)$, inclusive, a horizontal line may be drawn that must intersect the graph of $y = P(x)$ at least once in the interval $[a,b]$. Then, the x-coordinate of an intersection point is a number c in the interval $[a,b]$ such that $P(c) = d$.

The intermediate value theorem is used in many ways when analyzing polynomial functions, and a special case of this theorem, called the **location theorem,** may be used to approximate zeros of such functions. When $P(a)$ and $P(b)$ are opposite in sign, then 0 is between $P(a)$ and $P(b)$ so the

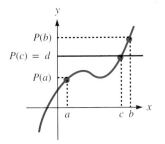

Figure 3.35

intermediate value theorem guarantees there is at least one number c in the interval (a,b) such that $P(c) = 0$. Thus, $y = P(x)$ has at least one real zero between a and b, which leads to the following result.

Location Theorem

> Let $y = P(x)$ be a polynomial function. If $P(a)$ and $P(b)$ have opposite signs, then $y = P(x)$ has at least one real zero between a and b.

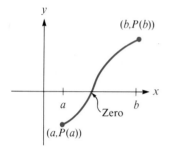

Figure 3.36

The geometric interpretation of this theorem is shown in Figure 3.36. Since the graph is on different sides of the x-axis at a and b and no breaks are possible, the graph must cross the x-axis between a and b, guaranteeing at least one real zero in this interval.

EXAMPLE 5   Verify that $P(x) = 3x^3 + 16x^2 - 8$ has a zero between -5 and -6.

Solution   By direct substitution or synthetic division, we determine that $P(-5) = 17$ and $P(-6) = -80$. Since $P(-5)$ and $P(-6)$ have opposite signs, the location theorem assures us of at least one real zero between -5 and -6.   ∎

Once we have determined an interval that contains a zero, the job is to keep narrowing the interval until we obtain a sufficient degree of accuracy. The most efficient methods for narrowing the interval are developed in calculus. A simple and relatively effective alternative is to basically halve the interval each time until we attain a specified accuracy.

EXAMPLE 6   To the nearest tenth, approximate the zero of the function $P(x) = 3x^3 + 16x^2 - 8$ that is between -5 and -6.

Solution   By using the location theorem and a calculator and by successively halving the intervals, we determine the following.

Calculation	Comment: A Zero Is Between
$P(-5) = 17$	
$P(-6) = -80$	-5 and -6
$P(-5.5) \approx -23$	-5 and -5.5
$P(-5.25) \approx -1.1$	-5 and -5.25
$P(-5.13) \approx 8.1$	-5.13 and -5.25
$P(-5.19) \approx 3.6$	-5.19 and -5.25

Thus, to the nearest tenth, a zero is -5.2. (*Note:* Exercise 14 in this section asks you to find the exact value of this zero by using the previous methods of this section.)   ∎

EXERCISES 3.5

In Exercises 1–4 list the possible rational zeros of the function.

1. $P(x) = x^3 + 8x^2 - 10x - 20$
2. $P(x) = x^4 - 3x^3 + x^2 - x - 6$
3. $P(x) = 4x^4 - 3x^3 + x^2 - x - 6$
4. $P(x) = 3x^3 + 7x^2 + 8$

In Exercises 5–8 use Descartes' rule of signs to determine the maximum number of positive and negative real zeros.

5. $P(x) = x^3 - 2x^2 + x - 3$
6. $P(x) = 5x^3 + x^2 + 4x + 1$
7. $P(x) = 2x^4 + 5x^3 - x^2 - 3x - 7$
8. $P(x) = x^8 + 1$

In Exercises 9–18 find the zeros of the function.

9. $P(x) = x^3 - x^2 - 10x - 8$
10. $P(x) = x^3 + 6x^2 + 11x + 6$
11. $P(x) = 3x^3 + x^2 + 15x + 5$
12. $P(x) = 4x^3 - x^2 - 28x + 7$
13. $P(x) = 8x^3 - 12x^2 + 6x - 1$
14. $P(x) = 3x^3 + 16x^2 - 8$
15. $P(x) = 4x^4 - 5x^3 - 2x^2 - 3x - 10$
16. $P(x) = x^4 - 6x^3 + 7x^2 + 12x - 18$
17. $P(x) = 2x^4 + 7x^3 + 25x^2 + 47x + 18$
18. $P(x) = 6x^4 + 11x^3 - 63x^2 - 7x + 5$

In Exercises 19–22 answer the following questions with respect to the given functions.

a. Find all zeros of the function.
b. How many zeros are rational numbers? Name them.
c. How many zeros are real numbers? Name them.
d. How many zeros are complex numbers? Name them.

19. $P(x) = x^3 - x^2 - 2x - 12$
20. $P(x) = x^3 + 7x^2 + 12x + 6$
21. $P(x) = 3x^4 + 5x^3 - 23x^2 - 35x + 14$
22. $P(x) = 10x^4 + 35x^3 - 73x^2 + 7x - 15$

In Exercises 23–26 use the location theorem and verify each statement.

23. $P(x) = 3x^3 + 16x^2 - 8$ has a zero between 0 and -1.

24. $P(x) = 4x^3 - x^2 - 28x + 7$ has a zero between 2 and 3.
25. $P(x) = x^4 - 6x^3 + 7x^2 + 12x - 18$ has a zero between 1 and 2 and another zero between -1 and -2.
26. $P(x) = 6x^4 + 11x^3 - 63x^2 - 7x + 5$ has a zero between 0 and 1 and another zero between -4 and -5.

In Exercises 27–32 approximate the zero of the function in the given interval to the nearest tenth.

27. $P(x) = x^3 + x - 1$; a zero is between 0.6 and 0.7.
28. $P(x) = x^3 + x^2 - 10x - 10$; a zero is between -3.1 and -3.2.
29. $P(x) = x^3 - 3x + 1$; a zero is between 0 and 1.
30. $P(x) = x^3 - 3x + 1$; a zero is between 1 and 2.
31. $P(x) = 2x^3 - 5x^2 - 3x + 9$; a zero is between -1 and -2.
32. $P(x) = 2x^3 - 5x^2 - 3x + 9$; a zero is between 2 and 3.
33. The volume of a rectangular box is 105 in.^{3}. If the width is 2 in. greater than the height and the length is 1 in. greater than twice the height, show that only one set of dimensions is possible for the given box. What are these dimensions?
34. The volume of a rectangular box is 20 ft^3. If the box has a square base and the height is 3 ft greater than the measure of the edges of the base, show that only one set of dimensions is possible for the given box. What are these dimensions?
35. A rectangular piece of cardboard 12 in. by 18 in. is to be used to make an open box by cutting a small square from each corner and bending up the sides. If the volume of the box is to be 216 in.^{3}, then find the length of the side of the square that is to be cut from each corner.
36. A rectangular sheet of tin 20 cm by 24 cm has identical squares cut out from each corner. The sides are then turned up to form an open box. Find the side length for each square cut out, given that the volume is to be 640 cm^3.

THINK ABOUT IT

1. Consider the following synthetic division:

$$\begin{array}{r|rrrr} 2\rfloor & 1 & -3 & 3 & -5 \\ & & 2 & -2 & 2 \\ \hline & 1 & -1 & 1 & -3. \end{array}$$

On the basis of this division and the upper bound theorem, can we conclude that 2 is an upper bound for the zeros of $P(x) = x^3 - 3x^2 + 3x - 5$? Justify your answer.

2. If $P(x) = x^3 + px + q$, determine the nature of the zeros when (a) p and q are both positive and (b) p is positive and q is negative.

3. Consider the function defined by $f(x) = x^n + a$.
   a. If n is a positive even integer and a is a negative real number, then find in terms of n the number of zeros of f that are not real numbers.
   b. Find in terms of n the number of zeros of f that are not real numbers, if n is a positive odd integer and a is a positive real number.

4. Use the rational-zero theorem and the function $P(x) = x^2 - 2$ to show that $\sqrt{2}$ is not a rational number.

5. Show that $\sqrt[5]{7}$ is not a rational number by using the rational-zero theorem.

REMEMBER THIS

1. Is the function $f(x) = 1/x^2$ a polynomial function?

2. Divide $\dfrac{x^2 - 8}{x + 2}$.

3. For what values of x does $\dfrac{x^2 - 4x}{x}$ equal $x - 4$?

4. a. Graph $y = -1$.
   b. Graph $x = 2$.

5. Solve $\dfrac{2x^2 - x - 10}{x^2} = 2$.

6. Solve $\dfrac{3x - 1}{2x + 1} = \dfrac{3}{2}$.

7. Find the equation that defines the linear function f if $f(-3) = -1$ and $f(1) = 5$.

8. Find the coordinates of the x- and y-intercepts of the graph of $y = 6x^2 - x - 2$.

9. What is the quotient when $4x^4 + 5x^3 - 8$ is divided by $x^2 + 1$?

10. If $P(x) = 2x(x + 3)^4$, find the zeros of the function and state the multiplicity of each zero.

3.6 Rational Functions

When we add, subtract, or multiply two polynomials, the result is another polynomial. However, a function of the form

$$y = \frac{P(x)}{Q(x)} \qquad Q(x) \neq 0,$$

where $P(x)$ and $Q(x)$ are polynomials, is a different type of function called a **rational function.** For example,

$$y = \frac{1}{x}, \qquad y = \frac{x - 4}{x^2 - 1}, \quad \text{and} \quad y = \frac{3x^5 + 2}{x^3}$$

are rational functions. In our statements about rational functions we will assume that $P(x)$ and $Q(x)$ have no common factors, so the rational function is in lowest terms.

The behavior of a rational function often differs dramatically from that of a polynomial function. This difference may be easily seen by comparing the graph of a rational function to the smooth unbroken curves that characterize a polynomial.

Consider the rational function $y = 1/x$. Since division by zero is undefined, x cannot equal zero. Thus, the graph of this function does not intersect the line $x = 0$ (the y-axis). We can, however, let x approach zero and consider x-values as close to zero as we wish. From the following tables, note that as the x-values squeeze in on zero, $|y|$ becomes larger.

x	1	0.5	0.1	0.01	0.001
$y = \dfrac{1}{x}$	1	2	10	100	1,000

x	-1	-0.5	-0.1	-0.01	-0.001
$y = \dfrac{1}{x}$	-1	-2	-10	-100	$-1,000$

Figure 3.37 shows the behavior of $y = 1/x$ in the interval $-1 \le x \le 1$. The vertical line $x = 0$ that the curve approaches, but never touches, is called a **vertical asymptote.** We may use the following rule to determine if the graph of a rational function has any vertical asymptotes.

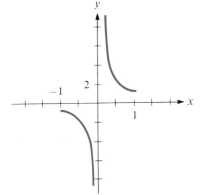

Figure 3.37

Vertical Asymptotes

If $P(x)$ and $Q(x)$ have no common factors, then the graph of the rational function $y = P(x)/Q(x)$ has as a vertical asymptote the line $x = a$ for each value a at which $Q(a) = 0$.

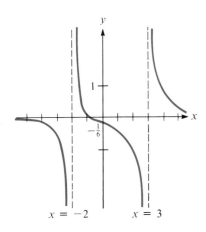

Figure 3.38

EXAMPLE 1   Find any vertical asymptotes of the function

$$y = \frac{x + 1}{(x - 3)(x + 2)}.$$

Solution   The polynomial in the denominator

$$Q(x) = (x - 3)(x + 2)$$

equals 0 when x is 3 or -2. Thus, there are two vertical asymptotes: $x = 3$ and $x = -2$. (*Note:* As a visual aid, the graph of this function is shown in Figure 3.38. By the end of this section, we will have developed sufficient techniques for you to be able to draw this graph.)   ∎

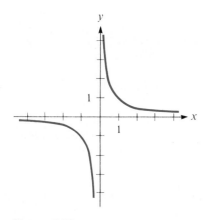

Figure 3.39

To complete the graph of $y = 1/x$, we must consider the behavior of the function as $|x|$ becomes larger. It is not difficult to see that $1/x$ squeezes in on zero from the positive side when x takes on larger positive values. Similarly, $1/x$ squeezes in on zero from the negative side when x increases in magnitude in the negative direction. Thus, the curve gets closer to the line $y = 0$ and the x-axis is a **horizontal asymptote**. Figure 3.39 shows the graph of $y = 1/x$.

Identifying any vertical and horizontal asymptotes is very important. These lines, together with plotting the intercepts and a few points, enable us to graph a rational function. A technique for determining horizontal asymptotes is included in the following examples.

EXAMPLE 2   If $y = \dfrac{x - 1}{x + 1}$, determine any vertical or horizontal asymptotes and graph the function.

Solution

a. The polynomial in the denominator $x + 1$ equals zero when x is -1. Thus, $x = -1$ is a vertical asymptote.

b. To determine any horizontal asymptotes, we change the form of the function by dividing each term in the numerator and the denominator by the highest power of x in the expression.

$$y = \frac{x - 1}{x + 1} = \frac{(x/x) - (1/x)}{(x/x) + (1/x)}$$

$$= \frac{1 - (1/x)}{1 + (1/x)} \qquad \text{(assuming that } x \neq 0)$$

Now as $|x|$ gets larger, $1/x$ approaches zero. Thus, y approaches $(1 - 0)/(1 + 0)$ and $y = 1$ is a horizontal asymptote.

c. By setting $x = 0$, we determine $(0, -1)$ is the y-intercept. Similarly, by setting $y = 0$, we determine that $(1, 0)$ is the x-intercept.

d. The vertical asymptote divides the x-axis into two regions. The intercepts are two points to the right of $x = -1$. If we plot a couple of points to the left of the vertical asymptote, say $(-2, 3)$ and $(-3, 2)$, we may complete the graph (see Figure 3.40).   ■

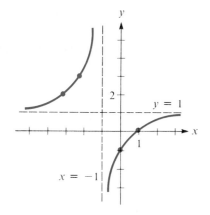

Figure 3.40

In Example 2 an important question is unanswered. We know that as $|x|$ becomes large, y approaches, but never quite reaches, 1. How do we know that y is not 1 when $|x|$ is small? We answer this question by determining if there is any value of x for which

$$\frac{x - 1}{x + 1} = 1.$$

Solving this equation, we have

$$x - 1 = x + 1$$
$$-1 = 1 \qquad \text{false.}$$

The equation has no solution, and we may conclude that the curve never crosses the horizontal asymptote. Example 3 illustrates why this possibility must be considered.

EXAMPLE 3 If

$$y = \frac{2x^2 + x - 3}{x^2},$$

determine any vertical and horizontal asymptotes and graph the function.

Solution

a. The polynomial in the denominator equals zero when x is zero. Thus, $x = 0$ is a vertical asymptote.

b. To determine any horizontal asymptotes, we use the procedure from Example 2. Dividing each term in the numerator and the denominator by x^2 gives

$$y = \frac{(2x^2/x^2) + (x/x^2) - (3/x^2)}{x^2/x^2}$$
$$= \frac{2 + (1/x) - (3/x^2)}{1} \qquad (x \neq 0).$$

As $|x|$ gets larger, $1/x$ and $-3/x^2$ approach 0. Thus, y approaches 2 and $y = 2$ is a horizontal asymptote.

c. To determine if the curve ever crosses the horizontal asymptote, find if there are any values of x for which

$$\frac{2x^2 + x - 3}{x^2} = 2.$$

Solving this equation for x, we have

$$2x^2 + x - 3 = 2x^2$$
$$x - 3 = 0$$
$$x = 3.$$

The curve crosses the asymptote at $(3,2)$.

d. Since x cannot be zero there is no y-intercept. After setting $y = 0$, we solve the equation $2x^2 + x - 3 = 0$ to find the x-intercepts $(1,0)$ and $(-\frac{3}{2},0)$.

e. To determine the behavior of the curve before it drops and starts approaching 2, we plot a couple of points to the right of $x = 3$, say $(4,\frac{33}{16})$ and $(5,\frac{52}{25})$. Additional points may always be plotted. Figure 3.41 shows the graph of the function. ∎

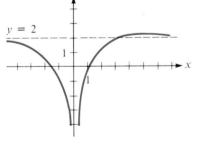

Figure 3.41

There are two theorems about asymptotes that can simplify our work. The first theorem indicates the behavior of the graph near any vertical asymptote and is based on the fact that even powers of positive and negative numbers have the same sign (positive) while odd powers of positive and negative numbers are opposite in sign.

Graph of $y = P(x)/Q(x)$ near Vertical Asymptotes

If $P(x)$ and $Q(x)$ have no common factors and if $(x - a)^n$ is a factor of $Q(x)$, where n is the largest positive integer for which this statement is true, then:

1. The graph of $y = P(x)/Q(x)$ goes in opposite directions about the vertical asymptote $x = a$ when n is odd.
2. The graph of $y = P(x)/Q(x)$ goes in the same direction about the vertical asymptote $x = a$ when n is even.

For instance, in Example 2 the graph goes to positive infinity on the left side of $x = -1$ and to negative infinity on the right side (see Figure 3.40). We can predict this type of behavior since $Q(x) = (x + 1)^1$ and the odd exponent indicates the graph goes in opposite directions about the vertical asymptote. In Example 3, however, $Q(x) = x^2$ and the even exponent indicates the graph goes in the same direction (to negative infinity) about the vertical asymptote ($x = 0$) as shown in Figure 3.41.

   The second theorem about asymptotes enables us to pick out horizontal asymptotes almost by inspection. This theorem may be derived by using the procedure shown in Examples 2 and 3 for determining a horizontal asymptote and applying it to the general form for a rational function given below.

Horizontal Asymptote Theorem

The graph of the rational function

$$y = \frac{P(x)}{Q(x)} = \frac{a_n x^n + a_{n-1} x^{n-1} + \cdots + a_0}{b_m x^m + b_{m-1} x^{m-1} + \cdots + b_0},$$

where $a_n, b_m \neq 0$, has

1. a horizontal asymptote at $y = 0$ (the x-axis) if $n < m$
2. a horizontal asymptote at $y = a_n/b_m$ if $n = m$
3. no horizontal asymptote if $n > m$

For instance, the graph of $y = 1/x$ satisfies the first case since the numerator is a zero-degree polynomial while the denominator is a first-degree polynomial. Because the higher degree polynomial is in the denominator, $y = 0$ is a horizontal asymptote. Examples 2 and 3 illustrate the second case in which both polynomials are of the same degree, so the horizontal asymptotes for

$$y = \frac{1x - 1}{1x + 1} \quad \text{and} \quad y = \frac{2x^2 + x - 3}{1x^2}$$

are $y = 1/1 = 1$ and $y = 2/1 = 2$, respectively. When there are no horizontal asymptotes as described in case 3, other techniques may be employed. In particular, if $n = m + 1$, that is, if the degree of the numerator is one more than the degree of the denominator, then the graph has a slant or oblique asymptote as discussed in the next example.

EXAMPLE 4   Graph the function $f(x) = \dfrac{x^2 - 1}{x + 2}$.

Solution

a. The polynomial in the denominator equals zero when x is -2. Thus, $x = -2$ is a vertical asymptote. Also, since $Q(x) = (x + 2)^1$, $Q(x)$ is an odd power of $x + 2$ so the graph of f goes in opposite directions about $x = -2$.

b. Because the higher degree polynomial is in the numerator, there is no horizontal asymptote. However, we can change the form of the equation by dividing $x^2 - 1$ by $x + 2$ as follows:

$$
\begin{array}{r}
x - 2 \\
x + 2 \overline{)x^2 - 1} \\
x^2 + 2x \\
\hline
-2x - 1 \\
-2x - 4 \\
\hline
3.
\end{array}
$$

Thus,

$$f(x) = \frac{x^2 - 1}{x + 2} = x - 2 + \frac{3}{x + 2}.$$

Now as $|x|$ gets larger, $3/(x + 2)$ approaches 0 and y approaches $x - 2$. Thus, the graph of f approaches the oblique (neither horizontal nor vertical) line $y = x - 2$, and we call this line an **oblique asymptote** of the graph of f.

c. By setting $x = 0$, we may determine that $(0, -\tfrac{1}{2})$ is the y-intercept. Similarly, by setting $f(x) = 0$, we may determine that $(1,0)$ and $(-1,0)$ are x-intercepts.

d. Using the above information and plotting one point to the left of the vertical asymptote, say $(-3, -8)$, we draw the graph in Figure 3.42.   ∎

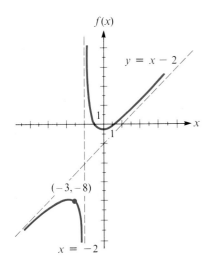

$f(x)$

$y = x - 2$

$(-3, -8)$

$x = -2$

Figure 3.42

In our statements about rational functions we have assumed that $P(x)$ and $Q(x)$ had no common factors. If the same factor does appear in the numerator and the denominator, we can divide it out as long as we keep track of all values that make the factor zero. The next example illustrates such a case.

EXAMPLE 5   Graph the function $f(x) = \dfrac{x^2 - 4}{x - 2}$.

Solution   By factoring the numerator, we rewrite the expression as

$$f(x) = \frac{x^2 - 4}{x - 2} = \frac{(x + 2)(x - 2)}{x - 2}.$$

If $x = 2$, the expression becomes $\tfrac{0}{0}$, which is undefined. If $x \neq 2$, we can divide out common factors and obtain $x + 2$. Thus, we can rewrite the expression as follows:

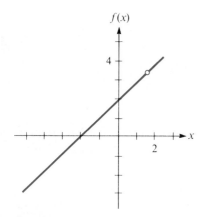

$f(x)$

4

2

x

Figure 3.43

$$f(x) = \frac{x^2 - 4}{x - 2} = \frac{(x + 2)(x - 2)}{x - 2} = \begin{cases} x + 2 & \text{if} \quad x \neq 2 \\ \text{undefined} & \text{if} \quad x = 2. \end{cases}$$

As shown in Figure 3.43, the graph is the set of all points on the line given by $f(x) = x + 2$ except $(2,4)$. ∎

Finally, here is an outline of the procedure for graphing rational functions with the form $y = P(x)/Q(x)$.

1. If $P(x)$ and $Q(x)$ have common factors, we can divide them out on the following conditions:
   a. If the degree of the common factor in the numerator is greater than or equal to the degree of the common factor in the denominator, then any value of x that makes the common factor zero produces a hole in the graph as shown in Example 5.
   b. If the degree of the common factor in the numerator is less than the degree of the common factor in the denominator, then any value of x that makes the common factor zero produces a vertical asymptote.
2. If $P(x)$ and $Q(x)$ have no common factors, find any vertical asymptotes by solving $Q(x) = 0$. Determine the behavior of the graph near any vertical asymptote by using the theorem given in this section.
3. Find any horizontal asymptotes by using the horizontal asymptote theorem or by dividing each term by the highest power of x and determining the value approached by y as $|x|$ gets larger. If the degree of the numerator is one more than the degree of the denominator, then determine an oblique asymptote by dividing $P(x)$ by $Q(x)$ and setting y equal to the quotient.
4. Check if the curve crosses the horizontal asymptote for small values of x by finding if there are any values of x for which $P(x)/Q(x)$ equals the value of the horizontal asymptote. There's no need to check for crossings of vertical asymptotes—they can't happen.
5. Find any x-intercepts by (setting $y = 0$ and) solving $P(x) = 0$. Find any y-intercepts by setting $x = 0$ and evaluating the expression.
6. Draw the graph. Plot additional points as needed (particularly if you have a calculator). You may wish to check for symmetries as discussed in Section 2.3.

EXERCISES 3.6

In Exercises 1–4 find any vertical asymptotes of the function.

1. $y = \dfrac{x + 7}{2x - 3}$

2. $y = \dfrac{x^2 + 1}{x(x + 4)}$

3. $y = \dfrac{2}{x^2 - 5x - 6}$

4. $y = \dfrac{x}{x^2 + x}$

In Exercises 5–24 determine any asymptotes and graph the function.

5. $y = \dfrac{-1}{x}$

6. $y = \dfrac{3}{x^2}$

7. $y = \dfrac{2}{x + 1}$

8. $y = \dfrac{-5}{3x + 2}$

9. $y = \dfrac{x - 2}{x + 2}$

10. $y = \dfrac{x + 3}{x - 5}$

11. $y = \dfrac{2x - 5}{3x + 5}$

12. $y = \dfrac{4x + 1}{2x + 7}$

13. $y = \dfrac{2}{(x - 1)^2}$

14. $y = \dfrac{2}{x^2 - 1}$

15. $y = \dfrac{1}{(x + 1)(x - 4)}$

16. $y = \dfrac{x - 1}{x^2 - 4}$

17. $y = \dfrac{x^2 - 1}{x^3}$

18. $y = \dfrac{x}{x^2 + 1}$

19. $y = \dfrac{x^2 - x - 6}{x^2}$

20. $y = \dfrac{3x^2 + x - 2}{2x^2}$

21. $y = \dfrac{x^2 + 1}{x}$

22. $y = \dfrac{(x - 1)^2}{x}$

23. $y = \dfrac{x^2 - 9}{x + 2}$

24. $y = \dfrac{x^2 - 2x - 3}{x + 3}$

In Exercises 25–33 graph the given function, with particular emphasis on any value of x for which $P(x) = Q(x) = 0$.

25. $h(x) = \dfrac{x(x + 1)}{x}$

26. $y = \dfrac{x(x + 1)}{x + 1}$

27. $y = \dfrac{x^2 - 4}{x + 2}$

28. $y = \dfrac{x^2 - x}{x}$

29. $g(x) = \dfrac{(2x + 1)(x - 3)(x + 2)}{(x - 3)(x + 2)}$

30. $f(x) = \dfrac{(x - 1)(x - 2)(x - 3)}{(x - 1)(x - 2)}$

31. $f(x) = \dfrac{x}{x(x + 1)}$

32. $y = \dfrac{x}{x^2 - 2x}$

33. $y = \dfrac{x - 1}{(x - 1)^2}$

THINK ABOUT IT

1. What is the difference between the graphs of
$y = \dfrac{x^2 - 1}{x - 1}$ and $y = x + 1$?

2. If $f(x) = \dfrac{4}{x^2 - 1}$, what is the range of f?

3. Graph $y = \dfrac{2x - 7}{5x + 10}$ and then solve the following inequalities by analyzing this graph.

a. $\dfrac{2x - 7}{5x + 10} \geq 0$ **b.** $\dfrac{2x - 7}{5x + 10} \leq 0$

4. Give two examples of an equation that defines a rational function such that the vertical asymptote is $x = -3$ and the horizontal asymptote is $y = \frac{5}{2}$.

5. a. Explain why the graph of a rational function cannot cross a vertical asymptote.
   b. We have seen that it is possible for the graph of a rational function to cross its horizontal asymptote. Do you think a graph may cross its oblique asymptote? Explain.
   c. Graph $y = \dfrac{x^3}{x^2 - 1}$ and check that your explanation in part b is in agreement with this graph.

REMEMBER THIS

1. Evaluate $(1.07)^8$ to the nearest hundredth using a calculator.

2. Evaluate using exponent definitions.
   a. $16^{-3/2}$ **b.** $(\frac{1}{4})^{-2}$ **c.** 10^0

3. Simplify using exponent properties.
   a. $2^x/2$ **b.** $b^x \cdot b^x$

4. Show that $(\frac{1}{3})^x$ is equivalent to 3^{-x}.

5. What is the relation between the graph of $y = f(x)$ and the graph of $y = f(x + 2)$?

6. If -1 is a zero of $P(x) = x^3 + x^2 + 12x + 12$, then find the other zeros for the function.

7. By Descartes' rule of signs what is the maximum number of negative real zeros of the function $P(x) = 3x^3 - 17x^2 + 28x - 12$?

8. Find the zeros of $P(x) = 3x^3 - 17x^2 + 28x - 12$.

9. Write an equation for the line passing through $(-7, 2)$ that is parallel to the line given by $6x - 2y = 1$.

10. Graph $y = (x + 3)(x - 1)^2$ based on information obtained from the intercepts.

CHAPTER OVERVIEW

Section	Key Concepts to Review
3.1	• Definitions of polynomial function and linear function • Slope formula: $m = \dfrac{\Delta y}{\Delta x} = \dfrac{y_2 - y_1}{x_2 - x_1}$ $(x_2 \neq x_1)$ • Point-slope equation: $y - y_1 = m(x - x_1)$ • Slope-intercept equation: $y = mx + b$ • For two lines with slopes m_1 and m_2: parallel lines: $m_1 = m_2$ perpendicular lines: $m_1 \cdot m_2 = -1$
3.2	• Definitions of quadratic function, quadratic inequality, and axis of symmetry • Methods to graph a quadratic function • Axis of symmetry formula: $x = -b/2a$ • Vertex formula: $(-b/2a, f(-b/2a))$ • The graph of $f(x) = a(x - h)^2 + k$ (with $a \neq 0$) is the graph of $y = ax^2$ (a parabola) shifted so the vertex is (h,k) and the axis of symmetry is $x = h$. • Methods to solve a quadratic inequality
3.3	• Long division procedure for the division of polynomials • Synthetic division procedure for the division of a polynomial by $x - b$ • Remainder theorem
3.4	• Definitions of a zero of a function and the multiplicity of a zero • Factor theorem • Fundamental theorem of algebra • Number of zeros theorem • Theorem about when complex zeros come in conjugate pairs • Theorem about when irrational zeros of the form $a \pm b\sqrt{c}$ come in conjugate pairs • Theorem about graphing polynomial functions by knowing the multiplicity of real-number zeros • Properties of polynomial graphs concerning continuity, turning points, and behavior for large $|x|$
3.5	• Rational-zero theorem • Descartes' rule of signs • Upper and lower bounds theorem • Intermediate value theorem • Location theorem
3.6	• Definition of a rational function • Methods to determine vertical, horizontal, or oblique asymptotes • Horizontal asymptote theorem • Theorem about the behavior of a graph near any vertical asymptote(s) • Outline of the procedure for graphing rational functions

CHAPTER REVIEW EXERCISES

In Exercises 1–6 indicate if the given function is a polynomial function.

1. $f(x) = (2x - 1)/3$
2. $y = (2x - 1)/x$
3. $y = x^2 + x^{-1} + 2$
4. $y = x^2 + x + 2^{-1}$
5. $y = \sqrt{x}$
6. $y = \sqrt{2}$

In Exercises 7–12 find the zeros of the function.

7. $y = 2x^2 - 3x$
8. $f(x) = 3 - 5x$
9. $g(x) = x^3 - 2x^2 + x - 2$
10. $y = x^2(x + 7)(x - 4)$
11. $f(x) = (2x^2 - 5x + 1)/x^3$
12. $y = 1/(x^2 - 1)$

In Exercises 13–20 graph each function.

13. $y = x - x^2$
14. $y = \dfrac{3}{2x - 3}$
15. $y = \dfrac{x + 1}{x}$
16. $y = \dfrac{x + 1}{2}$
17. $y = \dfrac{x^2 + 2x}{x}$
18. $y = \dfrac{5x^2 + 4}{4x^2 - 1}$
19. $y = x^3 - 3x^2 - 4x$
20. $y = x^2(x + 7)(x - 4)$

In Exercises 21 and 22 solve each inequality.

21. $x^2 + x \geq 6$
22. $x^2 > x - 1$

23. What is the domain of $y = \dfrac{3x - 1}{4x + 2}$?

24. What is the range of $y = 2x - x^2$?
25. Determine the vertex and axis of symmetry of the graph of $f(x) = 3(x + 5)^2 - 1$.
26. Determine the vertical asymptotes for the function $y = \dfrac{3x + 1}{x^2 + 2x}$.
27. Determine the horizontal asymptote for the function $y = \dfrac{2x^2 + 3}{3x^2 + x - 2}$. Does the graph cross this asymptote? If yes, find the point at which it crosses.
28. What is the difference between the graphs of $y = \dfrac{x^2 - 4}{x - 2}$ and $y = x + 2$?

29. Write the equation of the line with an x-intercept of 3 and a y-intercept of -2.
30. Find the equation that defines the linear function f if $f(-1) = 6$ and $f(2) = -3$.
31. Line L_1 passes through $(-2, -2)$ and $(0,3)$ and line L_2 passes through $(-4,k)$ and $(2,-1)$.
    a. Find k if L_1 is parallel to L_2.
    b. Find k if L_1 is perpendicular to L_2.

32. Write a polynomial function (in factored form) with rational coefficients of the lowest possible degree with zeros of $-i$, $\sqrt{2}$, and 0.
33. Write a polynomial function (in factored form) of degree 4 with zeros of 5, -2, 0, and 3.
34. If $f(x) = 2x^4 - x + 7$, find $f(-2)$ by (a) direct substitution and (b) the remainder theorem.
35. If the parabola $y = x^2 - 8x + k$ has its vertex on the x-axis, what is the value of k?
36. Divide $2x^4 - x + 7$ by $x^2 + 2$. Express the result in the form dividend = (divisor)(quotient) + remainder.
37. The dividend is x^2, the quotient is $x + 3$, and the remainder is 9. What is the divisor?
38. List the possible rational zeros of the function $P(x) = 3x^3 - 7x^2 + x - 2$.
39. Use Descartes' rule of signs to determine the maximum number of (a) positive and (b) negative real zeros of $P(x) = x^4 - 3x^3 + x^2 - x + 1$.
40. If 2 is a zero of $P(x) = 2x^3 - 4x^2 - 6x + 12$, find the other zeros.
41. If 1 is a root of $x^3 - 2x^2 - 5x + 6 = 0$, what is the quadratic equation that can be used to find the other two roots?
42. If b is a zero of $P(x) = x^3 + ax^2 + ax + 1$, show that $1/b$ is also a zero.
43. Find all the rational zeros of $y = 2x^{11} - x - 1$.
44. Verify that $P(x) = x^3 - 3x^2 - 1$ has a real zero between 3 and 4 by using the location theorem.
45. Approximate the zero of $P(x) = x^3 - 2x - 7$ that is between 2 and 3 to the nearest tenth.
46. Show that the largest rectangular field that can be enclosed by a given length of fence is a square.
47. A businessperson owns a parking lot that she wants to enclose with a fence. One side of the lot is bounded by a building, so no fencing is required. There is to be a 16-ft opening in the front of the fence, which is opposite the building. If side fencing costs $1 per foot and front fencing costs $1.50 per foot, what are the dimensions of the largest rectangular lot that can be fenced for $300?
48. A salesperson receives a monthly salary plus a commission on sales. She earned $1,050 during a month in which her sales totaled $5,000; the next month she sold $6,000 worth of merchandise and earned $1,150.
    a. Find the equation that defines this linear relationship.
    b. How much will she earn if she sells $9,000 worth of merchandise for the month?
    c. What is her monthly salary?
    d. What is her rate of commission?

49. A projectile shot vertically upward has a velocity of 192 ft/second at $t = 4$ seconds and 96 ft/second when $t = 7$ seconds.
  a. Find the equation that defines this linear relationship.
  b. Find the velocity of the projectile when $t = 9$ seconds.
  c. What is the initial velocity of the projectile (velocity when $t = 0$)?
  d. When is the velocity 0? What is the physical significance of this?

50. A computer manufacturer sells 6,000 units per week of its $3,000 model. Because of a backlog of computers a rebate program is introduced. It is estimated that each $100 decrease in price will result in the sale of 300 more units. The weekly revenue R is then given by the formula $R = (6,000 + 300x)(3,000 - 100x)$, where x is the number of $100 discounts offered. What rebate should be given to produce maximum revenue?

In Exercises 51–60 select the choice that answers the question or completes the statement.

51. The degree of $P(x) = x^2(x + 3)(x - 3)^4$ is
  a. 6   **b.** 7   **c.** 8   **d.** 16

52. If $P(x) = x^2(x + 3)(x - 3)^4$, what is the multiplicity of the zero of -3?
  a. 1   **b.** 2   **c.** 4   **d.** 5

53. A quadratic function with zeros of $2 \pm i$ is
  a. $y = x^2 - 4x + 5$ **b.** $y = x^2 - 5x + 4$
  c. $y = x^2 + 5x - 4$ **d.** $y = x^2 + 4x - 5$

54. The graph of $y = (x^2 + 1)/x^3$ lies in quadrants
  a. 1 and 2   **b.** 1 and 3
  c. 2 and 4   **d.** 3 and 4

55. If $x^3 - 2x^2 + ax + 9$ is exactly divisible by $x + 3$, then a equals
  a. -10 **b.** -12 **c.** 7   **d.** 16

56. The axis of symmetry for $y = 2x^2 + 4x - 1$ is
  a. $y = 1$ **b.** $x = 2$ **c.** $x = -1$ **d.** $y = 4$

57. The slope of the line $3x - 2y = 6$ is
  a. 3   **b.** -3 **c.** $\frac{3}{2}$ **d.** $-\frac{3}{2}$

58. The equation of the line that passes through the point $(-3,1)$ and is perpendicular to $y = \frac{1}{2}x + 4$ is
  a. $y = \frac{1}{2}x - \frac{1}{4}$ **b.** $y = \frac{1}{2}x + \frac{5}{2}$
  c. $y = 2x + 7$ **d.** $y = -2x - 5$

59. If $\frac{2}{5}$ is a zero of $y = ax^3 + bx^2 + cx + d$, in which $a, b, c,$ and d are integers with $a \neq 0$, then 5 must be a factor of
  a. a **b.** b **c.** c **d.** d

60. The function $y = x^3 - 3x - 3$ has exactly one positive real zero in which one of the following intervals?
  a. $[0,1]$ **b.** $[1,2]$ **c.** $[2,3]$ **d.** $[3,4]$

CHAPTER 3 TEST

1. Graph the line whose slope is $-\frac{3}{2}$ and whose y-intercept is $(0,2)$. Also, find an equation for the line.

2. Write an equation for the line passing through $(-4,1)$ that is perpendicular to the line given by $y = -\frac{1}{2}x + 3$.

3. Find the equation that defines the linear function f if $f(0) = -3$ and $f(5) = 0$.

4. Find the coordinates of the x-intercepts of the graph of $g(x) = 2x^2 - x - 3$.

5. Find the range of the function defined by $y = 6x - x^2$.

6. Solve $x^2 - 6 \leq 2x$.

7. The height (y) of a projectile shot vertically upward from the ground with an initial velocity of 128 ft/second is given by the formula $y = 128t - 16t^2$. What is the maximum height attained by the projectile?

8. What is the remainder when $5x^4 - 3x + 12$ is divided by $x^2 - 2$?

9. What is the quotient when $2x^4 - x^2 + 5$ is divided by $x + 3$?

10. If $P(x) = x^3 - 3x^2 + x + 9$, find $P(-2)$ by (a) direct substitution and (b) the remainder theorem.

11. If $P(x) = 4x^2(x + 1)^3$, find the zeros of the function and state the multiplicity of each zero.

12. Graph $y = x^4 - 4x^2$ based on information obtained from the intercepts.

13. If -5 is a zero of $P(x) = x^3 + 5x^2 - 6x - 30$, then find the other zeros for the function.

14. By the rational-zero theorem what are the possible rational zeros for $P(x) = 3x^3 + 2x^2 - 3x - 2$?

15. By Descartes' rule of signs what is the maximum number of negative real zeros of the function $P(x) = x^3 + 4x^2 - x + 5$?

16. Find the zeros of $P(x) = 2x^3 + 3x^2 + 6x + 9$.

17. Find any vertical asymptotes for the graph of
$$y = \frac{2x - 7}{x^2 - 6x}.$$

18. Find all points at which the graph of $y = \dfrac{x^2 - 4}{x^3}$ crosses its horizontal asymptote.

19. Find the oblique asymptote for the graph of
$$y = \frac{x^2 - 4x - 5}{x - 2}.$$

20. Graph $y = \dfrac{2x - 5}{x + 3}.$

CHAPTER 4

Exponential and Logarithmic Functions

If you win the state lottery and are offered $1 million or a penny that doubles in value on each day in the month of November, which option would you accept? What is the value of the penny by the end of the month? (See Example 2 of Section 4.1.)

Exponential functions and their inverses, the logarithmic functions, are important rules for analyzing a wide variety of relationships. Compound interest, population growth, radioactive decay, pH, decibels, and heat loss (from a home or a murder victim) are but some of the applications we consider in this chapter. Whereas a typical polynomial function looks like $y = x^2$, a typical exponential function looks like $y = 2^x$. The explosive growth of exponential functions like $y = 2^x$ can be quite amazing, and the introductory problem given above addresses this issue. Throughout this chapter, you will find that a scientific calculator is particularly useful, while a knowledge of exponents is indispensable.

4.1 Exponential Functions

In engineering, biology, economics, psychology, and other fields there are important phenomena whose behavior is described by functions in which the independent variable appears as the exponent of a fixed base. These functions are called exponential functions as shown in the following definition.

Exponential Function

The function f defined by

$$f(x) = b^x,$$

with $b > 0$ and $b \neq 1$, is called the **exponential function with base b.**

We eliminate a nonpositive base in the defining equation because such expressions as $(-1)^{1/2}$ and 0^{-2} are not real numbers. Since $1^x = 1$ for all values of x, we also eliminate 1.

EXAMPLE 1   In the biological reproduction of cells by division, one cell divides into two cells after a certain period of time. Each of these cells repeats the process and divides into two more, and so on. To be specific, assume that on a given day we start with one cell and the number doubles each day. Then there are two cells present after 1 day; $2 \cdot 2$, or 2^2, after 2 days; $2 \cdot 2^2$, or 2^3, after 3 days; and so on. Therefore, there are 2^x cells present after x days. This phenomenon of cell reproduction is described by the function

$$y = 2^x,$$

where x represents the number of complete days from the start of the experiment.   ∎

EXAMPLE 2   Solve the problem in the chapter introduction on page 211.

Solution   As in Example 1, the doubling process produces a base 2 exponential function. Since the initial value of a penny is $0.01, the value ($V$) of the penny after t days is given by

$$V = (0.01)2^t$$

and the penny's value increases as follows:

$$V = 0.01 \qquad\qquad\qquad\qquad \text{(when } t = 0)$$
$$V = (0.01)2 \quad\ = \$0.02 \qquad\qquad \text{(after 1 day)}$$
$$\vdots \qquad\qquad\qquad\qquad\qquad \vdots$$
$$V = (0.01)2^{10} = \$10.24 \qquad\qquad \text{(after 10 days)}$$
$$\vdots \qquad\qquad\qquad\qquad\qquad \vdots$$
$$V = (0.01)2^{20} = \$10{,}485.76 \qquad \text{(after 20 days)}$$
$$\vdots \qquad\qquad\qquad\qquad\qquad \vdots$$
$$V = (0.01)2^{25} = \$335{,}544.32 \qquad \text{(after 25 days)}$$
$$\vdots \qquad\qquad\qquad\qquad\qquad \vdots$$
$$V = (0.01)2^{30} = \$10{,}737{,}418 \qquad \text{(after 30 days).}$$

Thus, by the end of November the penny is worth $10,737,418 and is the better option. (*Note:* In this example we evaluated $(0.01)2^{30}$ by using the $\boxed{y^x}$ key on a calculator as follows:

$$0.01 \ \boxed{\times} \ 2 \ \boxed{y^x} \ 30 \ \boxed{=} \qquad 10{,}737{,}418.$$

Throughout this chapter, you will find that a scientific calculator is an essential aid. Exponential expressions as above [and later logarithmic expressions] are evaluated quickly and accurately on a calculator; otherwise, such evaluations require a cumbersome logarithmic approach that is discussed in Appendix A.3. So, at least for this chapter, try to obtain a calculator.) ∎

At present we are able to interpret exponents that are rational numbers. For example, if $f(x) = 2^x$, then

$$f(3) = 2^3 = 8$$
$$f(0) = 2^0 = 1$$
$$f(-2) = 2^{-2} = \frac{1}{2^2} = \frac{1}{4}$$
$$f(\tfrac{3}{2}) = 2^{3/2} = \sqrt{2^3} = \sqrt{8} \approx 2.83.$$

To interpret 2^x, where x is irrational, it is important to understand that real-number exponents are defined so that the following "betweenness" property applies.

Property of Real-Number Exponents

> For $b > 0$, $b \neq 1$, and any real number x, if x lies between the rational numbers r and s, then b^x lies between b^r and b^s.

For example, to interpret $2^{\sqrt{2}}$ by using this property, we first note that $\sqrt{2} = 1.4142135\ldots$ may be written to any desired accuracy. Then, since

$$1.4 < \sqrt{2} < 1.5 \text{ and } 2^{1.4} < 2^{1.5}, \text{ we know } 2^{1.4} < 2^{\sqrt{2}} < 2^{1.5},$$
$$\text{so } 2.639\ldots < 2^{\sqrt{2}} < 2.828.\ldots$$

Similarly,

$$1.41 < \sqrt{2} < 1.42 \text{ and } 2^{1.41} < 2^{1.42},$$
$$\text{so } 2^{1.41} < 2^{\sqrt{2}} < 2^{1.42}, \text{ or } 2.657\ldots < 2^{\sqrt{2}} < 2.675\ldots,$$
$$1.414 < \sqrt{2} < 1.415 \text{ and } 2^{1.414} < 2^{1.415},$$
$$\text{so } 2^{1.414} < 2^{\sqrt{2}} < 2^{1.415}, \text{ or } 2.664\ldots < 2^{\sqrt{2}} < 2.666\ldots,$$

and so on. By continuing in this manner, we can make the difference between the endpoints of the interval containing $2^{\sqrt{2}}$ as small as we wish and thereby approximate the unique real number $2^{\sqrt{2}}$ to any desired accuracy. In actual practice, we usually obtain a direct estimate for $2^{\sqrt{2}}$ by using the $\boxed{y^x}$ key on a calculator as follows:

$$2 \boxed{y^x} 2 \boxed{\sqrt{}} \boxed{=} \qquad 2.6651441.$$

Thus, although a precise definition for irrational exponents can only be given in more advanced mathematics, we are now able to interpret expressions like 2^x for all real numbers x, and it can be shown that all previous laws of exponents are valid for all real-number exponents. In the next example we use the results of our discussion to graph two typical exponential functions:

$$y = 2^x \qquad \text{in which } b > 1$$
$$y = (\tfrac{1}{2})^x \quad \text{in which } 0 < b < 1.$$

EXAMPLE 3   Sketch the graphs of $g(x) = 2^x$ and $h(x) = (\frac{1}{2})^x$. In each case indicate the domain and range of the function on the graph.

Solution   By picking convenient integer values for x and substituting them in the given equations, we generate the following table.

x	-3	-2	-1	0	1	2	3
2^x	$\frac{1}{8}$	$\frac{1}{4}$	$\frac{1}{2}$	1	2	4	8
$(\frac{1}{2})^x$	8	4	2	1	$\frac{1}{2}$	$\frac{1}{4}$	$\frac{1}{8}$

Based on these points and the "betweenness" property of real-number exponents, we then sketch the graphs of g and h as the smooth curves in Figure 4.1(a) and (b), respectively. In both cases the domain is the set of all real numbers and the range is the set of positive real numbers as shown on the graphs.

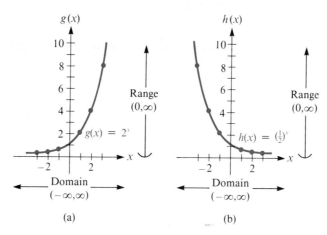

Figure 4.1

The graphs in Figure 4.1(a) and (b) illustrate the following important features of exponential functions of the form $f(x) = b^x$ (with $b > 0$, $b \neq 1$) and their graphs.

**Properties of $f(x) = b^x$
(with $b > 0$, $b \neq 1$)
and Its Graph**

1. The domain of f is $(-\infty,\infty)$ and the range of f is $(0,\infty)$.
2. The x-axis is a horizontal asymptote for the graph of f.
3. The y-intercept is always $(0,1)$ and there are no x-intercepts.
4. If $b > 1$, then as x increases, y increases.
5. If $0 < b < 1$, then as x increases, y decreases.

We can graph certain variations of the function $y = b^x$ by using the graphing techniques given in Section 2.3. Example 4 illustrates such a case.

EXAMPLE 4 Graph $y = 1 - 2^{-x}$. What is the range of this function?

Solution We begin by noting that 2^{-x} is equivalent to $(\frac{1}{2})^x$, so we start with the graph of $y = (\frac{1}{2})^x$ in Figure 4.1(b). We also note that $1 - 2^{-x} = -2^{-x} + 1$. To graph $y = -2^{-x} + 1$, we first reflect the graph of $y = 2^{-x}$ about the x-axis to obtain the graph of $y = -2^{-x}$ [see Figure 4.2(a)]. Then we raise the graph in Figure 4.2(a) up 1 unit, since the constant 1 is added after the exponential term. The completed graph is shown in Figure 4.2(b). From this graph we read that the range is the interval $(-\infty,1)$. ∎

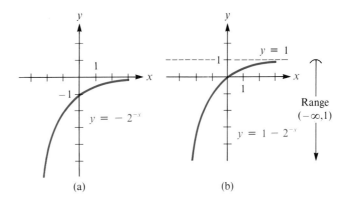

Figure 4.2

In this chapter we will frequently solve certain equations containing exponential expressions by using the principle that since exponential functions are one-to-one, if b is a positive number other than 1, then

$$b^x = b^y \text{ implies } x = y.$$

The next two examples show where we use this principle.

EXAMPLE 5 If $(x, \frac{1}{27})$ is a solution of the equation $y = 9^x$, then find x.

Solution We need to find the value of x that satisfies the equation $9^x = \frac{1}{27}$. In many cases it is hoped that through an understanding of exponents, you can determine the solution by inspection. If not, then try to rewrite the expressions in terms of a common base. In this case since $9 = 3^2$ and $27 = 3^3$ we solve the equation as follows:

$$9^x = \frac{1}{27}$$

$$(3^2)^x = \frac{1}{3^3}$$

$$3^{2x} = 3^{-3}$$

$$2x = -3 \qquad \text{Since } b^x = b^y \text{ implies } x = y$$

$$x = -\frac{3}{2}.$$

Thus, $-\frac{3}{2}$ is the x value in the ordered pair. (*Note:* In Section 4.4 we will consider equations like $3^x = 8$, in which it is difficult to rewrite the expressions in terms of a common base.) ∎

EXAMPLE 6 Solve $32^x = (\frac{1}{4})^{x+1}$.

Solution To rewrite this equation so both sides are expressed with the same base, recognize that $32 = 2^5$ and $\frac{1}{4} = \frac{1}{2}^2 = 2^{-2}$. Then,

$$32^x = (\tfrac{1}{4})^{x+1}$$

$$(2^5)^x = (2^{-2})^{x+1}$$

$$2^{5x} = 2^{-2x-2}$$

$$5x = -2x - 2 \qquad \text{Since } b^x = b^y \text{ implies } x = y.$$

$$7x = -2$$

$$x = -\tfrac{2}{7}$$

Thus, the solution set is $\{-\frac{2}{7}\}$. ∎

When working with exponential functions, it may be necessary to solve an equation that contains a rational exponent as shown in the next example.

EXAMPLE 7 Find the base of the exponential function $y = b^x$ that contains the point $(-\frac{1}{3}, \frac{1}{4})$.

Solution Replacing x by $-\frac{1}{3}$ and y by $\frac{1}{4}$ in the equation $y = b^x$ gives

$$\frac{1}{4} = b^{-1/3}.$$

To find b, use an extension of the principle of powers and raise both sides of this equation to the *reciprocal* power of $-\frac{1}{3}$, which is -3, and then simplify.

$$(\tfrac{1}{4})^{-3} = (b^{-1/3})^{-3} \qquad \text{Raise both sides to the reciprocal power.}$$
$$4^3 = b^1$$
$$64 = b$$

Since $64^{-1/3} = (\tfrac{1}{64})^{1/3} = \sqrt[3]{\tfrac{1}{64}} = \tfrac{1}{4}$, the solution checks. Thus, the base of the exponential function containing the given point is 64.   ∎

   Finally, let us consider two more applications involving exponential functions.

EXAMPLE 8   A company purchases a new machine for $3,000. The value of the machine depreciates at a rate of 10 percent each year.
   a. Find a formula (rule) showing the value (y) of the machine at the end of x years.
   b. What is the machine worth after 4 years?

Solution
   a. If the machine depreciates 10 percent each year, at the end of a year the machine is worth 90 percent of the value at which it began the year. Thus, the value is $3,000(0.9)$ after 1 year; $3,000(0.9)(0.9)$ or $3,000(0.9)^2$ after 2 years; and

$$y = 3,000(0.9)^x \qquad \text{after } x \text{ years.}$$

   b. Evaluating the function when $x = 4$, we have

$$y = 3,000(0.9)^4$$
$$= \$1,968.30.$$ ∎

EXAMPLE 9 $1,000 is invested at 6-percent interest compounded annually.
   a. Describe in functional notation the value of the investment at the end of t years.
   b. How much is the investment worth after 3 years?

Solution
   a. The $1,000 amounts to $1,000(1 + 0.06)$ or $1,060 at the end of 1 year. The principal during the second year is $1,000(1 + 0.06)$. Thus, the amount at the end of 2 years is $1,000(1 + 0.06)(1 + 0.06)$ or $1,000(1 + 0.06)^2$. In general, the value of the investment after t years is

$$f(t) = 1,000(1 + 0.06)^t.$$

b. Evaluating the function when $t = 3$, we have

$$f(3) = 1,000(1 + 0.06)^3$$
$$= \$1,191.02.$$

(*Note:* Compound interest, which produces exponential growth, is a veritable gold mine when compared to the linear growth of simple interest. For example, if $1 had been invested and compounded annually at 3-percent interest since the birth of Christ (0 A.D.), the compounded amount in 1993 would be about $38,000,000,000,000,000,000,000,000,000. At simple interest the same investment would amount to $60.79. We can generalize the procedure from this example to obtain the following formula for the compounded amount A when an original principal P is compounded annually for t years at annual interest rate r:

$$A = P(1 + r)^t.$$

Use this formula and the formula $A = P(1 + rt)$ associated with simple interest to verify that the investments described above grow to the stated amounts.)   ∎

In this section the applications describe situations in which no change is considered to have occurred until an entire time period has elapsed. Thus, the $1,000 investment in Example 9 is compounded annually and does not change until the end of the year. For this reason, the domain of these functions is the set of nonnegative integers. In Section 4.5 we discuss phenomena that change continuously. The domain of the function in these applications is usually the set of nonnegative real numbers.

EXERCISES 4.1

In Exercises 1–4 evaluate each function for the given elements of the domain.

1. $f(x) = 4^x$; $f(2), f(-1), f(\frac{3}{2})$
2. $g(x) = 9^x$; $g(0), g(-\frac{1}{2}), g(3)$
3. $h(x) = (\frac{2}{3})^x$; $h(4), h(-3), h(0)$
4. $f(x) = 8^{-x}$; $f(1), f(-2), f(\frac{2}{3})$

In Exercises 5–20 solve each equation.

5. $3^{x-1} = 3^5$
6. $2^{2x} = 2^6$
7. $9^x = 3$
8. $8^x = 2$
9. $2^x = \frac{1}{8}$
10. $6^x = \frac{1}{36}$
11. $(\frac{3}{4})^x = \frac{16}{9}$
12. $(\frac{2}{3})^x = \frac{27}{8}$
13. $3^{-x} = \frac{1}{81}$
14. $2^{3-x} = 1$
15. $4^{x+3} = \sqrt{2}$
16. $9^{2x-1} = \sqrt[3]{3}$
17. $4^x = 81^{-x}$
18. $27^x = 9^{x+2}$
19. $(\frac{1}{16})^{x-1} = 32^x$
20. $(\frac{1}{64})^{2-x} = (\frac{1}{32})^{2x}$

In Exercises 21–30 fill in the missing component of each of the ordered pairs that makes the pair a solution of the given equation. Also, choose a convenient scale for the vertical axis and graph each function.

21. $y = 2^x$; $(-3, \ ), (\ , 1), (\ , \frac{1}{2}), (2, \ )$
22. $y = (\frac{1}{2})^x$; $(0, \ ), (2, \ ), (\ , 2), (\ , \frac{1}{2})$
23. $y = (\frac{1}{10})^x$; $(\ , 0.01), (\ , 1), (-1, \ ), (-2, \ )$
24. $g(x) = 4^{-x}$; $(2, \ ), (\ , 0.5), (-0.5, \ ), (\ , 4)$
25. $y = 4^x$; $(\frac{1}{2}, \ ), (\ , 1), (1, \ ), (\ , \frac{1}{2})$
26. $y = -5^x$; $(0, \ ), (1, \ ), (-1, \ ), (\ , -0.04)$
27. $h(x) = (\sqrt{2})^x$; $(0, \ ), (-2, \ ), (1, \ ), (\ , 2)$
28. $f(x) = 3(2)^x$; $(-1, \ ), (0, \ ), (1, \ ), (2, \ )$
29. $f(x) = 100(1.06)^x$; $(-1, \ ), (0, \ ), (1, \ ), (2, \ )$
30. $y = 3^x + 3^{-x}$; $(-1, \ ), (0, \ ), (1, \ ), (2, \ )$

In Exercises 31–40 graph each function. Also, determine the range of the function in each case.

31. $f(x) = 3^{-x}$ **32.** $y = -3^x$

33. $y = 1 - 3^{-x}$ **34.** $y = 3^x - 1$

35. $y = 2^{x-2}$ **36.** $y = 2^{2-x}$

37. $h(x) = -(\frac{1}{2})^{-x}$ **38.** $y = (\frac{1}{2})^{x+3}$

39. $y = 2^x + 2^{-x}$ **40.** $f(x) = 2^{|x|}$

In Exercises 41–50 find the base of the exponential function $y = b^x$ that contains the given point.

41. $(1,4)$ **42.** $(2,9)$ **43.** $(-1,3)$

44. $(-2,\frac{1}{25})$ **45.** $(\frac{1}{2},4)$ **46.** $(0.5,10)$

47. $(\frac{2}{3},\frac{1}{4})$ **48.** $(\frac{3}{2},27)$ **49.** $(-\frac{1}{3},2)$

50. $(-\frac{2}{3},\frac{1}{9})$

51. Show that if $f(x) = b^x$ is the exponential function with base b, then $f(x_1 + x_2) = f(x_1) \cdot f(x_2)$.

52. A biologist has 500 cells in a culture at the start of an experiment. Hourly readings indicate that the number of cells is doubling every hour.
  a. Find a formula (rule) showing the number of cells present at the end of t hours.
  b. How many cells are present at the end of 4 hours?
  c. At the end of how many hours are 32,000 cells present?

53. $100 is invested at 5 percent compounded annually.
  a. Find a formula showing the value of the investment at the end of t years.
  b. How much is the investment worth at the end of 3 years?

54. The market value of a company's investment is $100,000. This value increases at a rate of 10 percent each year.
  a. Describe in functional notation the market value of the investment t years from now.
  b. What should be the market value of the investment 2 years from now?

55. A company purchases a machine for $10,000. The value of the machine depreciates at a rate of 20 percent each year.
  a. Find a formula to show the value of the machine after t years.
  b. What will be the value of the machine 3 years from now?

56. A biologist grows a colony of a certain kind of bacteria. It is found experimentally that $N = N_0 3^t$, where N represents the number of bacteria present at the end of t days. N_0 is the number of bacteria present at the start of the experiment. Suppose that there are 153,000 bacteria present at the end of 2 days.
  a. How many bacteria were present at the start of the experiment?
  b. How many bacteria are present at the end of 4 days?
  c. At the end of how many days are 459,000 bacteria present?

57. A new element has a half-life of 30 minutes; that is, if x oz of the element exist at a given time, $x/2$ oz exist 30 minutes later. The other half disintegrates into another element. Suppose that 1 g of the element is present 2 hours after the start of an experiment.
  a. Find a formula showing the amount of the element present t hours after the start of the experiment.
  b. How much of the element was present 1 hour after the start of the experiment?

58. You purchase a home for $150,000 and take out a 30-year mortgage. If the annual inflation rate remains fixed at 5 percent and the value of the home keeps pace with inflation, then what is the value of your home by the time the mortgage is paid off?

THINK ABOUT IT

1. The following figure shows the graph of $y = 2^x$ on the interval $[0,1]$. Use this graph to approximate the following numbers to the nearest tenth.
   a. $2^{0.7}$ **b.** $\sqrt{2}$ **c.** $\sqrt[5]{4}$

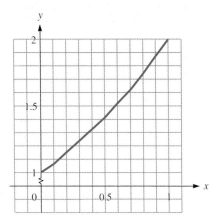

2. **a.** Without using a calculator, place the following numbers in their correct numerical order: $5^{\sqrt{2}}$, 5^{π}, 0.04, $\sqrt[3]{5}$, $(\frac{1}{5})^{1 \cdot 3}$. Justify your answer in terms of the graph of $y = 5^x$.
   b. Graph $y = (\frac{1}{5})^x$, $y = (\frac{1}{3})^x$, $y = 3^x$, and $y = 5^x$ on the same coordinate system. Use these graphs to explain why $a^x = b^x$ implies $a = b$, if a and b are positive real numbers with $x \neq 0$.

3. Determine an integer n such that $n < x < n + 1$.
   a. $10^x = 451$ **b.** $10^x = 45,100$
   c. $10^x = 0.451$ **d.** $10^x = 0.00451$

4. For the exponential function defined by $f(x) = b^x$, which of the given equations specifies a property of f?
   a. $f(u + v) = f(u) + f(v)$
   b. $f(uv) = f(u) + f(v)$
   c. $f(u + v) = f(u) \cdot f(v)$

5. **a.** Graph $f(x) = 10^x$ and then use this graph to explain why the inverse of f is a function.
   b. Use the graph of $y = f(x)$ in part a to graph $y = f^{-1}(x)$.
   c. Find the domain and range of f.
   d. Find the domain and range of f^{-1}.

REMEMBER THIS

1. Solve $x^{1/3} = 9$.
2. Solve $9^x = \frac{1}{3}$.
3. Express 0.00038 in scientific notation.
4. Express 4.21×10^{-6} in regular notation.
5. Write $\{x : x > -2\}$ in interval notation.
6. Graph $y = x$.

7. If $f = \{(0,1),(1,3),(2,9)\}$, find the inverse function f^{-1}.
8. If $f(x) = \sqrt[3]{x}$, find $f^{-1}(x)$.
9. If $f(x) = x^2$, is f a one-to-one function?
10. If the domain of f is $(-\infty, \infty)$, the range of f is $(0, \infty)$, and f is a one-to-one function, then find the domain and range of the inverse function f^{-1}.

4.2 Logarithmic Functions

The **logarithm** (or **log**) of a number is the exponent to which a fixed base is raised to obtain the number. In the statement $10^2 = 100$, we call 10 the base, 100 the number, and 2 the exponent or logarithm. We say 2 is the log to the base 10 of 100. In logarithmic form $10^2 = 100$ is written as $\log_{10} 100 = 2$. In general, the key relation between exponential form and logarithmic form is

$$\log_b N = L \text{ is equivalent to } b^L = N,$$

where b and N are positive numbers with $b \neq 1$. Examples 1 and 2 illustrate this definition.

EXAMPLE 1 Write $4^2 = 16$ and $b^r = s$ in logarithmic form.

Solution Since $b^L = N$ implies $\log_b N = L$,

$$4^2 = 16 \text{ may be written as } \log_4 16 = 2$$
$$b^r = s \text{ may be written as } \log_b s = r.$$ ∎

EXAMPLE 2 Write $\log_2 \frac{1}{8} = -3$ and $\log_a u = v$ in exponential form.

Solution Since $\log_b N = L$ implies $b^L = N$,

$$\log_2 \frac{1}{8} = -3 \text{ may be written as } 2^{-3} = \frac{1}{8}$$
$$\log_a u = v \text{ may be written as } a^v = u.$$ ∎

Now that we have defined a logarithm, we introduce the concept of a logarithmic function. To begin, we first need to recall three facts about inverse functions from Section 2.5.

1. Two functions with exactly reverse assignments are inverse functions.
2. The domain of a function f is the range of its inverse function, and the range of f is the domain of its inverse.
3. A function is one-to-one when each x-value in the domain is assigned a different y-value so that no two ordered pairs have the same second component. We define the inverse function of f, denoted f^{-1}, only when f is a one-to-one function.

The logarithmic function is defined by a rule that reverses the assignments of the exponential function, which is a one-to-one function. We start with the exponential function $y = b^x$ and obtain its inverse by interchanging the variables x and y. The resulting equation $x = b^y$ is a rule that assigns to each positive number exactly one exponent. Since $x = b^y$ is equivalent to $y = \log_b x$, this rule is written in logarithmic form as $y = \log_b x$ and we have the following definition.

Logarithmic Function

For $b > 0$, $b \neq 1$, and $x > 0$,
$$y = \log_b x \text{ if and only if } x = b^y.$$
The function f defined by
$$f(x) = \log_b x \quad \text{for } x > 0$$
is called the **logarithmic function with base b.**

Note that the fixed base (b), as in the exponential function, may be any positive real number except 1. Also, since a positive base raised to any power is positive, only positive numbers may be substituted for x, so the domain of $f(x) = \log_b x$ is $(0,\infty)$. We expect this result because the domain and range of inverse functions are interchanged. Thus, the domain of $y = \log_b x$ is the range of $y = b^x$ (set of positive real numbers), and the range of $y = \log_b x$ is the domain of $y = b^x$ (set of all real numbers).

To illustrate that a logarithmic function reverses the assignments of an exponential function, consider Figure 4.3 in which $y = 2^x$ and $y = \log_2 x$ are sketched on the same axes.

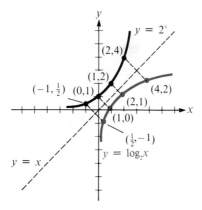

Figure 4.3

$y = 2^x$	x	-3	-2	-1	0	1	2	3	
	y	$\frac{1}{8}$	$\frac{1}{4}$	$\frac{1}{2}$	1	2	4	8	

$y = \log_2 x$ or $x = 2^y$	x	$\frac{1}{8}$	$\frac{1}{4}$	$\frac{1}{2}$	1	2	4	8
	y	-3	-2	-1	0	1	2	3

An examination of the tables and the two graphs shows that the functions have reverse assignments. Note in Figure 4.3 that the graphs are related in that each one is the reflection of the other across the line $y = x$ as expected.

Once again, we can graph variations of a familiar function $y = \log_b x$ by using the graphing techniques given in Section 2.3. Example 3 illustrates the case of a horizontal shift.

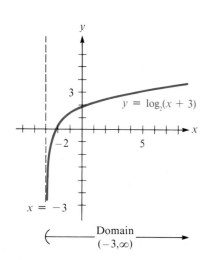

Figure 4.4

EXAMPLE 3   Graph $y = \log_2(x + 3)$. What is the domain of the function?

Solution   We start with the graph of $y = \log_2 x$ in Figure 4.3. To graph $y = \log_2(x + 3)$, we note that x has been replaced by $x + 3$, and the logarithmic rule is then applied to $x + 3$. Such a replacement means the graph of $y = \log_2(x + 3)$ is the graph of $y = \log_2 x$ shifted 3 units to the left. The completed graph is shown in Figure 4.4, and we read from this graph that the domain of $y = \log_2(x + 3)$ is $(-3,\infty)$. Without the graph, we may determine the domain by noting that logarithms are only defined for positive numbers. Thus, $x + 3 > 0$, so $x > -3$.   ∎

There are two specific bases associated with applications of logarithmic functions: base 10 or **common logarithms** and base e or **natural logarithms.** Logarithms to the base e are used extensively in calculus and will be considered in Section 4.5. Here we discuss common logarithms that are used to express many important formulas in science. As examples we have the formulas for pH (measure of the acidity of a solution), decibel (basic unit of loudness), and Richter scale ratings (measure of the magnitude of an earthquake). Before considering the pH formula, we first discuss how to evaluate common logarithm statements on a calculator.

It is standard notation to write a common logarithm (or $\log_{10} N$) as $\log N$ with base 10 being understood. Scientific calculators have a $\boxed{\log}$ key and to find $\log N$, simply enter N and press $\boxed{\log}$. Examples 4 and 5 illustrate this.

EXAMPLE 4 $\log 12.3$ $12.3 \boxed{\log}$ 1.0899051 ■

EXAMPLE 5 $\log(3.2 \times 10^{-5})$ $3.2 \boxed{\text{EE}} 5 \boxed{+/-} \boxed{\log}$ $\boxed{-4.4949\ 00}$ ■

To reverse this process and find N if we know $\log N$, we must remember that $\log_{10} x$ and 10^x are inverse functions. Therefore,

$$\boxed{\text{INV}} \ \boxed{\log} \text{ is equivalent to } \boxed{10^x}.$$

Some calculators write 10^x above the $\boxed{\log}$ key to specify that it is the inverse or second function associated with this key. We then find N as shown in Examples 6 and 7.

N equals

EXAMPLE 6 $\log N = 1.13$ $1.13 \boxed{\text{INV}} \boxed{\log}$ 13.489629 ■

EXAMPLE 7 $\log N = -7.4$ $7.4 \boxed{+/-} \boxed{\text{INV}} \boxed{\log}$ $\boxed{3.9811\ -08}$ ■

In summary, here are a few notes.

1. You must understand scientific notation and how it is displayed on a calculator, so you may need to study Appendix A.1. In the last example the display $\boxed{3.9811\ -08}$ means $N = 3.9811 \times 10^{-8}$.
2. In science it is common to refer to an inverse logarithm as an antilogarithm. Thus, antilog $x = $ INV log $x = 10^x$.
3. If you cannot obtain a calculator, logarithms and inverse logarithms can be determined as shown in Appendix A.3.
4. Remember that calculators replace log *evaluations,* not log *functions.* Analyzing relationships with logarithmic functions remains the major idea here.

We now consider the logarithmic function that defines pH.

You are probably aware of the concept of an acid, since it is associated with such familiar and diverse topics as indigestion, shampoo, swimming pools, and car batteries. In chemistry a logarithm is used to define pH (hydrogen potential), which is a convenient measure of the acidity of a solution. Briefly, let's see why pH is defined as a logarithm.

When an atom of hydrogen loses an electron, it is called a hydrogen ion. Since this atom is positively charged, the hydrogen ion is symbolized H^+. The electrical imbalance in an ion greatly affects chemical reactions, and the concentration of hydrogen ions (symbolized $[H^+]$) in a solution determines its acidity. However, hydrogen ion concentrations are very small numbers. For example, a weak acid solution might have a concentration (measured in moles/liter) of only 1 part H^+ in 10,000, or 1/10,000, or 10^{-4}. When writing such small numbers, the exponential form is obviously helpful. In 1909 a further simplification, pH notation, was introduced. With this notation only the exponent is considered, as follows:

$$\text{If } [H^+] = 10^{-4} \quad \text{then} \quad pH = 4$$
$$[H^+] = 10^{-8.3} \quad \text{then} \quad pH = 8.3$$
$$[H^+] = 10^{-x} \quad \text{then} \quad pH = x.$$

Since pH is defined as an exponent, we determine pH by finding a logarithm. That is, from

$$[H^+] = 10^{-pH}$$

we may write

$$-pH = \log_{10}[H^+]$$
$$pH = -\log[H^+].$$

Thus, pH is defined as the negative of the logarithm of the hydrogen ion concentration. A solution with $pH = 7$ is called neutral. In acids $pH < 7$; bases or alkalies have $pH > 7$.

EXAMPLE 8 The hydrogen ion concentration in a sample of human blood is 4.4×10^{-8}. Determine the pH of the sample.

Solution

$$pH = -\log[H^+]$$
$$= -\log(4.4 \times 10^{-8})$$

The calculation is as follows:

$$4.4 \boxed{\text{EE}} \, 8 \boxed{+/-} \, \boxed{\log} \qquad \boxed{-7.3565 \ 00}.$$

Thus, $pH \approx 7.4$. ∎

EXAMPLE 9   The pH of tomato juice is about 4.5. Determine $[H^+]$.

Solution

$$pH = -\log[H^+]$$

Replacing pH by 4.5 and solving for $\log[H^+]$, we have

$$\log[H^+] = -4.5.$$

We now compute an inverse logarithm as follows:

$$4.5 \boxed{+/-}\ \boxed{INV}\ \boxed{log} \qquad \boxed{3.1623\ -05}.$$

Thus, $[H^+] \approx 3.2 \times 10^{-5}$. ∎

EXERCISES 4.2

In Exercises 1–10 express in logarithmic form.

1. $3^2 = 9$ **2.** $2^3 = 8$ **3.** $(\frac{1}{2})^2 = \frac{1}{4}$
4. $(\frac{1}{3})^3 = \frac{1}{27}$ **5.** $4^{-2} = \frac{1}{16}$ **6.** $(\frac{1}{2})^{-3} = 8$
7. $25^{1/2} = 5$ **8.** $8^{-1/3} = \frac{1}{2}$ **9.** $7^0 = 1$
10. $10^{-4} = 0.0001$

In Exercises 11–20 express in exponential form.

11. $\log_5 5 = 1$ **12.** $\log_2 32 = 5$
13. $\log_{1/3} \frac{1}{9} = 2$ **14.** $\log_{1/2} \frac{1}{16} = 4$
15. $\log_2 \frac{1}{4} = -2$ **16.** $\log_{1/4} 4 = -1$
17. $\log_{49} 7 = \frac{1}{2}$ **18.** $\log_{27} \frac{1}{3} = -\frac{1}{3}$
19. $\log_{100} 1 = 0$ **20.** $\log_{10} 0.001 = -3$

In Exercises 21–30 find the value of each expression.

21. $\log_3 9$ **22.** $\log_2 16$ **23.** $\log_4 4$
24. $\log_5 1$ **25.** $\log_3 \frac{1}{3}$ **26.** $\log_{10} 0.01$
27. $\log_3 \sqrt{3}$ **28.** $\log_9 3$ **29.** $\log_{10}(\log_5 5)$
30. $\log_2(\log_4 2)$

In Exercises 31–40 determine the value of the unknown by inspection or by writing in exponential form.

31. $\log_2 8 = y$ **32.** $\log_4 2 = y$
33. $\log_5 x = -1$ **34.** $\log_{10} x = -5$
35. $\log_b 125 = 3$ **36.** $\log_b 10 = \frac{1}{2}$
37. $\log_{10} 10^3 = y$ **38.** $\log_{10} 10^{2.4} = y$
39. $\log_b b = 1$ **40.** $\log_b 1 = 0$

In Exercises 41–43 evaluate each function for the given elements of the domain.

41. $f(x) = \log_2 x$; $f(8)$, $f(0.5)$, $f(1)$
42. $g(x) = \log_9 x$; $g(3)$, $g(27)$, $g(\frac{1}{81})$
43. $h(x) = \log_{10} x$; $h(10{,}000)$, $h(1)$, $h(0.0000001)$

In Exercises 44–55 graph the function.

44. $y = \log_3 x$ **45.** $y = \log_{10} x$
46. $y = \log_{1/2} x$ **47.** $y = \log_{1/4} x$
48. $y = -\log_{1/4} x$ **49.** $y = -\log_4 x$
50. $y = \log_2(-x)$ **51.** $y = 1 - \log_2 x$
52. $y = 1 + \log_2 x$ **53.** $y = \log_2(1 + x)$
54. $y = \log_{10}|x|$ **55.** $y = |\log_{10} x|$

In Exercises 56–60 what is the domain of each function?

56. $f(x) = \log_{10}(2x - 1)$ **57.** $g(x) = \log_2(x^2 - 9)$
58. $h(x) = \log_5(-x)$ **59.** $f(x) = \log_{10}|x - 1|$
60. $f(x) = \log_{10}\sqrt{x + 1}$

In Exercises 61–64 determine the pH of the solution with the given hydrogen ion concentration.

61. $[H^+] = 10^{-7}$ (water)
62. $[H^+] = 4.2 \times 10^{-3}$ (5-percent vinegar)
63. $[H^+] = 2.5 \times 10^{-4}$ (orange juice)
64. $[H^+] = 3.3 \times 10^{-8}$ (swimming pool water)

In Exercises 65–68 determine the hydrogen ion concentration of the solution with the given pH.

65. pH = 0   (pure sulfuric acid)
66. pH = 8.9   (seawater)
67. pH = 2.1   (stomach juices)
68. pH = 11.3   (household ammonia)

   The most widely used unit in communications engineering is the **decibel**. This name is derived from the basic unit of loudness, the bel (in honor of Alexander Graham Bell), with decibel meaning one-tenth bel. The decibel is defined as a logarithm because the ear perceives changes in sound levels logarithmically. That is, the ear is very perceptive to changes in loudness at low sound levels; at high sound levels the intensity of the sound must be greatly increased for the ear to perceive change. The formula for decibels (dB) is

$$dB = 10 \log_{10} \frac{P_2}{P_1},$$

where P_1 and P_2 represent the powers required to produce sound levels 1 and 2. Use this formula in Exercises 69–72. (*Note:* Many other important measures are defined as the logarithm of a ratio. The Richter scale, which is a measure of the magnitude of an earthquake, and the formula for the magnitude or brightness of a star are common examples.)

69. Determine the difference in loudness (in decibels) between sounds 1 and 2 if the power ratio (P_2/P_1) between the sounds is
   a. 1   **b.** 10   **c.** 100   **d.** 1,000
   As the power ratio is progressively multiplied by 10, how does the loudness in decibels increase?

70. What is the decibel gain of an amplifier with a 3-watt input and a 30-watt output?

71. In a normal conversation the power ratio (P_2/P_1) between the highest and lowest sound volumes is about 300 to 1. What is the range of speech in decibels? (*Note:* The decibel scale for loudness ranges from 0 dB for a sound at the threshold of hearing to about 140 dB for an airplane engine. Normal conversation is rated at about 60 dB.)

72. When rating an amplifier from a radio or public address system, P_1 assumes the arbitrary reference level value of 6 milliwatts (or 0.006 watt). What is the gain in decibels of a 60-watt amplifier? That is, find dB when $P_2 = 60$ watts and $P_1 = 0.006$ watt.

THINK ABOUT IT

1. For the logarithmic function defined by $y = \log_b x$, explain why the restriction $b \neq 1$ is necessary for this equation to define a function.
2. If $\log_a 2 = r$ and $\log_b 5 = s$, then $a^r b^s$ equals
   a. $\log_{ab} 7$ **b.** $\log_{ab} 10$ **c.** 10   **d.** 7
3. **a.** If $\log_{10} n = x$, find $(0.001)^x$ in terms of n.
   b. If $\log_{10} x = k$, then express 10^{k+2} in terms of x.

4. The Richter scale rating for an earthquake of intensity I equals $\log_{10}(I/I_0)$, where I_0 is a standard reference level number. How many times stronger is a quake rated at 6 than a quake rated at 2?
5. For the logarithmic function defined by $f(x) = \log_b x$, which of the given equations specifies a property of f?
   a. $f(u + v) = f(u) + f(v)$
   b. $f(uv) = f(u) + f(v)$
   c. $f(u + v) = f(u) \cdot f(v)$

REMEMBER THIS

1. Find the product of 3^u and 3^v.
2. Simplify $5^{2x}/5^x$.
3. Simplify $(b^5)^3$.
4. Express $x^{2/3}$ in radical form.
5. Express $\sqrt{L}$ in exponential form.
6. Solve $32^x = 64$.
7. Solve $b^{3/2} = 64$.

8. If $f(x) = -4^x$, what is the range of f?
9. If $f(x) = 2x$ and $f^{-1}(x) = \frac{1}{2}x$, find $f[f^{-1}(x)]$ and $f^{-1}[f(x)]$.
10. You purchase a new car for $15,000 and take out a 5-year loan. If the car depreciates at a rate of 20 percent each year, then what is the value of the car by the time the loan is paid off?

4.3   **Properties of Logarithms**

Since a logarithm is an exponent, the properties of logarithms may be derived from the following properties of exponents.

1. $b^m \cdot b^n = b^{m+n}$
2. $b^m/b^n = b^{m-n}$ $(b \neq 0)$
3. $(b^m)^n = b^{mn}$
4. $b^1 = b$
5. $b^0 = 1$ $(b \neq 0)$

The logarithm properties that correspond to each of these laws are now given below.

Properties of Logarithms

> If b, x, and y are positive, with $b \neq 1$ and k any real number, then
>
> 1. $\log_b xy = \log_b x + \log_b y$
> 2. $\log_b \dfrac{x}{y} = \log_b x - \log_b y$
> 3. $\log_b(x)^k = k \log_b x$
> 4. $\log_b b = 1$
> 5. $\log_b 1 = 0$

To establish our first property, consider the product of two positive numbers x and y. Let

$$x = b^m \quad \text{or} \quad m = \log_b x$$

and

$$y = b^n \quad \text{or} \quad n = \log_b y,$$

then

$$x \cdot y = b^m \cdot b^n = b^{m+n}.$$

Writing the last equation in logarithmic form, we have

$$\log_b xy = m + n,$$

so

$$\log_b xy = \log_b x + \log_b y.$$

Through similar procedures we may establish properties 2 and 3 (and these proofs are requested in Exercise 45), while properties 4 and 5 are the direct result of the exponent laws $b^1 = b$ and $b^0 = 1$ and the definition of logarithm. An understanding of these properties is essential to the study of the theory and applications of logarithms, and you should consider carefully how these properties are used in the following examples.

EXAMPLE 1 Express each statement as the sum or difference of simpler logarithmic statements.

Statement	Solution
a. $\log_{10}(4 \cdot 5)$	$= \log_{10} 4 + \log_{10} 5$
b. $\log_2(\frac{3}{7})$	$= \log_2 3 - \log_2 7$
c. $\log_4 \sqrt{5}$	$= \log_4(5)^{1/2} = \frac{1}{2} \log_4 5$
d. $\log_{10}(xy/z)$	$= (\log_{10} x + \log_{10} y) - \log_{10} z$

$$\textbf{e. } \log_{10} 2\pi \sqrt{L/g} = \log_{10} 2 + \log_{10} \pi + \log_{10} \sqrt{L/g}$$
$$= \log_{10} 2 + \log_{10} \pi + \frac{1}{2} \log_{10}(L/g)$$
$$= \log_{10} 2 + \log_{10} \pi + \frac{1}{2}(\log_{10} L - \log_{10} g) \qquad \blacksquare$$

EXAMPLE 2 Express as a single logarithm with coefficient 1.

Statement	Solution
a. $\log_{10} 9 + \log_{10} 2$	$= \log_{10}(9 \cdot 2) = \log_{10} 18$
b. $\log_5 35 - \log_5 7$	$= \log_5(\frac{35}{7}) = \log_5 5 \qquad \text{(or 1)}$
c. $2 \log_4 5$	$= \log_4(5)^2 = \log_4 25$
d. $3 \log_b x + 2 \log_b y$	$= \log_b x^3 + \log_b y^2 = \log_b(x^3 \cdot y^2)$

$$\textbf{e. } \frac{1}{2}(\log_{10} x - 3 \log_{10} y) = \frac{1}{2}(\log_{10} x - \log_{10} y^3) = \frac{1}{2}(\log_{10}(x/y^3))$$
$$= \log_{10}(x/y^3)^{1/2} = \log_{10} \sqrt{x/y^3} \qquad \blacksquare$$

EXAMPLE 3 If $\log_{10} 2 = 0.3010$ and $\log_{10} 3 = 0.4771$, simplify the expressions given.

Expression	Solution
a. $\log_{10} 9$	$= \log_{10} 3^2 = 2 \log_{10} 3 = 2(0.4771) = 0.9542$

$$\textbf{b. } \log_{10} 12 = \log_{10}(2^2 \cdot 3) = \log_{10} 2^2 + \log_{10} 3$$
$$= 2 \log_{10} 2 + \log_{10} 3$$
$$= 2(0.3010) + 0.4771 = 1.0791$$

$$\textbf{c. } \log_{10} \sqrt{2} = \log_{10}(2)^{1/2} = \frac{1}{2} \log_{10} 2 = \frac{1}{2}(0.3010) = 0.1505 \qquad \blacksquare$$

There are two more important properties of logarithms that are often useful. We have been saying all along that a logarithm is an exponent. One way to express this idea is to say that

$$b^{\log_b x} = x$$

for all positive values of b and x. This property is merely a convenient way of restating the definition that $\log_b x$ is the exponent to which b is raised to obtain x. Similarly, since x is the exponent to which b is raised to obtain b^x, it follows that

$$\log_b b^x = x.$$

From a different viewpoint, these properties are a direct consequence of the inverse relation between exponential and logarithmic functions. That is, if we let $f(x) = b^x$, then $f^{-1}(x) = \log_b x$. From the definition of inverse functions in Section 2.5 we have

$$f[f^{-1}(x)] = f^{-1}[f(x)] = x.$$

Thus,

$$f[f^{-1}(x)] = b^{\log_b x} = x$$

and

$$f^{-1}[f(x)] = \log_b b^x = x.$$

The next example shows how these properties make it easy to simplify certain expressions that initially appear to be complicated.

EXAMPLE 4   Simplify.

a. $2^{\log_2 8}$

b. $\log_{10} 10^p$

Solution

a. As a direct result of the above property, $2^{\log_2 8} = 8$. We can verify this answer since $\log_2 8 = 3$ so that $2^{\log_2 8} = 2^3 = 8$.

b. Since $\log_b b^x = x$, we have $\log_{10} 10^p = p$. We also could reason $\log_{10} 10^p = p \log_{10} 10 = p(1) = p$.   ∎

EXERCISES 4.3

In Exercises 1–18 express each logarithm as the sum or difference of simpler logarithms.

1. $\log_{10}(7 \cdot 5)$
2. $\log_b xyz$
3. $\log_6(\frac{3}{5})$
4. $\log_b(x/y)$
5. $\log_5(11)^{16}$
6. $\log_b x^{16}$
7. $\log_2 \sqrt{3}$
8. $\log_b \sqrt{x}$
9. $\log_{10} \sqrt[5]{16}$
10. $\log_b \sqrt[5]{x}$
11. $\log_4(4^2 \cdot 3^3)$
12. $\log_b x^2 y^3$
13. $\log_b \sqrt{xy}$
14. $\log_b \sqrt{x/y}$
15. $\log_b \sqrt[3]{x^5}$
16. $\log_b(x)^\pi$
17. $\log_b \sqrt[4]{xy^2/z^3}$
18. $\log_{10} \sqrt{s(s-a)(s-b)(s-c)}$

In Exercises 19–32 express each statement as a single logarithm with coefficient 1.

19. $\log_2 3 + \log_2 4$
20. $\log_b x + \log_b y$
21. $\log_4 20 - \log_4 5$
22. $\log_b y - \log_b x$
23. $2 \log_7 3$
24. $3 \log_b z$
25. $\frac{1}{2} \log_{10} 9 - 3 \log_{10} 2$
26. $3 \log_b x - \frac{1}{2} \log_b z$
27. $\frac{1}{3} \log_b x + \frac{2}{3} \log_b y$
28. $2 \log_b x + \log_b(x + y)$

29. $\log_b(x^2 - 1) - \log_b(x + 1)$
30. $\log_b(y - 2) + \log_b y - 2 \log_b x$
31. $\frac{1}{2}[\log_b x - (5 \log_b y - 3 \log_b z)]$
32. $\frac{1}{2}[(\log_b x - 5 \log_b y) - 3 \log_b z]$

In Exercises 33–40 simplify each expression.

33. $\log_{10} 10^5$
34. $\log_2(4^{1/3})$
35. $\log_{10}(m \times 10^k)$
36. $3^{\log_3 2}$
37. $2^{\log_2 15 - \log_2 5}$
38. $b^{r \log_b N}$
39. $\dfrac{n}{x} b^{n \log_b x}$
40. $b^{\log_b x + \log_b y}$

41. If $\log_{10} 2 = 0.3010$ and $\log_{10} 3 = 0.4771$, find
    a. $\log_{10} 6$
    b. $\log_{10} 4$
    c. $\log_{10} \sqrt{3}$
    d. $\log_{10} \frac{2}{3}$
    e. $\log_{10}(0.5)$
    f. $\log_{10} 5$

42. If $\log_a 2 = x$ and $\log_a 3 = y$, express each of the following in terms of x and y.
    a. $\log_a 27$
    b. $\log_a 72$
    c. $\log_a \sqrt{\frac{1}{3}}$
    d. $\log_a \sqrt[5]{\frac{2}{3}}$

43. Show that $\log_b(1/a) = -\log_b a$.

44. If $f(x) = 10^x$ and $f^{-1}(x) = \log_{10} x$, verify that $f^{-1}[f(x)] = x$ for all x in the domain of f.

45. Prove the following properties of logarithms.

a. $\log_b \dfrac{x}{y} = \log_b x - \log_b y$

b. $\log_b(x)^k = k \log_b x$

46. What restrictions are placed on x, y, b, and k in Exercise 45?

THINK ABOUT IT

1. a. If $\log_{10} \sqrt{x} = a$, then express $\log_{10} 100x$ in terms of a.

   b. If $\log_{10} 10! = x$, find $\log_{10} 9!$ in terms of x. (*Note:* $n! = n(n - 1) \cdots 3 \cdot 2 \cdot 1$.)

2. If $y = \log_{10} N$, then $y + 3$ equals

   a. $\log_{10} 3N$ **b.** $\log_{10} 1{,}000N$

   c. $\log_{10}(N + 1{,}000)$ **d.** $\log_{10}(N + 3)$

3. Liquid X has 100 times the concentration of hydrogen ions as liquid Y. What is the relation of the pH of X to the pH of Y?

4. Give specific counterexamples to *disprove* each of the following statements.

   a. $\log_b xy = (\log_b x)(\log_b y)$

   b. $(\log_b x)(\log_b y) = \log_b x + \log_b y$

   c. $\log_b \dfrac{x}{y} = \dfrac{\log_b x}{\log_b y}$

   d. $\dfrac{\log_b x}{\log_b y} = \log_b x - \log_b y$

   e. $\log_b x^k = (\log_b x)^k$

   f. $(\log_b x)^k = k \log_b x$

5. Examine the following line of reasoning: $3 > 2$. If we multiply both sides of the inequality by $\log_{10}(\frac{1}{2})$, we have

$$3 \log_{10}(\tfrac{1}{2}) > 2 \log_{10}(\tfrac{1}{2})$$
$$\log_{10}(\tfrac{1}{2})^3 > \log_{10}(\tfrac{1}{2})^2$$
$$\log_{10}(\tfrac{1}{8}) > \log_{10}(\tfrac{1}{4}).$$

Thus,

$$\tfrac{1}{8} > \tfrac{1}{4}.$$

Our conclusion is incorrect. What went wrong?

REMEMBER THIS

1. Express in logarithmic form: $a^r = s$.

2. Express in exponential form: $\log x = -3$.

3. Solve $x(x - 4) = 5$.

4. Solve for x: $(x + 2)m = (2x - 1)n$.

5. True or false: $\dfrac{\log 25}{\log 8} = \log 25 - \log 8$.

6. If $f(x) = \log x$, what is the domain of f?

7. If $f(x) = \log_5 x$, find $f(1)$.

8. If $g(x) = 9^{-x}$, find $g(\frac{3}{2})$.

9. If $f(x) = 8^{x-1}$, for what value of x does $f(x) = \frac{1}{16}$?

10. The pH of a sample of human blood is 7.5. Use $\text{pH} = -\log_{10}[\text{H}^+]$ to determine the hydrogen ion concentration $[\text{H}^+]$ of this sample.

4.4   Exponential and Logarithmic Equations

In Section 4.1 we solved exponential equations in which it was not hard to rewrite the expressions in terms of a common base. For instance, $9^x = \frac{1}{27}$ was written as $3^{2x} = 3^{-3}$, so $x = -\frac{3}{2}$. Since this procedure is limited, we now consider a general approach that uses the following principle, which is based on the fact that a logarithmic correspondence is a one-to-one function.

Equation-Solving Principle

If x, y, and b are positive real numbers with $b \neq 1$, then

1. $x = y$ implies $\log_b x = \log_b y$; and conversely,
2. $\log_b x = \log_b y$ implies $x = y$.

We can therefore solve **exponential equations** (the unknown is in the exponent) by taking the logarithm of both sides of the equation, as shown in the following examples.

EXAMPLE 1   Solve for x: $3^x = 8$.

Solution

$$3^x = 8$$
$$\log 3^x = \log 8 \qquad \text{Apply common logarithms to each side.}$$
$$x \log 3 = \log 8 \qquad \text{Property of logarithms}$$
$$x = \frac{\log 8}{\log 3}$$

If an approximation is desired, compute

$$8 \boxed{\log} \div 3 \boxed{\log} \boxed{=} \qquad 1.8927893.$$

Thus, $x \approx 1.893$, and the solution set is $\{1.893\}$ when rounded to four significant digits.   ■

EXAMPLE 2   Solve for x: $3^{x+2} = 5^{2x-1}$.

Solution

$$3^{x+2} = 5^{2x-1}$$
$$\log 3^{x+2} = \log 5^{2x-1} \qquad \text{Apply common logarithms to each side.}$$
$$(x+2)\log 3 = (2x-1)\log 5 \qquad \text{Property of logarithms}$$
$$x \log 3 + 2 \log 3 = 2x \log 5 - \log 5 \qquad \text{Distributive property}$$
$$x \log 3 - 2x \log 5 = -2 \log 3 - \log 5 \qquad \text{Equivalent equation grouping } x \text{ terms}$$
$$x(\log 3 - 2 \log 5) = -2 \log 3 - \log 5 \qquad \text{Factoring}$$
$$x = \frac{-2 \log 3 - \log 5}{\log 3 - 2 \log 5} \approx 1.795$$

To four significant digits, the solution set is $\{1.795\}$.   ■

The next example is not an exponential equation, but it is difficult to solve because the exponent is not a rational number. When an exponent is troublesome, a logarithmic solution is not far behind.

EXAMPLE 3   Solve for x: $x^{\sqrt{2}} = 23$.

Solution

$$x^{\sqrt{2}} = 23$$
$$\log x^{\sqrt{2}} = \log 23$$
$$\sqrt{2} \log x = \log 23$$
$$\log x = \frac{\log 23}{\sqrt{2}}$$

To solve for x, evaluate $(\log 23)/\sqrt{2}$ and then compute an inverse logarithm.

$$23 \boxed{\log} \boxed{\div} 2 \boxed{\sqrt{}} \boxed{=} \boxed{\text{INV}} \boxed{\log} \qquad 9.1809366$$

Thus, $x \approx 9.181$, so to four significant digits, the solution set is $\{9.181\}$.   ■

We may change a logarithm in one base to a logarithm in another base by the above procedure. Suppose that $y = \log_b x$ and we wish to change to base a. This expression in exponential form is

$$b^y = x.$$

Taking the logarithm to the base a of both sides of the equation, we have

$$\log_a b^y = \log_a x$$
$$y \log_a b = \log_a x$$
$$y = \frac{\log_a x}{\log_a b}.$$

Since $y = \log_b x$, we have

$$\log_b x = \frac{\log_a x}{\log_a b}.$$

EXAMPLE 4   Use logarithms to the base 10 to determine

a. $\log_2 7$ **b.** $\log_5 0.043$

Solution

a. $\log_2 7 = \dfrac{\log 7}{\log 2} \approx 2.807$

b. $\log_5 0.043 = \dfrac{\log 0.043}{\log 5} \approx -1.955$ ■

A **logarithmic equation** (the unknown is in the log statement) is sometimes solved by using the principle that $\log_b x = \log_b y$ implies $x = y$. For instance, if $\log(3x - 8) = \log x$, then $3x - 8 = x$, and $x = 4$. If the equation contains only one log statement, we often solve by changing to exponential form. Thus, if $\log (x + 7) = 2$, then $x + 7 = 10^2$, and

$x = 93$. In some cases we obtain the single log statement by applying a property of logarithms as shown in the next two examples. One note of caution here: Always check your answers in the original equation and accept only solutions that result in the logarithms of positive numbers.

EXAMPLE 5 Solve for x: $\log_3(x^2 - 4) - \log_3(x + 2) = 2$.

Solution

$$\log_3(x^2 - 4) - \log_3(x + 2) = 2$$
$$\log_3\left(\frac{x^2 - 4}{x + 2}\right) = 2$$
$$\log_3(x - 2) = 2$$
$$3^2 = x - 2$$
$$11 = x$$

If we substitute 11 in our original equation, both log expressions are defined. Thus, the solution set is $\{11\}$. ∎

EXAMPLE 6 Solve for x: $\log_2(x - 3) = 2 - \log_2 x$.

Solution   First rewrite the given equation as

$$\log_2 x + \log_2(x - 3) = 2$$

so that we may apply a property of logarithms to obtain

$$\log_2[x(x - 3)] = 2.$$

Now change from logarithmic form to exponential form and solve.

$$2^2 = x(x - 3)$$
$$4 = x^2 - 3x$$
$$0 = x^2 - 3x - 4$$
$$0 = (x - 4)(x + 1)$$

$$x - 4 = 0, \quad \text{or} \quad x + 1 = 0$$
$$x = 4 \qquad\qquad x = -1$$

A check in the original equation shows that 4 is a solution, while -1 is extraneous because the domain of $\log_2 x$ is the set of positive real numbers. Thus, the solution set is $\{4\}$. ∎

EXERCISES 4.4

In Exercises 1–20 use logarithms to solve the equation.

1. $2^x = 10$
2. $5^x = 100$
3. $4^{x+1} = 17$
4. $8^{x-1} = 25$
5. $2^{-x} = 9$
6. $3^{2x} = 5$
7. $(0.7)^t = 0.3$
8. $(1.05)^t = 3$
9. $2^x = 3^{2x+1}$
10. $3^{2x-1} = 4^{x+2}$

11. $5^{x-2} = 6^{2x}$
12. $10^{1-x} = 5^{-x}$
13. $x^{-0.6} = 12$
14. $x^{-4.1} = 47.6$
15. $x^{\sqrt{2}} = 8$
16. $(\sqrt{2})^x = 8$
17. $(1 - r)^8 = \frac{1}{2}$
18. $\left(1 + \frac{r}{2}\right)^4 = 2$

19. $6{,}000 = 2{,}000(1.03)^{2t}$

20. $1{,}000 = 100\left(1 + \dfrac{0.05}{4}\right)^{4t}$

In Exercises 21–50 solve the logarithmic equation.

21. $\log x = 2$
22. $\log x = 0$
23. $5 \log x = 5$
24. $2 \log x = \log 2$
25. $\log(1 - x) = -1$
26. $\log(-x) = \frac{1}{2}$
27. $\log(2x - 5) = \log x$
28. $\log x = \log(1 - x)$
29. $2 \log x = \log 8$
30. $3 \log x = \log(3x)$
31. $1 + \log x = \log 5$
32. $\log 2 + \log 3 = \log x$
33. $\log_4 x + \log_4 2 = 1$
34. $\log_3 x - \log_3 4 = 2$
35. $\log_2(x^2 - 1) = 3$
36. $\log_5(x + 1) - \log_5 x = 1$
37. $\log_6(x + 1) + \log_6 x = 1$
38. $\log_2(x + 1) + \log_2(x + 4) = 2$
39. $\log_2(x - 2) + \log_2 x = 3$
40. $\log x = 1 - \log(x - 3)$
41. $\log_3(x - 4) = 2 - \log_3(x + 4)$
42. $\log_2(x + 1) = 3 - \log_2(x - 1)$

43. $\log(x + 6) - 2 \log x = 0$
44. $2 \log x - \log(x + 2) = 0$
45. $\log(x - 4) + \log(3x - 4) = \log 11$
46. $\log_5(5x - 6) + \log_5(x - 1) = \log_5 4$
47. $\log_4(x - 1) - \log_4(x + 3) = \log_4 x$
48. $\log(3x - 2) = 1 + \log(x + 4)$
49. $\log_b 2x = \log_b 4x - \log_b 2$
50. $\log_7(x^2 - x) = \log_7 x + \log_7(x - 1)$

In Exercises 51–58 use logarithms to the base 10 to determine each logarithm.

51. $\log_2 9$ **52.** $\log_4 31$ **53.** $\log_3 5$
54. $\log_5 3$ **55.** $\log_{1/2} 19$ **56.** $\log_{1/3} \frac{2}{3}$
57. $\log_4 0.012$ **58.** $\log_6 0.735$
59. If $\log_b a = 3$, find $\log_a b$.
60. Simplify $\log_x 5 \cdot \log_5 x$.
61. Simplify $(\log_2 10)(\log_{10} 12 - \log_{10} 3)$.
62. How long will it take for money invested at 5 percent compounded annually to double?
63. At what interest rate (compounded annually) must a sum of money be invested if it is to double in 4 years?

THINK ABOUT IT

1. If $A = P(1 + r)^t$, express t in terms of the common logarithms of A, P, and $1 + r$.
2. If $N = 3^\pi$, approximate N to the nearest thousandth using properties of logarithms. Check your answer using a calculator.
3. If $\log_b x = a$, find $\log_{1/b} x$ in terms of a.
4. Solve each equation.
   a. $x^2 10^x = 10^x$ **b.** $\log(\log x) = 1$
   c. $2^{2x} - 20 = 2^x$

5. If $b^x = a^{x-1}$ and a and b are positive numbers other than 1, then x equals
   a. $\dfrac{\log a}{\log(a/b)}$ **b.** $\dfrac{\log b}{\log(b/a)}$
   c. $\dfrac{\log(b/a)}{\log b}$ **d.** $\dfrac{\log(a/b)}{\log a}$

REMEMBER THIS

1. True or false: An irrational number is a real number that cannot be expressed as a repeating decimal.
2. If $f(x) = \log_{10} x$ and $g(x) = 10^x$, find $(f \circ g)(x)$.
3. If f and g are inverse functions, then how are their graphs related?
4. Graph $y = 3^x$ and $y = \log_3 x$ on the same coordinate system.
5. Simplify $4^{\log_4 2}$.
6. Simplify $\frac{1}{2} \log_3 36 - \log_3 2$.

7. Express $\frac{1}{3}(\log_b x + 2 \log_b y)$ as a single logarithm with coefficient 1.
8. Express $\log(x/y^4)$ as the sum or difference of simpler logarithms.
9. If $\log_b 2 = x$ and $\log_b 3 = y$, express $\log_b \sqrt{6b}$ in terms of x and y.
10. $8{,}000 is invested at 9.4 percent compounded annually. How much is the investment worth at the end of 7 years?

4.5 More Applications and the Number *e*

In Section 4.1 the applications describe situations in which no change is considered to have occurred until an entire time period has elapsed. We now wish to discuss phenomena that change continuously. To illustrate this type of change, we study more extensively the concept of compound interest.

EXAMPLE 1 $2,000 is invested at 6 percent compounded semiannually.

a. Find a formula showing the value of the investment at the end of *t* years.

b. How much is the investment worth after 2 years?

Solution

a. Since the amount is compounded semiannually, there are two conversions each year at a 3-percent interest rate. The $2,000 amounts to $2,000(1 + \frac{0.06}{2})$ after $\frac{1}{2}$ year; $2,000(1 + \frac{0.06}{2})^2$ after 1 year; and $2,000(1 + \frac{0.06}{2})^{2t}$ after *t* years. Thus,

$$A = 2,000(1.03)^{2t}.$$

b. Evaluating the function when $t = 2$, we have

$$A = 2,000(1.03)^{2(2)}$$
$$= 2,000(1.03)^4$$
$$= \$2,251.02.$$

■

We can generalize the procedure from Example 1 to obtain the following formula for the amount (including principal and interest) of an investment in terms of four items: (1) the original principal, *P;* (2) the rate of interest, *r;* (3) the number of conversions per year, *n;* and (4) the number of years, *t*.

$$A = P\left(1 + \frac{r}{n}\right)^{nt}$$

EXAMPLE 2 $100 is invested at 5-percent interest compounded quarterly. How much will it amount to in 3 years?

Solution   Substituting $P = 100$, $r = 0.05$, $n = 4$, and $t = 3$ in the formula above, we have

$$A = 100\left(1 + \frac{0.05}{4}\right)^{4(3)}$$
$$= 100(1.0125)^{12}$$
$$= \$116.08.$$

■

You should note that if the interest is computed frequently and added to the principal, the amount grows at a faster rate. Thus, it is more profitable to an investor for the conversion periods to be shorter. For example, $10,000 amounts to $10,600 when compounded annually at 6 percent; $10,609 when compounded semiannually; and $10,613.64 when compounded quarterly. An important consideration is to determine the effect of compounding more frequently, such as every day, or even every minute. Does the compounded amount become astronomical or is the growth limited?

To determine the result of continuous growth, let us examine the return on a $1 investment for 1 year at an interest rate of 100 percent. Progressively, we will compound our investment more frequently and observe the result. (*Note:* A $1 investment and a 100-percent interest rate are very unlikely, but we chose them to illustrate the situation with the easiest numbers.)

$P = \$1, r = 100\%$ or 1, and $t = 1$. Substituting in the formula

$$A = P\left(1 + \frac{r}{n}\right)^{nt},$$

we have

$$A = 1\left(1 + \frac{1}{n}\right)^{n(1)},$$

which simplifies to

$$A = \left(1 + \frac{1}{n}\right)^{n}.$$

If compounded

annually ($n = 1$):   then $A = (1 + \frac{1}{1})^1 = 2.00$

semiannually ($n = 2$):   then $A = (1 + \frac{1}{2})^2 = (1.5)^2 = 2.25$

quarterly ($n = 4$):   then $A = (1 + \frac{1}{4})^4 = (1.25)^4$
$\approx 2.441 \ldots$

monthly ($n = 12$):   then $A = (1 + \frac{1}{12})^{12} \approx (1.08333)^{12}$
$\approx 2.613 \ldots$

daily ($n = 365$):   then $A = (1 + \frac{1}{365})^{365} \approx (1.00274)^{365}$
$\approx 2.714 \ldots$

hourly ($n = 8,760$):   then $A = (1 + \frac{1}{8,760})^{8,760}$
$\approx (1.000114)^{8,760}$
$\approx 2.718. \ldots$

Note the small difference between compounding daily and compounding hourly. More frequent conversions result in even smaller changes in the amount, and it can be shown that as n gets larger $(1 + 1/n)^n$ gets closer to an irrational number that is denoted by the letter e (to honor the Swiss mathematician Leonhard Euler). This number is extremely important in mathematics and to six significant digits

$$e \approx 2.71828.$$

Thus, a \$1 investment compounded every instant (or continuously) for 1 year at an interest rate of 100 percent amounts to \$2.71828 . . . , or \$e. In the next year each dollar again grows to \$e, so at the end of 2 years the amount is \$$e^2$. We can therefore find the amount to which the \$1 has grown in t years by the formula

$$A = e^t,$$

where the base in the exponential function is the irrational number e.

This formula has to be generalized to handle more practical principals and interest rates as well as other physical situations that change in a similar manner. The generalized formula is

$$A = A_0 e^{kt},$$

where

$$A = \text{amount at time } t$$
$$A_0 = \text{initial amount}$$
$$k = \text{growth or decay constant.}$$

The important idea is that an exponential function with base e is used in a situation where a quantity is *continuously* either growing or decaying. The rate of increase or decrease is proportional to the amount present at any instant. k is positive if the quantity is increasing and negative if the quantity is decreasing.

EXAMPLE 3   \$2,000 is invested at 6 percent compounded continuously. How much is the investment worth in 10 years?

Solution $A_0 = \$2,000$, $k = 0.06$, and $t = 10$. Substituting in the formula

$$A = A_0 e^{kt},$$

we have

$$A = 2,000 e^{(0.06)(10)}$$
$$= 2,000 e^{0.6}.$$

We compute $2,000 e^{0.6}$ by pressing

$$2,000 \;\boxed{\times}\; 0.6 \;\boxed{e^x}\; \boxed{=} \qquad 3,644.2376,$$

although on some calculators you need the following sequence, which is explained shortly:

$$2,000 \;\boxed{\times}\; 0.6 \;\boxed{\text{INV}}\; \boxed{\ln}\; \boxed{=} \qquad 3,644.2376.$$

Thus, $A \approx \$3,644.24$. (*Note:* If we refer to Table 4 at the end of the book, we find that $e^{0.6} \approx 1.8221$ so that $A \approx 2,000(1.8221) \approx \$3,644.20$.) ■

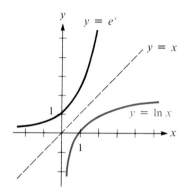

Figure 4.5

The calculator evaluation of $e^{0.6}$ in Example 3 once again forces us to consider the inverse relation between exponential and logarithmic functions. Just as 10^x and log x are inverse functions, so are e^x and $\log_e x$. Logarithms to the base e are called **natural logarithms** and $\log_e x$ is usually written as ln x. As with all inverse functions, the graphs of $y = e^x$ and $y = \ln x$ are symmetric about the line $y = x$ as shown in Figure 4.5. Thus,

$$\boxed{\text{INV}}\ \boxed{\text{ln}} \text{ is equivalent to } \boxed{e^x},$$

which explains the second keystroke sequence in Example 3. All the properties of logarithms and the equation-solving techniques from Section 4.4 apply to natural logarithms (or ln statements). In particular, note that the properties $\log_b b^x = x$ and $b^{\log_b x} = x$ (if $x > 0$) lead directly to

$$\ln e^x = x \text{ and } e^{\ln x} = x \text{ (if } x > 0)$$

when we use base e. Before continuing our discussion of exponential functions, it is useful to consider a few problems using these ideas.

EXAMPLE 4   Find x in decimal form.

a. $x = e^3$ **b.** $\ln x = 3$ **c.** $x = e^{\ln 3}$
d. $x = \ln e^3$ **e.** $e^x = 3$

Solution

a. Here we simply compute e^3 with one possible calculator sequence being

$$3 \ \boxed{\text{INV}}\ \boxed{\text{ln}} \qquad 20.085537.$$

b. $\ln x = 3$ is equivalent to $x = e^3$, so parts a and b have the same solution.
c. Although we could use a calculator, it is not necessary. Since $e^{\ln x} = x$ (if $x > 0$), we know $e^{\ln 3} = 3$.
d. It follows from $\ln e^x = x$ that $\ln e^3 = 3$.
e. As in Section 4.4, proceed as follows:

$$\begin{aligned}
e^x &= 3 \\
\ln e^x &= \ln 3 && \text{Apply natural logs to each side.} \\
x &= \ln 3 && \text{Property of logarithms} \\
x &\approx 1.0986123. && \text{Calculator}
\end{aligned}$$

(*Note:* If you cannot obtain a calculator, you can evaluate natural logarithms by using Table 3 at the end of the book. Appendix A.3 discusses the method.) ∎

EXAMPLE 5   Solve for x: $\ln x + \ln(x + 1) = 2$.

Solution   Applying logarithmic properties leads to a quadratic equation as shown below.

$$\ln x + \ln(x + 1) = 2$$
$$\ln[x(x + 1)] = 2$$
$$x(x + 1) = e^2$$
$$x^2 + x - e^2 = 0$$

Now using the quadratic formula with $a = 1$, $b = 1$, and $c = -e^2$ gives

$$x = \frac{-1 \pm \sqrt{1^2 - 4(1)(-e^2)}}{2(1)} = \frac{-1 \pm \sqrt{4e^2 + 1}}{2}.$$

A check in the original equation shows $(-1 + \sqrt{4e^2 + 1})/2$ is a solution (use your calculator), while $(-1 - \sqrt{4e^2 + 1})/2$ is extraneous (why?). Thus, the solution set is $\{(-1 + \sqrt{4e^2 + 1})/2\}$.   ∎

Now that we can solve equations involving e and natural logarithms, we continue our discussion of exponential functions.

EXAMPLE 6   An amount is invested at 7 percent compounded continuously. In how many years does the amount double?

Solution $A = 2A_0$ and $k = 0.07$. Substituting in the formula

$$A = A_0 e^{kt},$$

we have

$$2A_0 = A_0 e^{0.07t}$$
$$2 = e^{0.07t}$$
$$\ln 2 = \ln e^{0.07t} \qquad \text{Apply natural logs to each side.}$$
$$\ln 2 = 0.07t \qquad \text{Property of logarithms}$$
$$t = \frac{\ln 2}{0.07}.$$

To evaluate t, one possible calculator sequence is

$$2 \boxed{\ln} \boxed{\div} 0.07 \boxed{=} \qquad 9.9021026.$$

Thus, $t \approx 9.9$ years.   ∎

EXAMPLE 7   Physicists tell us that the half-life of radium is 1,620 years; that is, a given amount (A) of radium decomposes to the amount $A/2$ in 1,620 years. Radium (like all radioactive substances) decays at a rate proportional to the amount present. Thus, the formula $A = A_0 e^{kt}$ is appropriate. If a research center owns 1 g of radium, how much radium will it have in 100 years?

Solution   We know that $A_0 = 1$ g; thus, our formula

$$A = A_0 e^{kt}$$

becomes

$$A = 1e^{kt}.$$

We can find k since we know that $A = 0.5$ when $t = 1,620$.

$$0.5 = e^{k(1,620)}$$
$$\ln 0.5 = \ln e^{1,620k} \qquad \text{Apply natural logs to each side.}$$
$$\ln 0.5 = 1,620k \qquad \text{Property of logarithms}$$
$$k = \frac{\ln 0.5}{1,620} \approx -0.00043$$

The formula is then

$$A = e^{-0.00043t}.$$

Substituting $t = 100$, we have

$$A = e^{-0.00043(100)}$$
$$= e^{-0.043}$$
$$A \approx 0.96 \text{ g.} \qquad \blacksquare$$

EXAMPLE 8   If there is no restriction on food and living space, the rate of growth of a population of living organisms is proportional to the size of the population. Thus, the formula $A = A_0 e^{kt}$ is appropriate. Assume that in the absence of hunters, a certain animal population in New York State triples every 11 years.

a. Find the size of a population that initially numbers 500 after 5 years.
b. In how many years will the population number 5,000?

Solution

a. We know that $A_0 = 500$. Thus, our formula

$$A = A_0 e^{kt}$$

becomes

$$A = 500 e^{kt}.$$

We can find k since we know that $A = 1,500$ when $t = 11$.

$$1,500 = 500 e^{k(11)}$$
$$3 = e^{11k}$$
$$\ln 3 = \ln e^{11k}$$
$$\ln 3 = 11k$$
$$k = \frac{\ln 3}{11} \approx 0.10$$

The formula is then

$$A = 500e^{0.1t}.$$

Substituting $t = 5$, we have

$$A = 500e^{0.1(5)}$$
$$= 500e^{0.5}$$
$$= 500(1.6487)$$
$$A \approx 824.$$

b. Substituting $A = 5,000$ in our formula, we have

$$5,000 = 500e^{0.1t}$$
$$10 = e^{0.1t}$$
$$\ln 10 = \ln e^{0.1t}$$
$$\ln 10 = 0.1t$$
$$t = \frac{\ln 10}{0.1} \approx 23 \text{ years.}$$ ∎

EXAMPLE 9   Newton's law of cooling states that when a warm body is placed in colder surroundings at temperature t_α, the temperature (T) of the body at time t is given by

$$T - t_\alpha = D_0 e^{kt},$$

where D_0 is the initial difference in temperature and k is a constant. Because of an ice storm, there is a loss of power in a home heated to 68° F. If the outside temperature is 28° F and the temperature in the house drops from 68 to 64° F in 1 hour, when will the temperature in the house be down to 50° F?

Solution   We know that $t_\alpha = 28$ and $D_0 = 68 - 28 = 40$. Thus, our formula becomes

$$T - 28 = 40e^{kt}.$$

We can find k since we know that $T = 64$ when $t = 1$.

$$64 - 28 = 40e^{k(1)}$$
$$0.9 = e^k$$
$$\ln 0.9 = \ln e^k$$
$$\ln 0.9 = k$$
$$k \approx -0.105$$

(*Note:* k depends on the insulation in the house.) The formula is then

$$T - 28 = 40e^{-0.105t}.$$

Substituting $T = 50$, we have

$$50 - 28 = 40e^{-0.105t}$$
$$0.55 = e^{-0.105t}$$
$$\ln 0.55 = \ln e^{-0.105t}$$
$$\ln 0.55 = -0.105t$$
$$t = \frac{\ln 0.55}{-0.105}$$
$$t \approx 5.7 \text{ hours.}$$ ∎

In this section we discussed in detail only a few of the basic applications of the number e. However, keep in mind that the scope and power of the number e is immense. For example, consider the following relationships.

1. When the power is turned off, the electric current in a circuit does not vanish instantly but rather decreases exponentially.
2. The intensity of sunlight decreases exponentially as a diver descends further into ocean depths.
3. When a drug is injected into the bloodstream, the drug drains out of the body exponentially. Analyzing this relationship is important in medicine for it helps the doctor plan a series of injections that will maintain the prescribed level of some drug (such as an antibiotic) in a patient's bloodstream.
4. Atmospheric pressure decreases exponentially as altitude above sea level increases.

The list could easily be continued. In calculus you will probably consider some of these applications.

EXERCISES 4.5

In Exercises 1–20 write the indicated letter in decimal form or solve for the indicated letter.

1. $e^{0.16} = a$; for a
2. $e^{-0.35} = a$; for a
3. $e^a = 51$; for a
4. $e^a = 0.8906$; for a
5. $\ln x = -5.2$; for x
6. $\ln x = 2.18$; for x
7. $\ln e^x = 5$; for x
8. $\ln e^{4x} = 12$; for x
9. $e^{\ln 2} = k$; for k
10. $e^{2 \ln 3} = b$; for b
11. $10^x = e$; for x
12. $10^x = e^2$; for x
13. $\ln x = 2 \ln 3$; for x
14. $\ln x = 2 \log 3$; for x
15. $e^{2k} = 2$; for k
16. $e^{-k} = 0.25$; for k
17. $e^{-0.11t} = 0.45$; for t
18. $e^{5,600k} = 0.5$; for k
19. $e^{kt} = \frac{1}{2}$; for t
20. $y = ce^{kt}$; for t

21. Find the compounded amount for the following investments.
    a. $1,000 compounded semiannually for 3 years at 6 percent
    b. $5,000 compounded quarterly for 2 years at 12 percent
    c. $3,000 compounded monthly for 1 year at 5.4 percent

22. An amount is invested at 6 percent compounded monthly. In how many years does the amount triple?

23. How long will it take for $20,000 to grow to $30,000 at 10 percent compounded quarterly?

24. You are going to invest $10,000 for 3 years. Which offer is best?
    a. 10 percent compounded monthly
    b. 10.7 percent compounded annually
    c. 9.8 percent compounded continuously

25. Find the compounded amount for the following investments.
    a. $1,000 compounded continuously for 3 years at 6 percent

b. $30,000 compounded continuously for 5 years at 13 percent

c. $10,000 compounded continuously for 40 years at 8 percent

d. $200,000 compounded continuously for 9 months at 10.4 percent

26. An amount is invested at 6 percent compounded continuously. In how many years does the amount triple?

27. At what rate of interest compounded continuously must money be deposited if the amount is to double in 7 years?

28. How much would a person have to invest today at 6 percent compounded continuously if he wants to have $1,000 1 year from today? (*Note:* An important consideration of an economist is how much a dollar *x* years from now is worth today.)

29. Assume that in a specific culture the number of bacteria (*A*) present after *t* hours is given by $A = 1,000e^{0.15t}$.

    a. How many bacteria are present after $\frac{1}{2}$ hour?

    b. How long will it take to have 40,000 bacteria?

30. The number of bacteria in a culture at the start of an experiment is about 10^5. Four hours later the population has increased to about 10^7.

    a. Approximate the size of the population after 6 hours.

    b. When was the number of bacteria 250,000?

31. In 1960 the population of a certain town was 1,000, and in 1970 it was 4,000. What population can the city planning commission expect in 1994 if this growth rate continues?

32. The amount (*A*) of a certain radioactive element remaining after *t* minutes is given by $A = A_0e^{-0.03t}$. How much of the element would remain after 1 hour if 50 g were present initially?

33. Physicists tell us that the half-life of an isotope of strontium is 25 years. If a research center owns 10 g of this isotope, how much will it have in 10 years?

34. A radioactive substance decays from 10 g to 6 g in 5 days. Find the half-life of the substance.

35. Specimens in geology and archaeology are dated by considering the exponential decay of a particular radioactive element found in the specimen. This technique is very reliable because radioactive disintegration is not affected by conditions such as pressure and temperature, which change the rate of ordinary chemical reactions. For dating to an age of about 60,000 years, the researcher considers the concentration of carbon 14, an isotope of carbon. In the cells of all living plants and animals, there is a fixed ratio of carbon 14 to ordinary stable carbon. However, when the plant or animal dies, the carbon 14 decreases according to the law of exponential decay.

    a. The half-life of carbon 14 is about 5,600 years. Determine *k*, the decay constant, for this element.

    b. What percentage of the carbon 14 is left in a specimen of bone that is 2,000 years old?

(*Note:* The radiocarbon dating method was developed in 1947 by Dr. Willard Libby, who received the Nobel prize in chemistry for his discovery. For specimens older than 60,000 years, too much carbon 14 has disintegrated for an accurate measurement. In such cases the researcher considers the concentration of other radioactive substances. For example, long periods of geologic time are frequently determined by measuring the presence of uranium 238, which has a half-life of about 4.5 billion years.)

36. The temperature of a six-pack of beer (bought from a distributor on a hot summer day) is 90° F. The beer is placed in a refrigerator with a constant temperature of 40° F. If the beer cools to 60° F in 1 hour, when will the beer reach the more thirst-quenching temperature of 45° F?

37. A coroner examines the body of a murder victim at 9 A.M. and determines its temperature to be 88° F. An hour later the body temperature is down to 84° F. If the temperature of the room in which the body was found is 68° F and if the victim's body temperature was 98° F when she died, approximate the time of the murder.

38. The following formula (derived in calculus) is used to compute the values of e^x in Table 4 at the end of the book.

$$e^x = 1 + x + \frac{x^2}{2!} + \frac{x^3}{3!} + \cdots + \frac{x^n}{n!} + \cdots$$

where $n! = 1 \cdot 2 \cdot 3 \cdots n$ (for example, $4! = 1 \cdot 2 \cdot 3 \cdot 4$).

    a. Use the first four terms of this formula to approximate $e^{0.2}$. Compare your result with the value in Table 4.

    b. Use the first five terms of this formula to approximate *e* to the nearest hundredth.

In Exercises 39–44 express each logarithm as the sum or difference of simpler logarithms.

39. $\ln(1/x)$

40. $\ln\sqrt{x^2 + 1}$

41. $\ln 3\sqrt{e}$

42. $\ln(2/e^2)$

43. $\ln\dfrac{x(x - 1)}{x + 2}$

44. $\ln\sqrt[3]{x^2y}$

In Exercises 45–50 express each statement as a single logarithm with coefficient 1.

45. $2 \ln x$

46. $\frac{1}{4} \ln u$

47. $\ln x + \ln(x + 3)$

48. $\ln(x + h) - \ln x$

49. $\frac{1}{2} \ln x - \frac{1}{2} \ln a$

50. $\ln b - 2(\ln a + \ln x)$

In Exercises 51–58 solve the logarithmic equation.

51. $\ln(x + 3) = 2$

52. $\ln \sqrt{x + 3} = 1$

53. $\ln x = 3 \ln 2 - 2 \ln 3$

54. $\ln x - \ln(3x - 10) = 0$

55. $2 \ln(3x - 7) - \ln 4 = 0$

56. $\frac{1}{2} \ln x = 1 - \ln 2$

57. $\ln(x - 1) = 2 - \ln x$

58. $\ln x + \ln(x + 2) = 1$

In Exercises 59 and 60 find the value of k to the nearest thousandth that makes the equation a true statement.

59. $\ln x = k \log x$

60. $\log x = k \ln x$

61. Solve the formula for Newton's law of cooling (given in Example 9) for t: $T - t_\alpha = D_0 e^{kt}$.

62. Show that if $y = \dfrac{r}{1 + ce^{-at}}$, then $t = \dfrac{1}{a} \ln \dfrac{cy}{r - y}$.

THINK ABOUT IT

1. a. If $y = e^x$ and $x = \log_b y$, then find b.

  b. If $f(x) = -\ln x$, find $f^{-1}(x)$. Specify the domain and range of f^{-1}.

2. If an amount is invested at p percent compounded continuously, then a convenient rule of thumb for estimating the number of years required for the amount to double is

$$\text{doubling time} \approx \frac{70}{p}.$$

For example, an amount will double in about 10 years at 7 percent, and in about 5 years at 14 percent. Explain the mathematical basis for this rule of thumb.

3. Solve the following equations that are equations with quadratic form.

  a. $e^{2x} + 4e^x = 32$

  b. $(\ln x)^2 = \ln x^2$

  c. $(\ln x)^2 - 3 \ln x + 1 = 0$

4. Simplify each expression algebraically.

  a. $\left(\dfrac{e^x + e^{-x}}{2} \right)^2 - \left(\dfrac{e^x - e^{-x}}{2} \right)^2$

  b. $\left(\dfrac{e^2 - 1}{4e} + \dfrac{e^2 + 1}{2e} - e \right)$

  $- \left(\dfrac{e^2 - 1}{4e} - \dfrac{e^2 + 1}{2e} - \dfrac{1}{e} \right)$

5. From the following answers, identify *all* choices for k that result in the condition described.

  a. $0 < k < 1$ **b.** $k > 1$ **c.** $k = 0$

  d. $k = 1$ **e.** $k < 0$

  (i) $A = A_0 e^{kt}$ describes growth

  (ii) $A = A_0(k)^t$ describes growth

  (iii) $A = A_0 e^{kt}$ describes decay

REMEMBER THIS

In Exercises 1–7 solve each equation. Where necessary, approximate answers to four significant digits.

1. $x^{3/2} = 8$

2. $\left(\frac{1}{2} \right)^x = 8$

3. $3^{-x} = 6$

4. $e^{2x} = 6$

5. $2 \log_4 x = 6$

6. $\ln x = -0.5$

7. $\log(x - 9) = 1 - \log x$

8. Evaluate $\log_7 14$ to three significant digits.

9. Solve $y = a^t$ for t using common logarithms.

10. Solve $y = e^t$ for t using natural logarithms.

CHAPTER OVERVIEW

Section	Key Concepts to Review
4.1	• Definition of the exponential function with base b
	• For $b > 0$, $b \neq 1$, and any real number x, if x lies between the rational numbers r and s, then b^x lies between b^r and b^s.

Section	Key Concepts to Review
	• All previous laws of exponents hold for real-number exponents. • For $f(x) = b^x$ with $b > 0$ and $b \neq 1$: Domain: $(-\infty, \infty)$ Horizontal asymptote: x-axis Range: $(0, \infty)$ y-intercept: $(0,1)$ • If $b > 0$ and $b \neq 1$, then $b^x = b^y$ implies $x = y$. • Methods from Section 3.3 to graph variations of $f(x) = b^x$
4.2	• Definitions of a logarithm, a common logarithm, and the logarithmic function with base b • $\log_b N = L$ is equivalent to $b^L = N$. • For $b > 0$, $b \neq 1$, and $x > 0$, $y = \log_b x$ if and only if $x = b^y$; and for $f(x) = \log_b x$: Domain: $(0, \infty)$ Vertical asymptote: y-axis Range: $(-\infty, \infty)$ x-intercept: $(1,0)$ • The logarithmic function $y = \log_b x$ and the exponential function $y = b^x$ are inverse functions. Consequently, **a.** The functions interchange their domain and range. **b.** The graphs of the functions are symmetric about the line $y = x$. **c.** Calculator evaluations of exponential and log expressions often involve the $\boxed{\text{INV}}$ key. • Methods from Section 2.3 to graph variations of $f(x) = \log_b x$
4.3	• Properties of logarithms (for $b, x, y > 0$, $b \neq 1$, and k any real number): **1.** $\log_b xy = \log_b x + \log_b y$ **2.** $\log_b(x/y) = \log_b x - \log_b y$ **3.** $\log_b x^k = k \log_b x$ **4.** $\log_b b = 1$ **5.** $\log_b 1 = 0$ **6.** $\log_b b^x = x$ **7.** $b^{\log_b x} = x$
4.4	• If $x, y, b > 0$, with $b \neq 1$, then $x = y$ implies $\log_b x = \log_b y$; and $\log_b x = \log_b y$ implies $x = y$. • Change of base formula: $\log_b x = \dfrac{\log_a x}{\log_a b}$ • Methods to solve exponential and logarithmic equations
4.5	• Definition of a natural logarithm • As n gets larger $(1 + 1/n)^n$ gets closer to an irrational number that is denoted by the letter e. To six significant digits, $e \approx 2.71828$. • Compound interest formula: $A = P\left(1 + \dfrac{r}{n}\right)^{nt}$ • Continuous growth or decay formula: $A = A_0 e^{kt}$ • $y = \ln x$ and $y = e^x$ are inverse functions. • Graphs of $y = \ln x$ and $y = e^x$ • $\ln e^x = x$ and $e^{\ln x} = x$ (if $x > 0$)

CHAPTER REVIEW EXERCISES

In Exercises 1–10 solve each equation for x. Use a calculator or tables where necessary.

1. $\log_5 5^7 = x$ **2.** $2^{-x} = 3$

3. $\ln 8 = \ln x - \ln 4$ **4.** $\log_{10} 0.123 = x$

5. $x^{0.1} = 5.67$ **6.** $x = e^{\ln a}$

7. $\log_{10} x = -2.4157$ **8.** $5^{x-3} = 1$

9. $3^{\log_x 4} = 4$

10. $\log_{10}(3x - 2) - \log_{10}(x + 4) = 1$

In Exercises 11 and 12 express the sum or difference as a single logarithm with coefficient 1.

11. $2 \ln x + \frac{1}{3} \ln y$ **12.** $\log_a(x + h) - \log_a x$

In Exercises 13 and 14 express each as the sum or difference of simpler logarithms.

13. $\log_b \sqrt{x^3/y}$ **14.** $\ln \sqrt[3]{xy^2}$

15. What is the domain of $f(x) = \ln(3x - 7)$?

16. Express in logarithmic form $4^{-1} = \frac{1}{4}$.

17. Express in exponential form $\log_2 8 = 3$.

18. Simplify $\frac{n}{x} e^{n \ln x}$.

19. Sketch on the same axes the graphs of $f(x) = (\frac{1}{2})^x$ and $g(x) = \log_{1/2} x$. What is the domain and range of each function?

20. True or false: if false, give a specific counterexample.

$$\frac{\log_b x}{\log_b y} = \log_b x - \log_b y$$

21. If $y = (\frac{1}{4})^x$, fill in the missing component of each of the following ordered pairs:

$(3, \ ), (\ , 4), (\ , 1), (\frac{3}{2}, \ ), (-2, \ )$.

22. If $500 is invested at 6 percent compounded annually, find a formula for the value (y) of the investment at the end of t years.

23. $3,000 is invested at 5 percent compounded continuously. How much is the investment worth in 4 years?

24. A radioactive substance decays from 10 g to 8 g in 11 days. Find the half-life of the substance.

25. If $f = \{(3,-1), (4,-2)\}$, find f^{-1}. Determine the domain and range of each function.

26. If $\ln p^3 = b$, find $\ln p\sqrt{p}$ in terms of b.

27. If $\log_{10} M = K$, find 100^K in terms of M.

28. If $\ln 4 = 1.386$, solve $e^{3x} = 4$ for x.

29. Use $\log_{10} 2 = x$ and $\log_{10} 3 = y$ to find $\log_{10} \sqrt[3]{6}$ in terms of x and y.

30. If $f(x) = \log_b x$, find $f(1/b)$.

31. If $h(x) = \log_4 x$, find $h(1)$.

32. If $g(x) = 9^x$, find $g(-\frac{3}{2})$.

33. If $f(x) = 5^x - 2$, find the domain and range of f.

In Exercises 34–37 graph the given function.

34. $y = \ln x$ **35.** $y = 4^x$

36. $y = 10 - 10^{-x}$ **37.** $y = -\log_3(x - 2)$

38. Solve $9^x = \sqrt{3}$.

39. Solve $(\frac{1}{2})^x = 2^{x+3}$.

40. Solve $a^{x-4} = (1/a)^x$ for x if a is a positive real number other than 1.

41. If $\log_b(\frac{1}{10}) = -\frac{1}{2}$, find b.

42. Find the base of the exponential function $y = b^x$ that contains the point $(\frac{3}{2}, 8)$.

43. Evaluate $\log_9 3 - \frac{1}{2} \log_5 1 + 2 \log_3 27$.

44. Evaluate $\log_2 36 + \log_2 \frac{4}{9}$.

45. Solve $50 = 35 + 40e^{10k}$ for k by using natural logarithms.

46. Use logarithms to base 10 to evaluate $\log_4 9$.

47. Show that if $y = r + ce^{-kt}$, then $t = \frac{1}{k} \ln \frac{c}{y - r}$.

48. A common formula for pH is pH $= \log_{10} \frac{1}{[H^+]}$. Show that this formula is equivalent to the one given in this chapter. That is, show $\log_{10} \frac{1}{[H^+]} = -\log_{10}[H^+]$.

49. Liquid X has 10 times the concentration of hydrogen ions as liquid Y. What is the relation of the pH of X to the pH of Y?

50. If $\log_{10} 9! = x$, find $\log_{10} 10!$.

In Exercises 51–60 select the choice that answers the question or completes the statement.

51. Which function is an exponential function?

a. $y = x^{-3}$ **b.** $y = x^{1/3}$

c. $y = 3^{-1}$ **d.** $y = -3^x$

52. If $N = b^k$, then k equals

a. $\log N - \log b$ **b.** $\log b - \log N$

c. $\dfrac{\log N}{\log b}$ **d.** $\dfrac{\log b}{\log N}$

53. The graph of $f(x) = \log_3 x$ lies in quadrants

a. 1 and 2 **b.** 1 and 3

c. 1 and 4 **d.** 2 and 4

54. If $\log_{10} x^2 = a$, then $\log_{10} 10x$ equals

a. $(a + 2)/2$ **b.** $10a$ **c.** $a + 1$ **d.** a^2

55. An expression equivalent to $(\ln 27)/3$ is

  a. $\ln 9$ **b.** $-\ln 27$ **c.** $\ln 3$ **d.** $\ln 24$

56. An expression equivalent to $\dfrac{\log_{10} x}{\log_5 x}$ is

  a. $\log_2 x$ **b.** 2 **c.** $\log_5 10$ **d.** $\log_{10} 5$

57. If $4^x = \sqrt{8}$, then x equals

  a. $\frac{3}{4}$ **b.** $-\frac{2}{3}$ **c.** $\frac{3}{2}$ **d.** $\frac{1}{3}$

58. If $x^a = y^{a+1}$ and x and y are positive numbers other than 1, then a equals

  a. $\dfrac{\log x}{\log x - \log y}$ **b.** $\dfrac{\log y}{\log x - \log y}$

  c. $\dfrac{\log y - \log x}{\log x}$ **d.** $\dfrac{\log x + \log y}{\log y}$

59. If $f(x) = \ln x$, then $f(ab)$ equals

  a. $f(a) \cdot f(b)$ **b.** $f(a + b)$

  c. $f(a) + f(b)$ **d.** $f(a) - f(b)$

60. If $A/2 = Ae^{-kt}$, then k equals

  a. $\dfrac{\ln \frac{1}{2}}{t}$ **b.** $t \cdot \ln \frac{1}{2}$ **c.** $\ln \dfrac{2}{t}$ **d.** $\dfrac{\ln 2}{t}$

CHAPTER 4 TEST

1. If $(-2, a)$ is a solution of the equation $y = (\frac{1}{3})^x$, then find a.

2. Solve for x: $4^x = \frac{1}{8}$.

3. If $f(x) = 10^x - 3$, what is the range of f?

4. A company purchases a machine for $8,000. If the value of the machine depreciates at a rate of 30 percent each year, then what is the value of the machine 5 years after purchase?

5. **a.** Express in logarithmic form: $8^{2/3} = 4$.

  b. Express in exponential form: $\log_{10} 0.01 = -2$.

6. If $f(x) = \log_{10}(x + 2)$, what is the domain of f?

7. Graph $y = \log_2(x - 1)$.

8. The hydrogen ion concentration $[H^+]$ in a sample of grapefruit juice is 5.7×10^{-4}. Use $\text{pH} = -\log_{10}[H^+]$ to determine the pH of this sample.

9. Express $\log_b(5\sqrt{x})$ as the sum or difference of simpler logarithms.

10. If $\log_b 7 = x$ and $\log_b 3 = y$, express $\log_b(\frac{7}{9})$ in terms of x and y.

11. Express $3 \log_b 4 - \log_b 32$ as a single logarithm with coefficient 1.

12. If $f(x) = \log_b x$, find $f(\sqrt{b})$.

13. Solve $3^{x+2} = 72$.

14. Solve $\log_4 x + \log_4(x - 3) = 1$.

15. Evaluate $\log_4 25$ to four significant digits.

16. If $\ln x = -3.7$, find x to three significant digits.

17. Solve $\ln e^{3x} = 18$.

18. Solve for k to two significant digits: $e^{1.480k} = 0.5$.

19. A 2-year certificate of deposit (CD) is purchased for $4,000. If the CD pays 6 percent compounded monthly, then what is the value of the CD at maturity?

20. How long will it take for money invested at 9 percent compounded continuously to triple?

5 Trigonometric Functions

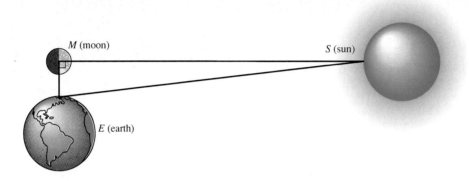

The first major attempt to measure the relative distances of the sun and the moon from the earth was made in about 280 B.C. by the Alexandrian astronomer Aristarchus. Basically he reasoned that the moon has no light of its own since we do not always see a "full moon." Therefore, the moon, like the earth, must receive its light from the sun and then reflect this light toward earth. As illustrated in the figure, he also correctly reasoned that at a quarter phase of the moon, when it is half light and half dark, angle *EMS* is a right angle. Using primitive instruments, Aristarchus measured angle *SEM* to be 29/30 of a right angle (or 87°). About how many times more distant is the sun than the moon in Aristarchus's estimate? (See Example 5 of Section 5.8.)

Historically, trigonometry involved the study of triangles for the purpose of measuring angles and distances in astronomy (as illustrated above). In early times this science was the primary concern of scholars who sought to understand God's design of the universe aesthetically and, practically, to obtain a more accurate system of navigation. As is often the case in mathematics, the ideas developed in this study later proved useful in analyzing a wide variety of situations, particularly physical phenomena that occur in cycles.

In this chapter we define and graph the trigonometric functions and their inverses. This topic can be approached from different viewpoints, and Sections 5.2 and 5.3 provide three definitions of the trigonometric functions. You must merge these definitions for a thorough understanding of trigonometry. Our primary viewpoint in this course will be to take the approach most useful in calculus. Thus, we begin by considering the radian measure of an angle. Radians are used extensively in calculus and are the link that makes a cohesive unit of trigonometry.

5.1 **Radian and Degree Measure**

A **ray** is a half line that begins at a point and extends indefinitely in some direction. Two rays that share a common endpoint (or vertex) form an angle. If we designate one ray as the **initial ray** and the other ray as the **terminal ray** (see Figure 5.1), the **measure of the angle** is the amount of rotation needed to make the initial ray coincide with the terminal ray. Notice that there are many rotations that will make the rays coincide, since there is no limitation on the number of revolutions made by the initial ray. In fact, it is useful to allow the initial ray to rotate through many revolutions, since the rotating initial ray will demonstrate a cyclic behavior that can serve as a model to simulate physical phenomena that occur in cycles. Also, the initial ray can rotate in two possible directions, as shown in Figure 5.1. To show the direction of the rotation, we define the measure of an angle to be positive if the rotation is counterclockwise and negative if the rotation is clockwise.

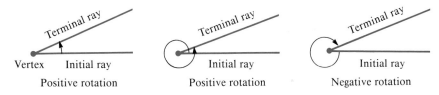

Figure 5.1

A common unit of the measure of an angle is a degree, written $1°$. We define **one degree** ($1°$) to be $\frac{1}{360}$ of a complete counterclockwise rotation. Equivalently, this means that there are $360°$ in a complete counterclockwise rotation. Figure 5.2 illustrates this case, a straight angle, and a right angle using the Greek letter θ (theta) to represent the angle measure. An angle is **acute** if it measures between $0°$ and $90°$ and **obtuse** if it measures between $90°$ and $180°$.

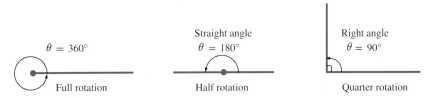

Figure 5.2

Because precise measurements are often needed, the degree is subdivided into 60 smaller units called minutes, with each minute being subdivided into 60 seconds. Thus, **one minute,** written $1'$, equals $\frac{1}{60}$ of one degree, and **one second,** written $1''$, equals $\frac{1}{60}$ of one minute or $\frac{1}{3600}$ of one degree. One degree may also be subdivided by using decimal degrees such as $18.5°$, which is a unit more convenient for calculator evaluation.

Radian Measure

Measuring angles in degrees is no doubt a familiar concept. However, in many applications and in calculus a different angle measure, called a **radian,** is more useful. We define the radian measure of an angle by first placing the vertex of the angle at the center of a circle. Let s be the length of the intercepted arc and let r be the radius. Then

$$\theta = \frac{s}{r}$$

is the radian measure of the angle (see Figure 5.3). Equivalently, an angle measuring 1 radian intercepts an arc equal in length to the radius of the circle (see Figure 5.4). In plane geometry it is shown that s *varies directly as r,* so that we can find the radian measure of θ by using a circle of *any* radius.

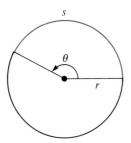

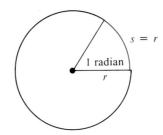

Figure 5.3   **Figure 5.4**

EXAMPLE 1   A central angle in a circle of radius 3 in. intercepts an arc of 12 in. Find the radian measure of the angle.

Solution   Substituting $r = 3$ in. and $s = 12$ in. in the formula

$$\theta = \frac{s}{r},$$

we have

$$\theta = \frac{12 \text{ in.}}{3 \text{ in.}} = 4.$$

Thus, the radian measure of the angle is 4. Note that r and s are measured in the same unit, which divides out in the ratio. Thus, the radian measure of an angle is a number without dimension. Although the word "radian" is often added, an angle measure with no units means radian measure.   ■

We may find the relation between degrees and radians by considering the measure of an angle that makes one complete rotation. In degrees the measure of the angle is 360. Since the circumference of a circle is $2\pi r$, the radian measure is

$$\theta = \frac{s}{r} = \frac{2\pi r}{r} = 2\pi.$$

Thus,

$$360° = 2\pi \text{ radians.}$$

From this relation we derive the following conversion rules between degrees and radians.

Degree–Radian Conversion Rules

> Degrees to radians formula:
>
> $$1° = \frac{\pi}{180} \text{ radian} \approx 0.0175 \text{ radians}$$
>
> Radians to degrees formula:
>
> $$1 \text{ radian} = \frac{180°}{\pi} \approx 57.3°$$

EXAMPLE 2 Express $30°$, $45°$, and $270°$ in terms of radians.

Solution Using the degrees to radians formula, we have

$$30° = 30 \cdot 1° = 30 \cdot \frac{\pi}{180} = \frac{\pi}{6}$$

$$45° = 45 \cdot 1° = 45 \cdot \frac{\pi}{180} = \frac{\pi}{4}$$

$$270° = 270 \cdot 1° = 270 \cdot \frac{\pi}{180} = \frac{3\pi}{2}.$$ ■

EXAMPLE 3 Express $\pi/3$, π, and $7\pi/5$ radians in terms of degrees.

Solution Using the radians to degrees formula, we have

$$\frac{\pi}{3} = \frac{\pi}{3} \cdot 1 = \frac{\pi}{3} \cdot \frac{180°}{\pi} = 60°$$

$$\pi = \pi \cdot 1 = \pi \cdot \frac{180°}{\pi} = 180°$$

$$\frac{7\pi}{5} = \frac{7\pi}{5} \cdot 1 = \frac{7\pi}{5} \cdot \frac{180°}{\pi} = 252°.$$ ■

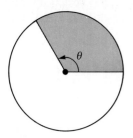

Figure 5.5

To illustrate an application of radians, consider the problem of determining the area of the shaded section in Figure 5.5. If you do not remember from geometry, notice intuitively that the area (A) varies directly as the central angle (θ). That is,

$$A = k\theta, \text{ where } k \text{ is a constant.}$$

We may find k since we know that $A = \pi r^2$ when the angle (θ) is a complete rotation of 2π radians. Thus,

$$\pi r^2 = k \cdot 2\pi$$
$$\tfrac{1}{2}r^2 = k.$$

The formula (when θ is expressed in radians) is then

$$A = \tfrac{1}{2}r^2\theta.$$

EXAMPLE 4   In a circle of radius 12 in., find the area of a sector whose central angle is 120°.

Solution   To use the formula we must first convert the angle to radians.

$$120° = 120 \cdot \frac{\pi}{180} = \frac{2\pi}{3}$$

Now, substituting $r = 12$ and $\theta = 2\pi/3$ in the formula

$$A = \tfrac{1}{2}r^2\theta,$$

we have

$$A = \frac{1}{2}(12)^2 \cdot \frac{2\pi}{3}$$
$$= 48\pi \text{ in.}^2.$$

One of the most important applications of radians concerns linear and angular velocity. If a point P moves along the circumference of a circle at a constant speed, then the linear velocity, v, is given by the formula

$$v = \frac{s}{t},$$

where s is the distance traveled by the point (or the length of the arc traversed) and t is the time required to travel this distance. During the same time interval, the radius of the circle connected to point P swings through θ angular units. Thus, the angular velocity, ω (Greek lowercase omega), is given by

$$\omega = \frac{\theta}{t}.$$

We can relate linear and angular velocity through the familiar formula

$$s = \theta r.$$

Dividing both sides of this equation by t, we obtain

$$\frac{s}{t} = \frac{\theta}{t} r.$$

Substituting v for s/t and ω for θ/t yields

$$v = \omega r.$$

Thus, the linear velocity is equal to the product of the angular velocity and the radius.

Since $s = \theta r$ is valid only when θ is measured in radians, the angular velocity ω must be expressed in radians per unit of time when using the formula $v = \omega r$. However, in many common applications the angular velocity is expressed in revolutions per minute. For these problems remember to convert ω to radians (rad) per unit of time through the relationship 1 rpm $= 2\pi$ rad/minute before using the formula.

EXAMPLE 5   A circular saw blade 12 in. in diameter rotates at 1,600 rpm. Find (a) the angular velocity in radians per second and (b) the linear velocity in inches per second at which the teeth would strike a piece of wood.

Solution

a. Since 1 rpm $= 2\pi$ rad/minute, we have

$$\omega = 1{,}600 \text{ rpm} = 1{,}600\left(2\pi\frac{\text{rad}}{\text{minute}}\right) = 3{,}200\pi\frac{\text{rad}}{\text{minute}}.$$

To convert to rad/second we use 1 minute $= 60$ seconds, as follows:

$$\omega = 3{,}200\pi\frac{\text{rad}}{\text{minute}}\frac{1 \text{ minute}}{60 \text{ seconds}} = \frac{160\pi}{3}\frac{\text{rad}}{\text{second}}.$$

b. Since ω is expressed in radians per unit time, we find v by the formula $v = \omega r$, with $r = 6$ in.

$$v = \omega r$$

$$= \frac{160\pi}{3}\frac{\text{rad}}{\text{second}} \cdot 6 \text{ in.}$$

$$= 320\pi\frac{\text{in.}}{\text{second}}$$

Remember that the radian measure of an angle is a number without dimension, so the word "radian" does not appear in the units for linear velocity. ■

EXERCISES 5.1

In Exercises 1–8 complete the table by replacing each question mark with the appropriate number.

The Radius Is	The Intercepted Arc Is	The Central Angle Is
1. 20 ft	100 ft	?
2. 8 in.	?	3
3. ?	56 yd	7
4. 5.2 meters	5.2 meters	?
5. ?	4.8 meters	$\frac{1}{2}$
6. 11 ft	?	3.2
7. 1 unit	5 units	?
8. 1 unit	π units	?

9. The radius of a wheel is 20 in. When the wheel moves 110 in., through how many radians does a point on the wheel turn? How many revolutions are made by the wheel?
10. The radius of a wheel is 16 in. Find the number of radians through which a point on the circumference turns when the wheel moves a distance of 2 ft. How many revolutions are made by the wheel?
11. Find the area of a sector of a circle whose central angle is $\pi/4$ and radius is 10 in.
12. Find the area of a sector of a circle whose central angle is 120° and radius is 5 in.
13. In a circle of radius 2 ft the arc length of a sector is 8 ft. Find the area of the sector.
14. In a circle of radius 5 ft the arc length of a sector is 3 ft. Find the area of the sector.

In Exercises 15–35 express each angle in radian measure.

15. 30°	16. 45°	17. 60°	18. 90°
19. 120°	20. 135°	21. 150°	22. 180°
23. 210°	24. 225°	25. 240°	26. 270°
27. 300°	28. 315°	29. 330°	30. 360°
31. 200°	32. 75°	33. 20°	34. 162°
35. 305°			

In Exercises 36–47 express each angle in degree measure.

36. $\dfrac{\pi}{4}$ 37. $\dfrac{\pi}{3}$ 38. $\dfrac{\pi}{6}$ 39. $\dfrac{2\pi}{9}$

40. $\dfrac{11\pi}{36}$ 41. $\dfrac{2\pi}{3}$ 42. $\dfrac{13\pi}{10}$ 43. $\dfrac{7\pi}{9}$

44. $\dfrac{2\pi}{15}$ 45. $\dfrac{12\pi}{5}$ 46. $\dfrac{\pi}{12}$ 47. $\dfrac{7\pi}{18}$

In Exercises 48–53 find, to the nearest degree, the number of degrees in each angle. (*Note:* Use $\pi \approx 3.14$.)

48. 1   49. 2   50. 6   51. 4   52. 3.5   53. 5.8

54. What is the linear velocity in inches/second of an object that moves along the circumference of a circle of radius 12 in. with an angular velocity of 7 rad/second?
55. What is the angular velocity in radians/second of an object that moves along the circumference of a circle of radius 9 in. with a linear velocity of 54 in./second?
56. What is the angular velocity in radians/second of the minute hand of a clock?
57. A phonograph record 12 in. in diameter is being played at $33\frac{1}{3}$ rpm.
    a. Find the angular velocity in radians/minute.
    b. Find the linear velocity in inches/minute of a point on the circumference of the record.
58. A pulley 20 in. in diameter makes 400 rpm.
    a. Find the angular velocity in radians/second.
    b. Find the speed in inches/second of the belt that drives the pulley.
    (*Note:* The speed of a point on the circumference of the pulley is the same as the speed of the belt.)
59. In a dynamo an armature 12 in. in diameter makes 1,500 rpm. Find the linear velocity in inches/second of the tip of the armature.
60. A car is traveling at 60 mi/hour (88 ft/second) on tires 30 in. in diameter. Find the angular velocity of the tires in radians/second.
61. Arcs of circles are used in many styles of molding that can be found in homes. Consider the illustration shown here. If $\overline{AB} = \overline{DE} = \frac{1}{2}\overline{BC} = \frac{1}{2}\overline{CD}$ and $\overline{AB} = \frac{1}{2}$ in., what is the length of the curved portion of the molding?

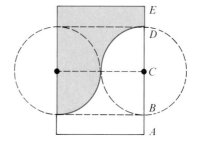

THINK ABOUT IT

1. **a.** If a central angle of α (Greek alpha) radians intercepts an arc of length a in a circle of radius b, then express b in terms of α and a.
    b. In a circle a central angle intercepts an arc equal in length to five eighths of the diameter of the circle. What is the measure of the central angle in radians?

2. Express $74°25'12''$ in radian measure to four significant digits.

3. Find, to the nearest $10'$, the latitude of Chicago, Illinois, which is 2,890 mi north of the equator. Assume that the earth is a sphere of radius 3,960 mi.

4. The formula $s = \theta r$ may be used to estimate certain distances when θ is a small angle (to about 10°). For example, consider the illustration below. An observer measures the angle from earth to opposite ends of the sun's diameter to be 0.53°. Since θ is a small angle, the arc length s is a good approximation for the

diameter d. If the distance from the earth to the sun is 93,000,000 mi, determine the diameter of the sun to two significant digits.

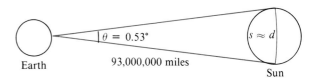

5. In architecture, arcs of circles are basic to the construction of many beautiful designs. For example, consider the quatrefoil in this illustration. If the side of the square measures 4 in., determine (a) the perimeter of the quatrefoil and (b) the area of the quatrefoil (the color portion of the figure).

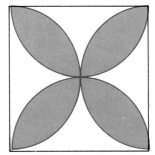

REMEMBER THIS

1. Rationalize the denominator: $1/\sqrt{2}$.
2. Find the reciprocal of $\sqrt{3}/2$.
3. Simplify $\dfrac{4\pi}{3} - \pi$.
4. The x-coordinate of a point is positive in which quadrants?
5. The y-coordinate of a point is negative in which quadrants?
6. If f is an even function and $f(a) = b$, find $f(-a)$.
7. If f is an odd function and $f(a) = b$, find $f(-a)$.
8. If $f(x) = x^2 - 3$, find the domain and range of f.
9. Write $\{x: -1 \le x \le 1\}$ in interval notation.
10. Write an equation for a circle with center at the origin and radius 1.

5.2 Trigonometric Functions of Real Numbers

Many natural phenomena repeat over definite periods of time. The waves broadcast by a radio station; the rhythmic motion of the heart; alternating electric current; weather-related issues such as air pollution; and the economic pattern of expansion, retrenchment, recession, and recovery—all these occur in cycles. The trigonometric functions, also repetitive, are very useful in analyzing such periodic phenomena. However, the independent variable in these applications is the time and not the angle. Thus, we need to define the trigonometric functions in terms of real numbers and not degrees. Since radians measure angles in terms of real numbers, they are the starting point for our discussion of the trigonometry of real numbers.

Consider Figure 5.6, in which the central angle θ is measured in radians. It is convenient to label the radius of the circle as 1 unit so that $r = 1$. Then

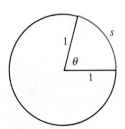

Figure 5.6

$$\theta = \frac{s}{r}$$
$$= \frac{s}{1}$$
$$\theta = s.$$

In a unit circle the same real number measures both the central angle θ and the intercepted arc s. Thus, we may base the definitions of the trigonometric functions on either an angle or an arc length. The results are valid for both interpretations. Since angles are meaningless in periodic phenomena, it is useful to emphasize the interpretation of a real number as the measure of an arc length s. We liberally interpret arc length as the distance traveled by a point as it moves around the unit circle repeating its behavior every 2π units.

To illustrate more forcefully the correspondence between real numbers and arc lengths, consider a unit circle with its center at the origin of the Cartesian coordinate system. The equation of this circle is $x^2 + y^2 = 1$. Through the point $(1,0)$ and parallel to the y-axis we draw a real number line, labeled s. The zero point of s coincides with the point $(1,0)$ on the circle. Units are marked off in the same scale as the y-axis (see Figure 5.7).

If the positive half of s is wrapped around the circle counterclockwise and the negative half of s is wrapped around the circle clockwise, we

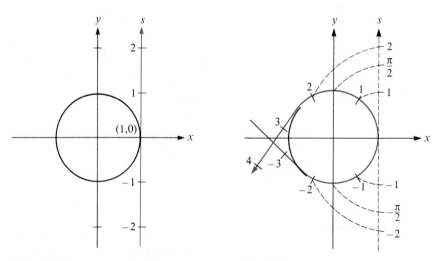

Figure 5.7   **Figure 5.8**

establish a one-to-one correspondence between real numbers and arc lengths of the circle (see Figure 5.8). Thus, each numbered point on s coincides with exactly one point on the circle, and the length of an arc may be read from this curved s-axis. We now relate this discussion to trigonometry.

Definition of the Sine and Cosine Functions

> Consider a point (x,y) on the unit circle $x^2 + y^2 = 1$ at arc length s from $(1,0)$. We define the cosine of s to be the x-coordinate of the point and the sine of s to be the y-coordinate.
>
> $$\cos s = x$$
> $$\sin s = y$$

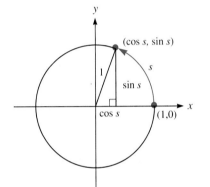

Figure 5.9

If you are familiar with right triangle trigonometry, note in Figure 5.9 that our definitions are consistent with the definitions of sine and cosine in terms of the sides and hypotenuse of a right triangle. The domain of both functions is the set of all real numbers, since the arc length s is determined by wrapping a real number line around the unit circle. The x- and y-coordinates in a unit circle vary between -1 and 1 (inclusive). Thus, for the range we have

$$-1 \le \cos s \le 1$$
$$-1 \le \sin s \le 1.$$

The remaining trigonometric functions are defined as follows.

Name of Function	Abbreviation	Ratio	
tangent of s	tan s	$y/x = \sin s/\cos s$	$(x \ne 0)$
cotangent of s	cot s	$x/y = \cos s/\sin s$	$(y \ne 0)$
secant of s	sec s	$1/x = 1/\cos s$	$(x \ne 0)$
cosecant of s	csc s	$1/y = 1/\sin s$	$(y \ne 0)$

Using the above definitions, we find the values of the trigonometric functions by determining the rectangular coordinates (x,y) of points on the unit circle. For certain real numbers these coordinates are easy to find. For example, let us determine the values of the trigonometric functions of zero. The coordinates for an arc length of zero are $(1,0)$ (see Figure 5.10). Thus,

$$\sin 0 = y = 0 \leftarrow \text{reciprocals} \rightarrow \csc 0 = \frac{1}{y} = \frac{1}{0} \text{ undefined}$$

$$\cos 0 = x = 1 \leftarrow \text{reciprocals} \rightarrow \sec 0 = \frac{1}{x} = \frac{1}{1} = 1$$

$$\tan 0 = \frac{y}{x} = \frac{0}{1} = 0 \leftarrow \text{reciprocals} \rightarrow \cot 0 = \frac{x}{y} = \frac{1}{0} \text{ undefined.}$$

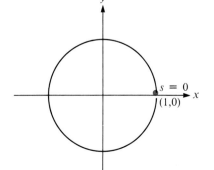

Figure 5.10

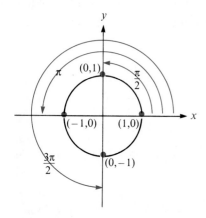

Figure 5.11

We can repeat this procedure for other arc lengths that terminate at one of the axes. Since the circumference of a circle of radius r is $2\pi r$, the circumference of a unit circle is 2π. The x- or y-axes may then intersect the unit circle at arc lengths of 0, $\pi/2$, π, and $3\pi/2$. Consider Figure 5.11, which illustrates the points of intersection for these arc lengths.

Using these coordinates, we may determine the trigonometric values of these numbers; the results are summarized in the following table and repeated on the endpaper at the back of the book.

s	$\sin s$	$\csc s$	$\cos s$	$\sec s$	$\tan s$	$\cot s$
0	0	undefined	1	1	0	undefined
$\pi/2$	1	1	0	undefined	undefined	0
π	0	undefined	-1	-1	0	undefined
$3\pi/2$	-1	-1	0	undefined	undefined	0

Other numbers terminate at one of the axes, but their trigonometric values are the same as one of the four listed. For example, the trigonometric values of 2π are the same as the trigonometric values of 0 since both numbers are assigned the point $(1,0)$ on the unit circle. The numbers 4π, 6π, -2π, and -4π are also assigned this point. A basic fact in our development is that in laying off a length 2π, we pass around the circle and return to our original point. Thus, the x- and y-coordinates repeat themselves at intervals of length 2π, and for any trigonometric function f we have

$$f(s + 2\pi k) = f(s), \text{ where } k \text{ is an integer.}$$

For example,

$$f(4\pi) = f(0 + 2\pi(2)) = f(0)$$
$$f(-2\pi) = f(0 + 2\pi(-1)) = f(0).$$

This observation is very important because it means that if we determine the values of the trigonometric functions in the interval $[0,2\pi)$, we know their values for all real s.

EXAMPLE 1   Find $\sin 7\pi$, $\cos 12\pi$, $\tan(11\pi/2)$, $\cot(17\pi/2)$, $\sec(-5\pi)$, and $\csc(-5\pi/2)$.

Solution   Since the values of the trigonometric functions repeat themselves at multiples of 2π, we have

$$\sin 7\pi = \sin(\pi + 6\pi) = \sin[\pi + 2\pi(3)] = \sin \pi = 0$$
$$\cos 12\pi = \cos(0 + 12\pi) = \cos[0 + 2\pi(6)] = \cos 0 = 1$$
$$\tan \frac{11\pi}{2} = \tan 5\tfrac{1}{2}\pi = \tan\left(\frac{3\pi}{2} + 4\pi\right) = \tan\left[\frac{3\pi}{2} + 2\pi(2)\right]$$
$$= \tan \frac{3\pi}{2} \quad \text{undefined}$$

$$\cot \frac{17\pi}{2} = \cot 8\tfrac{1}{2}\pi = \cot\left(\frac{\pi}{2} + 8\pi\right) = \cot\left[\frac{\pi}{2} + 2\pi(4)\right]$$

$$= \cot \frac{\pi}{2} = 0$$

$$\sec(-5\pi) = \sec[\pi + (-6\pi)] = \sec[\pi + 2\pi(-3)] = \sec \pi = -1$$

$$\csc\left(\frac{-5\pi}{2}\right) = \csc(-2\tfrac{1}{2}\pi) = \csc\left[\frac{3\pi}{2} + (-4\pi)\right]$$

$$= \csc\left[\frac{3\pi}{2} + 2\pi(-2)\right] = \csc \frac{3\pi}{2} = -1. \qquad \blacksquare$$

A **trigonometric identity** is a statement that is true for all real numbers for which the expressions are defined. We simplify our work by developing identities that relate a trigonometric function of a negative number to the same function of a positive number. The symmetry of the unit circle makes these identities easy to derive. Consider Figures 5.12 and 5.13, which illustrate the symmetry for two possible values of s. Note that the numbers s and $-s$ are assigned the same x-coordinate, so

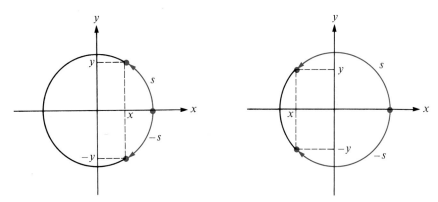

Figure 5.12   **Figure 5.13**

$$\cos(-s) = \cos s.$$

The y-coordinates differ only in their sign. Thus,

$$\sin(-s) = -\sin s.$$

The remaining functions are ratios of the sine and cosine, so it follows that

$$\tan(-s) = -\tan s$$
$$\cot(-s) = -\cot s$$
$$\sec(-s) = \sec s$$
$$\csc(-s) = -\csc s.$$

EXAMPLE 2　Find $\cos(-3\pi/2)$ and $\csc(-5\pi/2)$.

Solution

$$\cos\left(\frac{-3\pi}{2}\right) = \cos\frac{3\pi}{2} = 0$$

$$\csc\left(\frac{-5\pi}{2}\right) = -\csc\frac{5\pi}{2} = -\csc 2\tfrac{1}{2}\pi = -\csc\left(\frac{\pi}{2} + 2\pi\right)$$

$$= -\csc\frac{\pi}{2} = -1 \qquad \blacksquare$$

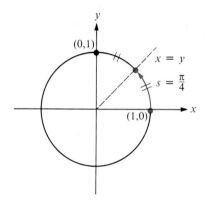

Figure 5.14

The coordinates on the unit circle for $\pi/4$, $\pi/3$, and $\pi/6$ may be found geometrically. First, consider $s = \pi/4$. Since $\pi/4$ is halfway between 0 and $\pi/2$, $\pi/4$ is assigned the midpoint of the arc joining the points $(1,0)$ and $(0,1)$ on the unit circle (see Figure 5.14). Thus, the x- and y-coordinates are equal (that is, $x = y$). Since the coordinates of any point on the circle satisfy the equation $x^2 + y^2 = 1$, we have

$$x^2 + x^2 = 1$$

$$x^2 = \frac{1}{2}$$

$$x = \frac{1}{\sqrt{2}} \quad \text{or} \quad -\frac{1}{\sqrt{2}}$$

and

$$y = \frac{1}{\sqrt{2}} \quad \text{or} \quad -\frac{1}{\sqrt{2}}.$$

Since x and y are positive in the first quadrant, we have

$$x = y = \frac{1}{\sqrt{2}}.$$

The trigonometric values of $\pi/4$ are then

$$\sin\frac{\pi}{4} = y = \frac{1}{\sqrt{2}} = \frac{\sqrt{2}}{2} \leftarrow \text{reciprocals} \rightarrow \csc\frac{\pi}{4} = \frac{1}{\sin(\pi/4)} = \sqrt{2}$$

$$\cos\frac{\pi}{4} = x = \frac{1}{\sqrt{2}} = \frac{\sqrt{2}}{2} \leftarrow \text{reciprocals} \rightarrow \sec\frac{\pi}{4} = \frac{1}{\cos(\pi/4)} = \sqrt{2}$$

$$\tan\frac{\pi}{4} = \frac{\sin(\pi/4)}{\cos(\pi/4)} = 1 \leftarrow \text{reciprocals} \rightarrow \cot\frac{\pi}{4} = \frac{\cos(\pi/4)}{\sin(\pi/4)} = 1.$$

Now consider $s = \pi/3$. Using the symmetry of the circle we can see that if $\pi/3$ is assigned the point (x,y), then $2\pi/3$ is assigned the point $(-x,y)$ (see Figure 5.15). Arcs AB and BC are both of length $\pi/3$. Since equal arcs in a circle subtend equal chords, the distance from $(1,0)$ to (x,y) is the same as the distance from (x,y) to $(-x,y)$. Therefore, we have

$$\sqrt{(x-1)^2 + (y-0)^2} = \sqrt{(x-(-x))^2 + (y-y)^2}$$

$$x^2 + y^2 - 2x + 1 = 4x^2.$$

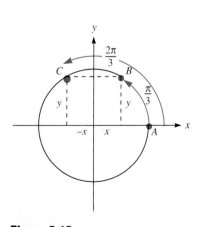

Figure 5.15

Since $x^2 + y^2 = 1$, by substitution we have

$$(1) - 2x + 1 = 4x^2$$
$$0 = 4x^2 + 2x - 2$$
$$= 2(2x - 1)(x + 1)$$
$$x = \tfrac{1}{2} \quad \text{or} \quad x = -1.$$

Because the point on the circle assigned to $\pi/3$ is in the first quadrant, we have

$$x = \tfrac{1}{2}.$$

To find y we replace x by $\tfrac{1}{2}$ in the equation $x^2 + y^2 = 1$.

$$(\tfrac{1}{2})^2 + y^2 = 1$$
$$y^2 = \sqrt{\frac{3}{4}}$$
$$y = \frac{\pm\sqrt{3}}{2}$$

Since the point is in quadrant one, we conclude that $y = \sqrt{3}/2$.

Using these coordinates, we may determine the trigonometric values of $\pi/3$. In the following table we list these values along with the trigonometric values of $\pi/4$ and $\pi/6$. As an exercise you are asked to determine the coordinates of the point on the circle from which the trigonometric values of $\pi/6$ were found.

s	$\sin s$	$\csc s$	$\cos s$	$\sec s$	$\tan s$	$\cot s$
$\dfrac{\pi}{6}$	$\dfrac{1}{2}$	2	$\dfrac{\sqrt{3}}{2}$	$\dfrac{2\sqrt{3}}{3}$	$\dfrac{\sqrt{3}}{3}$	$\sqrt{3}$
$\dfrac{\pi}{4}$	$\dfrac{\sqrt{2}}{2}$	$\sqrt{2}$	$\dfrac{\sqrt{2}}{2}$	$\sqrt{2}$	1	1
$\dfrac{\pi}{3}$	$\dfrac{\sqrt{3}}{2}$	$\dfrac{2\sqrt{3}}{3}$	$\dfrac{1}{2}$	2	$\sqrt{3}$	$\dfrac{\sqrt{3}}{3}$

Figure 5.16 shows a simple pattern you can use to generate the exact values for the sine and cosine of these numbers, 0, and $\pi/2$.

EXAMPLE 3   Find $\sin(-\pi/3)$, $\cos(-\pi/4)$, $\sin(9\pi/4)$, and $\tan(13\pi/3)$.

Solution   Using the table above and our previous procedure, we have

$$\sin\left(-\frac{\pi}{3}\right) = -\sin\frac{\pi}{3} = -\frac{\sqrt{3}}{2}$$

$$\cos\left(-\frac{\pi}{4}\right) = \cos\frac{\pi}{4} = \frac{\sqrt{2}}{2}$$

s	$\sin s$	$\cos s$
0	$\dfrac{\sqrt{0}}{2}$	$\dfrac{\sqrt{4}}{2}$
$\dfrac{\pi}{6}$	$\dfrac{\sqrt{1}}{2}$	$\dfrac{\sqrt{3}}{2}$
$\dfrac{\pi}{4}$	$\dfrac{\sqrt{2}}{2}$	$\dfrac{\sqrt{2}}{2}$
$\dfrac{\pi}{3}$	$\dfrac{\sqrt{3}}{2}$	$\dfrac{\sqrt{1}}{2}$
$\dfrac{\pi}{2}$	$\dfrac{\sqrt{4}}{2}$	$\dfrac{\sqrt{0}}{2}$

Figure 5.16

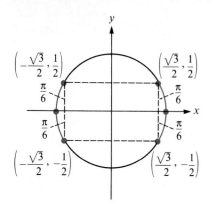

Figure 5.17

$$\sin \frac{9\pi}{4} = \sin 2\tfrac{1}{4}\pi = \sin\left(\frac{\pi}{4} + 2\pi\right) = \sin \frac{\pi}{4} = \frac{\sqrt{2}}{2}$$

$$\tan \frac{13\pi}{3} = \tan 4\tfrac{1}{3}\pi = \tan\left(\frac{\pi}{3} + 4\pi\right) = \tan \frac{\pi}{3} = \sqrt{3}. \qquad \blacksquare$$

A basic fact in our development has been that in laying off a length 2π, we pass around the unit circle and return to our original point. Thus, if we determine the values of the trigonometric functions in the interval $[0,2\pi)$, we know their values for all real s. Consider Figure 5.17, which illustrates how the symmetry of the circle may be used to further simplify the evaluation.

Note that the coordinates of the point assigned to $\pi/6$ differ only in sign from the coordinates assigned to $\pi - \pi/6 = 5\pi/6$, $\pi + \pi/6 = 7\pi/6$, and $2\pi - \pi/6 = 11\pi/6$. Thus, for a specific trigonometric function, say sine, we have

$$\sin \frac{\pi}{6} = \left|\sin \frac{5\pi}{6}\right| = \left|\sin \frac{7\pi}{6}\right| = \left|\sin \frac{11\pi}{6}\right| = \frac{1}{2}.$$

The number $\pi/6$ is called the **reference number** for $5\pi/6$, $7\pi/6$, and $11\pi/6$. In general, we determine the reference number for s (denoted by s_R) by finding the shortest positive arc length between the point on the circle assigned to s and the x-axis. This discussion indicates that the trigonometric values of s and s_R are related as follows.

Trigonometric Values of s and s_R

> Any trigonometric function of s is equal in absolute value to the same-named function of its reference number s_R.

We determine the correct sign by considering the function definitions together with the sign of x and y in the four quadrants. The following table indicates the signs of the functions in the various quadrants.

Function	quadrant$_1$ (Q$_1$)	quadrant$_2$ (Q$_2$)	quadrant$_3$ (Q$_3$)	quadrant$_4$ (Q$_4$)
$\sin s = y$	$+$	$+$	$-$	$-$
$\csc s = \dfrac{1}{\sin s}$	$+$	$+$	$-$	$-$
$\cos s = x$	$+$	$-$	$-$	$+$
$\sec s = \dfrac{1}{\cos s}$	$+$	$-$	$-$	$+$
$\tan s = \dfrac{\sin s}{\cos s}$	$\dfrac{+}{+} = +$	$\dfrac{+}{-} = -$	$\dfrac{-}{-} = +$	$\dfrac{-}{+} = -$
$\cot s = \dfrac{1}{\tan s}$	$+$	$-$	$+$	$-$

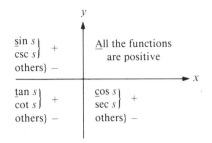

$\left.\begin{array}{l}\underline{\sin} s \\ \csc s\end{array}\right\}\ +$ $\underline{\text{All}}$ the functions
others} $-$ are positive

$\left.\begin{array}{l}\underline{\tan} s \\ \cot s\end{array}\right\}\ +$ $\left.\begin{array}{l}\underline{\cos} s \\ \sec s\end{array}\right\}\ +$
others} $-$ others} $-$

Figure 5.18

You may wish to remember this table by using the chart in Figure 5.18. Note that (1) in each quadrant, except for the first, the value of only two functions is positive, and (2) in each case the two functions are reciprocal functions. Thus, if we can remember in which quadrant sin s, tan s, and cos s are positive, we can generate the chart. A sentence that is useful in remembering this chart is: "$\underline{\text{A}}$ll $\underline{\text{s}}$tudents $\underline{\text{t}}$ake $\underline{\text{c}}$alculus." Note that the underlined first letter in each of the words corresponds to the underlined first letters in the chart.

EXAMPLE 4   Find $\sin(7\pi/4)$.

Solution   First, determine the reference number.

$$s_R = 2\pi - \frac{7\pi}{4} = \frac{\pi}{4} \qquad \text{(See Figure 5.19.)}$$

Second, determine $\sin(\pi/4)$.

$$\sin\frac{\pi}{4} = \frac{\sqrt{2}}{2}$$

Third, determine the correct sign. The point assigned to $7\pi/4$ is in Q_4, where the value of the sine function is negative. Therefore,

$$\sin\frac{7\pi}{4} = -\frac{\sqrt{2}}{2} \qquad ■$$

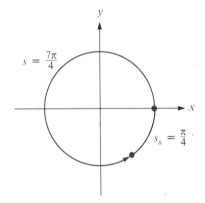

Figure 5.19

EXAMPLE 5   Find $\cos(-10\pi/3)$.

Solution   First, simplify the expression to the form cos s, where $0 \le s < 2\pi$.

$$\cos\left(\frac{-10\pi}{3}\right) = \cos\frac{10\pi}{3} = \cos 3\tfrac{1}{3}\pi = \cos\left(\frac{4\pi}{3} + 2\pi\right) = \cos\frac{4\pi}{3}$$

Second, determine s_R.

$$s_R = \frac{4\pi}{3} - \pi = \frac{\pi}{3} \qquad \text{(See Figure 5.20.)}$$

Third, determine $\cos(\pi/3)$.

$$\cos\frac{\pi}{3} = \frac{1}{2}$$

Fourth, determine the correct sign. The point assigned to $4\pi/3$ is in Q_3, where the value of the cosine function is negative. Therefore,

$$\cos\left(\frac{-10\pi}{3}\right) = -\frac{1}{2}. \qquad ■$$

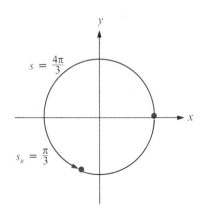

Figure 5.20

Until now we have been considering special numbers in that their trigonometric values may be found *exactly* by geometric means. To approximate the trigonometric values of other numbers we use the same procedure, together with Table 5 at the end of the book. This table lists the trigonometric values for numbers between 0 and 1.57 (or about $\pi/2$). These values are computed through the aid of calculus, and it should be understood that they are *approximations* that are rounded off to four significant digits. In these problems we use 3.14 as an approximation for π.

EXAMPLE 6   Approximate sin 2.

Solution   First, determine the reference number.

$$s_R = \pi - 2 \approx 3.14 - 2 = 1.14 \qquad \text{(See Figure 5.21.)}$$

Second, determine sin 1.14.

$$\sin 1.14 \approx 0.9086 \qquad \text{(Table 5)}$$

Third, determine the correct sign. The point assigned to 2 is in Q_2, where the value of the sine function is positive. Therefore,

$$\sin 2 \approx 0.9086. \qquad \blacksquare$$

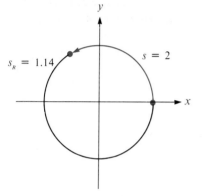

Figure 5.21

EXAMPLE 7   Approximate $\tan(-7.12)$.

Solution   First, simplify the expression to the form tan s, where $0 \le s < 6.28$ (or $0 \le s < 2\pi$).

$$\tan(-7.12) = -\tan 7.12 = -\tan(0.84 + 6.28) = -\tan 0.84$$

Second, determine tan 0.84.

$$\tan 0.84 \approx 1.116 \qquad \text{(Table 5)}$$

Therefore,

$$\tan(-7.12) = -\tan 0.84 \approx -1.116. \qquad \blacksquare$$

EXAMPLE 8   Approximate $\sec(8\pi/5)$.

Solution   First, determine the reference number.

$$s_R = 2\pi - \frac{8\pi}{5} = \frac{2\pi}{5} \approx \frac{2(3.14)}{5} = 1.26 \qquad \text{(See Figure 5.22.)}$$

Second, determine sec 1.26.

$$\sec 1.26 \approx 3.270 \qquad \text{(Table 5)}$$

Third, determine the correct sign. The point assigned to $8\pi/5$ is in Q_4, where the value of the secant function is positive. Therefore,

$$\sec \frac{8\pi}{5} \approx 3.270. \qquad \blacksquare$$

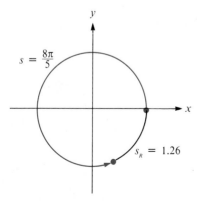

Figure 5.22

A scientific calculator provides a direct method for evaluating trigonometric functions of real numbers. For example, to find sin 4, simply set the calculator for radian mode, enter the number 4, and press $\boxed{\sin}$.

(radian mode)   4 $\boxed{\sin}$ -0.7568025

Thus, a calculator can easily give you a number associated with a trigonometric evaluation, and you should consider carefully the following notes on calculator usage.

1. Be sure to set the calculator for radian mode. Many calculators are in degree mode when first turned on, and forgetting this step is a very common error.
2. Some calculators limit the size of the number you can enter (under 8π, for instance, on Casio fx-910). If you exceed this input range, the calculator displays an error message. To evaluate such numbers, you must reduce the number to an acceptable range by subtracting a multiple of 2π, as discussed in this section.
3. Although a calculator is usually correct to at least eight decimal places, we will be displaying four decimal places in our solutions.
4. Cotangent, secant, and cosecant are usually computed by first evaluating either tangent, cosine, or sine, respectively, and then pressing the reciprocal key $\boxed{1/x}$. For instance, in Example 8 we determined $\sec(8\pi/5) \approx 3.270$. A calculator solution is

(radian mode)   8 $\boxed{\times}$ π $\boxed{\div}$ 5 $\boxed{=}$ $\boxed{\cos}$ $\boxed{1/x}$ 3.236068.

Note the calculator solution is much more accurate. The example solution is only accurate to two significant digits because 3.14 is used as an estimate for π.

EXERCISES 5.2

In Exercises 1–30 find the *exact* function value.

1. $\cos 4\pi$
2. $\sin 100\pi$
3. $\sin 5\pi$
4. $\cos 11\pi$
5. $\tan(-2\pi)$
6. $\cot(-36\pi)$
7. $\cos\left(-\dfrac{\pi}{2}\right)$
8. $\sin\left(-\dfrac{3\pi}{2}\right)$
9. $\sec\dfrac{5\pi}{2}$
10. $\tan\dfrac{19\pi}{2}$
11. $\csc\left(\dfrac{-7\pi}{2}\right)$
12. $\cos\left(\dfrac{-9\pi}{2}\right)$
13. $\sin\left(\dfrac{11\pi}{2}\right)$
14. $\sin\left(\dfrac{-97\pi}{2}\right)$
15. $\cos\left(-\dfrac{\pi}{3}\right)$
16. $\sin\left(-\dfrac{\pi}{4}\right)$
17. $\tan\dfrac{9\pi}{4}$
18. $\cos\dfrac{9\pi}{4}$
19. $\sin\dfrac{13\pi}{6}$
20. $\sec\dfrac{17\pi}{4}$
21. $\cot\dfrac{7\pi}{3}$
22. $\sin\dfrac{19\pi}{3}$
23. $\cos\left(\dfrac{-13\pi}{6}\right)$
24. $\sin\left(\dfrac{-13\pi}{3}\right)$

25. $\csc\left(\dfrac{-31\pi}{3}\right)$
26. $\cos\left(\dfrac{-25\pi}{4}\right)$
27. $\cot\dfrac{55\pi}{3}$
28. $\tan\left(\dfrac{-73\pi}{6}\right)$
29. $\sin\left(\pi - \dfrac{4\pi}{3}\right)$
30. $\cos\left(2\pi - \dfrac{13\pi}{6}\right)$

In Exercises 31–50 find the *exact* function value. Use reference numbers.

31. $\cos\dfrac{7\pi}{6}$
32. $\sin\dfrac{5\pi}{3}$
33. $\tan\dfrac{4\pi}{3}$
34. $\sec\dfrac{11\pi}{6}$
35. $\sin\dfrac{7\pi}{4}$
36. $\cos\dfrac{2\pi}{3}$
37. $\csc\dfrac{5\pi}{6}$
38. $\sin\dfrac{3\pi}{4}$
39. $\cos\dfrac{5\pi}{3}$

40. $\cot \dfrac{5\pi}{4}$ **41.** $\sin\left(\dfrac{-7\pi}{6}\right)$ **42.** $\cos\left(\dfrac{-4\pi}{3}\right)$

43. $\sec\left(\dfrac{-2\pi}{3}\right)$ **44.** $\tan\left(\dfrac{-7\pi}{4}\right)$ **45.** $\cos \dfrac{11\pi}{4}$

46. $\sin \dfrac{8\pi}{3}$ **47.** $\cot \dfrac{23\pi}{6}$ **48.** $\csc \dfrac{15\pi}{4}$

49. $\sin\left(\dfrac{-20\pi}{3}\right)$ **50.** $\cos\left(\dfrac{-23\pi}{6}\right)$

In Exercises 51–70 find the approximate function value. Use a calculator or reference numbers and Table 5.

51. $\cos 2$ **52.** $\sin 3$ **53.** $\tan 4$
54. $\sec 5$ **55.** $\sin 5.41$ **56.** $\csc 2.23$
57. $\cot 3.71$ **58.** $\cos 1.84$ **59.** $\sin(-6.07)$
60. $\tan(-1.69)$ **61.** $\cos 11.73$ **62.** $\cot 9.61$

63. $\csc(-7.57)$ **64.** $\sin(-10.25)$ **65.** $\sin \dfrac{2\pi}{5}$

66. $\tan \dfrac{9\pi}{5}$ **67.** $\cos \dfrac{3\pi}{7}$ **68.** $\sin \dfrac{4\pi}{9}$

69. $\sec\left(\dfrac{-7\pi}{8}\right)$ **70.** $\cos\left(\dfrac{-7\pi}{10}\right)$

THINK ABOUT IT

1. Explain why there is no real number s such that $\sin s > 1$.
2. Classify each of the six trigonometric functions as either an even function or an odd function. Give reasons for your answers.
3. In calculus it is important to determine the number approached by $(\sin h)/h$ as h approaches 0, where h is measured in radians. Complete the tables below and then state the number approached by this ratio that is suggested by these tables.

Table 1

h	0.3	0.2	0.1	0.01	0.001
$\dfrac{\sin h}{h}$					

Table 2

h	−0.3	−0.2	−0.1	−0.01	−0.001
$\dfrac{\sin h}{h}$					

4. The following formulas (derived from calculus) are used to compute the values of $\cos x$ and $\sin x$ in Table 5.

$$\cos x = 1 - \frac{x^2}{2!} + \frac{x^4}{4!} - \frac{x^6}{6!} + \cdots$$

$$\sin x = x - \frac{x^3}{3!} + \frac{x^5}{5!} - \frac{x^7}{7!} + \cdots,$$

where $n! = 1 \cdot 2 \cdots n$ (for example, $3! = 1 \cdot 2 \cdot 3$). Use the first three terms of these formulas to approximate the following expressions. Compare your results with the values in Table 5.

a. $\sin 1$ **b.** $\cos 1$ **c.** $\cos 0$ **d.** $\sin 0.5$

5. Determine the coordinates of the point on the unit circle assigned to $\pi/6$. (*Hint:* The method is similar to the one given for $\pi/3$. The arc in the circle from $-\pi/6$ to $\pi/6$ is equal to the arc from $\pi/6$ to $\pi/2$.)

REMEMBER THIS

1. Find the distance between $(0,0)$ and $(2,-5)$.
2. Rationalize the denominator: $2/\sqrt{29}$.
3. Find the length of the hypotenuse in a right triangle if the legs measure 2 ft and 5 ft.
4. Solve for a: $\dfrac{a}{39} = \dfrac{5}{13}$.
5. If the x-coordinate of a point is positive and the y-coordinate is negative, then in which quadrant is the point located?
6. Express $\pi/2$ radians in degree measure.
7. Express $\cot s$ in terms of $\tan s$.
8. True or false: $\sec(-s) = \sec s$ is an identity.
9. Evaluate $\sin(7\pi/6)$.
10. What is the range of the function $y = \sin x$?

5.3 **Trigonometric Ratios**

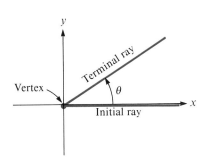

Figure 5.23

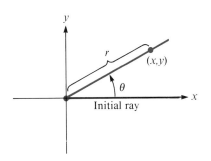

Figure 5.24

In Section 5.2 we viewed trigonometry in terms of a unit circle. In such a circle we saw that the same *real number* measures both the central angle θ and the intercepted arc s. Thus, we could base the definitions of the trigonometric functions either on an angle measured in radians or an arc length. The results are valid for both interpretations. However, regardless of the interpretation, *the distinguishing feature of this approach is that the domain elements for the trigonometric functions are real numbers.* In this section we take a different viewpoint. We now consider a general angular approach which includes the study of triangles as a special case.

The frame of reference that we use to define the trigonometric functions is the Cartesian coordinate system. First, we put an angle θ (theta) in **standard position** (see Figure 5.23).

1. We place the vertex of the angle at the origin (0,0).
2. We place the initial ray of the angle along the positive x-axis.

When we have the angle in standard position, we can define the trigonometric functions of an angle by considering any point on the terminal ray of θ [except (0,0)]. Three numbers can be associated with the location of this point (see Figure 5.24):

1. The x-coordinate of the point
2. The y-coordinate of the point
3. The distance r between the point and the origin

Since r represents the distance from the origin (0,0) to the point (x,y), we can find the relationship between x, y, and r by using the distance formula.

$$r = \sqrt{(x - 0)^2 + (y - 0)^2}$$

$$r = \sqrt{x^2 + y^2}$$

or

$$r^2 = x^2 + y^2$$

If we consider the number of ratios that can be obtained from the three variables x, y, and r, we find that there are six. It is these six ratios that define the six *trigonometric functions*.

Definition of the Trigonometric Functions

If θ is an angle in standard position and if (x,y) is any point on the terminal ray of θ [except $(0,0)$], then

Name of Function	Abbreviation	Ratio	
sine of angle θ	$\sin \theta$	$= \dfrac{y}{r}$	reciprocal functions
cosecant of angle θ	$\csc \theta$	$= \dfrac{r}{y} \ (y \neq 0)$	
cosine of angle θ	$\cos \theta$	$= \dfrac{x}{r}$	reciprocal functions
secant of angle θ	$\sec \theta$	$= \dfrac{r}{x} \ (x \neq 0)$	
tangent of angle θ	$\tan \theta$	$= \dfrac{y}{x} \ (x \neq 0)$	reciprocal functions
cotangent of angle θ	$\cot \theta$	$= \dfrac{x}{y} \ (y \neq 0)$	

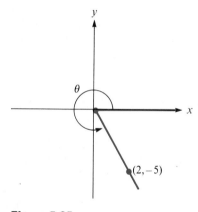

Figure 5.25

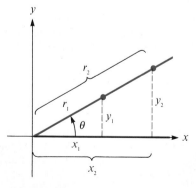

Figure 5.26

EXAMPLE 1   Find the value of the six trigonometric functions of an angle θ if $(2,-5)$ is a point on the terminal ray of θ (see Figure 5.25).

Solution

$$x = 2 \qquad y = -5$$
$$r = \sqrt{x^2 + y^2}$$
$$= \sqrt{(2)^2 + (-5)^2}$$
$$= \sqrt{4 + 25} = \sqrt{29} \qquad (\textit{Note: } \sqrt{4 + 25} \neq \sqrt{4} + \sqrt{25}.)$$

$$\sin \theta = \frac{y}{r} = \frac{-5}{\sqrt{29}} = \frac{-5\sqrt{29}}{29} \quad \leftarrow \text{reciprocals} \rightarrow \quad \csc \theta = \frac{r}{y} = \frac{\sqrt{29}}{-5}$$

$$\cos \theta = \frac{x}{r} = \frac{2}{\sqrt{29}} = \frac{2\sqrt{29}}{29} \quad \leftarrow \text{reciprocals} \rightarrow \quad \sec \theta = \frac{r}{x} = \frac{\sqrt{29}}{2}$$

$$\tan \theta = \frac{y}{x} = \frac{-5}{2} \quad \leftarrow \text{reciprocals} \rightarrow \quad \cot \theta = \frac{x}{y} = \frac{2}{-5} \qquad \blacksquare$$

Note in the definition above that (x,y) may be any point (except the origin) on the terminal ray of θ. Figure 5.26 shows that if we pick different points on the terminal ray of θ, say (x_1,y_1) and (x_2,y_2), they determine similar triangles. It follows that corresponding side lengths are proportional, so the trigonometric ratios are uniquely determined by the terminal ray of θ. This means that in addition to picking any point on the terminal ray of θ, we may also choose any of the angle measures associated with the angle. If we choose (x,y) so that $r = 1$, then we obtain the unit circle definitions of the trigonometric functions given in Section 5.2.

The coordinate system divides the plane into four quadrants. To determine the sign of the ratios for the trigonometric functions of an angle whose terminal ray is in a particular quadrant, we examine the sign of x and y in that quadrant. Since r is a distance represented by some positive number, r does not affect the sign of a ratio. Consider an angle whose terminal ray is in the second quadrant, where x is negative and y is positive.

$$\sin \theta = \frac{y}{r} = \frac{+}{+} = + \leftarrow \text{reciprocals} \rightarrow \csc \theta = \frac{r}{y} = \frac{+}{+} = +$$

$$\cos \theta = \frac{x}{r} = \frac{-}{+} = - \leftarrow \text{reciprocals} \rightarrow \sec \theta = \frac{r}{x} = \frac{+}{-} = -$$

$$\tan \theta = \frac{y}{x} = \frac{+}{-} = - \leftarrow \text{reciprocals} \rightarrow \cot \theta = \frac{x}{y} = \frac{-}{+} = -$$

Therefore, $\sin \theta$, and its reciprocal $\csc \theta$, will be positive if the terminal ray of θ is in the second quadrant. The values of the remaining four functions are negative.

If we repeat this procedure for all four quadrants, we obtain the results summarized in Figure 5.27. We shall use this important chart frequently, so see page 263 where we pointed out (with respect to an equivalent chart) how the sentence "All students take calculus" can help you remember it. Note that the chart does not mention what will happen if the terminal ray of θ coincides with one of the axes. This is a special case where θ is not contained in any quadrant and we will deal with it later.

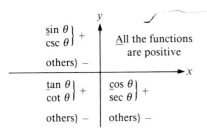

Figure 5.27

EXAMPLE 2   In which quadrant is the terminal ray of θ if $\sin \theta$ is positive and $\cos \theta$ is negative?

Solution $\sin \theta$ is positive in quadrant one (Q_1) or quadrant two (Q_2); $\cos \theta$ is negative in quadrant two (Q_2) or quadrant three (Q_3). Only quadrant two (Q_2) satisfies both conditions. Therefore, the terminal ray of θ is in quadrant two (Q_2). ∎

EXAMPLE 3   In which quadrant is the terminal ray of θ if $\csc \theta < 0$ and $\cot \theta > 0$?

Solution $\csc \theta < 0$ means that $\csc \theta$ is negative and θ is in Q_3 or Q_4; $\cot \theta > 0$ means that $\cot \theta$ is positive and θ is in Q_1 or Q_3. Only Q_3 satisfies both conditions. Therefore, the terminal ray of θ is in quadrant three (Q_3). ∎

EXAMPLE 4   Find the values of the remaining trigonometric functions if $\sin \theta = \frac{4}{5}$ and $\cos \theta$ is negative.

Solution   First, determine the quadrant that contains the terminal ray of θ. $\sin \theta = \frac{4}{5}$, which is positive. Therefore, θ is in Q_1 or Q_2; $\cos \theta$ is negative in Q_2 or Q_3. Only quadrant two (Q_2) satisfies both conditions. Therefore, the terminal ray of θ is in Q_2, where x is negative and y is positive.

Second, determine appropriate values for x, y, and r.

$$\sin \theta = \frac{y}{r} = \frac{4}{5}$$

Since y is positive in Q_2, let $y = 4$ and $r = 5$; then find x.

$$r^2 = x^2 + y^2$$
$$(5)^2 = x^2 + (4)^2$$
$$25 = x^2 + 16$$
$$9 = x^2$$
$$x = 3 \quad \text{or} \quad -3$$

Since x is negative in Q_2, let $x = -3$.

Third, find the values of the remaining trigonometric ratios. If $x = -3$, $y = 4$, and $r = 5$, then

$$\sin \theta = \frac{y}{r} = \frac{4}{5} \; \leftarrow\text{reciprocals}\rightarrow \; \csc \theta = \frac{r}{y} = \frac{5}{4}$$

$$\cos \theta = \frac{x}{r} = \frac{-3}{5} \; \leftarrow\text{reciprocals}\rightarrow \; \sec \theta = \frac{r}{x} = \frac{5}{-3}$$

$$\tan \theta = \frac{y}{x} = \frac{4}{-3} \; \leftarrow\text{reciprocals}\rightarrow \; \cot \theta = \frac{x}{y} = \frac{-3}{4}.$$

■

EXAMPLE 5   Find the value of the remaining trigonometric functions if $\tan \theta = \frac{2}{3}$ and $\sec \theta < 0$.

Solution   First, determine the quadrant that contains the terminal ray of θ. $\tan \theta = \frac{2}{3}$, which is positive. Therefore, θ is in Q_1 or Q_3; $\sec \theta < 0$ means that $\sec \theta$ is negative and θ is in Q_2 or Q_3. Only Q_3 satisfies both conditions. Therefore, the terminal ray of θ is in Q_3, where x is negative and y is negative.

Second, determine appropriate values for x, y, and r.

$$\tan \theta = \frac{y}{x} = \frac{2}{3}$$

Since both x and y are negative in Q_3, let $x = -3$ and $y = -2$; then find r.

$$r = \sqrt{x^2 + y^2}$$
$$= \sqrt{(-3)^2 + (-2)^2}$$
$$= \sqrt{9 + 4}$$
$$= \sqrt{13}$$

Third, calculate the values of the different trigonometric functions. If $x = -3$, $y = -2$, and $r = \sqrt{13}$, then

$$\sin \theta = \frac{y}{r} = \frac{-2}{\sqrt{13}} = \frac{-2\sqrt{13}}{13} \; \leftarrow \text{reciprocals} \rightarrow \; \csc \theta = \frac{r}{y} = \frac{\sqrt{13}}{-2}$$

$$\cos \theta = \frac{x}{r} = \frac{-3}{\sqrt{13}} = \frac{-3\sqrt{13}}{13} \quad \leftarrow \text{reciprocals} \rightarrow \quad \sec \theta = \frac{r}{x} = \frac{\sqrt{13}}{-3}$$

$$\tan \theta = \frac{y}{x} = \frac{-2}{-3} = \frac{2}{3} \quad \leftarrow \text{reciprocals} \rightarrow \quad \cot \theta = \frac{x}{y} = \frac{-3}{-2} = \frac{3}{2}.$$

Remember that the value of a trigonometric function is a ratio. $\tan \theta = \frac{2}{3}$ did not mean that $y = 2$ and $x = 3$. ∎

Right Triangle Trigonometry

Consider Figure 5.28 and note that if θ is a positive acute angle in standard position with the point (x,y) on its terminal ray, then a right triangle may be formed with opposite side of length y, adjacent side of length x, and hypotenuse of length r. Thus, right triangle trigonometry is a special case of the more general definition of the trigonometric functions for it arises when we are dealing with an acute angle whose terminal ray lies in the first quadrant, and the right triangle definitions of the trigonometric functions are as follows.

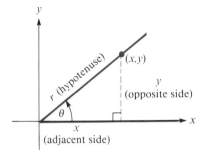

Figure 5.28

Definition of the Trigonometric Functions

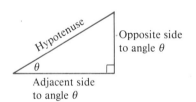

Figure 5.29

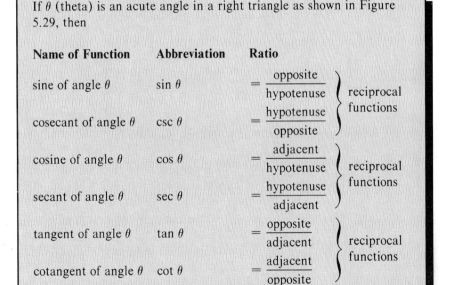

If θ (theta) is an acute angle in a right triangle as shown in Figure 5.29, then

Name of Function	Abbreviation	Ratio	
sine of angle θ	$\sin \theta$	$= \dfrac{\text{opposite}}{\text{hypotenuse}}$	reciprocal functions
cosecant of angle θ	$\csc \theta$	$= \dfrac{\text{hypotenuse}}{\text{opposite}}$	
cosine of angle θ	$\cos \theta$	$= \dfrac{\text{adjacent}}{\text{hypotenuse}}$	reciprocal functions
secant of angle θ	$\sec \theta$	$= \dfrac{\text{hypotenuse}}{\text{adjacent}}$	
tangent of angle θ	$\tan \theta$	$= \dfrac{\text{opposite}}{\text{adjacent}}$	reciprocal functions
cotangent of angle θ	$\cot \theta$	$= \dfrac{\text{adjacent}}{\text{opposite}}$	

EXAMPLE 6   Find the values of the six trigonometric functions of angle θ in Figure 5.30.

Solution   The length of the side opposite angle θ is 6, the adjacent side length is 8, and the length of the hypotenuse is 10. Substituting these numbers in the above definitions gives

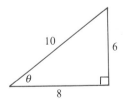

Figure 5.30

$$\sin \theta = \frac{\text{opp}}{\text{hyp}} = \frac{6}{10} = \frac{3}{5} \quad \leftarrow \text{reciprocals} \rightarrow \quad \csc \theta = \frac{\text{hyp}}{\text{opp}} = \frac{10}{6} = \frac{5}{3}$$

$$\cos \theta = \frac{\text{adj}}{\text{hyp}} = \frac{8}{10} = \frac{4}{5} \quad \leftarrow \text{reciprocals} \rightarrow \quad \sec \theta = \frac{\text{hyp}}{\text{adj}} = \frac{10}{8} = \frac{5}{4}$$

$$\tan \theta = \frac{\text{opp}}{\text{adj}} = \frac{6}{8} = \frac{3}{4} \quad \leftarrow \text{reciprocals} \rightarrow \quad \cot \theta = \frac{\text{adj}}{\text{opp}} = \frac{8}{6} = \frac{4}{3}.$$

Note that if we say $\tan \theta = \frac{3}{4}$, we do not necessarily mean that the opposite side length is 3 and the adjacent side length is 4. ∎

EXAMPLE 7   Find the values of the remaining trigonometric functions of acute angle θ if $\tan \theta = \frac{2}{5}$.

Solution   First, draw a right triangle as in Figure 5.31 and label one acute angle θ. Since $\tan \theta = \frac{2}{5}$, the ratio of the opposite side length to the adjacent side length is 2:5. Although many choices are possible, it is easiest to label the opposite side 2 and the adjacent side 5. The hypotenuse in the triangle is found by the Pythagorean relationship

$$(\text{hypotenuse})^2 = 2^2 + 5^2 = 29$$
$$\text{hypotenuse} = \sqrt{29}.$$

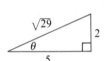

Figure 5.31

Then

$$\sin \theta = \frac{2}{\sqrt{29}} = \frac{2\sqrt{29}}{29} \quad \leftarrow \text{reciprocals} \rightarrow \quad \csc \theta = \frac{\sqrt{29}}{2}$$

$$\cos \theta = \frac{5}{\sqrt{29}} = \frac{5\sqrt{29}}{29} \quad \leftarrow \text{reciprocals} \rightarrow \quad \sec \theta = \frac{\sqrt{29}}{5}$$

$$\cot \theta = \frac{5}{2}.$$

Note that $\cot \theta = \frac{5}{2}$ can be determined from its reciprocal relation to $\tan \theta = \frac{2}{5}$ without constructing the triangle. ∎

It is common notation to label a right triangle as in Figure 5.32. Capital letters such as A, B, and C denote angles, while the lengths of the sides opposite these angles are labeled with the corresponding lowercase letters a, b, and c. The next example illustrates this notation.

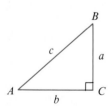

Figure 5.32

EXAMPLE 8   In right triangle ABC with $C = 90°$, if $\cos B = \frac{5}{13}$ and $c = 39$, find a.

Solution   From Figure 5.32 we have

$$\cos B = \frac{\text{adj}}{\text{hyp}} = \frac{a}{c} = \frac{5}{13}.$$

Substituting $c = 39$ gives

$$\frac{a}{39} = \frac{5}{13}.$$

Then

$$a = 39(\tfrac{5}{13})$$
$$= 15.$$

■

EXERCISES 5.3

In Exercises 1–10 find the values of the six trigonometric functions of the angle θ which is in standard position and satisfies the given condition.

1. $(3, -4)$ is on the terminal ray of θ.
2. $(-1, -1)$ is on the terminal ray of θ.
3. $(-12, 5)$ is on the terminal ray of θ.
4. $(2, 4)$ is on the terminal ray of θ.
5. $(9, -12)$ is on the terminal ray of θ.
6. $(-7, -13)$ is on the terminal ray of θ.
7. The terminal ray of θ lies in Q_1 on the line $y = x$.
8. The terminal ray of θ lies in Q_3 on the line $y = x$.
9. The terminal ray of θ lies on the line $y = -2x$ and θ is in Q_2.
10. The terminal ray of θ lies on the line $y = -2x$ and θ is in Q_4.

In Exercises 11–16 determine in which quadrant the terminal ray of θ lies.

11. $\sin \theta$ is negative, $\cos \theta$ is positive.
12. $\cot \theta$ is positive, $\csc \theta$ is positive.
13. $\tan \theta$ is positive, $\sec \theta$ is negative.
14. $\sec \theta > 0$, $\csc \theta < 0$.
15. $\tan \theta < 0$, $\sin \theta > 0$.
16. $\cos \theta < 0$, $\cot \theta < 0$.

In Exercises 17–24 find the values of the remaining trigonometric functions of θ.

17. $\sin \theta = -\tfrac{3}{5}$, terminal ray of θ is in Q_3.
18. $\sec \theta = \tfrac{13}{12}$, terminal ray of θ is in Q_4.
19. $\cos \theta = \tfrac{1}{2}$, $\tan \theta$ is negative.
20. $\tan \theta = \tfrac{2}{3}$, $\sin \theta$ is negative.
21. $\csc \theta = -\tfrac{4}{3}$, $\cot \theta$ is positive.
22. $\cot \theta = 1$, $\csc \theta < 0$.
23. $\sin \theta = \tfrac{1}{3}$, $\sec \theta > 0$.
24. $\csc \theta = 2$, $\cos \theta < 0$.

In Exercises 25–28 find the six trigonometric functions of the acute angles in the right triangle.

25.

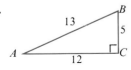

26.

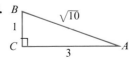

27.

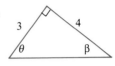

28.

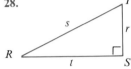

In Exercises 29–36 use these two triangles.

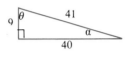

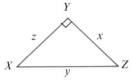

29. Find $\cos Z$.
30. Find $\sec \theta$.
31. Find $\cot \alpha$.
32. Find $\sin \alpha$.
33. Find $\csc X$.
34. Find $\tan \theta$.
35. Represent the ratio y/z as a trigonometric function of an acute angle.
36. Represent the ratio $9/40$ as a trigonometric function of an acute angle.

In Exercises 37–42 use the given trigonometric ratio to find the other trigonometric functions of acute angle θ. Use the definitions of the functions, not a calculator.

37. $\sin \theta = \tfrac{1}{2}$
38. $\csc \theta = 2$
39. $\cot \theta = \tfrac{3}{5}$
40. $\cos \theta = \tfrac{8}{17}$
41. $\tan \theta = 2$
42. $\sec \theta = 5$

In Exercises 43–50 refer to this triangle.

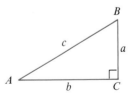

43. If $\cos B = \tfrac{4}{5}$ and $a = 16$, find c.
44. If $\sin A = \tfrac{12}{13}$ and $c = 39$, find a.
45. If $\tan A = \tfrac{7}{12}$ and $a = 21$, find b.
46. If $\cot B = \tfrac{5}{7}$ and $b = 35$, find a.
47. If $\sec A = 3$ and $c = 9$, find a.
48. If $\csc B = 2$ and $c = 8$, find a.
49. If $a = 5$, $b = 12$, and $c = 13$, find the cosine of the smaller acute angle.

50. If $c = 17$ and $a = 8$, find the tangent of the larger acute angle.

51. What is the value of $\sin \theta \cdot \csc \theta$ for all values of θ for which both functions are defined?

THINK ABOUT IT

1. Explain in terms of the general angle definition of the trigonometric functions why the values of $\csc \theta$ are always greater than 1 when θ is an acute angle.
2. Explain in terms of the right triangle definition of the trigonometric functions why the values of $\cos \theta$ range between 0 and 1 when θ is an acute angle.
3. If $\cos \theta = a$ and $270° < \theta < 360°$, then express the values of the other trigonometric functions in terms of a.
4. Consider the unit circle in the accompanying figure, in which each of the six trigonometric functions can be represented as a line segment. For example, since $\overline{OC} = 1$ in right triangle OAC, we have

$$\sin \theta = \frac{\overline{AC}}{\overline{OC}} = \frac{\overline{AC}}{1} = \overline{AC}.$$

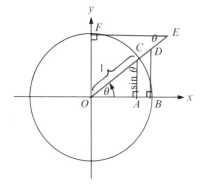

Notice we obtained the desired line segment by selecting a right triangle where the denominator in the defining ratio is 1. Determine the line segments representing the five remaining trigonometric functions in the figure.

5. A simple method to obtain a rough estimate of the values of the trigonometric functions is by construction. Start by using a compass to draw a circle with a radius of 20 spaces on a piece of graph paper. (*Note:* Do not be concerned if the circle goes slightly off the paper.) Label the graph so that a distance of 2 spaces corresponds to one tenth of a unit, and the radius of 20 spaces corresponds to 1 unit (thus $r = 1$). Now use a protractor to find the point on the circle that corresponds to the following rotations, and on the basis of your estimate for their x or y components, approximate to two significant digits the value of the given trigonometric functions.
   a. $\sin 30°$, $\sin 150°$, $\sin 210°$, $\sin 330°$
   b. $\cos 30°$, $\cos 150°$, $\cos 210°$, $\cos 330°$
   c. $\cos 45°$, $\cos 135°$, $\cos 225°$, $\cos 315°$
   d. $\sin 45°$, $\sin 135°$, $\sin 225°$, $\sin 315°$
   e. $\sin 62°$, $\sin 118°$, $\sin 242°$, $\sin 298°$
   f. $\cos 62°$, $\cos 118°$, $\cos 242°$, $\cos 298°$
   g. $\sec 15°$, $\sec 165°$, $\sec 195°$, $\sec 345°$
   h. $\csc 15°$, $\csc 165°$, $\csc 195°$, $\csc 345°$

REMEMBER THIS

1. If A and B are complementary angles and A measures $20°$, then find the measure of angle B.
2. How many minutes equal one degree?
3. Evaluate $\cos \pi$.
4. Evaluate $\sin \dfrac{\pi}{4}$.
5. Evaluate $\cot \dfrac{2\pi}{3}$.
6. Convert $\dfrac{\pi}{4}$ radians to degrees.
7. Find the length of the diagonal in a square with a side length of 1 unit.
8. In a right triangle the hypotenuse measures 2 units and one of the legs measures 1 unit. What is the measure of the other leg?
9. Does $\sin(-s) = -\sin s$ for all values of s?
10. In a circle of radius 6 ft the arc length of a sector is 3 ft. Find the area of the sector.

5.4   Evaluating Trigonometric Functions of Angles

We have determined the values of the trigonometric functions by applying their definitions when given a point on the terminal ray of θ or when given triangles whose side measures were known. A common problem is determining these values if we know the measure of angle θ. First, let us consider the special acute angles whose trigonometric values are known exactly. Figure 5.33 shows the acute angles 30°, 45°, and 60° contained in two right triangles. The diagonal in a square with a side length of 1 unit

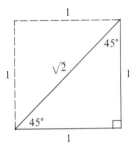
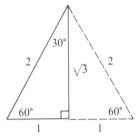

Figure 5.33

forms the right triangle with a 45° angle; the altitude in an equilateral triangle with a side length of 2 units forms the right triangle with angles of 30° and 60°. Using these two triangles, we may determine the value of any of their trigonometric functions. The results are tabulated below, where the ratios $1/\sqrt{2}$, $1/\sqrt{3}$, and $2/\sqrt{3}$ are written with rational denominators as $\sqrt{2}/2$, $\sqrt{3}/3$, and $2\sqrt{3}/3$, respectively. Note that this table is in agreement with the table for the exact trigonometric values of $\pi/6$, $\pi/4$, and $\pi/3$ from Section 5.2.

θ	$\sin \theta$	$\csc \theta$	$\cos \theta$	$\sec \theta$	$\tan \theta$	$\cot \theta$
30°	$\dfrac{1}{2}$	2	$\dfrac{\sqrt{3}}{2}$	$\dfrac{2\sqrt{3}}{3}$	$\dfrac{\sqrt{3}}{3}$	$\sqrt{3}$
60°	$\dfrac{\sqrt{3}}{2}$	$\dfrac{2\sqrt{3}}{3}$	$\dfrac{1}{2}$	2	$\sqrt{3}$	$\dfrac{\sqrt{3}}{3}$
45°	$\dfrac{\sqrt{2}}{2}$	$\sqrt{2}$	$\dfrac{\sqrt{2}}{2}$	$\sqrt{2}$	1	1

EXAMPLE 1   If the length of the short leg of a 30–60–90 triangle is 5, determine the lengths of the other two sides in the triangle.

Solution   The side lengths in a 30–60–90 triangle are in the ratio of $1:\sqrt{3}:2$. If the length of the short leg is 5, then the longer leg length is $5\sqrt{3}$ and the length of the hypotenuse is 10.   ∎

Note in the table of exact values for special angles the similarity between the values of the trigonometric functions for 30° and 60°. For example:

$$\sin 30° = \frac{1}{2} \qquad \text{and} \quad \cos 60° = \frac{1}{2}$$

$$\tan 30° = \frac{1}{\sqrt{3}} = \frac{\sqrt{3}}{3} \qquad \text{and} \quad \cot 60° = \frac{1}{\sqrt{3}} = \frac{\sqrt{3}}{3}$$

$$\sec 30° = \frac{2}{\sqrt{3}} = \frac{2\sqrt{3}}{3} \qquad \text{and} \quad \csc 60° = \frac{2}{\sqrt{3}} = \frac{2\sqrt{3}}{3}.$$

This similarity results from the fact that in the right triangle, the side opposite the 30° angle is adjacent to the 60° angle. Thus,

$$\sin 30° = \frac{\text{side opposite 30° angle}}{\text{hypotenuse}} = \frac{1}{2}$$

$$= \frac{\text{side adjacent to 60° angle}}{\text{hypotenuse}} = \cos 60°$$

$$\tan 30° = \frac{\text{side opposite 30° angle}}{\text{side adjacent to 30° angle}} = \frac{1}{\sqrt{3}}$$

$$= \frac{\text{side adjacent to 60° angle}}{\text{side opposite 60° angle}} = \cot 60°$$

$$\sec 30° = \frac{\text{hypotenuse}}{\text{side adjacent to 30° angle}} = \frac{2}{\sqrt{3}}$$

$$= \frac{\text{hypotenuse}}{\text{side opposite 60° angle}} = \csc 60°.$$

Observe that in each case a trigonometric function of 30° is equal to the corresponding cofunction of 60°. The corresponding cofunction is easy to remember since the *co*function of the sine is the *co*sine, the cofunction of the tangent is the *co*tangent, and the cofunction of the secant is the *co*secant. We can generalize from these examples concerning 30° and 60° to any two angles A and B that are **complementary** (that is, $A + B = 90°$), since in any right triangle the side opposite angle A is adjacent to angle B. Thus, a trigonometric function of any acute angle is equal to the corresponding cofunction of the complementary angle. This result may be stated as cofunction identities.

Cofunction Identities

For any acute angle θ,

$$\sin(90° - \theta) = \cos \theta \qquad \cos(90° - \theta) = \sin \theta$$
$$\tan(90° - \theta) = \cot \theta \qquad \cot(90° - \theta) = \tan \theta$$
$$\sec(90° - \theta) = \csc \theta \qquad \csc(90° - \theta) = \sec \theta.$$

EXAMPLE 2   Express tan 75° as a function of the angle complementary to 75°.

Solution   Using $\cot(90° - \theta) = \tan\theta$ with $\theta = 75°$ yields

$$\tan 75° = \cot(90° - 75°)$$
$$= \cot 15°.$$

In other words, tan 75° = cot 15° since 75° and 15° are complementary angles and the cofunction of the tangent is the cotangent.   ∎

Except for the special acute angles measuring 30°, 45°, and 60°, we evaluate trigonometric functions for acute angles by a calculator or a table. First consider Table 6 at the end of the book. Because of the relationship just mentioned between cofunctions of complementary angles, Table 6 lists angles from 0° to 45° on the left side of the page and their complementary angles from 45° to 90° on the right side. Functions listed at the top and bottom of each column are cofunctions. To use the table to find the approximate value of a particular trigonometric function for an acute angle, do the following.

To Evaluate Trigonometric Functions Using Table 6

> 1. Locate the row containing the angle measure.
> 2. Read at the intersection of this row and the column corresponding to the particular trigonometric function. The function names are taken from the top row if the angle is between 0° and 45°. Function names in the bottom row apply to angles between 45° and 90°.

Table 6 values are usually accurate to four significant digits, and greater accuracy may be obtained with a calculator.

EXAMPLE 3   Use Table 6 to approximate sin 22° and cos 68°.

Angle	sin	cos	tan	cot	sec	csc	
22°00′	.3746	.9272	.4040	2.475	1.079	2.669	68°00′
	cos	sin	cot	tan	csc	sec	Angle

Figure 5.34

Solution   Consider the excerpt from Table 6 in Figure 5.34. Thus,

$$\sin 22° \approx 0.3746$$
$$\cos 68° \approx 0.3746.$$ ∎

EXAMPLE 4   Use Table 6 to approximate cot 4°, cos 17°10′, tan 72°50′, sec 89°50′, and csc 0°.

Solution   From Table 6 we read the values

$$\cot 4° \approx 14.30$$
$$\cos 17°10′ \approx 0.9555$$
$$\tan 72°50′ \approx 3.237$$
$$\sec 89°50′ \approx 343.8$$
$$\csc 0° \text{ is undefined.}$$

A scientific calculator provides the most direct method for evaluating trigonometric functions. Often we need only set the calculator for degree mode, enter the angle, and press $\boxed{\sin}$, $\boxed{\cos}$, or $\boxed{\tan}$. For example,

Table 6	Calculator
$\sin 22° \approx 0.3746$	22 $\boxed{\sin}$ 0.3746066

Many calculators do not have a system for entering angle measure in minutes but instead express angles in decimal degrees. If the problem is stated in degrees-minutes, you can change to decimal degrees by dividing the minutes by 60. Thus,

Table 6	Calculator
$\cos 17°10′ \approx 0.9555$	17 $\boxed{+}$ 10 $\boxed{\div}$ 60 $\boxed{=}$ $\boxed{\cos}$ 0.9554502

Cotangent, secant, and cosecant are usually computed as reciprocal functions of tangent, cosine, and sine by using the $\boxed{1/x}$ key as follows:

Table 6	Calculator
$\cot 4° \approx 14.30$	4 $\boxed{\tan}$ $\boxed{1/x}$ 14.300666
$\sec 89°50′ \approx 343.8$	89 $\boxed{+}$ 50 $\boxed{\div}$ 60 $\boxed{=}$ $\boxed{\cos}$ $\boxed{1/x}$ 343.77516
$\csc 0°$ is undefined	0 $\boxed{\sin}$ $\boxed{1/x}$ Error

To find an angle when one of the trigonometric functions is known, we reverse the assignments of the trig functions and press $\boxed{\text{INV}}$ $\boxed{\sin}$ (or $\boxed{\sin^{-1}}$ or $\boxed{\text{Arcsin}}$) and so on. This sequence calculates the acute angle whose trig function is the displayed positive value.

Table 6	Calculator
$\sin 22° \approx 0.3746$	0.3746 $\boxed{\text{INV}}$ $\boxed{\sin}$ 21.999593

Once again, cotangent, secant, and cosecant problems are solved by using the $\boxed{1/x}$ key in conjunction with the corresponding reciprocal function. For example, the following sequence is used because $\cot \theta = 14.30$ is equivalent to $\tan \theta = 1/14.30$.

Table 6	Calculator
$\cot 4° \approx 14.30$	$14.30 \boxed{1/x} \boxed{INV} \boxed{\tan} 4.0001858$

The sine or cosine of an angle cannot be greater than 1; correspondingly, their reciprocals, cosecant and secant, cannot be positive numbers less than 1. An error message appears in the display when you attempt to solve such problems. The tangent and cotangent functions range over all real numbers.

At present we have discussed how a calculator can be used to approximate trigonometric values for acute angles. The extension to evaluating any angle on a scientific calculator is simple. Many calculators compute trigonometric values for all angles (positive or negative), so you need only enter the angle and press the appropriate function keys, as before. What is missing in this simple calculator method, however, is an understanding of some important concepts in trigonometry. Reference and coterminal angles, the signs of the functions in the various quadrants, and special angles with exact trigonometric values that are easily found and often used are just some of the ideas associated with a noncalculator approach. So keep in mind that a final numerical answer is only part of the objective of this section.

Quadrantal Angles

The definitions of the trigonometric functions may be used to determine the exact value of another special type of angle, as shown in the next example.

EXAMPLE 5   Find the six trigonometric functions of $90°$.

Solution   We have shown that we may pick any point on the terminal ray of the angle when applying the definition. The terminal ray of a $90°$ angle is the positive y-axis and a choice of $r = 1$ determines the point $(0,1)$ as shown in Figure 5.35. Then since $x = 0$, $y = 1$, and $r = 1$, we have

$$\sin 90° = \frac{y}{r} = \frac{1}{1} = 1 \leftarrow \text{reciprocals} \rightarrow \csc 90° = \frac{r}{y} = \frac{1}{1} = 1$$

$$\cos 90° = \frac{x}{r} = \frac{0}{1} = 0 \leftarrow \text{reciprocals} \rightarrow \sec 90° = \frac{r}{x} = \frac{1}{0} \text{ undefined}$$

$$\tan 90° = \frac{y}{x} = \frac{1}{0} \text{ undefined} \leftarrow \text{reciprocals} \rightarrow \cot 90° = \frac{x}{y} = \frac{0}{1} = 0.$$

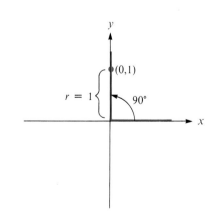

Figure 5.35

An angle of 90° is "special" because its terminal ray coincides with one of the axes. Such angles are called **quadrantal angles,** and any quadrantal angle can be expressed as the product of 90° and some integer. For example, 270° = 90° · 3 and −180° = 90° · (−2). We can repeat the procedure from Example 5 when evaluating functions for any quadrantal angle, and Figure 5.36 shows the four possible positions for the terminal ray of such angles. There are other quadrantal angles besides 0°, 90°, 180°,

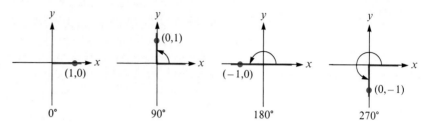

Figure 5.36

and 270° (which are shown in the figure), but their trigonometric values must be the same as one of the four listed. For example, the trigonometric values for 360° will be the same as the trigonometric values for 0°, since the terminal ray for both angles is the same (positive *x*-axis). In general, two angles have the same terminal ray, they are called **coterminal,** and the trigonometric functions of coterminal angles are equal. The following table summarizes the values of the trigonometric functions for quadrantal angles. Note that tabular values match our results for 0, $\pi/2$, π, and $3\pi/2$ from Section 5.2.

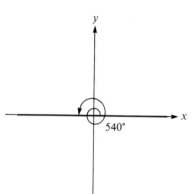

Figure 5.37

θ	$\sin\theta$	$\csc\theta$	$\cos\theta$	$\sec\theta$	$\tan\theta$	$\cot\theta$
0°	0	undefined	1	1	0	undefined
90°	1	1	0	undefined	undefined	0
180°	0	undefined	−1	−1	0	undefined
270°	−1	−1	0	undefined	undefined	0

EXAMPLE 6   Find the exact value of sin 540°.

Solution   540° is a quadrantal angle that is coterminal with 180° (see Figure 5.37). Therefore, sin 540° = sin 180° = 0.   ■

EXAMPLE 7   Find the exact value of cos(−270°).

Solution   −270° is a quadrantal angle that is coterminal with 90° (see Figure 5.38). Therefore, cos(−270°) = cos 90° = 0. (*Note:* Remember that a negative angle means a rotation in a clockwise direction.)   ■

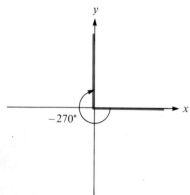

Figure 5.38

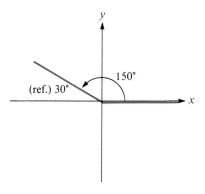

Figure 5.39

To evaluate trigonometric functions for angles that are not quadrantal angles, we introduce the concept of a **reference angle.** The reference angle for an angle θ is defined to be the positive acute angle formed by the terminal ray of θ and the horizontal axis. For example, the reference angle for 150° is 30°, since the closest segment of the horizontal axis (negative x-axis) may correspond to a rotation of 180° and $|180° - 150°| = 30°$, as shown in Figure 5.39.

Now consider the angles 30°, 150°, 210°, and 330°, where the reference angle for each of these angles is 30°. Note that the points on the terminal ray for each of these angles differ only in the sign of the x-coordinate or the y-coordinate. Therefore, the values of the trigonometric functions (which are ratios among x, y, and r) of these angles can differ only in their sign. For instance, by considering the appropriate points in Figure 5.40, we can determine $\sin 30° = \frac{1}{2}$, $\sin 150° = \frac{1}{2}$, $\sin 210° = -\frac{1}{2}$, and $\sin 330° = -\frac{1}{2}$. Thus, the reference angle provides us with a

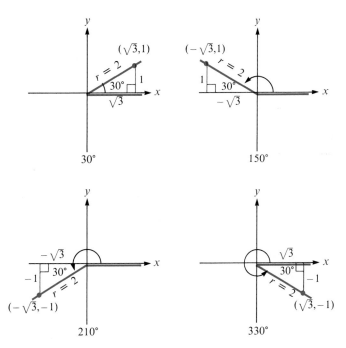

Figure 5.40

method for relating an angle in any quadrant to some acute angle whose trigonometric values can be found. Although the trigonometric values of an angle and its reference angle may differ in sign, we can determine the correct sign according to the quadrant containing the terminal ray of the angle and the chart in Figure 5.41.

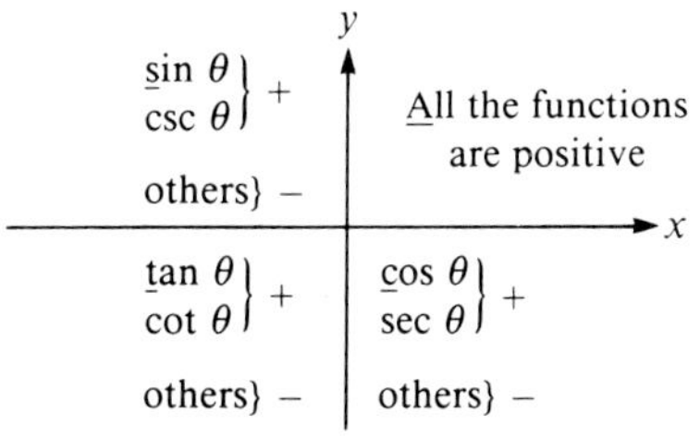

Figure 5.41

EXAMPLE 8   Find the value of tan 135°.

Solution   First, determine the reference angle.

$$|180° - 135°| = 45° \qquad \text{(See Figure 5.42.)}$$

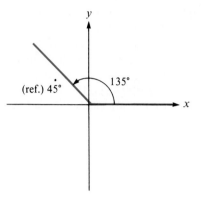

Figure 5.42   **Figure 5.43**

Second, determine tan 45°.

$$\tan 45° = 1 \qquad \text{(See Figure 5.43.)}$$

Third, determine the correct sign.

135° is in the second quadrant where the
value of the tangent function is negative.

Therefore, tan 135° = −1.   ■

In general, we can find the trigonometric value of nonquadrantal angles
by doing the following.

**To Evaluate Trigonometric
Functions**

For nonquadrantal angles:

1. Find the reference angle for the given angle.
2. Find the trigonometric value of the reference angle using the
   appropriate function. Since reference angles are always positive
   acute angles, this value can always be determined from Table 6.
   (*Note:* If the reference angle is 30°, 45°, or 60°, the exact
   answer is preferable.)
3. Determine the correct sign according to the terminal ray of the
   angle and the chart in Figure 5.41.

EXAMPLE 9   Find the exact value of cos 300°.

Solution   First, determine the reference angle.

$$|360° - 300°| = 60° \qquad \text{(See Figure 5.44.)}$$

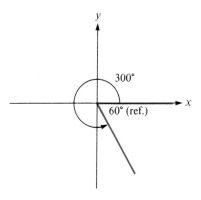

Figure 5.44   **Figure 5.45**

Second, determine cos 60°.

$$\cos 60° = \tfrac{1}{2} \qquad \text{(See Figure 5.45.)}$$

Third, determine the correct sign.

300° is in Q_4, where the cosine function is positive.

Therefore, cos 300° = cos 60° = $\tfrac{1}{2}$. ∎

EXAMPLE 10   Evaluate sec 510°.

Solution   First, determine the reference angle.

$$|540° - 510°| = 30° \qquad \text{(See Figure 5.46.)}$$

Second, determine sec 30°.

$$\sec 30° = \frac{2}{\sqrt{3}} = \frac{2\sqrt{3}}{3} \qquad \text{(See Figure 5.45.)}$$

Third, determine the correct sign.

510° is in Q_2, where the secant function is negative.

Therefore, sec 510° = $-$sec 30° = $-2\sqrt{3}/3$. ∎

EXAMPLE 11   Approximate cos 227°20'.

Solution   First, determine the reference angle.

$$|227°20' - 180°| = 47°20' \qquad \text{(See Figure 5.47.)}$$

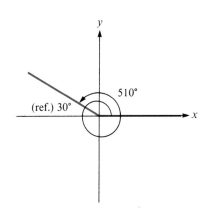

Figure 5.46

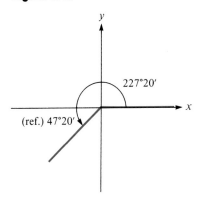

Figure 5.47

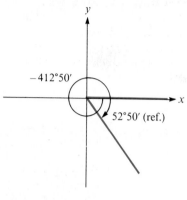

Figure 5.48

Second, determine cos 47° 20′.

$$\cos 47°20' \approx 0.6777 \qquad \text{(Table 6)}$$

Third, determine the correct sign.

227° 20′ is in Q₃, where the cosine function is negative.

Therefore, cos 227° 20′ = −cos 47° 20′ ≈ −0.6777.   ■

EXAMPLE 12 Approximate csc(−412° 50′).

Solution First, determine the reference angle.

$$\big|-360 - (-412°50')\big| = 52°50' \qquad \text{(See Figure 5.48.)}$$

Second, determine csc 52° 50′.

$$\csc 52°50' \approx 1.255 \qquad \text{(Table 6)}$$

Third, determine the correct sign.

−412° 50′ is in Q₄, where the cosecant function is negative.

Therefore, csc(−412° 50′) = −csc 52° 50′ ≈ −1.255.   ■

EXERCISES 5.4

In Exercises 1–8 use the exact trigonometric values of 30°, 45°, or 60°.

1. Does 2 sin 30° = sin 60°?
2. Does sin 60° csc 30° = tan 60°?
3. Does sin 45° cos 45° = 1?
4. Does $\sin 30° = \sqrt{\dfrac{1 - \cos 60°}{2}}$?
5. Does (tan 45°)² + 1 = (sec 45°)²?
6. Does cos 60° = 1 − 2(sin 30°)²?
7. If the length of the short leg of a 30–60–90 triangle is 2, determine the lengths of the other two sides of the triangle.
8. If the length of a leg of a 45–45–90 triangle is 5, determine the length of the hypotenuse of the triangle.

In Exercises 9–14 express each term as a function of the angle complementary to the given angle.

9. sin 17° **10.** cos 64° **11.** tan 81° 30′
12. sec 33° 58′ **13.** csc 68.1° **14.** cot 0.5°

In Exercises 15–24 approximate each using a calculator or Table 6. If the angle is given in decimal degrees, use a calculator.

15. cos 7° **16.** sin 42° **17.** cot 83°
18. csc 54° **19.** tan 79° 30′ **20.** sec 65° 10′
21. csc 51° 00′ **22.** cot 16° 40′ **23.** sin 12.3°
24. tan 47.6°

In Exercises 25–34 approximate the measure of angle θ. Write solutions to the nearest 10 minutes from Table 6 or to the nearest tenth of a degree by calculator.

25. sin θ ≈ 0.7071 **26.** cos θ ≈ 0.8660
27. tan θ ≈ 0.7907 **28.** cot θ ≈ 2.699
29. sec θ ≈ 1.781 **30.** csc θ ≈ 49.11
31. cot θ ≈ 0.7651 **32.** tan θ ≈ 0.0402
33. csc θ ≈ 16.00 **34.** sec θ ≈ 1.549

In Exercises 35–44 find the *exact* value of each expression.

35. sin 270° **36.** cos 180°
37. tan 90° **38.** cot 0°
39. cos 450° **40.** sin 630°
41. sin(−90°) **42.** sec(−180°)
43. csc(−630°) **44.** cot(−540°)

In Exercises 45–52 find the reference angle for the given angle.

45. 205° **46.** 181° **47.** 60° **48.** −30°
49. −97° **50.** 111.1° **51.** 385° 30′ **52.** 1,000°

In Exercises 53–82 find the *exact* value of each expression.

53. sin 210° **54.** tan 225°
55. sec 330° **56.** cos 135°
57. cot 315° **58.** csc 120°
59. sin 150° **60.** tan 300°
61. cos 225° **62.** cot 240°

63. cos 480°

64. sin 600°

65. sec 390°

66. tan 420°

67. cot 690°

68. cos 840°

69. sin 1,035°

70. csc 675°

71. cos 495°

72. sin 570°

73. sin(−60°)

74. cos(−45°)

75. tan(−120°)

76. sec(−225°)

77. cot(−315°)

78. csc(−210°)

79. sec(−330°)

80. cos(−480°)

81. cot(−495°)

82. sin(−1,050°)

In Exercises 83–112 find the approximate value of each expression. Use Table 6 or a calculator.

83. sin 212°

84. cos 307°

85. tan 254°

86. cot 115°

87. sec 301°20′

88. csc 163.4°

89. cos 148°50′

90. sin 354.5°

91. cot 298°10′

92. tan 190°40′

93. sin 177°10′

94. cos 252°20′

95. csc 672°40′

96. cot 392.1°

97. sin 626°40′

98. sin 531.5°

99. cos 952°20′

100. sec 452°20′

101. tan 738°30′

102. sin 521°50′

103. cos(−81°)

104. sin(−25°)

105. tan(−131°)

106. csc(−322°40′)

107. cos(−61°)

108. sin(−251.5°)

109. tan(−214°10′)

110. sec(−400°)

111. cot(−512°)

112. csc(−938°20′)

113. The formula for the horizontal distance traveled by a projectile neglecting air resistance, is

$$d = V^2 \frac{\sin A \sin B}{16}$$

where V is the initial velocity, A the angle of elevation, and $B = 90° − A$. In the 16-lb shot-putting event an athlete releases the ball at an angle of elevation of 42° with an initial velocity of 47 ft/second. Determine the distance of the throw. (*Note:* The maximum distance is attained when the release angle is 45°.)

114. A major principle in the theory of light is the **law of refraction**. Refraction is the bending of light as it passes from one medium to another. For example, consider the following diagram, which shows a ray of light bending toward the perpendicular as it passes

from air to water. The bending is caused by the change in speed of the light ray as it slows down in the water, which is the denser medium. The mathematical relation between the angle of incidence (*i*) and the angle of refraction (*r*) is a trigonometric ratio called **Snell's law.** The law is

$$\frac{\sin i}{\sin r} = \frac{v_i}{v_r},$$

where v_i/v_r is the ratio between the velocities of light in the two mediums.

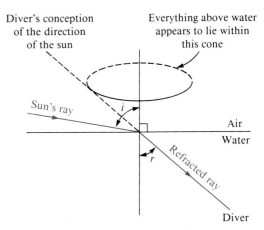

a. As light passes from air to water, v_i/v_r is about $\frac{4}{3}$. Find the angle of refraction if $i = 28.5°$.

b. When the sun is near the horizon, the angle of incidence approaches 90° and the angle at which the sun's rays penetrate the water approaches a limiting value. Determine this value of *r*. It is interesting to note that this restriction on the angle of refraction causes an optical illusion for a diver under water, as shown in the diagram. As he looks up, the world above the surface appears to be in the shape of a cone. This distorted perspective is called the "fish-eye view of the world."

THINK ABOUT IT

1. An engineer uses a calculator and computes sin 24° and tan(π/5) as follows:

   24 $\boxed{\text{sin}}$ 0.4067366

   π $\boxed{\div}$ 5 $\boxed{=}$ $\boxed{\text{tan}}$ 0.0109667

   This engineer made a common mistake. What is it?

2. **a.** Give two examples of specific values for θ_1 and θ_2 so that $\sin(\theta_1 + \theta_2) \neq \sin\theta_1 + \sin\theta_2$ is true. Verify your answers without using a calculator.

   b. Give two examples of specific values for θ_1 and θ_2 so that $\sin(\theta_1 + \theta_2) = \sin\theta_1 + \sin\theta_2$ is true. Verify your answers without using a calculator.

3. Evaluate $\sin 1° + \sin 2° + \cdots + \sin 360°$.

4. In the figure below find the exact values for m, n, p, and q.

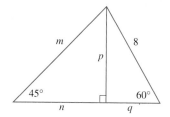

5. In the figure below find θ_1, θ_2, θ_3, θ_4, and θ_5. Which measures are exact and which are approximations?

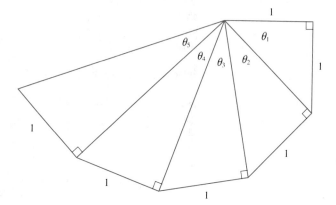

REMEMBER THIS

1. Find the exact value of $\sin(3\pi/2)$.

2. Evaluate $-4 \cos 10x$ for $x = 0$.

3. Solve $2x + \dfrac{\pi}{2} = 0$.

4. Simplify $\dfrac{2\pi}{1/2}$.

5. Evaluate $|-3|$.

6. Write the interval $[0, 2\pi)$ in set-builder notation using x as the variable.

7. In a circle a central angle intercepts an arc equal in length to the radius of the circle. What is the measure of the central angle in radians?

8. True or false: If $f(-x) = f(x)$, then the graph of f is symmetric with respect to the y-axis.

9. What is the relation between the graph of $y = f(x)$ and the graph of $y = -f(x)$?

10. What is the relation between the graph of $y = f(x)$ and the graph of $y = f(x - \pi/2)$?

5.5 Graphs of Sine and Cosine Functions

A picture or graph of the sine and cosine functions helps us understand their cyclic behavior. We use the Cartesian coordinate system and associate the arc length values with points on the horizontal or x-axis. This means we are now using x, instead of s, to represent the independent variable, which is an arc length. The vertical axis, labeled the y-axis, is used to represent the function values.

We begin by considering the values of the sine function as we lay off on the x-axis a length 2π and pass once around the unit circle. The following table indicates some values of the sine function on this typical interval.

$y = \sin x$													(for $0 \le x \le 2\pi$)
x	0	$\dfrac{\pi}{6}$	$\dfrac{\pi}{3}$	$\dfrac{\pi}{2}$	$\dfrac{2\pi}{3}$	$\dfrac{5\pi}{6}$	π	$\dfrac{7\pi}{6}$	$\dfrac{4\pi}{3}$	$\dfrac{3\pi}{2}$	$\dfrac{5\pi}{3}$	$\dfrac{11\pi}{6}$	2π
y	0	0.5	0.87	1	0.87	0.5	0	-0.5	-0.87	-1	-0.87	-0.5	0

If we plot these points and join them with a smooth curve, we obtain the graph in Figure 5.49, which describes the essential characteristics of the sine function during one cycle. The plot of $y = \sin x$ starts at the origin, attains a maximum at one fourth of the cycle length, returns to zero halfway through the cycle, attains a minimum at the three-quarter point, and returns to zero at the end of the cycle. Each time we lay off a length 2π we pass around the circle and repeat this behavior. Thus, the graph of $y = \sin x$ weaves continuously through cycles in both directions, as illustrated in Figure 5.50.

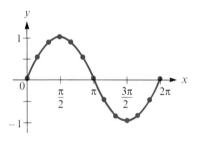

Figure 5.49

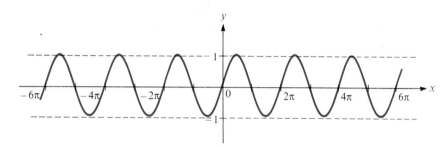

Figure 5.50

When we look at the graph of the sine function, we see that the function repeats its values on intervals of length 2π. For this reason, the sine function is said to be *periodic*. We define a periodic function as follows.

Periodic Function

A function f is periodic if

$$f(x) = f(x + p)$$

for all x in the domain of f. The smallest positive number p for which this is true is called the **period** of the function.

This definition applies to the sine function since

$$\sin x = \sin(x + 2\pi).$$

Thus, the sine function is periodic with period 2π.

We can use this information to sketch functions of the form $y = \sin bx$, where b represents a positive real number. These functions are similar to $y = \sin x$ in that they have the same basic shape, but they may differ by having different periods. For example, to obtain the graph of $y = \sin 2x$, we multiply x by 2 and take the sine of the resulting number. This means that when $x = \pi$, we evaluate the sine of $2 \cdot \pi$, or $\sin 2\pi$. Thus, when we substitute numbers from 0 to π for x, we evaluate the function from $\sin 0$ to $\sin 2\pi$ and complete one cycle. The period for $y = \sin 2x$ is therefore π. In general, the period of $y = \sin bx$ $(b > 0)$ is $2\pi/b$. Figure 5.51 compares the graphs of $y = \sin x$ and $y = \sin 2x$ for $0 \leq x \leq 2\pi$.

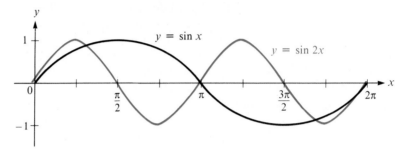

Figure 5.51

EXAMPLE 1 Sketch the graph of $y = \sin \frac{1}{2}x$ for $0 \leq x \leq 2\pi$.

Solution Determine the period by substituting $\frac{1}{2}$ for b.

$$\text{Period} = \frac{2\pi}{b} = \frac{2\pi}{1/2} = 4\pi$$

If the curve completes one cycle for $0 \leq x \leq 4\pi$, the curve completes one half of a sine wave for $0 \leq x \leq 2\pi$, as shown in Figure 5.52. ∎

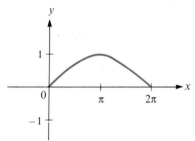

Figure 5.52

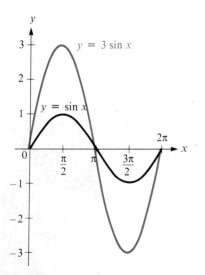

Figure 5.53

A second important characteristic of the graph of $y = \sin x$ is that the greatest y value is 1. When the graph of a periodic function is centered about the x-axis, the maximum y value is called the **amplitude** of the function. Thus, the amplitude of $y = \sin x$ is 1.

We can use this information to sketch functions of the form $y = a \sin bx$, where a represents a real number. These functions are similar to $y = \sin bx$ in that they have the same basic shape and period, but they may differ by having different amplitudes. For example, to sketch the graph of $y = 3 \sin x$ we obtain values for $\sin x$ and multiply these values by 3. Since the greatest y value that $\sin x$ attains is 1, the greatest y value that $3 \sin x$ attains is 3. Thus, the amplitude of $y = 3 \sin x$ is 3.

In general, since the greatest value that $\sin x$ attains is 1, the greatest value that $a \sin x$ attains is $|a|$. Thus, the amplitude of $y = a \sin bx$ is $|a|$. Figure 5.53 compares the graphs of $y = \sin x$ and $y = 3 \sin x$ on $[0, 2\pi]$.

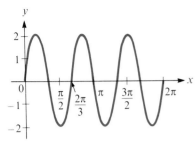

Figure 5.54

EXAMPLE 2   Sketch the graph of $y = 2 \sin 3x$ for $0 \le x \le 2\pi$.

Solution   First, determine the amplitude and period.

$$\text{Amplitude} = |a| = |2| = 2$$

$$\text{Period} = \frac{2\pi}{b} = \frac{2\pi}{3}$$

If the curve completes one cycle on the interval $[0, 2\pi/3]$, the curve completes three cycles on the given interval. Note that b gives the number of cycles on the interval $[0, 2\pi]$. Since the amplitude is 2, the curve oscillates between a maximum value of 2 and a minimum value of -2. This graph is sketched in Figure 5.54. ∎

EXAMPLE 3   Sketch the graph of $y = -\sin \pi x$ for $0 \le x \le 2\pi$.

Solution   First, determine the amplitude and period.

$$\text{Amplitude} = |a| = |-1| = 1$$

$$\text{Period} = \frac{2\pi}{b} = \frac{2\pi}{\pi} = 2$$

If the curve completes one cycle every 2 units, the curve completes slightly more than three cycles on the interval $[0, 2\pi]$. Since the amplitude is 1, the curve oscillates between a maximum value of 1 and a minimum value of -1. Because a is negative, we obtain the graph by reflecting the graph of $y = \sin \pi x$ about the x-axis, as shown in Figure 5.55.

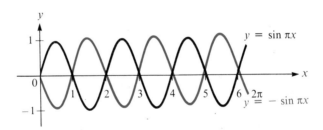

Figure 5.55 ∎

The graph of the cosine function has the same essential characteristics as the graph of the sine function. That is, the amplitude of the cosine function is 1 and the period is 2π. This is evidenced by the following table, which indicates some values of the cosine function on the interval $[0, 2\pi]$.

$y = \cos x$														
(for $0 \le x \le 2\pi$)														
x	0	$\frac{\pi}{6}$	$\frac{\pi}{3}$	$\frac{\pi}{2}$	$\frac{2\pi}{3}$	$\frac{5\pi}{6}$	π	$\frac{7\pi}{6}$	$\frac{4\pi}{3}$	$\frac{3\pi}{2}$	$\frac{5\pi}{3}$	$\frac{11\pi}{6}$	2π
y	1	0.87	0.5	0	-0.5	-0.87	-1	-0.87	-0.5	0	0.5	0.87	1

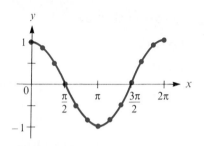

Figure 5.56

If we plot these points and join them with a smooth curve, we obtain the graph shown in Figure 5.56. This graph demonstrates that the cosine function completes one cycle on the interval $[0,2\pi]$ and attains a maximum value of 1. Like that of the sine function, this graph can be reproduced indefinitely in both directions to obtain as much of the graph of the cosine function as desired (see Figure 5.57).

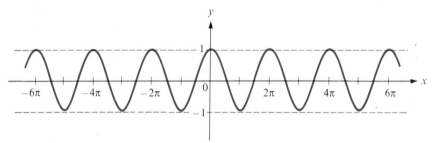

Figure 5.57

The great similarity between the graph of the sine and cosine functions should be apparent. In fact, if we shift the graph of the cosine function $\pi/2$ units to the right, the resulting graph is the sine function. Thus, the only difference between the two graphs is that one curve leads the other by $\pi/2$. That is, $\cos(x - \pi/2) = \sin x$.

We graph functions of the form $y = a \cos bx$ in a manner similar to graphing $y = a \sin bx$; that is, we find the amplitude by computing $|a|$ and the period by using $2\pi/b$. The difference is that the graph of $y = a \cos bx$ attains a maximum or minimum height at $x = 0$. A common error in these problems is to draw the graph of a sine wave, which has a height of 0 when $x = 0$.

EXAMPLE 4   Sketch the graph of $y = \frac{1}{2} \cos 2x$ for $0 \le x \le 2\pi$.

Solution   First, determine the amplitude and period.

$$\text{Amplitude} = |a| = \left|\frac{1}{2}\right| = \frac{1}{2}$$

$$\text{Period} = \frac{2\pi}{b} = \frac{2\pi}{2} = \pi$$

If the curve completes one cycle on $[0,\pi]$, the curve will complete two cycles on $[0,2\pi]$. Since the amplitude is $\frac{1}{2}$, the curve oscillates between a maximum value of $\frac{1}{2}$ and a minimum value of $-\frac{1}{2}$, as shown in Figure 5.58. ∎

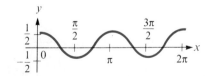

Figure 5.58

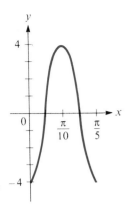

Figure 5.59

EXAMPLE 5   Sketch one cycle of the graph of $y = -4 \cos 10x$.

Solution   First, determine the amplitude and period.

$$\text{Amplitude} = |a| = |-4| = 4$$

$$\text{Period} = \frac{2\pi}{b} = \frac{2\pi}{10} = \frac{\pi}{5}$$

The curve completes one cycle on the interval $[0, \pi/5]$ and attains a maximum value of 4 and a minimum value of -4. Since a is a negative number, we obtain the graph shown in Figure 5.59 by starting and ending the graph at a minimum point. ■

Finally, let us determine the effect of the constant c in a function of the form $y = a \sin(bx + c)$. Consider $y = \sin(x - \pi/2)$. As $x - \pi/2$ ranges from 0 to 2π, the curve completes one sine wave.

$$x - \frac{\pi}{2} = 0 \quad \text{when} \quad x = \frac{\pi}{2}$$

$$x - \frac{\pi}{2} = 2\pi \quad \text{when} \quad x = \frac{5\pi}{2}$$

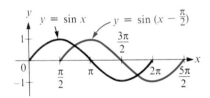

Figure 5.60

Thus, the function completes one cycle in the interval from $\pi/2$ to $5\pi/2$. The period of the function is 2π and the amplitude is 1. For comparison, one cycle of the graphs of $y = \sin x$ and $y = \sin(x - \pi/2)$ is given in Figure 5.60.

The complete graph of $y = \sin(x - \pi/2)$ is sketched by repeating the cycle shown in Figure 5.60 to the left and to the right. Notice that the graph of $y = \sin(x - \pi/2)$ may be obtained by shifting the graph of $y = \sin x$ to the right $\pi/2$ units. The sine wave then starts a cycle at $\pi/2$ instead of 0 and we call $\pi/2$ the **phase shift.**

In general, the constant c in the function $y = a \sin(bx + c)$ causes a shift of the graph of $y = a \sin bx$. The shift is of distance $|c/b|$ and is to the left if $c > 0$ and to the right if $c < 0$. The phase shift is $-c/b$. Similar remarks hold for functions of the form $y = a \cos(bx + c)$.

EXAMPLE 6   Graph one cycle of the function $y = 3 \cos(2x + \pi/2)$. Indicate the amplitude, period, and phase shift.

Solution   Determine the amplitude and period.

$$\text{Amplitude} = |a| = |3| = 3$$

$$\text{Period} = \frac{2\pi}{b} = \frac{2\pi}{2} = \pi$$

The function completes one cosine cycle as $2x + \pi/2$ varies from 0 to 2π.

$$2x + \frac{\pi}{2} = 0 \quad \text{when} \quad x = -\frac{\pi}{4}$$

$$2x + \frac{\pi}{2} = 2\pi \quad \text{when} \quad x = \frac{3\pi}{4}$$

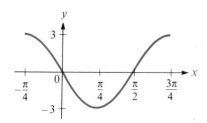

Figure 5.61

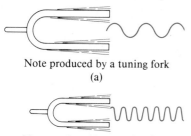

Note produced by a tuning fork
(a)

Note produced by a tuning fork
with a higher pitch
(b)

Sound produced by a piano
(c)

Figure 5.62

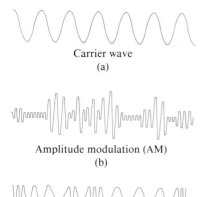

Carrier wave
(a)

Amplitude modulation (AM)
(b)

Frequency modulation (FM)
(c)

Figure 5.63

Thus, the function completes one cycle in the interval from $-\pi/4$ to $3\pi/4$ (see Figure 5.61). This interval checks with the computed period since $3\pi/4 - (-\pi/4) = \pi$. A cycle starts at $-\pi/4$, so $-\pi/4$ is the phase shift. We may verify the phase shift since

$$\frac{-c}{b} = \frac{-\pi/2}{2} = -\frac{\pi}{4}.$$

Amplitude, period, and phase shift are important considerations when analyzing any periodic phenomena. For example, let us briefly discuss two familiar concepts: musical sounds and radio waves. Musical sounds are caused by regular vibrations that have a definite period. On an electronic instrument called an **oscilloscope,** which changes sounds to electrical impulses and then to light waves, the sound from a tuning fork has the shape illustrated in Figure 5.62(a). The period of the wave depends on the pitch of the sound. With higher notes the pitch or frequency of the sound increases, producing a wave that has a smaller period [Figure 5.62(b)]. Humans can detect frequencies between about 50 and 15,000 vibrations (cycles) per second. However, some animals, such as the bat, can hear frequencies as high as 120,000 hertz (cycles per second). The amplitude of the wave depends on the intensity of the sound. Since humans hear best at a frequency of about 3,500 hertz, the loudness of a sound depends on both intensity and frequency. Although most musical sounds are very complex [such as the sound produced by the piano in Figure 5.62(c)], the French mathematician Joseph Fourier, in about 1800, showed that any periodic function is the sum of simple sine functions. Thus, all these sounds can be graphed and analyzed by some combination of sine waves. This analysis is indispensable in the design of sound recording and reproducing equipment.

Radio waves, used to transmit information at the speed of light (186,000 mi/second), are produced by oscillations in an electric current. Each current cycle produces a single radio wave. Special electronic equipment is needed to produce the alternating current since ordinary stations broadcast between 500,000 and 1,500,000 radio waves per second. Each station is licensed to broadcast a fixed number of radio waves per second, called its **frequency.** You receive the station's program by adjusting or tuning your set to accept this frequency.

Information is imposed on a radio wave as follows: A carrier wave [Figure 5.63(a)] is produced at a transmitting station, which makes the licensed number of cycles per second. This carrier wave is then modulated by the program current from the broadcasting site. In amplitude-modulated (AM) broadcasting, the amplitude of the carrier is made to vary according to the message; the wavelength remains constant [see Figure 5.63(b)]. With frequency modulation (FM), the amplitude of the wave remains constant and the wavelength varies [see Figure 5.63(c)].

FM broadcasting is superior to AM in that it produces better fidelity of sound and is relatively free from static and interference. However, AM stations are more numerous since they are less expensive and have a greater broadcasting range. In television an AM signal is used to transmit the picture but an FM signal carries the sound.

EXERCISES 5.5

In Exercises 1–10 state the amplitude and the period and sketch the curve for $-p \leq x \leq p$, where p is the period of the function.

1. $y = 2 \sin x$

2. $y = 3 \cos 2x$

3. $y = -3 \cos x$

4. $y = -4 \sin \frac{1}{3}x$

5. $y = 2 \sin 3x$

6. $y = \frac{1}{2} \cos 4x$

7. $y = -\cos 18x$

8. $y = -6 \sin \frac{x}{4}$

9. $y = 10 \cos \pi x$

10. $y = 110 \sin 120\pi x$

In Exercises 11–20 state the amplitude and the period and sketch the curve for $0 \leq x \leq 2\pi$.

11. $y = 3 \cos 4x$

12. $y = 2 \cos \frac{x}{4}$

13. $y = -\sin \frac{x}{2}$

14. $y = -3 \sin 2x$

15. $y = \frac{1}{2} \sin 3x$

16. $y = 1.5 \cos \frac{1}{3}x$

17. $y = \sin \frac{\pi}{2}x$

18. $y = 2 \cos \pi x$

19. $y = -2 \cos \frac{\pi}{3}x$

20. $y = -0.9 \sin \frac{\pi}{4}x$

In Exercises 21–30 state the amplitude, the period, and the phase shift and sketch one cycle of the function.

21. $y = \sin\left(x + \frac{\pi}{2}\right)$

22. $y = 2 \sin(x - \pi)$

23. $y = \cos\left(x - \frac{\pi}{4}\right)$

24. $y = 3 \cos\left(x + \frac{\pi}{3}\right)$

25. $y = \frac{1}{2} \cos\left(2x + \frac{\pi}{4}\right)$

26. $y = -\sin\left(\frac{x}{2} - \pi\right)$

27. $y = -\cos\left(\frac{x}{4} + \frac{\pi}{2}\right)$

28. $y = \sin(x - 1)$

29. $y = \sin(\pi x - \pi)$

30. $y = 1.2 \cos\left(2\pi x - \frac{\pi}{2}\right)$

In Exercises 31–40 find an equation for the curves with the given single cycle. The equations should be written in the form $y = a \sin bx$ or $y = a \cos bx$.

31.

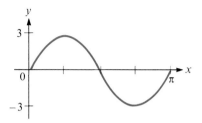

32.

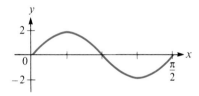

33.

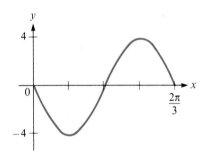

34.

35.

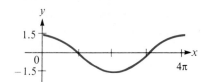

36.

37.

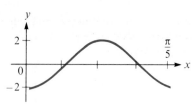

38.

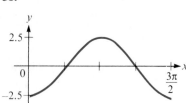

39.

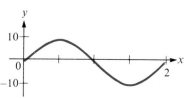

40.

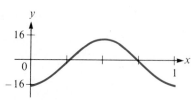

In Exercises 41–59 select the choice that completes the statement or answers the question.

41. The amplitude of the function $y = 4 \sin 3x$ is

    a. 1   **b.** 12   **c.** 3   **d.** 4

42. The maximum value of $y = 3 \cos \frac{1}{2}x$ is

    a. 1   **b.** $\frac{1}{2}$ **c.** 3   **d.** $\frac{3}{2}$

43. The minimum value of $y = 2 \cos 3x$ is

    a. 0   **b.** -3 **c.** 3   **d.** -2

44. The maximum value of $y = 2 + \sin x$ is

    a. 1   **b.** 2   **c.** 3   **d.** 4

45. The expression $y = 3 \sin \frac{1}{2}x$ reaches its maximum value when x equals

    a. 0   **b.** $\pi/2$ **c.** π **d.** $3\pi/2$

46. The expression $y = 2 \sin 3x$ reaches its maximum value when x equals

    a. 0   **b.** $\pi/6$ **c.** $\pi/4$ **d.** $2\pi/3$

47. The expression $y = 3 \cos \frac{1}{2}x$ reaches its minimum value when x equals

    a. 0   **b.** $\pi/2$ **c.** π **d.** 2π

48. The expression $y = 2 \cos 3x$ reaches its minimum value when x equals

    a. 0   **b.** $\pi/3$ **c.** $\pi/2$ **d.** $2\pi/3$

49. The period of the curve $y = 2 \sin x$ is

    a. 0   **b.** $\pi/2$ **c.** π **d.** 2π

50. The period of the curve $y = 3 \cos 2x$ is

    a. $\pi/2$ **b.** π **c.** 2π **d.** 4π

51. A function having the period π is

    a. $y = 2 \cos x$ **b.** $y = \cos \frac{1}{2}x$

    c. $y = \frac{1}{2} \cos x$ **d.** $y = \cos 2x$

52. A function having a period of $\pi/2$ is

    a. $y = \frac{1}{4} \sin x$ **b.** $y = 4 \sin 2x$

    c. $y = 2 \sin 4x$ **d.** $y = 4 \sin \frac{1}{4}x$

53. As x increases from $\pi/2$ to $3\pi/2$, then $y = \sin x$

    a. increases   **b.** decreases

    c. increases, then decreases

    d. decreases, then increases

54. As x increases from π to 2π, the cosine of x

    a. increases   **b.** decreases

    c. increases, then decreases

    d. decreases, then increases

55. As x increases in the interval from $\pi/2$ to $3\pi/2$, the value of $y = \cos x$ will

    a. increase   **b.** decrease

    c. increase, then decrease

    d. decrease, then increase

56. As x increases from $-\pi/2$ to 0, $\sin x$

    a. increases from 0 to 1

    b. decreases from 0 to -1

    c. increases from -1 to 0

    d. decreases from 1 to 0

57. $y = \sin x$ and $y = \cos x$ both increase in

    a. quadrant 1   **b.** quadrant 2

    c. quadrant 3   **d.** quadrant 4

58. Between $x = 0$ and $x = 2\pi$, the graphs of $y = \sin x$ and $y = \cos x$ have in common

    a. no points   **b.** one point

    c. two points   **d.** four points

59. When graphs of $y = \sin x$ and $y = \cos x$ are drawn on the same axes, how many times do they intersect on the interval $[\pi/2, \pi]$?

    a. 0   **b.** 1   **c.** 2   **d.** 3

THINK ABOUT IT

1. In each case describe how the graph of g may be obtained from the graph of f.
   a. $f(x) = \sin x$, $g(x) = -\sin x$
   b. $f(x) = \cos x$, $g(x) = 4 \cos x$
   c. $f(x) = 4 \cos x$, $g(x) = 4 \cos(x + \pi)$
   d. $f(x) = 3 \sin \pi x$, $g(x) = 3 \sin(\pi x - \pi)$

2. a. Give two examples of an equation of the form $y = \sin(bx + c)$ whose graph passes through $(\pi/4, 0)$.
   b. Give two examples of an equation of the form $y = \cos(bx + c)$ whose graph passes through $(1, 0)$.

3. Find the smallest nonnegative value of a for which the graph of $y = \sin x$ is symmetric about the line $x = a$. Confirm your answer graphically.

4. For any periodic function the amplitude is defined as $(M - m)/2$, where M is the maximum function value and m is the minimum function value. Use this definition to do the following problems.
   a. Show the amplitude of $y = \sin x$ is 1.
   b. Show the amplitude of $y = a \sin x$ is $|a|$.
   c. Show the amplitude of $y = 3 + \sin x$ is 1.
   d. What are the amplitude and the period of the function shown in this illustration?

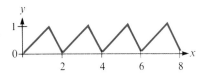

5. When graphs of $y = |\sin x|$ and $y = \frac{1}{6}x$ are drawn on the same coordinate system, how many times do they intersect? Confirm your answer graphically.

REMEMBER THIS

1. Evaluate $\tan(\pi/3)$ to two significant digits.
2. Evaluate $\sec 2\pi x$ at $x = \frac{1}{2}$.
3. Write the set of real numbers in interval notation.
4. Express $\csc x$ in terms of $\sin x$.
5. Express $\sec 2\pi x$ in terms of $\cos 2\pi x$.
6. True or false: If $f(-x) = -f(x)$, then the graph of f is symmetric with respect to the origin.
7. Find the reciprocal of -1.
8. Why is the reciprocal of 0 undefined?
9. If $f = \{(0,1),(1,-1)\}$ and $g = \{(0,2),(1,1)\}$, find $f + g$.
10. What is the relation between the graph of $y = f(x)$ and the graph of $y = f(x) + 3$?

5.6 Graphs of Other Trigonometric Functions

Although the tangent function is periodic, its behavior differs dramatically from that of the sine and cosine. To see this difference, compare the graph of $y = \tan x$ to the smooth weaving curves of these functions. We begin by constructing the following table, which lists some values of the tangent function on the interval $[0, 2\pi]$.

$y = \tan x = \sin x/\cos x$													(for $0 \leq x \leq 2\pi$)
x	0	$\frac{\pi}{6}$	$\frac{\pi}{3}$	$\frac{\pi}{2}$	$\frac{2\pi}{3}$	$\frac{5\pi}{6}$	π	$\frac{7\pi}{6}$	$\frac{4\pi}{3}$	$\frac{3\pi}{2}$	$\frac{5\pi}{3}$	$\frac{11\pi}{6}$	2π
y	0	0.6	1.7	und.	-1.7	-0.6	0	0.6	1.7	und.	-1.7	-0.6	0

Unlike the sine and cosine, the tangent function is not defined for all real numbers. That is, $y = \tan x = \sin x / \cos x$ is undefined if $\cos x = 0$. Thus, we must exclude from the domain of this function $\pi/2$, $3\pi/2$, and any x for which $x = (\pi/2) + k\pi$ (k any integer). As x approaches $\pi/2$, $\sin x$ approaches 1 and $\cos x$ approaches 0. This means that $|\tan x|$ becomes very large as x gets close to $\pi/2$. Consider Table 5 at the end of the book, which verifies this observation. On the basis of the preceding discussion and the fact that $\tan(-x) = -\tan x$, a portion of the graph of $y = \tan x$ is presented in Figure 5.64.

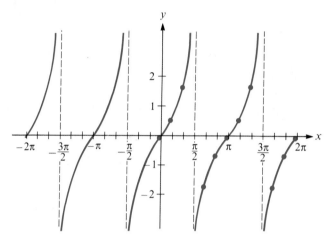

Figure 5.64

Using this graph we can see that the tangent function differs from the sine and cosine in the following important respects.

Comparison of the Sine, Cosine, and Tangent Functions

1. The domain of the sine and cosine functions is the set of all real numbers. The tangent function excludes $x = (\pi/2) + k\pi$ (k any integer).
2. The range of the sine and cosine functions is $[-1,1]$. The range of the tangent function is the set of all real numbers.
3. The sine and cosine are periodic with period 2π. Careful consideration of Figure 5.64 shows that the tangent function repeats its values every π units. Thus, the period of the tangent function is π, and the period of $y = \tan bx$ ($b > 0$) is π/b.

From this discussion, the value of graphs in analyzing the behavior of various functions should be clear.

The other three trigonometric functions may be graphed as the reciprocals of the sine, cosine, or tangent. That is,

$$\csc x = \frac{1}{\sin x}, \qquad \sec x = \frac{1}{\cos x}, \qquad \tan x = \frac{1}{\cot x}.$$

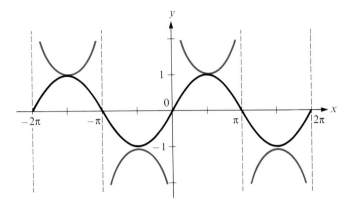

Figure 5.65 $y = \csc x$

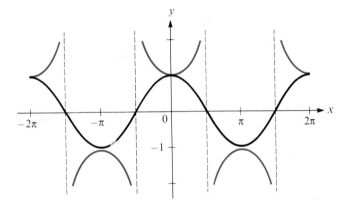

Figure 5.66 $y = \sec x$

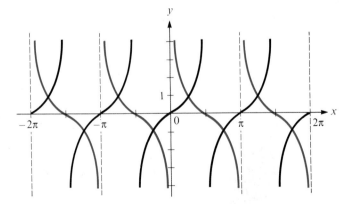

Figure 5.67 $y = \cot x$

In Figures 5.65–5.67 we first graph the sine, cosine, and tangent as black curves. By obtaining the reciprocals of various y values, we then graph in color the cosecant, secant, and cotangent. Note the following about a function and its reciprocal function.

1. As one function increases, the other decreases, and vice versa.
2. The two functions always have the same sign.
3. When one function is zero, the other is undefined.
4. When the value of the function is 1 or -1, the reciprocal function has the same value.

The dashed lines in these figures that the curves approach, but never touch, are called vertical asymptotes.

EXAMPLE 1   Graph one cycle of the function $y = \tan(x + \pi/4)$.

Solution   We start with the graph of $y = \tan x$ in Figure 5.64. To graph $y = \tan(x + \pi/4)$, we note that x has been replaced by $x + \pi/4$, so by our graphing techniques from Section 2.3, the graph of $y = \tan(x + \pi/4)$ is the graph of $y = \tan x$ shifted $\pi/4$ units to the left. Figure 5.68 shows the graph of one cycle on the interval that contains $x = 0$.

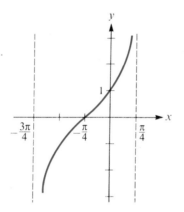

Figure 5.68

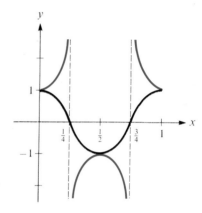

Figure 5.69

EXAMPLE 2   Graph one cycle of the function $y = \sec 2\pi x$.

Solution   Since $\sec 2\pi x = 1/\cos 2\pi x$, we first sketch one cycle of the graph of $y = \cos 2\pi x$ to use as a reference. The amplitude and period for $y = \cos 2\pi x$ are both 1, and the black curve in Figure 5.69 shows the graph of this function. Then by obtaining the reciprocals of the various y values, we graph one cycle of $y = \sec 2\pi x$ as shown in color in the figure. ■

Vertical Shifts

To this point we have used horizontal shifts (as in Example 1) to graph variations of a known curve. Vertical shifting is also useful. Recall from Section 2.3 that if $c > 0$, then the graph of $y = f(x) + c$ is the graph of f raised c units, while the graph of $y = f(x) - c$ is the graph of f lowered c units.

EXAMPLE 3   Graph one cycle of the function $y = 3 + 2 \sin 4x$.

Solution   First, by the methods of Section 5.5, we graph one cycle of $y = 2 \sin 4x$, as shown in black in Figure 5.70. Then we graph in color $y = 3 + 2 \sin 4x$ by raising the graph of $y = 2 \sin 4x$ up 3 units, since the constant 3 is added to $2 \sin 4x$.

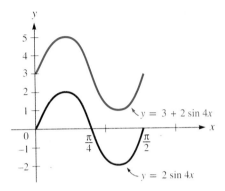

Figure 5.70

EXAMPLE 4   Graph one cycle of the function $y = -\tan x - 1$.

Solution   To graph $y = -\tan x - 1$, first reflect the graph of $y = \tan x$ about the x-axis to obtain the graph of $y = -\tan x$ [see Figure 5.71(a)]. Then we lower the graph in Figure 5.71(a) down 1 unit, because the constant 1 is subtracted from $-\tan x$. The completed graph is shown in Figure 5.71(b).

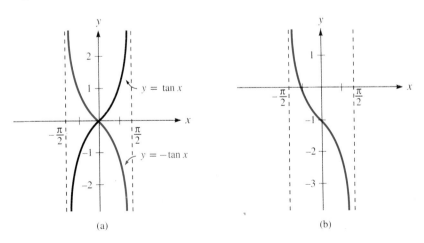

Figure 5.71

Addition of Ordinates

Graphing by vertical shifts is a special case of graphing equations of the form $f(x) = g(x) + h(x)$ by addition of ordinates (or y values) at each x value in the domain of f. In practice, if either g or h is a trigonometric function, then the graph is usually drawn by determining at least all points in the graph of f associated with intercepts, maximum points, or minimum points in the graphs of either g or h.

EXAMPLE 5   Graph $y = x + \sin x$.

Solution   Figure 5.72 shows the graphs of $g(x) = x$ and $h(x) = \sin x$ drawn on the same coordinate system. On $[0, 2\pi]$ the points in the graph of $y = x + \sin x$ associated with intercepts, maximum points, or minimum points in the graphs of g and h are determined below.

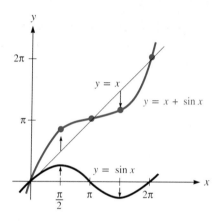

Figure 5.72

x	$g(x) = x$	$h(x) = \sin x$	$y = x + \sin x$
0	0	0	$0 + 0 = 0$
$\dfrac{\pi}{2}$	$\dfrac{\pi}{2}$	1	$\dfrac{\pi}{2} + 1 \approx 2.6$
π	π	0	$\pi + 0 = \pi$
$\dfrac{3\pi}{2}$	$\dfrac{3\pi}{2}$	-1	$\dfrac{3\pi}{2} - 1 \approx 3.6$
2π	2π	0	$2\pi + 0 = 2\pi$

By plotting $(0,0)$, $(\pi/2, \pi/2 + 1)$, (π, π), $(3\pi/2, 3\pi/2 - 1)$, and $(2\pi, 2\pi)$ and drawing a smooth curve through these points, we obtain the graph of $y = x + \sin x$ in Figure 5.72.   ∎

EXAMPLE 6   Graph $f(x) = \sin x - \cos x$.

Solution   Rewrite $f(x) = \sin x - \cos x$ as $f(x) = \sin x + (-\cos x)$ and then graph $g(x) = \sin x$ and $h(x) = -\cos x$ on the same coordinate system, as shown in Figure 5.73. On $[0,2\pi]$, the graphs of g and h have intercepts, maximum points, or minimum points at $x = 0, \pi/2, \pi, 3\pi/2$, and 2π; so evaluate $f(x) = \sin x + (-\cos x)$ at these points.

x	$g(x) = \sin x$	$h(x) = -\cos x$	$f(x) = \sin x + (-\cos x)$
0	0	-1	$0 + (-1) = -1$
$\pi/2$	1	0	$1 + 0 = 1$
π	0	1	$0 + 1 = 1$
$3\pi/2$	-1	0	$-1 + 0 = -1$
2π	0	-1	$0 + (-1) = -1$

Also, by adding ordinates (with the aid of a ruler) at the two points on $[0,2\pi]$ where the two curves intersect gives a maximum point and a minimum point in the graph of f. By connecting the seven derived points with a smooth curve, we obtain the graph of $f(x) = \sin x - \cos x$ in Figure 5.73.

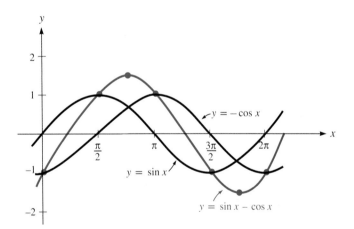

Figure 5.73

Note: In general, the equation $y = a \sin x + b \cos x$ may be written in the form $y = A \sin(x + c)$. Thus, you can anticipate that the graph of $y = a \sin x + b \cos x$ is a sine graph. For this example an identity derived in Section 6.4 can be used to show $y = \sin x - \cos x = \sqrt{2} \sin(x - \pi/4)$. ∎

EXERCISES 5.6

In Exercises 1–3 complete the table for the function and then sketch the curve from these points on the interval $[0,2\pi]$.

x	0	$\dfrac{\pi}{6}$	$\dfrac{\pi}{3}$	$\dfrac{\pi}{2}$	$\dfrac{2\pi}{3}$	$\dfrac{5\pi}{6}$	π	$\dfrac{7\pi}{6}$	$\dfrac{4\pi}{3}$	$\dfrac{3\pi}{2}$	$\dfrac{5\pi}{3}$	$\dfrac{11\pi}{6}$	2π
y													

1. $y = \cot x$　　　**2.** $y = \sec x$　　　**3.** $y = \csc x$

In Exercises 4–6 use Figures 5.65–5.67 to determine the domain, the range, and the period of the function.

4. $y = \cot x$　　　**5.** $y = \sec x$　　　**6.** $y = \csc x$

In Exercises 7–12 complete the table by determining if the function is increasing or decreasing in the interval. Use graphs to determine your answers.

	$\left(0,\dfrac{\pi}{2}\right)$	$\left(\dfrac{\pi}{2},\pi\right)$	$\left(\pi,\dfrac{3\pi}{2}\right)$	$\left(\dfrac{3\pi}{2},2\pi\right)$
7. $\sin x$				
8. $\cos x$				
9. $\tan x$				
10. $\cot x$				
11. $\sec x$				
12. $\csc x$				

13. Of the six trigonometric functions, only $y = \sin x$ and $y = \cos x$ have amplitudes. Why?

In Exercises 14–23 graph one cycle of the given functions.

14. $y = \tan\left(x - \dfrac{\pi}{4}\right)$　　　**15.** $y = -\cot\left(x + \dfrac{\pi}{2}\right)$

16. $y = \cot\frac{1}{2}x$　　　**17.** $y = \tan 2x$

18. $y = \sec 2x$　　　**19.** $y = 3\csc\dfrac{x}{2}$

20. $y = 2\csc \pi x$　　　**21.** $y = -\sec\left(x + \dfrac{\pi}{4}\right)$

22. $y = \sec\left(2\pi x - \dfrac{\pi}{2}\right)$　　　**23.** $y = \csc(\pi x - \pi)$

In Exercises 24–33 graph one cycle of the given function. Use vertical shifts.

24. $y = 2 + \sin x$　　　　**25.** $y = -3 + \cos x$
26. $y = -\cos x - 2$　　　**27.** $y = -\sin x - 1$
28. $y = 2 + 3\sin 4x$　　　**29.** $y = 1 - 2\cos 2x$
30. $y = 1 + \sec x$　　　　**31.** $y = -1 + \csc \pi x$
32. $y = -\cot x - 1$　　　**33.** $y = 2 - \tan x$

In Exercises 34–43 graph by the method of addition of ordinates.

34. $y = x + \cos x$　　　**35.** $y = x - \sin x$
36. $y = -x + \sin x$　　**37.** $y = \cos x - x$
38. $y = \cos x - \sin x$　**39.** $y = \sin x + \cos x$
40. $y = 2\sin x - \cos x$　**41.** $y = 2\cos x + \sin x$
42. $y = \cos x + \sin 2x$　**43.** $y = \sin x - \cos 2x$

THINK ABOUT IT

1.　a. Describe how we may translate the graph of $y = \csc x$ to obtain the graph of $y = \sec x$.
　　b. What is the value of c closest to zero such that
$$\csc(x + c) = \sec x$$
for all values of x for which the expressions are defined?

2. In each case find how many times the graphs of f and g intersect when drawn on the same coordinate system. Confirm your answer graphically.
　　a. $f(x) = \sec x,\ g(x) = \frac{1}{2}x$
　　b. $f(x) = \csc x,\ g(x) = \frac{1}{2}x$

3. Confirm graphically that $-\tan(x + \pi/2) = \cot x$ is an identity.

4. Solve each inequality.
　　a. $\tan x > 0$　　　　**b.** $\cot x < 0$

5.　a. Use properties of absolute value to show that if $f(x) = 2^{-x}\sin x$, then $-2^{-x} \le f(x) \le 2^{-x}$.
　　b. Graph $f(x) = 2^{-x}\sin x$. (*Note:* Analyzing this type of graph, which is called a **damped sine wave,** is important in engineering.)

REMEMBER THIS

1. If $f = \{(0,1),(\pi/2,0)\}$, find the inverse function f^{-1}.
2. If the domain of f is $[-\pi/2,\pi/2]$, the range of f is $[-1,1]$, and f is a one-to-one function, then find the domain and range of the inverse function f^{-1}.
3. If $f(x) = 3x - 2$, find $f^{-1}(x)$.
4. If $f(x) = 2x - 5$ and $f^{-1}(x) = \frac{1}{2}x + \frac{5}{2}$, find $f[f^{-1}(x)]$ and $f^{-1}[f(x)]$.

5. If f and g are inverse functions, then how are their graphs related?
6. What is the range of the function $y = \tan x$?
7. Evaluate $\sin 0$, $\sin \pi$, and $\sin(-\pi)$.
8. Write $\{x: -\pi/2 < x < \pi/2\}$ in interval notation.
9. Graph $y = \sin x$ for $-\pi/2 \le x \le \pi/2$.
10. Graph $y = \tan x$ for $-\pi/2 < x < \pi/2$.

5.7   Inverse Trigonometric Functions

The concept of an inverse function was introduced in Section 2.5. Recall these three facts.

1. Two functions with exactly reverse assignments are inverse functions.
2. The domain of a function f is the range of its inverse function and the range of f is the domain of its inverse.
3. A function is one-to-one when each x value in the domain is assigned a different y value so that no two ordered pairs have the same second component. We define the inverse of f, denoted f^{-1}, only when f is a one-to-one function.

Now consider the graph of $y = \sin x$ in Figure 5.74. Because the sine function is periodic, many x values are assigned the same y value. For example,

$$\sin 0 = \sin \pi = \sin 2\pi = \sin(-\pi) = 0.$$

Thus, the inverse of the sine function is not a function. The so-called **inverse sine function** is defined by restricting the domain of $y = \sin x$ to the interval $[-\pi/2,\pi/2]$. Note in Figure 5.74 that each x value is assigned a different y value in this limited version of the sine function.

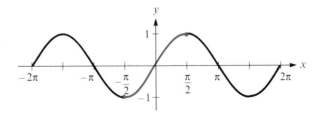

Figure 5.74

We find the rule for the inverse sine function by interchanging the x and y variables. Thus, the inverse of the function defined by

$$y = \sin x \qquad -\frac{\pi}{2} \le x \le \frac{\pi}{2}$$

is defined by

$$x = \sin y \qquad -\frac{\pi}{2} \le y \le \frac{\pi}{2},$$

which is written in inverse notation as

$$y = \arcsin x \quad \text{or} \quad y = \sin^{-1} x.$$

For example, since $\sin(\pi/2) = 1$, we write

$$\frac{\pi}{2} = \arcsin 1 \quad \text{or} \quad \frac{\pi}{2} = \sin^{-1} 1.$$

Both expressions are read "$\pi/2$ is the arc (or number) whose sin is 1," and note that the -1 in the function name $\sin^{-1}$ is not an exponent. Because a function and its inverse interchange their domain and range, for the inverse sine function the domain is the interval $[-1,1]$ and the range is $[-\pi/2,\pi/2]$. The following definition sums up the key aspects of our discussion.

Inverse Sine Function

> The inverse sine function, denoted by **arcsin** or **sin⁻¹**, is defined by
>
> $$y = \arcsin x \text{ if and only if } x = \sin y,$$
>
> where $-1 \le x \le 1$ and $-\pi/2 \le y \le \pi/2$.

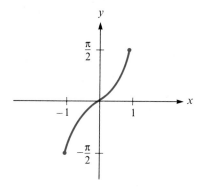

Figure 5.75

Consider Figure 5.75, which gives the graph of $y = \arcsin x$. Note that the rule in the inverse sine function basically selects the number *closest to zero* whose sine is x.

Through similar considerations we may define inverses for the other five trigonometric functions. For example, Figure 5.76 shows that $y = \cos x$ is one-to-one and assumes all its range values for $0 \le x \le \pi$. Thus, we may define the inverse function, $y = \arccos x$ as graphed in color, with domain $[-1,1]$ and range $[0,\pi]$ from this limited version of the cosine function. Similarly, Figure 5.77 shows that we use $y = \tan x$ for $-\pi/2 < x < \pi/2$ to define $y = \arctan x$ with the set of all real numbers for its domain and the interval $(-\pi/2,\pi/2)$ for its range. The remaining inverse functions are used less frequently, so we just list their respective domains and ranges in the following summary. Analysis of the graphs of $y = \cot x$, $y = \sec x$, and $y = \csc x$ would show the respective intervals to be suitable (though arbitrary) choices.

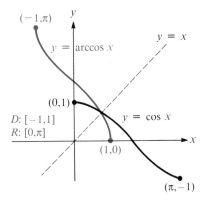

Figure 5.76

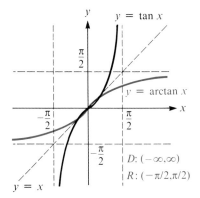

Figure 5.77

Inverse Trigonometric Functions

Function	Domain	Range
$y = \arcsin x$	$[-1,1]$	$[-\pi/2, \pi/2]$
$y = \arccos x$	$[-1,1]$	$[0, \pi]$
$y = \arctan x$	$(-\infty, \infty)$	$(-\pi/2, \pi/2)$
$y = \operatorname{arccot} x$	$(-\infty, \infty)$	$(0, \pi)$
$y = \operatorname{arcsec} x$	$(-\infty, -1] \cup [1, \infty)$	$[0, \pi/2) \cup [\pi, 3\pi/2)$
$y = \operatorname{arccsc} x$	$(-\infty, -1] \cup [1, \infty)$	$(0, \pi/2] \cup (-\pi, -\pi/2]$

There are two areas concerning inverse trigonometric functions in which there is no general agreement, so take note.

1. In terms of notation an inverse trig function is sometimes capitalized and written as Arcsin x or Sin^{-1} x. With this convention arcsin x and sin^{-1} x do not represent functions but represent the set of all numbers whose sine is x with no restriction placed on the range values.

2. The inverse secant and cosecant functions are sometimes assigned different range intervals depending on the application of the functions. The intervals chosen above are usually selected to obtain simpler derivative and integration formulas in calculus.

EXAMPLE 1   Find arcsin $\frac{1}{2}$.

Solution   Let $y = \arcsin \frac{1}{2}$; then $\frac{1}{2} = \sin y$. We solve this equation and find the number y in the interval $[-\pi/2, \pi/2]$ whose sin is $\frac{1}{2}$. Since $\sin \pi/6 = \frac{1}{2}$, arcsin $\frac{1}{2} = \pi/6$.   ∎

EXAMPLE 2   Find $\cos^{-1}(-\frac{1}{2})$.

Solution   Let $y = \cos^{-1}(-\frac{1}{2})$; then $-\frac{1}{2} = \cos y$. We solve this equation and find the number y in the interval $[0,\pi]$ whose cosine is $-\frac{1}{2}$. Since $\cos \pi/3 = \frac{1}{2}$, we have

$$\cos\left(\pi - \frac{\pi}{3}\right) = \cos\frac{2\pi}{3} = -\frac{1}{2}.$$

Thus, $\cos^{-1}(-\frac{1}{2}) = 2\pi/3$.   ∎

EXAMPLE 3   Find $\arctan(-1)$.

Solution   Let $y = \arctan(-1)$; then $-1 = \tan y$. We solve this equation and find the number y in the interval $(-\pi/2,\pi/2)$ whose tangent is -1. Since $\tan(-\pi/4) = -1$, $\arctan(-1) = -\pi/4$.   ∎

EXAMPLE 4   Find arccot 0.5312.

Solution   Let $y = $ arccot 0.5312; then $0.5312 = \cot y$. We solve this equation and find the number y in the interval $(0,\pi)$ whose cotangent is 0.5312. Table 5 at the end of the book indicates that $\cot 1.08 \approx 0.5312$. Thus, arccot $0.5312 \approx 1.08$.   ∎

EXAMPLE 5   Find $\sin(\arccos 0)$.

Solution   First, determine arccos 0. Since $\cos(\pi/2) = 0$, we have

$$\frac{\pi}{2} = \arccos 0.$$

Now replace arccos 0 by $\pi/2$ in the original expression.

$$\sin(\arccos 0) = \sin\frac{\pi}{2} = 1$$ ∎

EXAMPLE 6   Find $\tan(\sin^{-1}\frac{4}{5})$.

Solution   It is useful to interpret $\sin^{-1}\frac{4}{5}$ as the measure of an angle in a right triangle. Let $\theta = \sin^{-1}\frac{4}{5}$ and sketch the triangle in Figure 5.78. The length of the side opposite θ is 4 and the hypotenuse has length 5. The remaining side length is found by the Pythagorean relationship.

$$x = \sqrt{5^2 - 4^2}$$
$$= 3$$

Figure 5.78 $\theta = \sin^{-1}\frac{4}{5}$

Thus,

$$\tan \theta = \tan(\sin^{-1}\tfrac{4}{5}) = \tfrac{4}{3}.$$ ∎

The first six example problems from this section may all be solved easily on a calculator. Just set the calculator for radian mode and proceed as follows.

EXAMPLE 7   arcsin $\frac{1}{2}$ 1 $\boxed{\div}$ 2 $\boxed{=}$ $\boxed{\text{INV}}$ $\boxed{\sin}$ 0.5235988   ■

EXAMPLE 8 $\cos^{-1}(-\frac{1}{2})$ 1 $\boxed{+/-}$ $\boxed{\div}$ 2 $\boxed{=}$ $\boxed{\text{INV}}$ $\boxed{\cos}$ 2.0943951 ■

EXAMPLE 9 $\arctan(-1)$ 1 $\boxed{+/-}$ $\boxed{\text{INV}}$ $\boxed{\tan}$ -0.7853982 ■

EXAMPLE 10   arccot 0.5312   0.5312 $\boxed{1/x}$ $\boxed{\text{INV}}$ $\boxed{\tan}$ 1.0825014   ■

EXAMPLE 11 $\sin(\text{arccos } 0)$ 0 $\boxed{\text{INV}}$ $\boxed{\cos}$ $\boxed{\sin}$ 1   ■

EXAMPLE 12 $\tan(\sin^{-1} \frac{4}{5})$ 4 $\boxed{\div}$ 5 $\boxed{=}$ $\boxed{\text{INV}}$ $\boxed{\sin}$ $\boxed{\tan}$ 1.3333333   ■

Remember that, when possible, exact answers like $\pi/6$ and $2\pi/3$ should be given.

The right triangle method used to evaluate $\tan(\sin^{-1} \frac{4}{5})$ in Example 6 can be extended to solve a type of problem that occurs in calculus. The next two examples illustrate this important technique.

EXAMPLE 13   Simplify $\sin(\cos^{-1} x)$.

Solution   Let $\theta = \cos^{-1} x$. Then $\cos \theta = x$, where $0 \leq \theta \leq \pi$. First, assume x is positive and sketch the triangle in Figure 5.79. The length of the side adjacent to θ is x, and the hypotenuse has length 1. We find the opposite side length a to be $\sqrt{1 - x^2}$ using the Pythagorean relation.

$$x^2 + a^2 = 1^2$$
$$a^2 = 1 - x^2$$
$$a = \sqrt{1 - x^2}$$

Now sketch Figure 5.80 with the aid of this triangle, where θ may be in either Q_1 or Q_2 because $0 \leq \theta \leq \pi$.

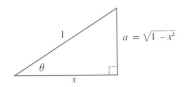

Figure 5.79 $\theta = \cos^{-1}x,\ x > 0$

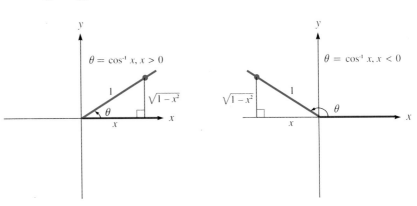

Figure 5.80

Then for all θ in $[0,\pi]$,

$$\sin \theta = \frac{\sqrt{1 - x^2}}{1},$$

so

$$\sin(\cos^{-1} x) = \sin \theta = \sqrt{1 - x^2}. \qquad \blacksquare$$

EXAMPLE 14   If $x = \sin \theta$, where $-\pi/2 \le \theta \le \pi/2$, then write $-\cot \theta - \theta$ as a function of x.

Solution   If $x = \sin \theta$, where $-\pi/2 \le \theta \le \pi/2$, then $\theta = \arcsin x$. We can represent $\cot \theta$ by first assuming $x > 0$ and sketching the triangle in Figure 5.81. The length of the side opposite θ is x, the hypotenuse has length 1, and the remaining side length is found by the Pythagorean relation (as in Example 13) to be $\sqrt{1 - x^2}$. Now sketch Figure 5.82 with the aid of this triangle, where θ may be in either Q_1 or Q_4 because $-\pi/2 \le \theta \le \pi/2$. Then, $\cot \theta = \sqrt{1 - x^2}/x$, so

$$-\cot \theta - \theta = \frac{-\sqrt{1 - x^2}}{x} - \arcsin x.$$

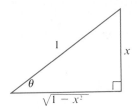

Figure 5.81

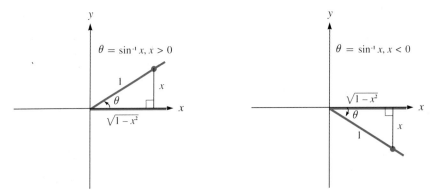

Figure 5.82 $\blacksquare$

Inverse Properties

As discussed in Section 2.5, inverse functions offset each other so that if f and g are inverses, then

$$f[g(x)] = x \text{ for all } x \text{ in the domain of } g$$

and

$$g[f(x)] = x \text{ for all } x \text{ in the domain of } f.$$

Applying this definition to the restricted sine, cosine, and tangent functions and their inverses yields the following properties.

$$
\begin{array}{lll}
\sin(\sin^{-1} x) = x & \text{for} & -1 \le x \le 1 \\
\sin^{-1}(\sin x) = x & \text{for} & -\pi/2 \le x \le \pi/2 \\
\cos(\cos^{-1} x) = x & \text{for} & -1 \le x \le 1 \\
\cos^{-1}(\cos x) = x & \text{for} & 0 \le x \le \pi \\
\tan(\tan^{-1} x) = x & \text{for} & \text{all } x \\
\tan^{-1}(\tan x) = x & \text{for} & -\pi/2 < x < \pi/2
\end{array}
$$

The next example shows that knowing the intervals for which these properties hold is essential to applying the properties.

EXAMPLE 15   Evaluate each expression if possible.

a. $\sin^{-1}\left(\sin \dfrac{\pi}{4}\right)$ **b.** $\cos^{-1}(\cos 2\pi)$ **c.** $\sin(\sin^{-1} 2)$

Solution

a. Because $\pi/4$ is in the interval $[-\pi/2, \pi/2]$, we use $\sin^{-1}(\sin x) = x$ to obtain $\sin^{-1}[\sin(\pi/4)] = \pi/4$. To check, note that $\sin(\pi/4) = \sqrt{2}/2$ and $\sin^{-1}(\sqrt{2}/2) = \pi/4$.

b. Because 2π is not in the interval $[0, \pi]$, we cannot use $\cos^{-1}(\cos x) = x$. Instead, since $\cos 2\pi = 1$, we have

$$\cos^{-1}(\cos 2\pi) = \cos^{-1} 1 = 0.$$

c. Because 2 is not in the interval $[-1, 1]$, we cannot use $\sin(\sin^{-1} x) = x$. There is no value for $\sin(\sin^{-1} 2)$ since 2 is not in the domain of $y = \sin^{-1} x$.   ∎

EXERCISES 5.7

In Exercises 1–30 evaluate the expression.

1. $\arccos \frac{1}{2}$

2. $\arccos(-\frac{1}{2})$

3. $\sin^{-1}\left(-\dfrac{\sqrt{3}}{2}\right)$

4. $\arcsin \dfrac{\sqrt{3}}{2}$

5. $\arctan 1$

6. $\operatorname{arccot}(-1)$

7. $\arcsin(-1)$

8. $\operatorname{arcsec}(-1)$

9. $\arcsin \dfrac{\sqrt{2}}{2}$

10. $\tan^{-1}\sqrt{3}$

11. $\tan^{-1}(-\sqrt{3})$

12. $\arccos\left(-\dfrac{\sqrt{2}}{2}\right)$

13. $\arcsin 0.3124$

14. $\sin^{-1}(-0.3124)$

15. $\cos^{-1}(-0.5509)$

16. $\arccos 0.5509$

17. $\arctan 1.758$

18. $\tan^{-1}(-1.758)$

19. $\sec^{-1} \frac{5}{4}$

20. $\operatorname{arccot} \frac{2}{3}$

21. $\cos(\arcsin 0)$

22. $\sin(\arccos 1)$

23. $\sin\left(\cos^{-1} \dfrac{\sqrt{3}}{2}\right)$

24. $\cos[\sin^{-1}(-\frac{1}{2})]$

25. $\cot(\arccos 0)$

26. $\tan^{-1}(\cos 0)$

27. $\arcsin(\tan 0)$

28. $\csc(\arcsin 1)$

29. $\tan[\arcsin(-0.5518)]$

30. $\csc(\arccos 0.0129)$

In Exercises 31–40 simplify the expression.

31. $\sin(\tan^{-1} \frac{3}{4})$

32. $\cos[\arcsin(-\frac{1}{3})]$

33. $\tan(\arcsin \frac{12}{13})$

34. $\cot(\cos^{-1} \frac{8}{17})$

35. $\cos(\tan^{-1} x)$

36. $\cos(\sin^{-1} x)$

37. $\cot(\arcsin x)$

38. $\sin(\operatorname{arcsec} x)$

39. $\tan\left(\arcsin \dfrac{x-2}{3}\right)$

40. $\sec\left(\tan^{-1} \dfrac{x}{2}\right)$

In Exercises 41–48 write the expression as a function of x. Use inverse functions and/or right triangles.

41. θ if $x = \sin \theta$, where $-\pi/2 \le \theta \le \pi/2$.

42. θ if $x = \tan \theta$, where $-\pi/2 < \theta < \pi/2$.

43. θ if $x - 2 = 3 \sec \theta$, where $0 \le \theta < \pi/2$ or $\pi \le \theta < 3\pi/2$.

44. θ if $x = a \sin \theta$, where $-\pi/2 \le \theta \le \pi/2$.

45. $-1/(4 \sin \theta)$ if $x = 2 \tan \theta$, where $-\pi/2 < \theta < \pi/2$.

46. $2 \sec \theta \tan \theta$ if $x = 2 \sec \theta$, where $0 \le \theta < \pi/2$ or $\pi \le \theta < 3\pi/2$.

47. $-\frac{1}{2} \csc \theta$ if $2(x + 2) = \sec \theta$, where $0 \le \theta < \pi/2$ or $\pi \le \theta < 3\pi/2$.

48. $\theta + \sin \theta \cos \theta$ if $x - 1 = \sin \theta$, where $-\pi/2 \le \theta \le \pi/2$.

In Exercises 49–52 solve the formula for θ.

49. $m \cdot \sin \theta = 1$, where $-\pi/2 \le \theta \le \pi/2$.

50. $\tan \theta = b/a$, where $-\pi/2 < \theta < \pi/2$.

51. $T = \dfrac{2V_0 \sin \theta}{g}$, where $-\dfrac{\pi}{2} \le \theta \le \dfrac{\pi}{2}$.

52. $V_x = V \cos \theta$, where $0 \le \theta \le \pi$.

In Exercises 53–58 for what values of x is each statement true?

53. $\sin^{-1}(\sin x) = x$ **54.** $\sin(\sin^{-1} x) = x$

55. $\cos^{-1}(\cos x) = x$ **56.** $\cos(\cos^{-1} x) = x$

57. $\tan(\arctan x) = x$ **58.** $\arctan(\tan x) = x$

In Exercises 59–68 evaluate each expression.

59. $\cos(\arccos 0)$ **60.** $\cos^{-1}(\cos 0)$

61. $\sin^{-1}[\sin(-1)]$ **62.** $\sin[\arcsin(-1)]$

63. $\cos^{-1}[\cos(3\pi/2)]$ **64.** $\sin^{-1}(\sin 2\pi)$

65. $\tan[\tan^{-1}(-1)]$ **66.** $\tan^{-1}[\tan(-\pi)]$

67. $\cos(\cos^{-1} 2)$ **68.** $\sin(\sin^{-1} 2\pi)$

69. Sketch the graph of $y = \text{arcsec } x$.

70. Sketch the graph of $y = \text{arccot } x$.

71. Sketch the graph of $y = \arcsin(x - 1)$.

72. Sketch the graph of $y = \arccos(x + 2)$.

73. Consider the illustration below.

   a. Write θ as a function of a.

   b. Write θ as a function of b.

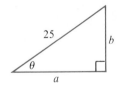

74. Consider the illustration below. Write θ as the difference between two inverse tangent expressions.

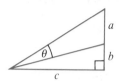

75. Consider the illustration below. The area of the region enclosed by the graphs of $y = 1/\sqrt{1 - x^2}$, $y = 0$, $x = 0$, and $x = \sqrt{2}/2$ is shown in calculus to be given by $\arcsin(\sqrt{2}/2) - \arcsin 0$. Find this area.

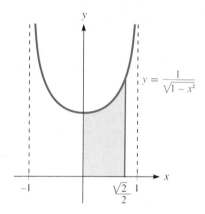

76. Consider the illustration below. The area of the region enclosed by the graphs of $y = 1/(x^2 + 1)$, $y = 0$, $x = -1$, and $x = 1$ is shown in calculus to be given by $\arctan 1 - \arctan(-1)$. Find this area.

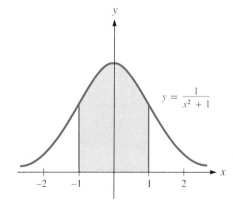

THINK ABOUT IT

1. Give an example of a value for x that shows the given equation is *not true* for all values of x for which the expressions are defined.

   a. $\tan^{-1} x = \dfrac{\sin^{-1} x}{\cos^{-1} x}$ b. $\sin^{-1} x = \dfrac{1}{\sin x}$

   c. $\text{arccot } x = \dfrac{1}{\arctan x}$ d. $\text{arccot } x = \arctan\left(\dfrac{1}{x}\right)$

   e. $\arccos(\cos x) = x$ f. $\cos(\cos^{-1} x) = x$

2. a. Consider the following keystroke sequence on a calculator set for radian mode:

   2 $\boxed{\sin}$ $\boxed{\text{INV}}$ $\boxed{\sin}$ 1.1415927

   Discuss why the output is 1.1415927, instead of 2.

   b. Give a keystroke sequence for approximating arcsec a on a calculator if $a \geq 1$.

   c. Give a keystroke sequence for approximating arcsec a on a calculator if $a \leq -1$.

3. Solve $\arcsin x = \arccos \frac{2}{3}$ without the aid of a calculator.

4. a. Show that $\arcsin(-x) = -\arcsin x$.

   b. If $f(x) = \arcsin x$, then is f an even function, an odd function, or neither?

5. Use the cofunction identity, $\cos(\pi/2 - \theta) = \sin \theta$, which is true for all values of θ, and show that

   $$\arcsin x + \arccos x = \frac{\pi}{2}$$

   for all x in $[-1,1]$. In terms of right triangles, explain the interpretation of this identity if $0 < x < 1$.

REMEMBER THIS

1. If $A = 65°30'$, find the measure of the angle complementary to A.

2. To three significant digits, find 225 tan 33.0°.

3. Evaluate sec 89°50' to four significant digits.

4. If $\tan \theta = 2.2$ and $0° < \theta < 90°$, find θ to the nearest 10 minutes.

5. Find the amplitude and period of the function $y = -2 \cos 10\pi x$.

6. Solve for b: $\dfrac{2\pi}{b} = \dfrac{\pi}{3}$.

7. If $(-5,6)$ is a point on the terminal ray of θ, find $\cos \theta$.

8. What is the reference angle for 240°?

9. Find the exact value of $\sin(7\pi/4)$.

10. If $f(x) = \csc x$, what is the range of f?

5.8 Right Triangle Applications and Harmonic Motion

In this chapter we have emphasized that the trigonometric functions may be applied to analyze both right triangles and physical phenomena that occur in cycles. This section discusses applications that illustrate both of these uses of trigonometry.

Right Triangle Applications

To solve a right triangle means to find the measures of the two acute angles and the lengths of the three sides of the triangle. To accomplish this, at least two of these five values must be known and one or more must be a side length. In this section we follow the standard practice of simplifying the notation by always labeling the angles of the triangle as A, B, and C, with C designating the right angle. Before attempting these problems, it is suggested that you study the topic of significant digits in the Appendix. (*Note:* As discussed in the Appendix, a bar above a zero, $\bar{0}$, is used to avoid ambiguity and indicates a zero that is a significant digit.)

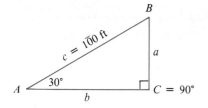

Figure 5.83

EXAMPLE 1 Solve the right triangle ABC in which $A = 30°$ and $c = 1\overline{0}0$ ft.

Solution First, sketch Figure 5.83. We can find angle B since angles A and B are complementary.

$$A + B = 90°$$
$$30° + B = 90°$$
$$B = 60°$$

Second, we can find side length a by using the sine function.

$$\sin A = \frac{a}{c}$$

$$\sin 30° = \frac{a}{100}$$

$$100(\sin 30°) = a$$
$$100(0.5000) = a$$
$$5\overline{0} \text{ ft} = a \quad \text{(two significant digits)}$$

Third, we can find b by using the cosine function.

$$\cos A = \frac{b}{c}$$

$$\cos 30° = \frac{b}{100}$$

$$100(\cos 30°) = b$$
$$100(0.8660) = b$$
$$87 \text{ ft} = b \quad \text{(two significant digits)}$$

To summarize, we found that $B = 60°$, $a = 5\overline{0}$ ft, and $b = 87$ ft. ∎

   In applications of trigonometry our values are only as accurate as the devices we use to measure the data. However, although all our answers are approximations, the symbol for equality ($=$) is generally used instead of the more precise symbol for approximation ($\approx$). Computed results should not be used to determine other parts of a triangle since the given data produce more accurate answers. The results are usually rounded off as follows.

Accuracy of Sides	Accuracy of Angle
Two significant digits	Nearest degree
Three significant digits	Nearest 10 minutes or tenth of a degree
Four significant digits	Nearest minute or hundredth of a degree

EXAMPLE 2 Solve the right triangle ABC in which $a = 11.0$ ft and $b = 5.00$ ft.

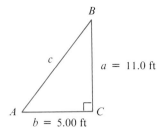

Figure 5.84

Solution   First, sketch Figure 5.84. Now find the length of the hypotenuse by the Pythagorean relation.

$$c^2 = a^2 + b^2$$
$$= 11^2 + 5^2$$
$$= 146$$
$$c = \sqrt{146}$$
$$= 12.1 \text{ ft} \qquad \text{(three significant digits)}$$

Second, we can find angle A by using the tangent function.

$$\tan A = \frac{a}{b}$$
$$= \frac{11}{5} = 2.2$$
$$A = 65°30' \qquad \text{(from Table 6 to the nearest 10 minutes)}$$

Third, we can find angle B since angle A and angle B are complementary.

$$A + B = 90°$$
$$65°30' + B = 90°$$
$$B = 24°30'$$

To summarize, we found that $c = 12.1$ ft, $A = 65°30'$, and $B = 24°30'$. ∎

In many practical applications of right triangles, an angle is measured with respect to a horizontal line. This measurement is accomplished by use of a transit. (By centering a bubble of air in a water chamber, the table of this instrument may be horizontally leveled.) The sighting tube of the transit is then tilted upward or downward until the desired object is sighted. This measuring technique will result in an angle that is described as either an angle of elevation or an angle of depression (see Figure 5.85). The measurement results in an **angle of elevation** if the object being sighted is above the observer, and the measurement results in an **angle of depression** if the object being sighted is below the observer.

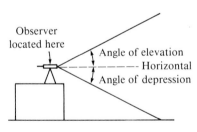

Figure 5.85

EXAMPLE 3   It is necessary to determine the height of a smokestack to estimate the cost of painting it. At a point 225 ft from the base of the stack, the angle of elevation is 33.0°. How high is the smokestack?

Solution   First, sketch Figure 5.86. In right triangle ABC we can find x, the height of the smokestack, by using the tangent function.

$$\tan A = \frac{\text{opposite}}{\text{adjacent}}$$
$$\tan 33.0° = \frac{x}{225}$$
$$225(\tan 33.0°) = x$$
$$225(0.6494) = x$$
$$146 \text{ ft} = x \qquad \text{(three significant digits)} \qquad ∎$$

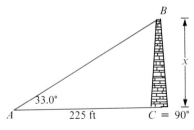

Figure 5.86

EXAMPLE 4   The measure of the angle of depression of a buoy from the platform of a radar tower that is 85 ft above the ocean is 15°. Find the distance of the buoy from the base of the tower. (See Figure 5.87.)

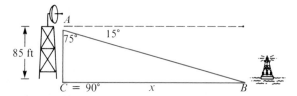

Figure 5.87

Solution   In right triangle ABC we can find the measure of angle A since angle A and the angle of depression are complementary.

$$A + 15° = 90°$$
$$A = 75°$$

We know the side length adjacent to angle A and we need to find x, which is the opposite side length. Thus,

$$\tan 75° = \frac{x}{85}$$
$$85(\tan 75°) = x$$
$$85(3.732) = x$$
$$320 \text{ ft} = x \qquad \text{(two significant digits)}. \qquad \blacksquare$$

EXAMPLE 5   Solve the problem in the chapter introduction on page 248.

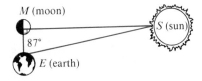

Figure 5.88

Solution   Consider Figure 5.88. The length of the hypotenuse $(\overline{ES})$ represents the distance from the earth to the sun, while the length of the adjacent side $(\overline{EM})$ represents the distance from the earth to the moon. The ratio $\overline{ES}/\overline{EM}$ represents how many times more distant is the sun than the moon from the earth. Then

$$\frac{\overline{ES}}{\overline{EM}} = \frac{\text{hypotenuse}}{\text{adjacent}} = \sec 87° \approx 19.11.$$

Thus, Aristarchus estimated that the sun is about 19 times farther away than the moon. (*Note:* Actually, Aristarchus computed his estimate by using Euclidean geometry. Trigonometry was developed later to solve precisely this type of problem. In reality, Aristarchus' estimate is not close to the true ratio. The error is caused by his measurement for angle SEM, which should be about 89°50′. On the basis of the correct measurement for angle SEM, about how many times farther away is the sun than the moon?) $\qquad \blacksquare$

Harmonic Motion

Motion that repeats itself in equal intervals of time is called **harmonic motion.** Such motion is both periodic and bounded, which are properties that are also possessed by the sine and cosine functions. A typical example of an object in harmonic motion is shown in Figure 5.89(a), which illustrates a block sliding horizontally on a frictionless surface back and forth between positions $x = x_1$ and $x = -x_1$.

If we record the position of the block as a function of time, as shown in Figure 5.89(b), then the graph is a sine or cosine curve, which implies motion that is called simple harmonic motion.

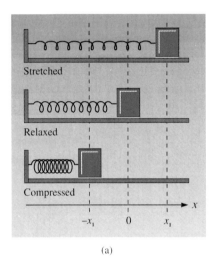
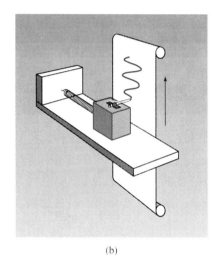

(a)   (b)

Figure 5.89

Simple Harmonic Motion

A particle that moves back and forth on a coordinate line is said to be in **simple harmonic motion** if its position x on the line at time t is given by

$$x = a \cos(\omega t + c) \qquad \text{or} \qquad x = a \sin(\omega t + c),$$

where a, ω, and c are constants with $\omega > 0$.

The magnitude of the maximum displacement is called the **amplitude** of the simple harmonic motion and is given by $|a|$. The **period** T is the time required to complete one round trip of motion and $T = 2\pi/\omega$. The **frequency** f is the number of cycles per unit of time, so the frequency is the reciprocal of the period. Thus, $f = 1/T = \omega/2\pi$.

EXAMPLE 6   A particle is moving in simple harmonic motion according to the equation $x = -2 \cos 10\pi t$. Find the amplitude, the period, and the frequency of this motion given that the amplitude is measured in centimeters and the period is measured in seconds.

Solution   In the given equation, $a = -2$ and $\omega = 10\pi$. Thus,

$$\text{Amplitude} = |a| = |-2| = 2 \text{ cm}$$

$$\text{Period} = \frac{2\pi}{\omega} = \frac{2\pi}{10\pi} = \frac{1}{5} \text{ second}$$

$$\text{Frequency} = \frac{1}{T} = \frac{1}{1/5} = 5 \text{ cycles per second.} \qquad \blacksquare$$

We have used the oscillatory motion of a mass attached to a spring as our prototype of simple harmonic motion. For such a problem, the period T may be determined from the mass m of the block and the force constant k of the spring according to the formula

$$T = 2\pi \sqrt{\frac{m}{k}}.$$

EXAMPLE 7   A block of mass $m = 1$ g is attached to a spring with a force constant $k = 36$ g/second². At $t = 0$ the displacement of the block is $x = 4$ cm and the block is then released.

a. What is the period of the simple harmonic motion?
b. What is an equation that relates displacement x and time t?

Solution

a. Replace m by 1 and k by 36 in the above formula for the period.

$$\text{Period} = T = 2\pi \sqrt{\frac{m}{k}} = 2\pi \sqrt{\frac{1}{36}} = 2\pi \cdot \frac{1}{6} = \frac{\pi}{3}$$

The period is $\pi/3$ seconds.
b. The block is at its maximum positive displacement at $t = 0$, so we may use $a = 4$ together with a cosine function and relate x and t with an equation of the form

$$x = 4 \cos \omega t.$$

To find ω, use $T = 2\pi/\omega$ and the result from part **a.**

$$\frac{2\pi}{\omega} = \frac{\pi}{3}, \qquad \text{so} \qquad \omega = 6.$$

Thus, an equation that describes the motion is

$$x = 4 \cos 6t. \qquad \blacksquare$$

EXERCISES 5.8

In Exercises 1–19 solve each right triangle ABC
($C = 90°$) for the given data. Round off answers using the
guidelines given in this section. (*Note:* As discussed in the
Appendix, a bar above a zero, $\bar{0}$, is used to avoid ambiguity
and indicates a zero that is a significant digit.)

1. $A = 30°$, $a = 5\bar{0}$ ft
2. $B = 45°$, $a = 85$ ft
3. $A = 60°$, $c = 15$ ft
4. $A = 22°$, $b = 62$ ft
5. $B = 71°$, $c = 25$ ft
6. $A = 19°$, $a = 17$ ft
7. $A = 55°$, $c = 25$ ft
8. $B = 10.3°$, $a = 24.5$ ft
9. $A = 45.5°$, $a = 86.6$ ft
10. $A = 84°50'$, $c = 12.4$ ft
11. $B = 52°40'$, $c = 625$ ft
12. $A = 31.5°$, $b = 29.7$ ft
13. $B = 88°10'$, $a = 31.2$ ft
14. $a = 5\bar{0}$ ft, $b = 120$ ft
15. $a = 6.0$ ft, $c = 15$ ft
16. $b = 1.0$ ft, $c = 2.0$ ft
17. $a = 1.0$ ft, $b = 1.0$ ft
18. $a = 7.00$ ft, $c = 11.0$ ft
19. $b = 12.0$ ft, $c = 26.0$ ft

In Exercises 20–40 solve each problem by making a careful
diagram and using right triangles.

20. A ladder leans against the side of a building and
makes an angle of 60° with the ground. If the foot of
the ladder is $1\bar{0}$ ft from the building, find the height
the ladder reaches on the building.
21. An escalator from the first floor to the second floor of
a building is $5\bar{0}$ ft long and makes an angle of 30°
with the floor. Find the vertical distance between the
floors.
22. If the angle of elevation of the sun at a certain time
is 40°, find the height of a tree that casts a shadow
of 45 ft.
23. A road has a uniform elevation of 6°. Find the
increase in elevation in driving $5\bar{0}0$ yd along the road.
24. The distance from ground level to the underside of a
cloud is called the "ceiling." At an airport, a ceiling
light projector throws a spotlight vertically on the
underside of a cloud. At a distance of 600 ft from the
projector, the angle of elevation of the spot of light on
the cloud is 58°. What is the ceiling?
25. To find the width of a river, a surveyor sets up her
transit at C and sights across the river to point B
(both B and C are at the water's edge, as shown
below). She then measures off $2\bar{0}0$ ft from C to A
such that C is a right angle. If she determines that
angle A is 24°, how wide is the river?

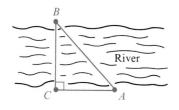

26. A pilot in an airplane at an altitude of $4,\bar{0}00$ ft
observes the angle of depression of an airport to be
12°. How far is the airport from the point on the
ground directly below the plane?
27. A lighthouse built at sea level is 180 ft high. From its
top, the angle of depression of a buoy is 24°. Find
the distance from the buoy to the foot of the
lighthouse.
28. A surveyor stands on a cliff $5\bar{0}$ ft above the water of
the river below. If the angle of depression to the
water's edge on the opposite bank is 10°, how wide is
the river at this point?
29. At an airport, cars drive down a ramp 85 ft long to
reach the lower baggage claim area 15 ft below the
main level. What angle does the ramp make with the
ground at the lower level?
30. For maximum safety the distance between the base
of a ladder and a building should be one fourth of the
length of the ladder. Find the angle that the ladder
makes with the ground when it is set up in the safest
position.
31. In building a warehouse, a carpenter checks the
drawings and finds the roof span to be $4\bar{0}$ ft, as
shown in the sketch. If the slope of the roof is 17°,
what length of 2- by 6-in. stock will he need to make
rafters if a 12-in. overhang is desired?

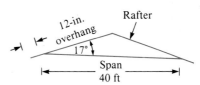

32. A machinist is given a 5.00-in.-diameter steel rod
with instructions to make a tapered pin 12.0 in. long.
The pin must have diameters of 4.00 in. and 2.00 in.
What angle of taper should he use to obtain the right
dimensions? (*Hint:* Make use of the dashed line
parallel to the center axis in the diagram.)

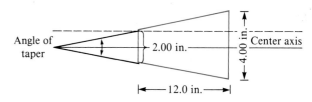

33. A circular disc 24.0 in. in diameter is to have five equally spaced holes as shown. Determine the correct setting (x) for the dividers to space these holes.

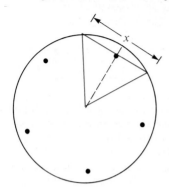

34. In an isosceles triangle each of the equal sides is $\overline{70}$ in. and the base is $\overline{80}$ in. Find the angles of the triangle.

35. The sides of a rectangle are 18 and 31 ft. Find the angle that the diagonal makes with the shorter side.

36. An observer on the third floor of a building determines that the angle of depression of the foot of a building across the street is 28° and the angle of elevation of the top of the same building is 51°. If the distance between the two buildings is $\overline{50}$ ft, find the height of the observed building.

37. An observer on the top of a hill 350 ft above the level of a road spots two cars due east of her. Find the distance between the cars if the angles of depression noted by the observer were 16° and 27°.

38. A right triangle, called an **impedance triangle,** is used to analyze alternating current (a.c.) circuits. Consider diagram (a), which shows a resistor and an inductor in series. As current flows through these components, it encounters some resistance, and the total effective resistance is called the **impedance.** We determine the impedance by making the resistances of the two circuit components the measures of the legs of a right triangle. As shown in diagram (b), the hypotenuse of the triangle then represents the impedance. The degree to which the voltage and current are in phase is given by angle θ, which is called the **phase angle.** From the data in this figure determine the impedance and the phase angle.

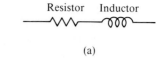

(a)

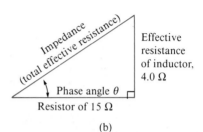

(b)

39. An important principle in the mathematical analysis of light is the **law of reflection.** This law states that a ray of light that strikes a reflecting surface is reflected so the angle of incidence (i) equals the angle of reflection (r). In the diagram below a photographer positions his flash at A. For a better lighting effect he aims this flash at position P on a reflecting surface to take a picture of a subject at B. If $\overline{AC}$ = 4.0 ft, $\overline{CD}$ = 12 ft, and i = 41°, find the length of $\overline{BD}$.

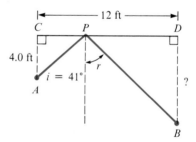

40. To determine the radius of the sun, an observer on earth at point O measures angle θ to be 16′, as illustrated by the diagram. If the distance from the earth to the sun is about 93,000,000 mi, what is the radius of the sun?

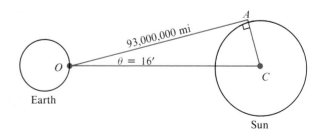

In Exercises 41–46 a particle is moving in simple harmonic motion according to the given equation. Find the amplitude, the period, and the frequency of this motion given that the amplitude is measured in centimeters and the period is measured in seconds.

41. $x = 3 \cos 8\pi t$ **42.** $x = 5 \sin 2\pi t$

43. $x = -1.5 \sin 6t$ **44.** $x = -\frac{1}{2}\cos 4t$

45. $x = 3 \cos(t - \pi/2)$ **46.** $x = \sqrt{2} \sin(4\pi t - \pi/6)$

47. A particle is moving in simple harmonic motion, passing back and forth through the origin an equal distance in each direction. Find an equation that relates displacement x and time t if the period is π seconds and at $t = 0$ the particle attains its maximum displacement in the positive direction $x = 1.7$ cm.

48. A particle is moving in simple harmonic motion, passing back and forth through the origin an equal distance in each direction. Find an equation that relates displacement x and time t if the period is 2 seconds and at $t = 0$ the particle attains its maximum displacement in the negative direction $x = -2.5$ cm.

49. A block of mass $m = 1$ g is attached to a spring with a force constant $k = 144$ g/second². At $t = 0$ the displacement of the block is $x = 3$ cm and the block is then released.

a. What is the period of the simple harmonic motion?

b. What is an equation that relates displacement x and time t?

50. A block of mass $m = 20$ g is attached to a spring with a force constant $k = 720$ g/second². At $t = 0$ the displacement of the block is $x = -1.8$ cm and the block is then released.

a. What is the period of the simple harmonic motion?

b. What is an equation that relates displacement x and time t?

THINK ABOUT IT

1. In the figure below show that $\tan \dfrac{\theta}{2} = \dfrac{\sin \theta}{1 + \cos \theta}$.

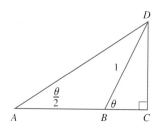

2. An artillery spotter in a plane at an altitude of 950 m observes two tanks in a line due east of the plane. If the angles of depressions of the two tanks measure 62° and 44°, then find the distance between the tanks.

3. In the figure below, derive a formula for h in terms of d, $\cot \theta$, and $\cot \alpha$.

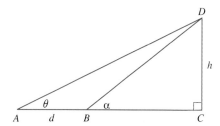

4. Create a word problem that may be solved by using the diagram and formula from Question 3. Then, find the solution to the problem you created.

5. For an experiment, a block of mass m is attached to a spring with a force constant k. For a second experiment, this spring is cut in half and the same block is attached to the spring. What is the relation between the frequencies of the simple harmonic motions in the two experiments?

REMEMBER THIS

1. Express $\sec x$ in terms of $\cos x$.

2. Express $\sin(-x)$ in terms of $\sin x$.

3. Factor $1 - s^2$.

4. Add $\dfrac{s^2}{c^2} + 1$.

5. Simplify $\left(\dfrac{s^2}{c^2} - 1\right) \div \left(\dfrac{s^2}{c^2} + 1\right)$.

6. What is the domain of $y = \arccos x$?

7. What is the range of $y = \cos x$?

8. Evaluate $\sin\left(\tan^{-1}\dfrac{15}{8}\right)$.

9. Evaluate $\cos(-90°)$.

10. In a unit circle what is the radian measure of a central angle that intercepts an arc that measures $\pi/4$ units?

CHAPTER OVERVIEW

Section	Key Concepts to Review
5.1	• Definitions of ray, measure of an angle, acute angle, obtuse angle, 1 degree, 1 minute, and 1 radian • Radian measure formula: $\theta = s/r$ • Degrees to radians formula: $1° = \pi/180$ radians • Radians to degrees formula: 1 radian $= 180°/\pi$ • Area formula: $A = \frac{1}{2}r^2\theta$ (θ in radians) • Formula relating linear velocity (v) and angular velocity (ω): $v = \omega r$
5.2	• Definition of the trigonometric functions for a unit circle • Definitions of trigonometric identity and reference number • In a unit circle the same real number measures both the central angle θ and the intercepted arc s. (That is, $\theta = s$.) • For the sine and cosine functions, the domain is the set of all real numbers, and the range is the set of real numbers between -1 and 1, inclusive. • Methods to determine (if defined) approximate trigonometric values for any number and exact values in special cases • For any trigonometric function f, we have $f(s + 2\pi k) = f(s)$, where k is an integer. • Negative angle identities • When using a scientific calculator, be sure to set the calculator for radian mode when required.
5.3	• Definitions of standard position of angle θ, and the trigonometric functions of an angle in standard position • Definition of the trigonometric functions of an acute angle θ of a right triangle • Chart summarizing the sign of the trigonometric ratios
5.4	• Definitions of quadrantal angle, coterminal angle, reference angle, complementary angles, and cofunctions • The side lengths in a 30–60–90 triangle are in the ratio $1:\sqrt{3}:2$. • The side lengths in a 45–45–90 triangle are in the ratio $1:1:\sqrt{2}$. • A trigonometric function of any angle is equal to the corresponding cofunction of the complementary angle. • Methods to determine (if defined) approximate trigonometric values for any angle and exact values in special cases
5.5	• Definitions of periodic function, period, amplitude, and phase shift • For $y = a \sin(bx + c)$ and $y = a \cos(bx + c)$, with $b > 0$, Amplitude $= |a|$, Period $= 2\pi/b$, Phase shift $= -c/b$

Section	Key Concepts to Review
5.6	• Graphs of $y = \tan x$, $y = \cot x$, $y = \sec x$, and $y = \csc x$ • Domain, range, and period for $y = \tan x$, $y = \cot x$, $y = \sec x$, and $y = \csc x$ • Relations between a function and its reciprocal function • Methods to graph trigonometric functions using horizontal shifting, vertical shifting, and addition of ordinates
5.7	• The inverse sine function is denoted by arcsin or $\sin^{-1}$. By definition, $\quad y = \arcsin x \quad$ if and only if $\quad x = \sin y$, where $-1 \le x \le 1$ and $-\pi/2 \le y \le \pi/2$. Similar remarks hold for the other inverse trigonometric functions. • Domain and range of the six inverse trigonometric functions • Graphs of $y = \arcsin x$, $y = \arccos x$, and $y = \arctan x$ • Right triangle method to simplify a trigonometric function of an inverse trigonometric expression • Inverse properties involving inverse trigonometric functions
5.8	• Definitions of angle of elevation, angle of depression, harmonic motion, and simple harmonic motion • Methods to solve a right triangle • Guidelines for accuracy in computed results • Methods to determine the amplitude, period, and frequency of a particle moving in simple harmonic motion • Formula relating period T, mass m, and spring force constant k: $T = 2\pi\sqrt{m/k}$

CHAPTER REVIEW EXERCISES

In Exercises 1–10 find the *exact* value of the given expression.

1. $\sin\left(\dfrac{\pi}{3}\right)$

2. $\cos\left(-\dfrac{\pi}{4}\right)$

3. $\sin 99\pi$

4. $\tan\left(\dfrac{56\pi}{3}\right)$

5. $\cot 300°$

6. $\cos 135°$

7. $\arctan(-1)$

8. $\arccos\left(-\dfrac{\sqrt{3}}{2}\right)$

9. $\sin[\cos^{-1}(-1)]$

10. $\tan(\sin^{-1} \frac{2}{3})$

In Exercises 11–20 find the value of the given expression to three significant digits. Use a calculator or tables.

11. $\sin 1°$

12. $\sin 1$

13. $\tan\left(\dfrac{8\pi}{5}\right)$

14. $\cot 400°$

15. $\sec 317°40'$

16. $\csc(-43.5°)$

17. $\arccos(-1.11)$

18. $\arcsin(-0.4439)$

19. $\cot(\text{arcsec } 1.238)$

20. $\tan[\sin^{-1}(-0.9563)]$

In Exercises 21–24 sketch the graph for $0 \le x \le 2\pi$.

21. $y = -\cos x$

22. $y = \sin 2x$

23. $y = \cot x$

24. $y = \sec x$

In Exercises 25–28 state the amplitude, the period, and the phase shift and sketch one cycle of the graph.

25. $y = \frac{1}{2}\cos 3x$

26. $y = 2 \sin \pi x$

27. $y = -\cos\left(2x + \dfrac{\pi}{2}\right)$

28. $y = \sin\left(\pi x - \dfrac{\pi}{4}\right)$

In Exercises 29–32 sketch one cycle of the function.

29. $y = 3 \sec\left(\dfrac{x}{2}\right)$

30. $y = \csc 2x$

31. $y = 1 - \tan x$

32. $y = \cot\left(x + \dfrac{\pi}{4}\right)$

In Exercises 33 and 34 graph by the method of addition of ordinates.

33. $y = \frac{1}{2}x + \sin x$ **34.** $y = 2 \cos x - \sin x$

35. Graph $y = \sin^{-1} x$.

36. Graph $y = \arccos(x - 1)$.

In Exercises 37–40 state the domain and range of the function.

37. $y = \tan x$ **38.** $y = \cos x$

39. $y = \csc x$ **40.** $y = \arcsin x$

41. In which quadrant do $y = \sin x$ and $y = \cos x$ both decrease?

42. Find an equation for the curve in the following illustration. The equation should be written in the form $y = a \sin bx$ or $y = a \cos bx$.

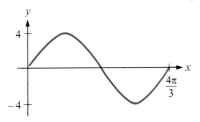

43. a. Change $210°$ to radians.

   b. Change $2\pi/5$ radians to degrees.

44. If $\tan \theta = \frac{1}{2}$ and $\cos \theta < 0$, find $\sin \theta$.

45. If the point $(1, -1)$ is on the terminal ray of θ, find $\cos \theta$.

46. Express $\cos 22°$ as a function of the angle complementary to $22°$.

47. For what values of x is $\sin(\sin^{-1} x) = x$ a true statement?

48. Find the area of a sector of a circle of radius 10 in. that is subtended by a central angle of $60°$.

49. In a circle a central angle intercepts an arc equal in length to the diameter of the circle. What is the measure of the central angle in radians?

50. If $y = \arccos(x/2)$, find x in terms of y.

51. If $x = 3 \sin \theta$, where $-\pi/2 \le \theta \le \pi/2$, write $\frac{9}{2}\theta - \frac{9}{4} \sin \theta \cos \theta$ as a function of x.

52. A car is traveling at 60 mi/hour (88 ft/second) on tires 32 in. in diameter. Find the angular velocity of the tires in radians/second.

53. A block of mass $m = 1$ g is attached to a spring with a force constant $k = 144$ g/second². At $t = 0$ the displacement of the block is $x = 2$ cm and the block is then released. What is an equation that relates displacement x and time t?

54. A road rises 25 ft in a horizontal distance of $4\overline{0}0$ ft. Find the angle that the road makes with the horizontal.

55. From the top of a building $3\overline{0}$ ft tall, the angle of depression to the foot of a building across the street is $60°$ and the angle of elevation to the top of the same building is $70°$. How tall is the building?

56. A carpenter has to build a stairway. The total rise is 8 ft 6 in. and the angle of rise is $30°$, as shown in the illustration. What is the shortest piece of 2- × -12-in. stock that can be used to make the stringer?

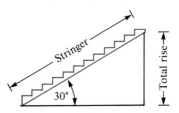

57. Consider the diagram below, which illustrates that twilight lasts until the sun is $18°$ below the horizon. From this, estimate the height (h) of the atmosphere.

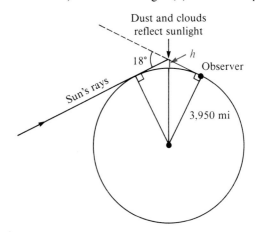

58. In a right triangle the length of the hypotenuse is 17 and the longer side length is 15. Determine the sine of the smaller acute angle.

59. Consider Snell's law (see Exercise 114 in Section 5.4). As light passes from air to water, v_i/v_r is about $\frac{4}{3}$. Find the angle of incidence of a ray of light if the angle of refraction is $38.6°$.

60. In right triangle ABC with $C = 90°$, if $b = 5.00$ and $c = 8.00$, determine angle B.

In Exercises 61–70 select the choice that completes the statement or answers the question.

61. If $x = \arccos(-\frac{1}{2})$, then x equals

    a. $\pi/3$ **b.** $\pi/6$ **c.** $5\pi/6$ **d.** $2\pi/3$

62. Which number is not in the range of $y = \sin x$?

    a. 1   **b.** $-\frac{1}{2}$ **c.** 2   **d.** 0

63. If $\sin x < 0$ and $\tan x > 0$, then the point assigned to x lies in quadrant

    a. 1   **b.** 2   **c.** 3   **d.** 4

64. The equation $\cos(-x) = -\cos x$ is true for

    a. all values of x **b.** only certain values of x

    c. no values of x

65. A function having the period π is

    a. $y = 2 \sin x$ **b.** $y = \frac{1}{2} \sin x$

    c. $y = \sin \frac{1}{2}x$ **d.** $y = \sin 2x$

66. To convert from radians to degrees, we multiply the number of radians by

    a. $180/\pi$ **b.** $\pi/90$ **c.** $90/\pi$ **d.** $\pi/180$

67. If a central angle of a radians intercepts an arc of length b in a circle of radius c, then

    a. $a = b/c$ **b.** $a = \pi/b$

    c. $a = c/b$ **d.** $a = bc$

68. If $f(x) = \cos 3x + \tan 2x$, then $f(\pi/6)$ equals

    a. $\sqrt{3}$ **b.** $1 + \sqrt{3}$

    c. $\sqrt{3}/3$ **d.** $(3 + \sqrt{3})/3$

69. Find $\sec R$ in the following right triangle.

    a. $3/5$ **b.** $4/3$ **c.** $\sqrt{41}/5$ **d.** $\sqrt{41}/4$

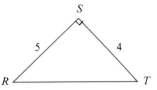

70. The value of $\sin 100°$ is equal to

    a. $\sin 10°$ **b.** $\sin 80°$

    c. $-\sin 80°$ **d.** $-\sin 10°$

CHAPTER 5 TEST

1. What is the range of the function $y = \arcsin x$?

2. What is the domain of the function $y = \tan x$?

3. Find the exact value of $\tan(2\pi/3)$.

4. To three significant digits evaluate $\sec 215°30'$.

5. Sketch one cycle of the graph of $y = -3 \cos 4\pi x$.

6. Sketch one cycle of the graph of $y = \tan(x - \pi/4)$.

7. Graph $y = \csc x$ for $0 \le x \le 2\pi$.

8. Find the amplitude, the period, and the phase shift for the graph of $y = -\sin(3x - \pi/2)$.

9. If $\sin \theta = \frac{1}{4}$ and $\tan \theta < 0$, find $\cos \theta$.

10.   **a.** Change $270°$ to radians.

      b. Change $7\pi/4$ radians to degrees.

11. Find the arc length of a sector of a circle of radius 10 cm that is subtended by a central angle of $150°$.

12. Find the exact value of $\arcsin(-\sqrt{3}/2)$.

13. Simplify $\tan(\cos^{-1} x)$.

14. Express $\sin 15°$ as a function of the angle complementary to $15°$.

15. In a right triangle the length of the hypotenuse is 15 and the longer side length is 12. Find the tangent of the smaller acute angle.

16. A road has a uniform elevation of $4.0°$. Find the increase in elevation in driving 95 m along this road.

17. If (x,y) is a point on the terminal ray of θ other than the origin, then what ratio defines $\cos \theta$?

18. For what values of x is $\sin^{-1}(\sin x) = x$ a true statement?

19. If $\cos s < 0$ and $\sin s > 0$, then find the quadrant in which the point assigned to s lies.

20. Give an example of an equation that defines a function with period $\pi/3$.

6 Trigonometric Identities and Equations

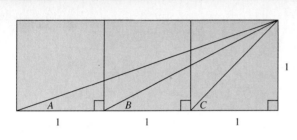

Show that angle C equals the sum of angles A and B in this diagram by showing $\sin C = \sin (A + B)$. (See Example 6 of Section 6.2.)

rigonometry is characterized by many formulas that may be used to simplify trigonometric expressions. For example, the formula

$$\cos(-x) = \cos x$$

made it easier to evaluate the cosine of a negative number. In Section 5.6 we used the formula

$$\csc x = \frac{1}{\sin x}$$

to study the behavior of the cosecant function. These formulas are examples of **trigonometric identities.** That is, they are true for all values of x for which the expressions are defined. In the first four sections of this chapter we develop many other useful identities that are an essential part of analyzing problems involving the trigonometric functions.

Conditional trigonometric equations differ from identities in that they are true only for certain values of x. The objective when working with a conditional equation like

$$\sin 4x = 1$$

is to find the solution set of the equation, and the last section of this chapter discusses methods for solving certain conditional trigonometric equations.

6.1   **Fundamental Identities**

Some basic trigonometric identities are easy to develop. Since all the trigonometric functions may be defined in terms of the sine and/or cosine, these functions are interrelated. This enables us to change the form of many trigonometric expressions to an expression that is either simpler or more useful for a particular problem. For convenience, we list below six fundamental identities that are used often. Except for identity 6, which is verified in Example 1, these statements follow directly from the definition of the trigonometric functions. Note in identity 6 that when using exponents with functions, it is common notation to avoid parentheses and to put the exponent between the function name and the symbol for the independent variable. Thus, $(\sin x)^2$, which means $(\sin x)(\sin x)$, is written $\sin^2 x$.

Six Fundamental Identities

Identity 1 $\csc x = \dfrac{1}{\sin x}$ or $\sin x = \dfrac{1}{\csc x}$ or $\sin x \csc x = 1$

Identity 2 $\sec x = \dfrac{1}{\cos x}$ or $\cos x = \dfrac{1}{\sec x}$ or $\cos x \sec x = 1$

Identity 3 $\cot x = \dfrac{1}{\tan x}$ or $\tan x = \dfrac{1}{\cot x}$ or $\tan x \cot x = 1$

Identity 4 $\tan x = \dfrac{\sin x}{\cos x}$

Identity 5 $\cot x = \dfrac{\cos x}{\sin x}$

Identity 6 $\sin^2 x + \cos^2 x = 1$

EXAMPLE 1   Verify the identity $\sin^2 x + \cos^2 x = 1$.

Solution   To avoid confusion, we return to the original notation of using s to represent the real numbers associated with arc lengths on the unit circle. Then by definition

$$x = \cos s \quad \text{and} \quad y = \sin s.$$

Any point (x,y) on the unit circle must satisfy the equation $x^2 + y^2 = 1$. Thus, by substitution we have

$$(\cos s)^2 + (\sin s)^2 = 1.$$

For convenience, this statement is usually written with x as the independent variable. Thus,

$$\sin^2 x + \cos^2 x = 1. \qquad \blacksquare$$

EXAMPLE 2 Show that the expression $\sin x \cot x \sec x$ is identical to 1.

Solution

$$\sin x \cot x \sec x = \sin x \left(\frac{\cos x}{\sin x} \right) \sec x \qquad \text{Identity 5}$$

$$= \sin x \, \frac{\cos x}{\sin x} \, \frac{1}{\cos x} \qquad \text{Identity 2}$$

$$= 1 \qquad \text{Simplify.}$$

Thus, $\sin x \cot x \sec x = 1$ for all values of x at which the expression is defined. ∎

EXAMPLE 3 Prove the identity $\tan^2 x + 1 = \sec^2 x$.

Solution

$$\tan^2 x + 1 = \frac{\sin^2 x}{\cos^2 x} + 1 \qquad \text{Identity 4}$$

$$= \frac{\sin^2 x + \cos^2 x}{\cos^2 x}$$

$$= \frac{1}{\cos^2 x} \qquad \text{Identity 6}$$

$$= \sec^2 x \qquad \text{Identity 2}$$

Thus, $\tan^2 x + 1 = \sec^2 x$ is an identity. ∎

In Example 3 we established the identity $\tan^2 x + 1 = \sec^2 x$, and in Exercise 16 you are asked to verify that $\cot^2 x + 1$ is identical to $\csc^2 x$. These two identities are used often and we need to add them to our list of basic identities.

> **Identity 7** $\quad \tan^2 x + 1 = \sec^2 x$
>
> **Identity 8** $\quad \cot^2 x + 1 = \csc^2 x$

The eight identities listed are commonly referred to as the **fundamental identities,** and among these, identities 6, 7, and 8 are called the **Pythagorean identities.** In the next example we include an alternative solution that effectively employs identity 7.

EXAMPLE 4 Show that the equation $\cos^2 x(1 + \tan^2 x) = 1$ is an identity.

Solution

$$\cos^2 x(1 + \tan^2 x) = \cos^2 x \left(1 + \frac{\sin^2 x}{\cos^2 x} \right) \qquad \text{Identity 4}$$

$$= \cos^2 x \cdot 1 + \cos^2 x \cdot \frac{\sin^2 x}{\cos^2 x} \qquad \text{Multiply.}$$

$$= \cos^2 x + \sin^2 x \qquad \text{Simplify.}$$

$$= 1 \qquad \text{Identity 6}$$

Alternative Solution

$$\cos^2 x(1 + \tan^2 x) = \cos^2 x(\sec^2 x) \qquad \text{Identity 7}$$

$$= \cos^2 x\left(\frac{1}{\cos^2 x}\right) \qquad \text{Identity 2}$$

$$= 1$$

Thus, $\cos^2 x(1 + \tan^2 x) = 1$ is an identity.   ■

In general, there is no standard procedure for working with identities. In fact, a given identity can usually be proved in several ways. However, these suggestions should be helpful.

Guidelines for Proving Identities

1. Change the more complicated expression in the identity to the same form as the less complicated expression. If both expressions are complicated, you might try to change them both to the same expression.
2. If you are having difficulty, change all functions to sines and cosines. This procedure might necessitate more algebra in some instances, but it will provide a direct approach to the problem. Gradually try to make use of the other trigonometric functions.
3. Do *not* attempt to prove an identity by treating it as an equation and using the associated techniques, for this involves assuming what you want to prove.

EXAMPLE 5   Show that the expression $(\tan^2 x - 1)/(\tan^2 x + 1)$ is identical to the expression $\sin^2 x - \cos^2 x$.

Solution

$$\frac{\tan^2 x - 1}{\tan^2 x + 1} = \frac{(\sin^2 x/\cos^2 x) - 1}{(\sin^2 x/\cos^2 x) + 1} \qquad \text{Identity 4}$$

$$= \frac{\cos^2 x[(\sin^2 x/\cos^2 x) - 1]}{\cos^2 x[(\sin^2 x/\cos^2 x) + 1]} \left.\vphantom{\frac{\frac{a}{b}}{\frac{a}{b}}}\right\} \quad \begin{array}{l}\text{Simplify the}\\\text{complex fraction.}\end{array}$$

$$= \frac{\sin^2 x - \cos^2 x}{\sin^2 x + \cos^2 x}$$

$$= \frac{\sin^2 x - \cos^2 x}{1} \qquad \text{Identity 6}$$

$$= \sin^2 x - \cos^2 x$$

Thus,

$$\frac{\tan^2 x - 1}{\tan^2 x + 1} = \sin^2 x - \cos^2 x$$

is an identity.   ■

EXAMPLE 6   Prove the identity $(\cos x \csc x)/\cot^2 x = \tan x$.

Solution

$$
\frac{\cos x \csc x}{\cot^2 x} = \frac{\cos x(1/\sin x)}{\cos^2 x/\sin^2 x} \qquad \text{Identities 1 and 5}
$$

$$
\left. \begin{aligned}
&= \frac{\sin^2 x(\cos x/\sin x)}{\sin^2 x(\cos^2 x/\sin^2 x)} \\[6pt]
&= \frac{\sin x \cos x}{\cos^2 x} \\[6pt]
&= \frac{\sin x}{\cos x}
\end{aligned} \right\} \quad \text{Simplify the complex fraction.}
$$

$$
= \tan x \qquad \text{Identity 4}
$$

Thus, the given equation is an identity.   ∎

The next example illustrates two additional guidelines for verifying identities. First, it may be helpful to factor the expression; and second, an identity like $\sin^2 x + \cos^2 x = 1$ may be expressed as $\sin^2 x = 1 - \cos^2 x$, so equivalent forms of the Pythagorean identities are often useful.

EXAMPLE 7   Verify the identity $\sin x - \cos^2 x \sin x = \sin^3 x$.

Solution

$$
\begin{aligned}
\sin x - \cos^2 x \sin x &= \sin x(1 - \cos^2 x) \qquad \text{Factor.} \\
&= \sin x(\sin^2 x) \qquad\qquad \text{Identity 6} \\
&= \sin^3 x
\end{aligned}
$$
                                                                            ∎

In calculus a trigonometric substitution may be used to simplify the square root of a quadratic expression. In such problems the Pythagorean identities are used in the forms $\cos^2 \theta = 1 - \sin^2 \theta$, $\sec^2 \theta = \tan^2 \theta + 1$, and $\tan^2 \theta = \sec^2 \theta - 1$.

Example 8   Eliminate the radical in the expression $\sqrt{9 - x^2}$ by substituting $3 \sin \theta$ for x, where $-\pi/2 \le \theta \le \pi/2$.

Solution   Replace x by $3 \sin \theta$ as directed and then simplify as follows.

$$
\begin{aligned}
\sqrt{9 - x^2} &= \sqrt{9 - (3 \sin \theta)^2} \\
&= \sqrt{9 - 9 \sin^2 \theta} \\
&= \sqrt{9(1 - \sin^2 \theta)} \\
&= \sqrt{9 \cos^2 \theta} \qquad \text{Identity 6} \\
&= 3 \cos \theta
\end{aligned}
$$

Note that the restriction $-\pi/2 \le \theta \le \pi/2$ allows us to replace $\sqrt{\cos^2 \theta}$ with $\cos \theta$ in the solution.   ∎

In Section 5.3 we considered a procedure for using a given trigonometric value to find other trigonometric values of that angle or number. We now show an alternative method for solving such problems that uses identities.

EXAMPLE 9   If $\sin x = \frac{3}{5}$ and $\pi/2 < x < \pi$, find

a. $\cos x$ b. $\tan x$ c. $\sec x$

Solution

a. Since $\pi/2 < x < \pi$, where $\cos x < 0$, the identity $\sin^2 x + \cos^2 x = 1$
when solved for $\cos x$ becomes

$$\cos x = -\sqrt{1 - \sin^2 x} = -\sqrt{1 - \left(\frac{3}{5}\right)^2} = -\sqrt{\frac{16}{25}} = -\frac{4}{5}.$$

b. $\tan x = \dfrac{\sin x}{\cos x} = \dfrac{3/5}{-4/5} = \dfrac{3}{-4}$

c. $\sec x = \dfrac{1}{\cos x} = \dfrac{1}{-4/5} = \dfrac{5}{-4}$ ■

EXERCISES 6.1

In Exercises 1–20 transform each first expression and show
that it is identical to the second expression.

1. $\sin x \sec x$; $\tan x$
2. $\tan x \csc x$; $\sec x$
3. $\sin^2 x \csc x$; $\sin x$
4. $\tan x \cot^2 x$; $\cot x$
5. $\cos x \tan x$; $1/\csc x$
6. $\sec x \cot x$; $1/\sin x$
7. $\cos x \tan x \csc x$; 1
8. $\cot x \sec x \sin x$; 1
9. $\sin^2 x \cot^2 x$; $\cos^2 x$
10. $\tan^2 x \csc^2 x$; $\sec^2 x$
11. $\sin^2 x \sec^2 x$; $\tan^2 x$
12. $\dfrac{\cot x}{\csc x}$; $\cos x$
13. $\dfrac{\cos^2 x}{\cot^2 x}$; $\sin^2 x$
14. $\dfrac{\sin x \sec x}{\tan x}$; 1
15. $\dfrac{\cot x \tan x}{\sec x}$; $\cos x$
16. $1 + \cot^2 x$; $\csc^2 x$
17. $(1 - \cos x)(1 + \cos x)$; $1/\csc^2 x$
18. $\sin x(\csc x - \sin x)$; $\cos^2 x$
19. $\dfrac{\cos^2 x}{1 + \sin x}$; $1 - \sin x$
20. $\sin^4 x - \cos^4 x$; $\sin^2 x - \cos^2 x$

In Exercises 21–40 prove that the equation is an identity.

21. $\dfrac{1}{\sin^2 x} - 1 = \cot^2 x$

22. $\csc x \sin x - \dfrac{1}{\sec^2 x} = \sin^2 x$

23. $\dfrac{1 + \tan^2 x}{\csc^2 x} = \tan^2 x$

24. $\dfrac{1 + \tan^2 x}{\tan^2 x} = \csc^2 x$

25. $\sin x \tan x + \cos x = \sec x$
26. $\tan x \csc^2 x - \tan x = \cot x$
27. $\dfrac{\tan^2 x - \sin^2 x}{\tan^2 x} = \sin^2 x$
28. $\dfrac{\sec^2 x + \csc^2 x}{\sec^2 x} = \csc^2 x$
29. $\dfrac{\csc x}{\tan x + \cot x} = \cos x$
30. $\dfrac{\sec x + \csc x}{1 + \tan x} = \csc x$
31. $(1 - \sin x)(\sec x + \tan x) = \cos x$
32. $\sec x \csc x - 2 \cos x \csc x + \cot x = \tan x$
33. $\sin x \cos^3 x + \cos x \sin^3 x = \sin x \cos x$
34. $\cos^4 x + 2 \cos^2 x \sin^2 x + \sin^4 x = 1$
35. $(\sin x + \cos x)^2 + (\sin x - \cos x)^2 = 2$
36. $1/(\sec^3 x \cos^4 x) = \sec x$
37. $\sec x - \cos x = \sin x \tan x$
38. $\sec^4 x - \tan^4 x = 2 \sec^2 x - 1$
39. $\dfrac{\sin x}{\csc x} + \dfrac{\cos x}{\sec x} = 1$ 40. $\dfrac{\csc x}{\sin x} - \dfrac{\cot x}{\tan x} = 1$

In Exercises 41–50 eliminate the radical in the expression
by using the given substitution, where θ is restricted to the
interval specified.

41. $\sqrt{1 - x^2}$; $x = \sin \theta$, $[-\pi/2,\pi/2]$
42. $\sqrt{1 + x^2}$; $x = \tan \theta$, $(-\pi/2,\pi/2)$
43. $\sqrt{x^2 + 9}$; $x = 3 \tan \theta$, $(-\pi/2,\pi/2)$
44. $\sqrt{25 - x^2}$; $x = 5 \sin \theta$, $[-\pi/2,\pi/2]$
45. $\sqrt{x^2 - 1}$; $x = \sec \theta$, $[0,\pi/2) \cup [\pi,3\pi/2)$
46. $\sqrt{(x - 3)^2 - 1}$; $x - 3 = \sec \theta$, $[0,\pi/2) \cup [\pi,3\pi/2)$

47. $\sqrt{1 - (x - 2)^2}$; $x - 2 = \sin\theta$, $[-\pi/2,\pi/2]$
48. $\sqrt{(16 + x^2)^3}$; $x = 4\tan\theta$, $(-\pi/2,\pi/2)$
49. $\sqrt{(4x^2 - 9)^3}$; $2x = 3\sec\theta$, $[0,\pi/2) \cup [\pi,3\pi/2)$
50. $\sqrt{4 - 9(x - 2)^2}$; $3(x - 2) = 2\sin\theta$, $[-\pi/2,\pi/2]$

In Exercises 51–54 use identities to find the given function value if $\sin x = \frac{12}{13}$ and $\pi/2 < x < \pi$.

51. $\csc x$ **52.** $\cos x$
53. $\tan x$ **54.** $\cot x$

In Exercises 55–58 use identities to find the given function value if $\cos x = a$ and $\pi < x < 3\pi/2$.

55. $\sin x$ **56.** $\csc x$
57. $\sec x$ **58.** $\tan x$

In Exercises 59–62 use identities to find the values of the remaining trigonometric functions of x.

59. $\sin x = \frac{8}{17}$, $\cos x > 0$
60. $\cos x = -\frac{3}{5}$, $\sin x < 0$
61. $\sec x = -\frac{4}{3}$, $\sin x > 0$
62. $\tan x = 2$, $\cos x < 0$

In Exercises 63–66 express the given function of x in terms of $\sin x$.

63. $\cos x$ **64.** $\tan x$
65. $\sec x$ **66.** $\csc x$

In Exercises 67–70 express the given function of x in terms of $\cos x$.

67. $\sin x$ **68.** $\cot x$
69. $\sec x$ **70.** $\csc x$

THINK ABOUT IT

1. Find all values of x in $[0,2\pi)$ that make the given equation a true statement.
   a. $\sin x = \sqrt{1 - \cos^2 x}$
   b. $\sin x = -\sqrt{1 - \cos^2 x}$
   c. $\sec x = -\sqrt{1 + \tan^2 x}$
   d. $\cot x = \sqrt{\csc^2 x - 1}$
2. Use the graph of $y = \tan^2 x$ that follows to graph $y = \sec^2 x$.

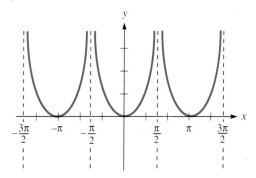

3. **a.** If $\log \sin a = b$, then express $\log \csc a$ in terms of b.
   b. Verify the identity:
   $$-\ln(\sec x - \tan x) = \ln(\sec x + \tan x).$$
4. Verify the identity: $\dfrac{\sec x + 1}{\tan x} = \dfrac{1}{\csc x - \cot x}$.

5. The figure below, called the **function hexagon,** arranges the six trigonometric functions of x so that

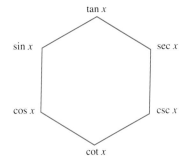

many basic identities are easily generated, as shown in the following exercises.
   a. Describe in geometric terms how reciprocal functions are positioned in the hexagon.
   b. Any function of x is identical to the product of the two functions of x on either side of it. For example, $\sin x = \cos x \tan x$ is an identity. Create another example of this type and verify the equation is an identity.
   c. The identities $\tan x = \sin x/\cos x$ and $\tan x = \sec x/\csc x$ illustrate quotient identities that may be read from the arrangement. How is this done? State and prove two identities of this type for $\sin x$.
   d. Another type of identity involves the product of three functions of x in alternate positions around the hexagon. Give two specific examples and state the general rule for this type of identity.

REMEMBER THIS

1. True or false: $\cos(60° + 30°) = \cos 60° + \cos 30°$.
2. Find the exact value of $\sin(\pi/2)$ and $\cos(\pi/2)$.
3. Find the exact value of $\tan 60°$.
4. If $f(x) = \sin x$, what is $f(x + h)$?
5. If θ is an acute angle, express $\cos \theta$ as a function of the angle complementary to θ.
6. Find the sine of the smaller acute angle in a right triangle with legs that measure 1 unit and 3 units.

7. Rationalize the denominator: $\dfrac{1}{1 + \sqrt{3}}$.
8. Simplify $-\sqrt{1 - (-\frac{1}{2})^2}$.
9. Simplify $\sin^2 s + \cos^2 s$.
10. Are $\sin(-x) = -\sin x$ and $\cos(-x) = \cos x$ identities?

6.2 Sum and Difference Formulas

Trigonometric expressions like $\sin(x + h)$ and $\cos[(\pi/2) - x]$ that involve the sum or difference of two numbers or angles occur often, so we now develop identities for analyzing such expressions. Examples 1 and 2 show derivations of the formulas for $\cos(x_1 + x_2)$ and $\cos(x_1 - x_2)$, and Example 3 shows how the formula for the cosine of the difference of two numbers may be used to simplify $\cos[(\pi/2) - x]$.

EXAMPLE 1   Verify the identity

$$\cos(x_1 + x_2) = \cos x_1 \cos x_2 - \sin x_1 \sin x_2.$$

Solution   As in Example 1 from Section 6.1, we first prove the identity by using s as the independent variable. The derivation is quite long, but once established, this formula yields many important results. Consider the coordinates of the points on the unit circle in Figure 6.1. The arc lengths from $(1,0)$ to $[\cos(s_1 + s_2), \sin(s_1 + s_2)]$ and from $[\cos(-s_1), \sin(-s_1)]$ to

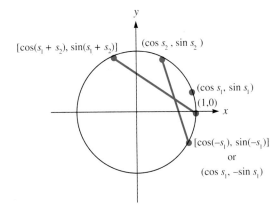

Figure 6.1

$(\cos s_2, \sin s_2)$ are both of length $s_1 + s_2$. Equal arcs in a circle subtend equal chords. Thus, by the distance formula we have

$$\sqrt{[\cos(s_1 + s_2) - 1]^2 + [\sin(s_1 + s_2) - 0]^2}$$
$$= \sqrt{(\cos s_2 - \cos s_1)^2 + [\sin s_2 - (-\sin s_1)]^2}.$$

After squaring both sides to remove radicals and performing the indicated operations, we obtain

$$\cos^2(s_1 + s_2) - 2\cos(s_1 + s_2) + 1 + \sin^2(s_1 + s_2)$$
$$= \cos^2 s_2 - 2\cos s_1 \cos s_2 + \cos^2 s_1 + \sin^2 s_2$$
$$+ 2\sin s_1 \sin s_2 + \sin^2 s_1.$$

Because $\sin^2 s + \cos^2 s = 1$, we regroup the terms as follows:

$$[\sin^2(s_1 + s_2) + \cos^2(s_1 + s_2)] - 2\cos(s_1 + s_2) + 1$$
$$= (\sin^2 s_2 + \cos^2 s_2) + (\sin^2 s_1 + \cos^2 s_1) - 2\cos s_1 \cos s_2$$
$$+ 2\sin s_1 \sin s_2.$$

Substituting 1 for the appropriate expressions, we have

$$1 - 2\cos(s_1 + s_2) + 1 = 1 + 1 - 2\cos s_1 \cos s_2 + 2\sin s_1 \sin s_2$$
$$-2\cos(s_1 + s_2) = -2\cos s_1 \cos s_2 + 2\sin s_1 \sin s_2$$
$$\cos(s_1 + s_2) = \cos s_1 \cos s_2 - \sin s_1 \sin s_2.$$

For convenience, we let x represent the independent variable. Thus,

$$\cos(x_1 + x_2) = \cos x_1 \cos x_2 - \sin x_1 \sin x_2.$$

Note that

$$\cos(x_1 + x_2) \neq \cos x_1 + \cos x_2. \qquad \blacksquare$$

EXAMPLE 2   Verify the identity

$$\mathbf{\cos(x_1 - x_2) = \cos x_1 \cos x_2 + \sin x_1 \sin x_2.}$$

Solution   Rewrite $\cos(x_1 - x_2)$ as $\cos[x_1 + (-x_2)]$ and use the formula from Example 1.

$$\cos[x_1 + (-x_2)] = \cos x_1 \cos(-x_2) - \sin x_1 \sin(-x_2)$$

Since $\cos(-x_2) = \cos x_2$ and $\sin(-x_2) = -\sin x_2$, we have

$$\cos[x_1 + (-x_2)] = \cos x_1 \cos x_2 - \sin x_1(-\sin x_2)$$

or

$$\cos(x_1 - x_2) = \cos x_1 \cos x_2 + \sin x_1 \sin x_2. \qquad \blacksquare$$

EXAMPLE 3   Verify the identity $\mathbf{\cos[(\pi/2) - x] = \sin x.}$

Solution   Use the formula for the cosine of the difference of two numbers (Example 2).

$$\cos\left(\frac{\pi}{2} - x\right) = \cos\frac{\pi}{2}\cos x + \sin\frac{\pi}{2}\sin x$$
$$= 0 \cdot \cos x + 1 \cdot \sin x$$
$$= \sin x$$

∎

To this point only identities involving the cosine function have been developed. We now show how the formula for the sine of the sum of two numbers is usually derived. This formula can then be used to derive a list of identities concerning the sine function, and the exercises ask you to verify many of these formulas.

In Example 3 we proved

$$\cos\left(\frac{\pi}{2} - x\right) = \sin x.$$

Replacing x by $x_1 + x_2$ gives

$$\sin(x_1 + x_2) = \cos\left[\frac{\pi}{2} - (x_1 + x_2)\right] = \cos\left[\left(\frac{\pi}{2} - x_1\right) - x_2\right].$$

If we now use the formula for the cosine of the difference of two numbers, we obtain

$$\sin(x_1 + x_2) = \cos\left[\left(\frac{\pi}{2} - x_1\right) - x_2\right]$$
$$= \cos\left(\frac{\pi}{2} - x_1\right)\cos x_2 + \sin\left(\frac{\pi}{2} - x_1\right)\sin x_2.$$

By the identity in Example 3 we know $\cos(\pi/2 - x_1) = \sin x_1$. In Exercise 6 of this section you are asked to verify $\sin(\pi/2 - x_1) = \cos x_1$. The formula for the sine of the sum of two numbers is then

$$\sin(x_1 + x_2) = \sin x_1 \cos x_2 + \cos x_1 \sin x_2.$$

This formula leads directly to the following identity whose verification is left to the exercises.

$$\sin(x_1 - x_2) = \sin x_1 \cos x_2 - \cos x_1 \sin x_2$$

Applications of these two formulas are shown in Examples 4–6, and Example 5 is an application from calculus.

EXAMPLE 4   Verify the identity $\sin(2\pi - x) = -\sin x$.

Solution   Use the formula for the sine of the difference of two numbers.

$$\sin(2\pi - x) = \sin 2\pi \cos x - \cos 2\pi \sin x$$
$$= 0 \cdot \cos x - 1 \cdot \sin x$$
$$= -\sin x$$

∎

EXAMPLE 5 If $f(x) = \sin x$, show that

$$\frac{f(x+h) - f(x)}{h} = \sin x \left(\frac{\cos h - 1}{h} \right) + \cos x \left(\frac{\sin h}{h} \right).$$

Solution If $f(x) = \sin x$, we have

$$f(x + h) = \sin(x + h) = \sin x \cos h + \cos x \sin h.$$

Then

$$\frac{f(x+h) - f(x)}{h} = \frac{\sin x \cos h + \cos x \sin h - \sin x}{h}$$

$$= \frac{\sin x (\cos h - 1) + \cos x \sin h}{h}$$

$$= \sin x \left(\frac{\cos h - 1}{h} \right) + \cos x \left(\frac{\sin h}{h} \right). \qquad \blacksquare$$

EXAMPLE 6 Solve the problem in the chapter introduction on page 324.

Solution Use the given figure and the Pythagorean relation to determine the three right triangles shown in Figure 6.2.

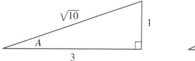 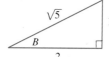

Figure 6.2

From the right triangle definitions of sine and cosine, $\sin A = \dfrac{1}{\sqrt{10}}$,

$\sin B = \dfrac{1}{\sqrt{5}}$, $\sin C = \dfrac{1}{\sqrt{2}}$, $\cos A = \dfrac{3}{\sqrt{10}}$, and $\cos B = \dfrac{2}{\sqrt{5}}$.

Then

$$\sin(A + B) = \sin A \cos B + \cos A \sin B$$

$$= \frac{1}{\sqrt{10}} \cdot \frac{2}{\sqrt{5}} + \frac{3}{\sqrt{10}} \cdot \frac{1}{\sqrt{5}}$$

$$= \frac{2}{\sqrt{50}} + \frac{3}{\sqrt{50}}$$

$$= \frac{5}{\sqrt{50}} \quad \text{or} \quad \frac{1}{\sqrt{2}}.$$

Because both $\sin C$ and $\sin(A + B)$ equal $1/\sqrt{2}$, and the sine function is one-to-one on the interval $(0°, 90°)$, we know $C = A + B$. (*Note:* Proving that $C = A + B$ may be done in many interesting ways, and Charles Trigg published 54 different proofs of the result in the *Journal of Recreational Mathematics*, Vol. 4, April 1971.) $\blacksquare$

Because $\tan x = \sin x / \cos x$, the identities for $\sin(x_1 + x_2)$ and $\cos(x_1 + x_2)$ may be used to derive an identity for $\tan(x_1 + x_2)$. The result and the formula for $\tan(x_1 - x_2)$ follow. The verification of these identities is requested in Exercises 47 and 50.

$$\tan(x_1 + x_2) = \frac{\tan x_1 + \tan x_2}{1 - \tan x_1 \tan x_2}$$

$$\tan(x_1 - x_2) = \frac{\tan x_1 - \tan x_2}{1 + \tan x_1 \tan x_2}$$

In our final examples we consider some numerical applications of the sum and difference identities.

EXAMPLE 7 Find the exact value of $\tan 15°$.

Solution The initial objective is to write $15°$ as a sum or difference of two special angles whose exact trigonometric values are known. By observing that $15° = 60° - 45°$ and using the formula for $\tan(x_1 - x_2)$, we have

$$\tan 15° = \tan(60° - 45°)$$
$$= \frac{\tan 60° - \tan 45°}{1 + \tan 60° \tan 45°}$$
$$= \frac{\sqrt{3} - 1}{1 + \sqrt{3}(1)}.$$

To simplify this result, rationalize the denominator.

$$\tan 15° = \frac{\sqrt{3} - 1}{1 + \sqrt{3}} \cdot \frac{1 - \sqrt{3}}{1 - \sqrt{3}}$$
$$= \frac{\sqrt{3} - 3 - 1 + \sqrt{3}}{1 - 3}$$
$$= 2 - \sqrt{3}$$ ∎

EXAMPLE 8 If $\sin x_1 = -\frac{1}{2}$, where $\pi < x_1 < 3\pi/2$, and $\cos x_2 = \sqrt{3}/2$, where $0 < x_2 < \pi/2$, find

a. $\cos x_1$ **b.** $\sin x_2$ **c.** $\cos(x_1 + x_2)$ **d.** $\sin(x_1 - x_2)$

Solution

a. Since $\pi < x_1 < 3\pi/2$, where $\cos x_1$ is negative, we have

$$\cos x_1 = -\sqrt{1 - \sin^2 x_1} = -\sqrt{1 - \left(-\frac{1}{2}\right)^2}$$
$$= -\sqrt{\frac{3}{4}} = -\frac{\sqrt{3}}{2}.$$

b. Since $0 < x_2 < \pi/2$, where $\sin x_2$ is positive, we have

$$\sin x_2 = \sqrt{1 - \cos^2 x_2} = \sqrt{1 - \left(\frac{\sqrt{3}}{2}\right)^2} = \sqrt{\frac{1}{4}} = \frac{1}{2}.$$

c. $\cos(x_1 + x_2) = \cos x_1 \cos x_2 - \sin x_1 \sin x_2$

$$= \left(-\frac{\sqrt{3}}{2}\right)\left(\frac{\sqrt{3}}{2}\right) - \left(-\frac{1}{2}\right)\left(\frac{1}{2}\right)$$

$$= -\frac{3}{4} - \left(-\frac{1}{4}\right)$$

$$= -\frac{1}{2}$$

d. $\sin(x_1 - x_2) = \sin x_1 \cos x_2 - \cos x_1 \sin x_2$

$$= \left(-\frac{1}{2}\right)\left(\frac{\sqrt{3}}{2}\right) - \left(-\frac{\sqrt{3}}{2}\right)\left(\frac{1}{2}\right)$$

$$= -\frac{\sqrt{3}}{4} - \left(-\frac{\sqrt{3}}{4}\right)$$

$$= 0$$

In summary, the six sum or difference identities developed in this section are listed below in consolidated form. Read the upper signs for the sum formula and the lower signs for the difference formula.

Sum or Difference Formulas

$$\sin(x_1 \pm x_2) = \sin x_1 \cos x_2 \pm \cos x_1 \sin x_2$$
$$\cos(x_1 \pm x_2) = \cos x_1 \cos x_2 \mp \sin x_1 \sin x_2$$
$$\tan(x_1 \pm x_2) = \frac{\tan x_1 \pm \tan x_2}{1 \mp \tan x_1 \tan x_2}$$

EXERCISES 6.2

In Exercises 1–6 verify the identity. Use the formulas for the cosine of either the sum or difference of two numbers.

1. $\cos\left(\frac{\pi}{2} + x\right) = -\sin x$

2. $\cos(\pi + x) = -\cos x$
3. $\cos(\pi - x) = -\cos x$

4. $\cos\left(\frac{3\pi}{2} - x\right) = -\sin x$

5. $\cos\left(\frac{3\pi}{2} + x\right) = \sin x$

6. Show $\sin(\pi/2 - x) = \cos x$. (*Hint:* Write cos x as $\cos[\pi/2 - (\pi/2 - x)]$ and use the formula for the cosine of the difference of two numbers.)

In Exercises 7–12 verify the identity. Use the formulas for the sine of either the sum or difference of two numbers.

7. $\sin\left(\frac{\pi}{2} - x\right) = \cos x$ 8. $\sin\left(\frac{\pi}{2} + x\right) = \cos x$

9. $\sin(\pi + x) = -\sin x$ 10. $\sin(\pi - x) = \sin x$

11. $\sin\left(\frac{3\pi}{2} - x\right) = -\cos x$

12. $\sin\left(\frac{3\pi}{2} + x\right) = -\cos x$

In Exercises 13–16 verify the identity. Use the formulas for the tangent of either the sum or difference of two numbers.

13. $\tan(\pi + x) = \tan x$ 14. $\tan(2\pi + x) = \tan x$
15. $\tan(2\pi - x) = -\tan x$ 16. $\tan(\pi - x) = -\tan x$

In Exercises 17–20 verify the cofunction identity.

17. $\sin(90° - \theta) = \cos\theta$ **18.** $\cos(90° - \theta) = \sin\theta$
19. $\sec(90° - \theta) = \csc\theta$ **20.** $\csc(90° - \theta) = \sec\theta$

In Exercises 21–26 verify the identity.

21. $\sqrt{2}\sin\left(x - \dfrac{\pi}{4}\right) = \sin x - \cos x$

22. $\sqrt{2}\sin\left(x + \dfrac{\pi}{4}\right) = \sin x + \cos x$

23. $\cos\left(x + \dfrac{\pi}{3}\right) = \dfrac{1}{2}(\cos x - \sqrt{3}\sin x)$

24. $\cos\left(x - \dfrac{\pi}{6}\right) = \dfrac{1}{2}(\sqrt{3}\cos x + \sin x)$

25. $\tan\left(x + \dfrac{\pi}{4}\right) = \dfrac{1 + \tan x}{1 - \tan x}$

26. $\tan\left(x - \dfrac{\pi}{4}\right) = \dfrac{\tan x - 1}{\tan x + 1}$

In Exercises 27–34 find exact function values.

27. $\sin 75°$ **28.** $\cos 15°$
29. $\cos 255°$ **30.** $\tan 105°$
31. $\sin(\pi/12)$ **32.** $\cos(5\pi/12)$
33. $\tan(13\pi/12)$ **34.** $\sin(11\pi/12)$
35. If $\sin x_1 = \dfrac{1}{2}$, where $0 < x_1 < \pi/2$, and
$\cos x_2 = -\sqrt{3}/2$, where $\pi/2 < x_2 < \pi$, find
    a. $\cos x_1$ **b.** $\sin x_2$
    c. $\cos(x_1 + x_2)$ **d.** $\sin(x_1 - x_2)$
36. If $\cos x_1 = \dfrac{3}{5}$, where $3\pi/2 < x_1 < 2\pi$, and
$\sin x_2 = -\dfrac{5}{13}$, where $3\pi/2 < x_2 < 2\pi$, find
    a. $\sin x_1$ **b.** $\cos x_2$
    c. $\cos(x_1 - x_2)$ **d.** $\sin(x_1 + x_2)$

In Exercises 37–42 find exact function values if $\sin\alpha = \dfrac{12}{13}$ and $\cos\beta = -\dfrac{7}{25}$, where α is a first-quadrant angle and β is a second-quadrant angle.

37. $\sin(\alpha + \beta)$ **38.** $\sin(\alpha - \beta)$
39. $\cos(\alpha - \beta)$ **40.** $\cos(\alpha + \beta)$
41. $\tan(\alpha + \beta)$ **42.** $\tan(\alpha - \beta)$

In Exercises 43–46 find exact function values. Use a sum or difference formula and right triangles.

43. $\cos(\sin^{-1}\frac{3}{5} - \cos^{-1}\frac{24}{25})$
44. $\cos(\arctan\frac{4}{3} + \arcsin\frac{3}{5})$
45. $\sin[\arctan(-\frac{4}{3}) - \arccos\frac{12}{13}]$
46. $\sin[\sin^{-1}\frac{8}{17} + \cos^{-1}(-\frac{12}{13})]$

In Exercises 47 and 48 use $\tan x = (\sin x)/(\cos x)$ to verify the identity.

47. $\tan(x_1 + x_2) = \dfrac{\tan x_1 + \tan x_2}{1 - \tan x_1 \tan x_2}$

48. $\tan\left(\dfrac{\pi}{2} + x\right) = -\cot x$

In Exercises 49 and 50 verify the identity. Use addition formulas, writing $x_1 - x_2$ as $x_1 + (-x_2)$.

49. $\sin(x_1 - x_2) = \sin x_1 \cos x_2 - \cos x_1 \sin x_2$

50. $\tan(x_1 - x_2) = \dfrac{\tan x_1 - \tan x_2}{1 + \tan x_1 \tan x_2}$

In Exercises 51 and 52 verify the identity. Use addition formulas, writing $2x$ as $x + x$.

51. $\cos 2x = \cos^2 x - \sin^2 x$
52. $\sin 2x = 2\sin x \cos x$
53. Show $\sin(x + y) + \sin(x - y) = 2\sin x \cos y$ is an identity.
54. Show

$$\sin a + \sin b = 2\sin\frac{a+b}{2}\cos\frac{a-b}{2}$$

is an identity. [*Hint:* Start with the identity in Exercise 53 and let $x = (a + b)/2$ and $y = (a - b)/2$.]
55. If $\tan(x + y) = 2$ and $\tan x = 1$, find $\tan y$.
56. Show that angle C equals the sum of angles A and B in this diagram by showing $\tan C = \tan(A + B)$.

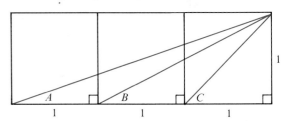

57. If $f(x) = \cos x$, show that

$$\frac{f(x + h) - f(x)}{h}$$

$$= \cos x\left(\frac{\cos h - 1}{h}\right) - \sin x\left(\frac{\sin h}{h}\right).$$

THINK ABOUT IT

1. **a.** Give an example of specific values for x_1 and x_2 so
        that $\cos(x_1 + x_2)$ does not equal $\cos x_1 + \cos x_2$.
        Verify your answer without using a calculator.
    b. Give an example of specific values for x_1 and x_2 so
        that $\cos(x_1 + x_2)$ does equal $\cos x_1 + \cos x_2$.
        Verify your answer without using a calculator.

2. Determine the exact value of the expression
    $\sin \frac{3}{10}\pi \cos \frac{1}{20}\pi - \cos \frac{3}{10}\pi \sin \frac{1}{20}\pi$.

3. Show that $\sin(A + B) = \sin C$ if A, B, and C are the
    angle measures in a triangle.

4. **a.** Show that $\arctan \frac{1}{2} + \arctan \frac{1}{3} = \pi/4$.
    b. Show that $\arccos(-x) = \pi - \arccos x$.

5. Use the diagram below and show that if α, β, and
    $\alpha + \beta$ are acute angles, then
    $$\sin(\alpha + \beta) = \sin \alpha \cos \beta + \cos \alpha \sin \beta.$$

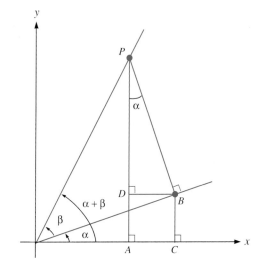

REMEMBER THIS

1. If $\sin x = -\frac{12}{13}$ and $\cos x = \frac{5}{13}$, find $\tan x$.
2. If $\sin x = \frac{7}{25}$ and $\pi/2 < x < \pi$, find $\cos x$.
3. Find the exact value of $\cos 330°$.
4. Express $\sin^2 x$ in terms of $\cos^2 x$.
5. If θ lies in quadrant 2, find the signs of $\sin \theta$, $\cos \theta$,
    and $\tan \theta$.
6. Show $\dfrac{\sqrt{2}/2}{1 + (\sqrt{2}/2)}$ simplifies to $\sqrt{2} - 1$.

7. Show $\tan 2x = 2 \tan x$ is not an identity by finding an
    admissible value of x for which the equation is not
    true.
8. Show $\cos x \cot x + \sin x$ is identical to $\csc x$.
9. Eliminate the radical in the expression $\sqrt{x^2 + 4}$ by
    substituting $2 \tan \theta$ for x, where $-\pi/2 < \theta < \pi/2$.
10. Verify the identity $2 \cos^2 x - 1 = 1 - 2 \sin^2 x$.

6.3 | Multiple-Angle Formulas

Multiple-angle formulas refer to identities for $\sin kx$, $\cos kx$, and $\tan kx$,
where k is a positive rational number. The most useful formulas of this type
result for $k = 2$ and are called the double-angle formulas.

Double-Angle Formulas

$$\sin 2x = 2 \sin x \cos x$$
$$\cos 2x = \cos^2 x - \sin^2 x = 2 \cos^2 x - 1 = 1 - 2 \sin^2 x$$
$$\tan 2x = \frac{2 \tan x}{1 - \tan^2 x}$$

To prove these formulas, we use the addition identities from the preceding section, rewriting $2x$ as $x + x$. For example, a formula for $\cos 2x$ may be derived as follows.

$$\cos 2x = \cos(x + x) = \cos x \cos x - \sin x \sin x$$
$$= \cos^2 x - \sin^2 x$$

There are two alternative forms that express $\cos 2x$ in terms of $\cos x$ or in terms of $\sin x$. By solving $\sin^2 x + \cos^2 x = 1$ for $\sin^2 x$ it follows that $\sin^2 x = 1 - \cos^2 x$. Thus,

$$\cos 2x = \cos^2 x - \sin^2 x$$
$$= \cos^2 x - (1 - \cos^2 x)$$
$$= 2 \cos^2 x - 1.$$

Similarly, because $\cos^2 x = 1 - \sin^2 x$, we have

$$\cos 2x = \cos^2 x - \sin^2 x$$
$$= (1 - \sin^2 x) - \sin^2 x$$
$$= 1 - 2 \sin^2 x.$$

Formulas for $\sin 2x$ and $\tan 2x$ may be derived by similar methods, and these derivations are requested in Exercises 73 and 74.

EXAMPLE 1   If $\sin x = \frac{3}{5}$ and $\pi/2 < x < \pi$, find

a. $\cos 2x$ **b.** $\sin 2x$ **c.** $\tan 2x$

Solution

a. Because $\sin x$ is known, use the formula that expresses $\cos 2x$ in terms of $\sin x$.

$$\cos 2x = 1 - 2 \sin^2 x = 1 - 2(\tfrac{3}{5})^2 = \tfrac{7}{25}$$

b. To find $\sin 2x$, both $\sin x$ and $\cos x$ must be known. First, determine $\cos x = -\frac{4}{5}$ as shown in Example 9 of Section 6.1 (or use the methods of Section 5.3). Then,

$$\sin 2x = 2 \sin x \cos x = 2(\tfrac{3}{5})(-\tfrac{4}{5}) = -\tfrac{24}{25}.$$

c. $\sin 2x$ and $\cos 2x$ are now known, so the easiest way to find $\tan 2x$ is to use a fundamental identity.

$$\tan 2x = \frac{\sin 2x}{\cos 2x} = \frac{-\frac{24}{25}}{\frac{7}{25}} = -\frac{24}{7}$$

If $\sin 2x$ or $\cos 2x$ were unknown, then we could use the double-angle formula for $\tan 2x$. From $\sin x = \frac{3}{5}$ and $\cos x = -\frac{4}{5}$, it follows that $\tan x = -\frac{3}{4}$. Then,

$$\tan 2x = \frac{2 \tan x}{1 - \tan^2 x} = \frac{2(-\frac{3}{4})}{1 - (-\frac{3}{4})^2} = \frac{-24}{16 - 9} = -\frac{24}{7}. \qquad\blacksquare$$

EXAMPLE 2 Verify the identity $\cos^2 x = \dfrac{1 + \cos 2x}{2}$.

Solution Because the left-hand member of the identity is $\cos^2 x$, use the formula that expresses $\cos 2x$ in terms of $\cos x$.

$$\frac{1 + \cos 2x}{2} = \frac{1 + (2 \cos^2 x - 1)}{2}$$

$$= \frac{2 \cos^2 x}{2}$$

$$= \cos^2 x \qquad \blacksquare$$

The identity in Example 2 may also be derived by solving for $\cos^2 x$ in the identity

$$\cos 2x = 2 \cos^2 x - 1.$$

Similarly, solving for $\sin^2 x$ in the identity $\cos 2x = 1 - 2 \sin^2 x$ yields

$$\sin^2 x = \frac{1 - \cos 2x}{2}.$$

Thus, the following identities, called power reduction formulas, are alternative forms of $\cos 2x$ formulas.

Power Reduction Formulas

$$\sin^2 x = \frac{1 - \cos 2x}{2}$$

$$\cos^2 x = \frac{1 + \cos 2x}{2}$$

An important application of these formulas in calculus is the simplification of powers of sine and cosine, as shown in the next example.

EXAMPLE 3 Verify the identity $\cos^4 x = \frac{3}{8} + \frac{1}{2} \cos 2x + \frac{1}{8} \cos 4x$.

Solution Rewrite $\cos^4 x$ in terms of $\cos^2 x$, use the power reduction formula for $\cos^2 x$, and then perform the indicated operations.

$$\cos^4 x = (\cos^2 x)^2$$

$$= \left(\frac{1 + \cos 2x}{2}\right)^2$$

$$= \tfrac{1}{4}(1 + 2 \cos 2x + \cos^2 2x)$$

Now $\cos^2 2x = (1 + \cos 4x)/2$ by the power reduction formula for $\cos^2 x$, with x replaced by $2x$. Thus,

$$\frac{1}{4}(1 + 2\cos 2x + \cos^2 2x) = \frac{1}{4}\left(1 + 2\cos 2x + \frac{1 + \cos 4x}{2}\right)$$

$$= \tfrac{1}{4} + \tfrac{1}{2}\cos 2x + \tfrac{1}{8} + \tfrac{1}{8}\cos 4x$$

$$= \tfrac{3}{8} + \tfrac{1}{2}\cos 2x + \tfrac{1}{8}\cos 4x.$$

Note that the right-hand member of the identity is simpler in the sense that the *fourth power* of cos x is expressed in terms of the *first powers* of the cosine of multiples of x. ∎

A second common type of multiple-angle identities, called the half-angle formulas, result for $k = \frac{1}{2}$. In these formulas which follow, the correct sign is determined by the quadrant in which $x/2$ lies for identities involving the plus or minus sign.

Half-Angle Formulas

$$\sin \frac{x}{2} = \pm \sqrt{\frac{1 - \cos x}{2}} \qquad \cos \frac{x}{2} = \pm \sqrt{\frac{1 + \cos x}{2}}$$

$$\tan \frac{x}{2} = \pm \sqrt{\frac{1 - \cos x}{1 + \cos x}} = \frac{1 - \cos x}{\sin x} = \frac{\sin x}{1 + \cos x}$$

The derivations of the identities for $\sin(x/2)$ and $\cos(x/2)$ are applications of the power reduction formulas, with x replaced by $x/2$. Thus,

$$\sin^2 \frac{x}{2} = \frac{1 - \cos 2(x/2)}{2} \qquad \text{implies} \qquad \sin \frac{x}{2} = \pm \sqrt{\frac{1 - \cos x}{2}}$$

$$\cos^2 \frac{x}{2} = \frac{1 + \cos 2(x/2)}{2} \qquad \text{implies} \qquad \cos \frac{x}{2} = \pm \sqrt{\frac{1 + \cos x}{2}}.$$

Question 5 in the "Think About It" exercises will lead you through derivations of the formulas for tan $(x/2)$.

EXAMPLE 4   Find the exact values of sin 165° and cos 165°.

Solution $165° = \frac{1}{2}(330°)$, and we know cos 330° = $\sqrt{3}/2$ by the methods of Section 5.4. Because 165° lies in quadrant 2, where sin $\theta > 0$ and cos $\theta < 0$, choose the positive square root to find sin 165° and the negative square root to find cos 165° in the half-angle formulas.

$$\sin 165° = \sin \frac{330°}{2} = \sqrt{\frac{1 - \cos 330°}{2}}$$

$$= \sqrt{\frac{1 - \sqrt{3}/2}{2}} = \frac{\sqrt{2 - \sqrt{3}}}{2}$$

$$\cos 165° = \cos \frac{330°}{2} = -\sqrt{\frac{1 + \cos 330°}{2}}$$

$$= -\sqrt{\frac{1 + \sqrt{3}/2}{2}} = \frac{-\sqrt{2 + \sqrt{3}}}{2} \qquad \blacksquare$$

EXAMPLE 5 Show that $\tan(\pi/8) = \sqrt{2} - 1$.

Solution $\pi/8 = \frac{1}{2}(\pi/4)$ and the exact trigonometric values of $\pi/4$ are known with $\sin(\pi/4) = \cos(\pi/4) = \sqrt{2}/2$. Choosing one of the formulas for $\tan(x/2)$, say

$$\tan \frac{x}{2} = \frac{\sin x}{1 + \cos x},$$

leads to

$$\tan \frac{\pi}{8} = \tan \frac{\pi/4}{2} = \frac{\sin(\pi/4)}{1 + \cos(\pi/4)} = \frac{\sqrt{2}/2}{1 + \sqrt{2}/2}.$$

Now simplify the complex fraction by multiplying the numerator and the denominator by 2. Then, rationalize the denominator.

$$
\begin{aligned}
\frac{\sqrt{2}/2}{1 + \sqrt{2}/2} &= \frac{\sqrt{2}}{2 + \sqrt{2}} \\
&= \frac{\sqrt{2}}{2 + \sqrt{2}} \cdot \frac{2 - \sqrt{2}}{2 - \sqrt{2}} \\
&= \frac{2\sqrt{2} - 2}{4 - 2} \\
&= \sqrt{2} - 1
\end{aligned}
$$

Thus, $\tan(\pi/8) = \sqrt{2} - 1$. Exercise 78 asks you to verify that the same result may be obtained by using the alternate formulas for $\tan(x/2)$. ■

EXAMPLE 6 Use the double-angle or half-angle identities to write each expression as a single function of a multiple angle.

a. $\dfrac{\sin 6t}{1 + \cos 6t}$

b. $\sin 5\theta \cos 5\theta$

c. $5 \cos^2 \dfrac{x}{8} - 5 \sin^2 \dfrac{x}{8}$

Solution

a. The given expression suggests the half-angle identity

$$\tan \frac{x}{2} = \frac{\sin x}{1 + \cos x}.$$

Replacing x with $6t$, we have

$$\frac{\sin 6t}{1 + \cos 6t} = \tan \frac{6t}{2} = \tan 3t.$$

b. Use the double-angle identity $\sin 2x = 2 \sin x \cos x$ with $x = 5\theta$, as follows.

$$\sin 5\theta \cos 5\theta = (\tfrac{1}{2} \cdot 2) \sin 5\theta \cos 5\theta$$
$$= \tfrac{1}{2}(2 \sin 5\theta \cos 5\theta)$$
$$= \tfrac{1}{2} \sin(2 \cdot 5\theta)$$
$$= \tfrac{1}{2} \sin 10\theta$$

c. Factor out the common factor 5, and then apply the double-angle identity $\cos 2x = \cos^2 x - \sin^2 x$, with $x/8$ in place of x.

$$5 \cos^2 \frac{x}{8} - 5 \sin^2 \frac{x}{8} = 5\left(\cos^2 \frac{x}{8} - \sin^2 \frac{x}{8} \right)$$
$$= 5 \cos\left(2 \cdot \frac{x}{8} \right)$$
$$= 5 \cos \frac{x}{4} \qquad \blacksquare$$

Besides the double-angle and half-angle formulas, there are other multiple-angle identities that are sometimes useful. The next example shows how we may derive a triple-angle formula.

EXAMPLE 7   Derive a formula for $\cos 3x$ in terms of $\cos x$.

Solution   Use the formula for the cosine of the sum of two angles, rewriting $3x$ as $2x + x$. Then express the result in terms of $\cos x$ by using the double-angle formulas for $\cos 2x$ and $\sin 2x$ and replacing $\sin^2 x$ with $1 - \cos^2 x$.

$$\cos 3x = \cos(2x + x)$$
$$= \cos 2x \cos x - \sin 2x \sin x$$
$$= (2 \cos^2 x - 1)\cos x - (2 \sin x \cos x)\sin x$$
$$= 2 \cos^3 x - \cos x - 2 \cos x \sin^2 x$$
$$= 2 \cos^3 x - \cos x - 2 \cos x(1 - \cos^2 x)$$
$$= 2 \cos^3 x - \cos x - 2 \cos x + 2 \cos^3 x$$
$$= 4 \cos^3 x - 3 \cos x$$

Thus, $\cos 3x = 4 \cos^3 x - 3 \cos x$ is a triple-angle formula for $\cos 3x$.   ■

EXERCISES 6.3

In Exercises 1–6 find $\sin 2x$, $\cos 2x$, and $\tan 2x$ by using the double-angle formulas.

1. $\sin x = \frac{4}{5}$, $0 < x < \pi/2$

2. $\cos x = \frac{12}{13}$, $3\pi/2 < x < 2\pi$

3. $\csc x = -\frac{4}{3}$, $3\pi/2 < x < 2\pi$

4. $\sec x = -3$, $\pi < x < 3\pi/2$

5. $\tan x = \frac{1}{2}$, $\pi < x < 3\pi/2$

6. $\cot x = -\sqrt{2}$, $\pi/2 < x < \pi$

In Exercises 7–16 show that the equation is an identity.

7. $\sin^2 x = \dfrac{1 - \cos 2x}{2}$

8. $\dfrac{\sin 2x}{2 \sin x} = \cos x$

9. $\csc x \sin 2x = 2 \cos x$

10. $\cos^4 x - \sin^4 x = \cos 2x$

11. $\cos 2x + 2 \sin^2 x = 1$

12. $\dfrac{2 \cot x}{\csc^2 x} = \sin 2x$

13. $\dfrac{1 + \cos 2x}{\sin 2x} = \cot x$

14. $\dfrac{\cos 2x}{\sin x} + \dfrac{\sin 2x}{\cos x} = \csc x$

15. $\dfrac{2 \tan x}{2 - \sec^2 x} = \tan 2x$

16. $\dfrac{2 \cot x}{\cot^2 x - 1} = \tan 2x$

In Exercises 17–20 find $\sin(x/2)$, $\cos(x/2)$, and $\tan(x/2)$ by using the half-angle formulas.

17. $\cos x = \frac{15}{17}$, $3\pi/2 < x < 2\pi$
18. $\sin x = \frac{3}{5}$, $0 < x < \pi/2$
19. $\csc x = -4$, $\pi < x < 3\pi/2$
20. $\tan x = -1$, $\pi/2 < x < \pi$

In Exercises 21–26 find the exact value of each expression by using the half-angle formulas.

21. $\sin 105°$
22. $\cos 105°$
23. $\cos(\pi/8)$
24. $\sin(\pi/8)$
25. $\tan(\pi/12)$
26. $\tan 165°$

27. Show that $\sin \dfrac{\pi}{12} = \dfrac{\sqrt{2 - \sqrt{3}}}{2}$.

28. Show that $\cos \dfrac{\pi}{12} = \dfrac{\sqrt{2 + \sqrt{3}}}{2}$.

29. Show that $\tan(7\pi/12) = -2 - \sqrt{3}$.
30. Show that $\tan 67.5° = \sqrt{2} + 1$.

In Exercises 31–34 use the figure below and write each expression in terms of a, b, and c.

31. $\cos 2\theta$
32. $\sin 2\theta$.
33. $\tan 2\theta$
34. $\tan(\theta/2)$

In Exercises 35–40 write each expression in terms of a if $\cos x = a$ and $\pi < x < 3\pi/2$. Do not simplify radicals.

35. $\sin 2x$
36. $\cos 2x$
37. $\tan 2x$
38. $\tan(x/2)$
39. $\cos(x/2)$
40. $\sin(x/2)$

In Exercises 41–50 use the double-angle or half-angle identities to write each expression as a single function of a multiple angle.

41. $2 \cos^2 6t - 1$
42. $2 \sin 8\theta \cos 8\theta$

43. $\dfrac{\sin 4x}{1 + \cos 4x}$
44. $\dfrac{1 - \cos 6x}{\sin 6x}$

45. $\pm \sqrt{\dfrac{1 + \cos 10\theta}{2}}$
46. $\dfrac{2 \tan 5t}{1 - \tan^2 5t}$

47. $\sin 2x \cos 2x$
48. $3 \cos^2 \dfrac{x}{4} - 3 \sin^2 \dfrac{x}{4}$

49. $4 - 8 \sin^2 9t$
50. $\dfrac{6 \sin 7\theta}{5 + 5 \cos 7\theta}$

In Exercises 51–62 show that the equation is an identity.

51. $\sin 6x = 2 \sin 3x \cos 3x$
52. $\cos 4x = \cos^2 2x - \sin^2 2x$

53. $10 \cos^2 \dfrac{x}{2} - 5 = 5 \cos x$

54. $6 \sin \dfrac{x}{2} \cos \dfrac{x}{2} = 3 \sin x$

55. $\csc 2x = \frac{1}{2}(\cot x + \tan x)$
56. $\csc 2x = \frac{1}{2} \sec x \csc x$
57. $\cot 2x = \frac{1}{2}(\cot x - \tan x)$

58. $\cot 2x = \dfrac{\cot^2 x - 1}{2 \cot x}$

59. $\sec 2x = \dfrac{1}{2 \cos^2 x - 1}$

60. $\sec 2x = \dfrac{\sec^2 x}{2 - \sec^2 x}$

61. $\tan \dfrac{x}{2} = \csc x - \cot x$

62. $\cot \dfrac{x}{2} = \csc x + \cot x$

In Exercises 63–72 use the power reduction formulas to verify each identity.

63. $2 \cos^2 4\theta = 1 + \cos 8\theta$
64. $\frac{1}{2}(1 - \sin \theta)^2 = \frac{3}{4} - \sin \theta - \frac{1}{4} \cos 2\theta$
65. $\sin^2 x \cos^2 x = \frac{1}{8}(1 - \cos 4x)$
66. $\sin^2 2x \cos^2 2x = \frac{1}{8}(1 - \cos 8x)$
67. $\sin^4 x = \frac{3}{8} - \frac{1}{2} \cos 2x + \frac{1}{8} \cos 4x$

68. $\cos^4 \dfrac{x}{2} = \dfrac{3}{8} + \dfrac{1}{2} \cos x + \dfrac{1}{8} \cos 2x$

69. $\sec^2 \dfrac{x}{2} = \dfrac{2}{1 + \cos x}$ **70.** $\csc^2 \dfrac{x}{2} = \dfrac{2}{1 - \cos x}$

71. $\cos^2 \dfrac{x}{2} = \dfrac{\sec x + 1}{2 \sec x}$ **72.** $\sin^2 \dfrac{x}{2} = \dfrac{\tan x - \sin x}{2 \tan x}$

73. Use the formula for $\sin(x_1 + x_2)$ to derive the double-angle formula for $\sin 2x$.
74. Use the formula for $\tan(x_1 + x_2)$ to derive the double-angle formula for $\tan 2x$.
75. Derive a formula for $\sin 3x$ in terms of $\sin x$.
76. Derive a formula for $\cos 4x$ in terms of $\cos x$.

77. Exercise 113 of Section 5.4 gave the formula

$$d = V^2 \frac{\sin A \sin B}{16}$$

for the horizontal distance traveled by a projectile neglecting air resistance, where V is the initial velocity, A the angle of elevation, and $B = 90° - A$. Show that an alternative form of this formula is

$$d = \tfrac{1}{32} V^2 \sin 2A.$$

78. Redo Example 5 of this section using the alternate forms of the formulas for $\tan(x/2)$ and show that all three formulas lead to $\tan(\pi/8) = \sqrt{2} - 1$.

THINK ABOUT IT

1. Use a power reduction formula to graph the given equation.
  a. $y = \sin^2 x$ **b.** $y = \cos^2 x$
2. Explain how the figure below can be used to show geometrically that

$$\tan \frac{\theta}{2} = \frac{\sin \theta}{1 + \cos \theta}$$

if θ is an acute angle.

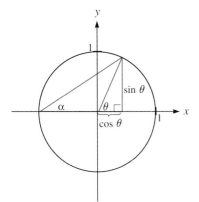

3. Is $\tan 2x$ less than or greater than $2 \tan x$, if $0 < x < \pi/4$? Explain.
4. In 1706, John Machin determined 100 correct decimal places for π with the aid of the relation

$$\frac{\pi}{4} = 4 \arctan\left(\frac{1}{5}\right) - \arctan\left(\frac{1}{239}\right).$$

Use identities to show this relation is true.
5. **a.** Use $\tan x = (\sin x)/(\cos x)$ to verify the identity

$$\tan \frac{x}{2} = \pm \sqrt{\frac{1 - \cos x}{1 + \cos x}}.$$

  b. Verify the identity

$$\tan \theta = \frac{\sin 2\theta}{1 + \cos 2\theta}.$$

  c. Replace θ by $x/2$ in the identity in part b to verify the identity

$$\tan \frac{x}{2} = \frac{\sin x}{1 + \cos x}.$$

  d. Multiply by $1 - \cos x$ in the numerator and the denominator of the right-hand member of the identity in part c to verify the identity

$$\tan \frac{x}{2} = \frac{1 - \cos x}{\sin x}.$$

REMEMBER THIS

1. Find the exact value of $\sin(2\pi/3)$.
2. Evaluate $\arctan(-\sqrt{3})$.
3. What are the amplitude, the period, and the phase shift of the graph of $y = -2 \sin(2x - \pi/3)$?
4. Simplify $\sin(x + y) - \sin(x - y)$.
5. Simplify $A^2 \cos^2 \theta + A^2 \sin^2 \theta$.
6. Simplify $\dfrac{2 \sin 2x \cos x}{2 \cos 2x \cos(-x)}$.

7. If $\sin x = b$ and $\pi/2 < x < \pi$, express $\cos x$ in terms of b.
8. Find the exact value of $\cos(\arcsin \tfrac{1}{2} + \arccos \tfrac{4}{5})$.
9. Show that $\sqrt{2} \sin(x + \pi/4) = \sin x + \cos x$ is an identity.
10. Show that $\dfrac{1 + \cot^2 x}{\cot^2 x} = \sec^2 x$ is an identity.

6.4   Product and Sum Formulas; Reduction Formula

The most commonly used trigonometric identities have been presented in Sections 6.1–6.3. We now consider some formulas that have occasional use in calculus and in higher mathematics. You should be able to apply these formulas by consulting a reference source as the need arises. First, consider the following *product formulas* that are used for converting certain products to sums (or differences).

Product Formulas

$$\sin x \cos y = \tfrac{1}{2}[\sin(x + y) + \sin(x - y)]$$
$$\cos x \sin y = \tfrac{1}{2}[\sin(x + y) - \sin(x - y)]$$
$$\cos x \cos y = \tfrac{1}{2}[\cos(x + y) + \cos(x - y)]$$
$$\sin x \sin y = \tfrac{1}{2}[\cos(x - y) - \cos(x + y)]$$

Each formula may be proved by starting with the right-hand member of the identity and using sum and difference formulas from Section 6.2. For example, the formula for $\cos x \cos y$ may be verified as follows.

$$\tfrac{1}{2}[\cos(x + y) + \cos(x - y)]$$
$$= \tfrac{1}{2}[(\cos x \cos y - \sin x \sin y) + (\cos x \cos y + \sin x \sin y)]$$
$$= \tfrac{1}{2}(2 \cos x \cos y)$$
$$= \cos x \cos y$$

Verifications of the other product formulas are requested in the exercises.

EXAMPLE 1   Express $\sin 3x \cos 2x$ as a sum or difference.

Solution   Use the product formula for $\sin x \cos y$, with x replaced by $3x$ and y replaced by $2x$.

$$\sin 3x \cos 2x = \tfrac{1}{2}[\sin(3x + 2x) + \sin(3x - 2x)]$$
$$= \tfrac{1}{2}(\sin 5x + \sin x)$$
$$= \tfrac{1}{2} \sin 5x + \tfrac{1}{2} \sin x \qquad ∎$$

EXAMPLE 2   Find the exact value of $\sin 75° \sin 15°$.

Solution   Substitute $75°$ for x and $15°$ for y in the product formula for $\sin x \sin y$ and simplify.

$$\sin 75° \sin 15° = \tfrac{1}{2}[\cos(75° - 15°) - \cos(75° + 15°)]$$
$$= \tfrac{1}{2}(\cos 60° - \cos 90°)$$
$$= \tfrac{1}{2}(\tfrac{1}{2} - 0)$$
$$= \tfrac{1}{4} \qquad ∎$$

The product formulas can be used to derive the following *sum formulas* that are useful for rewriting certain sums (or differences) as products.

Sum Formulas

$$\sin a + \sin b = 2 \sin \frac{a+b}{2} \cos \frac{a-b}{2}$$

$$\sin a - \sin b = 2 \cos \frac{a+b}{2} \sin \frac{a-b}{2}$$

$$\cos a + \cos b = 2 \cos \frac{a+b}{2} \cos \frac{a-b}{2}$$

$$\cos a - \cos b = -2 \sin \frac{a+b}{2} \sin \frac{a-b}{2}$$

The identity involving $\cos a + \cos b$ may be verified by using the product formula for $\cos x \cos y$, with x replaced by $(a+b)/2$ and y replaced by $(a-b)/2$.

$$\cos \frac{a+b}{2} \cos \frac{a-b}{2}$$
$$= \frac{1}{2} \left[\cos \left(\frac{a+b}{2} + \frac{a-b}{2} \right) + \cos \left(\frac{a+b}{2} - \frac{a-b}{2} \right) \right]$$
$$= \tfrac{1}{2} (\cos a + \cos b)$$

Multiplying both sides of this identity by 2 then results in the formula for $\cos a + \cos b$. The other sum formulas may be proved in a similar way, and these verifications are requested in the exercises.

EXAMPLE 3   Express $\cos 3x - \cos 5x$ as a product.

Solution   Use the formula for $\cos a - \cos b$, with a replaced by $3x$ and b replaced by $5x$. Note that the constant factor in the right-hand member of this formula is -2.

$$\cos 3x - \cos 5x = -2 \sin \frac{3x+5x}{2} \sin \frac{3x-5x}{2}$$
$$= -2 \sin 4x \sin(-x)$$
$$= -2 \sin 4x (-\sin x)$$
$$= 2 \sin 4x \sin x \qquad\blacksquare$$

EXAMPLE 4   Verify the identity $\dfrac{\sin t + \sin 3t}{\cos t + \cos 3t} = \tan 2t$.

Solution   First, rewrite the sum in the numerator using the formula for $\sin a + \sin b$, replacing a with t and b with $3t$.

$$\sin t + \sin 3t = 2 \sin \frac{t+3t}{2} \cos \frac{t-3t}{2}$$
$$= 2 \sin 2t \cos(-t)$$
$$= 2 \sin 2t \cos t$$

Next, rewrite the expression in the denominator using the formula for $\cos a + \cos b$, replacing a with t and b with $3t$.

$$\cos t + \cos 3t = 2 \cos \frac{t + 3t}{2} \cos \frac{t - 3t}{2}$$
$$= 2 \cos 2t \cos(-t)$$
$$= 2 \cos 2t \cos t$$

Then,

$$\frac{\sin t + \sin 3t}{\cos t + \cos 3t} = \frac{2 \sin 2t \cos t}{2 \cos 2t \cos t} = \frac{\sin 2t}{\cos 2t} = \tan 2t. \qquad \blacksquare$$

Sums of the form $a \sin Bx + b \cos Bx$ occur in many applied problems in physics, and such sums may be rewritten in the form $A \sin(Bx + C)$ by using the reduction formula that follows. Note that an immediate application of this identity is the result that any motion given by the equation $f(t) = a \sin \omega t + b \cos \omega t$ is simple harmonic motion (see Section 5.8) and that the amplitude, period, frequency, phase shift, and graph of the harmonic motion are easily determined by rewriting such equations in the form $f(t) = A \sin(\omega t + c)$.

Reduction Formula

If $a \neq 0$, then

$$a \sin Bx + b \cos Bx = A \sin(Bx + C),$$

where $C = \arctan(b/a)$, and $A = \sqrt{a^2 + b^2}$ if $a > 0$ or $A = -\sqrt{a^2 + b^2}$ if $a < 0$.

To prove this formula, begin with the right-hand member and use the formula for the sine of the sum of two numbers.

$$A \sin(Bx + C) = A(\sin Bx \cos C + \cos Bx \sin C)$$
$$= (A \cos C)\sin Bx + (A \sin C)\cos Bx$$

This result equals $a \sin Bx + b \cos Bx$ for every x if and only if

$$a = A \cos C \qquad \text{and} \qquad b = A \sin C.$$

Now squaring both sides of these two equations and adding the results gives

$$a^2 + b^2 = A^2 \cos^2 C + A^2 \sin^2 C$$
$$= A^2(\cos^2 C + \sin^2 C)$$
$$= A^2$$

Thus, $A = \pm\sqrt{a^2 + b^2}$. Furthermore, if $a \neq 0$,

$$\frac{b}{a} = \frac{A \sin C}{A \cos C} = \tan C,$$

and we choose $C = \arctan(b/a)$. Because C is in the interval $(-\pi/2, \pi/2)$, we know $\cos C > 0$; so the condition $a = A \cos C$ implies A and a have the same sign. Thus, $A = \sqrt{a^2 + b^2}$ if $a > 0$, and $A = -\sqrt{a^2 + b^2}$ if $a < 0$, which establishes the formula. Note that if $a = 0$, we may conclude $C = \pi/2$ and $A = b$.

EXAMPLE 5　Rewrite $-\sin 2x + \sqrt{3} \cos 2x$ using the reduction identity.

Solution　Matching $-\sin 2x + \sqrt{3} \cos 2x$ to the general form $a \sin Bx + b \cos Bx$ gives $a = -1$, $b = \sqrt{3}$, and $B = 2$. Then,

$$C = \arctan\frac{b}{a} = \arctan\frac{\sqrt{3}}{-1} = -\frac{\pi}{3}.$$

Because $a < 0$, use $A = -\sqrt{a^2 + b^2}$ to obtain

$$A = -\sqrt{(-1)^2 + (\sqrt{3})^2} = -\sqrt{4} = -2.$$

Thus, the reduction formula yields the identity

$$-\sin 2x + \sqrt{3} \cos 2x = -2 \sin\left(2x - \frac{\pi}{3}\right).$$ ∎

EXAMPLE 6　Graph $f(x) = \sin x - \cos x$ and give its amplitude, period, and phase shift. (*Note:* The graph of this function was also drawn in Example 6 of Section 5.6 by the method of addition of ordinates.)

Solution　Because $a = 1$ and $b = -1$, we have

$$C = \arctan\frac{b}{a} = \arctan\frac{-1}{1} = -\frac{\pi}{4}.$$

Since $a > 0$, use $A = \sqrt{a^2 + b^2}$ to determine

$$A = \sqrt{1^2 + (-1)^2} = \sqrt{2}.$$

Thus, the reduction formula with $B = 1$ gives

$$f(x) = \sin x - \cos x = \sqrt{2} \sin\left(x - \frac{\pi}{4}\right).$$

By the methods of Section 5.5, the amplitude is $\sqrt{2}$, the period is 2π, the phase shift is $\pi/4$, and Figure 6.3 gives the graph of f.

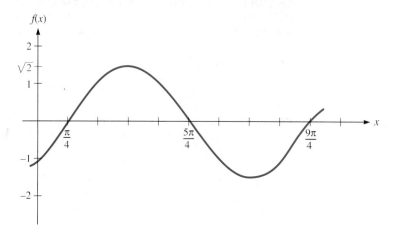

Figure 6.3

Finally, the major identities we have considered are summarized in the "Chapter Overview." You should ask your instructor which formulas must be memorized for your course.

EXERCISES 6.4

In Exercises 1–8 express each product as a sum or difference.

1. $\sin 3x \sin 2x$

2. $\cos 5x \cos 2x$

3. $4 \sin x \cos 4x$

4. $2 \cos x \sin 2x$

5. $\cos \frac{5}{2}\theta \sin \frac{1}{2}\theta$

6. $\sin 6t \sin(-4t)$

7. $\cos(A + B)\cos(A - B)$

8. $\sin(x + \pi)\cos(x - \pi)$

In Exercises 9–12 find the exact value of each expression.

9. $\cos 75° \cos 15°$

10. $\cos 105° \sin 15°$

11. $\sin \frac{5\pi}{12} \cos \frac{\pi}{12}$

12. $\sin \frac{7\pi}{8} \sin \frac{5\pi}{8}$

In Exercises 13–20 write each expression as a product.

13. $\sin 5x + \sin 3x$

14. $\cos 7x + \cos 5x$

15. $\cos x - \cos 3x$

16. $\sin 2x - \sin 4x$

17. $\cos \frac{3}{2}t + \cos \frac{1}{2}t$

18. $\cos 12\theta - \cos 3\theta$

19. $\sin(\theta + 90°) - \sin(\theta - 90°)$

20. $\sin(A + B) + \sin(A - B)$

In Exercises 21–24 find the exact value of each expression.

21. $\sin 105° + \sin 15°$

22. $\cos 15° + \cos 75°$

23. $\cos \frac{\pi}{12} - \cos \frac{5\pi}{12}$

24. $\sin \frac{13\pi}{12} - \sin \frac{5\pi}{12}$

In Exercises 25–32 verify the identity.

25. $\dfrac{\cos x - \cos 3x}{\sin x + \sin 3x} = \tan x$

26. $\dfrac{\cos 3x + \cos x}{\sin 3x - \sin x} = \cot x$

27. $\dfrac{\cos 3\theta + \cos 8\theta}{\sin 3\theta + \sin 8\theta} = \cot \dfrac{11\theta}{2}$

28. $\dfrac{\cos 3t - \cos 5t}{\sin 5t - \sin 3t} = \tan 4t$

29. $\dfrac{\sin A + \sin B}{\cos A + \cos B} = \tan \dfrac{A + B}{2}$

30. $\dfrac{\sin A + \sin B}{\cos A - \cos B} = -\cot \dfrac{A - B}{2}$

31. $\sin(x + y)\sin(x - y) = \sin^2 x - \sin^2 y$

32. $\sin(\theta + 45°)\sin(\theta - 45°) = -\frac{1}{2} \cos 2\theta$

In Exercises 33–40 rewrite the expression using the reduction identity.

33. $2 \sin x + 2 \cos x$

34. $-\sin x + \cos x$

35. $\sin x - \sqrt{3} \cos x$

36. $-\sqrt{3} \sin x - \cos x$

37. $\dfrac{\sqrt{3}}{2} \cos 2t - \dfrac{1}{2} \sin 2t$

38. $\dfrac{\sqrt{3}}{2} \cos \pi x + \dfrac{1}{2} \sin \pi x$

39. $4 \sin \pi x + 3 \cos \pi x$

40. $-5 \sin 2x - 12 \cos 2x$

In Exercises 41–44 graph each function and state the amplitude, the period, and the phase shift.

41. $y = \sin x + \cos x$

42. $y = -\sin x - \cos x$

43. $y = \dfrac{1}{2} \sin 2x - \dfrac{\sqrt{3}}{2} \cos 2x$

44. $y = \sqrt{3} \sin \pi x - \cos \pi x$

In Exercises 45–50 verify the formula that is given in this section for each expression.

45. $\sin x \cos y$

46. $\cos x \sin y$

47. $\sin x \sin y$

48. $\sin a + \sin b$

49. $\sin a - \sin b$

50. $\cos a - \cos b$

THINK ABOUT IT

1. In the figure below, show that
$$\cos A + \cos B = \sqrt{2} \cos \dfrac{A - B}{2}.$$

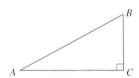

2. Explain why we may conclude $C = \pi/2$ and $A = b$, if $a = 0$ in the reduction formula
$$a \sin Bx + b \cos Bx = A \sin(Bx + C).$$

3. If the formula
$$a \sin x + b \cos x = A \sin(x + C_1)$$

is rewritten in the form
$$a \sin x + b \cos x = A \cos(x + C_2),$$
then find an expression for C_2 in terms of C_1.

4. Find all values of x at which $y = 3 \cos 2x - 3 \sin 2x$ attains a maximum value. What is this maximum value?

5. If a particle is moving in simple harmonic motion according to the given equation, then find the amplitude, the period, the frequency, and the phase shift of the motion given that the amplitude is measured in centimeters and the period is measured in seconds.

a. $f(t) = \sin \pi t + \cos \pi t$

b. $f(t) = \sin 8\pi t - \sqrt{3} \cos 8\pi t$

REMEMBER THIS

1. What is the reference number for $5\pi/6$?

2. What is the reference angle for $240°$?

3. Find the exact value of $\arcsin(\sqrt{2}/2)$.

4. What is the period of $y = \sin x$?

5. Does $\sin 3x = \sin x$ if $x = \pi/4$?

6. Does $\sin 3x = \sin x$ if $x = \pi/3$?

7. Is $-5(x - 2) = 5(2 - x)$ a conditional equation or an identity?

8. Solve the equation $2(1 - x^2) = 1 - x$.

9. Simplify $\tan^2 x - \sec^2 x$.

10. Verify the identity $\dfrac{1 - \cos 2x}{1 + \cos 2x} = \tan^2 x$.

6.5 Trigonometric Equations

In Sections 6.1–6.4 we considered **trigonometric identities,** which are equations like $\sin x \csc x = 1$ that are true for all values of x for which the expressions are defined. We now consider **conditional trigonometric equations** like $\sin x = -\frac{1}{2}$ that are true only for certain values of x. In solving such equations, we are looking for the set of all values of the variable that satisfy the given equation. Before we attempt to find a procedure for writing all the solutions to a particular equation, let us first establish a method for finding solutions between 0 and 2π.

EXAMPLE 1 Find the exact values of x ($0 \le x < 2\pi$) for which the equation $\sin x = -\frac{1}{2}$ is a true statement.

Solution First, determine the quadrant that contains the point assigned to x.

$$\sin x = -\frac{1}{2}, \qquad \text{which is a negative number.}$$

The point assigned to x could be in either Q3 or Q4, since the sine function is negative in both quadrants.

Second, determine the reference number (x_R).

$$\sin \frac{\pi}{6} = \frac{1}{2} \qquad \text{(from Section 5.2)}$$

Therefore, the reference number is $\pi/6$. (*Note:* When the reference number is $\pi/3$, $\pi/4$, or $\pi/6$, we may read exact values from the table in Section 5.2. Otherwise, we need Table 5 at the back of the book or a calculator.)

Third, determine the appropriate values of x (Figure 6.4).

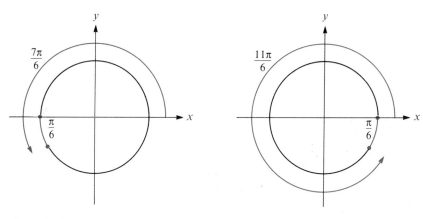

Figure 6.4

$\pi + \pi/6 = 7\pi/6$ is the number in Q3 with a reference number of $\pi/6$.

$2\pi - \pi/6 = 11\pi/6$ is the number in Q4 with a reference number of $\pi/6$.

Therefore, $7\pi/6$ and $11\pi/6$ make the equation a true statement, and the solution set in the interval $[0,2\pi)$ is $\{7\pi/6, 11\pi/6\}$.

Once we have the solutions of a trigonometric equation that are between 0 and 2π, we can determine all the solutions, since the trigonometric functions are periodic. In laying off a length 2π, we pass around the unit circle and return to our original point. Thus, we generate all the solutions to an equation by adding multiples of 2π to the solutions that are in the interval $[0,2\pi)$.

EXAMPLE 2   Approximate all the solutions to $4 \cos x + 1 = 0$.

Solution   First, solve the equation for $\cos x$.

$$4 \cos x + 1 = 0$$
$$\cos x = -\tfrac{1}{4} = -0.2500$$

Second, determine the quadrant that contains the point assigned to x.

$$\cos x = -0.2500, \text{ which is a negative number.}$$

The point assigned to x could be in either Q_2 or Q_3, since the cosine function is negative in both quadrants.

Third, determine the reference number.

$$\cos 1.32 \approx 0.2500$$

Therefore, the reference number is 1.32.

Fourth, determine the appropriate values of x (Figure 6.5).

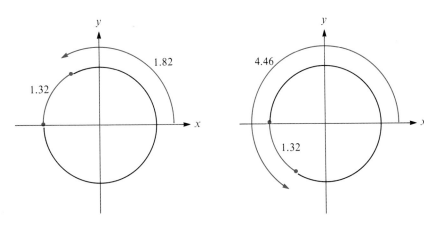

Figure 6.5

$3.14 - 1.32 = 1.82$ is the number in Q_2 with a reference number of 1.32.

$3.14 + 1.32 = 4.46$ is the number in Q_3 with a reference number of 1.32.

Thus, the formulas

$$1.82 + k2\pi \quad \text{and} \quad 4.46 + k2\pi,$$

where k is an integer, generate all the solutions to the equation, and the solution set is $\{x : x = 1.82 + k2\pi \text{ or } x = 4.46 + k2\pi, k \text{ any integer}\}$.   ∎

If the variable in a trigonometric equation represents an angle measure θ, then we may analyze the problem in terms of the terminal ray of θ and give answers in degree or radian measure.

EXAMPLE 3   Find the exact values of θ ($0° \leq \theta < 360°$) for which the equation $2 \sin \theta + \sqrt{3} = 0$ is a true statement.

Solution   First, solve the equation for $\sin \theta$.

$$2 \sin \theta + \sqrt{3} = 0$$
$$2 \sin \theta = -\sqrt{3}$$
$$\sin \theta = \frac{-\sqrt{3}}{2}$$

Second, determine the quadrant that contains the terminal ray of θ.

$$\sin \theta = \frac{-\sqrt{3}}{2}, \qquad \text{which is a negative number.}$$

The terminal ray of θ could be in either Q_3 or Q_4 since the sine function is negative in both quadrants.

Third, determine the reference angle.

$$\sin 60° = \frac{\sqrt{3}}{2} \qquad \text{(from Section 5.4)}$$

Therefore, 60° is the reference angle.

Fourth, determine the appropriate values of θ (see Figure 6.6).

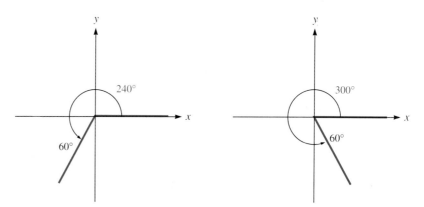

Figure 6.6

240° is the angle in Q_3 with a reference angle of 60°.

300° is the angle in Q_4 with a reference angle of 60°.

Therefore, 240° and 300° make the equation a true statement, and the solution set in the interval $[0°,360°)$ is $\{240°,300°\}$.

EXAMPLE 4   Approximate all the solutions to $10 \cot \theta = 3$.

Solution   First, solve the equation for $\cot \theta$.

$$10 \cot \theta = 3$$
$$\cot \theta = \tfrac{3}{10} = 0.3000$$

Second, determine which quadrant contains the terminal ray of θ.

$$\cot \theta \text{ is a positive number.}$$

The terminal ray of θ could be in either Q_1 or Q_3 since the cotangent function is positive in both quadrants.

Third, determine the reference angle.

$$\cot 73°20' \approx 0.3000$$

Therefore, $73°20'$ is the reference angle.

Fourth, determine the solutions between $0°$ and $360°$ (see Figure 6.7).

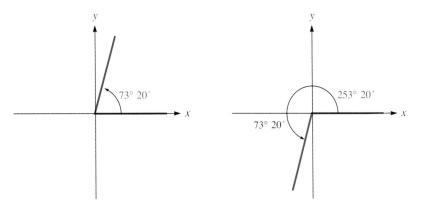

Figure 6.7

$73°20'$ is the angle in Q_1 with a reference angle of $73°20'$.

$253°20'$ is the angle in Q_3 with a reference angle of $73°20'$.

Therefore, $73°20'$ and $253°20'$ make the equation a true statement.

Fifth, indicate how angles coterminal to the above angles may be generated.

$$\left. \begin{array}{l} 73°20' + k360° \\ 253°20' + k360° \end{array} \right\} \quad \text{or equivalently} \quad 73°20' + k180°,$$

where k is an integer, generates all the solutions to the equation. Thus, the solution set is $\{\theta : \theta = 73°20' + k180°, k \text{ any integer}\}$. ∎

In the next two examples we show how to solve a trigonometric equation by factoring (Example 5) and how to solve a trigonometric equation that involves a multiple angle (Example 6).

EXAMPLE 5 Solve the equation $2 \sin x \cos x - \sin x = 0$ in the interval $[0,2\pi)$.

Solution First, factor out the common factor $\sin x$.

$$2 \sin x \cos x - \sin x = 0$$
$$\sin x(2 \cos x - 1) = 0$$

Note that we have found two factors whose product is zero. Hence, the original equation will be satisfied whenever either factor is zero, and we treat each factor separately from this point on.

First Factor $\sin x = 0$ The sine function is 0 when $x = 0$ and $x = \pi$.

Second Factor $2 \cos x - 1 = 0$
$$\cos x = \tfrac{1}{2}$$

1. Since $\cos x$ is positive, the point assigned to x could be in Q_1 or Q_4.
2. Since $\cos \pi/3 = \tfrac{1}{2}$, the reference number is $\pi/3$.
3. $\pi/3$ is the number in Q_1 with a reference number of $\pi/3$ (Figure 6.8). $2\pi - \pi/3 = 5\pi/3$ is the number in Q_4 with a reference number of $\pi/3$ (Figure 6.9).

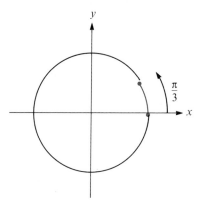

Figure 6.8   **Figure 6.9**

Thus, in the interval $[0,2\pi)$ the solution set is $\{0, \pi/3, \pi, 5\pi/3\}$.

EXAMPLE 6 Find the values of x in the interval $[0,2\pi)$ for which the equation $\sin 3x = \sqrt{2}/2$ is a true statement.

Solution Solve the equation for $3x$.

a. Since $\sin 3x$ is positive, the point assigned to $3x$ could be in Q_1 or Q_2.
b. Since $\sin \pi/4 = \sqrt{2}/2$, the reference number is $\pi/4$.
c. Determine the appropriate values of $3x$ and then of x (Figure 6.10).

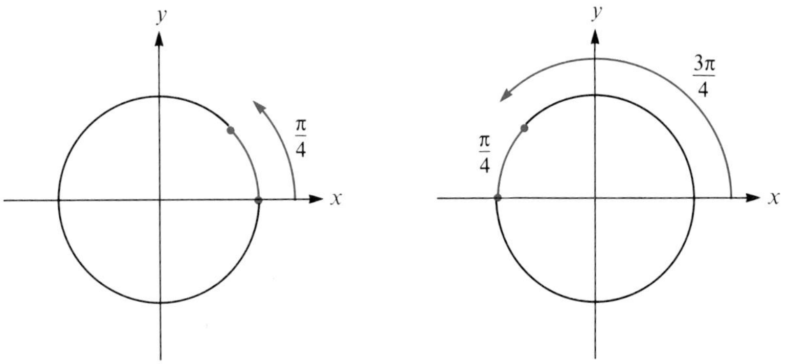

Figure 6.10

$\pi/4$ is the number in Q_1 with a reference number of $\pi/4$.

$\pi - \pi/4 = 3\pi/4$ is the number in Q_2 with a reference number of $\pi/4$.

Thus,

$$3x = \frac{\pi}{4} + k2\pi \quad \text{or} \quad 3x = \frac{3\pi}{4} + k2\pi,$$

from which we have

$$x = \frac{\pi}{12} + k\frac{2\pi}{3} \quad \text{or} \quad x = \frac{\pi}{4} + k\frac{2\pi}{3}.$$

d. To obtain solutions in the interval $[0,2\pi)$, use 0, 1, and 2 as replacements for k.

$$\frac{\pi}{12} + (0)\frac{2\pi}{3} = \frac{\pi}{12} \qquad \frac{\pi}{4} + (0)\frac{2\pi}{3} = \frac{\pi}{4}$$

$$\frac{\pi}{12} + (1)\frac{2\pi}{3} = \frac{3\pi}{4} \qquad \frac{\pi}{4} + (1)\frac{2\pi}{3} = \frac{11\pi}{12}$$

$$\frac{\pi}{12} + (2)\frac{2\pi}{3} = \frac{17\pi}{12} \qquad \frac{\pi}{4} + (2)\frac{2\pi}{3} = \frac{19\pi}{12}$$

Thus, $\{\pi/12, \pi/4, 3\pi/4, 11\pi/12, 17\pi/12, 19\pi/12\}$ is the solution set for $0 \le x < 2\pi$. ■

The following examples illustrate how the solution to a problem can depend on a knowledge of identities.

EXAMPLE 7 Solve the equation $\tan x + \cot x = 2 \csc x$ in the interval $[0,2\pi)$.

Solution $\tan x + \cot x = 2 \csc x$. Using *fundamental identities* 4, 5, and 1, we rewrite an equivalent equation that contains only $\sin x$ and $\cos x$.

$$\frac{\sin x}{\cos x} + \frac{\cos x}{\sin x} = \frac{2}{\sin x}$$

Simplifying the fractional equation, we have

$$\sin x \cos x \left(\frac{\sin x}{\cos x} + \frac{\cos x}{\sin x} \right) = \sin x \cos x \frac{2}{\sin x}$$

$$\sin^2 x + \cos^2 x = 2 \cos x$$

$$1 = 2 \cos x \qquad \text{Identity 6}$$

$$\tfrac{1}{2} = \cos x.$$

As illustrated in Example 5 of this section, $\cos x = \tfrac{1}{2}$ when $x = \pi/3$ and $x = 5\pi/3$, so the solution set in $[0,2\pi)$ is $\{\pi/3, 5\pi/3\}$. ∎

EXAMPLE 8 Solve the equation $2 \cos^2 x = 1 - \sin x$ in the interval $[0,2\pi)$.

Solution Use the identity $\cos^2 x = 1 - \sin^2 x$ to rewrite the equation in terms of a single trigonometric function of x. The resulting equation is quadratic in form and may be solved by using the factoring method considered in Example 5.

$$2 \cos^2 x = 1 - \sin x$$

$$2(1 - \sin^2 x) = 1 - \sin x$$

$$2 - 2 \sin^2 x = 1 - \sin x$$

$$0 = 2 \sin^2 x - \sin x - 1$$

$$0 = (2 \sin x + 1)(\sin x - 1)$$

$$2 \sin x + 1 = 0 \quad \text{or} \quad \sin x - 1 = 0$$

$$\sin x = -\tfrac{1}{2} \text{ or} \qquad \sin x = 1$$

In the interval $[0,2\pi)$, $\sin x = -\tfrac{1}{2}$ when $x = 7\pi/6$ and $x = 11\pi/6$ (see Example 1) while $\sin x = 1$ when $x = \pi/2$. Thus, the solution set in $[0,2\pi)$ is $\{\pi/2, 7\pi/6, 11\pi/6\}$. ∎

EXAMPLE 9 Solve $\sin 3x = \sin x$ in the interval $[0,2\pi)$.

Solution Use the formula for $\sin a - \sin b$ from Section 6.4 along with the methods of our previous examples.

$$\sin 3x = \sin x$$

$$\sin 3x - \sin x = 0$$

$$2 \cos \frac{3x + x}{2} \sin \frac{3x - x}{2} = 0$$

$$2 \cos 2x \sin x = 0$$

$$\cos 2x = 0 \quad \text{or} \quad \sin x = 0$$

The solutions of $\sin x = 0$ in the interval $[0,2\pi)$ are 0 and π. The solutions of $\cos 2x = 0$ may be expressed in consolidated form as

$$2x = \frac{\pi}{2} + k\pi$$

so

$$x = \frac{\pi}{4} + k\frac{\pi}{2}.$$

To find solutions in the interval $[0,2\pi)$, use 0, 1, 2, and 3 as replacements for k to obtain $\pi/4$, $3\pi/4$, $5\pi/4$, and $7\pi/4$. Thus the requested solution set is $\{0,\pi/4,3\pi/4,\pi,5\pi/4,7\pi/4\}$.   ∎

EXERCISES 6.5

In Exercises 1–16 solve for x in the interval $[0,2\pi)$.

1. $\sin x = \dfrac{\sqrt{3}}{2}$

2. $\sin x = -\dfrac{\sqrt{3}}{2}$

3. $\cos x = -\frac{1}{2}$

4. $\tan x = 1$

5. $\tan x = -1$

6. $\sec x = 2$

7. $\sin x = 0.1219$

8. $\sin x = -0.1219$

9. $\tan x = -3.145$

10. $\tan x = 3.145$

11. $\sqrt{2} \cos x = 1$

12. $\csc x - \sqrt{2} = 0$

13. $\sin x + 2 = 0$

14. $\cos x - 3 = 0$

15. $3 \tan x - 1 = 0$

16. $2 \tan x + 7 = 0$

In Exercises 17–20 find all real number solutions to the equation.

17. $4 \sin x - 1 = 1$

18. $10 \cos x - 2 = 0$

19. $4 \csc x + 9 = 0$

20. $5 \cot x + 3 = 0$

In Exercises 21–30 find the exact values of θ between $0°$ and $360°$ that make the equation a true statement.

21. $\cos \theta = \frac{1}{2}$

22. $\cos \theta = -\frac{1}{2}$

23. $2 \tan \theta = 2\sqrt{3}$

24. $-2 \cos \theta = \sqrt{3}$

25. $2 \sin \theta + \sqrt{3} = 0$

26. $\csc \theta - 2 = 0$

27. $4 \sin \theta + 3 = 1$

28. $3 \sec \theta - 7 = -1$

29. $2 \tan \theta + 5 = 7$

30. $3 \cot \theta + 4 = 1$

In Exercises 31–40 find all degree measures of θ for which the given trigonometric equation is a true statement. Where possible, find exact solutions.

31. $\sqrt{2} \sin \theta = 1$

32. $\sqrt{3} \csc \theta = 2$

33. $3 \sin \theta + 2 = 1$

34. $5 \tan \theta + 2 = -3$

35. $10 \cos \theta + 7 = 3$

36. $2 \cot \theta - 3 = 0$

37. $5 \sec \theta + 1 = 3$

38. $\sin \theta = 1$

39. $\cos \theta = -1$

40. $\cos \theta = 2$

In Exercises 41–56 solve for x in the interval $[0,2\pi)$.

41. $2 \sin^2 x - 1 = 0$

42. $3 \tan^2 x - 1 = 0$

43. $\cos^2 x = \cos x$

44. $\sin^3 x = \sin x$

45. $2 \cos x \sin x - \cos x = 0$

46. $\tan x \cos x = \tan x$

47. $\tan^2 x + 4 \tan x - 21 = 0$

48. $2 \sin^2 x - 5 \sin x = 3$

49. $\cos 3x = 0$

50. $\sin 4x = 1$

51. $\sin 2x = \frac{1}{2}$

52. $\sin \frac{1}{2}x = \frac{1}{2}$

53. $\tan \frac{1}{3}x = 1$

54. $\csc \frac{1}{4}x = 2$

55. $\sin^2 2x = \sin 2x$

56. $\cot^2 2x + 2 \cot 2x + 1 = 0$

In Exercises 57–60 solve for x in the interval $[0,2\pi)$. Use the quadratic formula.

57. $\cos^2 x + \cos x - 1 = 0$

58. $3 \sin^2 x - \sin x - 1 = 0$

59. $\sin^2 x + \sin x + 1 = 0$

60. $2 \tan^2 x + \tan x - 5 = 0$

In Exercises 61–74 solve each equation for $0 \le x < 2\pi$ through the aid of the *fundamental identities*.

61. $\sin x = \cos x$

62. $\sqrt{3} \sin x - \cos x = 0$

63. $\sin x = \tan x$

64. $2 \sin x - \tan x = 0$

65. $\sin^2 x + \cos^2 x = \tan x$

66. $\tan^2 x - \sec^2 x = \cos x$

67. $\sin x + \cos x \tan x = 1$

68. $\tan x + 4 \cot x = 5$

69. $2 \sin x - \csc x = 1$

70. $1 - \cos^2 x = \sin x$

71. $2 \cos^2 x = 3 \sin x + 3$

72. $2 \sin^2 x = \cos x + 1$

73. $\sec^2 x + 5 \tan x + 4 = 0$

74. $\tan^2 x - 3 \sec x = 9$

In Exercises 75–86 use identities to solve for x in the interval $[0,2\pi)$.

75. $\sin 2x + \sin x = 0$
76. $\sin 2x = \sin x$
77. $4 \sin x \cos x = -1$
78. $4 \sin x \cos x = \sqrt{3}$
79. $\cos 2x = \cos x$
80. $\sin x - \cos 2x = 0$
81. $\cos 2x = 2 \sin x \cos x$
82. $\sin 2x - \cos 2x = 1$
83. $\sin 3x + \sin x = 0$
84. $\cos 3x = \cos x$
85. $\cos 6x = \cos 2x$
86. $\sin 5x - \sin 3x = 0$

87. A formula for the horizontal distance traveled by a projectile, neglecting air resistance, is

$$d = \tfrac{1}{32}V^2 \sin 2\theta,$$

where θ and V measure the angle of elevation and the initial velocity in feet/second of the projectile, respectively. If a professional field goal kicker boots a football with an initial velocity of 76 ft/second and the ball travels 180 ft, find θ to the nearest degree. (*Note:* There are two possible solutions.)

88. A particle is moving in simple harmonic motion according to the equation $x = 5 \sin(6\pi t + \pi/2)$, where t represents time in seconds. Determine when the particle will first reach its maximum negative displacement.

THINK ABOUT IT

1. Explain, in terms of the unit circle definitions of the trigonometric functions, why the solution set of the equation $\sec t = -\tfrac{1}{2}$ is $\emptyset$.

2. Is $\sin(x - \pi/2) = \cos x$ a conditional equation or an identity? Explain why.

3. Find all solutions to $\sin(\ln x) = 0$.

In Questions 4 and 5 solve for x in the interval $[0,2\pi)$.

4. $\sin 3x = \cos x - \sin x$

5. $\dfrac{\cos 3x - \cos x}{\sin 3x - \sin x} = 1$

REMEMBER THIS

1. Simplify: $(\sin x + \cos x)^2 + (\sin x - \cos x)^2$.

2. True or false: $\cos^2 60° = (1 - \cos 120°)/2$.

3. Verify the cofunction identity $\tan(90° - \theta) = \cot \theta$.

4. Show $\dfrac{\csc^2 x}{\csc^2 x - 1}$, is identical to $\sec^2 x$.

5. Why is there no solution for θ if $\sin \theta > 1$?

6. Solve $\sin \theta = 0.4384$ for $0° < \theta < 180°$.

7. In the figure below express h in terms of b and $\sin A$.

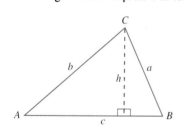

8. If (x,y) is a point on the terminal ray of θ other than the origin, then by definition $\sin \theta$ equals
  a. x/y **b.** r/x **c.** y/r **d.** x/r

9. To two significant digits find x: $\dfrac{\sin 35°}{18} = \dfrac{\sin 105°}{x}$.

10. Solve triangle ABC if $B = 25°$, $C = 90°$, and $b = 24$ ft.

CHAPTER OVERVIEW

Section	Key Concepts to Review
6.1	• Statements and applications of the **fundamental identities:**

$$\csc x = \frac{1}{\sin x} \qquad \sec x = \frac{1}{\cos x} \qquad \cot x = \frac{1}{\tan x}$$

$$\tan x = \frac{\sin x}{\cos x} \qquad \cot x = \frac{\cos x}{\sin x}$$

$$\left. \begin{array}{l} \sin^2 x + \cos^2 x = 1 \\ \tan^2 x + 1 = \sec^2 x \\ \cot^2 x + 1 = \csc^2 x \end{array} \right\} \begin{array}{l} \textbf{Pythagorean} \\ \textbf{identities} \end{array}$$

• Guidelines for proving identities

| **6.2** | • Statements and applications of the **sum and difference formulas:** |

$$\sin(x_1 \pm x_2) = \sin x_1 \cos x_2 \pm \cos x_1 \sin x_2$$

$$\cos(x_1 \pm x_2) = \cos x_1 \cos x_2 \mp \sin x_1 \sin x_2$$

$$\tan(x_1 \pm x_2) = \frac{\tan x_1 \pm \tan x_2}{1 \mp \tan x_1 \tan x_2}$$

(*Note:* Read the upper signs for the sum formula and the lower signs for the difference formula.)

| **6.3** | • Statements and applications of the **double-angle, half-angle,** and **power reduction formulas:** |

$$\sin 2x = 2 \sin x \cos x \qquad \cos 2x = \cos^2 x - \sin^2 x$$

$$\tan 2x = \frac{2 \tan x}{1 - \tan^2 x} \qquad\qquad = 2 \cos^2 x - 1$$

$$= 1 - 2 \sin^2 x$$

$$\sin \frac{x}{2} = \pm \sqrt{\frac{1 - \cos x}{2}} \qquad \cos \frac{x}{2} = \pm \sqrt{\frac{1 + \cos x}{2}}$$

$$\tan \frac{x}{2} = \pm \sqrt{\frac{1 - \cos x}{1 + \cos x}} = \frac{1 - \cos x}{\sin x} = \frac{\sin x}{1 + \cos x}$$

$$\sin^2 x = \frac{1 - \cos 2x}{2} \qquad \cos^2 x = \frac{1 + \cos 2x}{2}$$

| **6.4** | • Statements and applications of the **product to sum formulas** and the **sum to product formulas:** |

$$\sin x \cos y = \tfrac{1}{2}[\sin(x + y) + \sin(x - y)] \qquad \sin a + \sin b = 2 \sin \frac{a + b}{2} \cos \frac{a - b}{2}$$

$$\cos x \sin y = \tfrac{1}{2}[\sin(x + y) - \sin(x - y)] \qquad \sin a - \sin b = 2 \cos \frac{a + b}{2} \sin \frac{a - b}{2}$$

$$\cos x \cos y = \tfrac{1}{2}[\cos(x + y) + \cos(x - y)] \qquad \cos a + \cos b = 2 \cos \frac{a + b}{2} \cos \frac{a - b}{2}$$

$$\sin x \sin y = \tfrac{1}{2}[\cos(x - y) - \cos(x + y)] \qquad \cos a - \cos b = -2 \sin \frac{a + b}{2} \sin \frac{a - b}{2}$$

• Reduction formula: If $a \neq 0$, then

$$a \sin Bx + b \cos Bx = A \sin(Bx + C),$$

where $C = \arctan(b/a)$, and $A = \sqrt{a^2 + b^2}$ if $a > 0$ or $A = -\sqrt{a^2 + b^2}$ if $a < 0$.

Section	Key Concepts to Review
6.5	• Methods to solve certain trigonometric equations (Reference numbers, identities, and a scientific calculator may be involved.)
	• Determine solutions between 0 and 2π as follows.

Quadrant	Solution
1	Reference number
2	π − reference number
3	π + reference number
4	2π − reference number

• We generate all the solutions to a trigonometric equation by adding multiples of 2π to the solutions that are in the interval $[0,2\pi)$.

CHAPTER REVIEW EXERCISES

In Exercises 1–20 verify the given identity.

1. $\cot^2 x \sec^2 x = \csc^2 x$

2. $\dfrac{\sin^2 x}{1 + \cos x} = 1 - \cos x$

3. $\sin^2 x \tan^2 x = \tan^2 x - \sin^2 x$

4. $\sec \theta \csc \theta = \tan \theta + \cot \theta$

5. $\dfrac{\sin \theta + \tan \theta}{1 + \sec \theta} = \sin \theta$ **6.** $\dfrac{\sec x}{\cos x} - \dfrac{\tan x}{\cot x} = 1$

7. $(\sin x - \cos x)^2 = 1 - \sin 2x$

8. $\dfrac{1 - \cos 2\theta}{1 + \cos 2\theta} = \tan^2 \theta$

9. $\dfrac{1 - \cos 2x}{\sin 2x} = \tan x$

10. $\dfrac{2 \cot x}{\csc^2 x - 2} = \tan 2x$

11. $2 \sin^2 3x = 1 - \cos 6x$

12. $\tan(\theta/2) = \csc \theta - \cot \theta$

13. $\cot 2x = \cot x - \csc 2x$

14. $\sin(x - \pi) = -\sin x$

15. $\sin 3x = 3 \sin x \cos^2 x - \sin^3 x$

16. $\sqrt{3} \sin x + \cos x = 2 \sin\left(x + \dfrac{\pi}{6}\right)$

17. $\sin^4 2x = \dfrac{3}{8} - \dfrac{1}{2} \cos 4x + \dfrac{1}{8} \cos 8x$

18. $\cos(u + v) + \cos(u - v) = 2 \cos u \cos v$

19. $\cos(A + B)\cos(A - B) = \cos^2 A - \sin^2 B$

20. $\dfrac{\sin 3x + \sin x}{\cos 3x - \cos x} = -\cot x$

In Exercises 21–30 find exact function values if $\cos x = \frac{4}{5}$ and $3\pi/2 < x < 2\pi$.

21. $\sin x$ **22.** $\csc x$

23. $\sec x$ **24.** $\tan x$

25. $\cos 2x$ **26.** $\sin 2x$

27. $\tan 2x$ **28.** $\tan(x/2)$

29. $\sin(x/2)$ **30.** $\cos(x/2)$

31. Is $\sin[(\pi/2) + x] = -\cos x$ a conditional equation or an identity?

32. Find $\sin[(\pi/4) - x]$, if $\sin x = \frac{4}{5}$ and $0 < x < \pi/2$.

33. Find $\cos 2x$, if $\sin x = \frac{1}{4}$ and $\pi/2 < x < \pi$.

34. If $\cos x = a$, express $\cos 2x$ in terms of a.

35. If $\tan \alpha = a$ and $\tan \beta = 2a$, express $\tan(\beta - \alpha)$ in terms of a.

36. Find the exact value of $\cos 75°$ by using a sum formula.

37. Eliminate the radical in the expression $\sqrt{x^2 + 25}$ by substituting $5 \tan \theta$ for x, where $-\pi/2 < \theta < \pi/2$.

38. Express $8 \cos^2(x/4) - 8 \sin^2(x/4)$ as a single function of a multiple angle.

39. Simplify $\sin(3\pi - x)$.

40. Simplify $\tan 2x \cos 2x \csc x$.

41. Simplify $\dfrac{1}{\sin x \cos^2 x} - \dfrac{\tan^2 x}{\sin x}$.

42. Express $\sin 5x - \sin 3x$ as a product.
43. Express $\sin 3\theta \cos \theta$ as a sum.
44. Rewrite $4\sqrt{3} \sin 2x - 4 \cos 2x$ using the reduction identity.
45. Find exactly: $\cos(\sin^{-1} \frac{24}{25} - \cos^{-1} \frac{3}{5})$.
46. If $\sin x_1 = -\frac{1}{3}$, where $\pi < x_1 < 3\pi/2$, and $\cos x_2 = -\frac{1}{2}$, where $\pi/2 < x_2 < \pi$, find $\sin(x_1 - x_2)$.
47. Use the identity for $\sin(x_1 + x_2)$ to derive the identity for $\sin(x_1 - x_2)$.
48. Use the identity for $\cos(x_1 + x_2)$ to derive the identity for $\cos 2x$ in terms of $\sin x$.
49. Use the formula for the sine of the sum of two numbers and derive a formula for $\sin 4x$ in terms of $\sin x$ and $\cos x$.
50. In triangle ABC show that $A + B = 90°$ if
$$\frac{\sin A + \sin B}{\cos A + \cos B} = 1.$$

In Exercises 51–56 solve for x in the interval $[0,2\pi)$.

51. $\csc x = -2$
52. $4 \sin^2 x - \sin x - 2 = 0$
53. $2 \cos^2 x = 3 - 3 \sin x$ 54. $\tan 3x = 1$
55. $\sin 2x = \sin x$ 56. $\cos 3x + \cos x = 0$

In Exercises 57 and 58 solve for θ in the interval $[0°,360°)$.

57. $3 \tan^2 \theta - 1 = 0$
58. $2 \sin \theta \cos \theta - \cos \theta = 0$
59. Find all real number solutions: $\tan x = \cot x$.
60. Find all solutions to the nearest tenth of a degree: $4 \cos \theta - 3 = 0$.

In Exercises 61–70 select the choice that completes the statement or answers the question.

61. The expression $\sin x + (\cos^2 x/\sin x)$ is identical to
    a. $\sec x$ **b.** $\csc x$ **c.** $\cos x$ **d.** 1

62. Which one of the following is an identity?
    a. $\sin x + \cos x = 1$ **b.** $\sec x \cdot \csc x = 1$
    c. $\sin \frac{1}{2}x = \frac{1}{2} \sin x$ **d.** $\cos^2 x = 1 - \sin^2 x$
63. The equation $\cos(-x) = -\cos x$ is true for
    a. all values of x **b.** only certain values of x
    c. no values of x
64. The expression $\cos(x - \pi)$ simplifies to
    a. $\cos x$ **b.** $-\cos x$ **c.** $\sin x$ **d.** $-\sin x$
65. If $\log \tan x = a$, then $\log \cot x$ equals
    a. $1/a$ **b.** $1 - a$ **c.** $-a$ **d.** a
66. The exact value of $\sin(11\pi/12)$ is
    a. $\dfrac{\sqrt{6} - \sqrt{3}}{4}$ **b.** $\dfrac{\sqrt{3} - \sqrt{6}}{4}$
    c. $\dfrac{\sqrt{6} - \sqrt{2}}{4}$ **d.** $\dfrac{\sqrt{2} - \sqrt{6}}{4}$
67. To the nearest hundredth the solution of the equation $3 \sin^2 x + \sin x - 1 = 0$ on $[\pi/2,\pi]$ is
    a. 2.71   **b.** 2.00
    c. 2.02   **d.** 2.69
68. Which formula, where k is any integer, generates all solutions to $1 + \cos 2x = 0$?
    a. $\pi/2 + k\pi$ **b.** $\pi/4 + k\pi$
    c. $\pi/2 + k2\pi$ **d.** $\pi/4 + k2\pi$
69. The expression $\sin 3x + \sin x$ is identical to
    a. $\sin 4x \cos 2x$ **b.** $2 \sin 2x \cos x$
    c. $2 \sin 4x$ **d.** $\sin 4x$
70. If $3\pi/2 < x < 2\pi$, then
    a. both $\sin(x/2)$ and $\cos(x/2)$ are positive
    b. both $\sin(x/2)$ and $\cos(x/2)$ are negative
    c. $\sin(x/2)$ is positive and $\cos(x/2)$ is negative
    d. $\sin(x/2)$ is negative and $\cos(x/2)$ is positive

CHAPTER 6 TEST

1. Show $\dfrac{1 + \cot^2 x}{\cot^2 x} = \sec^2 x$ is an identity.

2. Simplify $\sec x \cos x - \dfrac{1}{\csc^2 x}$.

3. Use identities to find $\tan x$ if $\cos x = \frac{5}{13}$ and $3\pi/2 < x < 2\pi$.

4. Eliminate the radical in the expression $\sqrt{16 - x^2}$ by substituting $4 \sin \theta$ for x, where $-\pi/2 \le \theta \le \pi/2$.

5. Show $\cos(x - \pi) = -\cos x$ is an identity.

6. Rewrite $\tan\left(x + \dfrac{3\pi}{4}\right)$ in terms of $\tan x$.

7. Find the exact value of $\sin 15°$ by using a difference formula.

8. Find exactly: $\sin(\cos^{-1} \frac{4}{5} + \sin^{-1} \frac{7}{25})$.

9. Show $\dfrac{1 - \tan^2 x}{1 + \tan^2 x} = \cos 2x$ is an identity.

10. Express $\sin 3x \cos 3x$ as a single function of a multiple angle.

11. Show that $4 \sin^2 4\theta = 2 - 2 \cos 8\theta$ is an identity by using a power reduction formula.

12. If $\cos x = \frac{12}{13}$ and $3\pi/2 < x < 2\pi$, find $\cos(x/2)$.

13. Find the exact value of $\cos \dfrac{5\pi}{12} \sin \dfrac{\pi}{12}$.

14. Express $\sin 3x + \sin 2x$ as a product.

15. Express $\cos 3\theta \cos \theta$ as a sum.

16. Rewrite $2 \sin x - 2 \cos x$ by using the reduction identity.

17. Find all real number solutions: $4 \sin x + 3 = 1$.

18. Find all solutions to the nearest tenth of a degree: $3 \tan \theta - 4 = 0$.

19. Solve for x in the interval $[0,2\pi)$: $\cos 3x = \sqrt{2}/2$.

20. Solve for x in the interval $[0,2\pi)$: $2 \sin^2 x = 1 - \cos x$.

7 Further Applications of Trigonometry

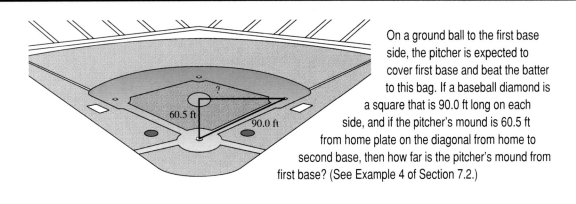

On a ground ball to the first base side, the pitcher is expected to cover first base and beat the batter to this bag. If a baseball diamond is a square that is 90.0 ft long on each side, and if the pitcher's mound is 60.5 ft from home plate on the diagonal from home to second base, then how far is the pitcher's mound from first base? (See Example 4 of Section 7.2.)

I n this chapter we continue our discussion of trigonometry by considering additional applications of the trigonometric functions with angle measures as domain elements. These applications involve the law of sines, the law of cosines, the area of a triangle, vectors, the trigonometric form of complex numbers, and polar coordinates.

7.1 Law of Sines

In Section 5.8 we learned to solve right triangles. We now wish to extend our ability to solve triangles by considering the solution of general triangles, which may or may not be right triangles. Remember that we "solve" a triangle by finding the measures of its three angles and three sides. To accomplish this, at least three of these six values must be known and one or more must be a side length.

The first technique that we use to solve general triangles is called the **law of sines.** We can derive this law by placing triangle ABC on a rectangular coordinate system so that angle A is in standard position. Figure 7.1(a) and (b) show the result when A is an acute angle and an obtuse angle, respectively. In both cases we draw the altitude of the triangle, CD, and note that its length is y.

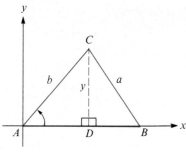

(a)

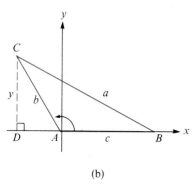

(b)

Figure 7.1 (a) A is an acute angle.
(b) A is an obtuse angle.

Then by the general definition of the sine function,

$$\sin A = \frac{y}{b}, \text{ so } y = b \sin A.$$

Also, in right triangle *BDC*,

$$\sin B = \frac{y}{a}, \text{ so } y = a \sin B.$$

Setting these two expressions for y equal to each other, we have

$$b \sin A = a \sin B, \text{ so } \frac{\sin A}{a} = \frac{\sin B}{b}.$$

Similarly, by placing angle C in standard position we can show that

$$\frac{\sin A}{a} = \frac{\sin C}{c}.$$

Combining these results, we have

$$\frac{\sin A}{a} = \frac{\sin B}{b} = \frac{\sin C}{c}.$$

This relationship, the **law of sines,** states the following.

Law of Sines

The sines of the angles in a triangle are proportional to the lengths of the opposite sides.

Note that if C is a right angle, $\sin C = \sin 90° = 1$, and the law of sines yields the right triangle relationships

$$\sin A = \frac{a}{c} \quad \text{and} \quad \sin B = \frac{b}{c}.$$

The law of sines can be used to solve any triangle in the following two cases:

1. If we know the measures for two angles and one side of the triangle
2. If we know the measures for two sides of the triangle and the angle opposite one of them

The following example illustrates how the law of sines can be used to solve a triangle in the first case in which the measures for two angles and one side of the triangle are known. Note that in computing results the symbol for equality (=) is generally used even though the symbol for approximation (≈) may be more appropriate.

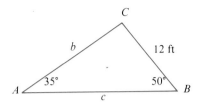

Figure 7.2

EXAMPLE 1   Approximate the missing parts of triangle ABC in which $A = 35°$, $B = 50°$, and $a = 12$ ft.

Solution   First, sketch Figure 7.2. We can find angle C since the sum of the angles in a triangle is $180°$.

$$A + B + C = 180°$$
$$35° + 50° + C = 180°$$
$$C = 95°$$

Second, we find side length b by applying the law of sines.

$$\frac{\sin A}{a} = \frac{\sin B}{b}$$

$$\frac{\sin 35°}{12} = \frac{\sin 50°}{b}$$

$$b = \frac{12 \sin 50°}{\sin 35°}$$

$$= \frac{12(0.7660)}{0.5736}$$

$$= 16 \text{ ft}$$

Third, we find c by applying the law of sines.

$$\frac{\sin A}{a} = \frac{\sin C}{c}$$

$$\frac{\sin 35°}{12} = \frac{\sin 95°}{c}$$

$$c = \frac{12 \sin 95°}{\sin 35°}$$

$$= \frac{12(0.9962)}{0.5736}$$

$$= 21 \text{ ft}$$

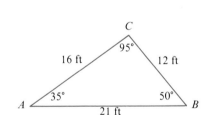

Figure 7.3

Thus, the solution to the triangle is as shown in Figure 7.3.   ∎

(*Note:* Remember that our computed results cannot be more accurate than the data that are given. Guidelines for the desired accuracy in a solution can be found in Section 5.8. The arithmetic involved in these calculations is quite tedious, so you are encouraged to use a calculator.)

EXAMPLE 2   Two observation towers, A and B, are located 10 mi apart. A fire is sighted at point C and the observer in tower A measures angle CAB to be $80°$. At the same time the observer in tower B measures angle CBA to be $40°$. How far is the fire from tower A?

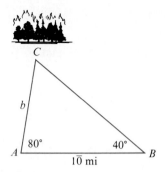

Figure 7.4

Solution   First, draw a diagram picturing the data (Figure 7.4). Second, find angle C.

$$A + B + C = 180°$$
$$80° + 40° + C = 180°$$
$$C = 60°$$

Third, we find b by applying the law of sines.

$$\frac{\sin B}{b} = \frac{\sin C}{c}$$

$$\frac{\sin 40°}{b} = \frac{\sin 60°}{10}$$

$$b = \frac{10 \sin 40°}{\sin 60°}$$

$$= \frac{10(0.6428)}{0.8660}$$

$$= 7.4$$

Thus, the fire is about 7.4 mi from station A.   ■

The following examples illustrate how the law of sines can be used to solve a triangle in the second case in which the measures for two sides of the triangle and the angle opposite one of them are known.

EXAMPLE 3   Solve the triangle ABC in which $B = 60°$, $b = 5\overline{0}$ ft, and $c = 3\overline{0}$ ft.

Solution   First, sketch Figure 7.5. We find angle C by applying the law of sines.

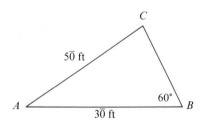

Figure 7.5

$$\frac{\sin B}{b} = \frac{\sin C}{c}$$

$$\frac{\sin 60°}{50} = \frac{\sin C}{30}$$

$$\frac{30(\sin 60°)}{50} = \sin C$$

$$\frac{30(0.8660)}{50} = \sin C$$

$$0.5196 = \sin C$$

We now have two possibilities for angle C since the sine of both first and second quadrant angles is positive.

Case 1   (acute angle in Q_1):   **Case 2**   (obtuse angle in Q_2):

$\sin C = 0.5196$ $\sin C = 0.5196$

reference angle $= 31.3°$ (by calculator)

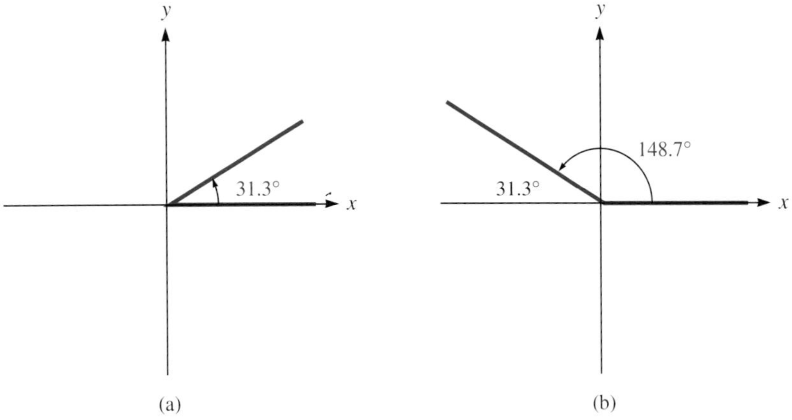

Figure 7.6

31.3° is the angle in Q_1 with a reference angle of 31.3° [Figure 7.6(a)].

148.7° is the angle in Q_2 with a reference angle of 31.3° [Figure 7.6(b)].

Therefore, $C = 31.3°$ or $C = 148.7°$.

Second, we find angle A in both of the above cases.

$$A + B + C = 180°$$
$$A + 60° + 31.3° = 180°$$
$$A = 88.7°$$

$$A + B + C = 180°$$
$$B = 60°\qquad C = 148.7°$$

Here we find $B + C = 208.7°$, so regardless of the value of A, the sum of the angles of the triangle exceeds 180°. Therefore, we reject $C = 148.7°$ as a solution.

Third, we find side length a by applying the law of sines.

$$\frac{\sin A}{a} = \frac{\sin B}{b}$$
$$\frac{\sin 88.7°}{a} = \frac{\sin 60°}{50}$$
$$\frac{50(\sin 88.7°)}{\sin 60°} = a$$
$$\frac{50(0.9997)}{0.8660} = a$$
$$58 \text{ ft} = a$$

When we round off the angle measures to the nearest degree, the solution to the triangle is as shown in Figure 7.7.   ∎

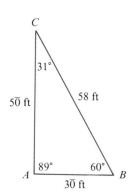

Figure 7.7

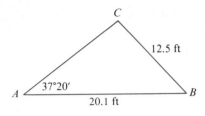

Figure 7.8

EXAMPLE 4   Approximate the missing parts of triangle ABC in which $A = 37°20'$, $a = 12.5$ ft, and $c = 20.1$ ft.

Solution   First, sketch Figure 7.8. We find angle C by applying the law of sines.

$$\frac{\sin A}{a} = \frac{\sin C}{c}$$

$$\frac{\sin 37°20'}{12.5} = \frac{\sin C}{20.1}$$

$$\frac{20.1(\sin 37°20')}{12.5} = \sin C$$

$$\frac{20.1(0.6065)}{12.5} = \sin C$$

$$0.9753 = \sin C$$

We now have two possibilities.

Case 1   (acute angle in Q_1):   **Case 2**   (obtuse angle in Q_2):

$\quad \sin C = 0.9753$ $\quad \sin C = 0.9753$

reference angle $= 77°10'$

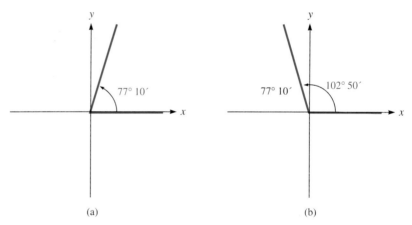

(a)   (b)

Figure 7.9

$77°10'$ is the angle in Q_1 with a reference angle of $77°10'$ [Figure 7.9(a)].

$102°50'$ is the angle in Q_2 with a reference angle of $77°10'$ [Figure 7.9(b)].

Therefore, $C = 77°10'$ or $C = 102°50'$.

Second, we find angle B in both of the above cases.

$$A + B + C = 180°$$
$$37°20' + B + 77°10' = 180°$$
$$B + 114°30' = 180°$$
$$B = 65°30'$$

$$A + B + C = 180°$$
$$37°20' + B + 102°50' = 180°$$
$$B + 140°10' = 180°$$
$$B = 39°50'$$

Third, we find side length b by applying the law of sines.

$$\frac{\sin A}{a} = \frac{\sin B}{b}$$

$$\frac{\sin 37°20'}{12.5} = \frac{\sin 65°30'}{b}$$

$$b = \frac{12.5(\sin 65°30')}{\sin 37°20'}$$

$$= \frac{12.5(0.9100)}{0.6065}$$

$$= 18.8 \text{ ft}$$

$$\frac{\sin A}{a} = \frac{\sin B}{b}$$

$$\frac{\sin 37°20'}{12.5} = \frac{\sin 39°50'}{b}$$

$$b = \frac{12.5(\sin 39°50')}{\sin 37°20'}$$

$$= \frac{12.5(0.6406)}{0.6065}$$

$$= 13.2 \text{ ft}$$

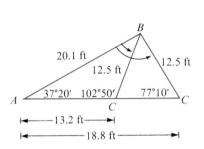

Figure 7.10

The two possible solutions from the given data are shown in Figure 7.10. ∎

Note that when we attempt to solve a triangle in which the measures for two sides of the triangle and the angle opposite one of them are given, there may be one triangle that fits the data (as in Example 3) or there may be two triangles that fit the data (as in Example 4). Consequently, this case is called the **ambiguous case** of the law of sines. It is also possible in the ambiguous case that no triangle can be constructed from the data; then we say the data is inconsistent. Figure 7.11 shows conditions that determine the various cases when given a, b, and acute angle A. In this diagram it is helpful to think that the side of length a can swing like a pendulum.

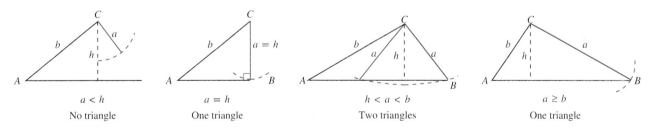

Figure 7.11 $\sin A = \dfrac{h}{b}$, so $h = b \sin A$.

If A is an obtuse angle, then the possibilities are more obvious, as shown in Figure 7.12.

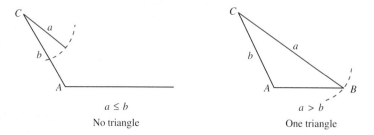

$a \leq b$
No triangle

$a > b$
One triangle

Figure 7.12

It is not recommended that the information in Figures 7.11 and 7.12 be memorized. Instead, in a given problem you should be aware of the various possibilities, make a careful sketch of the given information, and let an analysis based on the law of sines determine the case when the case is not obvious.

EXAMPLE 5   Use the law of sines to show that no triangle exists in the case in Figure 7.11 in which $a < b \sin A$.

Solution   Relate a, b, $\sin A$, and $\sin B$ by the law of sines and then solve for $\sin B$.

$$\frac{\sin B}{b} = \frac{\sin A}{a}$$

$$\sin B = \frac{b \sin A}{a}$$

Then $a < b \sin A$ implies $\sin B > 1$. Because the range of the sine function is $[-1,1]$, there is no solution for B and no triangle exists.   ■

For the next example we will assume all numbers are exact numbers, and because the angles given measure 45° and 30°, we may determine an exact solution.

EXAMPLE 6   In triangle RST, $r = 8$, $R = 45°$, and $S = 30°$. Find s.

Solution   First, sketch Figure 7.13. Then relate r, s, $\sin R$, and $\sin S$ by the law of sines and solve for s.

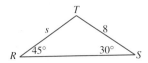

Figure 7.13

$$\frac{\sin R}{r} = \frac{\sin S}{s}$$

$$\frac{\sin 45°}{8} = \frac{\sin 30°}{s}$$

$$s = \frac{8 \sin 30°}{\sin 45°}$$

$$s = \frac{8 \cdot \frac{1}{2}}{\sqrt{2}/2}$$

$$s = 4\sqrt{2} \qquad \blacksquare$$

EXERCISES 7.1

(*Note:* There are more problems on the law of sines in Exercises 7.2.) In Exercises 1–12 approximate the remaining parts of the triangle for the data given.

1. $A = 30°$, $a = 25$ ft, and $B = 45°$
2. $C = 60°$, $c = 4\overline{0}$ ft, and $A = 80°$
3. $B = 120°$, $C = 40°$, and $a = 55$ ft
4. $C = 135°$, $c = 98$ ft, and $B = 15°$
5. $A = 62°10'$, $a = 31.5$ ft, and $B = 76°30'$
6. $A = 98°30'$, $B = 6°10'$, and $a = 415$ ft
7. $A = 45°$, $a = 8\overline{0}$ ft, and $b = 5\overline{0}$ ft
8. $C = 60°$, $c = 75$ ft, and $a = 45$ ft
9. $B = 30°$, $b = 3\overline{0}$ ft, and $a = 4\overline{0}$ ft
10. $B = 22°$, $b = 78$ ft, and $a = 86$ ft
11. $C = 150°$, $c = 92$ ft, and $b = 69$ ft
12. $C = 105°30'$, $c = 46.1$ ft, and $b = 75.2$ ft

In Exercises 13–22 assume exact numbers and find exact answers.

13. In triangle ABC, $a = 8$, $b = 12$, and $A = 30°$. Find $\sin B$.
14. In triangle ABC, $b = 8$, $c = 10$, and $C = 150°$. Find $\sin B$.
15. In triangle PQR, $p = 9$, $\sin P = \frac{3}{4}$, and $\sin Q = \frac{1}{2}$. Find q.
16. In triangle PQR, $\sin R = 0.6$, $\sin Q = 0.4$, and $q = 14$. Find r.
17. In triangle RST, $\sin R = \frac{1}{4}$ and $\sin S = \frac{7}{8}$. Find s/r.
18. In triangle RST, $S = 30°$ and $T = 45°$. Find s/t.
19. In triangle ABC, $b = 20$, $B = 45°$, and $C = 30°$. Find c.
20. In triangle ABC, $a = 10$, $A = 30°$, and $B = 60°$. Find b.

21. Find b and c in the figure below.
22. Find p and q in the figure below.

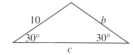 

In Exercises 23–28 determine if no triangle, one triangle, or two triangles are determined by the given conditions.

23. $A = 18°$, $a = 15$, $b = 28$
24. $A = 65°$, $a = 18$, $b = 24$
25. $C = 30°$, $b = 16$, $c = 32$
26. $B = 45°$, $a = 26$, $b = 21$
27. $A = 130°$, $a = 14$, $b = 18$
28. $A = 96°$, $a = 15$, $b = 11$
29. Two surveyors establish a baseline AB on a level field. The surveyor at point A is $20\overline{0}$ ft from the surveyor at point B. Each one sights a stake at point C. The surveyor at A measures angle CAB to be $72°30'$, while the surveyor at B measures angle CBA to be $81°20'$. Find the distance from B to C.
30. Engineers wish to build a bridge across a river to join point A on one side to either point B or point C on the other side. The distance from B to C is $40\overline{0}$ ft, angle ABC is $67°20'$, and angle ACB is $84°30'$. By how many feet does the distance from A to B exceed the distance from A to C?
31. Airport A is $30\overline{0}$ mi due north of airport B. Their radio stations receive a distress signal from a ship located at point C. It is determined that point C is located $54°$ south of east with respect to airport A and $76°$ north of east from airport B. How far is the ship from airport A?

32. Two engineers are located at points A and B on the opposite sides of a hill. They are both able to see a stake at point C, which is at a distance of $8\overline{0}0$ ft from A and $7\overline{0}0$ ft from B. If angle ABC is $25°$, find the distance $\overline{AB}$ through the hill.

33. In the illustration below find acute angles θ, α, and β to the nearest degree and x and y to two significant digits.

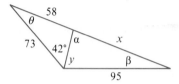

34. In triangle RST express $\sin T$ in terms of r, t, and $\sin R$.

35. If ABC is a right triangle ($C = 90°$), show that the law of sines simplifies to the right triangle relationships $\sin A = a/c$ and $\sin B = b/c$.

36. In a triangle if A, a, and b are given and $a = b \sin A$, then use the law of sines to show that the given conditions determine a right triangle.

37. Prove, by using the law of sines, that if the measures of two angles of a triangle are equal, then the lengths of the sides opposite these angles are equal.

38. In the figure below use the law of sines to show that if line segments AB and DE are parallel, then $e/b = d/a$. (This problem shows that if a line is drawn parallel to a side of a triangle, then the other two sides are divided proportionately.)

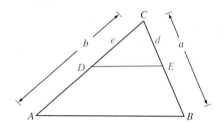

THINK ABOUT IT

1. In triangle ABC, show that $\dfrac{a - b}{b} = \dfrac{\sin A - \sin B}{\sin B}$.

$$\left(Hint: \frac{\sin A}{a} = \frac{\sin B}{b} \text{ implies } \frac{a}{b} = \frac{\sin A}{\sin B}. \text{ Then} \right.$$

subtract a constant on each side of this equation. $\Big)$

2. In triangle ABC, show that $\dfrac{a + b}{b} = \dfrac{\sin A + \sin B}{\sin B}$.

3. Use the results in Questions 1 and 2 and show that in triangle ABC, $\dfrac{a - b}{a + b} = \dfrac{\sin A - \sin B}{\sin A + \sin B}$.

4. Use the result in Question 3 and the sum (to product) formulas from Section 6.4 and show that in triangle ABC,

$$\frac{a - b}{a + b} = \frac{\tan \frac{1}{2}(A - B)}{\tan \frac{1}{2}(A + B)}.$$

This equation is one form of the **law of tangents** and gives an effective formula for finding A or B when a, b, and C are known.

5. **a.** The following equations are equivalent forms of an equation that is one of **Mollweide's check formulas.**

$$\frac{a - b}{c} = \frac{\sin \frac{1}{2}(A - B)}{\cos \frac{1}{2}C}$$

$$\frac{b - a}{c} = \frac{\sin \frac{1}{2}(B - A)}{\cos \frac{1}{2}C}$$

These equations relate all the parts in a triangle and are therefore useful for checking solutions when solving triangles. Check the solution given in Example 1 of this section by using the form from above that produces positive results on both sides of the equation. If the same result is not obtained on both sides (with minor allowances for round-off error), then the solution is incorrect.

b. Derive the Mollweide equation

$$\frac{a - b}{c} = \frac{\sin \frac{1}{2}(A - B)}{\cos \frac{1}{2}C}.$$

REMEMBER THIS

1. What may we conclude about triangle ABC when $\cos B = 0$?
2. Describe the measure of angle A in triangle ABC if (a) $\cos A < 0$ and (b) $\cos A > 0$.
3. In triangle ABC if $A = 60°$, $b = 25$ ft, and $a = 37$ ft, find B.
4. In triangle ABC if $A = 90°$, $b = 25$ ft, and $a = 37$ ft, find B.
5. Find the distance between $(2,6)$ and $(-1,4)$.

6. Find the area of the triangle with vertices at $(0,8)$, $(10,0)$, and $(0,0)$.
7. Find the exact value of $\frac{1}{2}(24)(14) \sin 45°$.
8. Solve $\cos B = -0.6157$ for $0° < B < 180°$.
9. If (x,y) is a point on the terminal ray of θ other than the origin, then by definition $\cos \theta$ equals
   a. x/y **b.** r/x **c.** y/r **d.** x/r
10. What do we call the case in which the measures of two sides of the triangle and the angle opposite one of them is given?

7.2   Law of Cosines and Area of Triangles

In Section 7.1 we found that the law of sines can be used to solve any triangle in the following two cases:

1. If we know the measures for **two angles** and **one side** of the triangle
2. If we know the measures for **two sides** of the triangle and the **angle opposite** one of them

However, there exist two other cases for which the law of sines cannot be applied. They are:

3. If we know the measures for **two sides** of the triangle and the **angle between** these two sides
4. If we know the measures for the **three sides** of the triangle

In these cases we use the **law of cosines,** which states the following.

Law of Cosines

> In any triangle, the square of any side length equals the sum of the squares of the other two side lengths, minus twice the product of these other two side lengths and the cosine of their included angle.

Thus, for triangle ABC in Figure 7.14 we have

$$a^2 = b^2 + c^2 - 2bc \cos A$$
$$b^2 = a^2 + c^2 - 2ac \cos B$$
$$c^2 = a^2 + b^2 - 2ab \cos C.$$

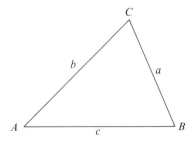

Figure 7.14

If we know the measures for two sides of the triangle and the included angle, we find the third side length by substituting in one of these formulas. After finding this part, we complete the solution by use of the law of sines. To obtain accuracy in the angle measures, the computed side length should be carried in the calculations to at least one more significant digit than stated in the solution.

If the three forms of the law of cosines are solved for the cosine of the angle, we have

$$\cos A = \frac{b^2 + c^2 - a^2}{2bc}$$

$$\cos B = \frac{a^2 + c^2 - b^2}{2ac}$$

$$\cos C = \frac{a^2 + b^2 - c^2}{2ab}.$$

These formulas are used to find the angle measures in a triangle when we know the three side lengths. In this case we do not use the law of sines to complete the solution because results are more accurate when they are computed from the data given.

Before starting the sample problems, let us first derive the law of cosines. Once again, we place triangle ABC on a rectangular coordinate system with angle A in standard position and consider both an acute and obtuse possibility for angle A as shown in Figure 7.15(a) and (b). In both cases vertex B obviously has coordinates $(c,0)$. Also, in general, $\cos A = x/b$ and $\sin A = y/b$, so the x-coordinate of vertex C is $b \cos A$ and the y-coordinate is $b \sin A$. If we now apply the distance formula to find the square of side length a, we have

$$
\begin{aligned}
a^2 &= (c - b \cos A)^2 + (0 - b \sin A)^2 \\
&= c^2 - 2bc \cos A + b^2 \cos^2 A + b^2 \sin^2 A \\
&= b^2(\sin^2 A + \cos^2 A) + c^2 - 2bc \cos A.
\end{aligned}
$$

Since $\sin^2 A + \cos^2 A = 1$ is an identity, the equation becomes

$$a^2 = b^2 + c^2 - 2bc \cos A,$$

which is one form of the law of cosines. A similar procedure with angles B and C, respectively, placed in standard position yields the other two forms.

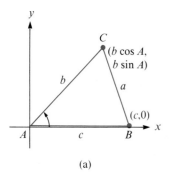

(a)

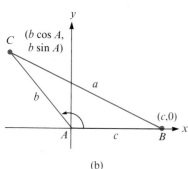

(b)

Figure 7.15

EXAMPLE 1   Approximate the missing parts of triangle ABC in which $A = 60°$, $b = 25$ ft, and $c = 42$ ft.

Solution   First, sketch Figure 7.16. We find side length a by applying the law of cosines.

$$
\begin{aligned}
a^2 &= b^2 + c^2 - 2bc \cos A \\
&= (25)^2 + (42)^2 - 2(25)(42)\cos 60° \\
&= 625 + 1{,}764 - 2{,}100(0.5000) \\
&= 2{,}389 - 1{,}050 \\
&= 1{,}339 \\
a &= \sqrt{1{,}339} \approx 36.6 \\
a &= 37 \text{ ft}
\end{aligned}
$$

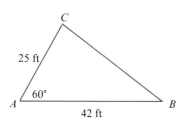

Figure 7.16

Second, we find the smaller of the remaining angles, angle B, by applying the law of sines. This angle must be acute. (Why?) We use 36.6 for a for better accuracy.

$$\frac{\sin A}{a} = \frac{\sin B}{b}$$

$$\frac{\sin 60°}{36.6} = \frac{\sin B}{25}$$

$$\frac{25 \sin 60°}{36.6} = \sin B$$

$$\frac{25(0.8660)}{36.6} = \sin B$$

$$0.5915 = \sin B$$

$$36° = B$$

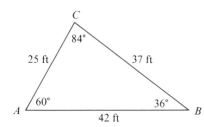

Figure 7.17

(*Note:* $\sin B = 0.5915$ is true if $B = 36°$ or if $B = 144°$. We eliminate $144°$ as a possible solution since we know that angle B must be acute.)

Third, we find angle C.

$$A + B + C = 180°$$
$$60° + 36° + C = 180°$$
$$C = 84°$$

Thus, the solution to the triangle is as shown in Figure 7.17.   ■

EXAMPLE 2   Approximate the missing parts of triangle ABC in which $a = 23.5$ ft, $b = 44.2$ ft, and $c = 30.1$ ft.

Solution   First, sketch Figure 7.18. We find angle A by applying the law of cosines.

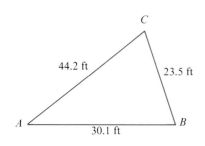

Figure 7.18

$$\cos A = \frac{b^2 + c^2 - a^2}{2bc}$$

$$= \frac{(44.2)^2 + (30.1)^2 - (23.5)^2}{2(44.2)(30.1)}$$

$$= 0.8672$$

$$A = 29.9°$$

(*Note:* Remember that if the cosine of the angle is positive, the angle is acute; if the cosine of the angle is negative, the angle is in Q_2 and is obtuse.)

Second, we find the smaller of the remaining angles, acute angle C, by applying the law of cosines.

$$\cos C = \frac{a^2 + b^2 - c^2}{2ab}$$

$$= \frac{(44.2)^2 + (23.5)^2 - (30.1)^2}{2(44.2)(23.5)}$$

$$= 0.7701$$

$$C = 39.6°$$

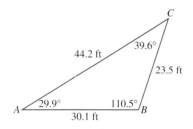

Figure 7.19

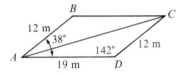

Figure 7.20

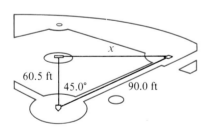

Figure 7.21

Third, we find angle B.

$$A + B + C = 180°$$
$$29.9° + B + 39.6° = 180°$$
$$B = 110.5°$$

Thus, the solution to the triangle is as shown in Figure 7.19. ∎

EXAMPLE 3 In parallelogram $ABCD$ the lengths of sides AB and AD are 12 m and 19 m, respectively. If $A = 38°$, find the length of the longer diagonal of the parallelogram.

Solution First, sketch Figure 7.20 and note that the longer diagonal is AC. In the parallelogram $\overline{AB} = \overline{DC} = 12$ and the sum of angles A and D is $180°$. Thus,

$$A + D = 180°$$
$$38° + D = 180°$$
$$D = 142°.$$

We find the length of the longer diagonal by applying the law of cosines.

$$(\overline{AC})^2 = (\overline{AD})^2 + (\overline{DC})^2 - 2(\overline{AD})(\overline{DC})\cos D$$
$$(\overline{AC})^2 = (19)^2 + (12)^2 - 2(19)(12)\cos 142°$$
$$(\overline{AC})^2 = (19)^2 + (12)^2 - 2(19)(12)(-0.7880)$$
$$(\overline{AC})^2 = 864.3$$
$$\overline{AC} = 29$$

Thus, the longer diagonal is about 29 m. ∎

EXAMPLE 4 Solve the problem in the chapter introduction on page 365.

Solution Consider the simplified sketch of the problem in Figure 7.21. We find x by applying the law of cosines to obtain

$$x^2 = (60.5)^2 + (90)^2 - 2(60.5)(90)\cos 45°$$
$$= 3,660.25 + 8,100 - 10,890(0.7071)$$
$$= 4,059.93$$
$$x = \sqrt{4,059.93} \approx 63.7.$$

Thus, the pitcher's mound is about 63.7 ft from first base, which should give the pitcher an adequate head start in a race to the bag. ∎

Area of a Triangle

The area of a triangle is equal to one half the product of its base and its altitude, which translates to the formula $A = \frac{1}{2}bh$. Because h is often unknown, it is useful to develop area formulas that do not require h. First, consider Figure 7.22 and note that in both cases by the general definition of the sine function

$$\sin A = \frac{h}{b}, \text{ so } h = b \sin A.$$

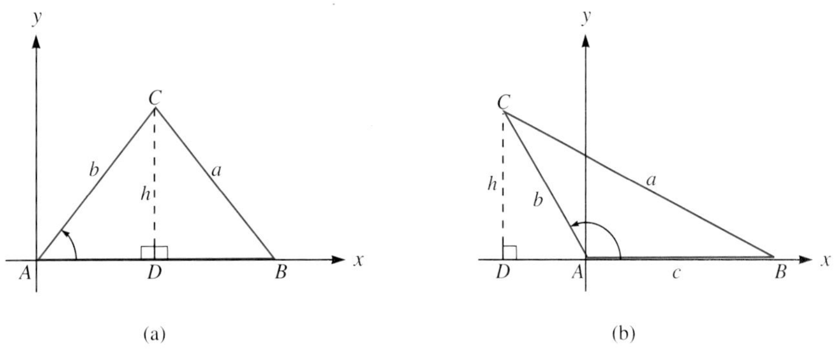

Figure 7.22 (a) *A* is an acute angle. (b) *A* is an obtuse angle.

Thus, if we represent the area by K (to avoid ambiguity with angle A), then

$$K = \tfrac{1}{2}(\text{base})(\text{height})$$
$$K = \tfrac{1}{2}c(b \sin A)$$

$$K = \tfrac{1}{2}bc \sin A.$$

Using similar reasoning with angle B and then angle C in standard position produces the formulas

$$K = \tfrac{1}{2}ac \sin B$$
$$K = \tfrac{1}{2}ab \sin C.$$

These formulas establish the following theorem.

Area of a Triangle

> The area of a triangle is given by one half the product of the lengths of two sides and the sine of the angle between these two sides.

EXAMPLE 5   Find the area of triangle ABC if $a = 32.4$ cm, $b = 49.2$ cm, and $C = 18.5°$.

Solution   Using the formula containing a, b, and C, we have

$$K = \tfrac{1}{2}ab \sin C$$
$$= \tfrac{1}{2}(32.4)(49.2)\sin 18.5°$$
$$= 253 \text{ cm}^2 \qquad \text{(three significant digits)}. \qquad ■$$

EXAMPLE 6   Show by using the law of sines that the area K of triangle ABC may be given by

$$K = \frac{c^2 \sin A \sin B}{2 \sin C}.$$

Solution Consideration of the above equation suggests that we start with an area formula that contains the factor $c \sin A$ or the factor $c \sin B$. We select $K = \frac{1}{2}bc \sin A$ and note that by the law of sines

$$\frac{\sin B}{b} = \frac{\sin C}{c}, \text{ so } b = \frac{c \sin B}{\sin C}.$$

Then,

$$K = \frac{1}{2}bc \sin A = \frac{1}{2}\left(\frac{c \sin B}{\sin C}\right)c \sin A = \frac{c^2 \sin A \sin B}{2 \sin C}. \quad \blacksquare$$

When the three side measures of a triangle are known, then the area may be determined by using the following formula, which is named after the mathematician Heron of Alexandria.

Heron's Area Formula

> The area K of the triangle with side lengths a, b, and c is
> $$K = \sqrt{s(s - a)(s - b)(s - c)}$$
> where s is the semiperimeter which is given by $\frac{1}{2}(a + b + c)$.

To derive this formula, start with one of the formulas for the area of a triangle in terms of two side lengths and the included angle, say $K = \frac{1}{2}bc \sin A$, and square both sides of the equation to obtain

$$K^2 = \frac{1}{4}b^2c^2 \sin^2 A = \frac{1}{4}b^2c^2 (1 - \cos^2 A)$$

$$= \frac{bc}{2}(1 + \cos A)\frac{bc}{2}(1 - \cos A).$$

Then by the law of cosines, $\cos A = \dfrac{b^2 + c^2 - a^2}{2bc}$, so

$$K^2 = \frac{bc}{2}\left(1 + \frac{b^2 + c^2 - a^2}{2bc}\right)\frac{bc}{2}\left(1 - \frac{b^2 + c^2 - a^2}{2bc}\right).$$

Through algebraic methods (see "Think About It" Question 5) this equation leads to

$$K^2 = \left(\frac{b + c + a}{2}\right)\left(\frac{b + c - a}{2}\right)\left(\frac{a - b + c}{2}\right)\left(\frac{a + b - c}{2}\right).$$

Finally, rewrite each factor so it contains the expression $(a + b + c)/2$, which represents the semiperimeter s of the triangle. Then, K^2 equals

$$\left(\frac{a + b + c}{2}\right)\left(\frac{a + b + c}{2} - a\right)\left(\frac{a + b + c}{2} - b\right)\left(\frac{a + b + c}{2} - c\right).$$

Thus, $K^2 = s(s - a)(s - b)(s - c)$, so $K = \sqrt{s(s - a)(s - b)(s - c)}$.

EXAMPLE 7 Find the area of triangle ABC if $a = 3.0$ ft, $b = 4.0$ ft, and $c = 6.0$ ft.

Solution First, find the semiperimeter s.

$$s = \tfrac{1}{2}(a + b + c) = \tfrac{1}{2}(3 + 4 + 6) = 6.5$$

Then, Heron's formula gives

$$\begin{aligned} K &= \sqrt{s(s - a)(s - b)(s - c)} \\ &= \sqrt{6.5(6.5 - 3)(6.5 - 4)(6.5 - 6)} \\ &= 5.3 \text{ ft}^2 \qquad \text{(two significant digits)}. \end{aligned}$$ ∎

EXERCISES 7.2

In Exercises 1–10 approximate the remaining parts of triangle ABC. The law of cosines will be needed in at least one of the steps.

1. $a = 12$ ft, $b = 15$ ft, and $C = 60°$
2. $a = 2\overline{0}$ ft, $c = 3\overline{0}$ ft, and $B = 30°$
3. $c = 19.2$ ft, $a = 46.1$ ft, and $B = 10°20'$
4. $b = 36$ ft, $c = 75$ ft, and $A = 98°$
5. $b = 11.1$ ft, $a = 19.2$ ft, and $C = 95°40'$
6. $a = 11$ ft, $b = 15$ ft, and $c = 19$ ft
7. $a = 12$ ft, $b = 5.2$ ft, and $c = 8.1$ ft
8. $a = 4.9$ ft, $b = 5.3$ ft, and $c = 2.6$ ft
9. $a = 34.4$ ft, $b = 56.1$ ft, and $c = 42.3$ ft
10. $a = 45.0$ ft, $b = 108$ ft, and $c = 117$ ft

In Exercises 11–14 use the law of cosines.

11. In parallelogram $ABCD$ the lengths of sides AB and AD are 21 m and 13 m, respectively. If $A = 52°$, find the length of the longer diagonal of the parallelogram.
12. In parallelogram $ABCD$ the length of sides AB and AD are 6.0 and 8.0 m, respectively. If the length of the shorter diagonal is 5.0 m, find angle A.
13. A surveyor at point C sights two points A and B on opposite sides of a lake. If C is 760 ft from A and 920 ft from B and angle ACB is 96°, how wide is the lake?
14. What is the distance between the two islands in the illustration below?

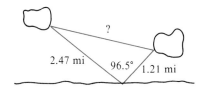

In Exercises 15–18 refer to the following triangle and complete each statement by using the law of cosines.

15. $r^2 = $ _____
16. $t = $ _____
17. $\cos T = $ _____
18. $\cos S = $ _____

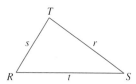

In Exercises 19–22 assume exact numbers and find exact answers.

19. In triangle ABC, $a = 7$, $b = 5$, and $c = 6$. Find $\cos A$.
20. Find the cosine of the largest angle of the triangle whose sides measure 3, 5, and 6 units.
21. In triangle RST, $R = 60°$, $s = 7$, and $t = 4$. Find r.
22. Find the perimeter in triangle ABC if $B = 120°$, $b = 16$, and $c = 11$.

In Exercises 23–34 the problems are mixed, that is, some use the law of sines, some the law of cosines, and some use both. Solve each triangle.

23. $A = 15°$, $C = 87°$, and $b = 42$ ft
24. $B = 68°$, $C = 72°$, and $a = 18$ ft
25. $a = 4\overline{0}$ ft, $b = 5\overline{0}$ ft, and $C = 120°$
26. $b = 126$ ft, $c = 92.1$ ft, and $A = 72°50'$
27. $B = 111°20'$, $C = 35°40'$, and $a = 142$ ft
28. $c = 127$ ft, $b = 315$ ft, and $A = 162°30'$
29. $a = 4.0$ ft, $b = 2.0$ ft, and $c = 3.0$ ft
30. $A = 95°$, $a = 54$ ft, and $c = 38$ ft
31. $B = 7°10'$, $b = 74.8$ ft, and $c = 92.4$ ft
32. $a = 84.8$ ft, $b = 36.8$ ft, and $c = 76.5$ ft
33. $a = 15\overline{0}$ ft, $b = 175$ ft, and $c = 20\overline{0}$ ft
34. $B = 152°50'$, $b = 13\overline{0}$ ft, and $c = 45.0$ ft

In Exercises 35–38 the solution will require a decision as to whether the law of sines or the law of cosines is appropriate.

35. A ship sails due east for $4\overline{0}$ mi and then changes direction and sails 20° north of east for $6\overline{0}$ mi. How far is the ship from its starting point?

36. One gun is located at point A, while a second gun at point B is located 5.0 mi directly east of A. From point A the direction to the target is 27° north of east. From point B the direction to the target is 72° north of east. For what firing range should the guns be set?

37. In a parallelogram the shorter diagonal makes angles of 25° and 72° with the sides. If the length of the shorter side is 15 m, what is the length of the longer side?

38. A and B are two points located on opposite edges of a lake. A third point C is located so that $\overline{AC}$ is 421 ft and $\overline{BC}$ is 376 ft. Angle ABC is measured to be 65.5°. Compute the distance $\overline{AB}$ across the lake.

In Exercises 39–50 find the area of the triangle satisfied by the given conditions.

39. $a = 7.0$ ft, $b = 4.0$ ft, and $C = 30°$
40. $a = 6.0$ ft, $b = 8.0$ ft, and $C = 150°$
41. $a = 12$ m, $c = 14$ m, and $B = 110°$
42. $a = 25$ m, $c = 19$ m, and $B = 70°$
43. $b = 4.74$ km, $c = 3.42$ km, and $A = 21.5°$
44. $b = 51.7$ cm, $c = 55.9$ cm, and $A = 16°50'$
45. $a = 5.0$ ft, $b = 4.0$ ft, and $c = 7.0$ ft
46. $a = 3.0$ m, $b = 4.0$ m, and $c = 3.0$ m
47. $a = 23.0$ m, $b = 14.0$ m, and $c = 18.0$ m
48. $a = 538$ ft, $b = 726$ ft, and $c = 981$ ft
49. $a = 2.51$ km, $b = 1.95$ km, and $c = 2.14$ km
50. $a = 42.56$ cm, $b = 37.83$ cm, and $c = 53.17$ cm

In Exercises 51–54 assume exact numbers and find exact answers.

51. Find the area of triangle ABC if $A = 120°$, $b = 16$, and $c = 11$.

52. Find the area of an isosceles triangle in which the vertex angle measures 30° and each leg measures 8 units.

53. Find the area of an equilateral triangle in which each side measures 6 units by using $K = \frac{1}{2}ab \sin C$.

54. Find the area of an equilateral triangle in which each side measures 6 units by using Heron's formula.

55. If ABC is a right triangle ($C = 90°$), show that the law of cosines simplifies to the relationship $c^2 = a^2 + b^2$.

56. Write a formula for the area K of an isosceles triangle in which s and θ measure the legs and vertex angle, respectively.

57. Three circles are tangent externally, as shown in the diagram below. If the diameters of the circles are given by $d_1 = 32.0$ mm, $d_2 = 16.0$ mm, and $d_3 = 18.0$ mm, then find the area of the triangle joining their centers.

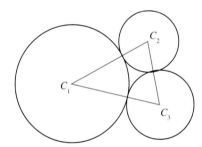

58. Show by using the law of sines that the area K of triangle ABC may be given by

$$K = \frac{a^2 \sin B \sin C}{2 \sin A}.$$

THINK ABOUT IT

1. Discuss why it is important to know how the law of cosines is stated in words.

2. a. Find the perimeter of the regular pentagon in the given illustration.

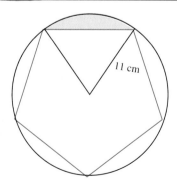

b. Find the area of the regular pentagon shown in the figure.

c. Find the area of the shaded region in the figure.

3. Show that the area K of an isosceles triangle in which s and β measure the legs and base angles, respectively, is given by

$$K = \tfrac{1}{2}s^2 \sin 2\beta.$$

4. Use Heron's formula to show that the area K of an isosceles triangle in which s and b measure the legs and base, respectively, is given by

$$K = \tfrac{1}{4}b\sqrt{4s^2 - b^2}.$$

5. In the derivation of Heron's formula show that

$$\frac{bc}{2}\left(1 + \frac{b^2 + c^2 - a^2}{2bc}\right)\frac{bc}{2}\left(1 - \frac{b^2 + c^2 - a^2}{2bc}\right)$$

$$= \left(\frac{b + c + a}{2}\right)\left(\frac{b + c - a}{2}\right)\left(\frac{a - b + c}{2}\right)\left(\frac{a + b - c}{2}\right).$$

REMEMBER THIS

1. Describe the two cases in which the law of cosines is used to solve a triangle.

2. Describe the two cases in which the law of sines is used to solve a triangle.

3. Find the measure of the larger acute angle in a right triangle whose legs measure 12 m and 16 m.

4. If the legs in a right triangle measure 27 ft and 36 ft, then what is the length of the hypotenuse?

5. Solve $\tan \theta = -3.0340$ for $90° < \theta < 180°$.

6. In triangle PQR, $P = 135°$ and $Q = 30°$. Find p/q.

7. In triangle ABC, if $B = 145°$, $a = 53$, and $c = 48$, then find b.

8. In triangle ACD, if $D = 145°$, $a = 47$, and $d = 95$, then find A.

9. In triangle ABC, if $\overline{AB} = 25$ and A and C measure $35°$ and $90°$, respectively, then find $\overline{AC}$.

10. In triangle OAC, if $\overline{OC} = 390$ and O and A measure $12°$ and $90°$, respectively, then find $\overline{AC}$.

7.3   **Introduction to Vectors**

An application of trigonometry occurs with the study of physical quantities that act in a definite direction. For example, when meteorologists describe wind they mention both the speed of the wind and the direction from which the wind is blowing. Similarly, quantities such as forces, weights, and velocities must be described in such a way that both the strength (magnitude) and the direction of the quantity can be determined. Mathematically, we represent such a quantity by a line segment with an arrowhead at one end; this directed line segment is called a **vector.** The direction in which the arrowhead is pointing represents the direction in which the quantity is acting, while the length of the line segment is proportional to the magnitude of the quantity.

EXAMPLE 1   Use a vector to represent graphically a wind that is blowing due north at a speed of 40 mi/hour.

Solution   If 1 unit represents 10 mi/hour, the directed line segment **OA,** which is 4 units long and pointing due north, is the appropriate vector

(Figure 7.23). (*Note:* A vector that starts at O and ends at A is labeled **OA**; a vector that starts at A and ends at O is a different vector and is labeled **AO.** In handwritten work a vector like **OA** is written $\overrightarrow{OA}$.)

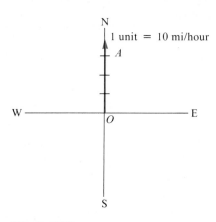

Figure 7.23 ▪

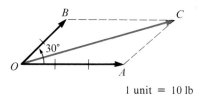

Figure 7.24

Frequently, there are two (or more) forces acting on a body from different directions, and their net effect is a third force with a new direction. This new force is called the **resultant,** or vector sum, of the given forces. For example, in Figure 7.24 two forces of 20 lb and 30 lb are both acting on a body with an angle of 30° between the two forces. It can be shown that the resultant of vectors **OA** and **OB** is vector **OC,** which is the diagonal of a parallelogram formed from the given vectors. The magnitude of the resultant **OC** can be determined by the length of the line segment from O to C. The direction of the resultant is the same as the direction of the arrowhead on vector **OC** and is described in terms of the original vectors by finding either angle AOC or angle BOC. The next three examples illustrate this procedure for finding the resultant of two forces. Examples 2 and 3 require only right triangle trigonometry, while Example 4 uses both the law of cosines and the law of sines.

EXAMPLE 2   A force of 3.0 lb and a force of 4.0 lb are acting on a body with an angle of 90° between the two forces. Find (a) the magnitude of the resultant and (b) the angle between the resultant and the larger force.

Solution   Let **OA** represent the 4.0-lb force and **OB** the 3.0-lb force. Then vector **OC** represents the resultant (see Figure 7.25).

a. We can find the length of **OC** by using the Pythagorean theorem in right triangle OAC. Notice that line segments OB and AC have the same length.

$$(\overline{OC})^2 = (4)^2 + (3)^2$$
$$(\overline{OC})^2 = 25$$
$$\overline{OC} = 5.0$$

Thus, the magnitude of the resultant is 5.0 lb.

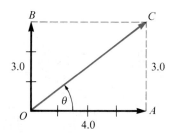

Figure 7.25

b. We can find the angle between the resultant and the 4.0-lb force by using the tangent function in right triangle *OAC*.

$$\tan \theta = \frac{3}{4}$$

$$\theta = 37° \qquad \text{(to the nearest degree)}$$

Thus, the angle between the resultant and the larger force is 37°.   ■

EXAMPLE 3   A ship is headed due east at $2\overline{0}$ knots (nautical miles per hour) while the current carries the ship due south at 5.0 knots. Find (a) the speed of the ship and (b) the direction (course) of the ship.

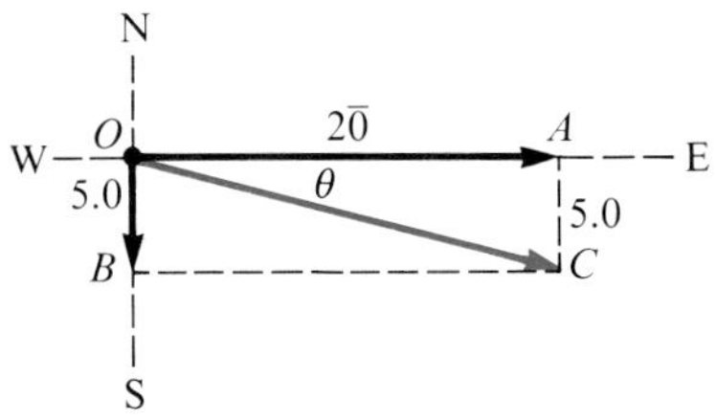

Solution   Let vector **OA** represent the velocity of the ship in the easterly direction. Let vector **OB** represent the velocity of the ship in the southerly direction due to the current. Let vector **OC** represent the actual velocity of the ship. (See Figure 7.26.)

a. We can find the length of **OC** by using the Pythagorean theorem in right triangle *OAC*.

$$(\overline{OC})^2 = (20)^2 + (5)^2$$
$$(\overline{OC})^2 = 425$$
$$\overline{OC} = \sqrt{425} \approx 21$$

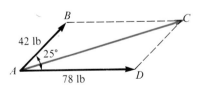

Figure 7.26

Thus, the speed of the ship is 21 knots.

b. We can find the angle between the resultant and vector **OA,** which is due east, by using the tangent function in right triangle *OAC*.

$$\tan \theta = \frac{5}{20}$$

$$\theta = 14° \qquad \text{(to the nearest degree)}$$

Thus, the ship is heading in a direction that is 14° south of east.   ■

EXAMPLE 4   A body is acted on by two forces with magnitudes of 78 lb and 42 lb that act at an angle of 25° with each other, as shown in Figure 7.27. Find (a) the magnitude of the resultant and (b) the angle between the resultant and the larger force.

Figure 7.27

Solution

a. The resultant of the two given forces is the diagonal of the parallelogram *ABCD* formed from the given vectors. In the parallelogram $\overline{AB} = \overline{DC} = 42$ and the sum of angle *A* and angle *D* equals 180°. Thus,

$$A + D = 180°$$
$$25° + D = 180°$$
$$D = 155°.$$

We find the magnitude of resultant **AC** by applying the law of cosines.

$$(\overline{AC})^2 = (78)^2 + (42)^2 - 2(78)(42) \cos 155°$$
$$(\overline{AC})^2 = (78)^2 + (42)^2 - 2(78)(42)(-0.9063)$$
$$\overline{AC} = \sqrt{13,786} \approx 117$$
$$\overline{AC} = 120$$

Thus, the magnitude of the resultant is 120 lb.

b. We find the angle between the resultant and the larger force, angle *CAD*, by applying the law of sines. Note that since angle *D* is greater than 90°, angle *CAD* must be acute. We use 117 for $\overline{AC}$ for better accuracy.

$$\frac{\sin CAD}{42} = \frac{\sin 155°}{117}$$
$$\sin CAD = \frac{42 \sin 155°}{117}$$
$$\sin CAD = \frac{42(0.4226)}{117}$$
$$= 0.1517$$
$$CAD = 9°$$

Thus, the angle between the resultant and the larger force is 9°.   ■

In the previous examples we considered how the net effect of having two forces acting on a body produced a third force, the resultant. However, the reverse situation often arises where we are given a single force that we think of as the resultant and we need to calculate two forces, which are called **components,** that produce the resultant. This process of expressing a single force in terms of two components, which are usually at right angles to each other, is called **resolving a vector.** The following examples illustrate the usefulness of resolving a vector into components.

EXAMPLE 5   An airplane, pointed due west, is traveling 10° north of west at a rate of $4\overline{0}0$ mi/hr. This resultant course is due to a wind blowing north. Find (a) the velocity of the plane if there were no wind (that is, the vector pointing due west) and (b) the velocity of the wind (that is, the vector pointing due north).

Solution   Let vector **OC** represent the resultant velocity of the plane. The resolution of vector **OC** results in a westerly component vector **OA** (the velocity of the plane if there were no wind) and a northerly component vector **OB** (the velocity of the wind) (see Figure 7.28).

a. We can find the velocity of the plane if there were no wind by using the cosine function in right triangle *OAC*.

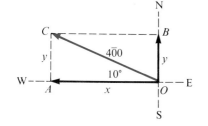

Figure 7.28

$$\cos 10° = \frac{x}{400}$$
$$400(\cos 10°) = x$$
$$400(0.9848) = x$$
$$390 = x \quad \text{(two significant digits)}$$

Thus, the velocity of the plane if there were no wind would be 390 mi/hour due west.

b. We can find the velocity of the wind by using the sine function in right triangle *OAC*.

$$\sin 10° = \frac{y}{400}$$
$$400(\sin 10°) = y$$
$$400(0.1736) = y$$
$$69 = y \quad \text{(two significant digits)}$$

Thus, the velocity of the wind would be 69 mi/hour due north. ∎

EXAMPLE 6 A man pulls with a force of $4\overline{0}$ lb on a window pole in an effort to lower a window. What part of the man's force lowers the window if the pole makes an angle of 20° with the window?

Solution Let vector **OA** represent the pull on the window pole of $4\overline{0}$ lb. The resolution of vector **OA** results in the vertical component vector **OC** (the force lowering the window) and the horizontal component vector **OB** (wasted force) (see Figure 7.29).

We can find the magnitude of vector **OC** by using the cosine function in right triangle *ACO*.

$$\cos 20° = \frac{\overline{OC}}{40}$$
$$40(\cos 20°) = \overline{OC}$$
$$40(0.9397) = \overline{OC}$$
$$38 = \overline{OC} \quad \text{(two significant digits)}$$

Thus, a force of 38 lb is lowering the window. ∎

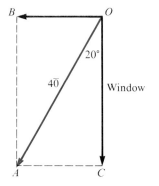

Figure 7.29

EXAMPLE 7 Find the horizontal and vertical components of a vector that has a magnitude of 25 lb and makes an angle of 40° with the positive *x*-axis.

Solution Let vector **OA** be the given vector. The resolution of vector **OA** results in the horizontal component, vector **OB**, and the vertical component, vector **OC** (see Figure 7.30).

We can find the magnitude of the horizontal component, vector **OB**, by using the cosine function in right triangle *OBA*.

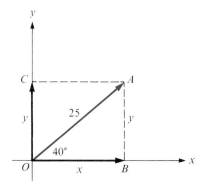

Figure 7.30

$$\cos 40° = \frac{x}{25}$$

$$25(\cos 40°) = x$$

$$25(0.7660) = x$$

$$19 = x \qquad \text{(two significant digits)}$$

We can find the magnitude of the vertical component, vector **OC**, by using the sine function in right triangle *OBA*.

$$\sin 40° = \frac{y}{25}$$

$$25(\sin 40°) = y$$

$$25(0.6428) = y$$

$$16 = y \qquad \text{(two significant digits)}$$

Thus, the magnitude of the horizontal component is 19 lb and the magnitude of the vertical component is 16 lb.

Example 7 shows how to resolve a vector into horizontal (or x) and vertical (or y) components. This breakdown can be very useful if we want to add two (or more) vectors. That is because the resultant of two (or more) vectors may be found by breaking each vector into its x and y components. The sum of the x components of each vector is the x component of the resultant (labeled R_x). Similarly, the sum of the y components of each vector is the y component of the resultant (labeled R_y). By the Pythagorean theorem the magnitude of the resultant is then

$$R = \sqrt{(R_x)^2 + (R_y)^2},$$

and the angle the resultant makes with the positive x-axis (labeled θ_R) is determined from the equation

$$\tan \theta_R = \frac{R_y}{R_x}.$$

Example 8 illustrates this approach to vector addition.

EXAMPLE 8 Vector **A** has a magnitude of 5.0 lb and makes an angle of 35° with the positive x-axis. Vector **B** has a magnitude of 9.0 lb and makes an angle of 160° with the positive x-axis. Find the resultant (or vector sum) of **A** and **B**.

Solution We organize the procedure described above with the following table.

Vector	Magnitude	Direction	x Component	y Component
A	5.0	35°	5.0 cos 35° = 4.096	5.0 sin 35° = 2.868
B	9.0	160°	9.0 cos 160° = −8.457	9.0 sin 160° = 3.078
R			$R_x = -4.361$	$R_y = 5.946$

The magnitude of the resultant is

$$R = \sqrt{(R_x)^2 + (R_y)^2}$$
$$= \sqrt{(-4.361)^2 + (5.946)^2}$$
$$= \sqrt{54.37}$$
$$= 7.4.$$

The angle the resultant makes with the positive x-axis is found by solving

$$\tan \theta_R = \frac{R_y}{R_x} = \frac{5.946}{-4.361} = -1.363.$$

Since the y component is positive and the x component is negative, θ_R is in Q_2. By calculator (or Table 6) we determine the reference angle is 54°. Then

$$\theta_R = 180° - 54°$$
$$= 126°.$$

Thus, the resultant has magnitude 7.4 lb and makes an angle of 126° with the positive x-axis. (See Figure 7.31.)   ∎

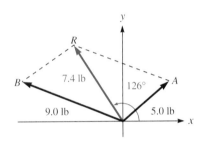

Figure 7.31

EXERCISES 7.3

In Exercises 1–10 find the magnitude of all forces to two significant digits and all angles to the nearest degree.

1. A force of 5.0 lb and a force of 12 lb are acting on a body with an angle of 90° between the two forces. Find
   a. the magnitude of the resultant
   b. the angle between the resultant and the larger force

2. A force of 12 lb and a force of 16 lb are acting on a body with an angle of 90° between the two forces. Find
   a. the magnitude of the resultant
   b. the angle between the resultant and the smaller force

3. Two forces, one $1\overline{0}$ lb and the other 15 lb, act on the same object at right angles to each other. Find
   a. the magnitude of the resultant
   b. the angle between the resultant and the larger force

4. Two velocities, one $2\overline{0}$ mi/hour north and the other $3\overline{0}$ mi/hour east, are acting on the same body. Find
   a. the speed of the resultant velocity
   b. the angle of the resultant velocity with respect to a direction of due east

5. Two velocities, one 5.0 mi/hour south and the other 15 mi/hour west, are acting on the same body. Find
   a. the speed of the resultant velocity
   b. the angle of the resultant velocity with respect to a direction of due west

6. A ship is headed due south at 18 knots (nautical miles per hour) while the current carries the ship due west at 5.0 knots. Find
   a. the speed of the ship
   b. the direction (course) of the ship

7. An airplane can fly $5\overline{0}0$ mi/hour in still air. If it is heading due north in a wind that is blowing due east at a rate of $8\overline{0}$ mi/hour, find
   a. the distance the plane can fly in 1 hour
   b. the angle that the resultant velocity will make with respect to due east

8. An object is dropped from a plane that is moving horizontally at a speed of $3\overline{0}0$ ft/second. If the vertical velocity of the object in terms of time is given by the formula $v = 32t$, 5 seconds later:
  a. What is the speed of the object?
  b. What angle does the direction of the object make with the horizontal?

9. A pilot wishes to fly due east at $3\overline{0}0$ mi/hour when a $7\overline{0}$ mi/hour wind is blowing due north.
  a. How many degrees south of east should the pilot point the plane to attain the desired course?
  b. What airspeed should the pilot maintain?

10. A ship wishes to travel due south at 25 knots in an easterly current of 7.0 knots.
  a. How many degrees west of south should the navigator direct the ship?
  b. What speed must the ship maintain?

In Exercises 11–14 find the horizontal and vertical component of each vector.

11. A magnitude of $1\overline{0}0$ lb; makes an angle of 30° with the positive *x*-axis.

12. A magnitude of 75 lb; makes an angle of 45° with the positive *x*-axis.

13. A magnitude of 18 lb; makes an angle of 27° with the positive *x*-axis.

14. A magnitude of 125 lb; makes an angle of 72° with the positive *x*-axis.

In Exercises 15–20 find the resultant (or vector sum) of the given vectors by resolving each vector into its *x* and *y* components.

	Vector	Magnitude	Direction (with respect to positive *x*-axis)
15.	**A**	5.0 lb	26°
	B	3.0 lb	84°
16.	**C**	6.0 lb	95°
	D	9.0 lb	15°
17.	**C**	12 lb	110°
	D	15 lb	180°
18.	**A**	4.0 lb	90°
	B	3.0 lb	190°
19.	**A**	$1\overline{0}$ lb	42°
	B	$2\overline{0}$ lb	140°
	C	$3\overline{0}$ lb	240°
20.	**A**	5.0 lb	60°
	B	2.0 lb	210°
	C	1.0 lb	270°

21. An airplane pointed due east is traveling 15° north of east at a rate of 350 mi/hour. The resultant course is due to a wind blowing north. Find
  a. the velocity of the plane if there were no wind (that is, the vector pointing due east)
  b. the velocity of the wind (that is, the vector pointing due north)

22. A man pulls with a force of 25 lb on a window pole in an effort to lower a window. How much force is wasted if the pole makes an angle of 15° with the window?

23. A woman pushes with a force of $4\overline{0}$ lb on the handle of a lawn mower that makes an angle of 33° with the ground. How much force pushes the lawn mower forward?

24. A car weighing 3,500 lb is parked in a driveway that makes an angle of 5° with the horizontal. Find the minimum brake force that is needed to keep the car from rolling down the driveway. (Use the illustration below and assume no friction.)

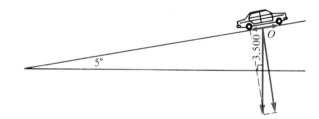

25. Find the force needed to keep a barrel weighing $1\overline{0}0$ lb from rolling down a ramp that makes an angle of 15° with the horizontal (assume no friction).

26. A box is resting on a ramp that makes an angle of 18° with the horizontal. What is the force of friction between the box and the ramp if the box weighs $8\overline{0}$ lb?

27. A body is acted on by two forces with magnitudes of 16 lb and 25 lb which act at an angle of 30° with each other. Find
  a. the magnitude of the resultant
  b. the angle between the resultant and the larger force

28. Two forces with magnitudes of 45 lb and $9\overline{0}$ lb are applied to the same point. If the angle between them measures 72°, find
  a. the magnitude of the resultant
  b. the angle between the resultant and the smaller force

29. Forces with magnitudes of 126 lb and 198 lb act simultaneously on a body in such a way that the angle between the forces is 14°50′. Find
  a. the magnitude of the resultant
  b. the angle between the resultant and the larger force

30. Two forces with magnitudes of 15 lb and $2\bar{0}$ lb act on a body in such a way that the magnitude of the resultant is 28 lb. Find the angles that the three forces make with each other.

31. Two forces act on a body to produce a resultant of 75 lb. If the angle between the two forces is 56° and one of the forces is $6\bar{0}$ lb, find the magnitude of the other force.

32. City B is located 50° north of east of city A. There is a $3\bar{0}$-mi/hour wind from the west and the pilot wishes to maintain an airspeed of 450 mi/hour. How many

degrees north of east should the pilot head the plane to arrive directly at city B from city A?

33. Two forces with magnitudes of 42 lb and 71 lb are applied to the same object. If the magnitude of the resultant is 85 lb, find the angles that the three forces make with each other.

34. A force $\mathbf{A}$ of $4\bar{0}0$ lb and a force $\mathbf{B}$ of $6\bar{0}0$ lb act at a point. Their resultant, $\mathbf{R}$, makes an angle of 42° with $\mathbf{A}$. Find the magnitude of $\mathbf{R}$.

THINK ABOUT IT

1. Two vectors $\mathbf{AB}$ and $\mathbf{CD}$ are said to be equal, and we write $\mathbf{AB} = \mathbf{CD}$ if and only if they have the same length and direction. Use this definition to explain why the two methods shown below give the same result for the vector sum, $\mathbf{A} + \mathbf{B}$.

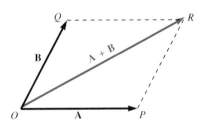

Parallelogram method

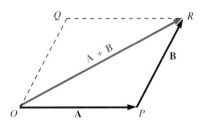

Tip-to-tail method

2. Use the tip-to-tail method of vector addition and show in a diagram that

$$\mathbf{A} + \mathbf{B} = \mathbf{B} + \mathbf{A} \qquad \text{(commutative property)}.$$

3. Use the tip-to-tail method of vector addition and show in a diagram that

$$(\mathbf{A} + \mathbf{B}) + \mathbf{C} = \mathbf{A} + (\mathbf{B} + \mathbf{C}) \qquad \text{(associative property)}.$$

4. If the magnitude of vector $\mathbf{A}$ is 6.0 lb and the magnitude of vector $\mathbf{B}$ is 8.0 lb, then show in a diagram how the vectors may be combined so the vector sum has the given magnitude.

  a. 14 lb   **b.** 2.0 lb
  c. $1\bar{0}$ lb   **d.** 7.2 lb

5. During a tournament, a golfer requires three putts to sink the ball in the hole. In respective order the three putts displaced the ball 18.00 ft due east, 6.00 ft in a direction 30.0° north of west, and 1.00 ft in a direction 45.0° south of east. Find the magnitude and the direction of the putt that was needed to sink the ball on the first putt.

REMEMBER THIS

1. To solve triangle ABC when given b, c, and B, do we first apply the law of sines or the law of cosines?

2. Solve $b^2 = a^2 + c^2 - 2ac \cos B$ for $\cos B$.

3. Find the area in triangle ABC if $a = 5.0$ m, $b = 6.0$ m, and $c = 5.0$ m.

4. Find the area in triangle PQR if $Q = 40°$, $p = 14$ ft, and $r = 19$ ft.

5. Name the property of real numbers illustrated by $a(b + c) = ab + ac$.

6. Find the exact values of 6 cos 150°, and 6 sin 150°.

7. Combine like terms: $(ax + by) - (cx + dy)$.

8. Find the distance from the origin to the point (a_1, a_2).

9. If the point (a, b) lies on the terminal ray of θ such that $a = 0$ and $b < 0$, then find θ in $[0°, 360°)$.

10. If the point $(-5, -1)$ lies on the terminal ray of θ and $0° \le \theta < 360°$, then find θ to the nearest degree.

7.4 Analytic Approach to Vectors

In this section we take an analytic approach to vectors and consider definitions and properties of vectors that enable us to analyze vectors using algebraic methods. We have seen that in a geometric approach, a vector is a directed line segment. Two vectors **AB** and **CD** are said to be **equal,** and we write **AB** = **CD,** if and only if they have the same length and direction (see Figure 7.32). This definition means that a vector may be shifted horizontally and vertically to different positions as long as its magnitude and direction are not changed. Thus, if we introduce a coordinate plane and place a vector **v** in this plane, then there are many representatives for **v,** as illustrated in Figure 7.33.

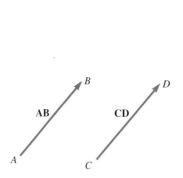

Figure 7.32 **AB** = **CD**

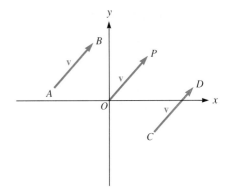

Figure 7.33 **v** = **OP** = **AB** = **CD**

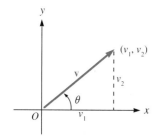

Figure 7.34 $\mathbf{v} = \langle v_1, v_2 \rangle$

It is usually most convenient to work with the representation of **v** that has its initial point at the origin. Such a vector is called a **position vector** and vector **OP** in Figure 7.33 is an example of a position vector. A position vector **v** that terminates at the point (v_1, v_2), as shown in Figure 7.34, is called the **vector** (v_1, v_2). This vector is written $\langle v_1, v_2 \rangle$ so that vectors and points are not confused, and the numbers v_1 and v_2 are called the **components** of $\langle v_1, v_2 \rangle$. A one-to-one correspondence has now been established between vectors in a plane and ordered pairs of real numbers, and to analyze vectors analytically means to work from the ordered-pair interpretation of a vector. Using the analytic approach, two vectors $\mathbf{v} = \langle v_1, v_2 \rangle$ and $\mathbf{w} = \langle w_1, w_2 \rangle$ are **equal** if and only if

$$v_1 = w_1 \quad \text{and} \quad v_2 = w_2.$$

The **magnitude, length,** or **norm** of **v,** symbolized $\|\mathbf{v}\|$, is given by the distance from the origin to the point (v_1, v_2), so

$$\|\mathbf{v}\| = \sqrt{v_1{}^2 + v_2{}^2}.$$

The **direction angle** θ of **v** is the smallest positive angle from the positive x-axis to the position vector, so if $v_1 \neq 0$, then

$$\tan \theta = \frac{v_2}{v_1}.$$

When $v_1 = 0$, then $\theta = 90°$ if $v_2 > 0$, and $\theta = 270°$ if $v_2 < 0$. As a special case the vector $\langle 0,0 \rangle$ is called the **zero vector** and denoted by a boldface **0**. The magnitude of **0** is 0 and the zero vector has no direction.

EXAMPLE 1   Find the magnitude and the direction angle of each vector.

a. v $= \langle -1, \sqrt{3} \rangle$ **b. w** $= \langle 0, -3 \rangle$

Solution   (See Figure 7.35.)

a. Replacing v_1 by -1 and v_2 by $\sqrt{3}$ in the magnitude formula gives

$$\|\mathbf{v}\| = \sqrt{(-1)^2 + (\sqrt{3})^2} = 2.$$

Now determine the direction angle θ.

$$\tan \theta = \frac{v_2}{v_1} = \frac{\sqrt{3}}{-1} = -\sqrt{3}$$

Since $(-1, \sqrt{3})$ is in Q_2 and the reference angle is $60°$,

$$\theta = 180° - 60° = 120°.$$

b. The magnitude is

$$\|\mathbf{w}\| = \sqrt{0^2 + (-3)^2} = 3.$$

By considering the sketch of **w** in Figure 7.35, we conclude the direction angle is $270°$. We may also reason that $\theta = 270°$ because $w_1 = 0$ and $w_2 < 0$.   ■

The basic arithmetic operations on vectors, called vector addition and scalar multiplication, are defined below in component form. Note that when discussing vectors, the term **scalar** means a real number. In the sciences scalars are quantities such as length or temperature that are completely described by the magnitudes of the quantities.

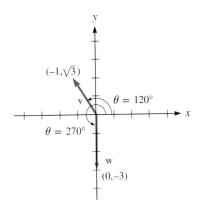

Figure 7.35

Vector Addition and Scalar Multiplication

For vectors **v** $= \langle v_1, v_2 \rangle$ and **w** $= \langle w_1, w_2 \rangle$ and scalar k:

$$\mathbf{v} + \mathbf{w} = \langle v_1 + w_1, v_2 + w_2 \rangle$$
$$k\mathbf{v} = \langle kv_1, kv_2 \rangle.$$

To reinforce these definitions geometrically, consider Figure 7.36. Note Figure 7.36(a) shows that the component definition of vector addition is in agreement with the parallelogram method of the previous section for finding a vector sum. This figure also shows that the vector sum $\mathbf{v} + \mathbf{w}$ may be found geometrically by placing the initial point of $\mathbf{w}$ at the endpoint of $\mathbf{v}$. Figure 7.36(b) shows the geometric interpretation of the vector $k\mathbf{v}$. The magnitude of $k\mathbf{v}$ is $|k|$ times the magnitude of $\mathbf{v}$, and $k\mathbf{v}$ has the same direction as $\mathbf{v}$ if $k > 0$ and the opposite direction of $\mathbf{v}$ if $k < 0$. Finally, Figure 7.36(c) shows the geometric interpretation of $k\mathbf{v}$ in component form if $k > 1$.

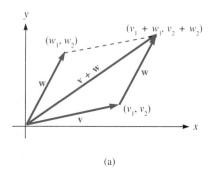

(a)

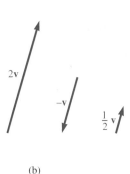

(b)

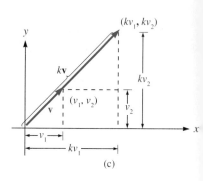

(c)

Figure 7.36

EXAMPLE 2   If $\mathbf{v} = \langle -3,4 \rangle$ and $\mathbf{w} = \langle 1,5 \rangle$, find the following vectors.

a. $\mathbf{v} + \mathbf{w}$ **b.** $5\mathbf{v}$ **c.** $-4\mathbf{w} + 3\mathbf{v}$

Solution   Apply the definitions of vector addition and scalar multiplication.

a. $\mathbf{v} + \mathbf{w} = \langle -3,4 \rangle + \langle 1,5 \rangle = \langle -3 + 1, 4 + 5 \rangle = \langle -2,9 \rangle$
b. $5\mathbf{v} = 5\langle -3,4 \rangle = \langle 5(-3), 5(4) \rangle = \langle -15,20 \rangle$
c. $-4\mathbf{w} + 3\mathbf{v} = -4\langle 1,5 \rangle + 3\langle -3,4 \rangle$
$= \langle -4,-20 \rangle + \langle -9,12 \rangle = \langle -13,-8 \rangle$ ■

Analogous to arithmetic with real numbers, the **negative** or **additive inverse** of $\mathbf{v} = \langle v_1,v_2 \rangle$ is defined by

$$-\mathbf{v} = \langle -v_1, -v_2 \rangle, \qquad \text{so} \qquad -\mathbf{v} = (-1)\mathbf{v},$$

and the **difference** of $\mathbf{w}$ and $\mathbf{v}$, denoted $\mathbf{w} - \mathbf{v}$, is defined by

$$\mathbf{w} - \mathbf{v} = \mathbf{w} + (-\mathbf{v}).$$

Thus, if $\mathbf{w} = \langle w_1,w_2 \rangle$ and $\mathbf{v} = \langle v_1,v_2 \rangle$, then

$$\mathbf{w} - \mathbf{v} = \langle w_1 - v_1, w_2 - v_2 \rangle.$$

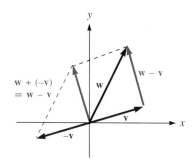

Figure 7.37

Figure 7.37 shows how to interpret a difference of two vectors geometrically. The difference $\mathbf{w} - \mathbf{v}$ is the vector from the endpoint of $\mathbf{v}$ to the endpoint of $\mathbf{w}$. We can check this interpretation by the methods for vector addition since $\mathbf{v} + (\mathbf{w} - \mathbf{v}) = \mathbf{w}$.

EXAMPLE 3   If $v = \langle 2, -6 \rangle$ and $w = \langle 1, -4 \rangle$, find the following vectors.

a. $-w$ **b.** $w - v$ **c.** $3v - 2w$

Solution

a. $-w = \langle -(1), -(-4) \rangle = \langle -1, 4 \rangle$

b. $w - v = \langle 1 - 2, -4 - (-6) \rangle = \langle -1, 2 \rangle$

c. Because $3v = \langle 6, -18 \rangle$ and $2w = \langle 2, -8 \rangle$,

$$3v - 2w = \langle 6, -18 \rangle - \langle 2, -8 \rangle = \langle 4, -10 \rangle.$$

Many of the familiar laws associated with real numbers carry over to working with vectors, and we list below the fundamental properties of vector addition and scalar multiplication.

Properties of Vector Addition and Scalar Multiplication

If **u**, **v**, and **w** are vectors, and c and d are scalars (real numbers), then

1. $u + v = v + u$
2. $(u + v) + w = u + (v + w)$
3. $u + 0 = u$
4. $u + (-u) = 0$
5. $c(u + v) = cu + cv$
6. $c(du) = (cd)u$
7. $(c + d)u = cu + du$
8. $1u = u.$

In Example 4 we prove analytically the fifth property listed above, and the proofs of the remaining properties are requested in the exercises.

EXAMPLE 4   Show that $c(u + v) = cu + cv$, where $u = \langle u_1, u_2 \rangle$, $v = \langle v_1, v_2 \rangle$, and c is a scalar.

Solution   The verification depends on the distributive property of real numbers that is used in the fourth step of the proof below.

$$
\begin{aligned}
c(u + v) &= c(\langle u_1, u_2 \rangle + \langle v_1, v_2 \rangle) \\
&= c\langle u_1 + v_1, u_2 + v_2 \rangle \\
&= \langle c(u_1 + v_1), c(u_2 + v_2) \rangle \\
&= \langle cu_1 + cv_1, cu_2 + cv_2 \rangle \\
&= \langle cu_1, cu_2 \rangle + \langle cv_1, cv_2 \rangle \\
&= c\langle u_1, u_2 \rangle + c\langle v_1, v_2 \rangle \\
&= cu + cv
\end{aligned}
$$

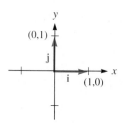

Figure 7.38

Unit Basis Vectors

A **unit vector** is a vector of length 1. Two special unit vectors, denoted by **i** and **j**, are shown in Figure 7.38 and are defined as follows.

$$\mathbf{i} = \langle 1,0 \rangle, \qquad \mathbf{j} = \langle 0,1 \rangle$$

These vectors are especially useful because for any vector $\mathbf{v} = \langle v_1, v_2 \rangle$, we have

$$\mathbf{v} = \langle v_1, v_2 \rangle = \langle v_1, 0 \rangle + \langle 0, v_2 \rangle$$
$$= v_1 \langle 1,0 \rangle + v_2 \langle 0,1 \rangle = v_1 \mathbf{i} + v_2 \mathbf{j}.$$

The vector sum $v_1 \mathbf{i} + v_2 \mathbf{j}$ is called a **linear combination** of **i** and **j**; and since every vector in a plane is expressible in this form, **i** and **j** are called **unit basis vectors.** To see a significant benefit of this notation, we restate below the rules for vector addition and scalar multiplication when vectors **v** and **w** are written as linear combinations of **i** and **j**.

$$(v_1 \mathbf{i} + v_2 \mathbf{j}) + (w_1 \mathbf{i} + w_2 \mathbf{j}) = (v_1 + w_1)\mathbf{i} + (v_2 + w_2)\mathbf{j}$$
$$(v_1 \mathbf{i} + v_2 \mathbf{j}) - (w_1 \mathbf{i} + w_2 \mathbf{j}) = (v_1 - w_1)\mathbf{i} + (v_2 - w_2)\mathbf{j}$$
$$c(v_1 \mathbf{i} + v_2 \mathbf{j}) = cv_1 \mathbf{i} + cv_2 \mathbf{j}$$

Note that operating on vectors in this form is like operating on ordinary algebraic expressions.

EXAMPLE 5   If $\mathbf{v} = 3\mathbf{i} - \mathbf{j}$ and $\mathbf{w} = -2\mathbf{i} + 5\mathbf{j}$, express the following vectors as linear combinations of **i** and **j**.

a. w + v   **b. v − w**   **c. 2w − 3v**

Solution

a. $\mathbf{w} + \mathbf{v} = (-2\mathbf{i} + 5\mathbf{j}) + (3\mathbf{i} - \mathbf{j})$
$$= \mathbf{i} + 4\mathbf{j}$$
b. $\mathbf{v} - \mathbf{w} = (3\mathbf{i} - \mathbf{j}) - (-2\mathbf{i} + 5\mathbf{j})$
$$= 5\mathbf{i} - 6\mathbf{j}$$
c. $2\mathbf{w} - 3\mathbf{v} = 2(-2\mathbf{i} + 5\mathbf{j}) - 3(3\mathbf{i} - \mathbf{j})$
$$= (-4\mathbf{i} + 10\mathbf{j}) - (9\mathbf{i} - 3\mathbf{j})$$
$$= -13\mathbf{i} + 13\mathbf{j}$$ ∎

To express a vector **v** as a linear combination of **i** and **j** when given the magnitude and direction angle of **v**, note in Figure 7.39 that

$$\cos \theta = \frac{v_1}{\|\mathbf{v}\|} \qquad \text{implies} \qquad v_1 = \|\mathbf{v}\| \cos \theta$$

and

$$\sin \theta = \frac{v_2}{\|\mathbf{v}\|} \qquad \text{implies} \qquad v_2 = \|\mathbf{v}\| \sin \theta.$$

Thus, $\mathbf{v} = v_1 \mathbf{i} + v_2 \mathbf{j}$ may be written as

$$\mathbf{v} = \|\mathbf{v}\|(\cos \theta)\mathbf{i} + \|\mathbf{v}\|(\sin \theta)\mathbf{j}$$

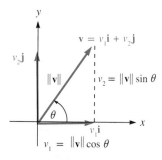

Figure 7.39

Consideration of Figure 7.39 also indicates why v_1 is called the **horizontal** or *x* **component** of $v_1\mathbf{i} + v_2\mathbf{j}$, while v_2 is called the **vertical** or *y* **component** of this vector.

EXAMPLE 6   Express **v** as a linear combination of **i** and **j** if the magnitude of **v** is 8 and the direction angle of **v** is 120°.

Solution   Using $\|\mathbf{v}\| = 8$ and $\theta = 120°$ in the above form of **v** gives

$$
\begin{aligned}
\mathbf{v} &= \|\mathbf{v}\|(\cos\theta)\mathbf{i} + \|\mathbf{v}\|(\sin\theta)\mathbf{j} \\
&= 8(\cos 120°)\mathbf{i} + 8(\sin 120°)\mathbf{j} \\
&= 8\left(-\frac{1}{2}\right)\mathbf{i} + 8\left(\frac{\sqrt{3}}{2}\right)\mathbf{j} \\
&= -4\mathbf{i} + 4\sqrt{3}\mathbf{j}.
\end{aligned}
$$ ■

In the next example we redo Example 8 from the previous section using our current notation and methods. Many applied problems may be solved in this way.

EXAMPLE 7   Vector **a** has a magnitude of 5.0 lb and a direction angle of 35°. Vector **b** has a magnitude of 9.0 lb and a direction angle of 160°. Find the vector sum or resultant **v** of vectors **a** and **b** and give the magnitude and direction angle of **v**.

Solution   First, sketch Figure 7.40. Then use the methods of Example 6 to write **a** and **b** as linear combinations of **i** and **j**.

$$
\begin{aligned}
\mathbf{a} &= 5(\cos 35°)\mathbf{i} + 5(\sin 35°)\mathbf{j} = 4.096\mathbf{i} + 2.868\mathbf{j} \\
\mathbf{b} &= 9(\cos 160°)\mathbf{i} + 9(\sin 160°)\mathbf{j} = -8.457\mathbf{i} + 3.078\mathbf{j}
\end{aligned}
$$

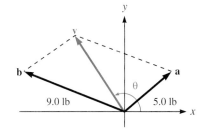

Figure 7.40

The vector sum **v** of **a** and **b** is given by

$$
\begin{aligned}
\mathbf{v} &= \mathbf{a} + \mathbf{b} \\
&= (4.096\mathbf{i} + 2.868\mathbf{j}) + (-8.457\mathbf{i} + 3.078\mathbf{j}) \\
&= -4.361\mathbf{i} + 5.946\mathbf{j},
\end{aligned}
$$

and the magnitude of the resultant is

$$
\|\mathbf{v}\| = \sqrt{v_1{}^2 + v_2{}^2} = \sqrt{(-4.361)^2 + (5.946)^2} = 7.4.
$$

Now determine the direction angle.

$$
\tan\theta = \frac{v_2}{v_1} = \frac{5.946}{-4.361} = -1.363
$$

Since $(-4.361, 5.946)$ is in Q_2 and the reference angle (by calculator) is 54°, we conclude

$$
\theta = 180° - 54° = 126°.
$$

In summary, the vector sum or resultant is $-4.361\mathbf{i} + 5.946\mathbf{j}$ and the magnitude and direction angle of the resultant are 7.4 lb and 126°, respectively.   ■

EXERCISES 7.4

In Exercises 1–10 find the magnitude and the direction angle of each vector.

1. $\langle -1,1 \rangle$ **2.** $\langle 2,2 \rangle$
3. $\langle -\sqrt{3},-1 \rangle$ **4.** $\langle 3,-3\sqrt{3} \rangle$
5. $\langle 0,1 \rangle$ **6.** $\langle -5,0 \rangle$
7. $\langle 5,-12 \rangle$ **8.** $\langle -4,-3 \rangle$
9. $\langle \frac{3}{5},\frac{4}{5} \rangle$ **10.** $\langle -\frac{1}{2}\sqrt{3},\frac{1}{2} \rangle$

In Exercises 11–22 let $v = \langle 4,-3 \rangle$ and $w = \langle -5,1 \rangle$ and find the vectors.

11. $v + w$ **12.** $w + v$
13. $3w$ **14.** $7v$
15. $-5v + 2w$ **16.** $2v + 3w$
17. $-w$ **18.** $-v$
19. $v - w$ **20.** $w - v$
21. $5w - 2v$ **22.** $3v - 4w$

In Exercises 23–26 sketch vectors corresponding to v, w, $v + w$, $2v$, $-v$, and $w - v$.

23. $v = \langle 1,0 \rangle$, $w = \langle 0,1 \rangle$
24. $v = \langle 0,-2 \rangle$, $w = \langle -2,0 \rangle$
25. $v = \langle 3,2 \rangle$, $w = \langle 1,4 \rangle$
26. $v = \langle -2,4 \rangle$, $w = \langle 5,-1 \rangle$

In Exercises 27–32 express (a) $v + w$, (b) $v - w$, and (c) $5w - 2v$ as linear combinations of i and j.

27. $v = i + j$, $w = 3i + 2j$
28. $v = i - j$, $w = -i - j$
29. $v = i - 3j$, $w = -5i + 2j$
30. $v = 4i - j$, $w = -i + 7j$
31. $v = i$, $w = 5j$
32. $v = -2i$, $w = j$

In Exercises 33–40 express the vectors as linear combinations of i and j.

33. $\langle 4,7 \rangle$ **34.** $\langle -9,-2 \rangle$
35. $\langle 0,-5 \rangle$ **36.** $\langle 3,0 \rangle$
37. magnitude 6, direction angle 180°
38. magnitude 3, direction angle 300°
39. magnitude 25, direction angle 120°
40. magnitude 11, direction angle 201°

In Exercises 41–46 find the horizontal and vertical component of each vector.

41. $2i - 8j$ **42.** $-i + 7j$
43. magnitude 5, direction angle 135°
44. magnitude 17, direction angle 270°
45. magnitude 8, direction angle 330°
46. magnitude 16, direction angle 303°

In Exercises 47–50 find the magnitude and direction angle of v.

47. $v = 3i - 3j$ **48.** $v = -i - j$
49. $v = 7j$ **50.** $v = -10i$

In Exercises 51–57 prove the given properties (which were stated in this section), where $u = \langle u_1,u_2 \rangle$, $v = \langle v_1,v_2 \rangle$, $w = \langle w_1,w_2 \rangle$, and c and d are scalars.

51. $u + 0 = u$ **52.** $u + (-u) = 0$
53. $u + v = v + u$ **54.** $1u = u$
55. $c(du) = (cd)u$ **56.** $(c + d)u = cu + du$
57. $(u + v) + w = u + (v + w)$

In Exercises 58–62 prove the following additional properties of vector addition and scalar multiplication.

58. $0u = 0$ **59.** $-u = (-1)u$
60. $c(u - v) = cu - cv$ **61.** $\|-2v\| = 2\|v\|$
62. The magnitude of cv is $|c|$ times the magnitude of v. That is, $\|cv\| = |c| \|v\|$.
63. Prove the vector i is a unit vector.
64. Prove the vector j is a unit vector.

The **dot product** $u \cdot v$ of two vectors $u = \langle u_1,u_2 \rangle$ and $v = \langle v_1,v_2 \rangle$ is defined by

$$u \cdot v = u_1v_1 + u_2v_2.$$

Note the dot product of two vectors is a real number (not a vector). For this reason the dot product is sometimes called the **scalar product**. In Exercises 65–68 find $u \cdot v$ for the given vectors.

65. $u = \langle 4,5 \rangle$, $v = \langle 3,-2 \rangle$
66. $u = \langle 3,-4 \rangle$, $v = \langle -4,-3 \rangle$
67. $u = 3i$, $v = 4j$
68. $u = i - j$, $v = 2i + 7j$

In Exercises 69–74 find the vector sum or resultant v of the given vectors and give the magnitude and direction of v.

	Vector	Magnitude	Direction
69.	a	5.0 lb	26°
	b	3.0 lb	84°
70.	c	6.0 lb	95°
	d	9.0 lb	15°
71.	c	12 lb	110°
	d	15 lb	180°
72.	a	4.0 lb	90°
	b	3.0 lb	190°
73.	a	$1\bar{0}$ lb	42°
	b	$2\bar{0}$ lb	140°
	c	$3\bar{0}$ lb	240°

74.

a	5.0 lb	60°
b	2.0 lb	210°
c	1.0 lb	270°

In Exercises 75–78 use the methods of this section to solve the problem. (*Note:* Alternative methods for solving these problems were shown in Examples 2, 3, 5, and 4 of the previous section.)

75. A force of 3.0 lb and a force of 4.0 lb are acting on a body with an angle of 90° between the two forces. Find (a) the magnitude of the resultant and (b) the angle between the resultant and the larger force.

76. A ship is headed due east at $2\bar{0}$ knots (nautical miles per hour) while the current carries the ship due south at 5.0 knots. Find (a) the speed of the ship and (b) the direction (course) of the ship.

77. An airplane, pointed due west, is traveling 10° north of west at a rate of $4\bar{0}0$ mi/hr. This resultant course is due to a wind blowing north. Find (a) the velocity of the plane if there were no wind (that is, the vector pointing due west) and (b) the velocity of the wind (that is, the vector pointing due north).

78. A body is acted on by two forces with magnitudes of 78 lb and 42 lb that act at an angle of 25° with each other, as was shown in Figure 7.27. Find (a) the magnitude of the resultant and (b) the angle between the resultant and the larger force.

THINK ABOUT IT

1. Show that if **v** is a nonzero vector and $\mathbf{v} = v_1\mathbf{i} + v_2\mathbf{j}$, then $\mathbf{u} = (v_1/\|\mathbf{v}\|)\mathbf{i} + (v_2/\|\mathbf{v}\|)\mathbf{j}$ is a unit vector having the same direction as **v**.
2. Use the result in Question 2 and find a unit vector having the same direction as the given vector.
   a. $\mathbf{v} = 3\mathbf{i} + 4\mathbf{j}$ **b.** $\mathbf{v} = \mathbf{i} - \mathbf{j}$
3. Use the law of cosines and the definition of dot product in Exercises 65–68 to show that if θ is the angle between two nonzero vectors $\mathbf{u} = \langle u_1,u_2\rangle$ and $\mathbf{v} = \langle v_1,v_2\rangle$, as shown below, then
$$\mathbf{u} \cdot \mathbf{v} = \|\mathbf{u}\|\,\|\mathbf{v}\|\cos\theta.$$

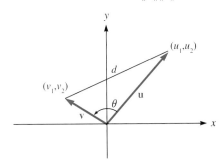

(*Note:* The **angle between two nonzero vectors u** and **v** is the angle θ, satisfying $0° \leq \theta \leq 180°$, that is determined by the position vectors of **u** and **v**.)

4. Two nonzero vectors are called **orthogonal** if and only if the angle between them measures 90°.
   a. Show that two nonzero vectors **u** and **v** are orthogonal if and only if $\mathbf{u} \cdot \mathbf{v} = 0$.
   b. Show that $\mathbf{u} = 2\mathbf{i} + 7\mathbf{j}$ and $\mathbf{v} = -7\mathbf{i} + 2\mathbf{j}$ are orthogonal.
5. Use the definition of dot product in Exercises 65–68 to prove the following properties of the dot product, where $\mathbf{u} = \langle u_1,u_2\rangle$, $\mathbf{v} = \langle v_1,v_2\rangle$, $\mathbf{w} = \langle w_1,w_2\rangle$, and c is a scalar.
   1. $\mathbf{u} \cdot \mathbf{v} = \mathbf{v} \cdot \mathbf{u}$
   2. $\mathbf{v} \cdot \mathbf{v} = \|\mathbf{v}\|^2$
   3. $c(\mathbf{u} \cdot \mathbf{v}) = (c\mathbf{u}) \cdot \mathbf{v}$
   4. $\mathbf{u} \cdot (\mathbf{v} + \mathbf{w}) = \mathbf{u} \cdot \mathbf{v} + \mathbf{u} \cdot \mathbf{w}$

REMEMBER THIS

1. To solve triangle ABC when given a, c, and B, do we first apply the law of sines or the law of cosines?
2. How many triangles satisfy the conditions that $C = 32°$, $b = 61$ ft, and $c = 47$ ft?
3. In triangle ABC if $a = 5.0$ m, $b = 8.0$ m, and $c = 5.0$ m, then find B.
4. In triangle RST if $R = 45°$, $S = 60°$, and $r = 16$ ft, then find t.

5. What is the horizontal component of a vector with a magnitude of 45 lb which makes an angle of 35° with the positive x-axis?
6. Explain geometrically why $|-3| = |3|$.
7. What is the conjugate of the complex number $3 + 4i$?
8. When does the complex number $a + bi$ equal the complex number $c + di$?
9. Write the formula for $\sin(\theta_1 + \theta_2)$.
10. Write the formula for $\cos(\theta_1 + \theta_2)$.

7.5   Trigonometric Form of Complex Numbers

Each complex number $a + bi$ involves a pair of real numbers, a and b. Graphically, this means that we may represent a complex number as a point in a rectangular coordinate system. The values for a are plotted on the horizontal axis (x-axis) and the values for b are plotted on the vertical axis (y-axis). Thus, the complex number $a + bi$ is represented by the point (a,b) with x value a and y value b. The plane on which complex numbers are graphed is called the **complex plane,** and in this context the horizontal axis is called the **real axis** and the vertical axis is called the **imaginary axis.** In Figure 7.41 the geometric representations of several complex numbers are shown.

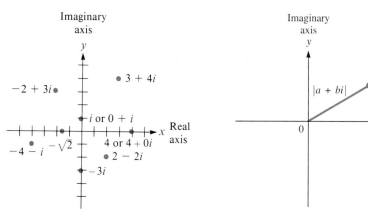

Figure 7.41   **Figure 7.42**

The **absolute value** or **modulus** of a complex number $a + bi$, denoted $|a + bi|$, is interpreted geometrically as the distance from the origin to the point (a,b), as shown in Figure 7.42. Thus, algebraically $|a + bi|$ is defined by

$$|a + bi| = \sqrt{a^2 + b^2}.$$

EXAMPLE 1   Graph $-3 + 2i$ and find $|-3 + 2i|$.

Solution   The complex number $-3 + 2i$, which is represented by the point $(-3,2)$, is graphed in Figure 7.43. Since $a = -3$ and $b = 2$,

$$|-3 + 2i| = \sqrt{(-3)^2 + 2^2} = \sqrt{13}.$$

Thus, the absolute value of $-3 + 2i$ is $\sqrt{13}$. ∎

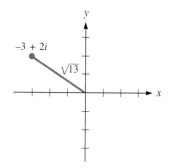

Figure 7.43

EXAMPLE 2   Show that if $z = a + bi$ is a real number, then $|z| = |a|$.

Solution   If $z = a + bi$ is a real number, then $b = 0$, so

$$|a + bi| = |a + 0i| = \sqrt{a^2 + 0^2} = \sqrt{a^2} = |a|.$$

Thus, $|z| = |a|$. This example shows that the definition of the absolute value of a complex number is in agreement with the definition of the absolute value of a real number.   ■

A useful new way of writing a complex number is to interpret the position of its geometric representation in terms of the trigonometric functions. Consider the point corresponding to the nonzero complex number $a + bi$ in Figure 7.44. When we use trigonometry, we obtain the following relationships.

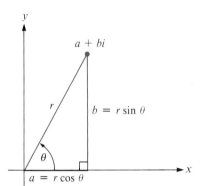

Figure 7.44

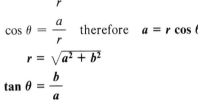

$$\sin \theta = \frac{b}{r} \quad \text{therefore} \quad b = r \sin \theta$$

$$\cos \theta = \frac{a}{r} \quad \text{therefore} \quad a = r \cos \theta$$

$$r = \sqrt{a^2 + b^2}$$

$$\tan \theta = \frac{b}{a}$$

Thus, we may change the form of the complex number as follows:

$$a + bi = r \cos \theta + (r \sin \theta)i = r(\cos \theta + i \sin \theta).$$

In this trigonometric form r is the absolute value of $a + bi$ and θ is called the **argument** of the complex number. Note that the argument is not unique, since all angles of the form $\theta + k \cdot 360°$, where k is an integer, are coterminal to θ and serve as suitable choices for the argument. In most cases we use the **principal argument** of the complex number, which is the unique choice for θ that satisfies $0° \le \theta < 360°$.

EXAMPLE 3   Write the number $3 - 3i$ in trigonometric form.

Solution   First, determine the absolute value, r. Since $a = 3$ and $b = -3$, we have

$$r = \sqrt{a^2 + b^2} = \sqrt{(3)^2 + (-3)^2} = \sqrt{18} \quad \text{or} \quad 3\sqrt{2}.$$

Now determine the argument.

$$\tan \theta = \frac{b}{a} = \frac{-3}{3} = -1$$

Since $(3, -3)$ is in Q_4 and the reference angle is $45°$, we conclude that

$$\theta = 315°.$$

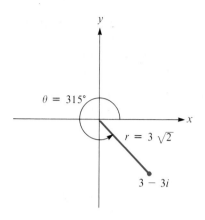

Figure 7.45

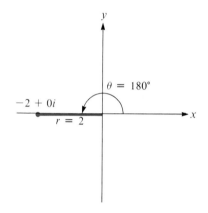

Figure 7.46

The number is written in trigonometric form as

$$3\sqrt{2}(\cos 315° + i \sin 315°).$$

See Figure 7.45. (*Note:* Any angle coterminal with 315° may also be used. For example, we may write $3\sqrt{2}[\cos(-45°) + i \sin(-45°)]$.) ■

EXAMPLE 4   Write the number -2 in trigonometric form.

Solution   First, determine the absolute value, r. Since $a = -2$ and $b = 0$, we have

$$r = \sqrt{a^2 + b^2} = \sqrt{(-2)^2 + 0^2} = \sqrt{4} = 2.$$

The argument is 180° since -2 is a point on the negative portion of the x-axis. The number in trigonometric form is

$$2(\cos 180° + i \sin 180°). \qquad \text{(See Figure 7.46.)} \qquad ■$$

EXAMPLE 5   Write the number $2(\cos 150° + i \sin 150°)$ in the form $a + bi$.

Solution   Since $\cos 150° = -\sqrt{3}/2$ and $\sin 150° = 1/2$, we have

$$2(\cos 150° + i \sin 150°) = 2\left[\frac{-\sqrt{3}}{2} + i\left(\frac{1}{2}\right)\right]$$

$$= -\sqrt{3} + i. \qquad ■$$

Multiplication and division of complex numbers is easy when the numbers are in trigonometric form. We derive the product rule as follows:

$$r_1(\cos \theta_1 + i \sin \theta_1) \cdot r_2(\cos \theta_2 + i \sin \theta_2)$$
$$= r_1r_2[\cos \theta_1 \cos \theta_2 + i \cos \theta_1 \sin \theta_2 + i \sin \theta_1 \cos \theta_2 + i^2 \sin \theta_1 \sin \theta_2].$$

Replacing i^2 by -1 and grouping the terms yields

$$r_1r_2[(\cos \theta_1 \cos \theta_2 - \sin \theta_1 \sin \theta_2) + i(\sin \theta_1 \cos \theta_2 + \cos \theta_1 \sin \theta_2)].$$

We now simplify the expressions in the parentheses by applying the formulas for the sine and cosine of the sum of two angles (see Section 6.2) and obtain our final result.

$$r_1(\cos \theta_1 + i \sin \theta_1) \cdot r_2(\cos \theta_2 + i \sin \theta_2)$$
$$= r_1r_2[\cos(\theta_1 + \theta_2) + i \sin(\theta_1 + \theta_2)]$$

In words, the product of two complex numbers is a third complex number. The absolute value is the product of the absolute values of the given numbers, and the argument is the sum of the arguments of the given numbers.

By a similar procedure we can obtain the following quotient rule.

$$\frac{r_1(\cos \theta_1 + i \sin \theta_1)}{r_2(\cos \theta_2 + i \sin \theta_2)} = \frac{r_1}{r_2}[\cos(\theta_1 - \theta_2) + i \sin(\theta_1 - \theta_2)], \text{ if } r_2 \neq 0$$

In words, the quotient of two complex numbers is a third complex number. The absolute value is the quotient of the absolute values of the given numbers, and the argument is the difference of the arguments of the given numbers.

EXAMPLE 6 Determine

a. the product $z_1 \cdot z_2$
b. the quotient z_1/z_2
c. the quotient z_2/z_1

of the following complex numbers.

$$z_1 = 3(\cos 72° + i \sin 72°), \qquad z_2 = 5(\cos 43° + i \sin 43°)$$

Solution

a. $3(\cos 72° + i \sin 72°) \cdot 5(\cos 43° + i \sin 43°)$
$= 3 \cdot 5[\cos(72° + 43°) + i \sin(72° + 43°)]$
$= 15(\cos 115° + i \sin 115°)$

b. $\dfrac{3(\cos 72° + i \sin 72°)}{5(\cos 43° + i \sin 43°)} = \dfrac{3}{5}[\cos(72° - 43°) + i \sin(72° - 43°)]$

$= \dfrac{3}{5}(\cos 29° + i \sin 29°)$

c. $\dfrac{5(\cos 43° + i \sin 43°)}{3(\cos 72° + i \sin 72°)} = \dfrac{5}{3}[\cos(43° - 72°) + i \sin(43° - 72°)]$

$= \dfrac{5}{3}[\cos(-29°) + i \sin(-29°)]$

or

$= \dfrac{5}{3}(\cos 331° + i \sin 331°)$ ∎

We often work with conjugate complex numbers, and Figure 7.47 shows relations between the complex number $z = a + bi$ and its conjugate $\bar{z} =$

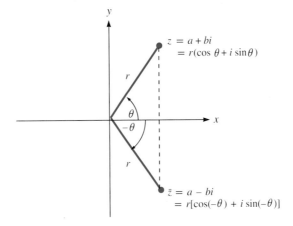

Figure 7.47

$a - bi$. Note that the graphs of z and $\bar{z}$ are symmetric about the real axis and that if $z = r(\cos \theta + i \sin \theta)$, then $\bar{z} = r[\cos(-\theta) + i \sin(-\theta)]$.

Example 7 Let $z = r(\cos \theta + i \sin \theta)$ and show $z \cdot \bar{z}$ is a real number.

Solution Multiplying z and $\bar{z}$ in trigonometric form gives

$$
\begin{aligned}
z \cdot \bar{z} &= r(\cos \theta + i \sin \theta) \cdot r[\cos(-\theta) + i \sin(-\theta)] \\
&= r \cdot r(\cos[\theta + (-\theta)] + i \sin[\theta + (-\theta)]) \\
&= r^2(\cos 0 + i \sin 0) \\
&= r^2(1 + 0i) \\
&= r^2
\end{aligned}
$$

Since r^2, which equals $a^2 + b^2$, is a real number, $z \cdot \bar{z}$ is a real number. ∎

In much of the literature dealing with complex numbers and their applications the expression

$$r(\cos \theta + i \sin \theta)$$

is written as

$$r \text{ cis } \theta \quad \text{or} \quad r \angle \theta.$$

These forms are merely convenient abbreviations and the expressions are interchangeable.

EXERCISES 7.5

In Exercises 1–10 graph each complex number and find its absolute value.

1. $4 + 3i$

2. $5 - 12i$

3. $-1 + 2i$

4. $-2 - i$

5. $\sqrt{3} - i$

6. $\sqrt{2} + \sqrt{2}i$

7. -3

8. 1

9. $4i$

10. $-2i$

In Exercises 11–24 write the number in trigonometric form.

11. 3 **12.** i **13.** $-2i$

14. -4 **15.** $1 - i$ **16.** $-1 + i$

17. $4 + 3i$ **18.** $-5 - 12i$ **19.** $-3 - 2i$

20. $7 - 4i$ **21.** $-1 + \sqrt{3}i$ **22.** $\sqrt{3} - i$

23. $\sqrt{2} - \sqrt{2}i$ **24.** $-2 - 2\sqrt{3}i$

In Exercises 25–34 write the number in the form $a + bi$.

25. $3(\cos 90° + i \sin 90°)$

26. $5(\cos 0° + i \sin 0°)$

27. $4(\cos 180° + i \sin 180°)$

28. $\sqrt{2}(\cos 270° + i \sin 270°)$

29. $\sqrt{3}(\cos 120° + i \sin 120°)$

30. $2(\cos 210° + i \sin 210°)$

31. $2(\cos 225° + i \sin 225°)$

32. $\sqrt{6}(\cos 315° + i \sin 315°)$

33. $\cos 52° + i \sin 52°$

34. $10(\cos 115° + i \sin 115°)$

In Exercises 35–44 find (a) the product $z_1 \cdot z_2$, (b) the quotient z_1/z_2, and (c) the quotient z_2/z_1 of the given complex numbers.

35. $z_1 = 2(\cos 52° + i \sin 52°)$,
$z_2 = 4(\cos 11° + i \sin 11°)$

36. $z_1 = 6(\cos 7° + i \sin 7°)$,
$z_2 = 9(\cos 90° + i \sin 90°)$

37. $z_1 = \cos 90° + i \sin 90°$,
$z_2 = \cos 180° + i \sin 180°$

38. $z_1 = \cos 33° + i \sin 33°$,
$z_2 = 4(\cos 63° + i \sin 63°)$

39. $z_1 = 3(\cos 131° + i \sin 131°)$,
$z_2 = 12(\cos 205° + i \sin 205°)$

40. $z_1 = \cos 270° + i \sin 270°$,
$z_2 = 5(\cos 3° + i \sin 3°)$

41. $z_1 = 2(\cos 300° + i \sin 300°)$,
$z_2 = \sqrt{2}(\cos 45° + i \sin 45°)$

42. $z_1 = 3\sqrt{2}(\cos 315° + i \sin 315°)$,
$z_2 = \cos 0° + i \sin 0°$

43. $z_1 = \cos(-20°) + i \sin(-20°)$,
$z_2 = \cos(-45°) + i \sin(-45°)$

44. $z_1 = 3(\cos 0° + i \sin 0°)$,
$z_2 = 4[\cos(-90°) + i \sin(-90°)]$

In Exercises 45 and 46 let $z = a + bi$ and establish each result.

45. $|\bar{z}| = |z|$ **46.** $|z| = \sqrt{z\bar{z}}$

In Exercises 47–50 let $z = r(\cos \theta + i \sin \theta)$ and establish each result.

47. $-z = r[\cos(\theta + \pi) + i \sin(\theta + \pi)]$, where
$-z = -1 \cdot z$

48. $z^{-1} = r^{-1}[\cos(-\theta) + i \sin(-\theta)]$, where $z^{-1} = 1/z$
and $z \neq 0$

49. $z^2 = r^2(\cos 2\theta + i \sin 2\theta)$

50. $z^3 = r^3(\cos 3\theta + i \sin 3\theta)$

51. If z and $\bar{z}$ are nonzero complex numbers, find $z/\bar{z}$ in trigonometric form. (*Note:* See Figure 7.47.)

THINK ABOUT IT

1. Graph all complex numbers z that satisfy the given condition.
 a. $|z| = 1$ **b.** $z = \bar{z}$
 c. The real part and the imaginary part of z are equal.

2. Multiply $z = r(\cos \theta + i \sin \theta)$ by i in trigonometric form. Use the result and the diagram below to describe what happens geometrically when a complex number is multiplied by i.

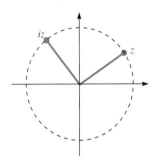

3. Let complex numbers $z_1 = r_1(\cos \theta_1 + i \sin \theta_1)$ and $z_2 = r_2(\cos \theta_2 + i \sin \theta_2) \neq 0$ and show that
$$\frac{z_1}{z_2} = \frac{r_1}{r_2}[\cos(\theta_1 - \theta_2) + i \sin(\theta_1 - \theta_2)].$$

4. Apply the product rule to the expression
$[\cos \theta + i \sin \theta][\cos(-\theta) + i \sin(-\theta)]$ and show that
$$(\cos \theta + i \sin \theta)^{-1} = \cos(-\theta) + i \sin(-\theta).$$

5. Use the result in Question 4 and the product rule to establish the quotient rule stated in Question 3.

REMEMBER THIS

1. In triangle PQR express $\cos R$ in terms of p, q, and r.

2. Find the magnitude and the direction angle of $\mathbf{v}$ if
$\mathbf{v} = -5\mathbf{i} + 5\mathbf{j}$.

3. If $\mathbf{v} = \langle 2, -5 \rangle$ and $\mathbf{w} = \langle -7, 1 \rangle$, find $2\mathbf{w} - \mathbf{v}$.

4. Assume exact numbers and find the exact area in an isosceles triangle in which the vertex angle measures 135° and each leg measures 6 units.

5. What is the absolute value of the complex number $4(\cos 150° + i \sin 150°)$?

6. Express $64^{1/3}$ in radical form and then simplify.

7. Find the exact value of $\sin 675°$.

8. Write the complex number $-4i$ in trigonometric form.

9. Find the product in the form $a + bi$:
$$[5(\cos 15° + i \sin 15°)][5(\cos 15° + i \sin 15°)].$$

10. Find all solutions of the equation $x^2 + 1 = 0$.

7.6 De Moivre's Theorem and *n*th Roots

When a complex number z is expressed in trigonometric form, it is easy to find powers of z. To establish a pattern that suggests a formula for z^n, we begin by considering Exercise 49 in the prevous section and we verify a formula for z^2 when $z = \cos \theta + i \sin \theta$.

$$\begin{aligned} z^2 &= [r(\cos \theta + i \sin \theta)][r(\cos \theta + i \sin \theta)] \\ &= r \cdot r[\cos(\theta + \theta) + i \sin(\theta + \theta)] \\ &= r^2(\cos 2\theta + i \sin 2\theta) \end{aligned}$$

Similarly, by writing z^3 as $z^2 \cdot z$ and applying the product rule, we obtain

$$z^3 = r^3(\cos 3\theta + i \sin 3\theta).$$

Continuing in this way suggests

$$z^n = r^n(\cos n\theta + i \sin n\theta)$$

is true for every positive integer n, and in fact, this pattern generalizes to the following statement, which is known as De Moivre's theorem.

De Moivre's Theorem

If $r(\cos \theta + i \sin \theta)$ is any complex number and n is any real number, then

$$[r(\cos \theta + i \sin \theta)]^n = r^n(\cos n\theta + i \sin n\theta).$$

Although advanced mathematics is required to prove the above result, De Moivre's formula may be proved for every positive integer n by using mathematical induction (see Section 10.4), and the result may be extended to all integral exponents, as discussed in "Think About It" Question 5.

EXAMPLE 1   Find $(1 + i)^{12}$.

Solution   First, write the number in trigonometric form.

$$r = \sqrt{a^2 + b^2} = \sqrt{(1)^2 + (1)^2} = \sqrt{2}$$

$$\tan \theta = \frac{b}{a} = \frac{1}{1} = 1 \qquad \theta = 45°$$

Thus, $1 + i = \sqrt{2}(\cos 45° + i \sin 45°)$. Now De Moivre's theorem tells us that

$$\begin{aligned} [\sqrt{2}(\cos 45° + i \sin 45°)]^{12} &= (\sqrt{2})^{12}[\cos(12 \cdot 45°) + i \sin(12 \cdot 45°)] \\ &= (2^{1/2})^{12}[\cos 540° + i \sin 540°] \\ &= 2^6[(-1) + i(0)] \\ &= -64. \end{aligned}$$

Analogous to roots of real numbers, an ***n*th root of a complex number** z is a complex number w such that

$$w^n = z,$$

where n is a positive integer. Any complex number has two square roots, three cube roots, four fourth roots, and so on. To find these roots we use De Moivre's theorem. However, to find all of the roots, you must remember that there are many trigonometric representations for the same complex number. For example, since the angle $\theta + k \cdot 360°$ (k any integer) is coterminal to θ, we have

$$1 + i = \sqrt{2}(\cos 45° + i \sin 45°) = \sqrt{2}(\cos 405° + i \sin 405°)$$
$$= \sqrt{2}(\cos 765° + i \sin 765°), \text{ and so on.}$$

Thus, the method is to find one root, add $360°$ to θ, use the new trigonometric representation to find another root, and repeat this procedure until all n roots are found.

EXAMPLE 2   Find and graph the five fifth roots of 32.

Solution   First, we write 32 in trigonometric form as

$$32(\cos 0° + i \sin 0°).$$

Now applying De Moivre's theorem, we have

$$[32(\cos 0° + i \sin 0°)]^{1/5} = 32^{1/5}\left(\cos \frac{0°}{5} + i \sin \frac{0°}{5}\right)$$
$$= 2(\cos 0° + i \sin 0°) = 2.$$

The first root is 2. To find another root, first add $360°$ to θ to obtain a different trigonometric representation for 32.

$$32 = 32[\cos(0° + 360°) + i \sin(0° + 360°)]$$

Then by De Moivre's theorem

$$\text{2nd root is } 32^{1/5}\left(\cos\frac{0° + 360°}{5} + i \sin\frac{0° + 360°}{5}\right)$$
$$= 2(\cos 72° + i \sin 72°).$$

Repeating this procedure, we have

$$\text{3rd root is } 32^{1/5}\left(\cos\frac{0° + 2 \cdot 360°}{5} + i \sin\frac{0° + 2 \cdot 360°}{5}\right)$$
$$= 2(\cos 144° + i \sin 144°)$$
$$\text{4th root is } 32^{1/5}\left(\cos\frac{0° + 3 \cdot 360°}{5} + i \sin\frac{0° + 3 \cdot 360°}{5}\right)$$
$$= 2(\cos 216° + i \sin 216°)$$
$$\text{5th root is } 32^{1/5}\left(\cos\frac{0° + 4 \cdot 360°}{5} + i \sin\frac{0° + 4 \cdot 360°}{5}\right)$$
$$= 2(\cos 288° + i \sin 288°).$$

The five roots are graphed in Figure 7.48, which illustrates that the roots all lie on a circle of radius 2 centered at O and that the arguments of consecutive roots differ by 72°. We do not have to consider arguments outside of the interval $[0°, 360°)$ because the associated graphs will repeat points already obtained.

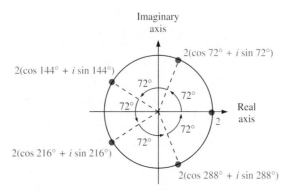

Figure 7.48

The type of problem considered in Example 2 may be solved more efficiently by generalizing our current methods. Because $\theta + k \cdot 360°$, where k is an integer, generates all angles coterminal to θ, the complex number $z = r(\cos \theta + i \sin \theta)$ may also be represented by

$$z = r[\cos(\theta + k \cdot 360°) + i \sin(\theta + k \cdot 360°)].$$

Then, by De Moivre's theorem this number may be written as

$$z = \left[r^{1/n}\left(\cos \frac{\theta + k \cdot 360°}{n} + i \sin \frac{\theta + k \cdot 360°}{n} \right) \right]^n.$$

Since $w^n = z$, where n is a positive integer, implies w is an nth root of z, every complex number of the form

$$r^{1/n}\left(\cos \frac{\theta + k \cdot 360°}{n} + i \sin \frac{\theta + k \cdot 360°}{n} \right)$$

is an nth root of z. Furthermore, we obtain n distinct nth roots of z for $k = 0, 1, 2, \ldots, n - 1$. Other integral values for k will lead to arguments that are coterminal to those previously obtained and no different nth roots will be derived. Thus, we may find all of the nth roots of a complex number by applying the following formula.

nth Root Formula

> The n distinct nth roots of the complex number $r(\cos\theta + i\sin\theta)$ are given by
>
> $$\sqrt[n]{r}\left(\cos\frac{\theta + k\cdot 360°}{n} + i\sin\frac{\theta + k\cdot 360°}{n}\right),$$
>
> where $k = 0,1,2,\ldots,n-1$.

The geometric interpretation of the above result is that the graphs of the nth roots all lie on a circle of radius $\sqrt[n]{r}$ that is centered at the origin. Also, because the arguments of consecutive roots differ by $360°/n$, the nth roots are equally spaced on this circle.

EXAMPLE 3   Find and graph the three cube roots of $-8i$.

Solution   First, we write $-8i$ in trigonometric form as

$$8(\cos 270° + i\sin 270°).$$

By the nth root formula, the three distinct cube roots are given by

$$\sqrt[3]{8}\left(\cos\frac{270° + k\cdot 360°}{3} + i\sin\frac{270° + k\cdot 360°}{3}\right), \quad\text{where}\quad k = 0,1,2,$$

which simplifies to

$$2[\cos(90° + k\cdot 120°) + i\sin(90° + k\cdot 120°)], \quad\text{where}\quad k = 0,1,2.$$

Finally, replacing k by 0, 1, and 2 yields the roots

$$2(\cos 90° + i\sin 90°) = 2i$$
$$2(\cos 210° + i\sin 210°) = -\sqrt{3} - i$$
$$2(\cos 330° + i\sin 330°) = \sqrt{3} - i.$$

The points determined by these roots are on a circle of radius 2, as shown in Figure 7.49.   ∎

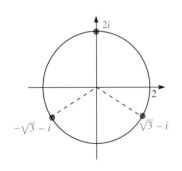

Figure 7.49

The nth root formula may be used to solve equations that may be written in the form

$$x^n = z, \quad\text{where n is a positive integer,}$$

since the roots of this equation are the n distinct nth roots of z. Note in the next example that the solutions are the sixth roots of 1, which is a special case. The nth roots of 1 are called the **nth roots of unity,** and such roots are especially useful in higher mathematics.

EXAMPLE 4   Find all solutions of the equation $x^6 = 1$.

Solution To solve $x^6 = 1$, we find the six 6th roots of 1. In trigonometric form 1 is written as $1(\cos 0° + i \sin 0°)$, so the nth root formula with $n = 6$ yields

$$\sqrt[6]{1}\left(\cos \frac{0° + k \cdot 360°}{6} + i \sin \frac{0° + k \cdot 360°}{6}\right), \quad \text{where} \quad k = 0,1,2,3,4,5,$$

which simplifies to

$$\cos(0° + k \cdot 60°) + i \sin(0° + k \cdot 60°), \quad \text{where} \quad k = 0,1,2,3,4,5.$$

Then, the six roots are

$$\cos 0° + i \sin 0° = 1 \qquad \text{(for } k = 0)$$

$$\cos 60° + i \sin 60° = \frac{1}{2} + \frac{\sqrt{3}}{2}i \qquad \text{(for } k = 1)$$

$$\cos 120° + i \sin 120° = -\frac{1}{2} + \frac{\sqrt{3}}{2}i \qquad \text{(for } k = 2)$$

$$\cos 180° + i \sin 180° = -1 \qquad \text{(for } k = 3)$$

$$\cos 240° + i \sin 240° = -\frac{1}{2} - \frac{\sqrt{3}}{2}i \qquad \text{(for } k = 4)$$

$$\cos 300° + i \sin 300° = \frac{1}{2} - \frac{\sqrt{3}}{2}i \qquad \text{(for } k = 5)$$

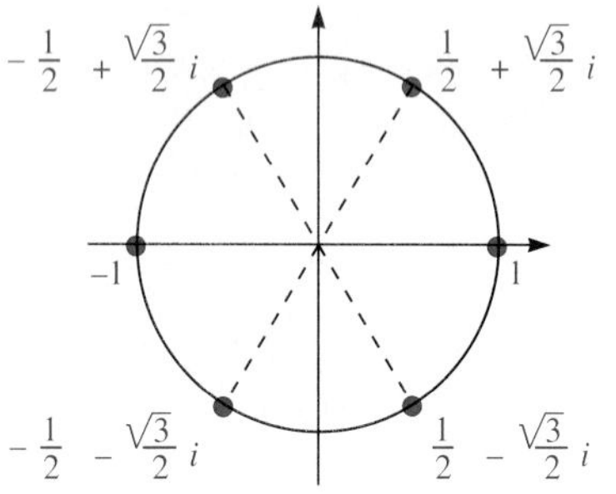

The six roots, which are the six 6th roots of unity, are graphed in Figure 7.50.

Figure 7.50

EXERCISES 7.6

In Exercises 1–14 use De Moivre's theorem and express the result in the form $a + bi$.

1. $[2(\cos 10° + i \sin 10°)]^3$
2. $[3(\cos 15° + i \sin 15°)]^4$
3. $[4(\cos 135° + i \sin 135°)]^5$
4. $[2(\cos 225° + i \sin 225°)]^6$
5. $(1 + i)^8$
6. $(-1 + i)^6$
7. $(1 - \sqrt{3}i)^5$
8. $(-\sqrt{3} - i)^7$
9. $\left(\frac{\sqrt{2}}{2} + \frac{\sqrt{2}}{2}i\right)^{16}$
10. $\left(\frac{\sqrt{2}}{2} - \frac{\sqrt{2}}{2}i\right)^{14}$
11. $(1 + \sqrt{3}i)^{-1}$
12. $(1 - i)^{-1}$
13. $(-\sqrt{2} + \sqrt{2}i)^{-3}$
14. $(-\sqrt{3} + i)^{-4}$

In Exercises 15–26, find and graph all roots. Express the answers in both trigonometric form and in the form $a + bi$.

15. the square roots of $25(\cos 60° + i \sin 60°)$
16. the cube roots of $8(\cos 135° + i \sin 135°)$
17. the fourth roots of $16(\cos 80° + i \sin 80°)$
18. the square roots of $9(\cos 70° + i \sin 70°)$
19. the cube roots of 1
20. the fourth roots of -1

21. the square roots of $-i$
22. the square roots of i
23. the fourth roots of $-16i$
24. the cube roots of $8i$
25. the fifth roots of $1 + \sqrt{3}i$
26. the fifth roots of $1 + i$

In Exercises 27–32 find all solutions of the equation. Leave the answers in trigonometric form except for answers that may be expressed exactly in $a + bi$ form.

27. $x^3 + 8 = 0$
28. $x^3 - 27i = 0$
29. $x^4 = 1$
30. $x^5 = 1$
31. $x^5 + 1 = 0$
32. $x^5 + 32 = 0$
33. Find and graph the fifth roots of unity.
34. Find and graph the fourth roots of unity.

THINK ABOUT IT

1. a. Solve $x^3 - 1 = 0$ using factoring and the quadratic formula. Verify that the methods of this section produce the same roots.

   b. Solve $x^3 + 1 = 0$ using factoring and the quadratic formula. Verify that the methods of this section produce the same roots.

2. a. Show that $\sin 2\theta = 2 \sin \theta \cos \theta$ by applying De Moivre's theorem to $(\cos \theta + i \sin \theta)^2$.

   b. Show that $\cos 2\theta = \cos^2 \theta - \sin^2 \theta$ by applying De Moivre's theorem to $(\cos \theta + i \sin \theta)^2$.

3. Show that $\cos 3\theta = 4 \cos^3 \theta - 3 \cos \theta$ by applying De Moivre's theorem to $(\cos \theta + i \sin \theta)^3$.

4. Show that $\sin 3\theta = 3 \sin \theta - 4 \sin^3 \theta$ by applying De Moivre's theorem to $(\cos \theta + i \sin \theta)^3$.

5. a. By definition, if z is a nonzero complex number, then $z^0 = 1 + 0i = 1$. Show that De Moivre's theorem holds in the case $n = 0$.

   b. By definition, if z is a nonzero complex number and n is a positive integer, then $z^{-n} = 1/z^n$. Show that if De Moivre's theorem holds for positive integers, then it holds for negative integers.

REMEMBER THIS

1. Assume exact numbers and find the exact area in an equilateral triangle in which each side measures 8 units.

2. In triangle PQR express $\sin Q$ in terms of p, q, and $\sin P$.

3. Find the magnitude and the direction angle of $\mathbf{v}$ if $\mathbf{v} = \langle 2, -2\sqrt{3} \rangle$.

4. Express $\mathbf{v}$ as a linear combination of $\mathbf{i}$ and $\mathbf{j}$ if the magnitude of $\mathbf{v}$ is 2 and the direction angle is 315°.

5. Two forces with magnitudes of 55 lb and 65 lb are applied to the same point. If the angle between them measures 78°, find the magnitude of the resultant.

6. Give the measures of three angles coterminal to 45°.

7. Simplify $\dfrac{11\pi}{6} - \pi$.

8. Evaluate $\sin 120°$ and $\sin 300°$ to two significant digits.

9. For what values of θ in the interval $[0°, 360°)$ does $y = \sin 2\theta$ reach its minimum value?

10. Show $\sin(180° - \theta) = \sin \theta$ is an identity.

7.7   **Polar Coordinates**

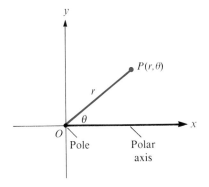

Figure 7.51

Up to now we have specified points in the plane by using the Cartesian (or rectangular) coordinate system. Another important system that can be used to obtain a picture of a relationship is the **polar coordinate system.** In this system we use a fixed point O, called the **pole** or **origin,** and a ray that extends from O and forms an axis, called the **polar axis.** The pole and polar axis correspond to the origin and positive x-axis of the rectangular coordinate system, as shown in Figure 7.51. Now any point P in the plane may be represented by an ordered pair of numbers (r, θ), called **polar coordinates,** where r gives the distance from O to P and θ gives an angle measure from the polar axis to ray OP. Example 1 graphs a few points that are given in polar coordinate form.

EXAMPLE 1   Plot the points $(3, 45°)$, $(1, 7\pi/6)$, $(2, -60°)$, and $(0, \pi)$.

Solution   The points are shown in Figure 7.52. Note that polar coordinates may use degree or radian angle measures that may be positive or negative.

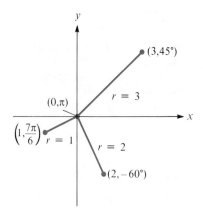

Figure 7.52

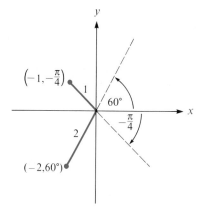

Figure 7.53

Also, the point $(0,\pi)$ is the pole. In general, $(0,\theta)$ is the pole for all values of θ.   ■

Negative values for r often arise when graphing equations like $r = \sin\theta$ in polar coordinates, since negative values for the trigonometric functions are common. If $r < 0$, then we plot (r,θ) by measuring $|r|$ units in a direction opposite to the terminal ray of θ. Example 2 illustrates this interpretation.

EXAMPLE 2   Plot the points $(-2,60°)$ and $(-1,-\pi/4)$.

Solution   The point $(-2,60°)$ is located 2 units in the opposite direction from the terminal ray of a $60°$ angle, while $(-1,-\pi/4)$ is located 1 unit along the extension of the terminal ray of $-\pi/4$. Both points are shown in Figure 7.53.   ■

There are many ways to represent a given point in polar coordinates, and we need to adjust to this. For instance, the point at $(1,30°)$ may also be described as

$$(1,390°), (1,750°), (1,-330°), (-1,210°), \text{ and so on.}$$

In general, different representations for a given point are obtained as follows:

$$(r,\theta) = (r,\theta + n \cdot 360°), \text{ where } n \text{ is an integer}$$
$$(-r,\theta) = (r,\theta + n \cdot 180°), \text{ where } n \text{ is an odd integer.}$$

Basically, these rules tell us that when the angle changes by $360°$, r remains the same; but when the angle changes by $180°$, r changes sign.

EXAMPLE 3   Express the points $(2,-45°)$, $(-1,2\pi/3)$, and $(-3,7\pi/6)$ in polar coordinates with $r \geq 0$ and $0° \leq \theta < 360°$ (or $0 \leq \theta < 2\pi$).

Solution   We satisfy the given conditions as follows:

$$(2,-45°) = (2,-45° + 360°) = (2,315°)$$
$$\left(-1,\frac{2\pi}{3}\right) = \left(-(-1),\frac{2\pi}{3} + \pi\right) = \left(1,\frac{5\pi}{3}\right)$$
$$\left(-3,\frac{7\pi}{6}\right) = \left(-(-3),\frac{7\pi}{6} - \pi\right) = \left(3,\frac{\pi}{6}\right).$$ ■

Polar equations relating r and θ, and graphed in polar coordinates, provide an efficient vehicle for analyzing some basic curves. For example, consider the problem of graphing $r = \sin\theta$ in polar coordinates. First, let's construct the following table and try point-by-point plotting.

θ	0°	30°	60°	90°	120°	150°	180°	210°	240°	270°	300°	330°	360°
$r = \sin\theta$	0	0.50	0.87	1	0.87	0.50	0	-0.50	-0.87	-1	-0.87	-0.50	0

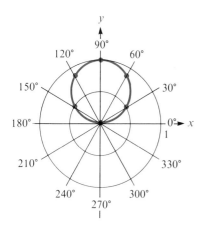

Figure 7.54

Note in Figure 7.54 that $r = \sin\theta$ graphs as a circle and that the complete circle is traced out for $0° \le \theta < 180°$. When $r < 0$, for $180° < \theta < 360°$ we merely obtain the same points a second time, and the sine function will continue this pattern if other values of θ are considered.

Point-by-point plotting is not very efficient and we usually try to draw graphs by recognizing that certain types of curves come from certain types of equations. In the case of the circle we have the following important special cases.

Equation	Graph	Comment
$r = a$		A circle of radius a and centered at the origin
$r = 2a \sin\theta$		A circle of radius a, centered on the positive y-axis, and passing through the origin
$r = -2a \sin\theta$		A circle of radius a, centered on the negative y-axis, and passing through the origin
$r = 2a \cos\theta$		A circle of radius a, centered on the positive x-axis, and passing through the origin
$r = -2a \cos\theta$		A circle of radius a, centered on the negative x-axis, and passing through the origin

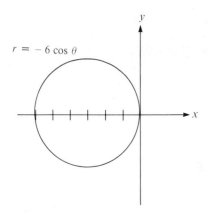

$r = -6 \cos \theta$

Figure 7.55

EXAMPLE 4   Graph $r = -6 \cos \theta$ in polar coordinates.

Solution   By matching $r = -6 \cos \theta$ to $r = -2a \cos \theta$ and using the above chart, we determine that the graph is a circle of radius 3, centered on the negative x-axis, and passing through the origin. From this information it is easy to sketch the graph in Figure 7.55.   ∎

Besides recognizing a curve from a given equation, we also want to make use of symmetry. We can test for x- and y-axis symmetry as follows:

Substitution	Symmetry
Replacing θ by $-\theta$ produces an equivalent equation.	x-axis
Replacing θ by $180° - \theta$ produces an equivalent equation.	y-axis

Since $\cos(-\theta) = \cos \theta$ and $\sin(180° - \theta) = \sin \theta$, we look for x-axis symmetry for polar equations involving cosine and y-axis symmetry for polar curves involving sine.

EXAMPLE 5   Graph $r = 2 - 2 \sin \theta$ in polar coordinates.

Solution   Since $r = 2 - 2 \sin(180° - \theta)$ and $r = 2 - 2 \sin \theta$ are equivalent equations, we have symmetry about the y-axis. Therefore, we first graph the curve to the right of the y-axis and then reflect this portion about the y-axis to obtain the entire graph. We can obtain the first portion by plotting the points in the following table, as shown in Figure 7.56(a).

θ	90°	60°	30°	0°	$-30°$	$-60°$	$-90°$
$r = 2 - 2 \sin \theta$	0	0.27	1	2	3	3.7	4

Now we just reflect the curve in Figure 7.56(a) about the y-axis to obtain the entire graph [Figure 7.56(b)].

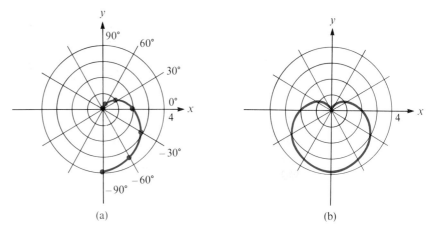

(a)   (b)

Figure 7.56   ∎

Because the curve in Figure 7.56(b) is heart shaped, it is called a **cardioid.** We can anticipate cardioids as graphs of equations of the form

$$r = a + a \sin \theta \qquad r = a - a \sin \theta$$
$$r = a + a \cos \theta \qquad r = a - a \cos \theta.$$

Each case produces the same heart-shaped figure but the orientation of the cardioid relative to the axes varies, with the sine versions being symmetric with respect to the *y*-axis and the cosine versions being symmetric with respect to the *x*-axis.

In our final illustration of graphing we wish to consider the graph of $r = \sin 2\theta$. We have seen that $r = 2 \sin \theta$ graphs as a circle, but doubling θ, instead of sin θ, is totally different. In general, equations of the form

$$r = a \sin n\theta$$
$$r = a \cos n\theta$$

graph as curves called **roses** (see Figure 7.57), because they contain loops (or petals) that produce a flower-shaped appearance. The rose has *n* loops when *n* is odd and 2*n* loops when *n* is even. Example 6 shows how to draw the rose corresponding to $r = \sin 2\theta$.

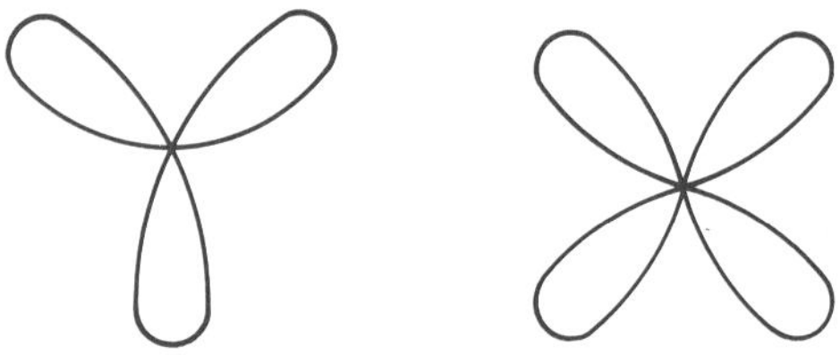

Three loops (*n* = 3)   Four loops (*n* = 2)

Figure 7.57

EXAMPLE 6   Graph $r = \sin 2\theta$ in polar coordinates.

Solution   The equation fits the form $r = a \sin n\theta$, with $n = 2$, so we know it graphs as a rose with 4 (or 2*n*) loops. To draw the rose, we need to determine the values of θ for which $|r|$ is greatest, since these points may be endpoints for the loops. By drawing the graph in rectangular coordinates of $r = \sin 2\theta$ in Figure 7.58, we see that 45°, 135°, 225°, and 315° lead

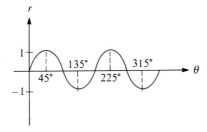

Figure 7.58

to the greatest values for $|r|$. We can now draw the rose for $r = \sin 2\theta$ as in Figure 7.59. The Q_1 and Q_3 loops are straightforward, but note that the Q_4 loop is formed for $90° < \theta < 180°$ and the Q_2 loop is formed for $270° < \theta < 360°$, because r is negative in these intervals. It is useful to know that when n is even, we have $2n$ loops with endpoints at *both* maximum and minimum r values. When n is odd, however, we need only consider maximum r values, because the minimum r values give no new points.

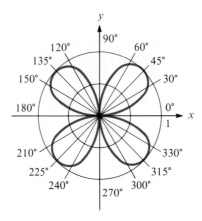

Figure 7.59

We often analyze a problem by using both polar and rectangular coordinates, so it is important we understand the basic relationships between the two systems. These may be found by considering Figure 7.60 and using our knowledge of trigonometry.

$$\sin \theta = \frac{y}{r}, \text{ so } y = r \sin \theta$$

$$\cos \theta = \frac{x}{r}, \text{ so } x = r \cos \theta$$

$$\tan \theta = \frac{y}{x}$$

$$x^2 + y^2 = r^2$$

The top two formulas show us how to switch from polar coordinates to rectangular coordinates, while the bottom two formulas let us go the other way.

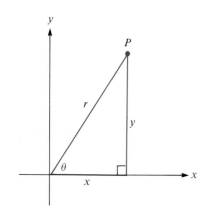

Figure 7.60

EXAMPLE 7   Express $(2, 210°)$ in rectanglar coordinates.

Solution   We substitute $r = 2$ and $\theta = 210°$ in the given formulas.

$$x = r \cos \theta = 2 \cos 210° = 2\left(\frac{-\sqrt{3}}{2}\right) = -\sqrt{3}$$

$$y = r \sin \theta = 2 \sin 210° = 2\left(-\frac{1}{2}\right) = -1$$

Thus, $(2, 210°)$ is written in rectangular coordinates as $(-\sqrt{3}, -1)$.   ■

EXAMPLE 8   Express $(-1, 1)$ in polar coordinates with $r \geq 0$ and $0° \leq \theta < 360°$.

Solution   We are given $x = -1$ and $y = 1$. Then

$$r^2 = x^2 + y^2 = (-1)^2 + (1)^2 = 2.$$

Since $r \geq 0$, we know $r = \sqrt{2}$. Also,

$$\tan \theta = \frac{y}{x} = \frac{1}{-1} = -1.$$

Since $(-1, 1)$ lies in Q_2 and $r \geq 0$, we conclude $\theta = 135°$. Thus, $(-1, 1)$ is written as $(\sqrt{2}, 135°)$ in polar coordinates. (*Note:* Without the stated restrictions on r and θ, many representations are possible.)   ■

The four basic relationships may also be used to change a polar equation to rectangular form and vice versa.

EXAMPLE 9   Transform $x^2 + y^2 + 6x = 0$ to an equation in polar form.

Solution   Since $x^2 + y^2 = r^2$ and $x = r \cos \theta$, we proceed as follows:

$$x^2 + y^2 + 6x = 0$$
$$r^2 + 6r \cos \theta = 0$$
$$r(r + 6 \cos \theta) = 0$$
$$r = 0 \qquad r + 6 \cos \theta = 0$$
$$r = -6 \cos \theta.$$

Since $r = -6 \cos \theta$ includes the pole, or $r = 0$, the complete relation is expressed by $r = -6 \cos \theta$. Note that both equations fit the form for circles in their respective systems.   ■

EXAMPLE 10   Transform $\theta = 3\pi/4$ to an equation in rectangular form.

Solution   When θ is a constant, we only need $\tan \theta = y/x$. Since $\theta = 3\pi/4$,

$$\tan \theta = \frac{y}{x}$$

$$\tan \frac{3\pi}{4} = \frac{y}{x}$$

$$-1 = \frac{y}{x}$$

$$y = -x.$$

Thus, $\theta = 3\pi/4$ may be written in rectangular form as $y = -x$.   ∎

EXERCISES 7.7

1. Plot the following points in polar coordinates.
   a. $(2,60°)$ **b.** $(3,\pi)$ **c.** $(1,-30°)$
   d. $(0,150°)$ **e.** $(0.5,-2\pi/3)$ **f.** $(2,500°)$
2. Plot the following points in polar coordinates.
   a. $(-1,90°)$ **b.** $(-2,-\pi/6)$ **c.** $(-\frac{3}{2},135°)$
   d. $(-2,210°)$ **e.** $(-3,-\pi)$ **f.** $(-1,2\pi/3)$
3. Express the following points in polar coordinates with $r \geq 0$ and $0° \leq \theta < 360°$ (or $0 \leq \theta < 2\pi$).
   a. $(1,-30°)$ **b.** $(-2,5\pi/6)$ **c.** $(-1,4\pi/3)$
   d. $(3,585°)$ **e.** $(-1,-7\pi/4)$ **f.** $(2,-180°)$
4. List three other polar representations for $(3,210°)$ with one having $r < 0$.

In Exercises 5–20 graph each equation in polar coordinates and give its name.

5. $r = 3$ 6. $\theta = \pi/4$
7. $r = \cos \theta$ 8. $r = 3 \sin \theta$
9. $r = -4 \sin \theta$ 10. $r = -2 \cos \theta$
11. $r = 1 + \sin \theta$ 12. $r = 1 - \sin \theta$
13. $r = \sin \theta - 1$ 14. $r = 1 + \cos \theta$
15. $r = 2(1 - \cos \theta)$ 16. $r = \cos \theta - 1$
17. $r = \cos 2\theta$ 18. $r = 3 \sin 2\theta$
19. $r = 2 \sin 3\theta$ 20. $r = \cos 4\theta$

In Exercises 21–28 graph each equation in polar coordinates.

21. $r = 2 + \cos \theta$ (limaçon)
22. $r = 3 - \sin \theta$ (limaçon)
23. $r = 1 - 2 \sin \theta$ (limaçon, inner loop)
24. $r = 1 + 2 \cos \theta$ (limaçon, inner loop)
25. $r^2 = 4 \cos 2\theta$ (lemniscate)
26. $r^2 = \sin 2\theta$ (lemniscate)
27. $r = \theta, \theta \geq 0$ (spiral)
28. $r = \theta, \theta \leq 0$ (spiral)

In Exercises 29–34 express each point in rectangular coordinates.

29. $(1,\pi)$ 30. $(2,60°)$ 31. $(-3,3\pi)$
32. $(-6,0°)$ 33. $(\sqrt{2},5\pi/4)$ 34. $(-\sqrt{3},-120°)$

In Exercises 35–40 express each point in polar coordinates with $r \geq 0$ and $0° \leq \theta < 360°$.

35. $(-1,0)$ 36. $(0,-2)$ 37. $(1,-1)$
38. $(-3,-3)$ 39. $(-\sqrt{3},1)$ 40. $(-2,2\sqrt{3})$

In Exercises 41–50 transform each equation to polar form.

41. $x = 2$ 42. $y = -2$
43. $x^2 + y^2 = 9$ 44. $y = x$
45. $x^2 + y^2 = x$ 46. $x^2 + y^2 - 4y = 0$
47. $y = x^2$ 48. $y = x^3$
49. $x^2 - y^2 = 4$ 50. $x^2 + y^2 + x = \sqrt{x^2 + y^2}$

In Exercises 51–60 transform each equation to rectangular form.

51. $r = 3$ 52. $\theta = -\pi/6$
53. $r \cos \theta = 2$ 54. $r \sin \theta = -3$
55. $r = -3 \sin \theta$ 56. $r = 3 \cos \theta$
57. $r = 2 \csc \theta$ 58. $r = \sec \theta$
59. $r(3 \sin \theta + 2 \cos \theta) = 1$ 60. $r = \sin 2\theta$

THINK ABOUT IT

1. Explain how the figure below can be used to interpret geometrically the graph of the equation $r = 2a \cos \theta$.

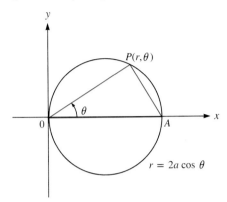

2. **a.** Convert $r = a \sin \theta + b \cos \theta$ to rectangular form. Name the graph of the resulting equation.
   b. Graph $r = 2 \sin \theta + 2 \cos \theta$.

3. Graph $r = \sin \theta$ and $r = \cos \theta$ on the same coordinate system and give a representation for all points of intersection of the two curves. Explain why geometric methods must be included when finding intersection points of two curves that are given in polar form.

4. **a.** Consider the graph of $r = \sin 2\theta$ in Example 6. Is the graph symmetric about the x-axis?
   b. If replacing θ by $-\theta$ produces an equivalent polar equation, then the graph of the equation is symmetric about the x-axis. Apply this test to the graph of $r = \sin 2\theta$. Is the test result in agreement with your answer to part a? Explain.

5. Use different polar representations for $(r, -\theta)$ and state two tests for determining symmetry about the x-axis that are different from the test in Question 4b. Can you conclude from either of these alternative tests that the graph of $r = \sin 2\theta$ is symmetric about the x-axis?

REMEMBER THIS

1. Find the absolute value of the complex number $7 - 24i$.
2. Find the magnitude of **v** if $\mathbf{v} = 7\mathbf{i} - 24\mathbf{j}$.
3. Find the argument of the complex number $-\sqrt{3} - i$ in $[0°, 360°)$.
4. Find the direction angle of **v** if $\mathbf{v} = -\sqrt{3}\mathbf{i} - \mathbf{j}$.
5. Express the complex number $2 - 2i$ in trigonometric form.

6. Express the point $(2, -2)$ in polar coordinates with $r \geq 0$ and $0° \leq \theta < 360°$.
7. Express $2(\cos 300° + i \sin 300°)$ in the form $a + bi$.
8. Express $(2, 300°)$ in rectangular coordinates.
9. If the magnitude of **v** is 2 and the direction angle of **v** is $300°$, then find the vertical component of **v**.
10. Find $(1 - \sqrt{3}i)^3$ in the form $a + bi$.

CHAPTER OVERVIEW

Section	Key Concepts to Review
7.1	• Law of sines: $\dfrac{\sin A}{a} = \dfrac{\sin B}{b} = \dfrac{\sin C}{c}$
	• Guidelines on when to use the law of sines
	• When given the measures for two sides of a triangle and the angle opposite one of them, there may be one triangle that fits the data (see Example 3) or there may be two triangles that fit the data (see Example 4). Sometimes no triangle can be constructed from the data; then we say the data is inconsistent.

Section	Key Concepts to Review
7.2	• Law of cosines: $a^2 = b^2 + c^2 - 2bc \cos A$ $\phantom{\text{Law of cosines: }}b^2 = a^2 + c^2 - 2ac \cos B$ $\phantom{\text{Law of cosines: }}c^2 = a^2 + b^2 - 2ab \cos C$ • Guidelines on when to use the law of cosines • The area of a triangle is given by one half the product of the lengths of two sides and the sine of the angle between these two sides. • Heron's area formula: $K = \sqrt{s(s-a)(s-b)(s-c)}$, where $s = \frac{1}{2}(a+b+c)$
7.3	• Definitions of vector, resultant of two (or more) vectors, and components of a vector • Methods to determine the resultant of two (or more) vectors • Methods to determine the horizontal and vertical components of a vector • Resultant formulas: $R = \sqrt{(R_x)^2 + (R_y)^2}$ and $\tan \theta = R_y/R_x$
7.4	• Definitions of vector equality, position vector, components of $\mathbf{v}$, magnitude and direction angle of $\mathbf{v}$, vector addition, scalar multiplication, vector subtraction, and unit vector • Magnitude (or length) of $\mathbf{v}$: $\|\mathbf{v}\| = \sqrt{v_1{}^2 + v_2{}^2}$ • The direction angle of $\mathbf{v}$ may be determined from $\tan \theta = v_2/v_1$ if $v_1 \neq 0$. • Every vector $\mathbf{v}$ in a plane is expressible in the form $\mathbf{v} = v_1\mathbf{i} + v_2\mathbf{j}$, where $\mathbf{i} = \langle 1,0 \rangle$ and $\mathbf{j} = \langle 0,1 \rangle$. The vector sum $v_1\mathbf{i} + v_2\mathbf{j}$ is called a linear combination of $\mathbf{i}$ and $\mathbf{j}$. • Use $\mathbf{v} = \|\mathbf{v}\|(\cos \theta)\mathbf{i} + \|\mathbf{v}\|(\sin \theta)\mathbf{j}$ to express $\mathbf{v}$ as a linear combination of $\mathbf{i}$ and $\mathbf{j}$ when given the magnitude and direction angle of $\mathbf{v}$.
7.5	• Definitions of complex plane, real axis, imaginary axis, and the argument and absolute value of a complex number • Graph of a complex number • In trigonometric form $a + bi$ is written as $r(\cos \theta + i \sin \theta)$. • The absolute value, r, is given by $r = |a + bi| = \sqrt{a^2 + b^2}$. • The argument, θ, is given by $\tan \theta = b/a$. • Methods to multiply and divide two complex numbers in trigonometric form
7.6	• Definitions of an nth root of a complex number and the nth roots of unity • De Moivre's theorem • The n distinct nth roots of the complex number $r(\cos \theta + i \sin \theta)$ are given by $$\sqrt[n]{r}\left(\cos \frac{\theta + k \cdot 360°}{n} + i \sin \frac{\theta + k \cdot 360°}{n} \right),$$ where $k = 0, 1, 2, \ldots, n-1$.

Section	Key Concepts to Review
7.7	• Definitions of pole, polar axis, and polar coordinates
	• If $r < 0$, we plot (r,θ) by measuring $\lvert r \rvert$ units in a direction opposite to the terminal ray of θ.
	• Formulas for different representations of a given point:
	$(r,\theta) = (r,\theta + n \cdot 360°)$, where n is an integer
	$(-r,\theta) = (r,\theta + n \cdot 180°)$, where n is an odd integer
	• Two tests for x- and y-axis symmetry
	• Basic graphs:
	$r = a, r = \pm 2a \sin \theta, r = \pm 2a \cos \theta$: circle
	$r = a \pm a \sin \theta, r = a \pm a \cos \theta$: cardioid
	$r = a \sin n\theta, r = a \cos n\theta$: rose $\begin{cases} n \text{ loops, if } n \text{ is odd} \\ 2n \text{ loops, if } n \text{ is even} \end{cases}$
	• Formulas relating rectangular coordinates and polar coordinates:
	$x = r \cos \theta, y = r \sin \theta, x^2 + y^2 = r^2, \tan \theta = y/x$

CHAPTER REVIEW EXERCISES

In Exercises 1–6 find the indicated part in triangle ABC.

1. Determine B if $a = 5.0$, $b = 9.0$, and $c = 6.0$.
2. Determine C if $A = 81°$, $a = 11$, and $c = 35$.
3. Determine A if $C = 44°50'$, $B = 86°20'$, and $a = 62.7$.
4. Determine B if $C = 90°$, $b = 5.00$, and $c = 8.00$.
5. Determine c if $C = 120°$, $a = 3.0$, and $b = 4.0$.
6. Determine b if $A = 40°$, $B = 60°$, and $c = 6.0$.

In Exercises 7–10 perform the indicated operations. Write the answers in the form $a + bi$ and in trigonometric form.

7. $8(\cos 90° + i \sin 90°) \div (\cos 0° + i \sin 0°)$
8. $2(\cos 73° + i \sin 73°) \cdot 3(\cos 107° + i \sin 107°)$
9. $\left(\dfrac{\sqrt{2}}{2} + \dfrac{\sqrt{2}}{2}i \right)^{10}$
10. $\sqrt[3]{-1}$ (all roots)

In Exercises 11–14 graph each equation in polar coordinates.

11. $r = -2 \sin \theta$
12. $r = 2$
13. $r = 2 + 2 \cos \theta$
14. $r = 2 \cos 3\theta$
15. Express the point $(-1,-1)$ in polar coordinates with $r \geq 0$ and $0° \leq \theta < 360°$.
16. Express the point $(-2,7\pi/4)$ in rectangular coordinates.
17. What is the vertical component of a vector with a magnitude of $5\overline{0}$ lb which makes an angle of $25°$ with the positive x-axis?

18. Find all solutions for the equation $x^4 + 4 = 0$.
19. Find the area of triangle ABC if $a = 16$ m, $c = 11$ m, and $B = 116°$.
20. In triangle ABC, $b = 5$, $c = 6$, and $\cos A = -\frac{1}{3}$. Find side length a.
21. Transform the equation $r(2 \cos \theta - \sin \theta) = 3$ to rectangular form.
22. Transform the equation $x^2 + y^2 = y$ to polar form.
23. Express the roots of $x^3 + 27 = 0$ in the form $r(\cos \theta + i \sin \theta)$.
24. Express the number $-1 - i$ in trigonometric form.

In Exercises 25–28 find the magnitude and direction angle of $\mathbf{v}$.

25. $\mathbf{v} = \langle 1,-\sqrt{3} \rangle$
26. $\mathbf{v} = \langle 3,0 \rangle$
27. $\mathbf{v} = -5\mathbf{j}$
28. $\mathbf{v} = -2\mathbf{i} + 2\mathbf{j}$
29. If $\mathbf{v} = \langle 8,-1 \rangle$ and $\mathbf{w} = \langle -5,-2 \rangle$, find $\mathbf{v} - 4\mathbf{w}$.
30. Express $\mathbf{v}$ as a linear combination of $\mathbf{i}$ and $\mathbf{j}$ if the magnitude of $\mathbf{v}$ is 6 and the direction angle of $\mathbf{v}$ is $150°$.
31. If $\mathbf{v} = 2\mathbf{i} - 7\mathbf{j}$ and $\mathbf{w} = 5\mathbf{i} + 3\mathbf{j}$, express $2\mathbf{v} - \mathbf{w}$ as a linear combination of $\mathbf{i}$ and $\mathbf{j}$.
32. Vector $\mathbf{a}$ has a magnitude of 7.0 lb and a direction angle of $74°$. Vector $\mathbf{b}$ has a magnitude of 4.0 lb and a direction angle of $195°$. Find the vector sum or resultant $\mathbf{v}$ of vectors $\mathbf{a}$ and $\mathbf{b}$, and give the magnitude and direction angle of $\mathbf{v}$.

33. If the outfielders are positioned as shown below, then how far is each of them from third base in case they need to throw out a runner at that base?

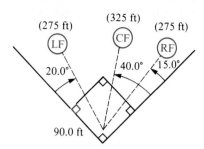

34. Two points, A and B, are $10\overline{0}$ yd apart. Point C across a canyon is located so that angle CAB is $70°$ and angle CBA is $80°$. Compute the distance $\overline{BC}$ across the canyon.

35. A draftsman drew to scale (1 in. = $5\overline{0}$ yd) a map of a development that includes a triangular recreation area with sides of lengths 75 yd, 125 yd, and 150 yd. What are the angles of the triangle representing the recreation area on the map?

36. A force of 12 lb and a force of 15 lb are acting on a body with an angle of $90°$ between the two forces. Find the magnitude of the resultant and the angle between the resultant and the larger force.

37. A force of $4\overline{0}$ lb and a force of $3\overline{0}$ lb act on a body so that their resultant is a force of 38 lb. Find the angle between the two original forces.

38. Vector **A** has a magnitude of 6.0 lb and makes an angle of $55°$ with the positive x-axis. Vector **B** has a magnitude of 2.0 lb and makes an angle of $110°$ with the positive x-axis. Find the resultant (or vector sum) of **A** and **B**.

39. Use the law of cosines to show that if p, q, and r are the side lengths of a triangle and $r^2 = p^2 + q^2$, then triangle PQR is a right triangle.

40. Use the law of sines to show that the area K of triangle ABC may be given by

$$K = \frac{b^2 \sin A \sin C}{2 \sin B}.$$

In Exercises 41–50 select the choice that completes the statement or answers the question.

41. The number of triangles satisfying the conditions that $B = 30°$, $a = 57$ ft, and $b = 39$ ft is
    a. two   **b.** one   **c.** none

42. We write $2(\cos 120° + i \sin 120°)$ in the form $a + bi$ as
    a. $1 - \sqrt{3}i$ **b.** $-\sqrt{3} + i$
    c. $\sqrt{3} - i$ **d.** $-1 + \sqrt{3}i$

43. The graph of $r = 2 + 2 \cos \theta$ is a
    a. circle   **b.** cardioid
    c. rose   **d.** line

44. The x component of a vector with a magnitude of 25 lb which makes an angle of $67°$ with the x-axis is
    a. 9.8 lb   **b.** 12 lb   **c.** 17 lb   **d.** 23 lb

45. An alternate polar representation for $(-1, -45°)$ is
    a. $(1, 45°)$ **b.** $(1, 135°)$
    c. $(-1, 135°)$ **d.** $(1, 315°)$

46. To solve a triangle when we know the measures for two sides of the triangle and the angle between these two sides, we first apply the
    a. law of sines   **b.** law of cosines

47. $(1 + i)^6$ simplifies to
    a. 8   **b.** -8 **c.** $-8i$ **d.** $8i$

48. In triangle ABC, if $\sin A = \frac{3}{4}$ and $\sin B = \frac{1}{2}$, then the ratio of side length a to side length b is
    a. 3:2   **b.** 8:3   **c.** 3:1   **d.** 4:3

49. Which illustration shows the graph of $r = \sin 5\theta$?

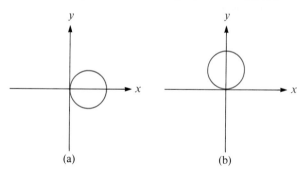

(a)   (b)

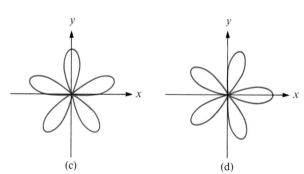

(c)   (d)

50. The absolute value of the complex number $-2 + 2i$ is
    a. $2\sqrt{2}$ **b.** 2   **c.** 4   **d.** $\sqrt{2}$

CHAPTER 7 TEST

1. True or false: To solve a triangle when given the measures for two sides of the triangle and the angle between these two sides, we first apply the law of cosines.

2. Assume exact numbers and find r in triangle RST if $R = 45°$, $T = 30°$, and $t = 24$.

3. In triangle ABC, $A = 34.5°$, $a = 11.6$ ft, and $c = 19.1$ ft. Find the two possible solutions for C.

4. Find the measure of the largest angle in a triangle with side measures of 18.5 m, 15.0 m, and 26.0 m.

5. Solve the triangle ABC in which $B = 76°$, $b = 45$ ft, and $c = 35$ ft.

6. An equilateral triangle is inscribed in a circle with radius 12.0 cm. What is the perimeter of the triangle?

7. Find the area of triangle PQR if $p = 5.6$ cm, $r = 4.1$ cm, and $Q = 48.0°$.

8. If $v = \langle 3\sqrt{3}, -3 \rangle$, find the magnitude and direction angle of v.

9. If $v = \langle 7, -1 \rangle$ and $w = \langle -3, 5 \rangle$, find $2v - 3w$.

10. Express v as a linear combination of i and j if the magnitude of v is 4 and the direction angle of v is 120°.

11. What is the vertical component of a vector with a magnitude of 56 lb which makes an angle of 37° with the positive x-axis?

12. A force of 28 lb and a force of 12 lb are acting on a body with an angle of 28° between the two forces. Find the magnitude of the resultant force.

13. Write the number $-15 + 8i$ in trigonometric form.

14. If complex numbers $z_1 = 6(\cos 44° + i \sin 44°)$ and $z_2 = 3(\cos 136° + i \sin 136°)$, write the product $z_1 z_2$ in the form $a + bi$.

15. Find $(1 + i)^{10}$.

16. Find the three cube roots of -125.

17. List two other polar representations for $(-3, 225°)$ with one having $r > 0$.

18. Graph $r = 5 + 5 \sin \theta$.

19. Express the point $(6, 2\pi/3)$ in rectangular coordinates.

20. Transform the equation $x^2 + y^2 - 9x = 0$ to polar form.

8 Analytic Geometry: Conic Sections

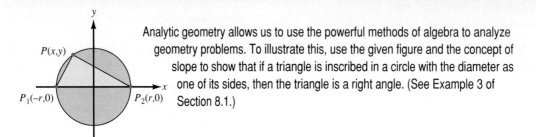

Analytic geometry allows us to use the powerful methods of algebra to analyze geometry problems. To illustrate this, use the given figure and the concept of slope to show that if a triangle is inscribed in a circle with the diameter as one of its sides, then the triangle is a right angle. (See Example 3 of Section 8.1.)

Analytic geometry bridges the gap between algebra and geometry. By representing the ordered pairs that satisfy some algebraic equation as points in the Cartesian coordinate system, we generate a geometric picture or graph. The fundamental relationship between an equation and its graph is as follows: Every ordered pair that satisfies the equation corresponds to a point in its graph, and every point in the graph corresponds to an ordered pair that satisfies the equation.

We have considered some techniques for sketching the graph of a given equation. However, the reverse question is of major importance. If we are given in geometric terms some object or motion from the physical world, can we describe it with some equation? If so, the powerful methods of algebra can be used to analyze them. Thus, we now consider the problem of determining an equation that corresponds to a given geometric condition.

8.1 Introduction to Analytic Geometry

Suppose the graph of an equation is the set of all points in a plane that are 3 units from the point $(2,-1)$. It is not hard to tell that the graph is a circle of radius 3. We now consider the more difficult problem of determining an equation for this circle.

We start by making the sketch in Figure 8.1, where we let $P(x,y)$ be any point on the given circle. By the distance formula, the distance from $(2,-1)$ to P is

$$d = \sqrt{(x-2)^2 + [y-(-1)]^2}.$$

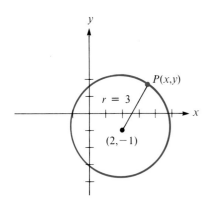

Figure 8.1

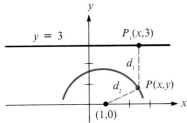

Figure 8.2

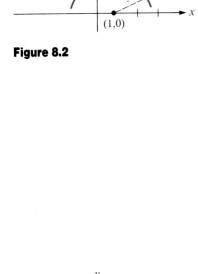

Figure 8.3

This distance is the radius of the circle, which is given as 3. Thus,

$$3 = \sqrt{(x - 2)^2 + [y - (-1)]^2}.$$

Squaring both sides of this equation yields

$$9 = (x - 2)^2 + (y + 1)^2.$$

We now simplify as follows:

$$9 = x^2 - 4x + 4 + y^2 + 2y + 1$$
$$0 = x^2 + y^2 - 4x + 2y - 4.$$

This equation defines the circle formed by the given geometric condition.

EXAMPLE 1   Find an equation for the graph that is the set of all points in a plane equidistant from the line $y = 3$ and the point $(1,0)$.

Solution   Let $P(x,y)$ represent any point satisfying the given geometric condition. Then draw the sketch in Figure 8.2 that includes a portion of the given graph. P_1 lies on the line $y = 3$, so we represent the point as $(x,3)$. Since P_1 and P have the same x-coordinate, the distance d_1 is given by

$$d_1 = |y - 3|.$$

The distance from P to $(1,0)$ is given by

$$d_2 = \sqrt{(x - 1)^2 + (y - 0)^2}.$$

The given geometric condition states that

$$d_1 = d_2.$$

Thus,

$$|y - 3| = \sqrt{(x - 1)^2 + (y - 0)^2}.$$

Squaring both sides of the equation yields

$$(y - 3)^2 = (x - 1)^2 + (y - 0)^2.$$

We simplify this equation as follows:

$$y^2 - 6y + 9 = x^2 - 2x + 1 + y^2$$
$$0 = x^2 - 2x + 6y - 8.$$

The graph of this equation satisfies the given geometric condition.   ■

EXAMPLE 2   Find an equation for the graph that is the set of all points in a plane the sum of whose distances from $(2,0)$ and $(-2,0)$ is 10.

Solution   Let $P(x,y)$ represent any point satisfying the given geometric condition. Then draw the sketch in Figure 8.3 that includes a portion of the given graph. By the distance formula, we have

$$d_1 = \sqrt{[x - (-2)]^2 + (y - 0)^2}$$
$$d_2 = \sqrt{(x - 2)^2 + (y - 0)^2}.$$

The given geometric condition states that

$$d_1 + d_2 = 10.$$

Thus,

$$\sqrt{(x + 2)^2 + y^2} + \sqrt{(x - 2)^2 + y^2} = 10$$

or

$$\sqrt{(x + 2)^2 + y^2} = 10 - \sqrt{(x - 2)^2 + y^2}.$$

Square both sides of the equation and simplify.

$$(x + 2)^2 + y^2 = 100 - 20\sqrt{(x - 2)^2 + y^2} + (x - 2)^2 + y^2$$
$$8x = 100 - 20\sqrt{(x - 2)^2 + y^2}$$
$$2x - 25 = -5\sqrt{(x - 2)^2 + y^2}$$

Square both sides of the equation and simplify.

$$4x^2 - 100x + 625 = 25[(x - 2)^2 + y^2]$$
$$4x^2 - 100x + 625 = 25x^2 - 100x + 100 + 25y^2$$
$$0 = 21x^2 + 25y^2 - 525$$

The graph of this equation satisfies the given geometric condition.   ■

Another aspect of analytic geometry is the proof of geometric theorems by using a coordinate system and algebraic methods as illustrated by the next example.

EXAMPLE 3   Solve the problem in the chapter introduction on page 424.

Solution   If the product of the slopes of line segments PP_1 and PP_2 is -1, then these sides are perpendicular and the triangle is a right triangle. Then

$$m_1m_2 = \frac{y - 0}{x - (-r)} \cdot \frac{y - 0}{x - r} = \frac{y^2}{x^2 - r^2}.$$

Since $x^2 + y^2 = r^2$ is an equation of a circle of radius r with center at the origin (as shown in Section 3.2), we know $y^2 = r^2 - x^2$, so

$$m_1m_2 = \frac{r^2 - x^2}{x^2 - r^2} = -1.$$

Thus, the product of the slopes is -1, so the triangle is a right triangle.   ■

Up to now we have relied on distance and slope formulas in our analysis. Another useful formula that is simple to apply is the formula for finding the coordinates of the midpoint of a line segment. To derive this formula, let $P_1(x_1,y_1)$ and $P_2(x_2,y_2)$ be two points in a plane with $x_1 < x_2$ and $y_1 < y_2$, and let $P(x,y)$ be any other point on line segment P_1P_2. We then construct similar triangles P_1RP and PSP_2 as shown in Figure 8.4, and since corresponding side lengths are proportional in such triangles, we know

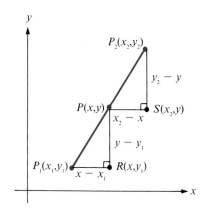

Figure 8.4

$$\frac{x - x_1}{x_2 - x} = \frac{y - y_1}{y_2 - y} = \frac{\overline{P_1P}}{\overline{PP_2}}.$$

If $P(x,y)$ is the midpoint of the segment, then $\overline{P_1P}/\overline{PP_2} = 1$, and we may solve for x and y as follows:

$$\frac{x - x_1}{x_2 - x} = 1 \qquad\qquad \frac{y - y_1}{y_2 - y} = 1$$

$$x - x_1 = x_2 - x \qquad\qquad y - y_1 = y_2 - y$$

$$2x = x_1 + x_2 \qquad\qquad 2y = y_1 + y_2$$

$$x = \frac{x_1 + x_2}{2} \qquad\qquad y = \frac{y_1 + y_2}{2}.$$

These results are valid regardless of the locations of P_1 and P_2 and we have the following result.

Midpoint Formula

The midpoint of the line segment joining (x_1,y_1) and (x_2,y_2) is

$$\left(\frac{x_1 + x_2}{2}, \frac{y_1 + y_2}{2} \right).$$

Note that the x-coordinate of the midpoint is simply the average of the x-coordinates of the two endpoints, and a similar relation applies for the y-coordinate of the midpoint. Also, in the above derivation of the midpoint formula, it is easy to extend our result by letting $\overline{P_1P}/\overline{PP_2}$ equal any positive number r and thereby develop a formula for the coordinates of the point that divides a line segment in any given ratio. We pursue this idea in the exercises.

EXAMPLE 4   Find the midpoint of the line segment joining $(-3,4)$ and $(7,-1)$.

Solution   Using the midpoint formula, the midpoint is

$$\left(\frac{-3 + 7}{2}, \frac{4 + (-1)}{2} \right) = \left(2, \frac{3}{2} \right).$$ ∎

EXAMPLE 5   Find an equation for the line that is the perpendicular bisector of the line segment whose endpoints are $(3,5)$ and $(11,-7)$.

Solution   (See Figure 8.5.)

a. The perpendicular bisector of a line segment passes through the midpoint of the segment. Substituting the coordinates of the endpoints $(3,5)$ and $(11,-7)$ in the midpoint formula gives

$$\left(\frac{3 + 11}{2}, \frac{5 + (-7)}{2} \right) = (7,-1)$$

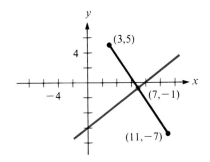

Figure 8.5

as the midpoint of the segment.

b. The slope of the given line segment is

$$m_1 = \frac{y_2 - y_1}{x_2 - x_1} = \frac{-7 - 5}{11 - 3} = \frac{-12}{8} = \frac{-3}{2}.$$

The slope m_2 of the perpendicular bisector must satisfy $-\frac{3}{2}m_2 = -1$, so $m_2 = \frac{2}{3}$.

c. We find an equation for the line through $(7, -1)$ with slope $\frac{2}{3}$ as follows:

$$y - y_1 = m(x - x_1)$$
$$y - (-1) = \tfrac{2}{3}(x - 7)$$
$$y = \tfrac{2}{3}x - \tfrac{17}{3}.$$

Thus, an equation for the specified line is $y = \frac{2}{3}x - \frac{17}{3}$. ∎

EXAMPLE 6 Use analytic geometry to prove that the midpoint of the hypotenuse of any right triangle is equidistant from each of the three vertices.

Solution We are free to place the specified figure in any convenient position on a coordinate system, so we start by placing the legs of a right triangle on the x- and y-axes as shown in Figure 8.6. The vertex of the right angle is at the origin, while the other vertices are at the points $A(a,0)$ and $B(0,b)$. By the midpoint formula, the midpoint M of the hypotenuse is

$$\left(\frac{a+0}{2}, \frac{0+b}{2}\right) = \left(\frac{a}{2}, \frac{b}{2}\right).$$

Then by the distance formula, the distance between the midpoint and each vertex is as follows:

$$\overline{AM} = \sqrt{\left(a - \frac{a}{2}\right)^2 + \left(0 - \frac{b}{2}\right)^3} = \sqrt{\frac{a^2}{4} + \frac{b^2}{4}}$$

$$\overline{BM} = \sqrt{\left(0 - \frac{a}{2}\right)^2 + \left(b - \frac{b}{2}\right)^2} = \sqrt{\frac{a^2}{4} + \frac{b^2}{4}}$$

$$\overline{CM} = \sqrt{\left(0 - \frac{a}{2}\right)^2 + \left(0 - \frac{b}{2}\right)^2} = \sqrt{\frac{a^2}{4} + \frac{b^2}{4}}.$$

Therefore, $\overline{AM} = \overline{BM} = \overline{CM}$, which establishes the given theorem. ∎

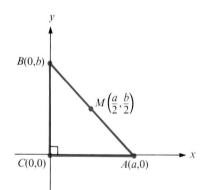

Figure 8.6

EXERCISES 8.1

In Exercises 1–10 find an equation for the graph in a plane that is defined by the given geometric condition.

1. The set of all points equidistant from $(0,2)$ and $(-1,5)$
2. The set of all points equidistant from $(1,-3)$ and $(-4,2)$
3. The set of all points that are 4 units from the point $(-2,-3)$
4. The set of all points that are 2 units from the point $(4,0)$
5. The set of all points equidistant from the line $x = 4$ and the point $(0,-2)$
6. The set of all points equidistant from the x-axis and the point $(3,-1)$
7. The set of all points the sum of whose distances from $(1,0)$ and $(-1,0)$ is 4
8. The set of all points the sum of whose distances from $(0,3)$ and $(0,-3)$ is 8
9. The set of all points the difference of whose distances from $(0,3)$ and $(0,-3)$ is 3

10. The set of all points the difference of whose distances from (4,0) and (−4,0) is 6

In Exercises 11–16 find the midpoint of the line segment joining the given pair of points.

11. (−2,3), (6,−1)
12. (5,−4), (−9,0)
13. (7,2), (−7,−9)
14. (−1,1), (1,−1)
15. (0,0), (0,3)
16. (8,−2), (5,−2)

In Exercises 17–24 find an equation for the line that is the perpendicular bisector of the line segment whose endpoints are given.

17. (1,2), (7,6)
18. (1,3), (−3,−1)
19. (−4,−9), (0,−5)
20. (−3,5), (1,−1)
21. (2,−3), (−3,4)
22. (1,0), (−2,7)
23. (6,−1), (−1,2)
24. (−4,4), (3,−5)

In Exercises 25–28 use the illustration below to prove the given theorem.

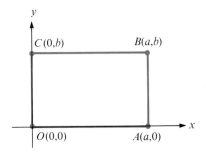

25. The diagonals of a rectangle are equal in length.
26. The diagonals of a rectangle bisect each other.
27. If the diagonals of a rectangle are perpendicular, then the rectangle is a square.
28. The midpoints of the sides of a rectangle are the vertices of a rhombus (that is, a parallelogram with all sides equal in length).

In Exercises 29–31 prove each theorem by the methods of analytic geometry.

29. The diagonals of a parallelogram bisect each other.
30. The line segments joining the midpoints of the opposite sides of any quadrilateral bisect each other.
31. The line segment joining the midpoints of two sides of a triangle is parallel to the third side and one-half its length.

THINK ABOUT IT

1. The points $A(−2,0)$ and $B(6,−4)$ are the endpoints of line segment AB. Find an equation for the line L that is the perpendicular bisector of AB and verify that $(3,0)$ is a point on L. Show an easier way to prove $(3,0)$ lies on line L without first finding an equation for L.

2. State a theorem from plane geometry concerning the diagonals of a rhombus. Then prove this theorem by the methods of analytic geometry.

3. Show by the methods of analytic geometry that the sum of the squares of the side lengths of a parallelogram is equal to the sum of the squares of the lengths of the diagonals.

4. **a.** Consider the derivation of the midpoint formula. If $P(x,y)$ is a point that divides the line segment from P_1 to P_2 so that $\overline{P_1P}/\overline{PP_2}$ equals a given positive ratio r, then show the coordinates of $P(x,y)$ are

$$\left(\frac{x_1 + rx_2}{1 + r}, \frac{y_1 + ry_2}{1 + r} \right).$$

This formula is called the **point-of-division formula.**

b. Derive the midpoint formula from the point-of-division formula.

5. Solve each problem using the point-of-division formula.

a. Find the coordinates of the point that divides the line segment from $P_1(1,3)$ to $P_2(8,−4)$ in the ratio $\frac{3}{4}$.

b. Determine the two points of trisection of the line segment joining $(0,−4)$ and $(6,4)$.

REMEMBER THIS

1. If $r = 2\sqrt{3}$, find r^2.
2. If $r^2 = 48$ and $r > 0$, find r in simplest radical form.
3. Find the distance between $(3,-4)$ and $(0,1)$.
4. What number should be added to $x^2 - 12x$ to make the expression a perfect square?
5. Factor $y^2 - 8y + 16$.
6. Write formulas for the diameter, the circumference, and the area of a circle of radius r.

7. Is the following graph the graph of a function?

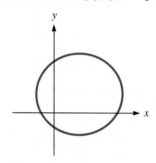

8. Graph $x^2 + y^2 = 4$.
9. Graph $y = \sqrt{4 - x^2}$.
10. Graph $y = -\sqrt{4 - x^2}$.

8.2   The Circle

Circles, ellipses, hyperbolas, and parabolas are referred to as **conic sections.** This name was first used by Greek mathematicians who discovered that these curves result from the intersections of a cone with an appropriate plane as shown in Figure 8.7. If the cutting plane does not contain the vertex, then the important conic sections are obtained as follows:

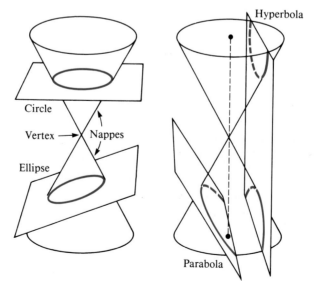

Figure 8.7

Circle   The intersecting plane is parallel to the base of the cone.

Ellipse   The intersecting plane is parallel to neither the base nor the side and intersects only one nappe of the cone.

Hyperbola   The intersecting plane cuts both nappes of the cone.

Parabola   The intersecting plane is parallel to the side of the cone.

Three degenerate cases occur when the cutting plane passes through the vertex. These degenerate conic sections are a point, a line, and a pair of intersecting lines.

The conic sections may be analyzed from different viewpoints, and in this chapter we take the analytic geometry approach in which we derive equations for the conic sections by starting with their geometric definitions. To illustrate, we begin with the definition of a circle that is given in plane geometry.

Definition of a Circle

A **circle** is the set of all points in a plane at a given distance from a fixed point.

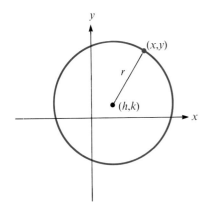

Figure 8.8

From this definition we derive an equation of a circle as follows (see also Figure 8.8):

Let (h,k) be the fixed point, the center of the circle.
Let r be the given distance, the radius of the circle.
Let (x,y) be any point on the circle.

Then by the distance formula, we have

$$\sqrt{(x - h)^2 + (y - k)^2} = r$$

or, after squaring both sides of the equation,

$$(x - h)^2 + (y - k)^2 = r^2.$$

This equation is the **standard form of the equation of a circle of radius r with center (h,k).**

EXAMPLE 1   Find the equation in standard form of the circle with center at $(2,-1)$ and radius 4.

Solution   Substituting $h = 2$, $k = -1$, and $r = 4$ in the equation

$$(x - h)^2 + (y - k)^2 = r^2,$$

we have

$$(x - 2)^2 + [y - (-1)]^2 = (4)^2$$
$$(x - 2)^2 + (y + 1)^2 = 16.$$

∎

EXAMPLE 2   Find the equation in standard form of the circle with center at (4,1) that passes through the origin.

Solution   Substituting $h = 4$ and $k = 1$ into the standard equation for a circle, we have

$$(x - 4)^2 + (y - 1)^2 = r^2.$$

Since the circle passes through the origin, we find r^2 by substituting the coordinates of the point (0,0) in the equation.

$$(0 - 4)^2 + (0 - 1)^2 = r^2$$
$$17 = r^2$$

The equation of the circle in standard form is then

$$(x - 4)^2 + (y - 1)^2 = 17 \quad \text{(see Figure 8.9).} \qquad \blacksquare$$

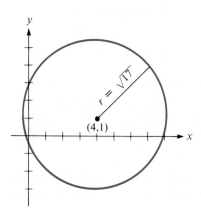

Figure 8.9

EXAMPLE 3   Find the center and radius of the circle given by

$$x^2 + y^2 + 4x - 6y = 12.$$

Solution   We must transform the given equation to the standard form.

$$(x - h)^2 + (y - k)^2 = r^2$$

We start by completing the square in the x terms and the y terms. To do this, first group the equation as

$$(x^2 + 4x \quad) + (y^2 - 6y \quad) = 12.$$

Now find one-half of the coefficient of x (2). Square it (4) and add the result to both sides of the equation. Similarly, find one-half of the coefficient of y (−3). Square it (9) and add the result to both sides of the equation. Thus, we have

$$(x^2 + 4x + 4) + (y^2 - 6y + 9) = 12 + 4 + 9$$
$$(x + 2)^2 + (y - 3)^2 = 25.$$

By comparing this equation to the standard form, we determine that the center of the circle is at (−2,3) and the radius is $\sqrt{25}$, or 5 (see Figure 8.10). $\qquad \blacksquare$

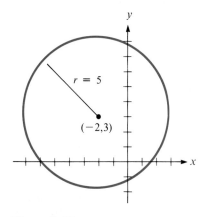

Figure 8.10

EXERCISES 8.2

In Exercises 1–10 find the equation in standard form for each of the circles from the given information.

1. Center at (−3,4), radius 2
2. Center at (−2,1), radius 3
3. Center at (0,0), radius 1
4. Center at (0,2), radius 5
5. Center at (−4,−1), tangent to the line $x = -1$
   (*Note:* A line is tangent to a circle if the two intersect at exactly one point. The tangent is perpendicular to the radius at the point of intersection.)
6. Center at (1,4), tangent to the line $y = -3$
7. Center at (3,−3), passes through the origin
8. Center at (2,−5), passes through (1,0)
9. Center at (−1,−1), passes through (6,2)
10. Center at (−4,−3), passes through (−1,−5)

In Exercises 11–24 find the center and radius of each circle.

11. $(x - 2)^2 + (y - 5)^2 = 16$
12. $(x + 1)^2 + (y + 4)^2 = 49$
13. $(x - 3)^2 + y^2 = 20$
14. $x^2 + (y + 2)^2 = 5$
15. $x^2 + y^2 = 9$
16. $x^2 + y^2 - 4 = 0$
17. $x^2 + y^2 - 10x - 6y = 15$

18. $x^2 + y^2 - 4x + 6y = 23$
19. $x^2 + y^2 + 8x - 2y - 1 = 0$
20. $x^2 + y^2 + 4x - 9 = 0$
21. $x^2 + y^2 + 7x + 3y + 4 = 0$
22. $x^2 + y^2 - 5x - y - 3 = 0$
23. $4x^2 + 4y^2 + 8x - 16y - 29 = 0$
24. $2x^2 + 2y^2 - 10x - 2y - 31 = 0$

THINK ABOUT IT

1. Find the equation in standard form of the circle with $P(-4,-3)$ and $Q(4,5)$ as endpoints of a diameter.
2. Write an equation for the circle that is tangent to both axes at $(6,0)$ and $(0,-6)$. Find the area and the circumference of this circle.

3. If the line $y = mx$ passes through the center of the circle $4x^2 + 4y^2 - 4x + 12y - 1 = 0$, then find the value of m.
4. Describe the graph of $x^2 + y^2 + 4x - 6y + 13 = 0$.
5. A graph is the set of all points in a plane that are twice as far from $(0,-4)$ as from $(4,-1)$. Show that this graph is a circle and find its center and radius.

REMEMBER THIS

1. If $c > 0$, solve for c: $36 = 9 + c^2$.
2. Solve $c^2 = a^2 - b^2$ for c if $a = 1$, $b = \frac{1}{2}$, and $c > 0$.
3. Find the distance between $(-6,-2)$ and $(10,-2)$.
4. If $a > 0$, find the distance between $(0,a)$ and $(0,-a)$.
5. Factor out 9 from $9y^2 - 54y$.
6. Factor $9x^2 + 36x + 36$.
7. Simplify $\dfrac{x^2}{\frac{1}{4}}$.

8. Find the midpoint of the line segment joining $(-6,-2)$ and $(10,-2)$.
9. Show that the diagonals of the parallelogram with vertices $A(0,0)$, $B(b,0)$, $C(b + c,d)$, and $D(c,d)$ bisect each other.
10. Find an equation for the graph that is the set of all points in a plane the sum of whose distances from $(2,0)$ and $(-2,0)$ is 8.

8.3 **The Ellipse**

The next conic section we consider is the ellipse, which is basically an oval-shaped curve. Once again, we begin our study with a geometric definition.

Definition of an Ellipse

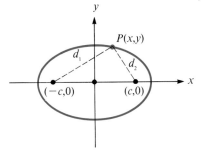

Figure 8.11

> An **ellipse** is the set of all points in a plane the sum of whose distances from two fixed points (**foci**) is a constant.

Equations for ellipses may be derived from this definition in the following way. For simplicity, let the foci be positioned at $(-c,0)$ and $(c,0)$ and let $2a$ represent the constant sum. If we let $P(x,y)$ be any point on the ellipse as shown in Figure 8.11, then by definition $d_1 + d_2 = 2a$, so it follows from the distance formula that

$$\sqrt{(x + c)^2 + (y - 0)^2} + \sqrt{(x - c)^2 + (y - 0)^2} = 2a$$

or

$$\sqrt{(x + c)^2 + y^2} = 2a - \sqrt{(x - c)^2 + y^2}.$$

Now square both sides of the equation and simplify.

$$(x + c)^2 + y^2 = 4a^2 - 4a\sqrt{(x - c)^2 + y^2} + (x - c)^2 + y^2$$
$$xc = a^2 - a\sqrt{(x - c)^2 + y^2}$$
$$a\sqrt{(x - c)^2 + y^2} = a^2 - cx$$

Again square both sides of the equation and simplify.

$$a^2[(x - c)^2 + y^2] = a^4 - 2a^2cx + c^2x^2$$
$$a^2x^2 - 2a^2cx + a^2c^2 + a^2y^2 = a^4 - 2a^2cx + c^2x^2$$
$$a^2x^2 - c^2x^2 + a^2y^2 = a^4 - a^2c^2$$
$$(a^2 - c^2)x^2 + a^2y^2 = a^2(a^2 - c^2)$$
$$\frac{x^2}{a^2} + \frac{y^2}{a^2 - c^2} = 1$$

Since $d_1 + d_2$ is greater than the distance between the foci, it follows that $a > c$ and $a^2 - c^2 > 0$. If we now define

$$b^2 = a^2 - c^2 \quad \text{or} \quad a^2 = b^2 + c^2,$$

we obtain the **standard form** of the equation of an ellipse with center at the origin and foci on the x-axis.

$$\frac{x^2}{a^2} + \frac{y^2}{b^2} = 1$$

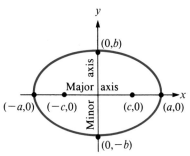

Foci: $(-c,0)$, $(c,0)$
Vertices: $(-a,0)$, $(a,0)$

Figure 8.12

Consider Figure 8.12. By setting $y = 0$, we find that the x-intercepts are $(-a,0)$ and $(a,0)$. By setting $x = 0$, we find that the y-intercepts are $(0,-b)$ and $(0,b)$. The larger segment from $(-a,0)$ to $(a,0)$ is called the **major axis;** the **minor axis** is the segment from $(0,-b)$ to $(0,b)$. The endpoints of the major axis are called the **vertices of the ellipse.**

Foci: $(-c,0)$, $(c,0)$
Vertices: $(-a,0)$, $(a,0)$

If the foci are placed on the y-axis at $(0,-c)$ and $(0,c)$, then the **standard form** of the equation of an ellipse is

$$\frac{x^2}{b^2} + \frac{y^2}{a^2} = 1,$$

where the larger denominator is denoted by a^2. The major axis is then along the y-axis as shown in Figure 8.13. Note that the foci always lie on the major axis.

Foci: $(0,c)$, $(0,-c)$
Vertices: $(0,a)$, $(0,-a)$

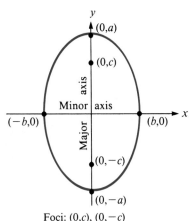

Foci: $(0,c)$, $(0,-c)$
Vertices: $(0,a)$, $(0,-a)$

Figure 8.13

EXAMPLE 1   Find the coordinates of the foci and the endpoints of the major and minor axes of the ellipse given by $4x^2 + y^2 = 36$.

Solution   First, divide both sides of the equation by 36 to obtain the standard form.

$$\frac{x^2}{9} + \frac{y^2}{36} = 1$$

Then

$$a^2 = 36 \qquad a = 6$$
$$b^2 = 9 \qquad b = 3.$$

To find c, we replace a^2 by 36 and b^2 by 9 in the formula

$$a^2 = b^2 + c^2$$
$$36 = 9 + c^2$$
$$27 = c^2 \quad \text{or} \quad c = \sqrt{27} = 3\sqrt{3}.$$

The major axis is along the y-axis since the larger denominator appears in the y term. Thus,

endpoints of major axis: $(0,-6)$, $(0,6)$
endpoints of minor axis: $(-3,0)$, $(3,0)$
coordinates of foci: $(0,-3\sqrt{3})$, $(0,3\sqrt{3})$.

These points are shown in the graph of the ellipse in Figure 8.14.   ∎

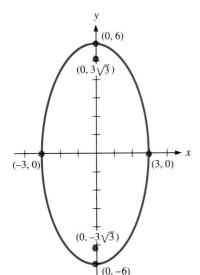

Figure 8.14

EXAMPLE 2   Find the equation in standard form of the ellipse with center at the origin, one focus at $(3,0)$, and one vertex at $(4,0)$.

Solution   Since c is the distance from the center to a focus, c is 3. Since a is the distance from the center to a vertex, a is 4. To find b^2, replace c by 3 and a by 4 in the formula

$$a^2 = b^2 + c^2$$
$$(4)^2 = b^2 + (3)^2$$
$$7 = b^2.$$

Since the major axis is along the x-axis, the standard form is

$$\frac{x^2}{a^2} + \frac{y^2}{b^2} = 1.$$

Thus,

$$\frac{x^2}{16} + \frac{y^2}{7} = 1$$

is the standard equation of the ellipse.   ∎

If the ellipse is centered at the point (h,k), the **standard form** is

$$\frac{(x - h)^2}{a^2} + \frac{(y - k)^2}{b^2} = 1$$

when the major axis is parallel to the x-axis and

$$\frac{(x - h)^2}{b^2} + \frac{(y - k)^2}{a^2} = 1$$

when the major axis is parallel to the y-axis. As in the case of the circle, the standard form is obtained by completing the square.

EXAMPLE 3   Find the coordinates of the foci and the endpoints of the major and minor axes of the ellipse given by

$$9x^2 + 4y^2 + 18x - 8y - 23 = 0.$$

Also, sketch the curve.

Solution   First, put the equation in standard form by completing the square.

$$(9x^2 + 18x \qquad) + (4y^2 - 8y \qquad) = 23$$
$$9(x^2 + 2x \qquad) + 4(y^2 - 2y \qquad) = 23$$
$$9(x^2 + 2x + 1) + 4(y^2 - 2y + 1) = 23 + 9(1) + 4(1) = 36$$

Be careful to add $9(1)$ and $4(1)$ to the right side of the equation. Now divide both sides of the equation by 36.

$$\frac{(x + 1)^2}{4} + \frac{(y - 1)^2}{9} = 1$$

From this equation in standard form we conclude that the center is at $(-1,1)$.

$$a^2 = 9 \qquad a = 3$$
$$b^2 = 4 \qquad b = 2$$

To find c, we replace a^2 by 9 and b^2 by 4 in the formula

$$a^2 = b^2 + c^2$$
$$9 = 4 + c^2$$
$$5 = c^2 \quad \text{or} \quad c = \sqrt{5}.$$

The major axis is parallel to the y-axis since the larger denominator appears in the y term; a is the distance from the center to the vertices. Thus,

endpoints of major axis: $(-1,-2)$, $(-1,4)$.

The distance from the center to the endpoints of the minor axis is b. Thus,

endpoints of minor axis: $(-3,1)$, $(1,1)$.

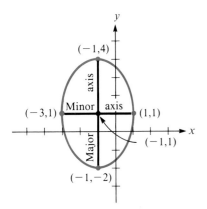

Figure 8.15

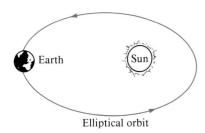

Elliptical orbit

Figure 8.16

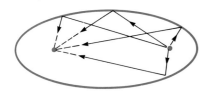

Figure 8.17

The distance from the center to the foci is c. Thus,

$$\text{coordinates of the foci:} \quad (-1, 1 - \sqrt{5}), (-1, 1 + \sqrt{5}).$$

The ellipse is graphed in Figure 8.15. ∎

Finally, let us consider a few of the ways the ellipse appears in the physical world.

1. In our solar system many bodies revolve in elliptical orbits around a larger body that is located at one focus. For instance, all planets and some comets travel in elliptical paths with the sun at one focus (see Figure 8.16). The moon and artificial satellites follow a similar path around the earth. The ellipse was first analyzed in detail by the Greek mathematician Apollonius in about 230 B.C. for its aesthetic, not practical, value. However, if this study had not been available, it is likely that Johannes Kepler (A.D. 1600) would have been unable to explain planetary motion scientifically, which would have caused a long delay in the development of astronomy and subsequent sciences. This episode is just one of many cases in which a purely mathematical investigation later proved to be of great practical value.

2. The ellipse has a reflection property that causes any ray or wave that originates at one focus to strike the ellipse and pass through the other focus. Acoustically, this property means that in a room with an elliptical ceiling, a slight noise made at one focus can be heard at the other focus (see Figure 8.17). However, if you are standing between the foci, you hear nothing. Such rooms are called whispering galleries, and a famous one is located in the Capitol in Washington, D.C.

3. Arches whose main purpose is beauty and not strength are often elliptical in shape. These ornamental arches are usually large structures made of masonry.

EXERCISES 8.3

In Exercises 1–14 find the coordinates of the foci and the endpoints of the major and minor axes. Sketch each ellipse.

1. $\dfrac{x^2}{25} + \dfrac{y^2}{16} = 1$

2. $\dfrac{x^2}{36} + \dfrac{y^2}{100} = 1$

3. $\dfrac{x^2}{1} + \dfrac{y^2}{4} = 1$

4. $\dfrac{x^2}{49} + \dfrac{y^2}{9} = 1$

5. $x^2 + 9y^2 = 36$
6. $9x^2 + 4y^2 = 144$
7. $4x^2 + y^2 = 1$
8. $x^2 + 9y^2 = 25$

9. $\dfrac{(x - 1)^2}{9} + \dfrac{(y + 3)^2}{25} = 1$

10. $\dfrac{(x + 2)^2}{100} + \dfrac{(y - 2)^2}{64} = 1$

11. $4(x - 3)^2 + 25(y + 1)^2 = 100$
12. $25(x + 4)^2 + 16y^2 = 400$
13. $4x^2 + 9y^2 + 8x - 54y + 49 = 0$
14. $9x^2 + y^2 - 36x + 2y + 1 = 0$

In Exercises 15–24 find the equation in standard form of the ellipse satisfying the conditions.

15. Center at origin, vertex $(0,5)$, focus $(0,3)$
16. Center at origin, vertex $(-10,0)$, focus $(-8,0)$

17. Center at origin, focus (4,0), length of major axis 12
18. Center at origin, focus (0,−2), length of minor axis 2
19. Center at origin, vertex (3,0), length of minor axis 4
20. Center at origin, vertex (0,3), passes through (2,1)
21. Center at (2,2), focus (2,−1), vertex (2,−3)
22. Center at (−1,3), vertex (3,3), length of minor axis 6
23. Foci at (5,3) and (−1,3), length of major axis 10
24. Vertices at (4,1) and (−8,1), focus (−4,1)

25. The earth travels in an elliptical orbit around the sun, which is located at one of the foci. The longest distance between the earth and the sun is 94,500,000 mi and the shortest distance is 91,500,000 mi.
    a. Find the length of the major axis of the ellipse.
    b. Find the distance between the foci of the orbit.
26. An elliptical arch has a height of 20 ft and a span of 50 ft. How high is the arch 15 ft each side of center?

THINK ABOUT IT

1. Describe the graph of the ellipse $(x^2/a^2) + (y^2/b^2) = 1$ if $a = b$. How are the foci related in this case?
2. An ellipse may be drawn using a pencil, string, and tacks as shown below. Explain the mathematical basis for this method.

3. If the area A of the ellipse enclosed by
$$\frac{(x - h)^2}{a^2} + \frac{(y - k)^2}{b^2} = 1$$
is given by $A = \pi ab$, find the area enclosed by $5x^2 + 6y^2 + 10x - 12y - 169 = 0$.
4. Use properties of the ellipse to find an equation for the graph that is the set of all points in a plane the sum of whose distances from (2,0) and (−2,0) is 10.
5. A line segment joining two points on an ellipse that contains a focus and that is perpendicular to the major axis is called a **focal chord** (or a **latus rectum**) of the ellipse. Find the endpoints for the two focal chords of an ellipse whose equation is $(x^2/a^2) + (y^2/b^2) = 1$. What is the length of these chords?

REMEMBER THIS

1. Solve $c^2 = a^2 + b^2$ for b^2 if $c = 10$ and $a = 7$.
2. Solve for b^2: $\dfrac{4^2}{4} - \dfrac{3^2}{b^2} = 1$.
3. Factor out -9 from $-9y^2 + 54y$.
4. What is the horizontal asymptote for $y = f(x) = 1/x$? What does it mean to say the line $y = b$ is a horizontal asymptote of the graph of f?
5. Graph the lines given by $y = \pm\frac{3}{2}x$.

6. Find an equation for the line that is the perpendicular bisector of the line segment whose endpoints are (0,6) and (−4,0).
7. Determine the center of the circle given by $(x + 6)^2 + (y - 3)^2 = 16$.
8. Write the equation in standard form for the circle with center (0,−5) and radius $\sqrt{7}$.
9. Write in standard form: $x^2 + y^2 - x - 3y + 1 = 0$.
10. Find an equation for the graph that is the set of all points in a plane the difference of whose distances from (2,0) and (−2,0) is 2.

8.4   The Hyperbola

The geometric definition of a hyperbola resembles the definition of an ellipse and is stated below.

Definition of a Hyperbola

A **hyperbola** is the set of all points in a plane the difference of whose distances from two fixed points (**foci**) is a positive constant.

Thus, the distances between the foci and a point on the figure maintain a *constant difference* for a hyperbola and a *constant sum* for an ellipse. To derive the standard form of the equation of a hyperbola with foci on the x-axis and center at the origin, we position the foci at $(-c,0)$ and $(c,0)$. Let $2a$ represent the positive constant of the definition and let $P(x,y)$ be any point on the hyperbola as shown in Figure 8.18.

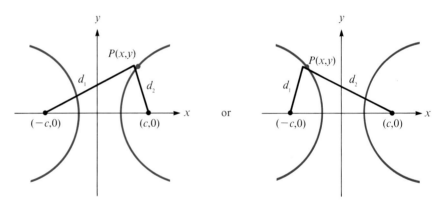

Figure 8.18

Since $d_1 < d_2$ or $d_1 > d_2$, by definition,

$$\left| d_1 - d_2 \right| = 2a \quad \text{or} \quad d_1 - d_2 = \pm 2a.$$

Then by the distance formula, we have

$$\sqrt{(x + c)^2 + (y - 0)^2} - \sqrt{(x - c)^2 + (y - 0)^2} = \pm 2a$$

or

$$\sqrt{(x + c)^2 + y^2} = \pm 2a + \sqrt{(x - c)^2 + y^2}.$$

Now square both sides of the equation and simplify.

$$(x + c)^2 + y^2 = 4a^2 \pm 4a\sqrt{(x - c)^2 + y^2} + (x - c)^2 + y^2$$
$$4cx = 4a^2 \pm 4a\sqrt{(x - c)^2 + y^2}$$
$$cx - a^2 = \pm a\sqrt{(x - c)^2 + y^2}$$

Again square both sides and simplify.

$$c^2x^2 - 2cxa^2 + a^4 = a^2[(x - c)^2 + y^2]$$
$$c^2x^2 - 2cxa^2 + a^4 = a^2x^2 - 2cxa^2 + a^2c^2 + a^2y^2$$
$$c^2x^2 - a^2x^2 - a^2y^2 = a^2c^2 - a^4$$
$$(c^2 - a^2)x^2 - a^2y^2 = a^2(c^2 - a^2)$$
$$\frac{x^2}{a^2} - \frac{y^2}{c^2 - a^2} = 1$$

If we now define

$$b^2 = c^2 - a^2 \quad \text{or} \quad c^2 = a^2 + b^2,$$

we obtain the **standard form** of a hyperbola with center at the origin and foci on the x-axis.

$$\frac{x^2}{a^2} - \frac{y^2}{b^2} = 1$$

Consider Figure 8.19. By setting $y = 0$, we find that the x-intercepts are $(-a,0)$ and $(a,0)$. The line segment joining these two points is called the **transverse axis.** The endpoints of the transverse axis are called the **vertices of the hyperbola.** By setting $x = 0$, we find that there are no y-intercepts. The line segment from $(0,b)$ to $(0,-b)$ is called the **conjugate axis.** To determine the significance of b, we rewrite $(x^2/a^2) - (y^2/b^2) = 1$ as

$$y = \frac{\pm bx}{a}\sqrt{1 - \frac{a^2}{x^2}}.$$

As $|x|$ gets very large, $1 - a^2/x^2$ approaches 1. Thus, the graph of the hyperbola approaches the lines

$$y = \pm\frac{b}{a}x.$$

These lines are called the **asymptotes of the hyperbola.** They are a great aid in sketching the curve. As shown in Figure 8.19, the asymptotes are the diagonals of a rectangle of dimensions $2a$ by $2b$.

EXAMPLE 1   For the hyperbola given by $4x^2 - y^2 = 36$, find the coordinates of the vertices and the foci. Also, determine the asymptotes and sketch the curve.

Solution   First, divide both sides of the equation by 36 to obtain the standard form.

$$\frac{x^2}{9} - \frac{y^2}{36} = 1$$

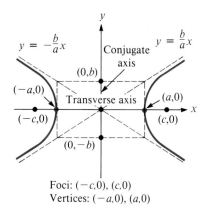

Foci: $(-c,0)$, $(c,0)$
Vertices: $(-a,0)$, $(a,0)$

Figure 8.19

Then

$$a^2 = 9 \qquad a = 3$$
$$b^2 = 36 \qquad b = 6.$$

Note that a^2 is not necessarily the larger denominator. Since a is the distance from the center to the vertices, we have

coordinates of the vertices: $(-3,0)$, $(3,0)$.

To find c, we replace a^2 by 9 and b^2 by 36 in the formula

$$c^2 = a^2 + b^2$$
$$= 9 + 36$$
$$c^2 = 45 \quad \text{or} \quad c = \sqrt{45} = 3\sqrt{5}.$$

The distance from the center to the foci is c. Thus,

coordinates of the foci: $(-3\sqrt{5},0)$, $(3\sqrt{5},0)$.

For the asymptotes we have

$$y = \pm\frac{b}{a}x = \pm\frac{6}{3}x = \pm 2x.$$

The hyperbola is sketched in Figure 8.20.   ∎

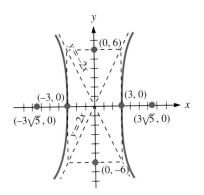

Figure 8.20

If the foci are positioned on the y-axis at $(0,-c)$ and $(0,c)$, the **standard form** of the equation of a hyperbola is

$$\frac{y^2}{a^2} - \frac{x^2}{b^2} = 1.$$

In this case the asymptotes are given by $y = \pm ax/b$, as shown in Figure 8.21.

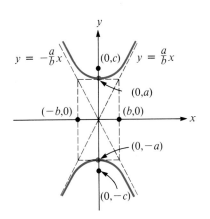

Figure 8.21

EXAMPLE 2   For the hyperbola given by $25y^2 - 4x^2 = 100$, find the coordinates of the vertices and the foci. Also, determine the asymptotes and sketch the curve.

Solution   First, divide both sides of the equation by 100 to obtain the standard form.

$$\frac{y^2}{4} - \frac{x^2}{25} = 1$$

Then

$$a^2 = 4 \qquad a = 2$$
$$b^2 = 25 \qquad b = 5.$$

Since a is the distance from the center to the vertices, we have

coordinates of the vertices: $(0,2)$, $(0,-2)$.

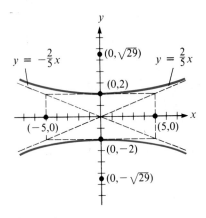

Figure 8.22

To find c, we replace a^2 by 4 and b^2 by 25 in the formula

$$c^2 = a^2 + b^2$$
$$= 4 + 25$$
$$c^2 = 29 \quad \text{or} \quad c = \sqrt{29}.$$

The distance from the center to the foci is c. Thus,

$$\text{coordinates of foci:} \quad (0, \sqrt{29}), (0, -\sqrt{29}).$$

For the asymptotes we have

$$y = \pm\frac{a}{b}x = \pm\frac{2}{5}x.$$

The hyperbola is sketched in Figure 8.22. ■

EXAMPLE 3   Find the equation in standard form of the hyperbola with center at the origin, one focus at (0,5), and one vertex at (0,3).

Solution   Since c is the distance from the center to a focus, c is 5. Since a is the distance from the center to a vertex, a is 3. To find b^2, we replace c by 5 and a by 3 in the formula

$$c^2 = a^2 + b^2$$
$$(5)^2 = (3)^2 + b^2$$
$$16 = b^2.$$

Since the foci are along the y-axis, the standard form is

$$\frac{y^2}{a^2} - \frac{x^2}{b^2} = 1.$$

Thus, the standard equation of the hyperbola is

$$\frac{y^2}{9} - \frac{x^2}{16} = 1.$$

■

If the hyperbola is centered at the point (h,k), the **standard form** is

$$\frac{(x - h)^2}{a^2} - \frac{(y - k)^2}{b^2} = 1$$

when the transverse axis is parallel to the x-axis. In this case the asymptotes are given by $y - k = (\pm b/a)(x - h)$. If the transverse axis is parallel to the y-axis, the **standard form** is

$$\frac{(y - k)^2}{a^2} - \frac{(x - h)^2}{b^2} = 1$$

and the asymptotes are given by $y - k = (\pm a/b)(x - h)$.

EXAMPLE 4 Find the coordinates of the vertices and the foci of the hyperbola given by $16y^2 - x^2 - 32y + 4x - 4 = 0$. Also, determine the asymptotes and sketch the curve.

Solution First, put the equation in standard form by completing the square.

$$(16y^2 - 32y \qquad) - (x^2 - 4x \qquad) = 4$$
$$16(y^2 - 2y \qquad) - (x^2 - 4x \qquad) = 4$$
$$16(y^2 - 2y + 1) - (x^2 - 4x + 4) = 4 + 16(1) + (-1)4$$
$$16(y - 1)^2 - (x - 2)^2 = 16$$

Dividing both sides of the equation by 16, we have

$$\frac{(y - 1)^2}{1} - \frac{(x - 2)^2}{16} = 1.$$

From this equation in standard form we conclude that the center is at $(2,1)$.

$$a^2 = 1 \qquad a = 1$$
$$b^2 = 16 \qquad b = 4$$

The transverse axis is parallel to the y-axis since the y term is positive; a is the distance from the center to the vertices. Thus,

coordinates of vertices: $(2,0)$, $(2,2)$.

To find c, we replace a^2 by 1 and b^2 by 16 in the formula

$$c^2 = a^2 + b^2$$
$$= 1 + 16$$
$$c^2 = 17 \quad \text{or} \quad c = \sqrt{17}.$$

The distance from the center to the foci is c. Thus,

coordinates of foci: $(2,1 - \sqrt{17})$, $(2,1 + \sqrt{17})$.

The asymptotes are the diagonals of the rectangle of dimensions $2a$ by $2b$ that is centered at $(2,1)$. The equations of these asymptotes are

$$y - k = \pm\frac{a}{b}(x - h)$$

$$y - 1 = \pm\frac{1}{4}(x - 2).$$

The hyperbola is graphed in Figure 8.23.   ■

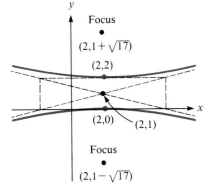

Figure 8.23

EXERCISES 8.4

In Exercises 1–14 find the coordinates of the vertices and the foci, determine the asymptotes, and sketch each curve.

1. $\dfrac{x^2}{16} - \dfrac{y^2}{9} = 1$

2. $\dfrac{x^2}{36} - \dfrac{y^2}{100} = 1$

3. $\dfrac{y^2}{25} - \dfrac{x^2}{16} = 1$

4. $\dfrac{y^2}{1} - \dfrac{x^2}{4} = 1$

5. $x^2 - 9y^2 = 36$

6. $9x^2 - 4y^2 = 144$

7. $4y^2 - x^2 = 1$

8. $y^2 - 9x^2 = 25$

9. $\dfrac{(x + 2)^2}{9} - \dfrac{(y - 3)^2}{25} = 1$

10. $\dfrac{(y - 1)^2}{64} - \dfrac{(x - 2)^2}{36} = 1$

11. $4(y + 1)^2 - 25(x + 3)^2 = 100$
12. $25(x + 4)^2 - 16y^2 = 400$
13. $4x^2 - 9y^2 + 8x - 54y - 113 = 0$
14. $9y^2 - x^2 - 36y + 2x - 1 = 0$

In Exercises 15–24 find the equation in standard form of the hyperbola satisfying the conditions.

15. Center at origin, focus $(0,5)$, vertex $(0,4)$
16. Center at origin, focus $(-10,0)$, vertex $(-8,0)$
17. Center at origin, focus $(4,0)$, length of transverse axis 6
18. Center at origin, focus $(0,-2)$, length of conjugate axis 2
19. Center at origin, vertex $(3,0)$, length of conjugate axis 10
20. Center at origin, vertex $(0,2)$, passes through $(3,4)$
21. Center at $(-3,-4)$, focus $(2,-4)$, vertex $(0,-4)$
22. Center at $(5,0)$, vertex $(5,6)$, length of conjugate axis 8
   Foci at $(3,4)$ and $(3,-2)$, length of transverse axis 4
24. Vertices at $(4,1)$ and $(-10,1)$, focus $(7,1)$

THINK ABOUT IT

1. Find equations in standard form for the two hyperbolas that satisfy the following conditions: The asymptotes are $y = \pm\frac{2}{3}x$ and the length of the transverse axis is 12.
2. If the asymptotes of the hyperbola $(x^2/a^2) - (y^2/b^2) = 1$ are perpendicular, then what is the relation between a and b? Establish the relation with a proof.
3. Describe the graph of $x^2 - y^2 + 4x + 4 = 0$.
4. Use properties of the hyperbola to find an equation for the graph that is the set of all points in a plane the difference of whose distances from $(4,0)$ and $(-4,0)$ is 6.
5. A line segment joining two points on a hyperbola that contains a focus and that is perpendicular to the transverse axis is called a **focal chord** (or **latus rectum**) of the hyperbola. Show that $2b^2/a$ is the length of both focal chords of a hyperbola whose equation is $(x^2/a^2) - (y^2/b^2) = 1$.

REMEMBER THIS

1. How are the graphs of $y = x^2$ and $y = -x^2$ related?
2. Graph $y = 4 - x^2$.
3. **a.** Graph $y = -1$.
   b. Graph $x = 2$.
4. Find the vertex of the graph of $y = 3(x - 2)^2 + 4$.
5. If $y = x^2 - 6x + 5$, what is the equation of the axis of symmetry?
6. Find the radius of the circle given by $x^2 + y^2 - 24y = 25$.
7. Find the standard equation of the circle with center $(2,4)$ that passes through $(-1,1)$.
8. Write the standard equation for the ellipse with vertices at $(6,2)$ and $(-10,2)$ and minor axis of length 6.
9. Find the length of the minor axis of the ellipse given by $x^2 + 9y^2 = 16$.
10. Find an equation for the graph that is the set of all points in a plane equidistant from the line $x = -4$ and the point $(4,0)$.

8.5 The Parabola

We have already done some work with parabolas in connection with graphing quadratic functions in Section 3.2. We now expand our coverage by considering some geometric properties of this curve and developing the standard forms of equations of parabolas with a horizontal or vertical axis of symmetry. As in the preceding sections, a geometric definition is our starting point.

Definition of a Parabola

A **parabola** is the set of all points in a plane equidistant from a fixed line (**directrix**) and a fixed point (**focus**) not on the line.

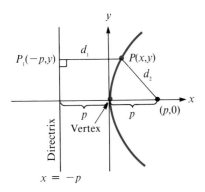

Figure 8.24

From this definition we derive a general equation for a parabola as follows. For simplicity, let the directrix be the line $x = -p$ and position the focus at $(p,0)$. $P(x,y)$ represents any point on the parabola (see Figure 8.24). Since P_1 and P have the same y-coordinate, the distance d_1 is given by

$$d_1 = |x - (-p)| = |x + p|.$$

The distance from P to $(p,0)$ is given by

$$d_2 = \sqrt{(x - p)^2 + (y - 0)^2}.$$

The geometric condition for a parabola states that

$$d_1 = d_2.$$

Thus,

$$|x + p| = \sqrt{(x - p)^2 + y^2}.$$

Squaring both sides of the equation yields

$$(x + p)^2 = (x - p)^2 + y^2.$$

Simplifying this equation, we have

$$x^2 + 2px + p^2 = x^2 - 2px + p^2 + y^2$$

$$4px = y^2.$$

This equation is the **standard form** of a parabola with directrix $x = -p$ and focus at $(p,0)$. The line through the focus that is perpendicular to the directrix is called the **axis of symmetry.** In this case the axis of symmetry is the x-axis. The point on the axis of symmetry that is midway between the focus and the directrix is called the **vertex.** The vertex is the turning point of the parabola. In this case the vertex is the origin. Note that p is the distance from the vertex to the focus and from the vertex to the directrix.

EXAMPLE 1   Find the focus and the directrix of the parabola given by $y^2 = 12x$.

Solution   Matching the equation $y^2 = 12x$ to the standard form

$$y^2 = 4px,$$

we write

$$y^2 = 12x = 4(3)x.$$

Thus,

$$p = 3.$$

The focus is on the axis of symmetry (the x-axis) p units to the right of the vertex (the origin). Thus,

$$\text{focus:} \quad (3,0).$$

The directrix is the line p units to the left of the vertex. Thus,

$$\text{directrix:} \quad x = -3. \qquad \blacksquare$$

There are three other possibilities for a parabola with a directrix that is parallel to the x- or y-axis and whose vertex is the origin. These cases are illustrated in Figures 8.25–8.27. In all cases $p > 0$.

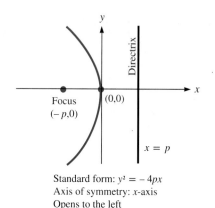

Standard form: $y^2 = -4px$
Axis of symmetry: x-axis
Opens to the left

Figure 8.25   Case 2

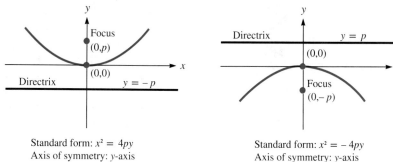

Standard form: $x^2 = 4py$
Axis of symmetry: y-axis
Opens upward

Figure 8.26   Case 3

Standard form: $x^2 = -4py$
Axis of symmetry: y-axis
Opens downward

Figure 8.27   Case 4

To summarize, the graphs of the equations $y^2 = \pm 4px$ and $x^2 = \pm 4py$ are parabolas. If the y term is squared, the axis of symmetry is the x-axis. The parabola opens to the right when the coefficient of x is positive and to the left when this coefficient is negative. When the x term is squared, the axis of symmetry is the y-axis. The parabola opens upward when the coefficient of y is positive and downward when this coefficient is negative. In all cases p gives the distance from the vertex to the focus and from the vertex to the directrix.

EXAMPLE 2   Find the focus and directrix of the parabola given by $x^2 = -6y$.

Solution   Matching the equation $x^2 = -6y$ to the standard form in case 4,

$$x^2 = -4py,$$

we write

$$x^2 = -6y = -4(\tfrac{3}{2})y.$$

Thus,

$$p = \tfrac{3}{2}.$$

The focus is on the axis of symmetry (the y-axis) p units down from the vertex (the origin). Thus,

$$\text{focus:} \quad (0, -\tfrac{3}{2}).$$

The directrix is the line p units up from the vertex. Thus,

$$\text{directrix:} \quad y = \tfrac{3}{2}. \qquad \blacksquare$$

EXAMPLE 3   A parabola with its vertex at the origin has its focus at $(-5,0)$. Find the directrix and the equation in standard form of the parabola and sketch the curve.

Solution   This is an example of case 2. The term p, which represents the distance from the vertex $(0,0)$ to the focus $(-5,0)$, is 5. Thus,

$$\text{directrix:} \quad x = 5.$$

To find the standard equation of the parabola, replace p by 5 in the form

$$y^2 = -4px$$

to obtain

$$y^2 = -20x.$$

The parabola opens to the left and is graphed in Figure 8.28. $\blacksquare$

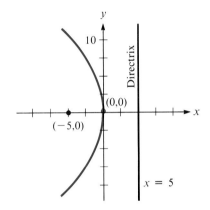

Figure 8.28

If the vertex of the parabola is at the point (h,k) and the directrix is parallel to the x- or y-axis, then there are four possible standard forms for the parabola. These cases are illustrated in Figure 8.29. In all cases $p > 0$.

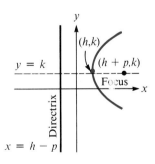

Case 1:
$(y - k)^2 = 4p(x - h)$

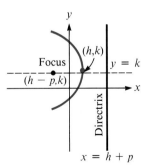

Case 2:
$(y - k)^2 = -4p(x - h)$

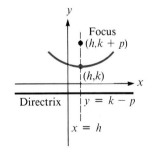

Case 3:
$(x - h)^2 = 4p(y - k)$

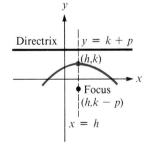

Case 4:
$(x - h)^2 = -4p(y - k)$

Figure 8.29

EXAMPLE 4   For the parabola given by $y = x^2 - 2x$, find the vertex, the focus, and the directrix. Also, sketch the curve.

Solution   First, put the equation in standard form by completing the square.

$$y = (x^2 - 2x \qquad)$$
$$y + 1 = (x^2 - 2x + 1)$$
$$y + 1 = (x - 1)^2$$

Matching this equation to the form in case 3,

$$(x - h)^2 = 4p(y - k),$$

we conclude that

$$h = 1, \qquad k = -1, \qquad 4p = 1, \quad \text{or} \quad p = \tfrac{1}{4}.$$

Thus,

$$\text{vertex:} \quad (1, -1).$$

The focus is on the axis of symmetry $(x = 1)$ p units up from the vertex $(1, -1)$. Thus,

$$\text{focus:} \quad (1, -\tfrac{3}{4}).$$

The directrix is the line p units down from the vertex. Thus,

$$\text{directrix:} \quad y = -\tfrac{5}{4}.$$

The parabola opens upward and is graphed in Figure 8.30.   ■

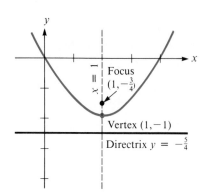

In conclusion, let us consider a few of the applications of the parabola.

1. The parabola has an important reflection property. Any ray or wave that originates at the focus and strikes the parabola is reflected parallel to the axis of symmetry (see Figure 8.31). For this reason, an instrument such as a flashlight or a searchlight uses a parabolic reflector with the bulb located at the focus. The reflector redirects light that would otherwise be wasted parallel to the axis so that a straight beam of light is formed. In an automobile headlight a bulb located at the focus produces the high beam. For the low beam the bulb is placed slightly ahead of and above the focus. In reverse, the reflection property of a parabola causes any ray or wave that comes into a parabolic reflector parallel to the axis of symmetry to be directed to the focus point. Radar, radio antennas, and reflecting telescopes work on this principle (see Figure 8.32).

2. Arches whose main purpose is strength are usually parabolic in shape and constructed of steel.

3. In a suspension bridge the cables are designed for engineering purposes to hang in a shape that is almost parabolic.

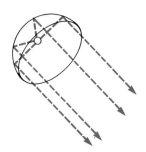

Figure 8.31

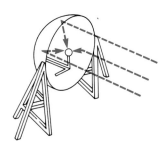

Figure 8.32

4. Every comet in our solar system travels in an orbit that is either an ellipse, a parabola, or a branch of a hyperbola. In each case the sun is located at one focus. Only comets with an elliptical orbit reappear.
5. Projectiles, such as a baseball or the water that streams from a fountain, follow a parabolic path.

EXERCISES 8.5

In Exercises 1–14 find the vertex, the focus, and the equation of the directrix and sketch each curve.

1. $y^2 = 4x$
2. $y^2 = -8x$
3. $x^2 = -2y$
4. $x^2 = 5y$
5. $y^2 + 12x = 0$
6. $y^2 - 10x = 0$
7. $4x^2 - 3y = 0$
8. $2x^2 + 3y = 0$
9. $(y + 2)^2 = 4(x + 1)$
10. $(x - 1)^2 = -6y$
11. $y = x^2 - 4x$
12. $x = y^2 + 2y$
13. $x^2 - 2x - 4y - 7 = 0$
14. $y^2 + 4y + 3x - 8 = 0$

In Exercises 15–24 find the equation in standard form of the parabola satisfying the conditions.

15. Vertex at origin, focus $(0,3)$
16. Vertex at origin, focus $(-2,0)$
17. Vertex at origin, directrix $x = -\frac{1}{2}$
18. Vertex at origin, directrix $y = \frac{3}{2}$
19. Focus $(0,4)$, directrix $y = -4$
20. Focus $(-\frac{5}{2},0)$, directrix $x = \frac{5}{2}$
21. Focus $(2,1)$, directrix $x = -4$
22. Focus $(-1,-3)$, directrix $y = 5$
23. Vertex $(3,0)$, focus $(3,3)$
24. Vertex $(1,4)$, directrix $x = \frac{7}{2}$
25. Consider the parabolic reflector shown. Choose axes in a convenient position and determine (a) an equation of the parabola and (b) the location of the focus point.

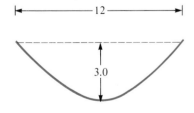

26. Answer the questions in Exercise 25 for the parabolic reflector shown below.

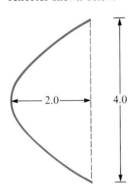

27. A parabolic arch has a height of 25 ft and a span of 40 ft. How high is the arch 8 ft each side of the center?

THINK ABOUT IT

1. Find the vertex of the parabola whose equation is $y = ax^2 + bx + c$ with $a \neq 0$.
2. Find the focus and the directrix of the parabola whose equation is $y = ax^2 + bx + c$ with $a \neq 0$.
3. Use properties of the parabola to find an equation for the graph that is the set of all points in a plane equidistant from the line $y = 4$ and the point $(-1,0)$.
4. Use the parabola with standard form $y^2 = 4px$ to explain why the point on a parabola closest to the focus is the vertex.

5. **a.** A line segment joining two points on a parabola that contains the focus and that is perpendicular to the axis of symmetry is called the **focal chord** (or **latus rectum**) of the parabola. Find the length of the focal chord of a parabola whose equation is $y^2 = 4px$.
   b. Use the result in part a to find equations in standard form for the two parabolas that satisfy the following conditions: The vertex is at the origin, the axis of symmetry is the x-axis, and the length of the focal chord is 8.

REMEMBER THIS

In Exercises 1–4 match each graph with the equation that illustrates that case.

Graph	Equation
1. Circle	a. $x^2 - y^2 = 4$
2. Ellipse	b. $x^2 + y^2 = 4$
3. Hyperbola	c. $2x^2 + y = 4$
4. Parabola	d. $2x^2 + y^2 = 4$

5. Explain why there is no ordered pair of real numbers that satisfies $x^2 + y^2 = -4$.
6. Find the standard equation of a circle with center $(-2,3)$ that is tangent to the line $y = -4$.
7. Write in standard form for an ellipse: $4x^2 + 9y^2 - 16x + 36y - 20 = 0$.
8. Write in standard form for a hyperbola: $25x^2 - 9y^2 + 150x + 36y - 36 = 0$.
9. Find the equations of the asymptotes of the graph of
$$\frac{(y-2)^2}{1} - \frac{(x-1)^2}{9} = 1.$$
10. Find an equation for the graph that is the set of all points in a plane equidistant from $(3,0)$ and $(-1,4)$. Identify the graph of this equation.

8.6 Classifying Conic Sections

All of the standard forms for conic sections that we have developed may be converted through basic algebra to an equation of the form

$$Ax^2 + Cy^2 + Dx + Ey + F = 0,$$

where A and C are not both zero. This equation is called the **general form** of an equation of a conic section with axis or axes parallel to the coordinate axes. If a graph is a circle, an ellipse, a hyperbola, or a parabola, then in general form A and C must satisfy certain conditions that are apparent from the standard forms of these conics. We summarize these conditions below, along with the degenerate possibilities for each case.

Conditions on A and C	Conic Section	Degenerate Possibilities
$A = C \neq 0$	Circle	A point or no graph at all
$A \neq C$, A and C have the same sign.	Ellipse	A point or no graph at all
A and C have opposite signs.	Hyperbola	Two intersecting straight lines
$A = 0$ or $C = 0$ (but not both)	Parabola	A line, a pair of parallel lines, or no graph at all

Example 1 illustrates the main application of this chart.

EXAMPLE 1   Identify the graph of the following equations if the graph is either a circle, an ellipse, a hyperbola, or a parabola.

a. $4x^2 - 9y^2 + 8x - 54y - 113 = 0$ b. $2x^2 - x = -3y^2$

Solution   Since the degenerate possibilities have been eliminated, we proceed as follows:

a. The equation $4x^2 - 9y^2 + 8x - 54y - 113 = 0$ is in general form with $A = 4$ and $C = -9$. Since A and C have different signs the graph of the equation is a hyperbola.

b. To determine the conic given by $2x^2 - x = -3y^2$, we first write the equation in general form as

$$2x^2 + 3y^2 - x = 0.$$

In this case A and C have the same sign with $A \neq C$. Thus, the graph of the equation is an ellipse.   ■

The next example shows the analysis of two equations that lead to degenerate cases.

EXAMPLE 2   Identify the graph of the following equations.

a. $2x^2 + y^2 + 1 = 0$ b. $x^2 + y^2 + 4x - 6y + 13 = 0$

Solution

a. The equation $2x^2 + y^2 + 1 = 0$ does not define an ellipse, even though A and C have the same sign with $A \neq C$. There is no ordered pair of real numbers that satisfies $2x^2 + y^2 = -1$ since $2x^2$ and y^2 are never negative, so their sum can never be -1. Thus, the equation has no graph.

b. On first inspection, this equation seems to define a circle. To be sure, we need to complete the square and convert the equation to standard form. By doing so, in this case we obtain

$$(x + 2)^2 + (y - 3)^2 = 0.$$

Only $x = -2$, $y = 3$ satisfies the equation, so the graph is the point $(-2,3)$.   ■

EXERCISES 8.6

In Exercises 1–10 identify the graph of each equation if the graph is either a circle, an ellipse, a hyperbola, or a parabola.

1. $y^2 - 3x + 2y + 1 = 0$
2. $9y^2 - x^2 - 36x + 2y - 1 = 0$
3. $4x^2 + 9y^2 + 8x - 54y - 49 = 0$
4. $x^2 + y^2 + 7x + 3y - 4 = 0$
5. $x^2 = y^2 - 4$
6. $3y^2 = 4x - x^2$
7. $3y = 4x - x^2$
8. $5x^2 = 1 - 5y^2$
9. $2y^2 = 9 - 2x^2$
10. $5x - 2y = y^2$

In Exercises 11–20 identify the graph of each equation. Degenerate cases are possible here.

11. $x^2 - y^2 - 4y - 4 = 0$
12. $x^2 + y^2 - 4y + 4 = 0$
13. $x^2 + 2x = 1 - y^2$
14. $x^2 + 2x = y^2 - 1$
15. $x^2 + 2x = 1 - y$
16. $x^2 + 2x = y^2 - 4$
17. $9x^2 + 4y^2 + 18x - 8y + 13 = 0$
18. $4x^2 + 9y^2 + 2x - 4 = 0$
19. $4x^2 + 9y^2 - 2x + 4 = 0$
20. $4y^2 + 3 = 8y$

THINK ABOUT IT

1. The distance an object falls due to gravity varies directly as the square of the time of fall. Identify the conic section that is the graph of this relation.
2. A graph is the set of all points in a plane whose distance from $(2,0)$ is always one-half their distance from $x = 8$. Find an equation for this graph and classify the graph.
3. a. If $x^2 + y^2 + Dx + Ey + F = 0$ is the graph of a circle, then find the center and the radius in terms of D, E, and F.
   b. State conditions involving D, E, and F that determine whether the equation in part a represents a circle, a point, or has no graph.

4. Match each graph with the equation that illustrates that case.

Graph	Equation
1. Two distinct lines through the origin	a. $x^2 + y^2 = 0$
2. Two distinct parallel lines	b. $x^2 + y^2 = -1$
3. One line through the origin	c. $x^2 - y^2 = 0$
4. A point (the origin)	d. $x^2 = 1$
5. No graph	e. $x^2 = 0$

5. Generalize from the result in the previous exercise and specify conditions for A, C, and k in the equation

$$Ax^2 + Cy^2 = k, \text{ with } A^2 + C^2 \neq 0,$$

that result in a graph that is a circle, an ellipse, a hyperbola, or each of the special cases listed above.

REMEMBER THIS

In Exercises 1–4 graph each equation and identify each graph.

1. $4y^2 - x^2 = 36$
2. $x^2 + 4y^2 = 36$
3. $y^2 + 9x = 0$
4. $x^2 + y^2 - 6y = 0$

In Exercises 5–8 find the focus or the foci of the graph of the equation.

5. $4y^2 - x^2 = 36$
6. $x^2 + 4y^2 = 36$
7. $y^2 + 2x = 0$
8. $(x - 2)^2 = -8y$
9. Find the directrix for the graph of $y^2 + 2x = 0$.
10. Find in standard form the equation for the graph that is the set of all points in a plane 7 units from the point $(5, -6)$. Identify the graph of this equation.

CHAPTER OVERVIEW

Section	Key Concepts to Review
8.1	• Fundamental relationship between an equation and its graph: Every ordered pair that satisfies the equation corresponds to a point in its graph, and every point in the graph corresponds to an ordered pair that satisfies the equation. • Procedures for finding an equation that corresponds to a given geometric condition • Procedures for proving certain theorems in plane geometry by using a coordinate system and algebraic methods • Midpoint formula: $\left(\dfrac{x_1 + x_2}{2}, \dfrac{y_1 + y_2}{2} \right)$
8.2	• Definition of circle • Standard form of a circle of radius r with center (h,k): $(x - h)^2 + (y - k)^2 = r^2$
8.3	• Definitions of ellipse, foci, major axis, minor axis, and vertices • Summary for an ellipse:

Foci	Center	Standard Form $(a > b)$
On x-axis at $(\pm c,0)$	Origin	$\dfrac{x^2}{a^2} + \dfrac{y^2}{b^2} = 1$
On y-axis at $(0, \pm c)$	Origin	$\dfrac{x^2}{b^2} + \dfrac{y^2}{a^2} = 1$
On major axis parallel to x-axis	(h,k)	$\dfrac{(x - h)^2}{a^2} + \dfrac{(y - k)^2}{b^2} = 1$
On major axis parallel to y-axis	(h,k)	$\dfrac{(x - h)^2}{b^2} + \dfrac{(y - k)^2}{a^2} = 1$

• In general, $a^2 = b^2 + c^2$, where a, b, and c represent the following distances:

a is the distance from the center to the endpoints on the major axis
b is the distance from the center to the endpoints on the minor axis
c is the distance from the center to the foci

Section	Key Concepts to Review
8.4	• Definitions of hyperbola, foci, transverse axis, vertices, conjugate axis, and asymptotes of the hyperbola • Summary for a hyperbola:

Foci	Center	Standard Form	Asymptotes
On x-axis at $(\pm c, 0)$	Origin	$\dfrac{x^2}{a^2} - \dfrac{y^2}{b^2} = 1$	$y = \pm\dfrac{b}{a}x$
On y-axis at $(0, \pm c)$	Origin	$\dfrac{y^2}{a^2} - \dfrac{x^2}{b^2} = 1$	$y = \pm\dfrac{a}{b}x$
On transverse axis parallel to x-axis	(h,k)	$\dfrac{(x-h)^2}{a^2} - \dfrac{(y-k)^2}{b^2} = 1$	$y - k = \pm\dfrac{b}{a}(x-h)$
On transverse axis parallel to y-axis	(h,k)	$\dfrac{(y-k)^2}{a^2} - \dfrac{(x-h)^2}{b^2} = 1$	$y - k = \pm\dfrac{a}{b}(x-h)$

• In general, $c^2 = a^2 + b^2$, where a, b, and c represent the following distances:

a is the distance from the center to the endpoints on the transverse axis
b is the distance from the center to the endpoints on the conjugate axis
c is the distance from the center to the foci

Section	Key Concepts to Review
8.5	• Definitions of parabola, directrix, focus, axis of symmetry, and vertex • Summary for a parabola with vertex at the origin: (In all cases p gives the distance from the vertex to the focus and from the vertex to the directrix. See Figures 8.24–8.27.)

Standard Form	Opens	Axis of Symmetry	Focus	Directrix
$y^2 = 4px$	Right	x-axis	$(p,0)$	$x = -p$
$y^2 = -4px$	Left	x-axis	$(-p,0)$	$x = p$
$x^2 = 4py$	Upward	y-axis	$(0,p)$	$y = -p$
$x^2 = -4py$	Downward	y-axis	$(0,-p)$	$y = p$

• Figure 8.29 summarizes the cases when the vertex of the parabola is at the point (h,k) and the directrix is parallel to the x- or y-axis.

Section	Key Concepts to Review
8.6	• The general form of an equation of a conic section with axis or axes parallel to the coordinate axes is $$Ax^2 + Cy^2 + Dx + Ey + F = 0,$$ where A and C are not both zero. • Chart summarizing the graphing possibilities for the above equation

CHAPTER REVIEW EXERCISES

1. Find an equation for the graph that is the set of all points in a plane equidistant from the line $y = 5$ and the point $(-1, -3)$.

2. Classify the graph defined by each of the following equations.
   a. $9x^2 = 4y^2 - 54x - 16y + 61$
   b. $x^2 + 4x + 8y = 4$
   c. $7x - 2y = 3$
   d. $16x^2 + 25y^2 - 32x - 284 = 0$
   e. $x^2 + 6x = 3 - y^2$

3. Determine the center and radius of the circle given by $x^2 + y^2 + 4x - 6y = 12$.

4. Determine the standard equation of the circle with center at $(4, -3)$ that passes through $(0, -1)$.

5. Find the coordinates of the foci of the ellipse given by $25x^2 + 16y^2 = 400$.

6. What is the equation in standard form of the ellipse with vertices at $(3, 2)$ and $(-7, 2)$ and a minor axis of length 6?

7. Find the equations of the asymptotes of the hyperbola given by $y^2/64 - x^2/100 = 1$.

8. Determine the standard equation of the hyperbola whose center is at the origin with one focus at $(-8, 0)$ and a transverse axis of length 14.

9. What are the coordinates of the focus point of the parabola given by $x^2 = -7y$?

10. Find the standard equation of the parabola with focus at $(5, 1)$ and directrix $x = -1$.

11. Find the standard equation of the circle whose diameter extends from $(-2, 2)$ to $(4, 2)$.

12. Show by the methods of analytic geometry that the midpoints of the sides of a rectangle are the vertices of a quadrilateral whose perimeter is equal to the sum of the lengths of the diagonals of the rectangle.

13. Graph $16(y - 2)^2 - 25(x + 3)^2 = 400$.

14. If $y = mx - 1$ passes through the center of the circle $(x - 2)^2 + (y + 3)^2 = 25$, then find the value of m.

15. What are the coordinates of the vertices of the hyperbola given by $9(x - 3)^2 - 4(y + 1)^2 = 144$?

In Exercises 16–20 select the choice that completes the statement or answers the question.

16. The graph of $x^2 = 1 - 2y^2$ is
    a. a circle   b. an ellipse
    c. a parabola   d. a hyperbola

17. Which conic section is defined as the set of all points in a plane equidistant from a fixed line and a fixed point not on the line?
    a. circle   b. ellipse
    c. parabola   d. hyperbola

18. If s varies directly as the square of t, then the graph of this relation is a
    a. line   b. circle
    c. parabola   d. hyperbola

19. Which of the following is the equation of a hyperbola?
    a. $2x^2 + y^2 = 5$ b. $x^2 = 1 - y^2$
    c. $x = y^2 - 4$ d. $x^2 = y^2 - 4$

20. The standard equation of the circle with center at $(1, -3)$ and radius 5 is
    a. $(x + 1)^2 + (y - 3)^2 = 5$
    b. $(x - 1)^2 + (y + 3)^2 = 25$
    c. $(x - 1)^2 + (y + 3)^2 = 5$
    d. $(x + 1)^2 + (y - 3)^2 = 25$

CHAPTER 8 TEST

1. Find an equation for the graph that is the set of all points in a plane that are 5 units from the point $(-4, 3)$.

2. The points $(8, -2)$ and $(0, 2)$ are the endpoints of line segment L. Find an equation for the line that is the perpendicular bisector of L.

3. Show by the methods of analytic geometry that the diagonals of a square are perpendicular to each other.

4. Find the standard equation of the circle with center $(1, -2)$ that passes through the origin.

5. Find the center and radius of the circle given by $(x + 5)^2 + (y - 1)^2 = 36$.

6. Determine the radius of the circle given by $x^2 + y^2 + 2x - 4y - 27 = 0$.

7. Find the coordinates of the foci of the ellipse given by $9x^2 + 4y^2 = 36$.

8. Write the standard equation of the ellipse with center $(-1, 1)$, one vertex at $(5, 1)$, and minor axis of length 4.

9. An elliptical arch has a height of 8 ft and a span of 20 ft. How high is the arch 5 ft each side of center?

10. Graph $\dfrac{(x + 3)^2}{25} + \dfrac{y^2}{9} = 1$.

11. Graph $9y^2 - x^2 = 36$.

12. Find the coordinates of the foci of the hyperbola given by $x^2 - 4y^2 = 1$.

13. Find the equation of the asymptotes of the graph of $x^2 - 4y^2 = 1$.

14. Find the standard equation of the hyperbola with vertices at $(0,0)$ and $(0,8)$ and one focus at $(0,10)$.

15. Determine the vertex of the parabola given by $y^2 + 4y + 5x - 11 = 0$.

16. Find the directrix for the graph of $x^2 + 10y = 0$.

17. Write the standard equation for the parabola with vertex at $(2,1)$ and focus at $(2,4)$.

18. Graph $(y + 1)^2 = 2(x - 2)$.

19. Identify the graph of $y^2 = x^2 - 4x + 5$ if the graph is either a circle, an ellipse, a hyperbola, or a parabola.

20. Identify the graph of $9x^2 + y^2 - 36x + 2y + 37 = 0$. Degenerate cases are possible here.

9 Systems of Equations and Inequalities

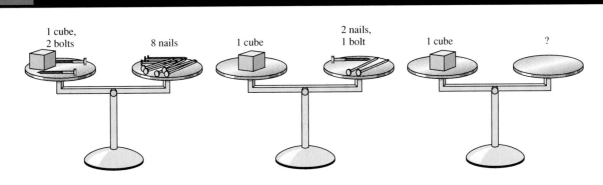

1 cube,
2 bolts

8 nails

1 cube

2 nails,
1 bolt

1 cube

?

Consider the figure. How many nails are needed to balance the cube? (See Example 8 of Section 9.1.)

In the analysis of a problem we often must take into account many variables and many relationships among the variables. This situation leads to the topic of setting up and solving systems of equations and inequalities. In most cases we deal with **linear equations in n variables,** which are equations of the form

$$a_1x_1 + a_2x_2 + \cdots + a_nx_n = c,$$

where $a_1, a_2, \ldots, a_n,$ and c are real numbers and $x_1, x_2, \ldots, x_n$ are variables. A set of linear equations is called a **linear system.** In this chapter we first review the substitution and addition-elimination methods for solving simple linear systems. Then we show how to solve more complicated linear systems by using Gaussian elimination, determinants, and matrix algebra. Finally, we discuss nonlinear systems of equations and systems of linear inequalities. Partial fractions and linear programming are considered as applications of systems of equations and inequalities, respectively.

9.1 Systems of Linear Equations in Two Variables

More than one linear equation is often needed to describe a situation adequately. For example, suppose that you have to decide between two positions as a salesperson. The first company offers a straight 20-percent commission, whereas the second company offers a salary of $70 per week plus a 10-percent commission. Before you choose your job, it is important to determine how much you would have to sell per week before the two job offers would produce the same income. For the purpose of comparison, we have graphed the two job offers in Figure 9.1.

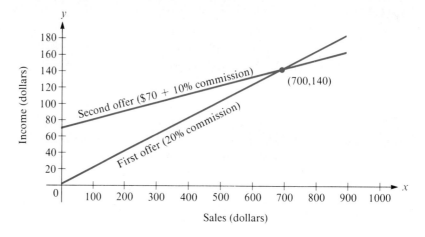

Figure 9.1

First offer (20% commission): $y = 0.2x$

Sales (in dollars) x	100	200	300	400	500	600	700	800	900	1,000
Income (in dollars) y	20	40	60	80	100	120	140	160	180	200

Second offer ($70 + 10% commission): $y = 70 + 0.1x$

Sales (in dollars) x	100	200	300	400	500	600	700	800	900	1,000
Income (in dollars) y	80	90	100	110	120	130	140	150	160	170

By examining the graph, we can see that the solution is the point (700,140), where the two lines intersect. That is, you earn an income of $140 from either job when you sell $700 worth of merchandise. Above $700 the larger commission is the better offer. Notice in our example that there are many ordered pairs that satisfy the first job offer and many ordered pairs that satisfy the second offer, but our solution is the only ordered pair common to the two equations. In general, the solution set of a system of linear equations is the set of all the ordered pairs that satisfy both equations. Graphically this corresponds to the collection of points where the lines intersect. Usually, as in our example, the two lines will intersect at only one point and, consequently, our solution is only one ordered pair. However, the next two examples illustrate two other possibilities.

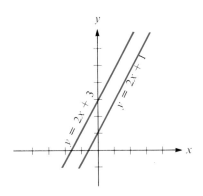

Figure 9.2

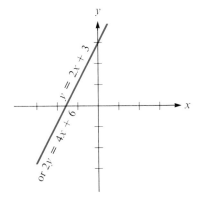

Figure 9.3

EXAMPLE 1   Find all the ordered pairs that satisfy the pair of equations

$$y = 2x + 1$$
$$y = 2x + 3.$$

Solution   These two lines are parallel and do not intersect, as shown in Figure 9.2. Thus, there is no ordered pair that satisfies both equations and the solution set is $\emptyset$. This situation will arise whenever two lines have the same slope and different y-intercepts. Such a system of equations is called **inconsistent.** ∎

EXAMPLE 2   Find all the ordered pairs that satisfy the pair of equations

$$y = 2x + 3$$
$$2y = 4x + 6.$$

Solution   When we attempt to graph both equations (see Figure 9.3), we find that any ordered pair that satisfies the first equation also satisfies the second equation. Thus, the two equations are equivalent. This means that the same line is the graph of both equations and that graphically the solution set is the set of all points on that line. This situation occurred because the second equation is obtained if we multiply both sides of the first equation by 2. A system containing equivalent equations is called **dependent.** ∎

The most useful situation is usually when the graphs of the two linear equations intersect at one point so there is only one ordered pair that satisfies both equations. We have been finding this solution graphically. This technique is good for illustrating the principle involved and for obtaining an approximate solution. However, if an exact solution is required, we must use algebraic methods. The first method we discuss makes use of the **substitution method,** as in the following example.

EXAMPLE 3   Find all the ordered pairs that satisfy the pair of equations

$$y = x + 2$$
$$y = 2x - 3.$$

Solution   We wish to find the value for x that makes both y values the same or, equivalently, we want to know the value for x that makes $x + 2$ equal to $2x - 3$.

$$x + 2 = 2x - 3$$
$$2 = x - 3$$
$$5 = x$$

Thus, the x-coordinate of the solution is 5. To find the y-coordinate, substitute 5 for x in either of the given equations.

$$
\begin{array}{ll}
y = x + 2 \quad\text{or}\quad & y = 2x - 3 \\
\;\; = (5) + 2 & \;\; = 2(5) - 3 \\
\;\; = 7 & \;\; = 7
\end{array}
$$

Thus, the solution is (5,7). (*Note:* A good check of your result is to find the *y*-coordinate by substituting the *x*-coordinate in both equations.)   ■

EXAMPLE 4   The total cost of producing gadgets consists of $300 per month for rent plus $4 per unit for material. If the selling price for a gadget is $9, how many units must be made and sold per month for the company to break even?

Solution   If *x* represents the number of gadgets made and sold, then

$$\text{Total cost} = 300 + 4x$$
$$\text{Total revenue} = 9x.$$

We wish to find the value for *x* that makes the total cost equal to the revenue or, equivalently, we want to know the value for *x* that makes $300 + 4x$ equal to $9x$.

$$300 + 4x = 9x$$
$$300 = 5x$$
$$60 = x$$

Thus, the company will break even when it makes and sells 60 gadgets. To check the result, substitute 60 in the original equations (see Figure 9.4).

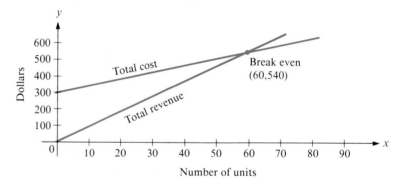

Figure 9.4

$$\begin{array}{lll}
\text{Total cost} = 300 + 4x & \text{or} & \text{Total revenue} = 9x \\
\qquad\quad\ = 300 + 4(60) & & \qquad\qquad\quad\ = 9(60) \\
\qquad\quad\ = 540 & & \qquad\qquad\quad\ = 540
\end{array}$$ ■

In these examples the substitution method has been appropriate since all the linear equations have been given in the form $y = mx + b$. However, if one of the linear equations is given as $3x - 5y - 4 = 0$, we would have to change the equation to $y = \frac{3}{5}x - \frac{4}{5}$ and continue with this equation. This procedure is awkward algebraically. Thus, a different method is used.

The following example illustrates the **addition-elimination method.** It is very important because it can be used to solve complicated systems of equations with more variables.

EXAMPLE 5   Find all the ordered pairs that satisfy the pair of equations

$$x + y = 9$$
$$x - y = 5.$$

Solution   This method attempts to eliminate one of the variables by adding the two equations together. In this example the coefficients for y are $+1$ and -1. If we add the equations, the result will be an equation that contains only x.

$$
\begin{array}{rl}
x + y = & 9 \\
x - y = & 5 \\
\hline
2x = & 14 \qquad \text{Add the equations.} \\
x = & 7
\end{array}
$$

Thus, the x-coordinate of the solution is 7.

To find the y-coordinate, substitute 7 for x in either of the given equations.

$$
\begin{array}{ll}
x + y = 9 \quad \text{or} & x - y = 5 \\
(7) + y = 9 & 7 - y = 5 \\
y = 2 & y = 2
\end{array}
$$

Thus, the solution is $(7,2)$. ■

EXAMPLE 6   Find all the ordered pairs that satisfy the pair of equations

$$3x - 2y = 27$$
$$2x + 5y = -1.$$

Solution   If we form equivalent equations by multiplying the top equation by -2 and the bottom equation by 3, we can eliminate the x variable.

$$
\begin{array}{rl}
-6x + 4y = & -54 \\
6x + 15y = & -3 \\
\hline
19y = & -57 \qquad \text{Add the equations.} \\
y = & -3
\end{array}
$$

Thus, the y-coordinate of the solution is -3.

To find the x-coordinate, substitute -3 for y in either of the given equations.

$$
\begin{array}{ll}
3x - 2y = 27 \quad \text{or} & 2x + 5y = -1 \\
3x - 2(-3) = 27 & 2x + 5(-3) = -1 \\
x = 7 & x = 7
\end{array}
$$

Thus, the solution is $(7,-3)$. ■

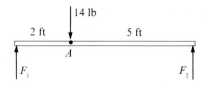

Figure 9.5

EXAMPLE 7   Find the forces F_1 and F_2 that achieve equilibrium for the beam in Figure 9.5.

Solution   If we consider two of the laws of physics, we can determine two equations that contain F_1 and F_2. First, since the system is in equilibrium, the sum of the forces pointing down must be equal to the sum of the forces pointing up.

$$F_1 + F_2 = 14$$

Second, the tendency of the lumber to rotate about point A in a clockwise direction is the product of F_1 and the distance (2 ft) between F_1 and the turning point A; the tendency of the lumber to rotate in a counterclockwise direction is the product of F_2 and the distance (5 ft) between F_2 and the turning point A. Since the system is in equilibrium, we have

$$\text{clockwise turning effect} = \text{counterclockwise turning effect}$$
$$2 \cdot F_1 \qquad = \qquad 5 \cdot F_2.$$

Thus, we can determine F_1 and F_2 by finding the ordered pair that satisfies these two equations.

$$F_1 + F_2 = 14$$
$$2F_1 - 5F_2 = 0 \quad \text{or} \quad (2F_1 = 5F_2)$$

We can eliminate F_2 by multiplying the top equation by 5 and then adding the two equations.

$$5F_1 + 5F_2 = 70$$
$$\underline{2F_1 - 5F_2 = 0}$$
$$7F_1 = 70$$
$$F_1 = 10$$

Thus, F_1 is 10 lb. To find F_2, substitute 10 for F_1 in either of the given equations.

$$F_1 + F_2 = 14 \quad \text{or} \quad 2F_1 = 5F_2$$
$$(10) + F_2 = 14 \qquad\quad 2(10) = 5F_2$$
$$F_2 = 4 \qquad\qquad\quad 4 = F_2$$

Thus,

$$F_1 = 10 \text{ lb and } F_2 = 4 \text{ lb.} \qquad\qquad \blacksquare$$

EXAMPLE 8   Solve the problem in the chapter introduction on page 457.

Solution   If we let b, c, and n represent the weights of a bolt, a cube, and a nail, respectively, then by inspection of the scales, we determine the following system of equations.

$$c + 2b = 8n$$
$$c = 2n + b$$

One way to find c in terms of n is to solve the second equation for b to determine that $b = c - 2n$ and then substitute the result in the first equation.

$$c + 2(c - 2n) = 8n$$
$$3c - 4n = 8n$$
$$3c = 12n$$
$$c = 4n$$

Thus, 4 nails are needed to balance the cube.   ■

In economics an analysis of the law of supply and demand involves the intersection of two curves. Basically, as the price for an item increases, the quantity of the product that is supplied increases while the quantity that is demanded decreases. The point at which the supply and demand curves intersect is called the **point of market equilibrium.** This principle is shown in Figure 9.6, where, for illustrative purposes, we assume the supply and demand equations to be linear.

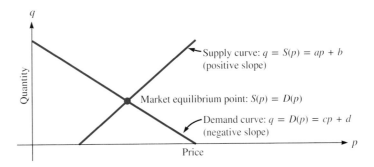

Figure 9.6

The equilibrium price is the value for p at which supply equals demand. The law of supply and demand states that in a system of free enterprise (a big assumption!) a product will sell near its equilibrium price. The theory asserts that if the price is above the equilibrium price, more of the product will be put on the market to capitalize on the higher price. A surplus results, causing the price of the product to lower. When the price is below the equilibrium price, the demand for the product will exceed the supply, so the sellers can raise their prices.

EXERCISES 9.1

In Exercises 1–20 find all the ordered pairs that satisfy the pair of equations. In 1–10 first estimate the solution graphically.

1. $y = x - 2$
   $y = 5x + 6$

2. $y = -x - 3$
   $y = 2x + 3$

3. $y = 3x + 4$
   $y = -x - 2$

4. $y = 4x + 7$
   $y = 3x + 5$

5. $y = x + 4$
   $y = 2x + 4$

6. $y = -5x + 2$
   $y = 4x - 7$

7. $y = x$
   $x - 2y = 6$

8. $y = -3x$
   $2x + 3y = -21$

9. $y = \frac{1}{3}x - 5$
   $x - 3y - 15 = 0$

10. $y = -\frac{5}{2}x + 3$
    $5x + 2y + 6 = 0$

11. $x + y = 25$
    $6x - y = 3$

12. $2x + 3y = 8$
    $2x - 7y = -32$

13. $5x + 3y = -2$
    $x - 2y = -3$

14. $-2x - y = -5$
    $5x + 2y = -17$

15. $3x - 2y = 1$
    $6x - 4y = 5$

16. $2x - 3y = 1$
    $6x - 9y = 3$

17. $2x - 5y = 5$
    $4x + 3y = 23$

18. $4x + 2y = 2$
    $6x - 5y = 27$

19. $7x - 2y - 19 = 0$
    $3x + 5y + 14 = 0$

20. $6x + 10y - 7 = 0$
    $15x - 4y - 3 = 0$

In Exercises 21–38 set up a system of linear equations and solve by an appropriate method illustrated in this section.

21. The sum of two numbers is 70; their difference is 22. What are the numbers?

22. Find two complementary angles whose difference is 20°.

23. Find two supplementary angles whose difference is 100°.

24. A piece of lumber is 120 in. long. Where must it be cut for one piece to be four times longer than the other piece?

25. A container holding a liquid weighs 500 g. If one-half the liquid is poured out, the weight is 350 g. What is the weight of the empty container?

26. If four black metal balls and one red metal ball are placed on a scale, they balance a weight of 100 g. A weight of 90 g will balance two black and three red balls. Find the weight of each kind of metal ball.

27. The velocity of a particle that accelerates at a uniform rate is linearly related to the elapsed time by the equation $v = v_0 + at$, where v_0 is the initial velocity and a the acceleration. If $v = 36$ ft/second when $t = 2$ seconds and $v = 4$ ft/second when $t = 3$ seconds, find values for v_0 and a.

28. In Exercise 27 find the values of v_0 and a if $v = 45$ ft/second when $t = 2$ seconds and if $v = 72$ ft/second when $t = 5$ seconds.

29. Find the forces F_1 and F_2 that achieve equilibrium for the beam in the following diagram.

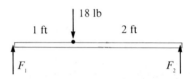

30. Find the forces F_1 and F_2 that achieve equilibrium for the beam in the following diagram.

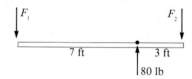

31. It takes a boat 1 hour to go 6 mi upstream (against the current); the return trip downstream (with the current) takes $\frac{1}{2}$ hour. What is the rate of the current? What would be the speed of the boat if there were no current?

32. A chemist has a 10-percent solution and a 25-percent solution of alcohol. How much of each should be mixed together to obtain 100 gal of a 20-percent solution?

33. A college wants to invest $10,000 to have an annual income of $700 for a scholarship. The college plans to invest part of the money in a bank that yields 6-percent interest and the remainder in a speculative fund that promises to yield 11-percent interest. How much should be invested in each to obtain the desired income?

34. You are trying to decide between two positions as a salesperson. The first offer is a straight 15-percent commission while the second offer pays a salary of $60 per week plus a 10-percent commission. How much must you sell each week for the two jobs to pay the same?

35. A manufacturer wants to know whether it will pay him to buy and install a special machine to turn out widgets that, until now, he has been purchasing from an outside supplier for $1 each. The machine will cost him $500 per year and will be able to produce widgets for 50 cents each. How many widgets would he have to make and use each year to justify purchasing this new machine?

36. A company is trying to decide between two machines for packaging its new product. Machine A will cost $5,000 per year plus $2 to package each unit; machine B will cost $8,000 per year plus $1 to package each unit.

   a. How many units must be produced for the cost of the two machines to be the same? If the company plans to produce more units, which machine should they purchase?

   b. If packaging can be subcontracted to another firm at a cost of $5 per unit, how many units would have to be made before the purchase of machine A would be worthwhile?

37. Consider Figure 9.6 and the accompanying discussion of the law of supply and demand. Determine the equilibrium price (p) in terms of the constants $a, b, c,$ and d. Which of these constants are positive? Which are negative?

38. In electronics the analysis of a circuit often utilizes several laws that apply at the same time. This

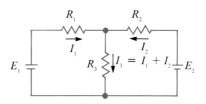

situation leads naturally to simultaneous equations. By applying Kirchhoff's laws to the circuit shown in the diagram, we obtain the following equations:

$$R_1 I_1 + R_3(I_1 + I_2) = E_1$$
$$R_2 I_2 + R_3(I_1 + I_2) = E_2.$$

Determine the value of the currents I_1 and I_2 if $E_1 = 8$ volts, $E_2 = 5$ volts, $R_1 = 6$ ohms, $R_2 = 10$ ohms, and $R_3 = 3$ ohms.

THINK ABOUT IT

1. Describe the graph and the solution set of a system of linear equations in two variables for each type of system.

   a. Inconsistent system   **b.** Dependent system

2. Which one of the figures would be used to graphically solve the system

$$5x + 3y + 2 = 0$$
$$x - 2y + 3 = 0?$$

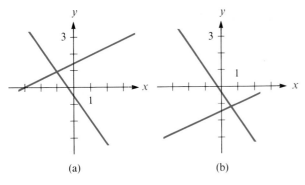

     (a)   (b)

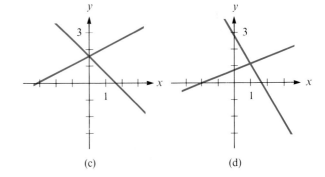

     (c)   (d)

3. Find x and y: $3^{x+y} = 9$
                      $4^{x-y} = 8$.

4. Create a word problem not considered in this section that may be solved by a system of linear equations in two variables. What is the solution to the problem you created?

5. The following problems are usually considered in the third semester of accounting. In these problems we find several important business considerations defined in terms of each other. Thus, they have interlocking solutions that lead naturally to simultaneous equations.

   a. A company operates in a state that levies a 10-percent tax on the income that remains after paying the federal tax. Meanwhile, the federal tax is 50 percent of the income that remains after paying the state tax. If during the current year a corporation has $400,000 in taxable income, determine the state and federal income taxes.

   b. A company is to give a performance bonus to one of its top managers. The income in the manager's division is $100,000 and the bonus rate is 10 percent. The bonus is based on profit, so it is based on income after both taxes and the bonus have been subtracted from the $100,000. The taxes amount to 30 percent of the taxable income, with the bonus counting as a tax deduction. What is the manager's bonus?

REMEMBER THIS

1. Simplify $-3(-5) + 4(-1) - 2(7)$.
2. Evaluate $2x - 3y - z$ if $x = -1$, $y = 2$, and $z = 1$.
3. Evaluate $3x - 2y - z$ if $x = \frac{4}{5}$, $y = -\frac{2}{5}$, and $z = \frac{1}{5}$.
4. Is $(-1,-2)$ a solution of the system
$$3x + 4y + 11 = 0$$
$$4x - 5y + 14 = 0?$$

5. Identify the coefficients of x, y, and z in the expression $-x - 2y + z$.
6. If $6x - 5y = 27$ and $y = -3$, find x.
7. If $a - b + c = -8$ with $b = 5$ and $c = -2$, find a.
8. Find the product of $-\frac{1}{2}$ and $-2y + 6z$.
9. Add -1 times $-7y + 15z$ to $-7y + 5z$.
10. Add -2 times $x + y - 3z$ to $3x - 2y - z$.

9.2 Triangular Form and Matrices

Systems of three linear equations with three variables, and more complicated linear systems, can always be solved by using a technique known as **Gaussian elimination.** This method is basically the addition-elimination method from Section 9.1, and its systematic nature can be programmed to allow effective computer solutions. To understand this method, first consider the following system of equations that is said to be in **triangular form.**

$$4x + 2y - z = 2$$
$$3y - 3z = 6$$
$$5z = 10$$

This system is easy to solve. The third equation, $5z = 10$, tells us $z = 2$, then back substitution yields

$$3y - 3(2) = 6 \qquad\qquad 4x + 2(4) - (2) = 2$$
$$3y = 12 \quad \text{and} \qquad 4x = -4$$
$$y = 4 \qquad\qquad\qquad x = -1.$$

Thus, the solution is $x = -1$, $y = 4$, and $z = 2$. However, systems of equations rarely start out in triangular form, so we need some procedure for obtaining this form for any given system. The three operations that follow are called the **elementary operations,** and they are used in Gaussian elimination to produce **equivalent systems** (ones with the same solution) until we reach triangular form.

Operations that Produce Equivalent Systems

1. Multiply both sides of an equation by a nonzero number.
2. Add a multiple of one equation to another equation.
3. Interchange the order in which two equations of a system are listed.

Note that we have already used the first two operations in Section 9.1, while the third operation clearly affects only the form of the system and not the solution. Now consider carefully Example 1, which shows how we can use these operations to change a system into triangular form.

EXAMPLE 1   Solve the system of equations

$$3x - y + 6z = 1$$
$$x + 2y - 3z = 0$$
$$2x - 3y - z = -9.$$

Solution   We want x to appear only in the first equation. It is easier to eliminate x from the other equations when the coefficient of x in the first equation is 1. So, first, we change the order of the equations to

$$x + 2y - 3z = 0$$
$$3x - y + 6z = 1$$
$$2x - 3y - z = -9.$$

We now add -3 times the first equation to the second equation, and we also add -2 times the first equation to the third equation. The result is

$$x + 2y - 3z = 0$$
$$-7y + 15z = 1$$
$$-7y + 5z = -9.$$

Finally, to eliminate y in the third equation, we add -1 times the second equation to the third equation to obtain

$$x + 2y - 3z = 0$$
$$-7y + 15z = 1$$
$$-10z = -10.$$

The system is now in triangular form. From the third equation we know $z = 1$, while back substitution gives

$$-7y + 15(1) = 1 \qquad\qquad x + 2(2) - 3(1) = 0$$
$$-7y = -14 \qquad \text{and} \qquad x = -1.$$
$$y = 2$$

Thus, the solution is $x = -1$, $y = 2$, and $z = 1$.   ∎

We can improve the procedure in our current method by keeping track of only the constants in the equations and not writing down the variables. The standard notation for such an abbreviation utilizes matrices. A **matrix** is a rectangular array of numbers that is enclosed in brackets (or parentheses) and commonly denoted by a capital letter such as A or B. Each number in the matrix is called an **entry** or **element** of the matrix. There are two matrices associated with the system

$$a_1x + b_1y + c_1z = d_1$$
$$a_2x + b_2y + c_2z = d_2$$
$$a_3x + b_3y + c_3z = d_3.$$

The **coefficient matrix** consists of the coefficients of x, y, and z and is written as

$$\begin{bmatrix} a_1 & b_1 & c_1 \\ a_2 & b_2 & c_2 \\ a_3 & b_3 & c_3 \end{bmatrix},$$

while the **augmented matrix** that follows includes these coefficients and an additional column (usually separated by a dashed line) that contains the constants on the right side of the equals sign.

$$\begin{bmatrix} a_1 & b_1 & c_1 & \vdots & d_1 \\ a_2 & b_2 & c_2 & \vdots & d_2 \\ a_3 & b_3 & c_3 & \vdots & d_3 \end{bmatrix}$$

We now restate in the language of matrices the operations that produce equivalent systems. These operations are called **elementary row operations** and they can be used to obtain a matrix solution to a system of equations. By analogy to *equivalent* systems of equations, two matrices that can be derived from each other by using one or more of these elementary row operations are called **equivalent matrices.**

Corresponding Operations for Solving a Linear System

Elementary Operations on Equations	**Elementary Row Operations on Matrices**
1. Multiply both sides of an equation by a nonzero number.	1. Multiply each entry in a row by a nonzero number.
2. Add a multiple of one equation to another.	2. Add a multiple of the entries in one row to another row.
3. Interchange two equations.	3. Interchange two rows.

We display both of these methods in the next example to reinforce the similarities in the methods.

EXAMPLE 2 Solve the system

$$4x - 2y + z = 11$$
$$x - y + 3z = 6$$
$$x + y + z = 2$$

and show both the matrix form of the system and the corresponding equations.

Solution   We use the elementary operations above and proceed as follows:

Equation Form	**Matrix Form**

$$\begin{aligned} 4x - 2y + z &= 11 \\ x - y + 3z &= 6 \\ x + y + z &= 2 \end{aligned} \qquad \left[\begin{array}{ccc|c} 4 & -2 & 1 & 11 \\ 1 & -1 & 3 & 6 \\ 1 & 1 & 1 & 2 \end{array}\right]$$

↓ Interchange equations 1 and 3.   ↓ Interchange rows 1 and 3.

$$\begin{aligned} x + y + z &= 2 \\ x - y + 3z &= 6 \\ 4x - 2y + z &= 11 \end{aligned} \qquad \left[\begin{array}{ccc|c} 1 & 1 & 1 & 2 \\ 1 & -1 & 3 & 6 \\ 4 & -2 & 1 & 11 \end{array}\right]$$

↓ Add -1 times the first equation   ↓ Add -1 times each entry in row 1
to the second equation.   to the corresponding entry in row 2.

$$\begin{aligned} x + y + z &= 2 \\ -2y + 2z &= 4 \\ 4x - 2y + z &= 11 \end{aligned} \qquad \left[\begin{array}{ccc|c} 1 & 1 & 1 & 2 \\ 0 & -2 & 2 & 4 \\ 4 & -2 & 1 & 11 \end{array}\right]$$

↓ Add -4 times the first equation   ↓ Add -4 times each entry in row 1
to the third equation.   to the corresponding entry in row 3.

$$\begin{aligned} x + y + z &= 2 \\ -2y + 2z &= 4 \\ -6y - 3z &= 3 \end{aligned} \qquad \left[\begin{array}{ccc|c} 1 & 1 & 1 & 2 \\ 0 & -2 & 2 & 4 \\ 0 & -6 & -3 & 3 \end{array}\right]$$

↓ Add -3 times the second equation   ↓ Add -3 times each entry in row 2
to the third equation.   to the corresponding entry in row 3.

$$\begin{aligned} x + y + z &= 2 \\ -2y + 2z &= 4 \\ -9z &= -9 \end{aligned} \qquad \left[\begin{array}{ccc|c} 1 & 1 & 1 & 2 \\ 0 & -2 & 2 & 4 \\ 0 & 0 & -9 & -9 \end{array}\right]$$

The last row or last equation tells us $-9z = -9$, so $z = 1$. Then

$$\begin{aligned} -2y + 2(1) &= 4 \\ y &= -1 \end{aligned} \qquad \text{and} \qquad \begin{aligned} x + (-1) + 1 &= 2 \\ x &= 2. \end{aligned}$$

Thus, the solution is $x = 2$, $y = -1$, and $z = 1$.   ∎

EXAMPLE 3   Use matrix form to solve the system

$$\begin{aligned} 6x + 10y &= 7 \\ 15x - 4y &= 3. \end{aligned}$$

Solution   The augmented matrix for the system is

$$\left[\begin{array}{cc|c} 6 & 10 & 7 \\ 15 & -4 & 3 \end{array}\right].$$

It is easier to obtain 0 for the second entry in column 1 when the first entry in column 1 is 1, so first multiply each entry in row 1 by $\frac{1}{6}$.

$$\begin{bmatrix} 1 & \frac{5}{3} & | & \frac{7}{6} \\ 15 & -4 & | & 3 \end{bmatrix}$$

Now add -15 times row 1 to row 2.

$$\begin{bmatrix} 1 & \frac{5}{3} & | & \frac{7}{6} \\ 0 & -29 & | & -\frac{29}{2} \end{bmatrix}$$

The last row tells us that $-29y = -\frac{29}{2}$, so $y = \frac{1}{2}$. Then

$$x + \frac{5}{3}\left(\frac{1}{2}\right) = \frac{7}{6}$$

$$x = \frac{7}{6} - \frac{5}{6}$$

$$x = \frac{1}{3}.$$

Thus, the solution is $(\frac{1}{3}, \frac{1}{2})$. ∎

For the remaining examples we will use the following convenient abbreviations for the elementary row operations.

Elementary row operation	**Abbreviation**
1. Replace row i by multiplying each entry in row i by k.	**1.** $R_i \rightarrow kR_i$
2. Replace row i by adding k times each entry in row j to the corresponding entry in row i.	**2.** $R_i \rightarrow kR_j + R_i$
3. Interchange row i and row j.	**3.** $R_i \leftrightarrow R_j$

EXAMPLE 4 Use matrix form to solve the system

$$\begin{aligned} a + b + c + d &= 1 \\ a - b + c + d &= 1 \\ b + c - d &= 1 \\ b \quad\quad + d &= 1. \end{aligned}$$

Solution The augmented matrix for the system is

$$\begin{bmatrix} 1 & 1 & 1 & 1 & | & 1 \\ 1 & -1 & 1 & 1 & | & 1 \\ 0 & 1 & 1 & -1 & | & 1 \\ 0 & 1 & 0 & 1 & | & 1 \end{bmatrix}.$$

To obtain 0's in the first column after row 1, we need only add -1 times the first row to the second row.

$$\begin{bmatrix} 1 & 1 & 1 & 1 & \vdots & 1 \\ 0 & -2 & 0 & 0 & \vdots & 0 \\ 0 & 1 & 1 & -1 & \vdots & 1 \\ 0 & 1 & 0 & 1 & \vdots & 1 \end{bmatrix} \qquad R_2 \rightarrow -1R_1 + R_2$$

The second row represents the equation $-2b = 0$, so we already know $b = 0$. To obtain 0's in the second column after row 2, we first multiply each entry in row 2 by $-\frac{1}{2}$ to make the coefficient of b in the second equation a 1.

$$\begin{bmatrix} 1 & 1 & 1 & 1 & \vdots & 1 \\ 0 & 1 & 0 & 0 & \vdots & 0 \\ 0 & 1 & 1 & -1 & \vdots & 1 \\ 0 & 1 & 0 & 1 & \vdots & 1 \end{bmatrix} \qquad R_2 \rightarrow -\frac{1}{2}R_2$$

Now add -1 times the second row to the third row as well as to the fourth row.

$$\begin{bmatrix} 1 & 1 & 1 & 1 & \vdots & 1 \\ 0 & 1 & 0 & 0 & \vdots & 0 \\ 0 & 0 & 1 & -1 & \vdots & 1 \\ 0 & 0 & 0 & 1 & \vdots & 1 \end{bmatrix} \qquad \begin{array}{l} R_3 \rightarrow -1R_2 + R_3 \\ R_4 \rightarrow -1R_2 + R_4 \end{array}$$

Row 2 and row 4 tell us that $b = 0$ and $d = 1$, while substitution of these values into the equations corresponding to row 3 and row 1 gives

$$\begin{array}{ccc} c - d = 1 & & a + b + c + d = 1 \\ c - (1) = 1 & \text{and} & a + 0 + 2 + 1 = 1 \\ c = 2 & & a = -2. \end{array}$$

Thus, the solution is $a = -2$, $b = 0$, $c = 2$, and $d = 1$. ∎

There are two additional ideas to consider about solving linear systems using row operations. First, if a row of 0's results in the coefficient portion in any matrix, then there is no unique solution to the problem. Such systems are either dependent (if the last column entry is also 0) or inconsistent (if the last column entry is not 0). Second, in our example problems to this point, we stopped when we reached triangular form and completed the solution by back substitution. An alternative method, called **Gauss-Jordon elimination,** is to continue to produce equivalent matrices until we reach a form like

$$\begin{bmatrix} 1 & 0 & 0 & \vdots & a \\ 0 & 1 & 0 & \vdots & b \\ 0 & 0 & 1 & \vdots & c \end{bmatrix}.$$

From this final form of the matrix, we directly read off that the solution is $x = a$, $y = b$, and $z = c$. The matrix above is an example of a **reduced row-echelon matrix,** which is defined as follows.

Reduced Row-Echelon Matrix

Matrix A is a reduced row-echelon matrix if and only if

1. Rows containing all 0's (if any exist) occur at the bottom of A.
2. The first nonzero entry in each nonzero row is 1, called a **leading 1**.
3. These leading 1's are positioned further to the right in succeeding lower rows of A.
4. Every column that contains a leading 1 has all 0's above and below that leading 1.

Example 5 shows a systematic procedure that may be used to obtain a reduced row-echelon matrix.

EXAMPLE 5   Use Gauss-Jordan elimination to solve the system

$$3x + y + z = 1$$
$$x - y + z = 3$$
$$2x + 2y - z = -3.$$

Solution   The augmented matrix for the system is

$$\left[\begin{array}{ccc|c} 3 & 1 & 1 & 1 \\ 1 & -1 & 1 & 3 \\ 2 & 2 & -1 & -3 \end{array}\right].$$

In the Gauss-Jordan elimination method, we transform this matrix to a reduced row-echelon matrix as shown next.

Row operation	Equivalent matrix	Objective
$R_1 \leftrightarrow R_2$	$\left[\begin{array}{ccc|c} 1 & -1 & 1 & 3 \\ 3 & 1 & 1 & 1 \\ 2 & 2 & -1 & -3 \end{array}\right]$	**1.** Obtain a leading 1 in row 1, column 1.
$R_2 \rightarrow -3R_1 + R_2$ $R_3 \rightarrow -2R_1 + R_3$	$\left[\begin{array}{ccc|c} 1 & -1 & 1 & 3 \\ 0 & 4 & -2 & -8 \\ 0 & 4 & -3 & -9 \end{array}\right]$	**2.** Use the leading 1 in row 1, column 1 to get other column 1 entries to be 0.
$R_2 \rightarrow \frac{1}{4}R_2$	$\left[\begin{array}{ccc|c} 1 & -1 & 1 & 3 \\ 0 & 1 & -\frac{1}{2} & -2 \\ 0 & 4 & -3 & -9 \end{array}\right]$	**3.** Obtain a leading 1 in row 2, column 2.
$R_1 \rightarrow R_2 + R_1$ $R_3 \rightarrow -4R_2 + R_3$	$\left[\begin{array}{ccc|c} 1 & 0 & \frac{1}{2} & 1 \\ 0 & 1 & -\frac{1}{2} & -2 \\ 0 & 0 & -1 & -1 \end{array}\right]$	**4.** Use the leading 1 in row 2, column 2 to get other column 2 entries to be 0.

$R_3 \rightarrow -1R_3$

$$\begin{bmatrix} 1 & 0 & \frac{1}{2} & \vdots & 1 \\ 0 & 1 & -\frac{1}{2} & \vdots & -2 \\ 0 & 0 & 1 & \vdots & 1 \end{bmatrix}$$

5. Obtain a leading 1 in row 3, column 3.

$R_1 \rightarrow -\frac{1}{2}R_3 + R_1$
$R_2 \rightarrow \frac{1}{2}R_3 + R_2$

$$\begin{bmatrix} 1 & 0 & 0 & \vdots & \frac{1}{2} \\ 0 & 1 & 0 & \vdots & -\frac{3}{2} \\ 0 & 0 & 1 & \vdots & 1 \end{bmatrix}$$

6. Use the leading 1 in row 3, column 3 to get other column 3 entries to be 0.

From the system of equations corresponding to the reduced row-echelon matrix at the bottom, we determine that the solution is $x = \frac{1}{2}$, $y = -\frac{3}{2}$, and $z = 1$. ∎

EXERCISES 9.2

In Exercises 1–4 solve the given systems of equations and show both the matrix form of the systems and the corresponding equations.

1. $x + y = 7$
$x - y = -2$

2. $2a + 5b = 1$
$3a - 4b = 13$

3. $a + b + c = 2$
$3a - b - c = -1$
$2a + 2b - c = 1$

4. $4x - 2y + 2z = 0$
$3x + 2z = 0$
$x - 2y = 0$

In Exercises 5–20 use matrix form to solve the given systems of equations. Use Gaussian elimination with back substitution or Gauss-Jordan elimination.

5. $5x + 2y + 17 = 0$
$-2x - y + 5 = 0$

6. $x - y = 3$
$-x - y = -3$

7. $2x - 3y = 1$
$-6x + 9y = -3$

8. $x - y = 3$
$-x + y = -5$

9. $x + y = 4$
$y + z = -8$
$x + z = 2$

10. $-a + b = -1$
$b - c = 3$
$a + c = -12$

11. $x + y + z = 1$
$x + y - z = 3$
$x - y - z = 5$

12. $x + y + 3z = 1$
$2x + 5y + 2z = 0$
$3x - 2y - z = 3$

13. $x_1 - x_2 + x_3 = 2$
$2x_1 - 3x_2 + 2x_3 = 6$
$3x_1 + x_2 + x_3 = 2$

14. $x_1 - x_2 + x_3 = 3$
$3x_1 + x_2 + x_3 = 1$
$2x_1 + 2x_2 - x_3 = -3$

15. $6A + 3B + 2C = 1$
$5A + 4B + 3C = 0$
$A + B + C = 0$

16. $2a + 3b - c = 3$
$a + b - 3c = -4$
$-a - b + 5c = 8$

17. $3x - 2y - z = -2$
$2x + 2y + 2z = 2$
$2x - 3y - 2z = -4$

18. $x + y = 2$
$y + z = 1$
$z + w = -1$
$x + w = 0$

19. $a + b + c + d = 0$
$a - b + 2c + d = 1$
$4a + b + 2c = 5$
$5a + 4c + 2d = 6$

20. $x_1 + x_2 + x_3 + x_4 = 1$
$x_1 - x_2 + x_3 - x_4 = -1$
$-2x_1 + x_2 - x_3 - 2x_4 = 1$
$2x_1 - 2x_2 + 2x_3 + x_4 = 1$

21. The points $(-1,-8)$, $(1,2)$, and $(2,4)$ lie on the parabola $y = ax^2 + bx + c$. Find a, b, and c.

22. Let $f(x) = ax^2 + bx + c$. Find values of a, b, and c such that $f(1) = 0$, $f(2) = 8$, and $f(3) = 22$.

23. If the graph of $y = ax^3 + bx^2 + cx + d$ passes through the points $(1,1)$, $(2,1)$, $(3,-11)$, and $(-1,-11)$, find a, b, c, and d.

24. The points $(3,3)$, $(-2,-2)$, and $(1,-1)$ lie on the circle $x^2 + y^2 + Dx + Ey + F = 0$. Find D, E, and F. What are the center and radius of this circle?

25. Find the radius of each circle in the diagram below.

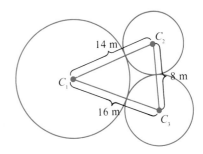

26. A 2,000-seat theater was sold out for a weekend concert. Total receipts on Friday night were $31,500 when seats in the orchestra, mezzanine, and balcony sold for $20, $15, and $10, respectively. On Saturday, prices were raised to $25, $20, and $10 and total receipts increased to $39,000. How many seats are in the mezzanine?

THINK ABOUT IT

1. Solve for x_1, x_2, and x_3 in terms of the constants a, b, and c.

$$\begin{aligned} x_1 + x_2 &= a \\ x_2 + x_3 &= b \\ x_1 + x_3 &= c \end{aligned}$$

2. Determine constants A, B, and C so that the following equation is an identity.

$$\frac{8x^2 - 4x + 5}{(x - 1)(x^2 + x + 1)} = \frac{A}{x - 1} + \frac{Bx + C}{x^2 + x + 1}$$

3. Find in terms of D an equation for all of the circles given by $x^2 + y^2 + Dx + Ey + F = 0$ that pass through the points $(1,3)$ and $(-7,-1)$. Choose arbitrarily two distinct values for D and write equations of two different circles passing through these two points.

4. One of the classic stories from the history of mathematics concerns Archimedes (250 B.C.), a great mathematician and scientist of ancient times. Archimedes lived in the Greek city-state of Syracuse, where the king, Hieron, suspected a goldsmith of giving him a gold crown that contained hidden silver. The king referred the problem to Archimedes and asked him to determine, without destroying the crown, the percentage of pure gold in the crown. While taking a bath, Archimedes found the principle needed to solve this problem. What Archimedes noticed in his bath was the obvious fact that when a body is immersed in water, it displaces a volume of water that is equal to the volume of the body. He also knew that bodies of the same weight do not necessarily have the same volume. Using these principles, Archimedes then filled a bucket of water to the brim. Suppose the crown weighed 10 lb and displaced 18 in.^{3} of water and that 10 lb of pure gold and 10 !b of pure silver displaced 15 and 30 in.^{3} of water, respectively. What percent of the crown would you tell King Hieron is made of each metal? (*Note:* Solve the linear system with two equations in two variables by a method of this section.)

5. Test your skills on the following problem from the *Mathematical Puzzles of Sam Loyd,* a two-volume series edited by Martin Gardner and published by Dover Publications, © 1960.

How much does each jar hold?

MRS. HUBBARD has invented a clever system for keeping tabs on her jars of blackberry jam. She has arranged the jars in her cupboard so that she has twenty quarts of jam on each shelf. The jars are in three sizes. Can you tell how much each size contains?

REMEMBER THIS

In Exercises 1–4 evaluate each expression.

1. $5(-10) - (-28)(4)$

2. $-1(-8) + 5(-10) + 2(-4)$

3. $\dfrac{8(-7) - 3(-32)}{2(-7) - 3(2)}$ **4.** $\dfrac{1(-9) - (-3)(3)}{2(-9) - (-3)(6)}$

In Exercises 5 and 6 evaluate $(-1)^{i+j}$ for the given conditions for i and j.

5. $i = 2, j = 3$ **6.** $i + j$ is even

7. Solve for y: $-acx - bcy = -pc$
$$acx + ady = qa.$$

In Exercises 8–10 solve the system

$$5x + 2y = 1$$
$$x - 3y = 7$$

by the method indicated.

8. Estimate the solution graphically.

9. Use the addition-elimination method.

10. Use the substitution method.

9.3   Determinants and Cramer's Rule

A linear system with n equations in n variables that is neither inconsistent nor dependent may be solved by using formulas known as Cramer's rule. To understand the derivation and application of these formulas, we start by finding the solution of the general system with two equations in two variables

$$a_1x + b_1y = c_1$$
$$a_2x + b_2y = c_2.$$

To solve for x, we multiply each member of the first equation by b_2 and each member of the second equation by $-b_1$ to obtain

$$a_1b_2x + b_1b_2y = c_1b_2$$
$$-b_1a_2x - b_1b_2y = -b_1c_2.$$

Now add the two equations to get

$$a_1b_2x - b_1a_2x = c_1b_2 - b_1c_2.$$

Factoring the left side of the equation, we have

$$(a_1b_2 - b_1a_2)x = c_1b_2 - b_1c_2.$$

Thus, if $a_1b_2 - b_1a_2 \neq 0$,

$$x = \frac{c_1b_2 - b_1c_2}{a_1b_2 - b_1a_2}.$$

In a similar manner, multiplying each member of the first equation by $-a_2$ and each member of the second equation by a_1, and then adding, leads to

$$y = \frac{a_1c_2 - c_1a_2}{a_1b_2 - b_1a_2}.$$

These formulas may now be used to find x and y whenever $a_1b_2 - b_1a_2 \neq 0$. If $a_1b_2 - b_1a_2 = 0$, there is no unique solution for x and y and the system is either dependent or inconsistent.

We do not memorize the formulas in this form since they may be obtained by defining what is called a determinant. Consider the expression $a_1b_2 - b_1a_2$, which is the denominator in the formulas for both x and y, and note that the coefficient matrix of the linear system is

$$A = \begin{bmatrix} a_1 & b_1 \\ a_2 & b_2 \end{bmatrix}.$$

This matrix is an example of a **square matrix,** which is a matrix having the same number of rows as columns. To each square matrix A there is assigned a unique real number called its **determinant** and denoted by $|A|$. As suggested above, when A has two rows and two columns, called a 2 by 2 matrix, then we define the value of $|A|$ to be $a_1b_2 - b_1a_2$.

Determinant of a 2 by 2 Matrix

If $A = \begin{bmatrix} a_1 & b_1 \\ a_2 & b_2 \end{bmatrix}$, then the determinant of A is given by

$$|A| = \begin{vmatrix} a_1 & b_1 \\ a_2 & b_2 \end{vmatrix} = a_1b_2 - b_1a_2.$$

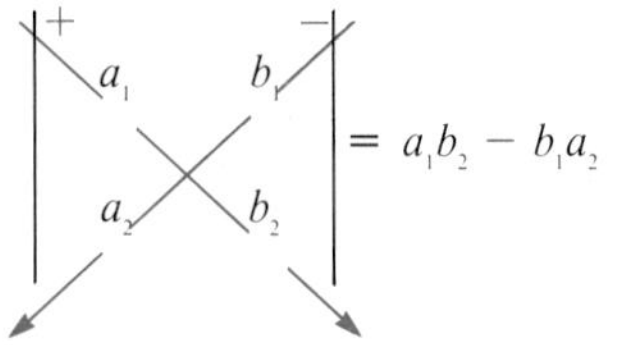

The numbers a_1 and b_2 are elements of the principal diagonal and the numbers b_1 and a_2 are elements of the secondary diagonal. Note in Figure 9.7 that the value of the determinant is the product of the elements of the principal diagonal minus the product of the elements of the secondary diagonal.

$$\begin{array}{cc} a_1 & b_1 \\ a_2 & b_2 \end{array} = a_1b_2 - b_1a_2$$

Figure 9.7

EXAMPLE 1 Evaluate $\begin{vmatrix} 1 & 2 \\ 3 & 4 \end{vmatrix}$.

Solution Using the definition above gives

$$\begin{vmatrix} 1 & 2 \\ 3 & 4 \end{vmatrix} = 1(4) - 2(3) = 4 - 6 = -2.$$

EXAMPLE 2 If $A = \begin{bmatrix} 3 & -5 \\ 4 & -7 \end{bmatrix}$, find $|A|$.

Solution The determinant of matrix A is given by

$$|A| = \begin{vmatrix} 3 & -5 \\ 4 & -7 \end{vmatrix} = 3(-7) - (-5)(4) = -21 + 20 = -1.$$

We now state Cramer's rule for two equations with two variables, which shows how determinants can be used to solve such a system of equations.

Cramer's Rule

(Two equations with two variables) The solution to the system

$$a_1x + b_1y = c_1$$
$$a_2x + b_2y = c_2$$

with $a_1b_2 - b_1a_2 \neq 0$ is

$$x = \frac{D_x}{D} = \frac{\begin{vmatrix} c_1 & b_1 \\ c_2 & b_2 \end{vmatrix}}{\begin{vmatrix} a_1 & b_1 \\ a_2 & b_2 \end{vmatrix}} = \frac{c_1b_2 - b_1c_2}{a_1b_2 - b_1a_2}$$

$$y = \frac{D_y}{D} = \frac{\begin{vmatrix} a_1 & c_1 \\ a_2 & c_2 \end{vmatrix}}{\begin{vmatrix} a_1 & b_1 \\ a_2 & b_2 \end{vmatrix}} = \frac{a_1c_2 - c_1a_2}{a_1b_2 - b_1a_2}.$$

In Cramer's rule the formulas for x and y are not difficult to remember. In both cases the determinant in the denominator is formed from the coefficients of x and y. Different determinants are used in the numerators. When solving for x, the coefficients of x, which are a_1 and a_2, are replaced by the constants c_1 and c_2. Similarly, when solving for y, the coefficients of y, which are b_1 and b_2, are replaced by the constants c_1 and c_2.

EXAMPLE 3   Use Cramer's rule to solve the system of equations

$$3x - 2y = 27$$
$$2x + 5y = -1.$$

Solution   First, evaluate the determinant in the denominator, which consists of the coefficients of x and y.

$$D = \begin{vmatrix} 3 & -2 \\ 2 & 5 \end{vmatrix} = 3(5) - (-2)(2) = 15 + 4 = 19$$

Second, evaluate the determinant in the numerator of the formula for x. Replace the column containing the coefficients of x by the column with the constants on the right side of the equations.

$$D_x = \begin{vmatrix} 27 & -2 \\ -1 & 5 \end{vmatrix} = 27(5) - (-2)(-1) = 135 - 2 = 133$$

Third, evaluate the determinant in the numerator of the formula for y. We replace the column containing the coefficients of y by the column with the constants on the right side of the equations.

$$D_y = \begin{vmatrix} 3 & 27 \\ 2 & -1 \end{vmatrix} = 3(-1) - 27(2) = -3 - 54 = -57$$

Fourth, substitute the values of the determinants in the formulas.

$$x = \frac{D_x}{D} = \frac{\begin{vmatrix} 27 & -2 \\ -1 & 5 \end{vmatrix}}{\begin{vmatrix} 3 & -2 \\ 2 & 5 \end{vmatrix}} = \frac{133}{19} = 7$$

$$y = \frac{D_y}{D} = \frac{\begin{vmatrix} 3 & 27 \\ 2 & -1 \end{vmatrix}}{\begin{vmatrix} 3 & -2 \\ 2 & 5 \end{vmatrix}} = \frac{-57}{19} = -3$$

Thus, the solution is $(7, -3)$.   ∎

EXAMPLE 4   Use determinants to solve the system of equations

$$5x + 2y - 1 = 0$$
$$x - 3y - 7 = 0.$$

Solution   First, change the equations to the form from which the formulas were derived.

$$5x + 2y = 1$$
$$x - 3y = 7$$

$$x = \frac{D_x}{D} = \frac{\begin{vmatrix} 1 & 2 \\ 7 & -3 \end{vmatrix}}{\begin{vmatrix} 5 & 2 \\ 1 & -3 \end{vmatrix}} = \frac{1(-3) - 2(7)}{5(-3) - 2(1)} = \frac{-17}{-17} = 1$$

$$y = \frac{D_y}{D} = \frac{\begin{vmatrix} 5 & 1 \\ 1 & 7 \end{vmatrix}}{\begin{vmatrix} 5 & 2 \\ 1 & -3 \end{vmatrix}} = \frac{5(7) - 1(1)}{-17} = \frac{34}{-17} = -2$$

Thus, the solution is $(1, -2)$.   ∎

Determinants may be used to solve more complicated systems, such as

$$a_1x + b_1y + c_1z = d_1$$
$$a_2x + b_2y + c_2z = d_2$$
$$a_3x + b_3y + c_3z = d_3.$$

This system involves three equations and three variables. Its solution is the collection of all values for x, y, and z that satisfy all three equations simultaneously. To solve this system, we must learn to evaluate determinants with three rows and three columns.

We first define the **minor** of an element to be the determinant formed by deleting the row and column containing the given element. For example,

$$\text{minor for } a_1 = \begin{vmatrix} a_1 & b_1 & c_1 \\ a_2 & b_2 & c_2 \\ a_3 & b_3 & c_3 \end{vmatrix} = \begin{vmatrix} b_2 & c_2 \\ b_3 & c_3 \end{vmatrix}$$

$$\text{minor for } b_2 = \begin{vmatrix} a_1 & b_1 & c_1 \\ a_2 & b_2 & c_2 \\ a_3 & b_3 & c_3 \end{vmatrix} = \begin{vmatrix} a_1 & c_1 \\ a_3 & c_3 \end{vmatrix}.$$

Now we define the **cofactor** of an element by considering the row and column position of that element. If an element is in the ith row and the jth column, then the cofactor of the element is given by the product

$$(-1)^{i+j} \cdot (\text{minor of the element}).$$

In other words, if $i + j$ is even, the cofactor of the element is the same as the minor of the element; but if $i + j$ is odd, then the cofactor equals the negative of the minor of the element. In the case of a 3 by 3 determinant, this definition means that the cofactor of an element is found by attaching the sign from the pattern in Figure 9.8 to the minor of that element. Now we can evaluate 3 by 3 (and more complicated) determinants by following these steps.

Figure 9.8

To Evaluate a Determinant

1. Pick any row or any column of the determinant.
2. Multiply each entry in that row or column by its cofactor.
3. Add the results. This sum is defined to be the value of the determinant.

Remember that to each determinant there corresponds exactly one number as the value of that determinant. This determinant value is independent of the row or column that is chosen in step 1.

EXAMPLE 5 Evaluate the following determinant by (a) expansion along the first row of the determinant and (b) expansion along the second column of the determinant.

$$\begin{vmatrix} 1 & 1 & 3 \\ 5 & 0 & 2 \\ -2 & 3 & -1 \end{vmatrix}$$

Solution

a. The sign pattern in Figure 9.8 for row 1 is $+, -, +$. Using row 1 and following the steps given above, we have

$$\begin{vmatrix} 1 & 1 & 3 \\ 5 & 0 & 2 \\ -2 & 3 & -1 \end{vmatrix}$$

$$= 1\begin{vmatrix} 1 & 1 & 3 \\ 5 & 0 & 2 \\ -2 & 3 & -1 \end{vmatrix} - 1\begin{vmatrix} 1 & 1 & 3 \\ 5 & 0 & 2 \\ -2 & 3 & -1 \end{vmatrix} + 3\begin{vmatrix} 1 & 1 & 3 \\ 5 & 0 & 2 \\ -2 & 3 & -1 \end{vmatrix}$$

$$= 1\begin{vmatrix} 0 & 2 \\ 3 & -1 \end{vmatrix} - 1\begin{vmatrix} 5 & 2 \\ -2 & -1 \end{vmatrix} + 3\begin{vmatrix} 5 & 0 \\ -2 & 3 \end{vmatrix}$$

$$= 1[0(-1) - 2(3)] - 1[5(-1) - 2(-2)] + 3[5(3) - 0(-2)]$$
$$= 1(-6) - 1(-1) + 3(15)$$
$$= -6 + 1 + 45$$
$$= 40.$$

b. The sign pattern in Figure 9.8 for column 2 is $-, +, -$. Expanding the determinant by this column gives

$$\begin{vmatrix} 1 & 1 & 3 \\ 5 & 0 & 2 \\ -2 & 3 & -1 \end{vmatrix}$$

$$= -1\begin{vmatrix} 1 & 1 & 3 \\ 5 & 0 & 2 \\ -2 & 3 & -1 \end{vmatrix} + 0\begin{vmatrix} 1 & 1 & 3 \\ 5 & 0 & 2 \\ -2 & 3 & -1 \end{vmatrix} - 3\begin{vmatrix} 1 & 1 & 3 \\ 5 & 0 & 2 \\ -2 & 3 & -1 \end{vmatrix}$$

$$= -1\begin{vmatrix} 5 & 2 \\ -2 & -1 \end{vmatrix} + 0\begin{vmatrix} 1 & 3 \\ -2 & -1 \end{vmatrix} - 3\begin{vmatrix} 1 & 3 \\ 5 & 2 \end{vmatrix}$$

$$= -1[5(-1) - 2(-2)] + 0[\text{not needed}] - 3[1(2) - 3(5)]$$
$$= -1(-1) + 0 - 3(-13)$$
$$= 1 + 39$$
$$= 40.$$

Thus, the value of the determinant is 40. As expected, the answer in parts a and b is the same. Since you have a choice, you may want to evaluate a determinant by picking the row or column with the most 0's. As shown in part b, it is unnecessary to evaluate the minor of 0 elements. ■

Cramer's rule extends to systems with three equations and three variables, and the formulas for x, y, and z follow an arrangement similar to that of the system with two equations and two variables. In each case the determinant in the denominator is formed from the coefficients of x, y, and z. Different determinants are used in the numerators. When solving for x, the coefficients of x are replaced by the constants d_1, d_2, and d_3. Similarly, when solving for y or z, the constants d_1, d_2, and d_3 replace the coefficients of the desired variable.

Cramer's Rule

(Three equations with three variables) The solution to the system

$$a_1 x + b_1 y + c_1 z = d_1$$
$$a_2 x + b_2 y + c_2 z = d_2 \quad \text{with } D = \begin{vmatrix} a_1 & b_1 & c_1 \\ a_2 & b_2 & c_2 \\ a_3 & b_3 & c_3 \end{vmatrix} \neq 0$$
$$a_3 x + b_3 y + c_3 z = d_3$$

is $x = D_x/D$, $y = D_y/D$, and $z = D_z/D$, where

$$D_x = \begin{vmatrix} d_1 & b_1 & c_1 \\ d_2 & b_2 & c_2 \\ d_3 & b_3 & c_3 \end{vmatrix}, D_y = \begin{vmatrix} a_1 & d_1 & c_1 \\ a_2 & d_2 & c_2 \\ a_3 & d_3 & c_3 \end{vmatrix}, \text{ and } D_z = \begin{vmatrix} a_1 & b_1 & d_1 \\ a_2 & b_2 & d_2 \\ a_3 & b_3 & d_3 \end{vmatrix}.$$

EXAMPLE 6   Use determinants to solve the system of equations

$$3x - y + 6z = 1$$
$$x + 2y - 3z = 0$$
$$2x - 3y - z = -9.$$

Solution   The determinant in the denominator is formed from the coefficients of x, y, and z. To find the determinant in the numerator for x, we replace the column containing the coefficients of x by the column containing the constants on the right side of the equations.

$$x = \frac{D_x}{D} = \frac{\begin{vmatrix} 1 & -1 & 6 \\ 0 & 2 & -3 \\ -9 & -3 & -1 \end{vmatrix}}{\begin{vmatrix} 3 & -1 & 6 \\ 1 & 2 & -3 \\ 2 & -3 & -1 \end{vmatrix}}$$

expansion by column 1 because of 0 element

$$= \frac{1\begin{vmatrix} 2 & -3 \\ -3 & -1 \end{vmatrix} - 0\begin{vmatrix} -1 & 6 \\ -3 & -1 \end{vmatrix} + (-9)\begin{vmatrix} -1 & 6 \\ 2 & -3 \end{vmatrix}}{3\begin{vmatrix} 2 & -3 \\ -3 & -1 \end{vmatrix} - \begin{vmatrix} -1 & 6 \\ -3 & -1 \end{vmatrix} + 2\begin{vmatrix} -1 & 6 \\ 2 & -3 \end{vmatrix}}$$

$$= \frac{70}{-70} = -1$$

The value of the determinant in the denominator for both y and z is -70. The column containing the constants on the right side of the equations first replaces the coefficients of y (to find y) and then the coefficients of z (to find z).

$$y = \frac{D_y}{D} = \frac{\begin{vmatrix} 3 & 1 & 6 \\ 1 & 0 & -3 \\ 2 & -9 & -1 \end{vmatrix}}{-70}$$

expansion by column 2 because of 0 element

$$= \frac{-1 \begin{vmatrix} 1 & -3 \\ 2 & -1 \end{vmatrix} + 0 \begin{vmatrix} 3 & 6 \\ 2 & -1 \end{vmatrix} - (-9) \begin{vmatrix} 3 & 6 \\ 1 & -3 \end{vmatrix}}{-70} = \frac{-140}{-70} = 2$$

$$z = \frac{D_z}{D} = \frac{\begin{vmatrix} 3 & -1 & 1 \\ 1 & 2 & 0 \\ 2 & -3 & -9 \end{vmatrix}}{-70}$$

expansion by column 3 because of 0 element

$$= \frac{1 \begin{vmatrix} 1 & 2 \\ 2 & -3 \end{vmatrix} - 0 \begin{vmatrix} 3 & -1 \\ 2 & -3 \end{vmatrix} + (-9) \begin{vmatrix} 3 & -1 \\ 1 & 2 \end{vmatrix}}{-70} = \frac{-70}{-70} = 1$$

The solution is $x = -1, y = 2, z = 1$. ∎

There are two additional points to mention about Cramer's rule. First, if the determinant in the denominator is zero, merely write that there is no unique solution to the problem. Such systems are either inconsistent or dependent. Second, although Cramer's rule extends to systems beyond three equations with three variables, the determinants associated with such systems are usually hard to evaluate. An easier approach in such cases is to use Gaussian elimination as discussed in the previous section.

EXERCISES 9.3

In Exercises 1–16 evaluate the determinant.

1. $\begin{vmatrix} 1 & 3 \\ 4 & 2 \end{vmatrix}$

2. $\begin{vmatrix} -2 & 5 \\ 1 & 3 \end{vmatrix}$

3. $\begin{vmatrix} 1 & -5 \\ 2 & -8 \end{vmatrix}$

4. $\begin{vmatrix} 2 & -3 \\ 2 & -5 \end{vmatrix}$

5. $\begin{vmatrix} 5 & -3 \\ 2 & 4 \end{vmatrix}$

6. $\begin{vmatrix} -2 & 14 \\ 3 & -3 \end{vmatrix}$

7. $\begin{vmatrix} -3 & 0 \\ 4 & 1 \end{vmatrix}$

8. $\begin{vmatrix} 5 & -28 \\ 4 & -10 \end{vmatrix}$

9. $\begin{vmatrix} 2 & 5 \\ -4 & -10 \end{vmatrix}$

10. $\begin{vmatrix} 4 & -8 \\ 6 & -12 \end{vmatrix}$

11. $\begin{vmatrix} -2 & 3 & 2 \\ 0 & 7 & 4 \\ 1 & -1 & 3 \end{vmatrix}$ **12.** $\begin{vmatrix} 0 & 2 & -1 \\ 5 & 1 & 0 \\ -3 & 0 & -4 \end{vmatrix}$

13. $\begin{vmatrix} 2 & 1 & 1 \\ 1 & -2 & -1 \\ 3 & 3 & -1 \end{vmatrix}$ **14.** $\begin{vmatrix} 2 & 3 & 2 \\ 1 & -3 & 3 \\ 5 & 1 & -4 \end{vmatrix}$

15. $\begin{vmatrix} 6 & 3 & 11 \\ -3 & 4 & 5 \\ -1 & -2 & 3 \end{vmatrix}$ **16.** $\begin{vmatrix} -1 & 2 & -1 \\ 5 & -2 & -3 \\ 2 & -4 & 2 \end{vmatrix}$

In Exercises 17–20, find $|A|$.

17. $A = \begin{bmatrix} 3 & 8 \\ -8 & 11 \end{bmatrix}$ **18.** $A = \begin{bmatrix} -7 & -9 \\ -10 & -6 \end{bmatrix}$

19. $A = \begin{bmatrix} 2 & -7 & 1 \\ 4 & 1 & 2 \\ 6 & -3 & 3 \end{bmatrix}$ **20.** $A = \begin{bmatrix} 9 & 10 & -2 \\ 4 & 1 & 8 \\ -1 & -7 & 3 \end{bmatrix}$

In Exercises 21–30 use determinants to solve the systems of equations in Exercises 11–20 of Section 7.1.

In Exercises 31–40 use determinants to solve the given systems of equations. If a variable is missing from an equation, enter zero in the determinant as its coefficient.

31. $x - y + z = -1$
$x + y - 3z = 3$
$2x + y - 2z = 4$

32. $x + y + z = 1$
$x + y - z = 3$
$x - y - z = 5$

33. $-2x + y - z = 3$
$2x + 3y + 2z = 5$
$3x + 2y + z = -3$

34. $x + y - 3z = -4$
$-x - y + 5z = 8$
$2x + 3y - z = 3$

35. $x - y + z = 3$
$3x + y + z = 1$
$2x + 2y - z = -3$

36. $x + y + 3z = 1$
$2x + 5y + 2z = 0$
$3x - 2y - z = 3$

37. $2x + 3y = 10$
$3x + 2z = 2$
$4y + z = 6$

38. $x + y + z = 0$
$2x - 5y = 3$
$-5y + 3z = 7$

39. $2x - y + 3z = 1$
$-x + y - z = -1$
$-4x + 2y - 6z = -2$

40. $x + y + z = 1$
$2x - 3y - 2z = -4$
$3x - 2y - z = -2$

In Exercises 41–44 solve the equations simultaneously to determine the value of the currents I_1, I_2, and I_3 shown in the diagram.

$$I_1 + I_2 + I_3 = 0$$
$$R_1 I_1 - R_3 I_3 = E_1$$
$$R_2 I_2 - R_3 I_3 = E_2$$

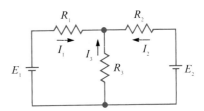

	E_1 (volts)	E_2 (volts)	R_1 (ohms)	R_2 (ohms)	R_3 (ohms)
41.	3	10	2	9	5
42.	8	5	6	10	3
43.	7	2	5	4	9
44.	10	20	50	12	100

THINK ABOUT IT

1. Consider the system of equations given by
$$x + y = b - 1$$
$$x + by = 5.$$

a. Use Cramer's rule to solve the system in terms of b.

b. How many solutions are there if $b = 1$?

c. For what value(s) of b does the system have a unique solution?

2. Consider the system of equations given by
$$ax + 2y + z = 0$$
$$2x - y + 2z = 0$$
$$x + y + 3z = 0.$$

a. For what value of a will the determinant D of the coefficients of x, y, and z equal 0?

b. When a has the value found in part a, how many solutions are there for the system?

c. For what value(s) of a does the system have a unique solution?

d. What is the unique solution?

3. a. Evaluate $\begin{vmatrix} a_1 & b_1 \\ 0 & b_2 \end{vmatrix}$.

b. Evaluate $\begin{vmatrix} a_1 & b_1 & c_1 \\ 0 & b_2 & c_2 \\ 0 & 0 & c_3 \end{vmatrix}$.

c. Evaluate $\begin{vmatrix} a_1 & b_1 & c_1 & d_1 \\ 0 & b_2 & c_2 & d_2 \\ 0 & 0 & c_3 & d_3 \\ 0 & 0 & 0 & d_4 \end{vmatrix}$.

d. Parts a–c show determinants of matrices that are called **upper triangular matrices.** How may the value of this type of matrix be found?

4. The following problems consider three useful properties of determinants that correspond to elementary row operations on matrices.

a. Verify the identity

$$\begin{vmatrix} ka_1 & kb_1 \\ a_2 & b_2 \end{vmatrix} = k \begin{vmatrix} a_1 & b_1 \\ a_2 & b_2 \end{vmatrix}.$$

If each element of a row of matrix A is multiplied by a nonzero real number to form matrix B, then how is $|B|$ related to $|A|$?

b. Verify the identity

$$\begin{vmatrix} a_1 & b_1 & c_1 \\ a_2 & b_2 & c_2 \\ a_3 & b_3 & c_3 \end{vmatrix} = - \begin{vmatrix} a_2 & b_2 & c_2 \\ a_1 & b_1 & c_1 \\ a_3 & b_3 & c_3 \end{vmatrix}.$$

If two rows of matrix A are interchanged to form matrix B, then how is $|B|$ related to $|A|$?

c. Create an example in which you start with a matrix, say A, and add a multiple of the entries in one row to another row to form matrix B. Then, find $|A|$ and $|B|$. What property of determinants is suggested by this example?

5. This section concentrated on the use of determinants to solve systems of linear equations. Determinants have many other applications as illustrated by the next two problems.

a. If the vertices of a triangle are at points (x_1,y_1), (x_2,y_2), and (x_3,y_3), then the area of the triangle is given by the absolute value of the following determinant.

$$\frac{1}{2} \begin{vmatrix} x_1 & y_1 & 1 \\ x_2 & y_2 & 1 \\ x_3 & y_3 & 1 \end{vmatrix}$$

Use this rule to find the area of the triangle whose vertices are at $(-4,-2)$, $(-3,2)$, and $(1,1)$.

b. An equation for the line that passes through distinct points (x_1,y_1) and (x_2,y_2) may be written in determinant form as

$$\begin{vmatrix} x & y & 1 \\ x_1 & y_1 & 1 \\ x_2 & y_2 & 1 \end{vmatrix} = 0.$$

Use this rule to find an equation for the line through $(-1,6)$ and $(2,-3)$.

REMEMBER THIS

In Exercises 1–5 let a and b be real numbers.

1. Name the property illustrated by $ab = ba$.

2. Express $a - b$ as a sum.

3. What is the additive inverse of a?

4. If $a \neq 0$, what is the product of a and its multiplicative inverse?

5. Solve $b + 4x = a$ for x without using division.

6. If $A = \begin{bmatrix} 2 & -1 \\ 4 & -2 \end{bmatrix}$, does the determinant of A equal 0?

7. Use Gauss-Jordan elimination to solve the system

$$5x + 2y = 1$$
$$x - 3y = 7.$$

8. Solve the system
$$x - 2y - z = 0$$
$$2y + z = -3$$
$$-3z = 12.$$

In Exercises 9 and 10 use the system
$$-x + y = 14$$
$$y - z = 11$$
$$x + z = -9.$$

9. Determine the augmented matrix for the system.

10. Use the answer to Exercise 9 and Gaussian elimination with back substitution to solve the system.

9.4 Solving Systems by Matrix Algebra

There is one other method for solving a system of linear equations that merits our attention. In Section 9.2 we used elementary row operations on matrices to solve the system. A different matrix method is to develop a matrix algebra that defines how to perform various operations with matrices and includes some properties that hold in this system. This approach is beneficial in two ways.

1. Matrices have many applications independent of systems of equations, and the ideas discussed here can serve as an introduction to this important topic. Because of this we develop a little more matrix algebra than is needed to solve a system of equations. More detailed coverage of this topic is usually given in a course in linear algebra.

2. Although the matrix algebra approach usually requires more work than our other methods to solve a given system, it is an efficient technique for solving systems such as

$$2x + 3y = b_1$$
$$3x + 5y = b_2,$$

where the coefficients of x and y remain fixed but different pairs of constants replace the b's. We consider this problem at the end of this section.

To begin, we first recall that a **matrix** is a rectangular array of numbers enclosed in brackets (or parentheses) and that each number in the matrix is called an **entry** or **element** of the matrix. We now consider some other basic definitions.

Matrix Definitions

1. A matrix with m rows and n columns is of **dimension $m \times n$.**
2. A **square matrix** is a matrix having the same number of rows as columns.
3. A **zero matrix** is a matrix containing only zero elements. We often denote a zero matrix by a boldface zero: **0.**
4. Two matrices are **equal** if and only if the elements in corresponding positions are equal.
5. To **add** two matrices of the same dimension, add elements in corresponding positions. Only matrices of the same dimension can be added.
6. To multiply a matrix by a real number, multiply each element in the matrix by that number. In this context we call the real number a **scalar,** and we call this operation **scalar multiplication.**
7. $A - B = A + (-B)$, where $-B = (-1)B$.

Now consider Example 1, which illustrates these definitions.

EXAMPLE 1   Answer the following questions given

$$A = \begin{bmatrix} 2 & -1 & 3 \\ 0 & 1 & -2 \end{bmatrix}, B = \begin{bmatrix} 0 & 0 \\ 0 & 0 \end{bmatrix},$$

$$C = \begin{bmatrix} 1 & 0 \\ 0 & 1 \end{bmatrix}, \text{ and } D = \begin{bmatrix} 1 & 3 & 6 \\ 2 & 5 & 7 \end{bmatrix}.$$

a. What is the dimension of each matrix?
b. Find (if possible) $A + D$.
c. Find (if possible) $A + C$.
d. Find $-A$.
e. If $A + 2X = D$, find X.

Solution

a. A and D are matrices of dimension 2×3 since both have 2 rows and 3 columns. B and C are square matrices with 2 rows and 2 columns and are of dimension 2×2. Matrix B is an example of a zero matrix.

b. A and D are both of dimension 2×3, so $A + D$ is defined and is found as follows:

$$\begin{bmatrix} 2 & -1 & 3 \\ 0 & 1 & -2 \end{bmatrix} + \begin{bmatrix} 1 & 3 & 6 \\ 2 & 5 & 7 \end{bmatrix} = \begin{bmatrix} 2+1 & -1+3 & 3+6 \\ 0+2 & 1+5 & -2+7 \end{bmatrix}$$

$$= \begin{bmatrix} 3 & 2 & 9 \\ 2 & 6 & 5 \end{bmatrix}.$$

c. A and C are not of the same dimension, so $A + C$ is undefined.

d. $-A = (-1)A$, so to find $-A$ we multiply each element in A by -1. Then

$$-A = -1 \begin{bmatrix} 2 & -1 & 3 \\ 0 & 1 & -2 \end{bmatrix} = \begin{bmatrix} -2 & 1 & -3 \\ 0 & -1 & 2 \end{bmatrix}.$$

e. Since A and D are of dimension 2×3, matrix X must also have 2 rows and 3 columns. If we let

$$X = \begin{bmatrix} a & b & c \\ d & e & f \end{bmatrix}, \text{ then } 2X = \begin{bmatrix} 2a & 2b & 2c \\ 2d & 2e & 2f \end{bmatrix}, \text{ so}$$

$$\begin{array}{ccccc} A & + & 2X & = & D \end{array}$$

$$\begin{bmatrix} 2 & -1 & 3 \\ 0 & 1 & -2 \end{bmatrix} + \begin{bmatrix} 2a & 2b & 2c \\ 2d & 2e & 2f \end{bmatrix} = \begin{bmatrix} 1 & 3 & 6 \\ 2 & 5 & 7 \end{bmatrix}$$

$$\begin{bmatrix} 2+2a & -1+2b & 3+2c \\ 0+2d & 1+2e & -2+2f \end{bmatrix} = \begin{bmatrix} 1 & 3 & 6 \\ 2 & 5 & 7 \end{bmatrix}.$$

For two matrices to be equal, elements in corresponding positions must be the same. Thus,

$$
\begin{array}{ll}
2 + 2a = 1, \text{ so } a = -\tfrac{1}{2}; & -1 + 2b = 3, \text{ so } b = 2; \\
3 + 2c = 6, \text{ so } c = \tfrac{3}{2}; & 0 + 2d = 2, \text{ so } d = 1; \\
1 + 2e = 5, \text{ so } e = 2; & -2 + 2f = 7, \text{ so } f = \tfrac{9}{2}.
\end{array}
$$

So

$$X = \begin{bmatrix} -\tfrac{1}{2} & 2 & \tfrac{3}{2} \\ 1 & 2 & \tfrac{9}{2} \end{bmatrix}.$$

■

Many of the properties from the algebra of real numbers carry over to the algebra of matrices. That is because we have defined our operations to this point in terms of operating on elements that are real numbers and that are in corresponding positions. Some of the more basic properties are as follows:

Properties of Matrices

If A, B, and C are matrices and c and k are scalars (real numbers), then

1. $(A + B) + C = A + (B + C)$
2. $A + B = B + A$
3. $A + 0 = 0 + A = A$
4. $A + (-A) = (-A) + A = 0$
5. $c(kA) = (ck)A$
6. $c(A + B) = cA + cB$

These properties provide us with an alternative method for solving the problem in Example 1e that is similar to our previous work with equations. That is, we start with $A + 2X = D$ and solve for X as follows:

$$A + 2X = D$$
$$A + (-A) + 2X = D + (-A) \quad \text{Add } -A \text{ to both sides.}$$
$$0 + 2X = D - A \quad \text{Property 4 and subtraction definition}$$
$$2X = D - A \quad \text{Property 3}$$
$$X = \tfrac{1}{2}(D - A). \quad \text{Multiply both sides by the scalar } \tfrac{1}{2}.$$

Then

$$X = \tfrac{1}{2}\left(\begin{bmatrix} 1 & 3 & 6 \\ 2 & 5 & 7 \end{bmatrix} - \begin{bmatrix} 2 & -1 & 3 \\ 0 & 1 & -2 \end{bmatrix}\right)$$

$$= \tfrac{1}{2}\begin{bmatrix} -1 & 4 & 3 \\ 2 & 4 & 9 \end{bmatrix}$$

$$= \begin{bmatrix} -\tfrac{1}{2} & 2 & \tfrac{3}{2} \\ 1 & 2 & \tfrac{9}{2} \end{bmatrix}.$$

Note that this answer agrees with our previous result.

To this point, our definitions and properties have been straightforward and predictable. Matrix multiplication, however, is unusual and we do *not* multiply matrices by simply multiplying elements in corresponding positions. To understand the usefulness of the different type of product that is defined in matrix multiplication, consider the following matrices.

$$\begin{array}{c} \\ \text{Plant A} \\ \text{Plant B} \\ \text{Plant C} \end{array} \begin{array}{cc} \text{Printers} & \text{Monitors} \\ \begin{bmatrix} 70 & 30 \\ 50 & 90 \\ 80 & 10 \end{bmatrix} & = A \end{array} \qquad \begin{array}{c} \\ \text{Printers} \\ \text{Monitors} \end{array} \begin{array}{c} \text{Value} \\ \begin{bmatrix} 300 \\ 200 \end{bmatrix} = B \end{array}$$

Matrix A summarizes information about the location of computer equipment, while matrix B indicates the value of that equipment. We can find the total value of the equipment at each plant as follows:

	$\begin{pmatrix} \text{number} \\ \text{of} \\ \text{printers} \end{pmatrix}$		$\begin{pmatrix} \text{value} \\ \text{per} \\ \text{printer} \end{pmatrix}$	$+$	$\begin{pmatrix} \text{number} \\ \text{of} \\ \text{monitors} \end{pmatrix}$		$\begin{pmatrix} \text{value} \\ \text{per} \\ \text{monitor} \end{pmatrix}$	$=$	$\begin{matrix} \text{total} \\ \text{value} \end{matrix}$
Plant A	70	$\cdot$	300	$+$	30	$\cdot$	200	$=$	27,000
Plant B	50	$\cdot$	300	$+$	90	$\cdot$	200	$=$	33,000
Plant C	80	$\cdot$	300	$+$	10	$\cdot$	200	$=$	26,000

and we can represent this result in the following matrix.

$$\begin{matrix} & \text{Value} \\ \begin{matrix} \text{Plant A} \\ \text{Plant B} \\ \text{Plant C} \end{matrix} & \begin{bmatrix} 27,000 \\ 33,000 \\ 26,000 \end{bmatrix} = C \end{matrix}$$

We wish to define matrix multiplication so that $AB = C$. First, note that for this product to make sense the columns in matrix A (type of computer equipment) must match the rows of matrix B. Thus, the product AB of two matrices is defined only when the number of columns in A is the same as the number of rows in B. Also note that the product AB (or matrix C) gets its row description from the rows of A (plant location) and its column description from the column(s) of matrix B (value). Thus, the product AB of two matrices has as many rows as A and as many columns as B. We summarize these results in this diagram.

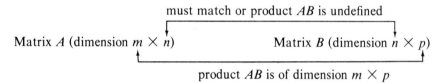

Finally, the element in row 1, column 1 in AB is found by multiplying the entries in the first row of A by the corresponding entries in the first column of B and adding the results. Similarly, the element in row 2, column 1 is found by multiplying the entries in the second row of A by the corresponding entries in the first column of B. This pattern continues and our observations suggest the following definition for matrix multiplication.

Matrix Multiplication

> If A is an $m \times n$ matrix and B is an $n \times p$ matrix, then the product AB is an $m \times p$ matrix in which the ith row, jth column element of AB is found by multiplying each element in the ith row of A by the corresponding element in the jth column of B and adding the results.

Examples 2–4 illustrate the definition.

EXAMPLE 2   Let

$$A = \begin{bmatrix} 3 & 0 & 1 \\ 4 & 4 & 2 \end{bmatrix} \quad \text{and} \quad B = \begin{bmatrix} 1 & 2 \\ 4 & 0 \\ 7 & 3 \end{bmatrix}.$$

Find (if possible) the product AB.

Solution A is of dimension 2×3 and B is of dimension 3×2. Thus, the product is defined and AB is a 2×2 matrix. Such a matrix has four elements that are determined as follows:

Step 1 (row 1, column 1)

$$\begin{bmatrix} 3 & 0 & 1 \\ 4 & 4 & 2 \end{bmatrix} \begin{bmatrix} 1 & 2 \\ 4 & 0 \\ 7 & 3 \end{bmatrix}$$

Computation

$3 \cdot 1 + 0 \cdot 4 + 1 \cdot 7 = 10$

Step 2 (row 1, column 2)

$$\begin{bmatrix} 3 & 0 & 1 \\ 4 & 4 & 2 \end{bmatrix} \begin{bmatrix} 1 & 2 \\ 4 & 0 \\ 7 & 3 \end{bmatrix}$$

$3 \cdot 2 + 0 \cdot 0 + 1 \cdot 3 = 9$

Step 3 (row 2, column 1)

$$\begin{bmatrix} 3 & 0 & 1 \\ 4 & 4 & 2 \end{bmatrix} \begin{bmatrix} 1 & 2 \\ 4 & 0 \\ 7 & 3 \end{bmatrix}$$

$4 \cdot 1 + 4 \cdot 4 + 2 \cdot 7 = 34$

Step 4 (row 2, column 2)

$$\begin{bmatrix} 3 & 0 & 1 \\ 4 & 4 & 2 \end{bmatrix} \begin{bmatrix} 1 & 2 \\ 4 & 0 \\ 7 & 3 \end{bmatrix}$$

$4 \cdot 2 + 4 \cdot 0 + 2 \cdot 3 = 14$

Thus,

$$AB = \begin{bmatrix} 10 & 9 \\ 34 & 14 \end{bmatrix}.$$

■

EXAMPLE 3   Let

$$A = [1 \quad 2 \quad 3 \quad 4] \quad \text{and} \quad B = \begin{bmatrix} -1 & 4 \\ 0 & 1 \\ 5 & -3 \\ -2 & 0 \end{bmatrix}.$$

Find (if possible) (a) the product AB and (b) the product BA.

Solution

a. A is a 1×4 matrix and B is a 4×2 matrix. Thus, the product is defined and is a 1×2 matrix. We determine these two elements as follows:

Entry in row 1, column 1: $1(-1) + 2(0) + 3(5) + 4(-2) = 6$
Entry in row 1, column 2: $1(4) + 2(1) + 3(-3) + 4(0) = -3$.

Thus,

$$AB = [6 \quad -3].$$

b. Since B is a 4×2 matrix and A is a 1×4 matrix, the number of columns in B does not equal the number of rows in A, so the product BA is undefined.

[*Note:* This example shows that matrix multiplication is not commutative (that is, in general $AB \neq BA$), so we must be careful about the order in which we write matrices when expressing a product. For square matrices of the same order we do, however, have an associative property $(AB)C = A(BC)$ and a distributive property $A(B + C) = AB + AC$.] ∎

EXAMPLE 4   Let

$$A = \begin{bmatrix} 2 & 3 \\ 3 & 5 \end{bmatrix}, \quad X = \begin{bmatrix} x \\ y \end{bmatrix}, \quad \text{and} \quad B = \begin{bmatrix} 5 \\ 9 \end{bmatrix}.$$

Write the system of linear equations represented by $AX = B$.

Solution A is a 2×2 matrix and X is a 2×1 matrix, so AX is defined and is a 2×1 matrix. The two elements in the product matrix are

Entry in row 1, column 1: $2x + 3y$
Entry in row 2, column 1: $3x + 5y$ so $AX = \begin{bmatrix} 2x + 3y \\ 3x + 5y \end{bmatrix}.$

Since $AX = B$, we know

$$\begin{bmatrix} 2x + 3y \\ 3x + 5y \end{bmatrix} = \begin{bmatrix} 5 \\ 9 \end{bmatrix},$$

so from the definition of equal matrices we obtain the system

$$2x + 3y = 5$$
$$3x + 5y = 9.$$ ∎

Example 4 illustrates that a system of linear equations can be represented as a matrix equation $AX = B$, so if we can solve this matrix equation for X, the result will give us the solution to the system of equations. To solve $AX = B$ for X, you might suggest we divide both sides by A, since that is one approach to solving equations. However, division of matrices is not defined, so this approach doesn't work here. A second approach is to multiply both sides of the equation by $1/A$ (or the multiplicative inverse of A), since in the algebra of real numbers

$$(1/A)Ax = (1/A)B$$
$$1 \cdot x = (1/A)B$$
$$x = (1/A)B.$$

This approach can be extended to matrix algebra if we define matrices that can play the parts of $1/A$ and 1 from the equations above. The necessary definitions for the part of 1 are as follows:

1. The **principal diagonal** of a square matrix consists of the elements in the diagonal extending from the upper left corner to the lower right corner.
2. The $n \times n$ **identity matrix** is the square matrix with n rows and n columns with 1's on the principal diagonal and 0's elsewhere. An identity matrix is symbolized by I.

An identity matrix is important because

$$AI = IA = A$$

for any square matrix A. Example 5 illustrates this property.

EXAMPLE 5   If $A = \begin{bmatrix} 3 & 1 \\ 2 & 4 \end{bmatrix}$, verify that $AI = IA = A$.

Solution   Since A is a 2×2 matrix, we use $I = \begin{bmatrix} 1 & 0 \\ 0 & 1 \end{bmatrix}$, which is the 2×2 identity matrix. Then

$$AI = \begin{bmatrix} 3 & 1 \\ 2 & 4 \end{bmatrix} \begin{bmatrix} 1 & 0 \\ 0 & 1 \end{bmatrix}$$

$$= \begin{bmatrix} 3 \cdot 1 + 1 \cdot 0 & 3 \cdot 0 + 1 \cdot 1 \\ 2 \cdot 1 + 4 \cdot 0 & 2 \cdot 0 + 4 \cdot 1 \end{bmatrix}$$

$$= \begin{bmatrix} 3 & 1 \\ 2 & 4 \end{bmatrix} = A.$$

Also,

$$IA = \begin{bmatrix} 1 & 0 \\ 0 & 1 \end{bmatrix} \begin{bmatrix} 3 & 1 \\ 2 & 4 \end{bmatrix}$$

$$= \begin{bmatrix} 1 \cdot 3 + 0 \cdot 2 & 1 \cdot 1 + 0 \cdot 4 \\ 0 \cdot 3 + 1 \cdot 2 & 0 \cdot 1 + 1 \cdot 4 \end{bmatrix}$$

$$= \begin{bmatrix} 3 & 1 \\ 2 & 4 \end{bmatrix} = A.$$

To play the part of $1/A$, we now wish to find a square matrix A^{-1}, called the **multiplicative inverse of A,** such that

$$A^{-1}A = AA^{-1} = I.$$

It can be shown that A has an inverse if and only if the determinant of A is not zero, and that when A^{-1} does exist, it is unique and can be obtained in the following way.

To Find A^{-1}

> 1. Write the augmented matrix $[A \mid I]$, where I is the identity matrix with the same dimension as A.
> 2. Use the elementary row operations shown in Section 9.2 to replace matrix $[A \mid I]$ with a matrix of the form $[I \mid B]$.
> 3. Then A^{-1} is matrix B.

When it is not possible to obtain $[I \mid B]$ (for instance, you may obtain all 0's in a row in the left portion of the matrix), then A has no inverse. Example 6 illustrates the procedure. Note that A^{-1} denotes the multiplicative inverse of matrix A and does not equal $1/A$ in matrix algebra.

EXAMPLE 6   If $A = \begin{bmatrix} 2 & 3 \\ 3 & 5 \end{bmatrix}$, find A^{-1} (if it exists).

Solution   First, we form the augmented matrix $[A \mid I]$.

$$\begin{bmatrix} 2 & 3 & \vdots & 1 & 0 \\ 3 & 5 & \vdots & 0 & 1 \end{bmatrix}$$

We then convert $[A \mid I]$ to the form $[I \mid B]$ using elementary row operations. To obtain 1 for the row 1, column 1 entry, we multiply each element in row 1 by $\frac{1}{2}$.

$$\begin{bmatrix} 1 & \frac{3}{2} & \vdots & \frac{1}{2} & 0 \\ 3 & 5 & \vdots & 0 & 1 \end{bmatrix}$$

To obtain 0 for the row 2, column 1 entry, we add -3 times each entry in row 1 to the corresponding entry in row 2.

$$\begin{bmatrix} 1 & \frac{3}{2} & \vdots & \frac{1}{2} & 0 \\ 0 & \frac{1}{2} & \vdots & -\frac{3}{2} & 1 \end{bmatrix}$$

Next, multiply each entry in row 2 by 2 so the row 2, column 2 entry is 1.

$$\begin{bmatrix} 1 & \frac{3}{2} & \vdots & \frac{1}{2} & 0 \\ 0 & 1 & \vdots & -3 & 2 \end{bmatrix}$$

Finally, add $-\frac{3}{2}$ times each entry in row 2 to the corresponding entry in row 1 to obtain 0 for the row 1, column 2 entry.

$$\begin{bmatrix} 1 & 0 & | & 5 & -3 \\ 0 & 1 & | & -3 & 2 \end{bmatrix}$$

The matrix is now in the form $[I \mid B]$, so by the given rule, B is the inverse of A. Thus,

$$A^{-1} = \begin{bmatrix} 5 & -3 \\ -3 & 2 \end{bmatrix}.$$

■

If A^{-1} exists, the matrix equation $AX = B$ can now be solved by multiplying both sides of the equation by A^{-1} and simplifying as follows:

$$
\begin{aligned}
A^{-1}AX &= A^{-1}B \\
IX &= A^{-1}B \qquad \text{(since } A^{-1}A = I) \\
X &= A^{-1}B \qquad \text{(since } IX = X).
\end{aligned}
$$

Thus, the product $A^{-1}B$ gives the solution to a system of linear equations. In practice, it is usually easier to use row operations to solve a particular linear system than to determine A^{-1}. However, Example 7 considers the problem mentioned in the introduction to this section in which the matrix algebra method is an efficient approach.

EXAMPLE 7   Solve the system

$$
\begin{aligned}
2x + 3y &= b_1 \\
3x + 5y &= b_2
\end{aligned}
$$

when (a) $b_1 = 5$, $b_2 = 9$ and (b) $b_1 = -3$, $b_2 = 0$.

Solution

a. The given system can be represented by $AX = B$, where

$$A = \begin{bmatrix} 2 & 3 \\ 3 & 5 \end{bmatrix}, \qquad X = \begin{bmatrix} x \\ y \end{bmatrix}, \qquad \text{and} \qquad B = \begin{bmatrix} 5 \\ 9 \end{bmatrix}.$$

If A^{-1} exists, the solution to $AX = B$ is $X = A^{-1}B$. We know from Example 6 that

$$A^{-1} = \begin{bmatrix} 5 & -3 \\ -3 & 2 \end{bmatrix},$$

so

$$\begin{bmatrix} x \\ y \end{bmatrix} = \begin{bmatrix} 5 & -3 \\ -3 & 2 \end{bmatrix}\begin{bmatrix} 5 \\ 9 \end{bmatrix} = \begin{bmatrix} 5(5) + (-3)(9) \\ -3(5) + 2(9) \end{bmatrix} = \begin{bmatrix} -2 \\ 3 \end{bmatrix}.$$

Thus, $x = -2$ and $y = 3$.

b. We simply proceed as before but with different elements in matrix B.

$$\begin{bmatrix} x \\ y \end{bmatrix} = \begin{bmatrix} 5 & -3 \\ -3 & 2 \end{bmatrix} \begin{bmatrix} -3 \\ 0 \end{bmatrix} = \begin{bmatrix} 5(-3) + (-3)(0) \\ -3(-3) + 2(0) \end{bmatrix} = \begin{bmatrix} -15 \\ 9 \end{bmatrix}$$

Thus, $x = -15$ and $y = 9$. Note that once we know A^{-1}, it is easy to solve the system for any pair of values for b_1 and b_2. ∎

EXERCISES 9.4

In Exercises 1–4 specify the dimension of each matrix.

1. $\begin{bmatrix} 3 & 0 & 1 & -5 \\ -2 & 8 & 4 & 7 \end{bmatrix}$　　**2.** $\begin{bmatrix} 0 & 1 \\ -1 & 0 \end{bmatrix}$

3. $\begin{bmatrix} 6 \\ 4 \end{bmatrix}$　　　　　　**4.** $[1 \ -3 \ 5]$

In Exercises 5–8 find the values of the variables.

5. $\begin{bmatrix} a & b \\ 5 & 4 \end{bmatrix} = \begin{bmatrix} 10 & -3 \\ c & d \end{bmatrix}$

6. $\begin{bmatrix} x \\ y \end{bmatrix} = \begin{bmatrix} -1 \\ 4 \end{bmatrix}$

7. $[1 + a \quad -1 - b \quad 3 + c] = [3 \ -1 \ 0]$

8. $\begin{bmatrix} 2 + 2a & -1 + 2b & 3 + 2c \\ 2d & 1 + 2e & -2 + 2f \end{bmatrix}$

$= \begin{bmatrix} 4 & 0 & 1 \\ -2 & 7 & 2 \end{bmatrix}$

In Exercises 9–14 perform the indicated operations.

9. $\begin{bmatrix} 1 & 2 \\ 3 & 1 \end{bmatrix} + \begin{bmatrix} -2 & -5 \\ 1 & 9 \end{bmatrix}$

10. $\begin{bmatrix} \frac{1}{8} \\ \frac{5}{8} \\ \frac{7}{8} \end{bmatrix} - \begin{bmatrix} \frac{3}{8} \\ -\frac{1}{8} \\ \frac{5}{8} \end{bmatrix}$

11. $\frac{1}{2}\begin{bmatrix} 0 & 4 & -2 \\ 8 & -6 & 2 \end{bmatrix}$

12. $-3\begin{bmatrix} 1 & 0 \\ 0 & 1 \end{bmatrix}$

13. $2\begin{bmatrix} 3 & 2 \\ 1 & 0 \end{bmatrix} - \begin{bmatrix} -4 & 2 \\ 3 & 5 \end{bmatrix}$

14. $-\frac{1}{2}[2 \ -4 \ 6] + \frac{1}{3}[-9 \ 3 \ 6]$

In Exercises 15–20 find the dimensions of the product AB and the product BA if they are defined.

15. A is 2×2, B is 2×2.
16. A is 2×3, B is 2×3.
17. A is 2×3, B is 3×2.
18. A is 4×1, B is 1×4.
19. A is 4×1, B is 3×2.
20. A is 3×3, B is 1×3.

In Exercises 21–26 determine the given product.

21. $\begin{bmatrix} 1 & 3 \\ -1 & 0 \end{bmatrix} \begin{bmatrix} 2 & -3 \\ 3 & 1 \end{bmatrix}$

22. $\begin{bmatrix} 2 & -3 \\ 3 & 1 \end{bmatrix} \begin{bmatrix} 1 & 3 \\ -1 & 0 \end{bmatrix}$

23. $\begin{bmatrix} 1 & 3 \\ -1 & 0 \end{bmatrix} \begin{bmatrix} 1 & 0 \\ 0 & 1 \end{bmatrix}$

24. $\begin{bmatrix} 1 & 3 \\ -1 & 0 \end{bmatrix} \begin{bmatrix} 2 \\ 3 \end{bmatrix}$

25. $\begin{bmatrix} 3 & -4 & 1 \\ 0 & 5 & -3 \\ 2 & -1 & 4 \end{bmatrix} \begin{bmatrix} -1 \\ 2 \\ 1 \end{bmatrix}$

26. $\begin{bmatrix} 1 & 3 & -2 \\ -1 & 0 & 2 \end{bmatrix} \begin{bmatrix} -1 & 3 \\ 2 & 0 \\ 1 & 4 \end{bmatrix}$

In Exercises 27–38 perform the indicated operations (if they are defined) for matrices A, B, C, D, and I as defined below.

$$A = \begin{bmatrix} 2 & -1 & 3 \\ 0 & 1 & -2 \end{bmatrix}, \ B = \begin{bmatrix} 1 & 2 \\ -3 & -4 \end{bmatrix},$$

$$C = \begin{bmatrix} -1 & 0 \\ 0 & -1 \end{bmatrix}, \ D = \begin{bmatrix} 1 & 3 & 6 \\ 2 & 5 & 7 \end{bmatrix},$$

$$I = \begin{bmatrix} 1 & 0 \\ 0 & 1 \end{bmatrix}.$$

27. $A - D$　　**28.** $A - B$　　**29.** AB
30. BA　　　**31.** BI　　　**32.** IB
33. $2(B + C)$　**34.** $3D - 2A$　**35.** II
36. $(BC)A$　　**37.** $C(A + D)$　**38.** $(A + D)C$

In Exercises 39 and 40 write the system of linear equations represented by $AX = B$.

39. $A = \begin{bmatrix} 1 & 1 \\ 5 & -2 \end{bmatrix}$, $X = \begin{bmatrix} x \\ y \end{bmatrix}$, and $B = \begin{bmatrix} 0 \\ 3 \end{bmatrix}$.

40. $A = \begin{bmatrix} 3 & -1 & 5 \\ 1 & 0 & -2 \\ -1 & 1 & -1 \end{bmatrix}$,

$X = \begin{bmatrix} x \\ y \\ z \end{bmatrix}$, and $B = \begin{bmatrix} 1 \\ -1 \\ 1 \end{bmatrix}$.

In Exercises 41–49 find the inverse, if it exists, for each matrix.

41. $\begin{bmatrix} -2 & -1 \\ 7 & 3 \end{bmatrix}$

42. $\begin{bmatrix} 1 & 1 \\ 6 & -1 \end{bmatrix}$

43. $\begin{bmatrix} 5 & 3 \\ 1 & -2 \end{bmatrix}$

44. $\begin{bmatrix} -2 & -1 \\ 5 & 2 \end{bmatrix}$

45. $\begin{bmatrix} 3 & -2 \\ 6 & -4 \end{bmatrix}$

46. $\begin{bmatrix} 7 & -2 \\ 3 & 5 \end{bmatrix}$

47. $\begin{bmatrix} 1 & -1 & 1 \\ -1 & 1 & 0 \\ 0 & -1 & 1 \end{bmatrix}$

48. $\begin{bmatrix} 2 & 3 & 0 \\ 3 & 0 & 2 \\ 0 & 4 & 1 \end{bmatrix}$

49. $\begin{bmatrix} 2 & -1 & 3 \\ -1 & 1 & -1 \\ -4 & 2 & -6 \end{bmatrix}$

In Exercises 50–58 write the given system in the form $AX = B$ and use A^{-1} (if it exists) to solve the system. (*Note:* The inverses for these problems are found in Exercises 41–49.)

50. $-2x - y = 2$
$7x + 3y = -1$

51. $x + y = 25$
$6x - y = 3$

52. $5x + 3y = -2$
$x - 2y = -3$

53. $-2x - y = -5$
$5x + 2y = -17$

54. $3x - 2y = 1$
$6x - 4y = 5$

55. $7x - 2y = 19$
$3x + 5y = -14$

56. $\begin{aligned} x - y + z &= 1 \\ -x + y &= -1 \\ -y + z &= 2 \end{aligned}$

57. $\begin{aligned} 2x + 3y &= 10 \\ 3x + 2z &= 2 \\ 4y + z &= 6 \end{aligned}$

58. $\begin{aligned} 2x - y + 3z &= 1 \\ -x + y - z &= -1 \\ -4x + 2y - 6z &= -2 \end{aligned}$

59. If the inverse of $\begin{bmatrix} 3 & -5 \\ -4 & 7 \end{bmatrix}$ is $\begin{bmatrix} 7 & 5 \\ 4 & 3 \end{bmatrix}$, then solve

$3x - 5y = b_1$
$-4x + 7y = b_2$

when (a) $b_1 = 2$, $b_2 = 3$ and (b) $b_1 = 0$, $b_2 = -5$.

60. If the inverse of $\begin{bmatrix} 6 & -1 & -5 \\ -7 & 1 & 5 \\ -10 & 2 & 11 \end{bmatrix}$

is $\begin{bmatrix} -1 & -1 & 0 \\ -27 & -16 & -5 \\ 4 & 2 & 1 \end{bmatrix}$, then solve

$6x - y - 5z = b_1$
$-7x + y + 5z = b_2$
$-10x + 2y + 11z = b_3$

when (a) $b_1 = 1$, $b_2 = 0$, $b_3 = 0$ and (b) $b_1 = 2$, $b_2 = -1$, $b_3 = -2$.

THINK ABOUT IT

1. Give an example of two matrices A and B of dimension 2×2 for which $AB = 0$ but neither A nor B is the zero matrix.
2. The equation $(A + B)^2 = A^2 + 2AB + B^2$ is an identity if A and B are real numbers. However, in matrix algebra if the square of a matrix, say A, is defined by $A^2 = AA$, then this equation is not necessarily true for square matrices A and B even though the distributive properties $A(B + C) = AB + AC$ and $(B + C)A = BA + CA$ are true. Explain why.
3. Consider the following matrices.

	Midterm	Final
Jennifer	80	96
David	75	83
Tom	94	86
Kelly	85	67

$= A$

	System 1	System 2
Midterm	0.5	0.4
Final	0.5	0.6

$= B$

Find AB and discuss what it means.

4. If $A = \begin{bmatrix} a & 0 & 0 \\ 0 & b & 0 \\ 0 & 0 & c \end{bmatrix}$, state conditions on a, b, and c so that A^{-1} exists. Then, find A^{-1}.

5. Consider the following theorem:

If $A = \begin{bmatrix} a & b \\ c & d \end{bmatrix}$ with $ad - bc \neq 0$, then

$$A^{-1} = \frac{1}{|A|} \begin{bmatrix} d & -b \\ -c & a \end{bmatrix}.$$

a. Use this theorem to find A^{-1} for

$$A = \begin{bmatrix} 5 & 3 \\ 1 & -2 \end{bmatrix}.$$

Check that the answer agrees with the result for Exercise 43 of this section.

b. For A and A^{-1} as specified in the theorem, show $A^{-1}A = I$.

REMEMBER THIS

In Exercises 1–3 add and simplify.

1. $\dfrac{5}{x-2} + \dfrac{3}{x+1}$

2. $\dfrac{-2}{x+3} + \dfrac{9}{(x+3)^2}$

3. $\dfrac{1}{x} + \dfrac{-x+4}{x^2+3}$

In Exercises 4 and 5 factor completely.

4. $2Bx + Cx$

5. $x^4 - 6x^3 + 9x^2$

In Exercises 6 and 7 divide and express the answer in the form

$$\frac{\text{dividend}}{\text{divisor}} = \text{quotient} + \frac{\text{remainder}}{\text{divisor}}.$$

6. $\dfrac{x^2+6}{x^2+3x}$

7. $\dfrac{x^5 - x^3 + x - 1}{x^4 + x^2}$

In Exercises 8 and 9 solve each system.

8. $\begin{array}{rcl} A + B &=& 1 \\ -A - 4B &=& 2 \end{array}$

9. $\begin{array}{rcl} A + B &=& 0 \\ 2B + C &=& 0 \\ 4A + 2C &=& 16 \end{array}$

10. Use Cramer's rule to find the solution for y of the system

$$\begin{array}{rcl} 3x + 3y - z &=& 0 \\ 3x - y - z &=& 1 \\ -x - 4y - 2z &=& 1. \end{array}$$

9.5 Partial Fractions

Solving a system of linear equations is a useful component for determining algebraic identities that enable us to analyze rational functions. Recall from Section 3.6 that rational functions fit the form

$$y = \frac{P(x)}{Q(x)} \qquad Q(x) \neq 0,$$

where $P(x)$ and $Q(x)$ are polynomials. In certain situations (especially in calculus), if $P(x)/Q(x)$ is complicated, we try to write this expression as the sum of simpler fractions that are easier to analyze. For example, from the usual addition of fractions we know that

$$\frac{2}{x+2} + \frac{3}{x-1}$$

is identical to

$$\frac{5x+4}{(x+2)(x-1)}.$$

The question is how to work this problem backward and split the more complicated fraction into the sum of the simpler fractions. This sum is called the **partial fraction decomposition** of the expression, and each term in the sum is called a **partial fraction.**

In the above problem, determining the partial fraction decomposition (assume it is unknown) amounts to finding constants A and B such that

$$\frac{5x + 4}{(x + 2)(x - 1)} = \frac{A}{x + 2} + \frac{B}{x - 1}$$

is an identity. To do this, we first clear the expression of fractions by multiplying both sides of the equation by $(x + 2)(x - 1)$ to obtain

$$5x + 4 = A(x - 1) + B(x + 2).$$

Now since this equation is an identity, the coefficients of like powers of x must be equal on both sides of the equation. By rewriting the above equation as

$$5x + 4 = (A + B)x - A + 2B$$

and equating coefficients, we obtain the linear system

$$A + B = 5$$
$$-A + 2B = 4.$$

Adding these two equations yields $3B = 9$, so $B = 3$. Then it is easy to determine that $A = 2$ and that our results check with the original problem.

In this example we have illustrated the method of equating coefficients to determine the constants A and B because it is a powerful method that extends to more complicated cases, and it also gives us additional exposure to systems of linear equations. A shortcut to finding A and B in this particular example is to realize that

$$5x + 4 = A(x - 1) + B(x + 2)$$

is an identity and therefore true when $x = 1$ and $x = -2$. Substitution of $x = 1$ readily yields $9 = 3B$, so $B = 3$, while replacing x by -2 gives $-6 = -3A$, so $A = 2$. However, in the long run this method is limited, and it is most useful only when A and B are coupled with factors that are linear expressions.

EXAMPLE 1   Determine the constants A, B, and C so that

$$\frac{16}{(x + 2)(x^2 + 4)} = \frac{A}{x + 2} + \frac{Bx + C}{x^2 + 4}$$

is an identity.

Solution   First, multiply both sides of the given equation by $(x + 2)(x^2 + 4)$ to obtain

$$16 = A(x^2 + 4) + (Bx + C)(x + 2).$$

Then expand the right side of this equation.

$$16 = Ax^2 + 4A + Bx^2 + 2Bx + Cx + 2C$$

Next, group together like powers of x and then factor.

$$16 = (A + B)x^2 + (2B + C)x + 4A + 2C$$

Now equate coefficients of like powers of x and set up a system of linear equations. (Note that $16 = 0x^2 + 0x + 16$.)

Equating coefficients of x^2:	$A + B = 0$
Equating coefficients of x:	$2B + C = 0$
Equating constant terms:	$4A + 2C = 16$

Finally, we solve this system by the methods in this chapter to determine that $A = 2$, $B = -2$, and $C = 4$. ∎

To this point, the form of the partial fraction decomposition has been given, and our job has been to determine constants so the given equation is an identity. In practice, however, you will start only with a rational function

$$y = \frac{P(x)}{Q(x)} \qquad Q(x) \neq 0$$

and you need to supply the form of the decomposition. To accomplish this, first divide out any common factors of $P(x)$ and $Q(x)$ and then (if necessary) do the following.

1. The degree of $P(x)$ must be less than the degree of $Q(x)$. If it is not, then through long division rewrite the fraction as

$$\frac{P(x)}{Q(x)} = \text{quotient} + \frac{\text{remainder}}{Q(x)},$$

   where the remainder term is in proper form.

2. The denominator, $Q(x)$, must be factored into linear factors and/or quadratic factors. The quadratic factors should be **irreducible,** which means they cannot be factored into linear factors (with real coefficients). Factoring $Q(x)$ in this manner is always possible (theoretically), but it may be very difficult.

Now consider Example 1, which illustrates the form used for distinct linear and quadratic factors. For a linear factor like $x + 2$, we merely put a constant, say A, in the numerator of the partial fraction. For an irreducible quadratic factor like $x^2 + 4$, we just put a linear expression, say $Bx + C$, in the numerator of the partial fraction. In general, we state these rules as follows:

Nonrepeated Linear or Quadratic Factors

Distinct Linear Factors Each linear factor $ax + b$ of $Q(x)$ that is not repeated produces in the decomposition a term of the form

$$\frac{A}{ax + b}.$$

Distinct Quadratic Factors Each irreducible quadratic factor $ax^2 + bx + c$ of $Q(x)$ that is not repeated produces in the decomposition a term of the form

$$\frac{Ax + B}{ax^2 + bx + c}.$$

We illustrate our procedures to this point in the following example.

EXAMPLE 2   Express $\dfrac{x^3 + 8}{x^3 + 4x}$ as a sum of partial fractions.

Solution   Since the degree of the numerator is not less than the degree of the denominator, we need long division to determine

$$\frac{x^3 + 8}{x^3 + 4x} = 1 + \frac{-4x + 8}{x^3 + 4x}.$$

Now we work with the remainder term. First, factor the denominator and then write the partial fraction decomposition in the form determined by the above rules.

$$\frac{-4x + 8}{x^3 + 4x} = \frac{-4x + 8}{x(x^2 + 4)} = \frac{A}{x} + \frac{Bx + C}{x^2 + 4}$$

Then multiplying both sides of this equation by $x(x^2 + 4)$ gives

$$\begin{aligned}
-4x + 8 &= A(x^2 + 4) + (Bx + C)x \\
&= Ax^2 + 4A + Bx^2 + Cx \\
&= (A + B)x^2 + Cx + 4A.
\end{aligned}$$

Equating the coefficients, we have

$$\begin{aligned}
A + B &= 0 \\
C &= -4 \\
4A &= 8
\end{aligned}$$

so

$$A = 2, B = -2, \text{ and } C = -4.$$

Thus,

$$\frac{x^3 + 8}{x^3 + 4x} = 1 + \frac{2}{x} + \frac{-2x - 4}{x^2 + 4}. \qquad \blacksquare$$

To handle the cases in which linear or quadratic factors may be repeated in the denominator, we need a more general version of our methods for partial fraction decomposition. These more powerful procedures are as follows:

Partial Fraction Decomposition of $P(x)/Q(x)$

> **Linear Factors** Each linear factor of $Q(x)$ of the form $(ax + b)^n$ produces in the decomposition a sum of n terms of the form
>
> $$\frac{A_1}{ax + b} + \frac{A_2}{(ax + b)^2} + \cdots + \frac{A_n}{(ax + b)^n}.$$
>
> **Quadratic Factors** Each irreducible quadratic factor of $Q(x)$ of the form $(ax^2 + bx + c)^n$ produces in the decomposition a sum of n terms of the form
>
> $$\frac{A_1x + B_1}{ax^2 + bx + c} + \frac{A_2x + B_2}{(ax^2 + bx + c)^2} + \cdots + \frac{A_nx + B_n}{(ax^2 + bx + c)^n}.$$

Note that when a linear or quadratic factor appears just once, then $n = 1$ and the above procedures simplify to our previous methods. Also, in practice, we usually denote the constants in the numerators as $A, B, C,$ and so on, instead of using subscript notation.

EXAMPLE 3   Express $\dfrac{1}{x^4 - 2x^3 + x^2}$ as a sum of partial fractions.

Solution   No division is necessary here, so first factor the denominator.

$$x^4 - 2x^3 + x^2 = x^2(x^2 - 2x + 1) = x^2(x - 1)^2$$

Now $(x - 1)^2$ is a repeated linear factor, and x^2 is also analyzed with the repeated linear factor procedure. Each of these factors produces two terms in the decomposition as follows:

$$\frac{1}{x^2(x - 1)^2} = \frac{A}{x} + \frac{B}{x^2} + \frac{C}{x - 1} + \frac{D}{(x - 1)^2}.$$

Multiplying both sides of this equation by $x^2(x - 1)^2$ gives

$$\begin{aligned}
1 &= Ax(x - 1)^2 + B(x - 1)^2 + Cx^2(x - 1) + Dx^2 \\
&= Ax^3 - 2Ax^2 + Ax + Bx^2 - 2Bx + B + Cx^3 - Cx^2 + Dx^2 \\
&= (A + C)x^3 + (-2A + B - C + D)x^2 + (A - 2B)x + B.
\end{aligned}$$

Equating the coefficients, we have

$$\begin{aligned}
A + C &= 0 \\
-2A + B - C + D &= 0 \\
A - 2B &= 0 \\
B &= 1,
\end{aligned}$$

so

$$A = 2, B = 1, C = -2, \text{ and } D = 1.$$

Thus,

$$\frac{1}{x^4 - 2x^3 + x^2} = \frac{2}{x} + \frac{1}{x^2} + \frac{-2}{x - 1} + \frac{1}{(x - 1)^2}. \qquad \blacksquare$$

EXERCISES 9.5

In Exercises 1–10 determine the constants A, B, C, and/or D so that the equation is an identity.

1. $\dfrac{1}{(x + 2)(x - 2)} = \dfrac{A}{x + 2} + \dfrac{B}{x - 2}$

2. $\dfrac{1}{(x - 1)(x + 2)} = \dfrac{A}{x - 1} + \dfrac{B}{x + 2}$

3. $\dfrac{6}{(x + 1)(x^2 + 1)} = \dfrac{A}{x + 1} + \dfrac{Bx + C}{x^2 + 1}$

4. $\dfrac{2x^3 + 5x + 2}{(x^2 + 1)(x^2 + 2)} = \dfrac{Ax + B}{x^2 + 1} + \dfrac{Cx + D}{x^2 + 2}$

5. $\dfrac{2x}{(x + 2)^2} = \dfrac{A}{x + 2} + \dfrac{B}{(x + 2)^2}$

6. $\dfrac{3x - 10}{(x - 1)^2} = \dfrac{A}{x - 1} + \dfrac{B}{(x - 1)^2}$

7. $\dfrac{x^3 - 2x^2 - 4x + 3}{(x^2 + 1)^2} = \dfrac{Ax + B}{x^2 + 1} + \dfrac{Cx + D}{(x^2 + 1)^2}$

8. $\dfrac{4x^2 + 3x - 1}{(x^2 + x + 1)^2} = \dfrac{Ax + B}{x^2 + x + 1} + \dfrac{Cx + D}{(x^2 + x + 1)^2}$

9. $\dfrac{-2x + 4}{(x - 1)^2(x^2 + 1)} = \dfrac{A}{x - 1} + \dfrac{B}{(x - 1)^2} + \dfrac{Cx + D}{x^2 + 1}$

10. $\dfrac{2x^3 - 5x^2 + 4x - 3}{x^2(x^2 + 1)} = \dfrac{A}{x} + \dfrac{B}{x^2} + \dfrac{Cx + D}{x^2 + 1}$

In Exercises 11–24 find the partial fraction decomposition of the given expression.

11. $\dfrac{1}{1 - x^2}$

12. $\dfrac{x + 2}{x^2 - 5x + 4}$

13. $\dfrac{x}{x^2 + 2x + 1}$

14. $\dfrac{5x + 6}{x^3 - 2x^2}$

15. $\dfrac{x^3}{x^2 - 4}$

16. $\dfrac{x^2}{x^2 - 2x + 1}$

17. $\dfrac{-12x - 6}{x^3 + x^2 - 6x}$

18. $\dfrac{1}{(x + 1)(x + 2)(x + 3)}$

19. $\dfrac{1}{x^3 + x}$

20. $\dfrac{x^5 - x^3 + x - 1}{x^4 + x^2}$

21. $\dfrac{4x}{x^4 - 1}$

22. $\dfrac{1}{x^3 - 1}$

23. $\dfrac{4x^2 + x + 8}{x^4 + 3x^2 + 2}$

24. $\dfrac{4x^3 - 7x^2 + 5x - 2}{x^4 + 2x^2 + 1}$

THINK ABOUT IT

1. **a.** Choose arbitrary values for A, B, and a in the expression

$$\frac{A}{x + a} + \frac{B}{x - a}$$

and then add the fractions.

b. Find the partial fraction decomposition of the answer in part a by the methods of this section.

2. Use the shortcut method discussed in the paragraph preceding Example 1 to express

$$\frac{x + 2}{x^2 - 5x + 4}$$

as a sum of partial fractions.

3. If x^2 is a factor of the denominator, then decomposition using the repeated linear factor form

$$\frac{A}{x} + \frac{B}{x^2}$$

is preferred to using the distinct quadratic factor form

$$\frac{Ax + B}{x^2}.$$

Start with the second form and show these two forms are equivalent. Discuss why the first form is preferred in terms of the basic goal of partial fraction decomposition.

4. If $f(x) = \dfrac{1}{a^2 - x^2}$ where a is a constant, then express

$f(x)$ as a sum of partial fractions.

5. Find the partial fraction decomposition of

$$\dfrac{bx + c}{(x - a)^2}$$

where a, b, and c are constants.

REMEMBER THIS

In Exercises 1–4 solve each equation.

1. $x^2 + 2(-4)^2 = 40$ **2.** $x^2 - 2x - 3 = 2x + 2$
3. $y^2 + y - 5 = 0$ **4.** $2y^4 - 9y^2 + 4 = 0$

In Exercises 5 and 6 graph each equation.

5. $g(x) = 10 - x^2$ **6.** $x^2 + y^2 = 10$

In Exercises 7–10 let $A = \begin{bmatrix} -7 & 9 \\ 4 & -5 \end{bmatrix}$.

7. What is the dimension of A?
8. Find the determinant of A.
9. Find $-A$.
10. Find A^{-1}, if it exists.

9.6   Nonlinear Systems of Equations

We started this chapter by showing how to determine any intersection point for two straight lines. A more general problem is to find intersection points for any two curves. For instance, in calculus it is often necessary to find any intersection points of one of the common conic sections (circle, ellipse, hyperbola, or parabola) with a line or another conic section. The substitution method and the addition-elimination method from Section 9.1 are sufficient in many cases to solve systems even when one or more of the equations is not linear. The following examples illustrate a few such cases.

EXAMPLE 1   Find all the ordered pairs that satisfy the pair of equations

$$y = x^2 - 6x + 8$$
$$y = x + 2.$$

Solution   We wish to find the value(s) for x that make both y-values the same or, equivalently, we want to know the value(s) for x that make $x^2 - 6x + 8$ equal to $x + 2$.

$$x^2 - 6x + 8 = x + 2$$
$$x^2 - 7x + 6 = 0$$
$$(x - 6)(x - 1) = 0$$
$$x - 6 = 0 \qquad x - 1 = 0$$
$$x = 6 \qquad x = 1$$

Thus, the parabola and the line intersect at $x = 1$ and $x = 6$. To find the y-coordinates, substitute these numbers in the simpler equation $y = x + 2$.

$$
\begin{array}{lcl}
y = x + 2 & \text{and} & y = x + 2 \\
\quad = (1) + 2 & & \quad = (6) + 2 \\
\quad = 3 & & \quad = 8
\end{array}
$$

Thus, the solutions are $(1,3)$ and $(6,8)$. Consider Figure 9.9, which shows the parabola and the line intersecting at these points. ■

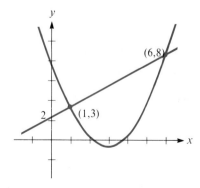

Figure 9.9

EXAMPLE 2　Find all intersection points of the ellipse $2x^2 + 3y^2 = 5$ and the hyperbola $4y^2 - 3x^2 = 1$.

Solution　The system we need to solve is

$$2x^2 + 3y^2 = 5$$
$$-3x^2 + 4y^2 = 1.$$

If we form equivalent equations by multiplying the top equation by 3 and the bottom equation by 2, we can eliminate the x variable.

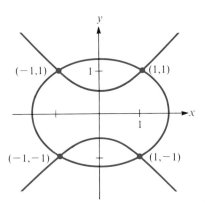

$$
\begin{array}{l}
6x^2 + 9y^2 = 15 \\
\underline{-6x^2 + 8y^2 = \;\; 2} \\
\qquad\quad 17y^2 = 17 \qquad \text{Add the equations.} \\
\qquad\quad\;\; y^2 = \;\; 1 \\
\qquad\quad\;\;\; y = \pm 1
\end{array}
$$

Now substituting $y = \pm 1$ into $2x^2 + 3y^2 = 5$ gives

$$2x^2 + 3(\pm 1)^2 = 5$$
$$x^2 = 1$$
$$x = \pm 1.$$

Thus, the solutions are $(1,1)$, $(1,-1)$, $(-1,1)$, and $(-1,-1)$. Figure 9.10 shows the ellipse and the hyperbola intersecting at these points.　■

Figure 9.10

EXAMPLE 3　Solve the system

$$\frac{5}{x} + \frac{2}{y} - 1 = 0$$

$$\frac{1}{x} - \frac{3}{y} - 7 = 0.$$

Solution　If we rewrite the above system as

$$5\left(\frac{1}{x}\right) + 2\left(\frac{1}{y}\right) = 1$$

$$\left(\frac{1}{x}\right) - 3\left(\frac{1}{y}\right) = 7$$

and let $a = 1/x$ and $b = 1/y$, we obtain

$$5a + 2b = 1$$
$$a - 3b = 7,$$

which is a relatively simple system of linear equations. If we now add -5 times the second equation to the first equation, we find

$$17b = -34 \quad \text{so} \quad b = -2.$$

Then substituting $b = -2$ into equation 2 gives

$$a - 3(-2) = 7$$
$$a = 1.$$

Finally, from the definition of a and b we have

$$\frac{1}{x} = 1 \qquad \text{and} \qquad \frac{1}{y} = -2$$

$$x = 1 \qquad \text{and} \qquad y = -\tfrac{1}{2}.$$

Thus, the solution is $(1, -\tfrac{1}{2})$.

EXAMPLE 4 Solve the system

$$x^2 + y^2 = 1$$
$$y = x^2.$$

Solution Substituting $y = x^2$ into the top equation gives

$$x^2 + (x^2)^2 = 1 \quad \text{or} \quad (x^2)^2 + x^2 - 1 = 0,$$

which is an equation with quadratic form. As discussed in Section 1.7, if we let $t = x^2$, the equation becomes

$$t^2 + t - 1 = 0.$$

By the quadratic formula,

$$t = \frac{-1 \pm \sqrt{(1)^2 - 4(1)(-1)}}{2(1)} = \frac{-1 \pm \sqrt{5}}{2},$$

so

$$x^2 = \frac{-1 \pm \sqrt{5}}{2} \quad \text{and} \quad x = \pm\sqrt{\frac{-1 \pm \sqrt{5}}{2}}.$$

Since x must be a real number we eliminate $\pm\sqrt{\dfrac{-1 - \sqrt{5}}{2}}$ and conclude that

$$x = \pm\sqrt{\frac{-1 + \sqrt{5}}{2}}.$$

Finally, substituting these two values of x into $y = x^2$ gives us solutions of

$$\left(\sqrt{\frac{-1 + \sqrt{5}}{2}}, \frac{-1 + \sqrt{5}}{2}\right) \quad \text{and} \quad \left(-\sqrt{\frac{-1 + \sqrt{5}}{2}}, \frac{-1 + \sqrt{5}}{2}\right).$$

Figure 9.11 shows the circle $x^2 + y^2 = 1$ and the parabola $y = x^2$ intersecting at these points, which are approximately $(0.8, 0.6)$ and $(-0.8, 0.6)$.

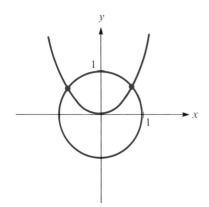

Figure 9.11

EXERCISES 9.6

In Exercises 1–30 find all the ordered pairs that satisfy the pairs of equations.

1. $y = x^2$
   $y = x$

2. $f(x) = 2 - x^2$
   $g(x) = -x$

3. $g(x) = 10 - x^2$
   $f(x) = 1$

4. $y = x^2 - 2x - 3$
   $y = 2x + 2$

5. $y = x^2 - 4$
   $y = 4 - x^2$

6. $f(x) = x^2$
   $g(x) = 2 - x^2$

7. $x^2 + y^2 = 100$
   $3x + y = 10$

8. $x^2 + y^2 = 1$
   $y = x + 1$

9. $x^2 + y^2 = 8$
   $2x^2 - y^2 = 4$

10. $x^2 - y^2 = 8$
    $2x^2 + y^2 = 19$

11. $x^2 + 2y^2 = 40$
    $2x^2 + y^2 = 32$

12. $2x^2 + 3y^2 = 5$
    $3x^2 = 4y^2 - 1$

13. $xy = 15$
    $x + y = 8$

14. $xy = 1$
    $x - 2y = 1$

15. $xy = 2$
    $x^2 + 2y^2 = 9$

16. $xy = 4$
    $x^2 + y^2 = 8$

17. $x^2 + y^2 = 2$
    $x = y^2$

18. $x^2 + y^2 = 3$
    $x = y^2$

19. $x - y = 2$
    $x = y^2 - 4$

20. $y^2 = x + 9$
    $x = 3 - 2y - y^2$

21. $3x + y = 7$
    $x^2 y = 4$

22. $x = y^3 - 3y^2$
    $x - y = -3$

23. $x^2 + y^2 = 5$
    $y = x^2$

24. $(x - 1)^2 + (y + 2)^2 = 5$
    $y + 2x = 0$

25. $\dfrac{x^2}{4} + \dfrac{y^2}{9} = 1$
    $\dfrac{x^2}{4} - \dfrac{y^2}{9} = 1$

26. $\dfrac{x^2}{4} + y^2 = 1$
    $\dfrac{x^2}{4} - \dfrac{y^2}{2} = 1$

27. $\dfrac{3}{x} - \dfrac{2}{y} = 27$
    $\dfrac{2}{x} + \dfrac{5}{y} = -1$

28. $\dfrac{5}{x} + \dfrac{3}{y} - 5 = 0$
    $\dfrac{4}{x} + \dfrac{1}{y} - 11 = 0$

29. $\dfrac{2}{x} + \dfrac{2}{y} - \dfrac{1}{z} = 0$
    $\dfrac{2}{x} - \dfrac{1}{y} + \dfrac{2}{z} = 3$
    $\dfrac{1}{x} - \dfrac{2}{y} - \dfrac{2}{z} = 12$

30. $\dfrac{1}{x} + \dfrac{2}{y} - \dfrac{3}{z} = 0$
    $\dfrac{3}{x} - \dfrac{1}{y} + \dfrac{6}{z} = 1$
    $\dfrac{2}{x} - \dfrac{3}{y} - \dfrac{1}{z} = -9$

31. Find all intersection points of the ellipse $x^2 + 2y^2 = 3$ and the hyperbola $3x^2 - y^2 = 2$.

32. Find all intersection points of the semicircle $y = -\sqrt{36 - x^2}$ and the line $y = -x$.

33. Find to the nearest tenth all points where the circle $x^2 + y^2 = 9$ and the ellipse $(x^2/25) + (y^2/4) = 1$ intersect.

34. Find to the nearest hundredth all intersection points of the circle $x^2 + y^2 = 1$ and the parabola $x = y^2$.

35. Find the length of the legs of a right triangle if the area is 36 ft² and the hypotenuse is $5\sqrt{6}$ ft.

36. Find the radius of each circle in the illustration below if the combined areas of the two circles is 48.5π in.²

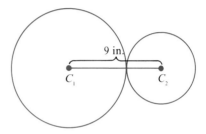

37. Both the parallelogram and the triangle in the illustration below have areas of 36 m². Find b and h.

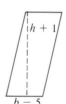

38. If the area and perimeter of a rectangle are denoted by a and p, respectively, then show the length is given by $(p + \sqrt{p^2 - 16a})/4$.

39. Three pipes A, B, and C can fill a tank in 2 hours. If only A and B are used it takes 3 hours, while B and C can do the job in 4 hours. How long does it take each pipe alone to fill the tank?

40. Pumps A and B working together can fill a tank in 7 hours and 30 minutes. This same tank may also be filled by running pumps A and C for 6 hours or pumps B and C for 10 hours. How long does it take each pump alone to fill the tank? How long does it take to fill the tank using all three pumps?

THINK ABOUT IT

1. Show that a hyperbola defined by the equation $(x^2/a^2) - (y^2/b^2) = 1$ does not intersect its asymptotes.
2. Find all intersection points of the graphs of $y = \log_{10}(x - 4)$ and $y = -1 + \log_{10}(6x + 16)$.
3. A rectangle is inscribed in a circle. If the areas of the rectangle and the circle are 4 m² and $\frac{5}{2}\pi$ m², respectively, then find the dimensions of the rectangle.

4. Find in terms of r all intersection points of the circle $x^2 + y^2 = r^2$ and the hyperbola $y^2 - x^2 = 2$. For what values of r do the curves intersect?
5. Find the slope-intercept equation of the line through $(1,1)$ that intersects the parabola $y = x^2$ in exactly one point.

REMEMBER THIS

1. True or false: $-2(0) + 0 \leq 5$.
2. Graph $-2x + y = 5$.

In Exercises 3–5 find the intersection point of each pair of lines.

3. $-2x + y = 5$
   $x = 0$
5. $-2x + y = 5$
   $x + y = 8$

4. $15x + 3y = 12,000$
   $y = 600$

In Exercises 6–9 let

$$A = \begin{bmatrix} -1 & 3 & 1 \\ 1 & 0 & -1 \\ 2 & 5 & -2 \end{bmatrix} \text{ and } B = \begin{bmatrix} 30 \\ 10 \\ 20 \end{bmatrix}.$$

6. What is the dimension of AB?
7. Find AB.
8. Find the determinant of A.
9. Find A^{-1}, if it exists.
10. Express $\dfrac{7x - 4}{x^2 + 2x}$ as a sum of partial fractions.

9.7 Systems of Linear Inequalities and Linear Programming

The analysis of certain problems often requires a system of inequalities instead of a system of equations. Example 1 illustrates this point.

EXAMPLE 1 A company manufactures two products, A and B. Product A costs $20 per unit to produce and product B costs $60 per unit. The company has $90,000 to spend on the production. If x represents the number of units produced of product A and y represents the number of units produced of product B, write an inequality that expresses the restriction placed on the company because of available funds.

Solution x represents the number of units produced of product A and each unit costs $20 to produce. Thus, the company spends $20x$ dollars to produce x units of product A. Similarly, the company spends $60y$ dollars to produce y units of product B. Since the total cost cannot exceed $90,000, we have

$$20x + 60y \leq 90,000.$$

Since the number of units produced cannot be negative, we also have

$$x \geq 0 \text{ and } y \geq 0.$$

Thus, the restriction placed on the company because of available funds is given by

$$20x + 60y \leq 90{,}000 \text{ with } x \geq 0 \text{ and } y \geq 0.$$ ∎

The question in Example 1 is designed to show the usefulness of inequalities in describing practical situations. Capital is only one of many restrictions that a company must consider because of limited resources. For example, machine time, labor hours, and storage space are other considerations. Subject to these restrictions, the company must determine the number of units of each product that should be manufactured for maximum profit. The solution to such problems is found by techniques from a branch of applied mathematics called **linear programming.** To explore this topic further, we first need to develop methods for graphing the solution set to a system of inequalities.

An inequality in two variables such as $y < x$ has infinitely many solutions. To illustrate these solutions graphically, we note that the line $y = x$ separates the plane into two regions. Each region consists of the set of points on one side of the line and is called a **half-plane.** The solution set to $y < x$ is given by the half-plane *below* the line $y = x$, while the solution set to $y > x$ is given by the half-plane *above* $y = x$. We usually indicate the half-plane in the solution set through shading as shown in Figure 9.12, and the sketch of the solution set gives the graph of the inequality. Note that we use either a solid line or a dashed line in the graph, depending on whether the line is included in the solution set.

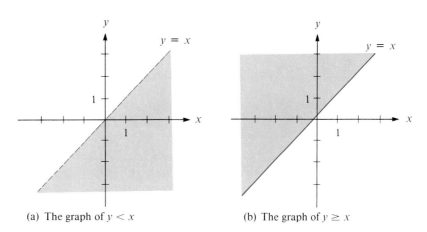

(a) The graph of $y < x$ (b) The graph of $y \geq x$

Figure 9.12

EXAMPLE 2   Graph each inequality.

a. $y \leq 2x - 1$ **b.** $y > -3x$

Solution

a. See Figure 9.13.
b. See Figure 9.14.

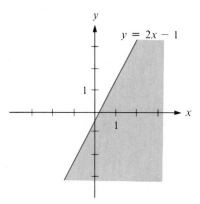

Figure 9.13

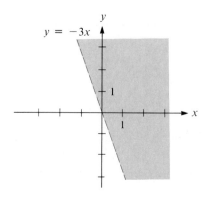

Figure 9.14 ■

When the inequality is not solved for y, you may have difficulty choosing which half-plane to shade in your answer. The easiest way to decide in such cases is usually to pick some **test point** that is not on the line and substitute the coordinates of this point into the inequality. If the resulting statement is true, then shade the half-plane containing the test point. Otherwise, shade the half-plane on the other side of the line from the test point. The origin $(0,0)$ is often a convenient point to use in this test.

EXAMPLE 3   Graph $2x - 3y \geq 6$.

Solution   The origin is not on the line $2x - 3y = 6$, so substitute 0 for x and y in the above inequality.

$$2(0) - 3(0) \geq 6$$
$$0 \geq 6$$

Since this statement is false we shade the half-plane not containing the origin, as shown in Figure 9.15. If you feel uncomfortable with the test point method, an alternative is to solve the given inequality for y. The result, $y \leq \frac{2}{3}x - 2$, tells us to shade the half-plane below the line. ■

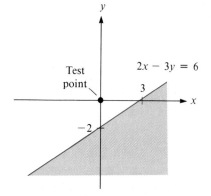

Figure 9.15

The solution set of a system of inequalities is the intersection of the solution sets of all the individual inequalities in the system. The best way to give the solution set is in a graph. To do this, we graph the solutions to each inequality in the system and then shade in the overlap (intersection) of these half-planes.

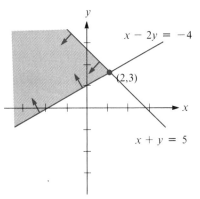

Figure 9.16

EXAMPLE 4   Graph the solution set of the system

$$x + y \leq 5$$
$$x - 2y \leq -4.$$

Solution   By using the origin in both cases as a test point, we determine that $x + y \leq 5$ is satisfied by the points on or below the line $x + y = 5$, while $x - 2y \leq -4$ is satisfied by the points on or above the line given by $x - 2y = -4$. The intersection of these two half-planes is shown in Figure 9.16.   ∎

The **vertices** or **corners** of a region are the intersection points of the bounding sides in the region. For example, the corner in the region defined by the system of inequalities in Example 4 is (2,3) as shown in Figure 9.16. Such corners are important in the solution to linear programming problems, and they can be found by any of the methods we have considered for finding the intersection point for two lines. The next example includes a discussion of this concept.

EXAMPLE 5   Graph the solution set of the system

$$x + y \leq 6, \qquad x - 2y \leq 3,$$
$$y - x \leq 2, \qquad x \geq 0, \qquad y \geq 0.$$

Specify the coordinates of any corner in the graph.

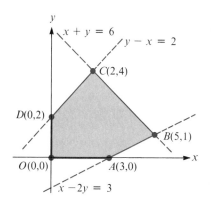

Figure 9.17

Solution   The inequalities $x \geq 0$ and $y \geq 0$ restrict the graph of the solution set to the first quadrant. By using the origin as a test point, we determine that the inequalities specify the region on or below the lines $x + y = 6$ and $y - x = 2$ as well as the region on or above the line $x - 2y = 3$. Figure 9.17 shows the graph of the solution set. To determine the coordinates of the corners O, A, B, C, and D, we note that O is the origin $(0,0)$; A is the x-intercept of $x - 2y = 3$, which is $(3,0)$; and D is the y-intercept of $y - x = 2$, which is $(0,2)$. To find point B, we solve the system

$$x + y = 6$$
$$x - 2y = 3$$

by adding -1 times the bottom equation to the top equation to get $3y = 3$. Then $y = 1$, so $x = 5$, and point B is $(5,1)$. Finally, the system

$$x + y = 6$$
$$y - x = 2$$

is solved by adding the two equations together to get $2y = 8$. Then $y = 4$, so $x = 2$, and point C is $(2,4)$.   ∎

We now return to the topic of linear programming. Basically, linear programming involves finding the maximum or minimum value of a linear function $F = ax + by$, which is called the **objective function.** The variables x and y must satisfy certain inequalities, called **constraints,** that impose restrictions on the solution. Only an ordered pair (x,y) that satisfies all the constraints is a possible or **feasible solution,** and the set of all feasible solutions is given by the graph of the given system of inequalities (or constraints). Although the set of feasible solutions may seem overwhelming, the corners are the crucial points as shown in the following theorem.

Corner Point Theorem

> If a linear function $F = ax + by$ assumes a maximum or minimum value subject to a system of linear inequalities, then it does so at the corners or vertices of that system of constraints.

Furthermore, we are guaranteed that a maximum or minimum exists as long as the set of feasible solutions is a closed convex polygon. The condition that a polygon be closed merely means that the region includes its boundary lines, while a region is said to be **convex** provided the region contains the line segment joining *any* two points in the region. Figure 9.18(a) shows a set of points that is convex, while Figure 9.18(b) shows a nonconvex region.

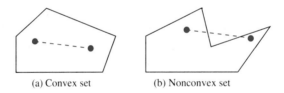

(a) Convex set   (b) Nonconvex set

Figure 9.18

Since the solution set of a system of linear inequalities is always a convex region, this condition poses no problem. Thus, the only time we may not be able to find a maximum or minimum value occurs when the set of feasible solutions is unbounded and does not form a polygon.

We can now consider a typical problem in linear programming.

EXAMPLE 6   Find the maximum value of $F = 3x - 2y$ subject to the constraints

$$x + y \leq 6$$
$$y - x \leq 2$$
$$x - 2y \leq 3$$
$$x \geq 0$$
$$y \geq 0.$$

Solution   The constraints are the system of inequalities in Example 5, so the set of all feasible solutions including all vertices is shown in Figure 9.17. Since the set of feasible solutions is a closed convex polygon, we know that a maximum value exists at one of the corners. We need only substitute the coordinates of these vertices into the objective function to determine the correct corner.

Vertices	$F = 3x - 2y$
O (0,0)	$3(0) - 2(0) = 0$
A (3,0)	$3(3) - 2(0) = 9$
B (5,1)	$3(5) - 2(1) = 13$
C (2,4)	$3(2) - 2(4) = -2$
D (0,2)	$3(0) - 2(2) = -4$

From the table we see that when $x = 5$ and $y = 1$, the function attains a maximum value of 13. Note that we can also determine from the table that the minimum function value of -4 occurs at (0,2).   ■

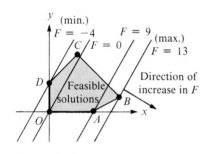

Figure 9.19

If we look a little closer at Example 6, we can see why any maximum or minimum occurs at a corner. Consider Figure 9.19, which plots the line $F = 3x - 2y$ for various values of F. Note that as F increases, the y-intercepts of the line change but the slopes remain fixed at $\frac{3}{2}$. The result is a family of parallel lines. The line with the maximum or minimum value of F will then intersect the set of feasible solutions at a corner as shown in the figure.

The next example considers a linear programming problem in a more practical setting.

EXAMPLE 7   A manufacturer of personal computers makes two types of printers, A and B. To comply with contracts, the company must produce at least 100 type A printers and 200 type B printers each week. A total of 4 labor hours is required to assemble printer A and a total of 2 labor hours is required to assemble printer B with a maximum of 1,000 labor hours available each week. If the company can sell all its printers for a profit of $70 on printer A and $40 on printer B, then how many of each type should be produced weekly for the maximum profit?

Solution   If we let x and y represent the number of assembled type A and type B printers, respectively, then the profit function is

$$P = 70x + 40y$$

and the constraints are

$$x \geq 100$$
$$y \geq 200$$
$$4x + 2y \leq 1,000.$$

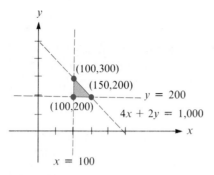

Figure 9.20

The constraints determine the set of feasible solutions shown in Figure 9.20 with vertices (100,200), (150,200), and (100,300). Since the set of feasible solutions is a closed convex polygon, we know that a maximum value exists at one of these corners. The value of the profit function at each of these vertices is as follows:

Vertices	$P = 70x + 40y$
(100,200)	$70(100) + 40(200) = 15,000$
(150,200)	$70(150) + 40(200) = 18,500$
(100,300)	$70(100) + 40(300) = 19,000$

Thus, the profit function reaches a maximum value of $19,000 when 100 type A printers and 300 type B printers are produced each week.   ■

The geometrical method we have discussed is intended as an introduction to linear programming. In most cases the situation being analyzed is quite complicated and an algebraic method, called the **Simplex method,** is used to solve the problem using computers. Although we have only considered this topic in the context of maximizing profit, other applications are common. Some of these include:

1. Determining the most economical mixture of ingredients that will result in a product with certain minimum requirements.
2. Determining the quickest and most economical route in distributing a product.
3. Determining the most efficient use of industrial machinery.
4. Determining the most effective production schedule.

EXERCISES 9.7

In Exercises 1–8 graph the given inequalities. Use shading to indicate the graph.

1. $y \leq 2x$
2. $y > x - 2$
3. $y > 1 - 2x$
4. $y \geq 2$
5. $x \leq 4$
6. $x - y > 0$
7. $2x - 4y \geq 8$
8. $x + 2y \leq 5$

In Exercises 9–16 graph the solution set of the given system of inequalities. Use shading to indicate the graph and specify the coordinates of any corner in the graph.

9. $x + y \leq 7$
  $x - 2y \leq -8$
10. $x + 2y \geq -12$
  $2x - y \geq 1$
11. $y \leq 3 - 4x$
  $x - y \geq 0$
12. $y \geq 3x + 4$
  $x + y \leq -2$
13. $x + y \leq 4$
  $x \geq 0$
  $y \geq 0$
14. $x + y \geq 1$
  $x - y \leq 1$
  $x + 2y \leq 2$
15. $y \leq x + 1$
  $y \geq -x + 1$
  $y \geq x - 1$
  $y \leq -x + 3$
16. $x + y \leq 8$
  $2x + y \leq 11$
  $-2x + y \leq 5$
  $x \geq 0$
  $y \geq 0$

17. Find the maximum and minimum values of $F = 2x - 3y$ subject to the constraints given in Exercise 13.

18. Find the maximum and minimum values of $F = 6x - 12y$ subject to the constraints given in Exercise 14.

19. Find the values of x and y that yield the maximum value of $F = \frac{1}{2}x + \frac{1}{4}y$ subject to the constraints given in Exercise 15.

20. Find the values of x and y that yield the minimum value of $F = 5x - y$ subject to the constraints given in Exercise 16.

In Exercises 21–22 find the maximum value of P subject to the given constraints.

21. $P = 60x + 40y$
  $4x + 2y \leq 2,000$
  $x \geq 100$
  $y \geq 200$
22. $P = 120x + 20y$
  $15x + 3y \leq 12,000$
  $x \geq 300$
  $y \geq 600$

23. Minimize $C = -2x - 3y + 200$ subject to
  $•3x + 4y \leq 180$
  $15 \leq x \leq 40$
  $10 \leq y \leq 30.$

24. Maximize $P = 15x + 8y + 9$ subject to
  $x - y \leq 7$
  $x - y \geq 1$
  $8 \leq x \leq 13.$

25. A woman has $20,000 to invest in the stock market. Stock A sells for $90 per share, stock B costs $50 per share, and stock C costs $30 per share. If x, y, and z represent the number of shares that she bought of each stock, respectively, find an inequality that expresses this relationship.

26. A TV manufacturer has available 10,000 labor hours per month. A console requires 7 labor hours and a portable requires 5 labor hours. If x represents the number of consoles produced and y represents the number of portables produced, find an inequality that expresses this relationship.

27. Government regulations limit the amount of a pollutant that a company can discharge into the water to A gal/day. If product 1 produces B gal of pollutant per unit and product 2 produces C gal of pollutant per unit, find an inequality that expresses the restriction placed on this company because of government regulations.

28. Sad Sam has been informed by his doctor that he would be less sad if he obtained at least the minimum adult requirements of thiamine (2 mg) and of niacin (25 mg) per day. A trip to the supermarket reveals the following facts about his favorite cereals.

Cereal	Thiamine per Ounce (mg)	Niacin per Ounce (mg)
A	1.0	10.0
B	0.7	12.0

If x represents the number of ounces he eats of cereal A and y represents the number of ounces he eats of cereal B, find two inequalities that express what must be done if he is to satisfy his minimum daily requirements of thiamine and niacin.

29. A manufacturer of personal computers makes two types of disc drives, A and B. To comply with contracts, the company must produce at least 80 type A and 150 type B disc drives each week. A total of 5 labor hours is required to assemble drive A and a total of 4 labor hours is required to assemble drive B with a maximum of 2,000 labor hours available each week. If the company can sell all its disc drives for a profit of $90 on type A and $75 on type B, then how many of each should be produced weekly for the maximum profit?

30. Chris needs at least 9 units of vitamin A and 12 units of vitamin C per day. The table below summarizes information about two of her favorite foods.

Food	Vitamin A per Ounce	Vitamin C per Ounce	Cost per Ounce
A	3 units	3 units	25 cents
B	2 units	4 units	20 cents

How many ounces of each food should Chris eat to satisfy at least cost her minimum requirements for vitamins A and C?

31. Repeat Exercise 30 with food A costing 28 cents per ounce and food B costing 18 cents per ounce.

32. Use the information given in Exercise 28 and determine the combination of cereals Sam should eat to satisfy his doctor's recommendation at least cost if cereal A costs 15 cents per ounce and cereal B costs 13 cents per ounce.

THINK ABOUT IT

1. Graph $x^2 + y^2 < 9$.

2. Graph the solution set of the system
$$y \le 4 - x^2$$
$$y \ge 3x.$$

In Exercises 3 and 4 find a system of linear inequalities with the given solution set.

3.

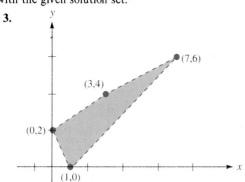

4.

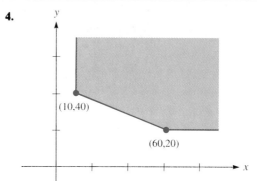

5. Redo the problem in Example 7 of this section assuming the company can sell all its printers for a profit of $80 on printer A and $40 on printer B. Explain geometrically why the solution is not unique.

REMEMBER THIS

In Exercises 1–7 answer true or false.

1. $\begin{vmatrix} 1 & 2 \\ 3 & 4 \end{vmatrix} = \begin{vmatrix} 1 & 3 \\ 2 & 4 \end{vmatrix}$

2. $\begin{bmatrix} 1 & 2 \\ 3 & 4 \end{bmatrix} = \begin{bmatrix} 1 & 3 \\ 2 & 4 \end{bmatrix}$

3. If matrix B and matrix C have the same dimension, then BC is defined.

4. If A is a square matrix with all positive elements, then $|A|$ is positive.

5. A linear system with no solution is called inconsistent.

6. If $A = \begin{bmatrix} 7 & -3 & -3 \\ -1 & 1 & 0 \\ -1 & 0 & 1 \end{bmatrix}$, then

$A^{-1} = \begin{bmatrix} 1 & 3 & 3 \\ 1 & 4 & 3 \\ 1 & 3 & 4 \end{bmatrix}$.

7. If A^{-1} exists, then the solution to the matrix equation $AX = B$ is given by $X = BA^{-1}$.

8. Find all intersection points of the circle $x^2 + y^2 = 64$ and the hyperbola $2x^2 - y^2 = 17$.

9. Express $\dfrac{3x}{(x + 3)^2}$ as a sum of partial fractions.

10. If $A = \begin{bmatrix} 1 & 3 & 2 \\ -1 & 0 & 4 \end{bmatrix}$,

$B = \begin{bmatrix} a + b & 3 & b + c \\ a - c & 0 & 4 \end{bmatrix}$, and $A = B$, then find a, b, and c.

CHAPTER OVERVIEW

Section	Key Concepts to Review
9.1	• Definitions of linear equations with n variables, linear system, inconsistent system, and dependent system
	• The solution set of a system of linear equations in two variables is the set of all the ordered pairs that satisfy both equations. Graphically this corresponds to the collection of points where the lines intersect.
	• Methods to solve a system of equations by the substitution method and by the addition-elimination method
9.2	• Definitions of triangular form, equivalent systems, matrix, entry or element of a matrix, coefficient matrix, augmented matrix, equivalent matrices, and reduced row-echelon matrix
	• The following operations are used in Gaussian elimination and Gauss-Jordan elimination to solve a linear system.

Elementary Operations on Equations	Elementary Row Operations on Matrices
1. Multiply both sides of an equation by a nonzero number.	**1.** Multiply each entry in a row by a nonzero number.
2. Add a multiple of one equation to another.	**2.** Add a multiple of the entries in one row to another row.
3. Interchange two equations.	**3.** Interchange two rows.

Section	Key Concepts to Review
9.3	• Definitions of square matrix, determinant, minor of an element, and cofactor of an element
	• To evaluate a 2 by 2 determinant, use $\begin{vmatrix} a_1 & b_1 \\ a_2 & b_2 \end{vmatrix} = a_1b_2 - b_1a_2$.
	• Cramer's rule
	• Method to evaluate 3 by 3 (and more complicated) determinants
9.4	• Definitions of dimension, square matrix, zero matrix, equal matrices, scalar, $n \times n$ identity matrix, and the multiplicative inverse of A (symbolized A^{-1})
	• Definitions for scalar multiplication and addition, subtraction, and multiplication of matrices
	• Six basic properties of matrices
	• Only matrices of the same dimension can be added or subtracted.
	• The product AB of two matrices is defined only when the number of columns in A matches the number of rows in B.
	• Methods to determine if A^{-1} exists, and if it does, how to find it
	• If A^{-1} exists, the solution to the linear system $AX = B$ is given by $X = A^{-1}B$.
9.5	• Definitions of partial fraction decomposition, partial fraction, and irreducible quadratic factor
	• Methods to determine the partial fraction decomposition of $P(x)/Q(x)$, where $P(x)$ and $Q(x)$ are (nonzero) polynomials (Note that the denominator in each partial fraction in the decomposition is either a linear factor, a repeated linear factor, an irreducible quadratic factor, or a repeated irreducible quadratic factor.)

Section	Key Concepts to Review
9.6	• Methods to solve nonlinear systems of equations by the substitution method and by the addition-elimination method
9.7	• Definitions of half-plane, vertices or corners of a region, objective function, constraints, feasible solution, and convex region
	• Methods to graph inequalities in two variables
	• Methods to solve a system of inequalities
	• The solution set of a system of inequalities is the intersection of the solution sets of all the individual inequalities in the system.
	• Methods to find the maximum and/or minimum value of an objective function subject to given constraints
	• Corner point theorem

CHAPTER REVIEW EXERCISES

In Exercises 1–10 solve the given system of linear equations using the method indicated.

1. $2x + 3y = -17$ (addition-elimination)
$3x - y = 2$

2. $y = -3x + 2$ (substitution)
$y = 2x - 3$

3. $y = \frac{1}{2}x - 3$ (graphing)
$3x - 7y - 18 = 0$

4. $3x + y = 2$ (Cramer's rule)
$-3x + y = -2$

5. $x + y = 10$ (row operations on
$-2x - 5y = -11$ matrices)

6. $5x + 3y = 1$ (matrix algebra)
$3x + 2y = 0$

7. $x - y + z = 5$ (substitution)
$y - z = 0$
$z = -1$

8. $x + y + z = 2$ (row operations on
$x + y - z = 4$ matrices)
$2x - y - 4z = -5$

9. $x + z = 10$ (Cramer's rule)
$y + 2z = 3$
$-x + y = -2$

10. $x + 3y + 3z = 5$ (matrix algebra)
$x + 4y + 3z = 2$
$x + 3y + 4z = 8$

In Exercises 11–22 solve each system of equations.

11. $x = 3y - 5$
$y = 4x - 2$

12. $5x - 4y - 11 = 0$
$2x + 7y + 13 = 0$

13. $\dfrac{2}{x} - \dfrac{1}{y} = 1$
$\dfrac{3}{x} + \dfrac{2}{y} = 12$

14. $x^2 + y^2 = 10$
$2x^2 - y^2 = 17$

15. $x + y = 6$
$xy = 4$

16. $3^{x+y} = 27$
$2^{x-y} = 4$

17. $x + y = 6$
$y = \sqrt{x}$

18. $y = x^3 - 6x^2 + 8x$
$y = x^2 - 4x$

19. $2x - y + z = 3$
$x - 3y + 5z = 4$
$x + y - 2z = 0$

20. $x + y - z = 2$
$2x - y + 5z = 3$
$-x - y + z = -1$

21. $-x + y + z = 0$
$x - y + z = 0$
$x + y - z = 0$

22. $\dfrac{1}{x} - \dfrac{1}{y} + \dfrac{1}{z} = 2$
$\dfrac{1}{x} + \dfrac{2}{y} - \dfrac{1}{z} = 7$
$\dfrac{2}{x} - \dfrac{6}{y} + \dfrac{1}{z} = 1$

In Exercises 23–26 evaluate each determinant.

23. $\begin{vmatrix} -2 & 3 \\ 1 & 4 \end{vmatrix}$

24. $\begin{vmatrix} 1 & -6 & 2 \\ 2 & 2 & -3 \\ 3 & -4 & 1 \end{vmatrix}$

25. $\begin{vmatrix} 3 & -3 & 1 \\ -1 & 9 & -3 \\ 5 & -6 & 2 \end{vmatrix}$ **26.** $\begin{vmatrix} 1 & 0 & 1 & 2 \\ 0 & 0 & -1 & 0 \\ -2 & 3 & 0 & -1 \\ 1 & 2 & 3 & 0 \end{vmatrix}$

In Exercises 27 and 28 verify the given identity.

27. $\begin{vmatrix} a & b \\ c & d \end{vmatrix} = -\begin{vmatrix} c & d \\ a & b \end{vmatrix}$

28. $\begin{vmatrix} ka & b \\ kc & d \end{vmatrix} = k\begin{vmatrix} a & b \\ c & d \end{vmatrix}$

In Exercises 29–36 find (if defined) the specified matrix given

$$A = \begin{bmatrix} 1 & -2 & 3 \\ -1 & 0 & 4 \end{bmatrix}, B = \begin{bmatrix} -3 & 0 \\ 0 & -3 \end{bmatrix},$$

$$C = \begin{bmatrix} 4 & 5 \\ 3 & 4 \end{bmatrix}, \text{ and } D = \begin{bmatrix} 3 & -1 \\ 0 & 1 \\ 2 & 5 \end{bmatrix}.$$

29. $B - C$ **30.** BC **31.** CD
32. DC **33.** $3C$ **34.** C^{-1}
35. $AD + C$ **36.** $2A - 3D$
37. Graph $2x - y \le 7$. Use shading to indicate the graph.
38. Graph the solution set of the system

$$x + 4y \le 17$$
$$4x + 3y \ge 16.$$

39. Maximize $P = 50x + 30y$ subject to

$$5x + 2y \le 1{,}000$$
$$x \ge 50$$
$$y \ge 100.$$

40. Minimize $C = 3x - 2y + 20$ subject to

$$x + y \le 8$$
$$y - x \le 6$$
$$x - 2y \le 2$$
$$x \ge 0$$
$$y \ge 0.$$

41. Find a, b, and c so the parabola $y = ax^2 + bx + c$ passes through the points $(1,3)$, $(-1,9)$, and $(2,6)$.
42. Find the intersection point of the line $y = \frac{3}{4}x - \frac{1}{4}$ and the line whose x- and y-intercepts are $(6,0)$ and $(0,4)$.

43. Find in terms of r all intersection points of the circle $x^2 + y^2 = r^2$ and the hyperbola $x^2 - y^2 = 5$. For what values of r do the curves intersect?
44. The area and perimeter of a rectangle are 341 m² and 84 m, respectively. Find the length and width of the rectangle.
45. If the vertices of a triangle are at points (x_1,y_1), (x_2,y_2), and (x_3,y_3), then the area of the triangle is given by the absolute value of the following determinant.

$$\frac{1}{2}\begin{vmatrix} x_1 & y_1 & 1 \\ x_2 & y_2 & 1 \\ x_3 & y_3 & 1 \end{vmatrix}$$

Use this rule to find the area of the triangle whose vertices are at $(0,-3)$, $(-2,-1)$, and $(1,4)$.
46. A company manufactures two types of fertilizer, A and B. A bag of fertilizer A is made from 4 units of nitrogen and 1 unit of potassium, while a bag of fertilizer B is made from 2 units of nitrogen and 3 units of potassium. Raw materials in stock are limited to 600 units of nitrogen and 260 units of potassium. If the company can sell all its fertilizer for profits of \$5 per bag of fertilizer A and \$4 per bag of fertilizer B, then how many of each type should be produced for the maximum profit?
47. True or false: $\begin{bmatrix} 1 & 2 \\ 3 & 4 \end{bmatrix} = \begin{bmatrix} 3 & 4 \\ 1 & 2 \end{bmatrix}$.

48. True or false: $\begin{vmatrix} 1 & 2 \\ 3 & 4 \end{vmatrix} = -\begin{vmatrix} 3 & 4 \\ 1 & 2 \end{vmatrix}$.

In Exercises 49 and 50 determine the constants A, B, C, and/or D so that the equation is an identity.

49. $\dfrac{-4x - 5}{(x + 3)^2} = \dfrac{A}{x + 3} + \dfrac{B}{(x + 3)^2}$

50. $\dfrac{-x^3 - x^2 - 8x + 2}{(x^2 + 3)(x^2 - 2)} = \dfrac{Ax + B}{x^2 + 3} + \dfrac{Cx + D}{x^2 - 2}$

In Exercises 51 and 52 find the partial fraction decomposition of the given expression.

51. $\dfrac{x^2}{x^2 - 9}$

52. $\dfrac{2x^3 + 8x}{x^4 + 4x^2 + 4}$

CHAPTER 9 TEST

1. Evaluate $\begin{vmatrix} 2 & -3 & 1 \\ -1 & 4 & -7 \\ 3 & 0 & 2 \end{vmatrix}$.

2. If $A = \begin{bmatrix} -3 & 4 \\ -5 & 7 \end{bmatrix}$, find A^{-1} if it exists.

3. Find BA if

$$B = \begin{bmatrix} 3 & 0 & -5 \\ -4 & 2 & 6 \end{bmatrix} \text{ and } A = \begin{bmatrix} 8 & 1 \\ 10 & -2 \\ 7 & 0 \end{bmatrix}.$$

4. Find all intersection points of the parabola $y = x^2 + 5$ and the line $x + y = 7$.

5. Express $\dfrac{4x + 6}{x^2 - 4}$ as a sum of partial fractions.

6. Use Cramer's rule to find the solution for x of the system

$$\begin{aligned} x - 2y - z &= 1 \\ 2x + y - z &= 2 \\ -x + 3y + z &= 1. \end{aligned}$$

7. Solve the system in Exercise 6 using row operations on matrices.

8. A manager invests a total of $150,000. The investment is split between a bank CD yielding 6-percent interest and a stock that pays a 9-percent dividend. If the total annual income from the investment is $12,300, then how much is invested in the bank CD?

9. The points $(1,-1)$, $(2,3)$, and $(-1,9)$ lie on the parabola given by $y = ax^2 + bx + c$. Find a, b, and c.

10. Find the maximum value of $P = 7x + 4y$ subject to the constraints

$$\begin{aligned} x + y &\leq 18 \\ 2x + y &\leq 24 \\ x &\geq 0 \\ y &\geq 0. \end{aligned}$$

10 Discrete Algebra and Probability

If 512 players enter a single elimination tennis tournament (one loss and you're out), determine the number of matches needed to declare a winner by considering the following two approaches:
a. The number of matches from each round form a geometric progression. Write down this sequence and determine its sum.
b. Instead of reasoning from the first round to the end of the tournament, try a favorite technique in problem solving—reason from the end result. Write an equivalent question to "How many matches were played?" that transforms this problem into one that is really simple. Now, how many matches are needed in this tournament? (See Example 3 of Section 10.2.)

Up to now we have dealt primarily with situations that are continuous in nature and that are analyzed by using the set of real numbers. However, in this chapter we discuss topics in which the positive integers play an important defining role because of the step-by-step or discrete nature of the processes in these problems. Throughout this chapter, as we discuss sequences, series, mathematical induction, the binomial theorem, and counting techniques, it is important to note that in our formulas, *n* is restricted to the positive integers. Finally, we conclude this chapter with an introduction to probability, since it is a natural outgrowth of our discussion of counting techniques.

10.1

A **sequence** is a set of numbers arranged in a definite order. For example, the collection of even positive integers

$$2, 4, 6, 8, \ldots$$

is a sequence. Each number in the sequence is called a **term** of the sequence. We usually write a sequence as

$$a_1, a_2, a_3, \ldots, a_n, \ldots,$$

519

where the subscript gives the term number and a_n represents the general or nth term. We also denote a sequence by $\{a_n\}$. In the above sequence, 2 is the first term, a_1, 4 is the second term, a_2, and so on. A sequence with a first and last term is called a **finite sequence;** a sequence with an infinite number of terms is called an **infinite sequence.**

Any term in a sequence is an assignment of a_n to n. Thus, we can think of a sequence as a function. For example, the sequence of even positive integers is given by

$$a(n) = 2n, \qquad n = 1, 2, 3, \ldots .$$

Note that a is a function name (just like f) and the domain of a is the set of positive integers. We can convert from this form and list the terms of the sequence by substituting the positive integers for n in the given rule.

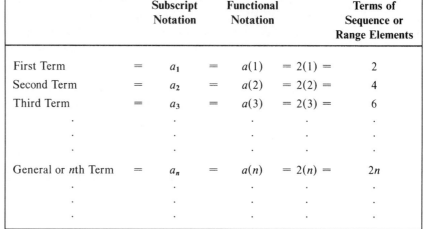

$a_n = a(n)$

		Subscript Notation		Functional Notation		Terms of Sequence or Range Elements
First Term	$=$	a_1	$=$	$a(1)$	$= 2(1) =$	2
Second Term	$=$	a_2	$=$	$a(2)$	$= 2(2) =$	4
Third Term	$=$	a_3	$=$	$a(3)$	$= 2(3) =$	6
.		.		.	.	.
.		.		.	.	.
General or nth Term	$=$	a_n	$=$	$a(n)$	$= 2(n) =$	$2n$
.		.		.	.	.
.		.		.	.	.

Figure 10.1

The graph of this function is given in Figure 10.1 and consists of the points

$$(1,2), (2,4), (3,6), \ldots .$$

Note that we do not connect the points and that we are limited to plotting a finite number of points that suggest the behavior of the function.

The functional interpretation of a sequence leads to the following definitions.

Definition of a Sequence

> A **sequence** is a function whose domain is the set of positive integers 1, 2, 3, ... The functional values or range elements are called the **terms** of the sequence.

The word *sequence* normally means infinite sequence and this definition is consistent with that understanding. In the examples and exercises that follow we stick with the custom of using subscript notation, instead of functional notation, when giving a sequence.

EXAMPLE 1   Write the first four terms of the sequence given by $a_n = 3n - 1$; also find a_{25}.

Solution   Substituting $n = 1, 2, 3, 4$ in the rule for a_n gives

$$a_1 = 3(1) - 1 = 2, \qquad a_2 = 3(2) - 1 = 5,$$
$$a_3 = 3(3) - 1 = 8, \qquad a_4 = 3(4) - 1 = 11.$$

For the 25th term, a_{25}, we have

$$a_{25} = 3(25) - 1 = 74. \qquad \blacksquare$$

In this book we discuss in detail two specific kinds of sequences: arithmetic progressions and geometric progressions. A famous prediction involving these sequences appeared in an essay on population written by the English economist Thomas Malthus in 1798. In this essay he was the first to express concern about the rapid growth of the earth's population and claimed that population increases in a geometric sequence while subsistence (food) increases in an arithmetic sequence. As we investigate these sequences, consider the consequences of this theory for the future. We begin with the definition of an arithmetic progression.

Definition of an Arithmetic Progression

> An **arithmetic progression** is a sequence of numbers in which each term after the first is found by adding a constant to the preceding term. This constant is called the **common difference** and is symbolized by d.

For example, the following sequences are arithmetic progressions.

$$7, \underset{3}{\quad} 10, \underset{3}{\quad} 13, \underset{3}{\quad} 16, \ldots \qquad \text{common difference: } 3$$

$$5, \underset{-4}{\quad} 1, \underset{-4}{\quad} -3, \underset{-4}{\quad} -7, \ldots \qquad \text{common difference: } -4$$

$$1, \quad \tfrac{3}{2}, \quad 2, \quad \tfrac{5}{2}, \ldots \qquad \text{common difference: } \tfrac{1}{2}$$

We find the rule for the general term in an arithmetic progression with first term a_1 and common difference d by observing the following pattern in such a sequence.

1st term, 2nd term, 3rd term, . . . , nth term

$$a_1, \quad a_1 + 1 \cdot d, \quad a_1 + 2 \cdot d, \ldots, a_1 + (n - 1)d$$

Thus, the formula for the nth term is

$$a_n = a_1 + (n - 1)d.$$

EXAMPLE 2   Find the formula for the general term a_n in the arithmetic progression

$$3, 7, 11, \ldots .$$

What is the 17th term in the sequence?

Solution   The first term is $a_1 = 3$ and the common difference is $d = 7 - 3 = 4$. Substituting these numbers in the above formula gives

$$a_n = 3 + (n - 1)4$$
$$= 4n - 1.$$

To find the 17th term, a_{17}, replace n by 17. Thus,

$$a_{17} = 4(17) - 1 = 67. \qquad \blacksquare$$

Next we consider geometric progressions.

Definition of a Geometric Progression

> A **geometric progression** is a sequence of numbers in which each number after the first is found by multiplying the preceding term by a constant. This constant is called the **common ratio** and is symbolized by r.

For example, the following sequences are geometric progressions.

$$2, 4, 8, 16, \ldots \qquad\qquad \text{common ratio: } \frac{4}{2} = \frac{8}{4} = \frac{16}{8} = 2$$

$$\frac{1}{2}, -\frac{1}{4}, \frac{1}{8}, -\frac{1}{16}, \ldots \qquad \text{common ratio: } -\frac{1}{2}$$

$$1, 1.06, (1.06)^2, (1.06)^3, \ldots \qquad \text{common ratio: } 1.06$$

We find the rule for the general term in a geometric progression with first term a_1 and common ratio r by observing the following pattern in such a sequence.

1st term, 2nd term, 3rd term, . . . , nth term

$$a_1, \qquad a_1 r^1, \qquad a_1 r^2, \ldots, a_1 r^{n-1}$$

Thus, the formula for the nth term is

$$a_n = a_1 r^{n-1}.$$

EXAMPLE 3   Find the formula for the general term a_n in the geometric progression

$$24, -12, 6, \ldots .$$

What is the 7th term in the sequence?

Solution   The first term is $a_1 = 24$ and the common ratio is   .
$r = -\frac{12}{24} = -\frac{1}{2}$. Substituting these values in the above formula gives

$$a_n = 24(-\tfrac{1}{2})^{n-1}.$$

To find the 7th term a_7, replace n by 7. Thus,

$$\begin{aligned}
a_7 &= 24(-\tfrac{1}{2})^{7-1} \\
&= 24(-\tfrac{1}{2})^6 \\
&= 24(\tfrac{1}{64}) = \tfrac{3}{8}.
\end{aligned}$$

Arithmetic and geometric progressions by no means exhaust the list of important sequences. As an example consider the set of numbers

$$1, 1, 2, 3, 5, 8, 13, 21, 34, 55, 89, \ldots,$$

which is called the **Fibonacci sequence.** The rule that generates these numbers is as follows: Start with 1 and 1 and thereafter each term is the sum of the two previous terms. In symbols we write

$$a_1 = 1, \quad a_2 = 1, \quad a_n = a_{n-1} + a_{n-2} \qquad \text{for } n \geq 3.$$

Leonardo Fibonacci, who was a mathematician in the Middle Ages, introduced this sequence innocently enough in his book on arithmetic and algebra as the answer to the following problem (paraphrased): A pair of rabbits produces a pair of baby rabbits by the end of their second month and every month thereafter. Each new pair of rabbits does the same. If none of the rabbits die, how many pairs of rabbits are there at the beginning of each month? The answer for the first seven months (and take the time to analyze why) is

Beginning of month number	1	2	3	4	5	6	7
Numbers of pairs	1	1	2	3	5	8	13

What is special about this sequence? Why is a mathematics journal, *The Fibonacci Quarterly,* devoted exclusively to issues connected with these numbers? Well, the numbers in the Fibonacci sequence keep appearing in the analysis of topics in such diverse areas as mathematics (naturally), physics, chemistry, art, and music, as well as appearing with surprising regularity in nature. Most of these applications are a bit involved, but it is not difficult to consider some of the fascinating ways in which Fibonacci numbers occur in nature.

On pine cones, pineapples, sunflowers, and other growths that appear on the tip of a stem there are two distinct sets of spirals. As shown in the diagram of a small sunflower in Figure 10.2, one set radiates clockwise; the other, counterclockwise. The number of spirals in each set is a Fibonacci number. In a small sunflower (as in this figure) there are 21 clockwise and 34 counterclockwise spirals. Some common counts for such growths are

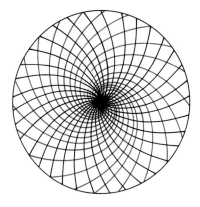

Figure 10.2

pine cone	5, 8	small sunflower	21, 34
pineapple	8, 13	large sunflower	34, 55
large pineapple	13, 21	giant sunflower	55, 89.

Note that each pair of numbers gives consecutive terms in the Fibonacci sequence.

These numbers also occur when considering the spiral leaf growth pattern of many trees. For example, consider the cyclic arrangements in the positioning of the leaf buds on the sample from an oak tree in Figure 10.3. Note that there are 5 buds to a complete cycle. In addition, each cycle contains 2 revolutions of the spiral. Both numbers are terms of the Fibonacci sequence. Similar remarks are true of other trees (such as elm and birch trees) and certain types of bushes. These examples just begin to scratch the surface of the immense volume of material dealing with the Fibonacci sequence.

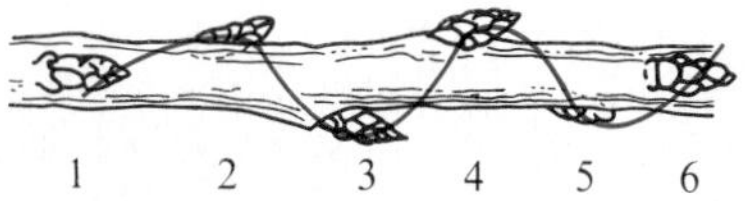

Figure 10.3

EXERCISES 10.1

In Exercises 1–10 write the first four terms in the sequence with general terms as given; also find the indicated term.

1. $a_n = 2n - 1$; a_{50}

2. $a_n = 7n + 2$; a_{25}

3. $a_n = n^2$; a_7

4. $a_n = 1/n$; a_{33}

5. $a_n = \dfrac{(-1)^{n+1}}{n}$; a_{10}

6. $a_n = \dfrac{(-1)^n}{n}$; a_{10}

7. $a_n = \dfrac{1}{2^n}$; a_5

8. $a_n = \dfrac{3^n - 1}{4^n - 1}$; a_5

9. $a_n = \dfrac{n(n+1)}{2}$; a_8

10. $a_n = \frac{1}{6}n(n+1)(2n+1)$; a_6

In Exercises 11–20 state whether the sequence is an arithmetic progression, geometric progression, or neither. For any arithmetic progression state the common difference and write the next two terms. For any geometric progression state the common ratio and write the next two terms.

11. 2, 7, 12, 17, 22, . . .

12. $-2, 4, -8, 16, -32, . . .$

13. $1, \frac{1}{2}, \frac{1}{3}, \frac{1}{4}, \frac{1}{5}, . . .$

14. $1, -4, -9, -14, -19, . . .$

15. $1, 1.02, (1.02)^2, (1.02)^3, (1.02)^4, . . .$

16. $\frac{1}{2}, \frac{2}{3}, \frac{3}{4}, \frac{4}{5}, \frac{5}{6}, . . .$

17. 3, 0.3, 0.03, 0.003, 0.0003, . . .

18. 1.01, 1.02, 1.03, 1.04, 1.05, . . .

19. $1, \dfrac{1}{2^2}, \dfrac{1}{3^2}, \dfrac{1}{4^2}, \dfrac{1}{5^2}, . . .$

20. $1, \dfrac{1}{2}, \dfrac{1}{2^2}, \dfrac{1}{2^3}, \dfrac{1}{2^4}, . . .$

In Exercises 21–30 the sequences are arithmetic progressions. Find the formula for the general term a_n and the value of the indicated term in the sequence.

21. 3, 7, 11, . . . ; 40th term

22. 2, 9, 16, . . . ; 50th term

23. $9, 4, -1, . . . ;$ 5th term

24. $2, -2, -6, . . . ;$ 10th term

25. 1.1, 1.4, 1.7, . . . ; a_{25}

26. 3.15, 3.10, 3.05, . . . ; a_{18}

27. $\frac{1}{3}, \frac{5}{3}, 3, . . . ;$ a_{30}

28. $\frac{10}{7}, \frac{1}{7}, -\frac{8}{7}, . . . ;$ a_9

29. $4, \frac{35}{8}, \frac{19}{4}, . . . ;$ a_8

30. $\frac{13}{4}, \frac{9}{2}, \frac{23}{4}, . . . ;$ a_{20}

In Exercises 31–40 the sequences are geometric progressions. Find the formula for the general term a_n and the value of the indicated term in the sequence.

31. 2, 6, 18, . . . ; 7th term

32. $2, -6, 18, . . . ;$ 7th term

33. $1, -\frac{1}{2}, \frac{1}{4}, . . . ;$ 6th term

34. $1, \frac{1}{2}, \frac{1}{4}, . . . ;$ 6th term

35. $\sqrt{2}, 2, 2\sqrt{2}, . . . ;$ a_8

36. $1, 1.04, (1.04)^2, . . . ;$ a_{10}

37. $6, 4, \frac{8}{3}, . . . ;$ a_7

38. $1, 0.1, 0.01, . . . ;$ a_{10}

39. $1, -\frac{1}{3}, \frac{1}{9}, . . . ;$ a_6

40. $1, \frac{3}{4}, \frac{9}{16}, . . . ;$ a_5

41. Write the first four terms in the arithmetic progression in which $a_1 = 4$ and $d = \frac{1}{2}$.

42. Write the first four terms in the geometric progression in which $a_1 = 4$ and $r = \frac{1}{2}$.

43. Which term of the geometric progression 18, 9, . . . is $\frac{9}{16}$?

44. Which term of the arithmetic progression 7, 4, . . . is -68?

45. Find the first term in an arithmetic progression in which $a_{17} = 39$ and $d = 2$.

46. The first term of a geometric progression is 8 and the fourth term is 27. Find the common ratio.

47. Find two different values of x so that $-\frac{2}{3}$, x, $-\frac{25}{54}$ are in geometric progression.

48. If, in an arithmetic progression, $a_1 = 80$, $d = -3$, and $a_n = 26$, find n.

49. If, in an arithmetic progression, $a_6 = 15$ and $a_{12} = 24$, find a_1 and d.

50. If, in an arithmetic progression, $a_{15} = \frac{22}{3}$ and $a_{30} = \frac{91}{6}$, find a_{40}.

THINK ABOUT IT

1. Is 5,000 a term in the arithmetic sequence 1,4,7, . . . ? Explain your answer.

2. **a.** If a, b, and c are the first three terms in an arithmetic progression, express b in terms of a and c.

  b. If a, b, and c are the first three terms in a geometric progression, express b in terms of a and c.

3. **a.** Find three numbers in arithmetic progression whose sum is 60 and whose product is 5,120.

  b. Find three numbers in geometric progression whose sum is 52 and whose product is 1,728.

4. State whether the sequence is an arithmetic progression, geometric progression, or neither. Find the common difference for any arithmetic progression and the common ratio for any geometric progression.

  a. log 2, log 4, log 8, log 16, . . .

  b. log 2, log 4, log 6, log 8, . . .

  c. log 2, log 4, log 16, log 256, . . .

5. **a.** Write the first 15 terms of the Fibonacci sequence.

b. The Fibonacci sequence appears in the family tree of a male bee. A male bee develops from an unfertilized egg and has only one parent—his mother. A female bee has both a mother and a father. The number of ancestors of a male bee in each successive generation is therefore the numbers of the Fibonacci sequence as shown below.

Number of Ancestors

1

2

3

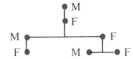

Continue the diagram for the next three generations and verify that the number of ancestors in these generations is given by the next three terms in the Fibonacci sequence.

REMEMBER THIS

In Exercises 1–4 find the formula for the general term a_n of the sequence.

1. 2,3,4,5, . . .

2. 2,12,22,32, . . .

3. 1,4,9,16, . . .

4. 2,4,8,16, . . .

In Exercises 5–8 evaluate each expression.

5. $\dfrac{8}{2}[2(1.1) + (8 - 1)(0.3)]$

6. $\dfrac{1 - \frac{1}{64}}{\frac{1}{2}}$

7. $\dfrac{1 - (\frac{3}{4})^6}{1 - \frac{3}{4}}$

8. $\dfrac{100[(1.08)^5 - 1]}{1.08 - 1}$

9. $500 is invested at 7 percent compounded annually. How much is the investment worth at the end of 4 years?

10. Simplify $\dfrac{a_n}{r^{n-1}} \cdot r^n$.

10.2   Series

Associated with any sequence

$$a_1, a_2, \ldots, a_n, \ldots$$

is a **series**

$$a_1 + a_2 + \cdots + a_n + \cdots,$$

which is the sum of all the terms in the sequence. The series associated with arithmetic and geometric progressions are called, respectively, an **arithmetic**

series and a **geometric series.** In this section we consider only series with a *finite* number of terms. The meaning of a series with an infinite number of terms is much more complex. Infinite series are discussed briefly in the next section and in detail in calculus.

We begin with a classic story from the history of mathematics that suggests the procedure for finding the sum of an arithmetic series. Carl Friedrich Gauss dominated mathematics in the nineteenth century and is considered, along with Archimedes and Newton, in a special class among the great mathematicians. He contributed profoundly to astronomy, physics (especially electromagnetic theory), non-Euclidean geometry, number theory, probability and statistics, and function theory. Reportedly, he demonstrated his imaginative insight to problem solving at the early age of ten when his class at school was assigned some "busy work." Their task—add all the numbers from 1 to 100. To the teacher's amazement Gauss quickly scribbled the correct result 5,050. How did he do it? Well, as is often the case when an ingenious method is used, the answer is easily obtained. Gauss noticed that the 100 numbers could be arranged in 50 pairs that all add up to 101 as follows:

$$\begin{array}{c} 1 + \quad 2 + \quad 3 + \quad 4 + \cdots + \quad 48 + \quad 49 + \quad 50 + \\ \underline{100 + 99 + 98 + 97 + \cdots + 53 + 52 + 51} \\ 101 + 101 + 101 + 101 + \cdots + 101 + 101 + 101. \end{array}$$

Thus, the result is simply 50(101) or 5,050.

By a similar method of adding pairs of terms with the same sum, we can develop the formula for the sum, denoted S_n, of the first n terms of an arithmetic progression. In general form an arithmetic series can be written as

$$S_n = a_1 + (a_1 + d) + (a_1 + 2d) + \cdots + (a_n - 2d) + (a_n - d) + a_n.$$

By reversing the order of the terms, we can also write this series as

$$S_n = a_n + (a_n - d) + (a_n - 2d) + \cdots + (a_1 + 2d) + (a_1 + d) + a_1.$$

If we now add term by term the equivalent expressions for S_n, we have n pairs, which all add up to $a_1 + a_n$. Thus,

$$2S_n = n(a_1 + a_n)$$

so that the general formula is

$$S_n = \frac{n}{2}(a_1 + a_n).$$

This formula is used when we know a_1 and a_n. However, a_n is often not given. If we replace a_n by $a_1 + (n - 1)d$, the above formula becomes

$$S_n = \frac{n}{2}\{a_1 + [a_1 + (n - 1)d]\}$$

$$= \frac{n}{2}[2a_1 + (n - 1)d].$$

We summarize our results as follows.

Arithmetic Series Formulas

The sum of the first n terms of an arithmetic series is given by

$$S_n = \frac{n}{2}(a_1 + a_n)$$

or

$$S_n = \frac{n}{2}[2a_1 + (n - 1)d].$$

EXAMPLE 1   Find the sum of the first 21 terms of the arithmetic progression 4, 7, 10, ...

Solution   Here $a_1 = 4$, $d = 3$, and $n = 21$. Thus,

$$S_n = \frac{n}{2}[2a_1 + (n - 1)d]$$

$$= \frac{21}{2}[2(4) + (21 - 1)3]$$

$$= \frac{21}{2}(68) = 714. \qquad \blacksquare$$

We now develop a formula for the sum of a given number of terms in a geometric progression. In general form a geometric series with n terms can be written as

$$S_n = a_1 + a_1r + a_1r^2 + \cdots + a_1r^{n-1}.$$

Multiplying both sides of this equation by r gives

$$rS_n = a_1r + a_1r^2 + \cdots + a_1r^{n-1} + a_1r^n.$$

Subtracting the second equation from the first, we obtain

$$S_n - rS_n = a_1 - a_1r^n$$
$$S_n(1 - r) = a_1 - a_1r^n$$
$$S_n = \frac{a_1 - a_1r^n}{1 - r}, \text{ for } r \neq 1.$$

This formula is used when we know a_1, r, and n. If we know the nth or last term in the series, a different formula is used. Since $a_n = a_1r^{n-1}$, we have $a_1 = a_n/r^{n-1}$. This replacement in the above formula gives

$$S_n = \frac{a_1 - (a_n/r^{n-1})r^n}{1 - r} = \frac{a_1 - a_nr}{1 - r}.$$

To summarize, we have derived the following formulas.

Geometric Series Formulas

> The sum of the first n terms of a geometric series with $r \neq 1$ is given by
>
> $$S_n = \frac{a_1 - a_1 r^n}{1 - r}$$
>
> or
>
> $$S_n = \frac{a_1 - a_n r}{1 - r} .$$

EXAMPLE 2   Find the sum of the first seven terms of the geometric progression $1, \frac{1}{2}, \frac{1}{4}, \ldots$.

Solution   Here $a_1 = 1$, $r = \frac{1}{2}$, and $n = 7$. Thus,

$$
\begin{aligned}
S_n &= \frac{a_1 - a_1 r^n}{1 - r} \\[2mm]
&= \frac{1 - 1\left(\frac{1}{2}\right)^7}{1 - \frac{1}{2}} \\[2mm]
&= \frac{1 - \frac{1}{128}}{\frac{1}{2}} = \frac{\frac{127}{128}}{\frac{1}{2}} = \frac{127}{64} .
\end{aligned}
$$ ∎

EXAMPLE 3   Solve the problem in the chapter introduction on page 519.

Solution

a. In the first round the 512 players are paired up two at a time, so 256 matches are needed. The 256 winners then advance to the second round, which will require 128 matches. Each subsequent round continues to require half the number of matches as the preceding round, so the number of matches from each round forms the following geometric progression.

$$256, \ 128, \ 64, \ 32, \ 16, \ 8, \ 4, \ 2, \ 1$$

It is easy enough to just add up these numbers, but we can also find the sum by using the second of our formulas for the sum of a geometric series.

$$S_n = \frac{a_1 - a_n r}{1 - r} = \frac{256 - 1\left(\frac{1}{2}\right)}{1 - \frac{1}{2}} = \frac{512 - 1}{2 - 1} = 511$$

Thus, 511 matches are needed in the tournament.

b. When the tournament is *over* there are 511 losers and 1 winner. Each loser lost exactly one match so the question "How many matches are played?" can be found by answering "How many losers are there?" With this approach it is easy to see that 511 matches are needed. ∎

A series is a sum, so it is convenient to use the Greek letter Σ, read "sigma," to mean *add*. By this convention, called **sigma notation,** we write

$$S_n = a_1 + a_2 + \cdots + a_n$$

in compact form as

$$S_n = \sum_{i=1}^{n} a_i$$

so that $\sum_{i=1}^{n} a_i$ means to add the terms that result from replacing i by 1, then 2, . . . , then n. The use of the letter i in this notation is arbitrary so that

$$\sum_{i=1}^{n} a_i, \qquad \sum_{j=1}^{n} a_j, \qquad \sum_{k=1}^{n} a_k$$

all represent the same series.

EXAMPLE 4   Write the series $\sum_{i=1}^{4} (2i + 3)$ in expanded form and determine the sum.

Solution

$$\sum_{i=1}^{4} (2i + 3) = (2 \cdot 1 + 3) + (2 \cdot 2 + 3) + (2 \cdot 3 + 3) + (2 \cdot 4 + 3)$$
$$= \quad 5 \quad + \quad 7 \quad + \quad 9 \quad + \quad 11$$
$$= 32 \qquad\qquad\qquad\qquad\qquad\blacksquare$$

EXAMPLE 5   Write the series $\sum_{j=2}^{7} 3^j$ in expanded form and determine the sum.

Solution

$$\sum_{j=2}^{7} 3^j = 3^2 + 3^3 + 3^4 + 3^5 + 3^6 + 3^7$$

This series is a geometric series with $a_1 = 3^2$, $a_n = 3^7$, and $r = 3$. Thus,

$$S_n = \frac{a_1 - a_n r}{1 - r}$$
$$= \frac{3^2 - (3^7)3}{1 - 3}$$
$$= \frac{9 - 6{,}561}{-2} = 3{,}276.$$

Therefore $\sum_{j=2}^{7} 3^j = 3{,}276.$ $\blacksquare$

EXAMPLE 6   Find an expression for the general term and write the series $2 + 5 + 8 + 11 + 14 + 17$ in sigma notation.

Solution   In most cases it is hoped that you can guess the general term by inspection. For this series it is

$$3i - 1$$

and since there are six terms, we have

$$\sum_{i=1}^{6} (3i - 1) = 2 + 5 + 8 + 11 + 14 + 17.$$

If you could not guess the general term in this case, you could note the series is an arithmetic series with $a_1 = 2$ and $d = 3$ so that

$$\begin{aligned} a_n &= a_1 + (n - 1)d \\ &= 2 + (n - 1)3 \\ &= 3n - 1. \end{aligned}$$

We now consider one of the important applications of geometric series. Sooner than you might imagine one (or more) of your former classmates will probably call you up and ask you to invest in your future. The plan—purchase an annuity. An **annuity** is any series of equal payments made at equal time intervals. There are different types of annuities. Here we discuss an **ordinary annuity,** in which payments are made at the end of each time period that coincides with a point at which interest is converted. As an example, suppose that at the end of each year for 5 years you invest $100, as shown below, in an account that pays 6-percent interest compounded annually.

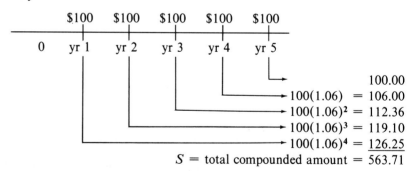

Each deposit earns interest for a different period of time. Thus, we compute separately the compounded amount of each $100 investment by the methods from Section 4.1 and add the results to obtain the total compounded amount. This sum is called the amount of the annuity and denoted S. The amount of this annuity (or the value of this series of equal periodic payments) is $563.71.

Treating each $100 investment separately is cumbersome. It is more efficient to note that

$$S = 100 + 100(1.06) + 100(1.06)^2 + 100(1.06)^3 + 100(1.06)^4$$

is a geometric series with $a_1 = 100$, $r = 1.06$, and $n = 5$. The formula for the sum of a geometric series

$$S = \frac{a_1 - a_1 r^n}{1 - r}$$

is usually written for these problems as

$$S = \frac{a_1(r^n - 1)}{r - 1}$$

so that for this series we have

$$S = \frac{100[(1.06)^5 - 1]}{1.06 - 1}$$

$$= \frac{100[1.338226 - 1]}{0.06}$$

$$= \frac{33.8226}{0.06} = \$563.71.$$

This answer coincides with our previous result and is easier to obtain.

Finally, let us generalize our discussion and derive the formula for the amount of an annuity—the most efficient way to determine S. Refer to the specific example just given to clarify any steps you do not understand. If we let

$$p = \text{periodic payment of the annuity}$$
$$i = \text{interest rate per period}$$
$$n = \text{number of payment periods}$$
$$S = \text{amount of the annuity}$$

then

$$S = p + p(1 + i) + p(1 + i)^2 + \cdots + p(1 + i)^{n-1}.$$

This series is a geometric series with n terms in which $r = 1 + i$ and $a_1 = p$. Substituting these values in the formula

$$S = \frac{a_1(r^n - 1)}{r - 1}$$

gives

$$S = \frac{p[(1 + i)^n - 1]}{(1 + i) - 1} = \frac{p[(1 + i)^n - 1]}{i}.$$

This formula is usually written as

$$S = p\left[\frac{(1 + i)^n - 1}{i}\right].$$

To simplify the evaluation of this expression, tables (which may be found in most business math books) give the values of

$$\frac{(1 + i)^n - 1}{i}$$

for common values of i and n. Then we need only multiply the appropriate table value by p (the periodic payment) to determine S (the amount of the annuity).

EXERCISES 10.2

In Exercises 1–10 find the sum of the indicated number of terms of the following arithmetic and geometric progressions.

1. $3, 7, 11, \ldots$; 10 terms
2. $9, 4, -1, \ldots$; 12 terms
3. $2, 6, 18, \ldots$; 7 terms
4. $2, -6, 18, \ldots$; 7 terms
5. $1, 1.04, (1.04)^2, \ldots$; 5 terms
6. $1, 0.1, 0.01, \ldots$; 9 terms
7. $1, \frac{3}{4}, \frac{9}{16}, \ldots$; 6 terms
8. $1.1, 1.4, 1.7, \ldots$; 8 terms
9. $3.15, 3.10, 3.05, \ldots$; 11 terms
10. $\frac{13}{4}, \frac{9}{2}, \frac{23}{4}, \ldots$; 10 terms

In Exercises 11–20 write the series in expanded form and determine the sum.

11. $\sum\limits_{i=1}^{10} i$

12. $\sum\limits_{i=1}^{4} i^2$

13. $\sum\limits_{j=1}^{6} (-1)^j$

14. $\sum\limits_{j=1}^{3} \frac{(-1)^{j+1}}{j}$

15. $\sum\limits_{k=1}^{12} (2k - 1)$

16. $\sum\limits_{k=1}^{20} (6k + 1)$

17. $\sum\limits_{i=2}^{5} (\frac{1}{2}i + 3)$

18. $\sum\limits_{j=2}^{8} 2^j$

19. $\sum\limits_{k=1}^{5} (\frac{1}{3})^{k-1}$

20. $\sum\limits_{i=1}^{4} (2 + 3^i)$

In Exercises 21–30 write the series in sigma notation. (*Note:* A given series can be expressed in sigma notation in more than one way.)

21. $3 + 4 + 5 + \cdots + 10$
22. $1 + 4 + 9 + 16 + 25$
23. $\frac{1}{2} + \frac{1}{4} + \frac{1}{8} + \frac{1}{16}$
24. $\frac{1}{2} - \frac{1}{4} + \frac{1}{8} - \frac{1}{16}$
25. $\frac{1}{2} + \frac{2}{3} + \frac{3}{4} + \frac{4}{5} + \frac{5}{6}$
26. $\frac{2}{3} + \frac{3}{5} + \frac{4}{7} + \frac{5}{9} + \frac{6}{11}$
27. $x + x^2 + x^3 + x^4 + x^5 + x^6$

28. $x + x^3 + x^5 + x^7$
29. $1 + 2^3 + 3^3 + \cdots + n^3$
30. $1 + 2^3 + 3^3 + \cdots + (n - 1)^3$
31. What is the sum of the first n even positive integers?
32. A grandfather clock chimes as many times as the hour. How many times does it chime in striking the hours 1 through 12?
33. A free-falling body that starts from rest drops about 16 ft the first second, 48 ft the second second, 80 ft the third second, and so on. How many feet does the object fall in 10 seconds?
34. You accept a position at a salary of $10,000 for the first year with an increase of $1,000 per year each year thereafter. How many years will you have to work for your total earnings to equal $231,000?
35. What is the total number of your ancestors in the seven generations that immediately precede you? (*Note:* One generation back is your parents, two generations back is your grandparents, and so on.)
36. A certain ball always rebounds $\frac{1}{2}$ as far as it falls. If the ball is thrown 10 ft into the air, how far up and down has it traveled when it hits the ground for the sixth time?
37. A certain ball always rebounds $\frac{2}{3}$ as far as it falls. If the ball is dropped from a height of 9 ft, how far up and down has it traveled when it hits the ground for the seventh time?
38. At the end of each year a woman deposits $1,000 in an account that pays 6-percent interest compounded annually. How much is in the account after 8 deposits have been made?
39. The principle of an annuity might help one to stop smoking. Suppose one saves the price of a package of cigarettes (let's say $1.50) each day of the year (use 365 days). At the end of the year the money is deposited in an account paying 9-percent interest compounded annually. How much is in the account after 10 deposits have been made?

THINK ABOUT IT

1. Find the sum of the first 250 positive integers in each of the following ways.
   a. Use the method attributed to Gauss in the beginning of this section.
   b. Use the method that was applied when deriving the formula for S_n.
   c. Use the formula $S_n = \dfrac{n}{2}(a_1 + a_n)$.

2. Prove each property of series.
   a. $\displaystyle\sum_{i=1}^{n} ca_i = c \sum_{i=1}^{n} a_i$
   b. $\displaystyle\sum_{i=1}^{n} (a_i - b_i) = \sum_{i=1}^{n} a_i - \sum_{i=1}^{n} b_i$

3. Find the arithmetic progression in which the sum of the first n terms is given by $S_n = n^2 + 3n$. What is the 50th term in this arithmetic sequence?

4. At the beginning of each month you deposit $100 in an account that pays 6-percent interest compounded monthly. Explain why the amount in this account at the end of 1 year is given by
$$100\left(1 + \frac{0.06}{12}\right)^1 + 100\left(1 + \frac{0.06}{12}\right)^2$$
$$+ \cdots + 100\left(1 + \frac{0.06}{12}\right)^{12}.$$
Use the methods of this section to find this sum.

5. Let $a_1, a_2, \ldots$ be the Fibonacci sequence. For successive values of n, find $a_1 + a_2 + \cdots + a_n$. Compare the sums with the terms of the sequence and determine the formula for the sum of the first n terms in the Fibonacci sequence.

REMEMBER THIS

In Exercises 1 and 2 find the common ratio in each geometric progression.

1. $\frac{1}{2}, -\frac{1}{3}, \frac{2}{9}, \ldots$ 2. $0.45, 0.0045, 0.000045, \ldots$

In Exercises 3 and 4 evaluate each expression.

3. $\dfrac{\frac{3}{2}}{1 - \left(-\frac{2}{3}\right)}$ 4. $2.14 + \dfrac{0.003}{1 - 0.1}$

5. Express $\frac{7}{11}$ as a repeating decimal.

6. Express the repeating decimal $0.\overline{45}$ as the ratio of two integers using the methods of Section 1.1.

7. Solve $|r| < 1$.

8. If $a_n = \dfrac{(-1)^{n+1}}{2^n}$, find a_6.

In Exercises 9 and 10 find the 11th term in the sequence.

9. $\sqrt{3}, 3, 3\sqrt{3}, \ldots$ 10. $3.75, 3.60, 3.45, \ldots$

10.3 Infinite Geometric Series

Consider the series

$$\tfrac{1}{2} + \tfrac{1}{4} + \tfrac{1}{8} + \tfrac{1}{16} + \cdots + \left(\tfrac{1}{2}\right)^n + \cdots.$$

This series contains infinitely many terms and raises a fundamental question: Can we ever add infinitely many numbers? In the usual sense the operation of addition has meaning only with respect to a finite number of terms so, in the usual sense, the answer to the question quite simply is no. However, instead of abandoning this train of thought let us look at the behavior of S_n, the sum of the first n terms in the series, as n increases from 1 to 6.

$$S_1 = \tfrac{1}{2}$$
$$S_2 = \tfrac{1}{2} + \tfrac{1}{4} = \tfrac{3}{4}$$
$$S_3 = \tfrac{1}{2} + \tfrac{1}{4} + \tfrac{1}{8} = \tfrac{7}{8}$$
$$S_4 = \tfrac{1}{2} + \tfrac{1}{4} + \tfrac{1}{8} + \tfrac{1}{16} = \tfrac{15}{16}$$
$$S_5 = \tfrac{1}{2} + \tfrac{1}{4} + \tfrac{1}{8} + \tfrac{1}{16} + \tfrac{1}{32} = \tfrac{31}{32}$$
$$S_6 = \tfrac{1}{2} + \tfrac{1}{4} + \tfrac{1}{8} + \tfrac{1}{16} + \tfrac{1}{32} + \tfrac{1}{64} = \tfrac{63}{64}$$

It appears that as n gets larger, S_n approaches, but never equals, 1. In fact, we can get S_n as close to 1 as we wish merely by taking a sufficiently large value for n. It is in this sense that we are going to assign the number 1 as the "sum" of the series. Note we are bending the usual meaning of sum to suit our purposes. To say that 1 is the sum of the above infinite series does not mean we add up all the numbers and get 1; it means that as n gets larger, S_n converges or closes in on 1.

Even with our new meaning for sum, many infinite series are still meaningless. Infinite arithmetic series and many infinite geometric series do not close in on some number. Intuitively, you can probably see that if the series is to converge to some number, then the terms in the series as n gets larger must approach 0. This condition occurs in infinite geometric series with $|r| < 1$ (or, equivalently, with $-1 < r < 1$), so we investigate this type of series further.

The formula for the sum of the first n terms of a geometric series is

$$S_n = \frac{a_1 - a_1 r^n}{1 - r}.$$

Now if $|r| < 1$, as n gets larger $a_1 r^n$ approaches 0, so that S_n closes in on $a_1/(1 - r)$. Thus, we have the following important result.

Sum of an Infinite Geometric Series

> An infinite geometric series with $|r| < 1$ converges to the value or sum
>
> $$S = \frac{a_1}{1 - r}.$$

EXAMPLE 1   Find the sum of the infinite geometric series

$$1 + \tfrac{1}{3} + \tfrac{1}{9} + \tfrac{1}{27} + \cdots.$$

Solution   Here $a_1 = 1$ and $r = \tfrac{1}{3}$. Since the common ratio is between -1 and 1 we can assign a sum to the series. Substituting in the formula

$$S = \frac{a_1}{1 - r},$$

we have

$$S = \frac{1}{1 - \tfrac{1}{3}} = \frac{1}{\tfrac{2}{3}} = \frac{3}{2}.$$

∎

In Section 1.1 we saw that every repeating decimal is a rational number and can be expressed as the quotient of two integers. One method for finding the appropriate fraction is to use an infinite geometric series. For example, the repeating decimal $0.\overline{3}$ (or equivalently 0.333 . . .) can be written as

$$0.3 + 0.03 + 0.003 + \cdots .$$

This series is an infinite geometric series with $r = (0.03/0.3) = 0.1$ and $a_1 = 0.3$, so

$$S = \frac{a_1}{1 - r} = \frac{0.3}{1 - (0.1)} = \frac{0.3}{0.9} = \frac{1}{3} .$$

Thus, $0.\overline{3}$ is equivalent to $\frac{1}{3}$.

EXAMPLE 2   Express the repeating decimal $7.\overline{54}$ as the ratio of two integers.

Solution   The repeating decimal $7.\overline{54}$ can be written as

$$7 + 0.54 + 0.0054 + 0.000054 + \cdots .$$

The series

$$0.54 + 0.0054 + 0.000054 + \cdots$$

is an infinite geometric series with $a_1 = 0.54$ and $r = 0.01$, so

$$S = \frac{a_1}{1 - r} = \frac{0.54}{1 - (0.01)} = \frac{0.54}{0.99} = \frac{54}{99} = \frac{6}{11} .$$

Adding the first term in the original series, 7, to $\frac{6}{11}$ gives

$$7 + \tfrac{6}{11} = \tfrac{77}{11} + \tfrac{6}{11} = \tfrac{83}{11} .$$

Thus, $7.\overline{54}$ is equivalent to $\frac{83}{11}$.   ∎

EXERCISES 10.3

In Exercises 1–10 find the sum of the infinite geometric series.

1. $1 + \frac{1}{2} + \frac{1}{4} + \frac{1}{8} + \cdots$

2. $1 + \frac{1}{5} + \frac{1}{25} + \frac{1}{125} + \cdots$

3. $\frac{1}{3} - \frac{1}{9} + \frac{1}{27} - \frac{1}{81} + \cdots$

4. $2 - 1 + \frac{1}{2} - \frac{1}{4} + \cdots$

5. $3 + \frac{3}{10} + \frac{3}{100} + \frac{3}{1000} + \cdots$

6. $5 + \frac{5}{3} + \frac{5}{9} + \frac{5}{27} + \cdots$

7. $\frac{3}{2} - 1 + \frac{2}{3} - \frac{4}{9} + \cdots$

8. $\frac{1}{2} - \frac{1}{3} + \frac{2}{9} - \frac{4}{27} + \cdots$

9. $5 + 0.5 + 0.05 + 0.005 + \cdots$

10. $11 + 1.1 + 0.11 + \cdots$

In Exercises 11–20 express each repeating decimal as the ratio of two integers.

11. $0.\overline{2}$ 12. $0.\overline{9}$ 13. $0.\overline{07}$

14. $0.0\overline{7}$ 15. $0.3\overline{21}$ 16. $0.6\overline{332}$

17. $5.\overline{9}$ 18. $4.8\overline{1}$ 19. $2.1\overline{43}$

20. $2.1\overline{43}$

21. A certain ball always rebounds $\frac{2}{3}$ as far as it falls. If the ball is dropped from a height of 9 ft, how far up and down has it traveled before it comes to rest?

22. For what values of x does the series

$$1 + x + x^2 + x^3 + \cdots$$

converge to a sum? What is this sum?

23. For what values of x does the series

$$(x + 1) + 2(x + 1)^2 + 4(x + 1)^3 + \cdots$$

converge to a sum? What is this sum?

24. Evaluate $\displaystyle\sum_{k=1}^{\infty} (-\tfrac{2}{3})^{k-1}$.

THINK ABOUT IT

1. Explain in terms of infinite geometric series why

$$\sum_{n=1}^{\infty} \frac{1}{x^n}$$ converges to a sum if $|x| > 1$. What is this sum?

2. The figure below shows the start of an infinite sequence of squares that is generated by using the midpoints of the sides of the first square as the vertices of the second square, and so on. If the side length of the initial square is 4 cm, then find the sum of the areas of all the squares. What is the sum of all the perimeters?

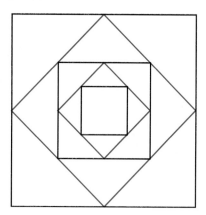

3. If $a_1 + a_2 + \cdots + a_n + \cdots$ is an infinite geometric series with $|r| < 1$, then what is the common

difference in the arithmetic sequence $\log |a_1|$, $\log |a_2|, \ldots, \log |a_n|, \ldots$?

4. **a.** If S denotes the sum of an infinite geometric series with $|r| < 1$, and S_n denotes the sum of the first n terms in this series, then write a formula for $S - S_n$.

   b. Use the result in part a and find the least number of terms in the series $1 + \tfrac{1}{3} + \tfrac{1}{9} + \cdots$ that should be summed so that the difference between S and S_n is less than $\tfrac{1}{1,000}$.

5. Consider the series

$$1 - 1 + 1 - 1 + 1 - 1 + \cdots.$$

What is the sum of this series? The problem can be attacked from three viewpoints.

Method 1   The sum is 0 since

$$(1 - 1) + (1 - 1) + (1 - 1) + \cdots$$
$$= 0 + 0 + 0 + \cdots = 0.$$

Method 2   The sum is 1 since

$$1 + (-1 + 1) + (-1 + 1) + \cdots$$
$$= 1 + 0 + 0 + \cdots = 1.$$

Method 3   The sum is $\tfrac{1}{2}$ since the series is a geometric series with $r = -1$, so

$$S = \frac{a_1}{1 - r} = \frac{1}{1 - (-1)} = \frac{1}{2}.$$

Therefore $0 = 1 = \tfrac{1}{2}$. What is wrong?

REMEMBER THIS

In Exercises 1–3 factor completely each expression.

1. $2k^2 + 4k + 2$ 2. $xa^k - a^{k+1}$
3. $2k(k + 1) + 4(k + 1)$
4. Multiply $(1 + r)^k(1 + r)$.
5. Add $\dfrac{k(2k + 1)}{6} + (k + 1)$.

6. State whether the sequence is an arithmetic progression, geometric progression, or neither:

$$1, \frac{1}{\sqrt{2}}, \frac{1}{\sqrt{3}}, \ldots$$

7. Find the formula for the general term a_n of the arithmetic progression $2, 6, 10, \ldots$.

8. Find the formula for the sum of the first n terms of the arithmetic progression $2, 6, 10, \ldots$.

9. Write the series $\displaystyle\sum_{i=1}^{5} (3i - 2)$ in expanded form and determine the sum.

10. Find the sum of the first 7 terms of the sequence $50, 50(1.04), 50(1.04)^2, \ldots$.

10.4 **Mathematical Induction**

Certain statements are true for every positive integer, and these results are quite useful in mathematics. We usually prove these statements about positive integers by a method called *mathematical induction*. The word *induction* is used because in many cases we can guess the general formula by looking at some specific cases. For example, consider the following sums of odd integers.

$$
\begin{array}{rcll}
1 & & \text{or} & 1^2 \\
1 + 3 = & 4 & \text{or} & 2^2 \\
1 + 3 + 5 = & 9 & \text{or} & 3^2 \\
1 + 3 + 5 + 7 = & 16 & \text{or} & 4^2 \\
1 + 3 + 5 + 7 + 9 = & 25 & \text{or} & 5^2
\end{array}
$$

From these additions we strongly suspect (or inductively conclude) that the sum of the first n odd integers always equals n^2 or, equivalently, that

$$ 1 + 3 + 5 + \cdots + (2n - 1) = n^2 $$

for every positive integer value of n. However, the validity of this formula in some specific cases does not guarantee that the rule will not break down for some higher number of odd integers. We can never prove the principle for *all* positive integers by checking a finite number of cases. Thus, mathematicians developed a technique called "proof by mathematical induction" that is based on the following principle.

Principle of Mathematical Induction

> If a given statement S_n concerning positive integer n is true for $n = 1$, and if its truth for $n = k$ implies its truth for $n = k + 1$, then S_n is true for every positive integer n.

According to this principle, we can prove that a statement (or formula) is true for every positive integer value of n by showing two things.

1. Prove the statement is true for $n = 1$.
2. Prove that if the statement is true for any positive integer k, then it is also true for the next highest integer $k + 1$.

Proof by mathematical induction is similar to lining up a row of dominoes so that when the first is knocked over it causes the second domino to fall, the second knocks down the third, and so on. In step 1 we knock down the first domino, and by proving step 2 we show the desired chain reaction then takes place.

By using mathematical induction, we can now continue our discussion concerning the sum of the first n odd integers, and we can prove that this sum *always* equals n^2. We show how in Example 1.

EXAMPLE 1   Use mathematical induction to prove that

$$1 + 3 + 5 + \cdots + (2n - 1) = n^2$$

is true for every positive integer n.

Solution   We prove the statement by mathematical induction with the following two steps.

Step 1   Replacing n by 1 in the formula gives

$$2(1) - 1 = 1^2$$

which is true. Step 1 is the easy part of the proof.

Step 2   Replace n by k and write what it means for the formula to work for any positive integer k. Call this equation (1). It is the induction assumption.

$$1 + 3 + 5 + \cdots + (2k - 1) = k^2 \qquad (1)$$

Now replace n by $k + 1$ and write what it means for the formula to work for the next highest integer $k + 1$. Call this equation (2).

$$1 + 3 + 5 + \cdots + (2k - 1) + [2(k + 1) - 1] = (k + 1)^2 \quad (2)$$

We now show that equation (1) implies equation (2). That is, we assume equation (1) is true and use this information to prove equation (2) is true. We accomplish this in either of the following ways.

Method 1   We start with equation (1) and add the $k + 1$ term to both sides of this equation. Then

$$\begin{aligned}
1 + 3 + \cdots + (2k - 1) + [2(k + 1) - 1] &= k^2 + [2(k + 1) - 1] \\
&= k^2 + 2k + 1 \\
&= (k + 1)^2
\end{aligned}$$

which is equation (2). Thus, equation (1) implies equation (2).

Method 2   We start with the left side of equation (2) and derive the other side with the aid of equation (1). Then

$$\begin{aligned}
&1 + 3 + \cdots + (2k - 1) + [2(k + 1) - 1] \\
&= [1 + 3 + \cdots + (2k - 1)] + [2(k + 1) - 1] \\
&\qquad \underbrace{}_{\text{induction assumption}} \\
&= \qquad k^2 \qquad\qquad + [2(k + 1) - 1] \\
&= k^2 + 2k + 1 \\
&= (k + 1)^2
\end{aligned}$$

which shows equation (1) implies equation (2).

The proof is complete. Step 1 shows the statement is true for $n = 1$. Step 2 guarantees the statement is true for the next highest integer 2, then 3, and so on, with no positive integer escaping the chain reaction.   ∎

EXAMPLE 2   Prove that if a and b are real numbers, then $(ab)^n = a^n b^n$ for every positive integer n.

Solution   We prove the statement by mathematical induction with the following two steps.

Step 1   Replacing n by 1 in the statement gives

$$(ab)^1 = a^1 b^1$$

which is true.

Step 2   When $n = k$, the statement becomes

$$(ab)^k = a^k b^k. \tag{1}$$

When $n = k + 1$, the statement becomes

$$(ab)^{k+1} = a^{k+1} b^{k+1}. \tag{2}$$

We now use equation (1) to establish equation (2) as follows:

$$\begin{aligned}
(ab)^{k+1} &= (ab)^k \cdot (ab) \qquad \text{First exponent law: } x^m \cdot x^n = x^{m+n} \\
&= (a^k b^k) \cdot (ab) \qquad \text{By equation (1)} \\
&= (a^k \cdot a) \cdot (b^k \cdot b) \\
&= a^{k+1} \cdot b^{k+1}.
\end{aligned}$$

The proof is complete. Note that it is important to state clearly what you need to prove (the "$k + 1$" statement) and what you assume (the "k" statement) so you can prove it.   ∎

Some statements are not true for *all* positive integers but instead are valid only from some particular integer on up. We prove a statement for all integers greater than or equal to some particular integer (say q) as follows:

1. By direct substitution show the statement is true for $n = q$.
2. Show that if the statement is true for any positive integer $k \geq q$, then it is also true for the next highest integer $k + 1$.

Example 3 illustrates this case.

EXAMPLE 3   Prove that $2^n > n + 1$ for every positive integer $n \geq 2$.

Solution   To prove the statement by mathematical induction, we need two steps. In step 1, however, we cannot start at $n = 1$ because 2^1 is not greater than $1 + 1$. Thus, as indicated in the problem, we start at $n = 2$.

Step 1   Replacing n by 2 in the inequality gives

$$2^2 > 2 + 1$$

which is true.

Step 2   We need to show that

$$2^k > k + 1 \text{ implies } 2^{k+1} > (k + 1) + 1$$

or, equivalently, that

$$2^k > k + 1 \text{ implies } 2^{k+1} > k + 2.$$

Since $2^k \cdot 2 = 2^{k+1}$, we start with $2^k > k + 1$ and multiply both sides by 2 as follows:

$$2^k > k + 1$$
$$2 \cdot 2^k > 2(k + 1)$$
$$2^{k+1} > 2k + 2.$$

Since $k \geq 2$, it follows that $2k + 2$ is greater than $k + 2$. Thus,

$$2^{k+1} > 2k + 2 > k + 2$$

and the proof is complete. ∎

The principle of mathematical induction is a very powerful method of proof that can be applied to a wide variety of mathematical statements. The main difficulty is that some ingenuity is often needed in deciding how to establish the "$k + 1$" statement from the *assumption* of the "k" statement. Note that we accomplished this with different approaches in the example problems after illustrating these approaches in Example 1. To be specific, in Example 2 we started with one side of the "$k + 1$" statement and derived the other side with the aid of the "k" statement. However, in Example 3 we started with the "k" statement and did something to it (we multiplied both sides by 2) to establish the "$k + 1$" statement. It will be easier for you to decide how to proceed in step 2 if you are aware of both methods. Eventually, step 2 becomes manageable through the experience gained in considering many problems.

EXERCISES 10.4

In Exercises 1–10 prove by mathematical induction that the following formulas are true for every positive integer value of n.

1. $1 + 2 + 3 + \cdots + n = \dfrac{n(n + 1)}{2}$

2. $\dfrac{1}{2} + \dfrac{2}{2} + \dfrac{3}{2} + \cdots + \dfrac{n}{2} = \dfrac{n(n + 1)}{4}$

3. $2 + 4 + 6 + \cdots + 2n = n(n + 1)$

4. $3 + 6 + 9 + \cdots + 3n = \dfrac{3n(n + 1)}{2}$

5. $4 + 8 + 12 + \cdots + 4n = 2n(n + 1)$

6. $2 + 6 + 10 + \cdots + (4n - 2) = 2n^2$

7. $2 + 2^2 + 2^3 + \cdots + 2^n = 2(2^n - 1)$

8. $3 + 3^2 + 3^3 + \cdots + 3^n = \dfrac{3(3^n - 1)}{2}$

9. $1^3 + 2^3 + 3^3 + \cdots + n^3 = \dfrac{n^2(n + 1)^2}{4}$

10. $1^2 + 2^2 + 3^2 + \cdots + n^2 = \dfrac{n(n + 1)(2n + 1)}{6}$

In Exercises 11–20 prove the statement by mathematical induction.

11. $2^n > n$ for all positive integers n.

12. $3^n < 3^{n+1}$ for all positive integers n.

13. $n^2 < 2^n$ for all positive integers $n \geq 5$.

14. If a is a real number, then $(a^2)^n = a^{2n}$ for all positive integers n.

15. If the x's are positive real numbers, then
$$\log(x_1 x_2 \cdots x_n) = \log x_1 + \log x_2 + \cdots + \log x_n$$
for all positive integers $n \geq 2$.

16. If P dollars are invested at r percent compounded annually, then at the end of n years the compounded amount is given by $A_n = P(1 + r)^n$.

17. If a is a real number such that $0 < a < 1$, then $0 < a^n < 1$ for all positive integers n.

18. $n^2 + n$ is divisible by 2 for all positive integers n.

19. If $x \neq a$, then $x^n - a^n$ is divisible by $x - a$ for all positive integers n. (*Hint:* Add and subtract xa^k to $x^{k+1} - a^{k+1}$, then group terms and factor.)

20. If $x \neq y$, then $x^{2n} - y^{2n}$ is divisible by $x - y$ for all positive integers n. (*Note:* See Exercise 19.)

THINK ABOUT IT

1. Show by mathematical induction that the sum of n even integers is an even integer for $n \geq 2$.

2. Show by mathematical induction that

$$\sum_{i=1}^{n} ca_i = c \sum_{i=1}^{n} a_i.$$

3. The following sequence is an example of one that is defined *recursively:* The first term is 1; thereafter, each term is one more than twice the preceding term. This translates to $a_1 = 1$ and $a_n = 2a_{n-1} + 1$ for $n \geq 2$. Find a_2, a_3, a_4, and a_5, and then make a conjecture about a formula for a_n in terms of n. Verify your formula is correct using mathematical induction.

4. **a.** Simplify each expression.

$$1 - \frac{1}{2}; \left(1 - \frac{1}{2}\right)\left(1 - \frac{1}{3}\right);$$
$$\left(1 - \frac{1}{2}\right)\left(1 - \frac{1}{3}\right)\left(1 - \frac{1}{4}\right)$$

b. Make a conjecture about the simplification of the following product, and then prove it using mathematical induction.

$$\left(1 - \frac{1}{2}\right)\left(1 - \frac{1}{3}\right) \cdots \left(1 - \frac{1}{n}\right)$$

5. The following problems analyze false statements to reinforce that proof by mathematical induction is a two-step process and that neither step by itself is sufficient in such proofs.

a. Let S_n be the statement that for any positive integer n,

$$n^2 + n + 11 \text{ is a prime number.}$$

Verify that S_1 is true. Show S_n is true for $n = 1, 2, \ldots, 9$ but false for $n = 10$.

b. Let S_n be the statement that for any positive integer n,

$$n = n + 5.$$

Show that if S_n is true for any positive integer k, then it is also true for the next highest integer $k + 1$. Then show S_n is a false statement by showing S_1 is false.

REMEMBER THIS

In Exercises 1 and 2 multiply and simplify.

1. $(5x - 2)^2$

2. $(a + b)^3$

3. How many terms are in the expansion of $(a + b)^3$?

4. For what value(s) of x does $(x + 3)^2 = x^2 + 3^2$?

5. Simplify $10(3x)^2(-2y)^3$.

6. If in an arithmetic sequence $a_{20} = 18$ and $a_{30} = 23$, then find a_6.

7. Write the series $1 - \frac{1}{3} + \frac{1}{9} - \frac{1}{27}$ in sigma notation.

In Exercises 8–10 find the sum of each series.

8. $\sum_{n=1}^{20} (7n - 5)$

9. $\sum_{n=1}^{5} \left(\frac{2}{5}\right)^{n-1}$

10. $\sum_{n=1}^{\infty} \frac{2}{5^{n-1}}$

10.5 Binomial Theorem

A leading candidate for the most common mistake in algebra appears when students are asked to find the product or expansion of $(x + y)^2$. Too frequently they write the expansion as $x^2 + y^2$, which leaves out the middle term $2xy$. The problem recurs when expanding other expressions of the form $(x + y)^n$, where n is a positive integer, since there are usually many more terms than just x^n and y^n. We now discuss a method for finding these middle terms that is much easier than repeated multiplication.

We start by trying to find some patterns in the following expansions of the powers of $a + b$. You can verify each expansion by direct multiplication.

$(a + b)^1 = a + b$	(2 terms)
$(a + b)^2 = a^2 + 2ab + b^2$	(3 terms)
$(a + b)^3 = a^3 + 3a^2b + 3ab^2 + b^3$	(4 terms)
$(a + b)^4 = a^4 + 4a^3b + 6a^2b^2 + 4ab^3 + b^4$	(5 terms)
$(a + b)^5 = a^5 + 5a^4b + 10a^3b^2 + 10a^2b^3 + 5ab^4 + b^5$	(6 terms)

Observe that in all the above cases the expansion of $(a + b)^n$ behaved as follows:

1. The number of terms in the expansion is $n + 1$. For example, in the expression $(a + b)^5$ we have $n = 5$ and there are six terms in the expansion.
2. The first term is a^n and the last term is b^n.
3. The second term is $na^{n-1}b$ and the nth term is nab^{n-1}.
4. The exponent of a decreases by 1 in each successive term while the exponent of b increases by 1.
5. The sum of the exponents of a and b in any term is n.

Thus, if we assume that this pattern continues when n is a positive integer greater than 5 (and it can be proved that it does), we need only find a method for determining the constant coefficients if we wish to expand such expressions. Our observations to this point may be summarized in the following expansion formula for positive integral powers of $a + b$.

$$(a + b)^n = a^n + na^{n-1}b + (\text{constant})a^{n-2}b^2$$
$$+ (\text{constant})a^{n-3}b^3 + \cdots + nab^{n-1} + b^n$$

You may think of this expansion formula as an informal version of the binomial theorem. A complete statement of the binomial theorem includes a method for obtaining the constant coefficients, and we will do this later. However, for our present purpose, we have arranged the constant coefficients in the above expansions of $(a + b)^n$ for $n \leq 5$ in the triangular array shown in the following chart.

Powers of $a + b$	Constant Coefficients (Pascal's Triangle)
$(a + b)^0$	Row 0
$(a + b)^1$	Row 1
$(a + b)^2$	Row 2
$(a + b)^3$	Row 3
$(a + b)^4$	Row 4
$(a + b)^5$	Row 5

The triangular array of numbers that specifies the constant coefficients in the expansions of $(a + b)^n$ for $n = 0, 1, 2, \ldots$ is called **Pascal's triangle.** Note that except for the 1's, each entry in Pascal's triangle is the sum of the two numbers on either side of it in the preceding row as diagrammed in the chart. We are now ready to try some problems.

EXAMPLE 1   Expand $(x + y)^8$ using Pascal's triangle.

Solution   In our expansion formula we substitute x for a, y for b, and 8 for n. To determine the constant coefficients, we must continue the pattern in the above chart until we have specified the entries in row 8 of Pascal's triangle. To do this, we start at row 5 and proceed as follows:

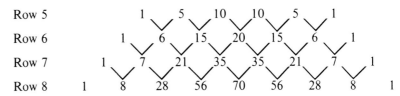

Row 5			1	5	10	10	5	1	
Row 6		1	6	15	20	15	6	1	
Row 7	1	7	21	35	35	21	7	1	
Row 8	1	8	28	56	70	56	28	8	1

Thus, the expansion of $(x + y)^8$ is

$$(x + y)^8 = x^8 + 8x^7y + 28x^6y^2 + 56x^5y^3$$
$$+ 70x^4y^4 + 56x^3y^5 + 28x^2y^6 + 8xy^7 + y^8.$$ ∎

EXAMPLE 2   Expand $(x + 2y)^4$ using Pascal's triangle.

Solution   We substitute x for a, $2y$ for b, and 4 for n in our expansion formula. From Pascal's triangle we determine the coefficients of the five terms in the expansion as 1, 4, 6, 4, and 1, respectively. Thus, we have

$$(x + 2y)^4 = x^4 + 4x^3(2y) + 6x^2(2y)^2 + 4x(2y)^3 + (2y)^4$$
$$= x^4 + 8x^3y + 24x^2y^2 + 32xy^3 + 16y^4.$$ ∎

EXAMPLE 3   Expand $(2x - y)^3$ using Pascal's triangle.

Solution   First, rewrite $(2x - y)^3$ as $[2x + (-y)]^3$. In our expansion formula we now substitute $2x$ for a, $-y$ for b, and 3 for n. From Pascal's triangle we determine the coefficients of the four terms in the expansion as 1, 3, 3, and 1, respectively. Thus, we have

$$(2x - y)^3 = (2x)^3 + 3(2x)^2(-y) + 3(2x)(-y)^2 + (-y)^3$$
$$= 8x^3 - 12x^2y + 6xy^2 - y^3.$$

Note that when the binomial is the difference of two terms, the terms in the expansion alternate in sign.   ∎

By now you should have the feel of the basic patterns in the binomial expansion, so let's add the final step—a formula for the constant coefficients. We have been using Pascal's triangle to generate these constants. However, this approach is often impractical since constructing the triangle can be time consuming, particularly for large values of n.

The formula for the coefficients can be quite efficient and follows a nice pattern. First, though, it is helpful to introduce **factorial** notation. For any positive integer n, the symbol $n!$ (read "n factorial") means the product $n \cdot (n - 1) \cdot (n - 2) \cdots \cdots 3 \cdot 2 \cdot 1$. For example,

$$6! = 6 \cdot 5 \cdot 4 \cdot 3 \cdot 2 \cdot 1 = 720.$$

We also define $0! = 1$. Using factorials we now state the binomial theorem.

Binomial Theorem

For any positive integer n,

$$(a + b)^n = a^n + \frac{n}{1!}a^{n-1}b + \frac{n(n-1)}{2!}a^{n-2}b^2 + \cdots$$

$$+ \frac{n(n-1)(n-2)\cdots(n-r+1)a^{n-r}b^r}{r!}$$

$$+ \cdots + b^n.$$

The binomial theorem is proved by mathematical induction, and we consider such a proof in the exercises. You will find the application of this theorem to be very systematic if you consider the following example carefully.

EXAMPLE 4   Expand $(x + y)^5$ by the binomial theorem.

Solution   In our statement of the binomial theorem we substitute x for a and y for b. We also note that in this case $n = 5$. The expansion goes as follows:

$$(x + y)^5 = x^5 + \frac{5}{1!}x^4y^1 + \frac{5 \cdot 4}{2!}x^3y^2 + \frac{5 \cdot 4 \cdot 3}{3!}x^2y^3$$

$$+ \frac{5 \cdot 4 \cdot 3 \cdot 2}{4!}xy^4 + y^5$$

$$= x^5 + 5x^4y + 10x^3y^2 + 10x^2y^3 + 5xy^4 + y^5. \qquad \blacksquare$$

In Example 4 there are a few patterns in determining the coefficients that you should note.

1. The coefficients 1, 5, 10, 10, 5, 1 are symmetrical; that is, they increase and then decrease in the same manner. This symmetry cuts our job in half since we need determine the coefficients only as far as the middle term to obtain the entire list.
2. The coefficients go

$$1, \frac{5}{1!}, \frac{5 \cdot 4}{2!}, \frac{5 \cdot 4 \cdot 3}{3!}, \quad \text{etc.}$$

or, in general, as

$$1, \frac{n}{1!}, \frac{n(n-1)}{2!}, \frac{n(n-1)(n-2)}{3!}, \quad \text{etc.}$$

Each new coefficient (after n) can be generated from the previous coefficient. Just insert the next lowest integer as a factor in the numerator and use the next highest factorial in the denominator. It might also help if you note that in each coefficient the denominator is the factorial of the exponent for y in that term.

3. When computing each coefficient it is possible to divide out common factors so that the expression is easy to evaluate. For example,

$$\frac{5 \cdot 4 \cdot 3 \cdot 2}{4!} = \frac{5 \cdot 4 \cdot 3 \cdot 2}{4 \cdot 3 \cdot 2 \cdot 1} = \frac{5}{1} = 5.$$

With a little practice you will find this method to be quite straightforward. The idea is to balance the different benefits that Pascal's triangle and the formula method have to offer. Here is a problem in which the formula method is a lot easier to use than constructing Pascal's triangle.

EXAMPLE 5   Write the first four terms in the expansion of $(x - 2)^{14}$.

Solution   First, rewrite $(x - 2)^{14}$ as $[x + (-2)]^{14}$. Here $n = 14$, $a = x$, and $b = -2$. Then by the binomial theorem the first four terms are

$$x^{14} + \frac{14}{1!}x^{13}(-2)^1 + \frac{14 \cdot 13}{2!}x^{12}(-2)^2 + \frac{14 \cdot 13 \cdot 12}{3!}x^{11}(-2)^3$$

which simplifies to

$$x^{14} - 28x^{13} + 364x^{12} - 2{,}912x^{11}. \qquad \blacksquare$$

To state the binomial theorem in its most efficient form, we need one more refinement in our methods. Note in the statement of the binomial theorem that the constant coefficient of the term containing b^r for $0 < r \le n$ is

$$\frac{n(n - 1) \cdots (n - r + 1)}{r!}.$$

For instance, in the expansion of $(x + y)^5$ in Example 4 the constant coefficient of the term containing y^4 is

$$\frac{5 \cdot 4 \cdot 3 \cdot 2}{4!}$$

or

$$\frac{5(5 - 1)(5 - 2)(5 - 4 + 1)}{4!}.$$

In its present form this formula is awkward to use and we may restate the formula more efficiently as follows:

$$\frac{n(n - 1) \cdots (n - r + 1)}{r!} = \frac{n(n - 1) \cdots (n - r + 1)}{r!} \cdot \frac{(n - r)!}{(n - r)!}$$

$$= \frac{n!}{r!(n - r)!}.$$

If we now follow customary notation and denote the constant coefficient of the term containing b^r by $\binom{n}{r}$, then we have developed that a formula for $\binom{n}{r}$ is

$$\binom{n}{r} = \frac{n!}{r!(n-r)!}.$$

In addition, $\binom{n}{0} = \binom{n}{n} = 1$, and we may obtain these results with our formula since by definition $0! = 1$. Our discussion has led us to the definition of a binomial coefficient.

Binomial Coefficient

Let r and n be nonnegative integers with $r \le n$. Then the symbol $\binom{n}{r}$ is defined by

$$\binom{n}{r} = \frac{n!}{r!(n-r)!}.$$

Each of the numbers $\binom{n}{r}$ is called a **binomial coefficient.**

As you might expect, binomial coefficients are entries in Pascal's triangle, and the next example illustrates this point.

EXAMPLE 6 Evaluate the following binomial coefficients and compare the results to the entries in Pascal's triangle:

$$\binom{4}{0}, \binom{4}{1}, \binom{4}{2}, \binom{4}{3}, \binom{4}{4}.$$

Solution By applying the formula in the definition, we have

$$\binom{4}{0} = \frac{4!}{0!(4-0)!} = \frac{4!}{0!4!} = 1$$

$$\binom{4}{1} = \frac{4!}{1!(4-1)!} = \frac{4!}{1!3!} = \frac{4 \cdot 3 \cdot 2 \cdot 1}{1 \cdot (3 \cdot 2 \cdot 1)} = 4$$

$$\binom{4}{2} = \frac{4!}{2!(4-2)!} = \frac{4!}{2!2!} = \frac{4 \cdot 3 \cdot 2 \cdot 1}{(2 \cdot 1)(2 \cdot 1)} = 6$$

$$\binom{4}{3} = \frac{4!}{3!(4-3)!} = \frac{4!}{3!1!} = \frac{4 \cdot 3 \cdot 2 \cdot 1}{(3 \cdot 2 \cdot 1) \cdot 1} = 4$$

$$\binom{4}{4} = \frac{4!}{4!(4-4)!} = \frac{4!}{4!0!} = 1.$$

By direct comparison, we note that the binomial coefficients 1, 4, 6, 4, and 1 match the entries in row 4 of Pascal's triangle. This example illustrates the general result that the binomial coefficients

$$\binom{n}{0}, \binom{n}{1}, \binom{n}{2}, \cdots, \binom{n}{n}$$

are precisely the numbers in the nth row of Pascal's triangle. Note also that by using a scientific calculator with a factorial key $\boxed{x!}$, it is easy to evaluate binomial coefficients. For instance, since $\binom{18}{7} = \dfrac{18!}{7!11!}$, we may evaluate $\binom{18}{7}$ by calculator as follows:

$$18 \boxed{x!} \div 7 \boxed{x!} \div 11 \boxed{x!} = 31{,}824. \qquad \blacksquare$$

By using binomial coefficients and sigma notation, the binomial theorem may be stated in a very efficient way.

Binomial Theorem

For any positive integer n,

$$(a + b)^n = \sum_{r=0}^{n} \binom{n}{r} a^{n-r} b^r$$

$$= \binom{n}{0} a^n + \binom{n}{1} a^{n-1} b + \binom{n}{2} a^{n-2} b^2 + \cdots + \binom{n}{n} b^n.$$

We now redo the expansion from Example 1 using this version of the theorem.

EXAMPLE 7   Expand $(x + y)^8$ by the binomial theorem.

Solution   Applying the binomial theorem with $x = a$, $y = b$, and $n = 8$ yields

$$(x + y)^8 = \binom{8}{0} x^8 + \binom{8}{1} x^7 y + \binom{8}{2} x^6 y^2 + \binom{8}{3} x^5 y^3 + \binom{8}{4} x^4 y^4$$

$$+ \binom{8}{5} x^3 y^5 + \binom{8}{6} x^2 y^6 + \binom{8}{7} x y^7 + \binom{8}{8} y^8.$$

To evaluate the binomial coefficients, we first note that $\binom{8}{0} = \binom{8}{8} = 1$. Then we use the symmetry in the coefficients and the given formula to obtain

$$\binom{8}{1} = \binom{8}{7} = \frac{8!}{1!7!} = 8, \qquad \binom{8}{2} = \binom{8}{6} = \frac{8!}{2!6!} = 28,$$

$$\binom{8}{3} = \binom{8}{5} = \frac{8!}{3!5!} = 56, \qquad \binom{8}{4} = \frac{8!}{4!4!} = 70.$$

Thus, the desired expansion is

$$(x + y)^8 = x^8 + 8x^7y + 28x^6y^2 + 56x^5y^3 + 70x^4y^4$$
$$+ 56x^3y^5 + 28x^2y^6 + 8xy^7 + y^8.$$

(*Note:* The symmetry in the binomial coefficients follows from the identity

$$\binom{n}{r} = \binom{n}{n-r}.$$

You are asked to verify this property in Exercise 48.)   ■

With our current methods it is now easy to write any single term of a binomial expansion without producing the rest of the terms. Note in the binomial theorem that the exponent of b is 1 in the second term, 2 in the third term, and, in general, $r - 1$ in the rth term. Since the sum of the exponents of a and b is always n, the exponent above a in the rth term must be $n - (r - 1)$. Finally, the binomial coefficient of the term containing b^{r-1} is $\binom{n}{r-1}$. These observations lead to the following formula.

rth Term of the Binomial Expansion

The rth term of the binomial expansion of $(a + b)^n$ is

$$\binom{n}{r-1}a^{n-(r-1)}b^{r-1}.$$

EXAMPLE 8   Find the 14th term in the expansion of $(3x - y)^{15}$.

Solution   In this case $n = 15$, $r = 14$, $a = 3x$, and $b = -y$. By the above formula, the 14th term is

$$\binom{15}{13}(3x)^2(-y)^{13} = 105(9x^2)(-y)^{13}$$
$$= -945x^2y^{13}.$$ ■

EXERCISES 10.5

In Exercises 1–12 expand each expression by the binomial theorem. Use both Pascal's triangle and the formula methods in determining the constant coefficients.

1. $(x + y)^6$ 2. $(x + y)^7$ 3. $(x - y)^5$
4. $(x - y)^4$ 5. $(x + h)^4$ 6. $(x + h)^3$
7. $(x - 1)^7$ 8. $(y + 1)^5$ 9. $(2x + y)^3$
10. $(x - 2y)^6$ 11. $(3c - 4d)^4$ 12. $(4c + 3d)^3$

In Exercises 13–16 write the first four terms in the expansion of the given expression.

13. $(x + y)^{15}$ 14. $(x - y)^{10}$
15. $(x - 3y)^{12}$ 16. $(x + 3)^{17}$

In Exercises 17 and 18 evaluate each of the binomial coefficients and compare the results to the entries in Pascal's triangle.

17. $\binom{3}{0}, \binom{3}{1}, \binom{3}{2}, \binom{3}{3}$

18. $\binom{5}{0}, \binom{5}{1}, \binom{5}{2}, \binom{5}{3}, \binom{5}{4}, \binom{5}{5}$

In Exercises 19 and 20 express the entries in the given row of Pascal's triangle using the notation $\binom{n}{r}$.

19. Row 6   20. Row 2

In Exercises 21–30 evaluate each of the binomial coefficients.

21. $\dbinom{5}{3}$ **22.** $\dbinom{6}{4}$ **23.** $\dbinom{9}{4}$

24. $\dbinom{10}{3}$ **25.** $\dbinom{10}{0}$ **26.** $\dbinom{9}{9}$

27. $\dbinom{20}{2}$ **28.** $\dbinom{25}{24}$ **29.** $\dbinom{18}{8}$

30. $\dbinom{15}{5}$

In Exercises 31–40 write the indicated term of the expansion.

31. 2nd term of $(x + y)^{18}$
32. 5th term of $(x - y)^7$
33. 12th term of $(3x - y)^{13}$
34. 6th term of $(x + 3y)^{10}$
35. 6th term of $(x^2 + 1)^9$
36. 11th term of $(2x^3 - 1)^{14}$
37. 3rd term of $(2 - \sqrt{x})^5$

38. 4th term of $(\sqrt{x} + 3)^6$
39. Middle term of $(3x^2 + 2y^3)^6$
40. Middle term of $(x^{-1} + y^{-1})^{10}$
41. Find the coefficient of x^7y^4 in the expansion of $(x + y)^{11}$.
42. Find the coefficient of $x^{15}y^5$ in the expansion of $(x - y)^{20}$.
43. Find the coefficient of the term containing x^7 in the expansion of $(x - 2y)^{10}$.
44. Find the term with the variable factor x^5 in the expansion of $(2x - y)^{15}$.
45. Find the value of $(1.2)^4$ by writing it in the form $(1 + 0.2)^4$ and using the binomial theorem.
46. Find the value of $(0.97)^3$ by writing it in the form $(1 - 0.03)^3$ and using the binomial theorem.

47. Show that $\dbinom{n}{n-1} = n$.

48. Show that $\dbinom{n}{r} = \dbinom{n}{n-r}$.

THINK ABOUT IT

1. Expand $(x^2 + x - 1)^4$ using the binomial theorem.
2. **a.** Use the binomial expansion of $(1 + 1)^n$ to show that

$$\binom{n}{0} + \binom{n}{1} + \binom{n}{2} + \cdots + \binom{n}{n} = 2^n.$$

   b. What is the sum of the entries in row 12 of Pascal's triangle?
3. If $f(x) = x^n$, where n is a positive integer, show that

$$\frac{f(x + h) - f(x)}{h}$$

$$= nx^{n-1} + \frac{n(n-1)}{2!}x^{n-2}h + \cdots + h^{n-1}.$$

4. **a.** Show that $\dbinom{4}{3} + \dbinom{4}{2} = \dbinom{5}{3}$.

   b. Consider the positions in Pascal's triangle of the binomial coefficients in part a. What feature of Pascal's triangle does the statement illustrate?

   c. The property we have been considering states that for all positive integers k and r with $r \le k$,

$$\binom{k}{r} + \binom{k}{r-1} = \binom{k+1}{r}.$$

   Prove this property by using the binomial coefficient formula.

5. Below we use mathematical induction and the property shown in Exercise 4 to prove the binomial theorem (that is, $(a + b)^n = \sum\limits_{r=0}^{n} \dbinom{n}{r} a^{n-r}b^r$). Fill in the missing steps (denoted by the letters a–i) along with appropriate justifications.

Step 1   Replacing n by 1 gives a, which is true.

Step 2   When $n = k$ the statement becomes

$$(a + b)^k = \binom{k}{0}a^k + \binom{k}{1}a^{k-1}b + \cdots \tag{1}$$
$$+ \binom{k}{k-1}ab^{k-1} + \binom{k}{k}b^k.$$

When $n = k + 1$ the statement becomes

   b. $\qquad\qquad\qquad\qquad\qquad\qquad$ (2)

We now use equation (1) to establish equation (2). To begin, multiply both sides of equation (1) by $(a + b)$ and then continue as follows:

$$(a + b)^{k+1} = a\left[\binom{k}{0}a^k + \binom{k}{1}a^{k-1}b + \cdots\right.$$
$$\left. + \binom{k}{k-1}ab^{k-1} + \binom{k}{k}b^k\right] + b[c]$$
$$= \binom{k}{0}a^{k+1} + \binom{k}{1}a^kb + \cdots$$
$$+ \binom{k}{k-1}a^2b^{k-1} + \binom{k}{k}ab^k + d$$

$$= \binom{k}{0}a^{k+1} + \left[\binom{k}{1} + \binom{k}{0}\right]a^kb$$
$$+ \cdots + [e]ab^k + \binom{k}{k}b^{k+1}$$
$$= (f)a^{k+1} + (g)a^kb + \cdots$$
$$+ (h)ab^k + (i)b^{k+1}$$

which is equation (2). Thus, equation (1) implies equation (2) and the proof by induction is complete.

REMEMBER THIS

In Exercises 1 and 2 evaluate each expression.

1. $6!$

2. $\dfrac{7!}{4!3!}$

3. Express $12!$ in scientific notation.

4. Simplify $\dfrac{n!}{(n-2)!}$.

5. Does $\binom{5}{2} = \binom{5}{3}$?

6. Find two different values of x so that $2, x, 16$ are in geometric progression.

7. Find x so that a^2, a^x, a^{16} are in geometric progression.

8. What is the sum of the first n positive integers?

9. Express $1.\overline{75}$ as the ratio of two integers. (Use infinite geometric series.)

10. Prove by mathematical induction: For every positive integer n,

$$3 + 9 + 15 + \cdots + (6n - 3) = 3n^2.$$

10.6   Counting Techniques

When making certain decisions, it is often important to analyze the possibilities associated with the situation. For example, if your college offers 2 sections of a required math course and 3 sections of a required English course, then how many schedules are possible in enrolling for these courses? We can answer this question by constructing a **tree diagram** as shown in Figure 10.4.

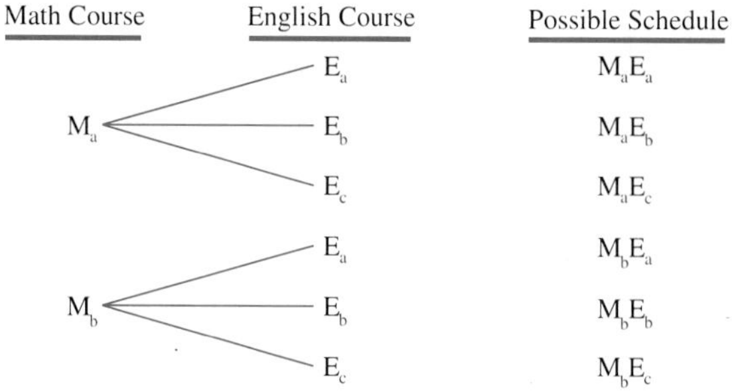

Math Course	English Course	Possible Schedule
	E_a	$M_a E_a$
M_a	E_b	$M_a E_b$
	E_c	$M_a E_c$
	E_a	$M_b E_a$
M_b	E_b	$M_b E_b$
	E_c	$M_b E_c$

Figure 10.4

We start by listing the 2 math sections and then we branch out from these points to the 3 English sections. Reading each branch from left to right, we determine there are 6 schedules as shown in Figure 10.4. Note that the number of possible schedules is the product of the number of math choices and the number of English choices. This example illustrates the following key principle in counting.

Fundamental Counting Principle

> If event A can occur in a ways, and following this, event B can occur in b ways, then the two events taken together can occur in $a \cdot b$ ways.

Furthermore, this principle extends to three or more events as considered in Example 1.

EXAMPLE 1   In how many ways can a three-item true–false test be answered? Include a tree diagram listing the possibilities.

Solution   There are three questions, so we start by making three boxes.

☐ ☐ ☐

Since there are two choices (true or false) for each question, we now enter a 2 in each box and then multiply as indicated in the fundamental counting principle.

$$2 \cdot 2 \cdot 2 = 8$$

Thus, there are 8 ways to answer the test. As shown in Figure 10.5, we can systematically list these choices by writing T and F under the first question and then branching to T and F for each successive question. Following each branch from left to right then gives the possibilities shown.

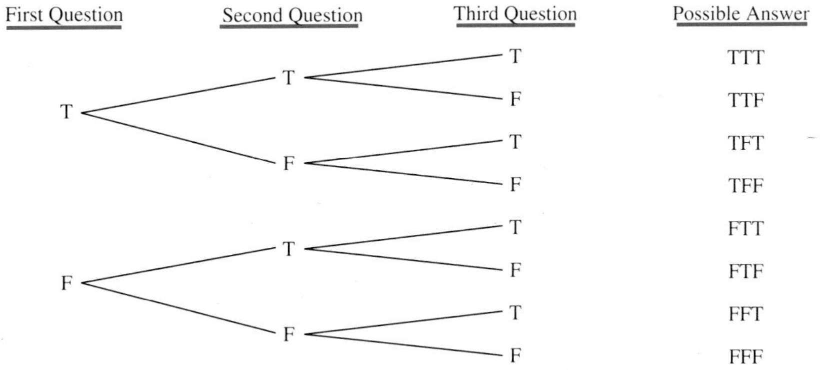

First Question	Second Question	Third Question	Possible Answer
		T	TTT
	T	F	TTF
T		T	TFT
	F	F	TFF
		T	FTT
	T	F	FTF
F		T	FFT
	F	F	FFF

Figure 10.5

EXAMPLE 2   Find the number of different batting orders that are possible for a baseball team that starts 9 players. (Assume the team has only 9 players.)

Solution   The leadoff batter may be any of the 9 players. However, there are only 8 choices for the second position (since the leadoff batter cannot be used again), 7 choices for the third position, and so on. By the fundamental counting principle, this gives

$$9 \cdot 8 \cdot 7 \cdot 6 \cdot 5 \cdot 4 \cdot 3 \cdot 2 \cdot 1 = 362{,}880$$

different batting orders.   ■

The question in Example 2 illustrates an important type of counting problem. A **permutation** of a collection of objects or symbols is an arrangement without repetition of these objects in which order is important. Thus, in Example 2 we say there are 362,880 permutations or orderings of the 9 players on a baseball team. In general, when ordering n objects we have n choices for the first position, $n - 1$ choices for the second position, $n - 2$ choices for the third position, and so on. By using the fundamental counting principle and recalling factorial notation (see Section 10.5), we then conclude the following:

Permutations of a Set of n Objects

> The number of permutations of n different objects using all of them is
>
> $$n(n - 1)(n - 2) \cdots 3 \cdot 2 \cdot 1 = n!$$

EXAMPLE 3   In how many ways can four candidates be arranged on a ballot?

Solution   Since each candidate is listed only once and the order or arrangement is significant, we need to determine the number of permutations of the four candidates. From the rule above, this number is

$$4! = 4 \cdot 3 \cdot 2 \cdot 1 = 24.$$ ■

In many cases it is useful to compute the number of permutations of n objects using only some of the objects in each permutation. For instance, by the fundamental counting principle, the number of four-letter permutations of the 26 letters in the alphabet is $26 \cdot 25 \cdot 24 \cdot 23 = 358{,}800$. We refer to this problem as the number of permutations of 26 objects taken 4 at a time and symbolize this as $_{26}P_4$. Note that

$$_{26}P_4 = \underbrace{26 \cdot 25 \cdot 24 \cdot 23}_{4 \text{ factors}}$$

so that to evaluate $_{26}P_4$ we multiply four consecutive integers that start at 26 and then decrease. This example suggests the following rule.

Permutation Formula

If $_nP_r$ represents the number of permutations of n objects taken r at a time, then

$$_nP_r = \underbrace{n(n-1)(n-2)\cdots(n-r+1)}_{r \text{ factors}} \qquad r \le n.$$

Note in this formula that n is the starting factor while r indicates the number of factors. When the number of factors is large, it is convenient to evaluate $_nP_r$ by applying the formula derived below and using a scientific calculator with a factorial key.

$$_nP_r = n(n-1)(n-2)\cdots(n-r+1)$$
$$= \frac{n(n-1)(n-2)\cdots(n-r+1)[(n-r)\cdots2\cdot1]}{[(n-r)\cdots2\cdot1]}$$

so

$$_nP_r = \frac{n!}{(n-r)!}$$

EXAMPLE 4   In how many ways can nine members from a hockey team be assigned to six different starting positions?

Solution   Since position is important and repetition cannot occur, the number is

$$_9P_6 = \underbrace{9\cdot8\cdot7\cdot6\cdot5\cdot4}_{6 \text{ factors}} = 60{,}480.$$

Using our alternative formula, the evaluation is

$$_9P_6 = \frac{9!}{(9-6)!} = \frac{9!}{3!} = \frac{9\cdot8\cdot7\cdot6\cdot5\cdot4\cdot3\cdot2\cdot1}{3\cdot2\cdot1} = 60{,}480. \qquad \blacksquare$$

When we cannot distinguish between certain members in our collection of objects, we must revise our permutation formula. For instance, how many distinct permutations can be made from the letters of the word ERROR? If we could distinguish among the R's by calling them say R_1, R_2, and R_3, then there would be 5! or 120 permutations of the five letters. Since there are 3! or 6 orderings of the three R's, we could group the 120 possibilities into 20 groups in which the 6 arrangements in each group are not distinguishable when we drop the subscripts. Thus, the number of distinct arrangements of the letters in ERROR is 20, which is 5! ÷ 3!. This example suggests the following rule for analyzing this type of problem.

Distinguishable Permutations Formula

> The number of distinguishable permutations of n objects taken all at a time when n_1 are of one kind, n_2 are of a second kind, . . . , and n_k are of a kth kind, where $n_1 + n_2 + \cdots + n_k = n$, is
>
> $$\frac{n!}{n_1! n_2! \cdots n_k!}.$$

EXAMPLE 5 How many distinct permutations can be made from the letters of the word BEGINNING?

Solution The word BEGINNING has 9 letters with duplications of 3 N's, 2 I's, and 2 G's. By the given theorem the number of distinct permutations is

$$\frac{9!}{3!2!2!} = 15,120.$$ ∎

We now consider one last, but very important, type of counting problem. A distinguishing characteristic of a permutation is that the order of the objects is significant. However, in many situations order does not matter. For example, if we want to know the number of ways in which 2 people can be hired from 5 job applicants, then selecting individuals A and B is the same as selecting B and A. What is important here is only who gets the jobs. A **combination** of a collection of objects or symbols is a selection without repetition in which the order of selection does not matter. In our example problem we therefore need to determine the number of combinations of 5 applicants when taken 2 at a time. To answer this, we first note that if order does matter, there are

$$_5P_2 = 5 \cdot 4 = 20$$

possibilities. However, since the two positions can be ordered in 2! or 2 ways, only half of these possibilities are different when we disregard order. Thus, there are 20 ÷ 2, or 10, combinations of applicants who can be hired. This example illustrates that we analyze combinations as follows:

Combination Formula

> If $_nC_r$ represents the number of combinations of n objects taken r at a time, then
>
> $$_nC_r = \frac{_nP_r}{r!} = \frac{n!}{(n-r)!r!}.$$

Note that the formula for $_nC_r$ matches the formula for the binomial coefficient $\binom{n}{r}$, which we considered in Section 10.5. For this reason, $\binom{n}{r}$ is often used in place of $_nC_r$ to denote the number of combinations of n objects taken r at a time.

EXAMPLE 6   From a class of 16 students an instructor asks 3 students to write their homework on the board. How many student selections are possible?

Solution   Order does not matter and repetition is not possible, so we need to determine the number of combinations of 16 students taken 3 at a time.

$$_{16}C_3 = \frac{_{16}P_3}{3!} = \frac{\overbrace{16 \cdot 15 \cdot 14}^{3 \text{ factors}}}{3!} = 560$$

An alternative evaluation is

$$_{16}C_3 = \frac{16!}{(16-3)!3!} = \frac{16!}{13!3!} = \frac{16 \cdot 15 \cdot 14}{3!} = 560. \qquad \blacksquare$$

EXAMPLE 7   How many different 5-card poker hands can be dealt from a deck of 52 cards?

Solution   The number of combinations of 52 cards taken 5 at a time is

$$_{52}C_5 = \frac{_{52}P_5}{5!} = \frac{52 \cdot 51 \cdot 50 \cdot 49 \cdot 48}{5 \cdot 4 \cdot 3 \cdot 2 \cdot 1} = 2,598,960. \qquad \blacksquare$$

If you use the formula $_nC_r = {}_nP_r/r!$ and r is a large number, it is useful to know that

$$_nC_r = {}_nC_{n-r}.$$

For example, this formula enables us to evaluate $_{16}C_{13}$ by replacing it with $_{16}C_3$ and proceeding as in Example 6. If you consider the alternative evaluation given in this example, you should be able to see why these two combinations are equal.

EXERCISES 10.6

In Exercises 1–10 evaluate the expression.

1. $6!$

2. $(6-3)!$

3. $\dfrac{9!}{5!4!}$

4. $\dfrac{100!}{98!}$

5. $_5P_4$

6. $_5C_4$

7. $_8P_5$

8. $_4P_2$

9. $_{10}C_1$

10. $_{25}C_{22}$

11. One reason 0! is defined as 1 is that this definition helps validate our permutation and combination formulas when r equals n or 0. Use $0! = 1$ when evaluating each of the following.

    a. $_5P_5$

    b. $_5C_5$

    c. $_5C_0$

In Exercises 12–15 answer the question using the fundamental counting principle.

12. In how many ways can a 2-item true–false test be answered? Include a tree diagram listing the possibilities.

13. How many head–tail orderings are possible when a coin is flipped 3 times? Include a tree diagram listing the possibilities.

14. In how many ways can a student choose 1 of 4 math courses and 1 of 3 business courses?

15. In how many ways can a student choose 2 of 3 computer courses and 3 of 4 elective courses?

In Exercises 16–19 answer the question by using the permutation theorems.

16. How many distinct permutations can be made from the letters of the word NUMBER?
17. How many distinct permutations can be made from the letters of the word MISSISSIPPI?
18. In how many ways can 8 members from a basketball team be assigned to 5 different starting positions?
19. How many orders of finish (excluding ties) are possible in a race among 6 people?

In Exercises 20–23 answer the question by using the combination theorems.

20. In how many ways can we obtain 4 volunteers from 10 people?
21. In how many ways can 3 elective courses be selected from 8 choices?
22. How many different 13-card bridge hands can be dealt from a deck of 52 cards?
23. How many subcommittees of 4 can be made from a 12-person committee?

In Exercises 24–40 consider carefully whether order is significant and repetition is possible. Then answer the question using an appropriate theorem.

24. In how many ways can the 12 members of a jury be seated in the jury box?
25. How many 12-member juries can be selected from 18 potential jurors?
26. In how many ways can 2 cards be selected from a deck of 52 cards if the first card is replaced before the second card is chosen? How many ways are possible if the first card is not replaced?
27. In how many ways can 4 red flags and 2 white flags be arranged one above the other?
28. Find the number of different batting orders that are possible for a baseball team that starts 9 players if the pitcher must bat last. (Assume the team has only 9 players.)
29. In a group of 10 people how many handshakes are possible if each person shakes hands once with the other members of the group?
30. In how many ways can the 4 quadrants in the Cartesian coordinate system be colored if 4 different colors are used?

31. In how many ways can 3 of the same math book and 2 of the same English book be arranged on a shelf?
32. In how many ways can a 5-item multiple-choice test with choices a, b, c, and d be answered?
33. How many subcommittees of 3 Republicans and 3 Democrats can be selected from a 10-member committee that includes 6 Republicans and 4 Democrats?
34. In how many ways can we divide 10 people into 2 groups of 5?
35. In how many ways can we divide 10 people into 5 groups of 2?
36. On a test you are asked to answer 4 problems out of 7. How many choices are possible?
37. How many 3-digit odd numbers can be written from the digits 1, 2, 3, 4, and 5 without repetition in the digits?
38. A soccer league consists of 6 teams. If each team plays 2 games with each of the other teams, then find the total number of games played.
39. In how many ways can we introduce, one by one, the 5 nominees for an award?
40. How many line segments can be drawn joining 7 points if no 3 of the points lie on the same line?
41. Solve for n: $_nP_2 = 20$.
42. Solve for n: $_nC_2 = 66$.
43. Use $_nC_r = \dfrac{n!}{(n - r)!r!}$ and prove that $_nC_r = {_nC_{n-r}}$.
44. Use $_nC_r = {_nC_{n-r}}$ and solve $_nC_{16} = {_nC_4}$ for n.

In Exercises 45 and 46 select the choice that completes the statement.

45. If $_nC_a = {_nC_b}$, where $a \neq b$ and a, b, and n are positive integers, then

    a. $n = a + b$ b. $n = a - b$

    c. $n = ab$ d. $n = a/b$

46. If n is a positive integer greater than 1, then

    a. $_nC_n > {_nC_{n-1}}$

    b. $_nC_n = {_nC_{n-1}}$

    c. $_nC_n < {_nC_{n-1}}$

THINK ABOUT IT

1. Create a word problem not discussed in this section that may be solved by computing the given expression.

   a. $6!$ **b.** $_6P_3$ **c.** $_6C_3$

2. In how many different orders may 4 people stand in line? How about a circle? In general, what is the number of distinguishable circular arrangements of n people?

3. By using sigma notation and combinations the binomial theorem may be efficiently stated as follows:

   For any positive integer n,

$$(a + b)^n = \sum_{r=0}^{n} {}_nC_r\, a^{n-r}b^r.$$

   Expand $(a + b)^4$ using this formula.

4. Expand $(x - 2y)^5$ using the formula in Exercise 3.

5. **a.** The rth term in the expansion of $(a + b)^n$ is $_nC_{r-1}a^{n-(r-1)}b^{r-1}$. Use this formula and find the third term in the expansion of $(x + y)^6$.

   b. Find in simplest form the fifth term in the expansion of $(x^2 + x^{-1})^8$ by using the formula in part a.

REMEMBER THIS

1. Express 40 percent in decimal form and in fraction form.

In Exercises 2–5 evaluate each expression.

2. $1 - \dfrac{4}{13}$

3. $\dfrac{1}{2} + \dfrac{1}{6} - \dfrac{1}{12}$

4. $\dfrac{_3C_3}{_{10}C_3}$

5. $\dbinom{10}{4}$

6. Expand $(5x - y)^4$ by the binomial theorem.

7. Write the first four terms in the expansion of $(x + h)^n$, where n is a positive integer.

8. Prove by mathematical induction: For every integer $n \geq 0$,

$$\log a^n = n \log a, \text{ where } a > 0.$$

In Exercises 9 and 10 a ball that always rebounds $\frac{3}{4}$ as far as it falls is dropped from a height of 8 ft.

9. How far up and down has it traveled when it hits the ground for the fifth time?

10. How far up and down has it traveled before it comes to rest?

10.7 Probability

Once we can analyze the number of ways in which certain events can occur, a natural and important question is to ask what the chances are of these events taking place. Probability theory is used to assign a number between 0 and 1, inclusive, that indicates how certain we are that an event will occur. On this scale, impossible events are assigned a probability of 0 and events that must take place are assigned a probability of 1. Before we begin to develop some of the laws for determining probabilities, it is important to understand that there are two varieties of probability statements. To illustrate this, consider the following predictions.

There is an 80-percent chance interest rates will go down.

There is a 60-percent chance the Bears will win.

There is a 50-percent chance a flipped coin will land heads.

There is a 25-percent chance of correctly guessing both answers on a two-item true–false test.

The first two statements are examples of subjective statements in that the stated number is a reflection of (hopefully) an expert's opinion and cannot really be verified. In such cases a different expert will often predict a different number. However, the last two statements are examples of **theoretical probability** in that the numbers are determined by the laws of probability and everyone should obtain the same theoretical result. When we say there is a 50-percent chance a flipped coin will land heads, we note that in practice we rarely obtain 50 heads in 100 flips of a coin, but theoretically

$$\frac{\text{number of heads}}{n \text{ tosses}} \xrightarrow{\text{approaches}} 50 \text{ percent}$$

as n gets larger. This idea that probability is a ratio suggests the following definition for the probability of an event.

Definition of the Probability of Event E

> If an event E can occur in s ways out of a total of T equally likely outcomes, then the probability of E, denoted $P(E)$, is
>
> $$P(E) = \frac{s}{T} = \frac{\text{number of ways } E \text{ can occur}}{\text{total number of equally likely outcomes}}.$$

Note that this definition assures us that if E cannot occur, $P(E) = 0$, while if E is certain, then $P(E) = 1$. Now consider Examples 1–4, which illustrate this definition.

EXAMPLE 1   Find the probability of selecting a spade from a regular deck of 52 cards.

Solution   There are 52 cards with the same chance of being picked. Among these are 13 spades. Thus,

$$P(\text{spade}) = \frac{13}{52}, \text{ or } \frac{1}{4}, \text{ or } 0.25, \text{ or } 25 \text{ percent.}$$

Note that fractional answers are simplified (if possible) and that probabilities often take the form of decimals and percents.   ■

For a given experiment, the set of all possible outcomes is called the **sample space** of the experiment. For some probability problems it is useful to specify the sample space as shown in Example 2. In other cases it is not practical to list all the possibilities, and Examples 3 and 4 consider such problems.

EXAMPLE 2   Two dice are rolled. What is the probability that the sum of the outcomes on the dice is 7?

1,1	1,2	1,3	1,4	1,5	(1,6)
2,1	2,2	2,3	2,4	(2,5)	2,6
3,1	3,2	3,3	(3,4)	3,5	3,6
4,1	4,2	(4,3)	4,4	4,5	4,6
5,1	(5,2)	5,3	5,4	5,5	5,6
(6,1)	6,2	6,3	6,4	6,5	6,6

Figure 10.6

Solution   There are 6 equally likely outcomes on each die, so by the fundamental counting principle, there are 6 × 6 or 36 outcomes to consider. By using a tree diagram and associating each outcome on one die with each possible outcome on the other, we obtain the sample space in Figure 10.6. As encircled in Figure 10.6, there are 6 outcomes that produce a sum of 7, so

$$P(\text{sum } 7) = \frac{6}{36}, \text{ or } \frac{1}{6}.$$ ∎

EXAMPLE 3   From a group of 8 students, 3 are selected randomly for prizes. If Ellen, Michael, and Ilene are among the group, then what is the probability they are the winners?

Solution   The prizes may be awarded to $_8C_3$ or 56 equally likely groups. Only one of these combinations is Ellen, Michael, and Ilene. Thus, the chances are $\frac{1}{56}$.   ∎

EXAMPLE 4   From a committee of 6 men and 6 women, a 3-person subcommittee is randomly chosen. What is the probability that the subcommittee contains 2 men and 1 woman?

Solution   There are $_{12}C_3$ or 220 equally likely subcommittees. Among these are $_6C_2 \cdot {}_6C_1$ or 90 subcommittees with 2 of 6 men and 1 of 6 women. Thus, the chances are $\frac{90}{220}$, or $\frac{9}{22}$.   ∎

To determine probabilities involving more than one event, it is useful to add to our current methods probability formulas that enable us to analyze "and," "or," and "not" statements. We first illustrate the "not" formula. On one roll of a die it is easy to determine that

$$P(4) = \frac{1}{6} \text{ while } P(\text{not } 4) = \frac{5}{6}.$$

Note that $P(4) + P(\text{not } 4) = 1$ or, equivalently, $P(\text{not } 4) = 1 - P(4)$. We generalize this observation as follows:

Formula for P(not E)

> The probability that event E will not occur is given by
>
> $$P(\text{not } E) = 1 - P(E).$$

EXAMPLE 5   If you guess randomly at the answers on a two-item true–false test, what is the probability you do not get both answers right?

Solution   From a tree diagram the possibilities in the sample space are

$$R,R \quad\quad R,W \quad\quad W,R \quad\quad W,W,$$

where the question 1 outcome is listed as the first component in each pair. Since there are only two answer choices, $P(R) = P(W)$, so these outcomes are equally likely. Then

$$P(\text{both right}) = \frac{1}{4} \text{ while } P(\text{not both right}) = 1 - \frac{1}{4} = \frac{3}{4}. \quad \blacksquare$$

In Example 5 we can determine the chances of being right on both questions without listing all the outcomes. Note that

$$P(\text{question 1 right}) = \frac{1}{2}, \text{ and } P(\text{question 2 right}) = \frac{1}{2},$$

$$\text{while } P(\text{both right}) = \frac{1}{4}.$$

Thus,

$$P(\text{both right}) = P(\text{question 1 right}) \cdot P(\text{question 2 right}).$$

In general, we say two events are **independent** if the outcome of one does not affect the outcome of the other, and we have the following theorem for independent events.

Probability of Independent Events

If A and B are independent events, then

$$P(A \text{ and } B) = P(A) \cdot P(B).$$

Furthermore, this rule extends to three or more independent events.

EXAMPLE 6   Three pills and five capsules are in a box. Two of the pills cause drowsiness, while four of the capsules remedy this side effect. What is the probability that a person who swallows a pill and a capsule will be drowsy? What are the chances of not being drowsy?

Solution   For a person to be drowsy, two independent events must occur as follows:

$$P(\text{drowsy}) = P(\text{pill causes drowsiness } and \text{ capsule is not a remedy})$$

$$= \frac{2}{3} \times \frac{1}{5}$$

$$= \frac{2}{15}.$$

Also,

$$P(\text{not drowsy}) = 1 - P(\text{drowsy}) = 1 - \frac{2}{15} = \frac{13}{15}. \quad \blacksquare$$

Finally, we need a probability formula for "or" statements. Two events that cannot occur at the same time are called **mutually exclusive,** and for such events the "or" formula is simply

$$P(A \text{ or } B) = P(A) + P(B).$$

For example, on one selection from a regular deck of cards

$$P(\text{jack or queen}) = P(\text{jack}) \quad + P(\text{queen})$$

$$= \frac{4}{52} \quad + \frac{4}{52}$$

$$= \frac{8}{52}, \text{ or } \frac{2}{13}.$$

When A and B have outcomes in common, we must subtract this overlapping from the sum so we don't count these outcomes twice. Thus, the general "or" formula is

$$P(A \text{ or } B) = P(A) + P(B) - P(A \text{ and } B).$$

For example,

$$\overbrace{\phantom{P(\text{jack and spade})}}^{\text{jack of spades}}$$

$$P(\text{jack or spade}) = P(\text{jack}) + P(\text{spade}) - P(\text{jack and spade})$$

$$= \frac{4}{52} \quad + \frac{13}{52} \quad - \frac{1}{52}$$

$$= \frac{16}{52}, \quad \text{or} \quad \frac{4}{13}.$$

To summarize, the "or" formulas are as follows:

Formulas for $P(A \text{ or } B)$

For any events A and B,

$$P(A \text{ or } B) = P(A) + P(B) - P(A \text{ and } B).$$

When A and B are mutually exclusive, this formula simplifies to

$$P(A \text{ or } B) = P(A) + P(B).$$

EXAMPLE 7   We flip a coin and roll a die. Find the following probabilities by both listing the equally likely outcomes and also using the "and" and "or" formulas.

a. tails on the coin *and* 5 on the die
b. tails on the coin *or* 5 on the die

Solution

a. There are 2 equally likely outcomes for the coin and 6 for the die. So by the fundamental counting principle we must list 2 × 6 or 12 outcomes. By using a tree diagram these outcomes are

$H1$	$H2$	$H3$	$H4$	$H5$	$H6$
$T1$	$T2$	$T3$	$T4$	$T5$	$T6$.

The outcome $T5$ represents tails on the coin *and* 5 on the die, so the chances for this event are $\frac{1}{12}$. We obtain this answer by formula as follows:

$$P(T \text{ and } 5) = P(T) \cdot P(5) = \frac{1}{2} \cdot \frac{1}{6} = \frac{1}{12}.$$

b. The outcomes $T1$, $T2$, $T3$, $T4$, $T5$, $T6$, and $H5$ all contain either a T or a 5, so the probability of tails on the coin *or* 5 on the die is $\frac{7}{12}$. This answer is obtained by formula as follows:

$$P(T \text{ or } 5) = P(T) + P(5) - P(T \text{ and } 5)$$
$$= \frac{1}{2} + \frac{1}{6} - \frac{1}{12}$$
$$= \frac{7}{12}.$$ ■

EXERCISES 10.7

1. A committee consists of 7 men and 5 women. If the chairperson is selected at random, find the probability the chairperson is a woman.

2. If a fair die is rolled once, find the probability of obtaining each of the following.
   a. a 3   **b.** an outcome less than 3
   c. a 7   **d.** an outcome less than 7

3. A card is drawn from a regular deck of 52 cards. Find the probability that the card drawn is
   a. the queen of spades **b.** a heart
   c. a picture card   **d.** not a red queen
   e. an ace or a king   **f.** neither a jack nor a queen
   g. a spade or a 7   **h.** neither a spade nor a 7

4. A student is selected at random from a group in which 30 percent are freshmen, 25 percent are sophomores, 35 percent are juniors, and 10 percent are seniors. Find the probability of selecting each of the following.
   a. a senior
   b. not a senior
   c. a freshman or a sophomore
   d. neither a freshman nor a junior

5. If a woman's chance of winning a prize is $\frac{2}{19}$, then what is her chance of not winning?

6. If the probability of playing a game 3 times and winning all 3 times is $\frac{27}{64}$, then what is the probability of playing the game 3 times and not winning at least once?

7. Two dice are rolled. What is the probability that the sum of the outcomes on the dice is 5?

8. On two rolls of a pair of dice, what is the probability that at least once the sum of the outcomes on the dice is 5?

9. A boy and a girl are playing a game in which both simultaneously call out a number from 1 through 3. Find the probability for each of the following.
   a. Both call an odd number.
   b. Both call the same number.
   c. One of them calls 2, while the other does not.

10. What is the probability that on 4 flips of a coin the result will be 4 tails?

11. Four pills and three capsules are in a box. One of the pills causes drowsiness, while two of the capsules remedy this side effect. What is the probability that a person who swallows a pill and a capsule will be drowsy?

12. From the letters *a*, *b*, *c*, and *d*, three letters are selected at random without replacement. What is the probability that a person will draw the word *bad* in the correct order of spelling?

13. From 10 tickets, 3 different winners are to be chosen. If Chris, Amy, and Pat each holds a ticket, then what is the probability they are the 3 winners?

14. If 5 nominees with different last names are introduced randomly, then what is the probability they are introduced in alphabetical order?

15. What is the probability that a family with 3 children contains 2 boys and 1 girl? (Assume boy and girl are equally likely.)

16. From a committee of 6 men and 4 women, a 3-person subcommittee is randomly chosen. What is the probability that the committee contains 2 women and 1 man?

17. If your pocket contains 7 nickels and 5 quarters and you randomly select three coins, then what is the probability of obtaining exactly 35 cents to pay for a newspaper?
18. On a multiple-choice test with 4 choices, what is the probability you guess randomly at the answers on 3 questions and get 2 out of 3 right?
19. On two independent events, A and B, the probability that A will occur is $\frac{1}{4}$ while the probability that B will occur is $\frac{1}{3}$. What is the probability that neither event occurs?
20. If 40 percent of the population has type A blood, then find the following probabilities for two blood donors chosen at random.
    a. Both are type A.
    b. Neither is type A.
    c. 1 out of 2 is type A.

THINK ABOUT IT

1. Which of the following numbers *cannot* be a probability? Explain why.
$$0, 1, -1, \tfrac{4}{3}, 30\%, 300\%, 1.01, \tfrac{31}{365}$$
2. True or false: If events A and B are mutually exclusive, then A and B are independent events. Explain your answer.
3. Create a word problem that may be solved by computing $3(\frac{1}{6})(\frac{5}{6})^2$.

4. The *odds against an event E* is given by the ratio $P(\text{not } E):P(E)$, or the fraction $P(\text{not } E)/P(E)$. Answer each question using this definition.
   a. When two dice are rolled, find the odds against the sum of the outcomes on the dice being 7.
   b. If the odds against winning in a certain game are $b:a$, then what is the probability of winning?
5. What is the probability of being dealt a poker hand that is any type of flush (that is, five cards all of the same suit)?

REMEMBER THIS

In Exercises 1–6 evaluate each expression.

1. $_{12}P_4$
2. $_{12}C_4$
3. $\binom{12}{4}$
4. $\sum\limits_{n=1}^{10} 3$
5. $1 - \dfrac{3}{5} + \dfrac{9}{25} - \dfrac{27}{125} + \cdots$
6. $3 + 6 + 9 + \cdots + 600$

7. In how many ways can a 10-item true–false test be answered?
8. How many distinct permutations can be made from the letters in the given word?
   a. CODE
   b. PARALLEL
9. Find the middle term in the expansion of $(2x - 3)^6$.
10. Prove by mathematical induction that $n^3 + 2n$ is divisible by 3 for all positive integers n.

CHAPTER OVERVIEW

Section	Key Concepts to Review
10.1	• Definitions of sequence, terms (of the sequence), arithmetic progression, common difference, geometric progression, common ratio, and Fibonacci sequence • Formulas for the nth term: arithmetic progression: $a_n = a_1 + (n-1)d$ geometric progression: $a_n = a_1 r^{n-1}$ Fibonacci sequence: $a_1 = 1$, $a_2 = 1$, $a_n = a_{n-1} + a_{n-2}$ for $n \geq 3$
10.2	• Definitions of series, arithmetic series, and geometric series • Formulas for the sum of the first n terms: arithmetic series: $S_n = \dfrac{n}{2}(a_1 + a_n)$ or $S_n = \dfrac{n}{2}[2a_1 + (n-1)d]$

Section	Key Concepts to Review

geometric series: $S_n = \dfrac{a_1 - a_1 r^n}{1 - r} = \dfrac{a_1(r^n - 1)}{r - 1}$ or $S_n = \dfrac{a_1 - a_n r}{1 - r}$

- The Greek letter Σ (read "sigma") is used to simplify the notation involved with series. We define sigma notation as follows:

$$\sum_{i=1}^{n} a_i = a_1 + a_2 + \cdots + a_n.$$

10.3

- An infinite geometric series with $|r| < 1$ converges to the value or sum S given by $S = \dfrac{a_1}{1 - r}$.

10.4

- Principle of mathematical induction
- We prove by mathematical induction that a statement is true for all positive integers by doing the following.
  1. Show by direct substitution that the statement (or formula) is true for $n = 1$.
  2. Show that if the statement is true for any positive integer k, then it is also true for the next highest integer $k + 1$.
- Two methods for establishing the "$k + 1$" statement from the assumption of the "k" statement

---

10.5

- Binomial theorem: For any positive integer n,

$$(a + b)^n = a^n + \frac{n}{1!}a^{n-1}b + \frac{n(n-1)}{2!}a^{n-2}b^2 + \cdots$$

$$+ \frac{n(n-1)(n-2)\cdots(n-r+1)}{r!}a^{n-r}b^r + \cdots + b^n,$$

  or an alternate form is

$$(a + b)^n = \sum_{r=0}^{n} \binom{n}{r}a^{n-r}b^r.$$

- The binomial coefficient $\binom{n}{r}$ is defined by

$$\binom{n}{r} = \frac{n!}{r!(n-r)!}.$$

- The rth term of the binomial expansion of $(a + b)^n$ is

$$\binom{n}{r-1}a^{n-(r-1)}b^{r-1}.$$

- The symbol $n!$ (read "n factorial") means the product $n \cdot (n - 1) \cdot \cdots \cdot 3 \cdot 2 \cdot 1$.
- Pascal's triangle

10.6

- Definitions of permutation and combination
- Method to construct a tree diagram
- Fundamental counting principle
- Permutation formulas:
  n objects, n at a time: $_nP_n = n!$

Section	Key Concepts to Review

n objects, r at a time: $_nP_r = \underbrace{n(n-1)\cdots(n-r+1)}_{r \text{ factors}} = \dfrac{n!}{(n-r)!}$

- Formula for distinguishable permutations (if certain objects are all alike)
- Combination formulas:

  n objects, r at a time: $_nC_r = \dfrac{_nP_r}{r!} = \dfrac{n!}{(n-r)!r!}$

- The order of the objects is important in a permutation but does not matter in a combination.

10.7

- Definitions of probability of an event, sample space, independent events, and mutually exclusive events
- The probability of an event is a number between 0 and 1, inclusive, with impossible events assigned a probability of 0 and events that must take place assigned a probability of 1.
- Probability formulas
  1. $P(\text{not } E) = 1 - P(E)$
  2. $P(A \text{ and } B) = P(A) \cdot P(B)$ (if A and B are independent events)
  3. a. $P(A \text{ or } B) = P(A) + P(B) - P(A \text{ and } B)$ (for any events A and B)
     b. $P(A \text{ or } B) = P(A) + P(B)$ (if A and B are mutually exclusive)

CHAPTER REVIEW EXERCISES

1. Write the first four terms of the sequence given by $a_n = (-1)^n 2^{n-1}$; also find a_6.

2. Determine a_2, a_4, and a_n so that the following sequence is (a) an arithmetic progression and (b) a geometric progression.

   $5, a_2, 45, a_4, \ldots, a_n, \ldots$

3. Find the sum of the first 70 positive integers.

4. Write the series $\displaystyle\sum_{i=2}^{7} 2^{i-1}$ in expanded form and determine the sum.

5. Write the series $1 + (\frac{1}{2})^2 + (\frac{1}{3})^2 + (\frac{1}{4})^2 + (\frac{1}{5})^2$ in sigma notation.

6. Find the sum of the infinite geometric series

   $1 - \frac{1}{4} + \frac{1}{16} - \frac{1}{64} + \cdots$.

7. By mathematical induction, prove that the sum of the first n terms of an arithmetic series is given by

   $a + (a + d) + (a + 2d) + \cdots$

   $+ [a + (n-1)d] = \dfrac{n}{2}[2a + (n-1)d]$.

8. Express the repeating decimal $3.\overline{4}$ as the ratio of two integers. (Use infinite geometric series.)

9. a. If a, b, and c are the first three terms in an arithmetic progression, express c in terms of a and b.

   b. If a, b, and c are the first three terms in a geometric progression, express c in terms of a and b.

10. $2,000 is split up into 8 prizes by a lottery system in which each award is $10 less than the preceding award. How much money is awarded as first prize?

11. Evaluate $\displaystyle\sum_{k=1}^{\infty} \frac{1}{3^k}$.

12. Find the 20th term in the arithmetic progression 3, 7, 11, ...

13. Evaluate $\displaystyle\sum_{k=1}^{5} k^2$.

14. What is the sum of the first n odd positive integers?

15. In how many different orders may 7 people stand in a line? How about a circle? In how many different orders can 4 men and 3 women stand in a line so each woman is between two men?

16. How many line segments are needed to connect in all possible ways the 9 points on the circle below?

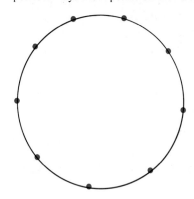

17. The first term in an arithmetic progression is a_1, the nth term is a_n, the common difference is d, and the number of terms is n. Express d in terms of a_1, a_n, and n.

18. Find the sum of the infinite geometric series $3^{-1} + 3^{-2} + 3^{-3} + \cdots$.

19. Is $\sum_{i=0}^{n-1} (2i + 1)$ equal to $\sum_{k=1}^{n} (2k - 1)$?

20. Simplify $_nC_{n-1}$.

21. If $_nC_{12} = {_nC_3}$, find n.

22. Expand $(x + h)^6$ by the binomial theorem.

23. A jack, a queen, and a king are face down on a table. If two of the cards are turned over at random, what is the probability that one of them is a king?

24. If the probability of winning a game is a/b, what is the probability of not winning the game?

25. How many permutations are there of 7 things taken 3 at a time?

26. How many combinations are there of 7 things taken 3 at a time?

27. If $_nP_r = 720$ and $_nC_r = 120$, find n and r.

28. Write the 10th term in the expansion of $(2x - y)^{11}$.

29. Write the fourth term in the expansion of $(\sqrt{x} + 1)^7$.

30. Prove by mathematical induction that the following formula is true for all positive integer values of n.
$$1 + 5 + 9 + \cdots + (4n - 3) = n(2n - 1)$$

31. If 10 percent of the population has type B blood, then what is the probability that two blood donors chosen at random are type B?

32. If 1 card is selected from a regular deck of 52 cards, then what is the probability this card is a club or a picture card?

33. Write the first four terms in the expansion of $(1 - x)^{15}$.

34. A slot machine contains 3 independent wheels. On each wheel a star, a lemon, a grape, or a cherry may appear. How many outcomes are possible on the slot machine?

In Exercises 35–44 select the choice that completes the statement or answers the question.

35. The sequence $\sqrt{2}, \sqrt{3}, \sqrt{4}, \sqrt{5}$ is
    a. an arithmetic progression
    b. a geometric progression
    c. neither an arithmetic nor geometric progression

36. If the first three terms in a geometric progression are b^2, b^x, and b^8, then x equals
    a. 4   b. $4\sqrt{2}$ c. $2\sqrt{2}$ d. 5

37. The next term in the geometric progression $\sqrt[4]{2}, \sqrt{2}, \sqrt[4]{8}, \ldots$ is
    a. 2   b. $2\sqrt{2}$ c. 4   d. 8

38. Which statement is true?
    a. $_{50}C_{10} < {_{50}C_{40}}$ b. $_{50}C_{10} = {_{50}C_{40}}$ c. $_{50}C_{10} > {_{50}C_{40}}$

39. Which number cannot be a probability?
    a. 1   b. 0   c. 2   d. $\frac{1}{2}$

40. How many telegraphic characters can be made using 3 dots and 2 dashes in each character?
    a. 10   b. 8   c. 32   d. 25

41. The coefficient of the x^9y^2 term in the binomial expansion of $(x + y)^{11}$ is
    a. 55   b. 9   c. 11   d. 165

42. The expression $\sum_{n=1}^{5} c$ (where c is a constant) simplifies to
    a. c b. $5c$ c. cn d. $c + 5$

43. The sum of the infinite geometric series
$$5 + \frac{5}{3} + \frac{5}{3^2} + \cdots \text{ is}$$
    a. $\frac{25}{3}$ b. $\frac{45}{8}$ c. $\frac{15}{2}$ d. $\frac{125}{9}$

44. The fifth term in the binomial expansion of $(a - b)^7$ is
    a. $35a^3b^4$ b. $-35a^3b^4$ c. $-21a^2b^5$ d. $21a^2b^5$

CHAPTER 10 TEST

1. State whether the sequence is an arithmetic progression, geometric progression, or neither: $1, 0.1, 0.01, 0.001, \ldots$.

2. What is the 70th term of the arithmetic progression $5, 9, 13, \ldots$?

3. Find the formula for the general term a_n of the geometric progression $27, -18, 12, \ldots$.

4. Find the sum of the first 30 terms of the arithmetic sequence $2, 7, 12, \ldots$.

5. Evaluate $\displaystyle\sum_{n=1}^{6} 100(1.08)^{n-1}$.

6. A certain ball always rebounds $\frac{1}{2}$ as far as it falls. If the ball is dropped from a height of 8 ft, how far up and down has it traveled when it hits the ground for the eighth time?

7. Find the sum of the series $7 - \dfrac{7}{4} + \dfrac{7}{16} - \dfrac{7}{64} + \cdots$.

8. Express the repeating decimal $2.\overline{16}$ as the ratio of two integers. (Use infinite geometric series.)

9. Expand $(x - 3y)^4$ by the binomial theorem.

10. Write the first four terms in the expansion of $(y + 2)^{11}$.

11. Find the sixth term in the expansion of $(x^2 - 1)^7$.

12. How many 7-digit telephone numbers are possible if the first digit cannot be 0?

13. In how many ways can 4 taxi drivers be assigned to 6 taxicabs?

14. On a test you are asked to answer 5 problems out of 8. How many choices are possible?

15. Two dice are rolled. What is the probability that the sum of the outcomes on the dice is 11?

16. What is the probability that a family with 3 children contains at least 1 girl?

17. A card is drawn at random from a regular deck of 52 cards. Find the probability that the card drawn is either a black card or a face card.

18. Prove by mathematical induction: For every positive integer n,
$$10 + 20 + 30 + \cdots + 10n = 5n(n + 1).$$

Appendix

Scientific Notation

Many numbers that appear in scientific work are either very large or very small. For example, the average distance from the earth to the sun is approximately 93,000,000 mi, and the mass of an atom of hydrogen is approximately

$$0.00000000000000000000000017 \text{ g.}$$

To work conveniently with these numbers, we often write them in a form called **scientific notation.**

A positive number is expressed in scientific notation when it is written in the form

$$N = m \times 10^k, \quad \text{where} \quad 1 \leq m < 10 \text{ and } k \text{ is some integer.}$$

For example,

$$93,000,000 = 9.3(10,000,000) = 9.3 \times 10^7$$

$$0.00103 = 1.03\frac{1}{1,000} = 1.03 \times 10^{-3}.$$

To convert a number from regular notation to scientific notation, use the following procedure.

To Convert to Scientific Notation

1. Immediately after the first nonzero digit of the number, place an apostrophe (').
2. Starting at the apostrophe, count the number of places to the decimal point. If you move to the right, your count is expressed as a positive number; if you move to the left, the count is negative.
3. The apostrophe indicates the position of the decimal in the factor between 1 and 10; the count represents the exponent to be used in the factor, which is a power of 10.

The following examples illustrate how this procedure is used. (*Note:* The arrow indicates the direction of the counting.)

Number	=	Number from 1 to 10	×	Power of 10
9'3000000.	=	9.3	×	10^7
1'36	=	1.36	×	10^2
0.0001'36	=	1.36	×	10^{-4}
0.6'2	=	6.2	×	10^{-1}
6'.2	=	6.2	×	10^0

Note from the above examples that a number can be changed from scientific notation to regular notation by adding zeros and moving the decimal point the appropriate number of spaces to the right if the exponent is positive, to the left if the exponent is negative.

Scientific calculators are programmed to work with scientific notation. To enter a number in scientific notation, first enter the significant digits of the number from 1 to 10, press $\boxed{\text{EE}}$ or $\boxed{\text{EXP}}$, and finally enter the exponent of the power of 10. For example, to enter 1.36×10^{-4} press

$$1.36 \ \boxed{\text{EE}} \ 4 \ \boxed{+/-} .$$

The display looks like

$$\boxed{1.36 \ - \ 04}$$

with the exponent appearing on the right in the display. Read the owner's manual to your calculator to learn its scientific notation capabilities. Here are a few features to check for.

1. If a calculation results in an answer with too many digits for display in ordinary notation, the calculator automatically displays the answer in scientific notation.

   Example: 47^5 $47 \ \boxed{y^x} \ 5 \ \boxed{=}$ $\boxed{2.2935 \ 08}$

   Note here that a calculator uses two display places in scientific notation to indicate the exponent and one display place to separate the significant digits from the power of 10. Thus, the display is accurate in this case to only five significant digits. However, $47^5 = 229,345,007$ and the calculator internally carries this answer. Check your owner's manual to determine the accuracy you can expect in displayed and internal values.

2. Entries in ordinary and scientific notation may be mixed in the same problem.

   Example: $(7.6 \times 10^3) + 127$

   $$7.6 \ \boxed{\text{EE}} \ 3 \ \boxed{+} \ 127 \ \boxed{=} \ \boxed{7.727 \ 03}$$

3. On some calculators (notably Texas Instruments models) pressing $\boxed{\text{INV}}$ $\boxed{\text{EE}}$ converts back to ordinary notation.

Example: $(7.6 \times 10^3) + 127$

7.6 $\boxed{\text{EE}}$ 3 $\boxed{+}$ 127 $\boxed{=}$ $\boxed{\text{INV}}$ $\boxed{\text{EE}}$ 7,727

EXERCISES A.1

In Exercises 1–10 write the number in scientific notation by replacing the question marks in the following table with the appropriate number.

Number (N)	=	m	×	10^k
1. 740,000	=	?	×	10^5
2. 0.0005	=	?	×	10^{-4}
3. 0.024	=	2.4	×	$10^?$
4. 240	=	2.4	×	$10^?$
5. 12,300,000	=	?	×	?
6. 0.00000123	=	?	×	?
7. ?	=	4.5	×	10^1
8. ?	=	4.5	×	10^{-1}
9. ?	=	9.08	×	10^{-7}
10. ?	=	9.08	×	10^7

In Exercises 11–20 express each number in scientific notation.

11. 42   **12.** 0.6   **13.** 34,251   **14.** 7.21

15. A light year (that is, the distance light travels in 1 year) is about 5,900,000,000,000 mi.

16. A single red cell of human blood contains about 270,000,000 hemoglobin molecules.

17. A certain radio station broadcasts at a frequency of about 1,260,000 hertz (cycles per second).

18. A certain computer can perform an addition in about 0.000014 second.

19. The weight of an oxygen molecule is approximately 0.000000000000000000000053 g.

20. The wavelength of yellow light is about 0.000023 in.

In Exercises 21–30 express each number in regular notation.

21. 9.2×10^4 **22.** 3×10^{-1}

23. 4.21×10^1 **24.** 6.3×10^0

25. The earth travels about 5.8×10^8 mi in its trip around the sun each year.

26. The number of atoms in 1 oz of gold is approximately 8.65×10^{21}.

27. One coulomb is equal to about 6.28×10^{18} electrons.

28. An atom is about 5×10^{-9} in. in diameter.

29. The mass of a molecule of water is about 3×10^{-23} g.

30. The wavelength of red light is approximately 6.6×10^{-5} cm.

31. If light travels about 186,000 mi/second, about how far will it travel in 5 minutes?

32. If one red blood cell contains about 270,000,000 hemoglobin molecules, about how many molecules of hemoglobin are there in 2 million red cells?

33. If the mass of one electron is about 0.0000000000000000000000000009 g, what is, approximately, the mass of 400 electrons?

34. If a certain computer can perform an addition in about 1.4×10^{-5} second, about how long will it take the computer to perform 50 additions?

A.2   Approximate Numbers

Measurements of one kind or another are essential to both scientific and nontechnical work. Weight, distance, time, volume, and temperature are only a few of the quantities for which measurements are required. We measure through the medium of numbers and arbitrary units, such as the foot (British system) or the meter (metric system). These numbers are approximations and can only be as accurate as the measuring instruments allow. For example, if we measure the height of a man to be 68 in., we are saying that his height to the nearest inch is 68 in. This means that his exact height (h) is somewhere in the interval 67.5 in. $\leq h < 68.5$ in. Using a different measuring device, we might record the height to be 68.2 in., so that 68.15 in. $\leq h < 68.25$ in.

The exact height is contained in these intervals because the measurements are estimated by rounding off to some decimal place. We round off by considering the digit in the next place to the right of the desired decimal place. If this digit is less than 5, the digit in the desired decimal place remains the same; if the digit is 5 or greater, the digit in the desired decimal place is increased by one. The digits to the right of the desired decimal place are then dropped and replaced by zeros if the zeros are needed to maintain the position of the decimal point. Rounding off is best understood by considering the following table, which indicates how we round off a measurement of 265.307 ft to various places.

Precision Desired	Measurement of 265.307 Ft Is Given as
Nearest 100 ft	300 ft
Nearest 10 ft	270 ft
Nearest foot	265 ft
Nearest tenth of a foot	265.3 ft
Nearest hundredth of a foot	265.31 ft

The **precision** of a number refers to the place at which we are rounding off. A number rounded off to the nearest hundredth is more precise than a number rounded off to the nearest tenth, and so on.

Another important consideration is the number of **significant digits** in our approximation. All digits, except the zeros that are required to indicate the position of the decimal point, are significant. The following table illustrates this concept.

Approximate Number	Number of Significant Digits
15.13	Four
0.44	Two
0.003	One; the number is *three* thousandths and only the 3 is significant. The zeros are needed to indicate the position of the decimal point.
0.0196	Three
307	Three; the zero is *not* used to indicate the position of the decimal point. Zeros between nonzero digits are always significant.
50.007	Five
8.0	Two; the zero is significant because it shows that the number is rounded off to the nearest tenth. The zero is not required to write the number 8 and should not be written unless it is significant.
0.2600	Four

230	Two; we do not have sufficient information to determine if the number is rounded off in the tens place or to the nearest unit. Unless stated otherwise, we assume the number to be rounded off in the tens place so the zero is needed to indicate the position of the decimal point.
2,000	One
2,00$\overline{0}$	Four; the bar is used to avoid ambiguity by indicating that the number is rounded off at the digit below the bar.
2,$\overline{0}$00	Two

The **accuracy** of a number is determined by the number of significant digits in the number. Thus, 21 (two significant digits) is a more accurate number than 0.07 (one significant digit). It is important to distinguish between the accuracy and the precision of a number. For example, consider the following pairs of numbers.

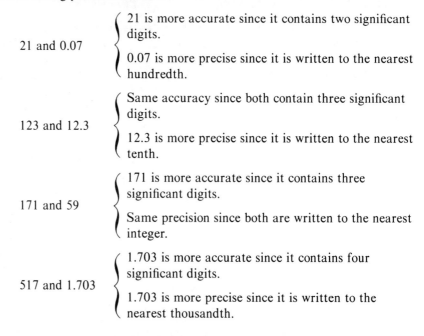

21 and 0.07
- 21 is more accurate since it contains two significant digits.
- 0.07 is more precise since it is written to the nearest hundredth.

123 and 12.3
- Same accuracy since both contain three significant digits.
- 12.3 is more precise since it is written to the nearest tenth.

171 and 59
- 171 is more accurate since it contains three significant digits.
- Same precision since both are written to the nearest integer.

517 and 1.703
- 1.703 is more accurate since it contains four significant digits.
- 1.703 is more precise since it is written to the nearest thousandth.

Operations with Approximate Numbers

When computing with approximate numbers, we should express our results to a precision or accuracy that is appropriate to the data. We establish guidelines for the result by considering the error involved in computing with such numbers. For example, suppose that $x \approx 123$ and $y \approx 4.27$. Adding these numbers, we have $123 + 4.27 = 127.27$. However, because of rounding off, we know that the exact values of x and y are in the intervals

$$122.5 \leq x < 123.5$$
$$4.265 \leq y < 4.275.$$

Adding the minimum values to find the minimum possible sum, and adding the maximum values to find the maximum possible sum, we have

$$122.5 + 4.265 \le x + y < 123.5 + 4.275$$
$$126.765 \le x + y < 127.775.$$

To the nearest integer, the minimum sum is 127 and the maximum sum is 128. Thus, the precision in the result may not be given to more than the nearest integer, and there is even a chance of error if we are that precise with our answer. This example motivates us to the following guideline.

Rounding Off in Addition or Subtraction

> When **adding** or **subtracting** approximate numbers, perform the operation and then round off the result to the **precision** of the least precise number.

Thus, if $x \approx 123$ and $y \approx 4.27$, then $x + y \approx 123 + 4.27 = 127.27 \approx 127$. We round off to the nearest integer because 123, the less precise number, is rounded off to the nearest integer.

Before adding or subtracting, it is permissible to round off the more precise numbers since the extra digits are not considered when we round off our result. If this is done, the numbers should be rounded off to one place beyond that of the least precise number.

EXAMPLE 1   Add the approximate numbers: 12.79, 4.3131, 46.2, 9.618.

Solution   46.2 is the least precise number, so before adding we may round off the other numbers to the nearest hundredth.

$$
\begin{array}{r}
12.79 \\
4.31 \\
46.2 \\
\underline{9.62} \\
72.92
\end{array}
$$

Rounding off to the nearest tenth, the result is 72.9.   ■

EXAMPLE 2   Perform the indicated operation for the approximate numbers: $0.396 - 17.6 + 150$.

Solution   150 is the least precise number. We assume that 150 is rounded off in the tens place, so before performing the operations, we round off the other numbers to the nearest integer. Thus, we have $0 - 18 + 150 = 132$. Rounding off in the tens place, the result is 130.   ■

To determine the possible error when multiplying or dividing approximate numbers, consider the product of $x \approx 12$ and $y \approx 4.27$.

$$x \cdot y \approx 12(4.27) = 51.24$$

Because of rounding off, we know that the exact values of x and y are in the intervals

$$11.5 \leq x < 12.5$$
$$4.265 \leq y < 4.275.$$

Multiplying the minimum values to find the minimum possible product and the maximum values to find the maximum possible product, we have

$$(11.5)(4.265) \leq xy < (12.5)(4.275)$$
$$49.0475 \leq xy < 53.4375.$$

To two significant digits, the minimum product is 49 and the maximum product is 53. Thus, the accuracy of the result may not be given to more than two significant digits, and even that accuracy may be wrong. This example motivates us to the following guideline.

Rounding Off in Multiplication or Division

When **multiplying** or **dividing** approximate numbers, perform the operation and then round off the result to the **accuracy** of the least accurate number.

Thus, if $x \approx 12$ and $y \approx 4.27$, then $x \cdot y \approx 12(4.27) = 51.24 \approx 51$. We round off to two significant digits because 12, the less accurate number, has two significant digits.

Before multiplying or dividing, it is permissible to round off the more accurate numbers to one more significant digit than the least accurate number. If you are using a calculator, you may find it easier to key in the number without rounding off. Either way, you will usually obtain the same final result.

Since finding a power or root of a number involves multiplication, we are concerned with accuracy in such problems. The result should contain the same number of significant digits as the given approximate number.

EXAMPLE 3   Multiply the approximate numbers: $(11.5)(0.042649)$.

Solution   11.5 is the less accurate number, so before multiplying we may round off the other number to four significant digits:

$$(11.5)(0.04265) = 0.490475.$$

Rounding off to three significant digits, the result is 0.490.   ∎

EXAMPLE 4   Perform the indicated operations for the approximate numbers.

$$\frac{(215.1)(7.63)}{0.0002}$$

Solution   0.0002 is the least accurate number. We round off the other numbers to two significant digits and then perform the operation.

$$\frac{220(7.6)}{0.0002} = 8,360,000$$

Rounding off to one significant digit, the result is 8,000,000.   ■

EXAMPLE 5   Find the approximate value of $\sqrt{79.1}$.

Solution   Using a calculator, the initial result is $\sqrt{79.1} \approx 8.8938181$. Since 79.1 contains three significant digits, the final result is 8.89.   ■

Exact Numbers

Although most of the numbers in scientific work are approximate, there are some numbers that are exact. Exact numbers are often the result of a counting process or a definition and are commonly seen in formulas. The following are examples of situations in which the numbers are exact.

1. There are 25 students in the class (counting process).
2. Of 100 light bulbs tested, 4 were defective (counting process).
3. There are 60 minutes in 1 degree (definition).
4. A bank pays 6-percent interest (definition).
5. $P = 4s$ (formula for the perimeter of a square).
6. $A = \frac{1}{2}ab$ (formula for the area of a triangle).

Since these numbers are exact, no error is involved when using them in computations. Only approximate numbers must be considered in determining the accuracy or precision of a result.

EXERCISES A.2

In Exercises 1–20 determine the number of significant digits in the approximate number.

1. 123.14
2. 0.59
3. 0.02
4. 0.000491
5. 904
6. 30.03
7. 22.0
8. 0.78000
9. 670
10. 80,000
11. 5.40
12. 540
13. 0.054
14. 54.0
15. 504
16. 540.0
17. 5,40$\overline{0}$
18. 5,4$\overline{0}$0
19. 0.010
20. 0.100

In Exercises 21–32 round off the numbers (a) to four significant digits and (b) to two significant digits.

21. 12.3456
22. 46.054
23. 59,372
24. 72,488
25. 29,697
26. 59,999
27. 84.096
28. 10.998
29. 0.068547
30. 0.0034581
31. 0.0020006
32. 0.39999

In Exercises 33–44 determine the approximate number in each pair that is (a) more accurate and (b) more precise.

33. 423, 0.004

34. 18, 0.700

35. 402, 40.2

36. 312, 98

37. 6.430, 0.304

38. 0.9, 12.0

39. 2,000, 0.0002

40. 1,230, 3,200

41. 9,080, 48.2

42. $4,\overline{0}00, 4\overline{0}$

43. $3\overline{0}0, 3\overline{0}0$

44. $8,88\overline{0}, 8.008$

In Exercises 45–64 perform the indicated operations for the approximate numbers.

45. $16.27 + 2.1515 - 4.3$

46. $15 - 3.043 + 103.1$

47. $0.002 - 11.0 + 3.59$

48. $0.200 + 4.6 + 120$

49. $0.3047 + 0.8 + 0.092 + 0.69$

50. $214.32 + 1,000 + 0.75 + 3.2$

51. $(2.1)(0.57892)$

52. $(0.08)(489.0)$

53. $(16.4)(2,001)$

54. $(7.00)(0.094265)$

55. $408 \div 0.002$

56. $602 \div 0.200$

57. $\dfrac{(799.4)(8.6)}{0.04}$

58. $\dfrac{(0.460)(126.1)}{2,200}$

59. $9.714 + (2.3)(0.812)$

60. $(4.50)(0.1234) - 5.8$

61. $\dfrac{4.07}{0.06} + (0.8715)(20)$

62. $5(0.0861) - \dfrac{0.90}{123}$

63. $\sqrt{95.0}$

64. $\sqrt{0.040}$

A.3 Logarithmic Computations and Tables

To determine by table the common logarithm of a positive number N (or, equivalently, the exponent above 10 that produces N), we first express N in the scientific notation form

$$N = m \times 10^k, \quad \text{where} \quad 1 \le m < 10 \text{ and } k \text{ is some integer.}$$

Finding the logarithm to the base 10 of N, we get

$$\log_{10} N = \log_{10}(m \times 10^k) = \log_{10} m + \log_{10} 10^k$$
$$= \log_{10} m + k \log_{10} 10.$$

Since $\log_{10} 10 = 1$, we have

$$\log_{10} N = (\log_{10} m) + k.$$

Thus, $\log_{10} N$ is the sum of two numbers: an integer, k, called the **characteristic,** and $\log_{10} m$, called the **mantissa.** Since $1 \le m < 10$, we find the mantissa by consulting Table 2 at the end of the book, which gives approximations of the logarithms (to the base 10) of numbers between 1 and 10. Consider the excerpt from this table shown in Figure A.1. The first column on the left contains the first two significant digits of m, and the top row contains the third significant digit of m. For example, to find $\log_{10} 5.68$ we look at the intersection of the row opposite 5.6 and the column headed by 8. Thus,

$$\log_{10} 5.68 \approx 0.7543.$$

m	0	1	2	3	4	5	6	7	8	9
5.5	.7404	.7412	.7419	.7427	.7435	.7443	.7451	.7459	.7466	.7474
5.6	.7482	.7490	.7497	.7505	.7513	.7520	.7528	.7536	.7543	.7551
5.7	.7559	.7566	.7574	.7582	.7589	.7597	.7604	.7612	.7619	.7627
5.8	.7634	.7642	.7649	.7657	.7664	.7672	.7679	.7686	.7694	.7701
5.9	.7709	.7716	.7723	.7731	.7738	.7745	.7752	.7760	.7767	.7774

Figure A.1

The following examples should clarify the procedure for finding the logarithm to the base 10 of any positive number. In most cases the equal sign will be used to relate logarithms, with the understanding that these logarithms are only approximations.

EXAMPLE 1   Find $\log_{10} 427$.

Solution

$$\log_{10} 427 = \log_{10}(4.27 \times 10^2)$$
$$= \log_{10} 4.27 + \log_{10} 10^2$$

Since $\log_{10} 10^2 = 2$ and $\log_{10} 4.27 \approx 0.6304$ (Table 2), we have

$$\log_{10} 427 = 0.6304 + 2 = 2.6304.$$ ■

EXAMPLE 2   Find $\log_{10} 80$.

Solution

$$\log_{10} 80 = \log_{10}(8.00 \times 10^1)$$
$$= \log_{10} 8.00 + \log_{10} 10^1$$

Since $\log_{10} 10^1 = 1$ and $\log_{10} 8.00 \approx 0.9031$ (Table 2), we have

$$\log_{10} 80 = 0.9031 + 1 = 1.9031.$$ ■

EXAMPLE 3   Find $\log_{10} 0.485$.

Solution $\log_{10} 0.485 = \log_{10}(4.85 \times 10^{-1}) = \log_{10} 4.85 + \log_{10} 10^{-1}$
Since $\log_{10} 10^{-1} = -1$ and $\log_{10} 4.85 \approx 0.6857$ (Table 2), we have

$$\log_{10} 0.485 = 0.6857 + (-1) \text{ or } -0.3143.$$ ■

EXAMPLE 4   Find $\log_{10} 0.00123$.

Solution

$$\log_{10} 0.00123 = \log_{10}(1.23 \times 10^{-3})$$
$$= \log_{10} 1.23 + \log_{10} 10^{-3}$$

Since $\log_{10} 10^{-3} = -3$ and $\log_{10} 1.23 \approx 0.0899$ (Table 2), we have

$$\log_{10} 0.00123 = 0.0899 + (-3) \text{ or } -2.9101.$$ ■

To perform computations, it is necessary to be able to reverse the above procedure and find N if we know $\log_{10} N$. The following examples illustrate this procedure.

EXAMPLE 5   If $\log_{10} N = 2.5250$, find N.

Solution   We know $N = m \times 10^k$, where $1 \leq m < 10$ and k is the characteristic.

$$\log_{10} N = 2.5250 = 0.5250 + 2$$

Since $\log_{10} 3.35 \approx 0.5250$ (Table 2), we have $m = 3.35$. Since 2 is the characteristic, we have $k = 2$. Thus,

$$N = 3.35 \times 10^2 = 335.$$ ∎

EXAMPLE 6   If $\log_{10} N = 0.9990 + (-3)$, find N.

Solution   We know that $N = m \times 10^k$, where $1 \le m < 10$ and k is the characteristic.

$$\log_{10} N = 0.9990 + (-3)$$

Since $\log_{10} 9.98 \approx 0.9990$ (Table 2), we have $m = 9.98$. Since -3 is the characteristic, we have $k = -3$. Thus,

$$N = 9.98 \times 10^{-3} = 0.00998.$$ ∎

EXAMPLE 7   If $\log_{10} N = -1.4567$, find N.

Solution   We must be careful because -1.4567 *does not equal* $0.4567 + (-1)$. We must have a positive mantissa to use Table 2, so we change the form of -1.4567 as follows:

$$-1.4567 = (-1.4567 + 2) + (-2)$$
$$= 0.5433 + (-2).$$

We know that $N = m \times 10^k$, where $1 \le m < 10$ and k is the characteristic. Since $\log_{10} 3.49 \approx 0.5433$ (Table 2), $m = 3.49$. Since -2 is the characteristic, $k = -2$. Thus,

$$N = 3.49 \times 10^{-2} = 0.0349.$$ ∎

The following examples illustrate the use of logarithms in difficult computations. The properties of logarithms discussed in Section 4.3 are essential to this work.

EXAMPLE 8   Compute $[(343)(0.678)]^{10}$.

Solution

$$N = [(343)(0.678)]^{10}$$
$$\log_{10} N = \log_{10}[(343)(0.678)]^{10}$$
$$= 10(\log_{10} 343 + \log_{10} 0.678)$$
$$= 10\{2.5353 + [0.8312 + (-1)]\}$$
$$= 10(2.3665)$$
$$= 23.665$$

We know that $N = m \times 10^k$, where $1 \le m < 10$ and k is the characteristic. Since $\log_{10} 4.62 \approx 0.6650$ (Table 2), $m = 4.62$. Since 23 is the characteristic, $k = 23$. Thus,

$$N = 4.62 \times 10^{23}$$
$$= 462,000,000,000,000,000,000,000.$$ ∎

EXAMPLE 9   Find an approximate value for $2^{\sqrt{2}}$. (*Note:* $\sqrt{2} \approx 1.41$.)

Solution

$$N = 2^{\sqrt{2}}$$
$$\log_{10} N = \log_{10} 2^{\sqrt{2}} = \sqrt{2} \log_{10} 2$$

Referring to Table 2, we have $\log_{10} 2 \approx 0.3010$.

$$\log_{10} N = \sqrt{2}(0.3010) = (1.41)(0.3010)$$
$$= 0.4244$$

We know that $N = m \times 10^k$, where $1 \le m < 10$ and k is the characteristic. Since $\log_{10} 2.66 \approx 0.4244$ (Table 2), $m = 2.66$. Since 0 is the characteristic, $k = 0$. Thus,

$$N = 2.66 \times 10^0 = 2.66.$$ ∎

EXAMPLE 10   Compute $\sqrt[5]{7.13/22.1}$.

Solution

$$N = \sqrt[5]{\frac{7.13}{22.1}}$$

$$\log_{10} N = \log_{10} \left(\frac{7.13}{22.1}\right)^{1/5}$$
$$= \tfrac{1}{5}(\log_{10} 7.13 - \log_{10} 22.1)$$
$$= \tfrac{1}{5}(0.8531 - 1.3444)$$
$$= \tfrac{1}{5}(-0.4913)$$
$$= -0.0983$$

To find N, we must write the logarithm in a form in which the mantissa is positive.

$$\log_{10} N = (-0.0983 + 1) + (-1)$$
$$= 0.9017 + (-1)$$

We know that $N = m \times 10^k$, where $1 \le m < 10$ and k is the characteristic. Since $\log_{10} 7.97 \approx 0.9017$ (Table 2), $m = 7.97$. Since -1 is the characteristic, $k = -1$. Thus,

$$N = 7.97 \times 10^{-1} = 0.797.$$ ∎

One method of determining the natural log of a number is to change the expression to base 10 logarithms.

EXAMPLE 11   Use logarithms to the base 10 to determine $\ln 170$.

Solution

$$\ln 170 = \frac{\log_{10} 170}{\log_{10} e} = \frac{\log_{10} 170}{\log_{10} 2.72} = \frac{2.2304}{0.4346}$$

$$\ln 170 \approx 5.132$$ ∎

A second method, illustrated in the following example, is more accurate and uses a table of natural logarithms (Table 3 at the end of the book).

EXAMPLE 12   Find ln 170.

Solution

$$\ln 170 = \ln(1.70 \times 10^2)$$
$$= \ln 1.7 + \ln 10^2$$
$$= \ln 1.7 + (2) \ln 10$$

Referring to Table 3, we have ln 1.7 ≈ 0.5306 and ln 10 ≈ 2.3026.

$$\ln 170 = 0.5306 + 2(2.3026)$$
$$\ln 170 \approx 5.136$$

EXAMPLE 13   Find ln 0.45.

Solution

$$\ln 0.45 = \ln(4.5 \times 10^{-1})$$
$$= \ln 4.5 + \ln 10^{-1}$$
$$= \ln 4.5 + (-1) \ln 10$$

Referring to Table 3, we have ln 4.5 ≈ 1.5041 and ln 10 ≈ 2.3026. Thus,

$$\ln 0.45 = 1.5041 + (-1)(2.3026)$$
$$= -0.7985.$$

EXAMPLE 14   If ln x = 0.95, find x.

Solution   ln x = 0.95. Writing this statement in exponential form, we have

$$e^{0.95} = x$$
$$2.5857 = x. \quad \text{(Table 4)}$$

EXAMPLE 15   If $e^{0.07t}$ = 2, find t by using natural logarithms.

Solution   If $e^{0.07t}$ = 2, then ln 2 = 0.07t. By Table 3, ln 2 = 0.6931. Thus, t = 0.6931/0.07 ≈ 9.9.

EXERCISES A.3

In Exercises 1–10 find the logarithm to the base 10 of the number. Use Table 2.

1. 2.34   **2.** 23.4   **3.** 0.234
4. 0.0234   **5.** 2,340   **6.** 187,000
7. 90   **8.** 0.9   **9.** 0.111
10. 0.000717

In Exercises 11–20 find N if $\log_{10} N$ equals the number. Use Table 2.

11. 0.8189   **12.** 2.8189
13. 0.8189 + (−1)   **14.** 0.8189 + (−3)

15. 1.5559   **16.** 3.9803
17. −1.4123   **18.** −2.6712
19. −0.5173   **20.** −4.2990

In Exercises 21–40 compute by means of logarithms.

21. (413)(2.74)   **22.** (0.0123)(12.7)
23. $\dfrac{908}{14.3}$ **24.** $\dfrac{19.2}{604}$
25. $(3.2)^5$ **26.** $(0.000491)^4$
27. $\sqrt{97}$ **28.** $\sqrt[5]{0.4}$

29. $\dfrac{(514)(29.6)}{0.02}$

30. $\dfrac{17.9}{(49.6)(0.0821)}$

31. $\sqrt{(16.1)(2.5)}$

32. $\sqrt[3]{(123)(0.22)^2}$

33. $\sqrt[6]{\dfrac{195}{7.6}}$

34. $\sqrt{\dfrac{22.2}{459}}$

35. $\dfrac{(1.8)^2(5.6)^3}{\sqrt{104}}$

36. $\dfrac{\sqrt[3]{17}}{(671)(0.08)^5}$

37. $3^{\sqrt{3}}$ (*Note:* $\sqrt{3} \approx 1.73$.)

38. 2^{π} (*Note:* $\pi \approx 3.14$.)

39. $(12.1)^{-5}$

40. $(1 + 0.01)^{-24}$

In Exercises 41–50 use Table 3 to determine the natural logarithm.

41. ln 7 **42.** ln 3.8 **43.** ln 42
44. ln 123 **45.** ln 6,500 **46.** ln 100
47. ln 0.56 **48.** ln 0.009 **49.** ln 0.00044
50. ln 0.292

In Exercises 51–58 find x if ln x equals the number.

51. 0.19 **52.** 4.6 **53.** 2.3 **54.** 0.55
55. −0.24 **56.** −7.5 **57.** −3.7 **58.** −0.09

A.4   Graphs on Logarithmic Paper

Sometimes we need to analyze data that range over a large collection of values. In addition, the analysis often requires us to find a formula that fits a set of data. Here we study a scheme that is useful in such situations.

On logarithmic graph paper the axes are laid off at distances proportional to the logarithms of numbers. Consider Figure A.2, in which we construct an axis by marking on the paper the following numbers:

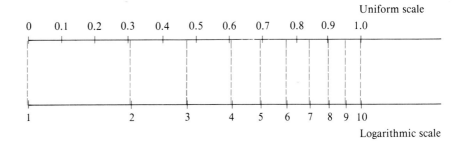

Figure A.2

$$\log_{10} 1 = 0 \qquad \log_{10} 6 = 0.778$$
$$\log_{10} 2 = 0.301 \qquad \log_{10} 7 = 0.845$$
$$\log_{10} 3 = 0.477 \qquad \log_{10} 8 = 0.903$$
$$\log_{10} 4 = 0.602 \qquad \log_{10} 9 = 0.954$$
$$\log_{10} 5 = 0.699 \qquad \log_{10} 10 = 1 \qquad \text{etc.}$$

On the logarithmic scale we use 1 as shorthand for $\log_{10}$ 1. Since $\log_{10} x$ is undefined for $x \leq 0$, only positive numbers may be marked off on logarithmic paper. Using this paper, we may obtain accuracy in a graph in which the variables range over a large collection of values.

Logarithmic paper is especially useful when graphing functions of the form $y = ax^m$, which are called **power functions.** By taking the logarithm to the base 10 of a power function, we have

$$\log y = \log ax^m$$
$$= \log a + m \log x.$$

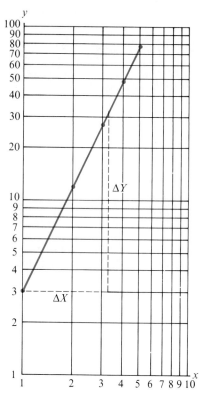

Figure A.3 $y = 3x^2$

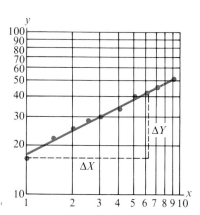

Figure A.4 $y = 17\sqrt{x}$

If we let log $y = Y$, log $x = X$, and log $a = b$, the equation becomes

$$Y = b + mX.$$

Thus, on logarithmic paper the graph of a power function is a straight line with slope m and Y-intercept b. It also follows that if the graph of experimental data is a straight line on logarithmic paper, the variables are related by an equation of the form $y = ax^m$. The slope m is measured with a ruler and the coefficient a is the intercept on the vertical axis at $x = 1$.

EXAMPLE 1   Use logarithmic paper to graph $y = 3x^2$.

Solution   First, construct a table of corresponding values of x and y.

x	1	2	3	4	5
y	3	12	27	48	75

As shown in Figure A.3, the graph is a straight line. Using a ruler you may verify that $m = \Delta Y/\Delta X = 2$. The intercept at $x = 1$ is 3. These readings agree with the constants in the equation $y = 3x^2$. In this example the y-axis is scaled in two cycles from log 1 = 0 to log 10 = 1 to log 100 = 2. Logarithmic paper may contain any number of cycles. A cycle ranges between any two numbers whose logarithms are consecutive integers. ■

EXAMPLE 2   Determine the equation relating the variables in the following table.

x	1.0	1.5	2.0	2.5	3.0	4.0	5.0	6.0	7.0	9.0
y	16	21	24	27	29	33	38	41	44	50

Solution   Consider Figure A.4. The vertical axis requires one cycle that ranges from log 10 = 1 to log 100 = 2. Since the graph is approximately a straight line, the equation relating the variables is a power function. Using a ruler, $m = \Delta Y/\Delta X \approx 0.5$. The intercept at $x = 1$ is about 17. Thus, the power function is $y \approx 17x^{0.5}$ or $y \approx 17\sqrt{x}$. ■

On semilogarithmic graph paper one axis is scaled as on logarithmic paper and the other axis is scaled as on Cartesian coordinate paper. This paper is used when only one variable ranges over a large collection of values. Semilogarithmic paper is especially useful when graphing exponential functions since the graph is a straight line.

EXAMPLE 3   Use semilogarithmic paper to graph $y = 2^x$.

Solution   First, construct a table of corresponding values of x and y.

x	-2	-1	0	1	2	3
y	0.25	0.5	1	2	4	8

The vertical axis requires two logarithmic cycles. The horizontal axis is equally spaced and is scaled with positive and negative numbers. As shown in Figure A.5, the graph is a straight line.

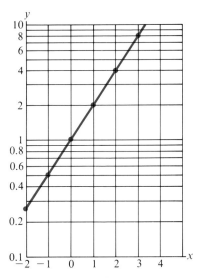

Figure A.5 $y = 2^x$ ■

Although the graphs in the examples are all straight lines, other curves may result. We obtain a straight line only when graphing a power function on logarithmic paper and an exponential function on semilogarithmic paper.

EXERCISES A.4

In Exercises 1–6 graph the functions on logarithmic paper. Also, find the slope and the intercept at $x = 1$.

1. $y = x^2$ **2.** $y = 2x^3$ **3.** $y = 4\sqrt{x}$
4. $y = 5x^{2/3}$ **5.** $xy = 1$ **6.** $x^2 y = 10$

In Exercises 7–10 use logarithmic paper to determine the equation relating the variables in the given tables.

7.

x	1	2	3	4	5	6	7
y	7.0	5.5	4.8	4.3	4.0	3.7	3.5

8.

x	1.0	2.0	3.0	4.0	5.0	6.0	7.0
y	5.0	1.4	.66	.39	.26	.19	.14

9.

x	1	3	5	7	9	25	30	70
y	3.0	6.0	8.6	11	13	25	28	49

10.

x	1.5	4.5	7.5	11	16	20	30
y	2.1	8.4	16	25	40	53	88

In Exercises 11–14 graph the functions on semilogarithmic paper.

11. $y = 2^{-x}$ **12.** $y = 5^x$
13. $y = 0.5(3)^x$ **14.** $y = 2(3^{-x})$

Tables

Table 1 **Squares, Square Roots, and Prime Factors**

No.	Sq.	Sq. rt.	Factors	No.	Sq.	Sq. rt.	Factors
1	1	1.000		51	2,601	7.141	$3 \cdot 17$
2	4	1.414	2	52	2,704	7.211	$2^2 \cdot 13$
3	9	1.732	3	53	2,809	7.280	53
4	16	2.000	2^2	54	2,916	7.348	$2 \cdot 3^3$
5	25	2.236	5	55	3,025	7.416	$5 \cdot 11$
6	36	2.449	$2 \cdot 3$	56	3,136	7.483	$2^3 \cdot 7$
7	49	2.646	7	57	3,249	7.550	$3 \cdot 19$
8	64	2.828	2^3	58	3,364	7.616	$2 \cdot 29$
9	81	3.000	3^2	59	3,481	7.681	59
10	100	3.162	$2 \cdot 5$	60	3,600	7.746	$2^2 \cdot 3 \cdot 5$
11	121	3.317	11	61	3,721	7.810	61
12	144	3.464	$2^2 \cdot 3$	62	3,844	7.874	$2 \cdot 31$
13	169	3.606	13	63	3,969	7.937	$3^2 \cdot 7$
14	196	3.742	$2 \cdot 7$	64	4,096	8.000	2^6
15	225	3.873	$3 \cdot 5$	65	4,225	8.062	$5 \cdot 13$
16	256	4.000	2^4	66	4,356	8.124	$2 \cdot 3 \cdot 11$
17	289	4.123	17	67	4,489	8.185	67
18	324	4.243	$2 \cdot 3^2$	68	4,624	8.246	$2^2 \cdot 17$
19	361	4.359	19	69	4,761	8.307	$3 \cdot 23$
20	400	4.472	$2^2 \cdot 5$	70	4,900	8.367	$2 \cdot 5 \cdot 7$
21	441	4.583	$3 \cdot 7$	71	5,041	8.426	71
22	484	4.690	$2 \cdot 11$	72	5,184	8.485	$2^3 \cdot 3^2$
23	529	4.796	23	73	5,329	8.544	73
24	576	4.899	$2^3 \cdot 3$	74	5,476	8.602	$2 \cdot 37$
25	625	5.000	5^2	75	5,625	8.660	$3 \cdot 5^2$
26	676	5.099	$2 \cdot 13$	76	5,776	8.718	$2^2 \cdot 19$
27	729	5.196	3^3	77	5,929	8.775	$7 \cdot 11$
28	784	5.292	$2^2 \cdot 7$	78	6,084	8.832	$2 \cdot 3 \cdot 13$
29	841	5.385	29	79	6,241	8.888	79
30	900	5.477	$2 \cdot 3 \cdot 5$	80	6,400	8.944	$2^4 \cdot 5$
31	961	5.568	31	81	6,561	9.000	3^4
32	1,024	5.657	2^5	82	6,724	9.055	$2 \cdot 41$
33	1,089	5.745	$3 \cdot 11$	83	6,889	9.110	83
34	1,156	5.831	$2 \cdot 17$	84	7,056	9.165	$2^2 \cdot 3 \cdot 7$
35	1,225	5.916	$5 \cdot 7$	85	7,225	9.220	$5 \cdot 17$
36	1,296	6.000	$2^2 \cdot 3^2$	86	7,396	9.274	$2 \cdot 43$
37	1,369	6.083	37	87	7,569	9.327	$3 \cdot 29$
38	1,444	6.164	$2 \cdot 19$	88	7,744	9.381	$2^3 \cdot 11$
39	1,521	6.245	$3 \cdot 13$	89	7,921	9.434	89
40	1,600	6.325	$2^3 \cdot 5$	90	8,100	9.487	$2 \cdot 3^2 \cdot 5$
41	1,681	6.403	41	91	8,281	9.539	$7 \cdot 13$
42	1,764	6.481	$2 \cdot 3 \cdot 7$	92	8,464	9.592	$2^2 \cdot 23$
43	1,849	6.557	43	93	8,649	9.644	$3 \cdot 31$
44	1,936	6.633	$2^2 \cdot 11$	94	8,836	9.695	$2 \cdot 47$
45	2,025	6.708	$3^2 \cdot 5$	95	9,025	9.747	$5 \cdot 19$
46	2,116	6.782	$2 \cdot 23$	96	9,216	9.798	$2^5 \cdot 3$
47	2,209	6.856	47	97	9,409	9.849	97
48	2,304	6.928	$2^4 \cdot 3$	98	9,604	9.899	$2 \cdot 7^2$
49	2,401	7.000	7^2	99	9,801	9.950	$3^2 \cdot 11$
50	2,500	7.071	$2 \cdot 5^2$	100	10,000	10.000	$2^2 \cdot 5^2$

Table 2   *Common Logarithms*

m	0	1	2	3	4	5	6	7	8	9
1.0	.0000	.0043	.0086	.0128	.0170	.0212	.0253	.0294	.0334	.0374
1.1	.0414	.0453	.0492	.0531	.0569	.0607	.0645	.0682	.0719	.0755
1.2	.0792	.0828	.0864	.0899	.0934	.0969	.1004	.1038	.1072	.1106
1.3	.1139	.1173	.1206	.1239	.1271	.1303	.1335	.1367	.1399	.1430
1.4	.1461	.1492	.1523	.1553	.1584	.1614	.1644	.1673	.1703	.1732
1.5	.1761	.1790	.1818	.1847	.1875	.1903	.1931	.1959	.1987	.2014
1.6	.2041	.2068	.2095	.2122	.2148	.2175	.2201	.2227	.2253	.2279
1.7	.2304	.2330	.2355	.2380	.2405	.2430	.2455	.2480	.2504	.2529
1.8	.2553	.2577	.2601	.2625	.2648	.2672	.2695	.2718	.2742	.2765
1.9	.2788	.2810	.2833	.2856	.2878	.2900	.2923	.2945	.2967	.2989
2.0	.3010	.3032	.3054	.3075	.3096	.3118	.3139	.3160	.3181	.3201
2.1	.3222	.3243	.3263	.3284	.3304	.3324	.3345	.3365	.3385	.3404
2.2	.3424	.3444	.3464	.3483	.3502	.3522	.3541	.3560	.3579	.3598
2.3	.3617	.3636	.3655	.3674	.3692	.3711	.3729	.3747	.3766	.3784
2.4	.3802	.3820	.3838	.3856	.3874	.3892	.3909	.3927	.3945	.3962
2.5	.3979	.3997	.4014	.4031	.4048	.4065	.4082	.4099	.4116	.4133
2.6	.4150	.4166	.4183	.4200	.4216	.4232	.4249	.4265	.4281	.4298
2.7	.4314	.4330	.4346	.4362	.4378	.4393	.4409	.4425	.4440	.4456
2.8	.4472	.4487	.4502	.4518	.4533	.4548	.4564	.4579	.4594	.4609
2.9	.4624	.4639	.4654	.4669	.4683	.4698	.4713	.4728	.4742	.4757
3.0	.4771	.4786	.4800	.4814	.4829	.4843	.4857	.4871	.4886	.4900
3.1	.4914	.4928	.4942	.4955	.4969	.4983	.4997	.5011	.5024	.5038
3.2	.5051	.5065	.5079	.5092	.5105	.5119	.5132	.5145	.5159	.5172
3.3	.5185	.5198	.5211	.5224	.5237	.5250	.5263	.5276	.5289	.5302
3.4	.5315	.5328	.5340	.5353	.5366	.5378	.5391	.5403	.5416	.5428
3.5	.5441	.5453	.5465	.5478	.5490	.5502	.5514	.5527	.5539	.5551
3.6	.5563	.5575	.5587	.5599	.5611	.5623	.5635	.5647	.5658	.5670
3.7	.5682	.5694	.5705	.5717	.5729	.5740	.5752	.5763	.5775	.5786
3.8	.5798	.5809	.5821	.5832	.5843	.5855	.5866	.5877	.5888	.5899
3.9	.5911	.5922	.5933	.5944	.5955	.5966	.5977	.5988	.5999	.6010
4.0	.6021	.6031	.6042	.6053	.6064	.6075	.6085	.6096	.6107	.6117
4.1	.6128	.6138	.6149	.6160	.6170	.6180	.6191	.6201	.6212	.6222
4.2	.6232	.6243	.6253	.6263	.6274	.6284	.6294	.6304	.6314	.6325
4.3	.6335	.6345	.6355	.6365	.6375	.6385	.6395	.6405	.6415	.6425
4.4	.6435	.6444	.6454	.6464	.6474	.6484	.6493	.6503	.6513	.6522
4.5	.6532	.6542	.6551	.6561	.6571	.6580	.6590	.6599	.6609	.6618
4.6	.6628	.6637	.6646	.6656	.6665	.6675	.6684	.6693	.6702	.6712
4.7	.6721	.6730	.6739	.6749	.6758	.6767	.6776	.6785	.6794	.6803
4.8	.6812	.6821	.6830	.6839	.6848	.6857	.6866	.6875	.6884	.6893
4.9	.6902	.6911	.6920	.6928	.6937	.6946	.6955	.6964	.6972	.6981
5.0	.6990	.6998	.7007	.7016	.7024	.7033	.7042	.7050	.7059	.7067
5.1	.7076	.7084	.7093	.7101	.7110	.7118	.7126	.7135	.7143	.7152
5.2	.7160	.7168	.7177	.7185	.7193	.7202	.7210	.7218	.7226	.7235
5.3	.7243	.7251	.7259	.7267	.7275	.7284	.7292	.7300	.7308	.7316
5.4	.7324	.7332	.7340	.7348	.7356	.7364	.7372	.7380	.7388	.7396

Table 2 (Continued)

m	0	1	2	3	4	5	6	7	8	9
5.5	.7404	.7412	.7419	.7427	.7435	.7443	.7451	.7459	.7466	.7474
5.6	.7482	.7490	.7497	.7505	.7513	.7520	.7528	.7536	.7543	.7551
5.7	.7559	.7566	.7574	.7582	.7589	.7597	.7604	.7612	.7619	.7627
5.8	.7634	.7642	.7649	.7657	.7664	.7672	.7679	.7686	.7694	.7701
5.9	.7709	.7716	.7723	.7731	.7738	.7745	.7752	.7760	.7767	.7774
6.0	.7782	.7789	.7796	.7803	.7810	.7818	.7825	.7832	.7839	.7846
6.1	.7853	.7860	.7868	.7875	.7882	.7889	.7896	.7903	.7910	.7917
6.2	.7924	.7931	.7938	.7945	.7952	.7959	.7966	.7973	.7980	.7987
6.3	.7993	.8000	.8007	.8014	.8021	.8028	.8035	.8041	.8048	.8055
6.4	.8062	.8069	.8075	.8082	.8089	.8096	.8102	.8109	.8116	.8122
6.5	.8129	.8136	.8142	.8149	.8156	.8162	.8169	.8176	.8182	.8189
6.6	.8195	.8202	.8209	.8215	.8222	.8228	.8235	.8241	.8248	.8254
6.7	.8261	.8267	.8274	.8280	.8287	.8293	.8299	.8306	.8312	.8319
6.8	.8325	.8331	.8338	.8344	.8351	.8357	.8363	.8370	.8376	.8382
6.9	.8388	.8395	.8401	.8407	.8414	.8420	.8426	.8432	.8439	.8445
7.0	.8451	.8457	.8463	.8470	.8476	.8482	.8488	.8494	.8500	.8506
7.1	.8513	.8519	.8525	.8531	.8537	.8543	.8549	.8555	.8561	.8567
7.2	.8573	.8579	.8585	.8591	.8597	.8603	.8609	.8615	.8621	.8627
7.3	.8633	.8639	.8645	.8651	.8657	.8663	.8669	.8675	.8681	.8686
7.4	.8692	.8698	.8704	.8710	.8716	.8722	.8727	.8733	.8739	.8745
7.5	.8751	.8756	.8762	.8768	.8774	.8779	.8785	.8791	.8797	.8802
7.6	.8808	.8814	.8820	.8825	.8831	.8837	.8842	.8848	.8854	.8859
7.7	.8865	.8871	.8876	.8882	.8887	.8893	.8899	.8904	.8910	.8915
7.8	.8921	.8927	.8932	.8938	.8943	.8949	.8954	.8960	.8965	.8971
7.9	.8976	.8982	.8987	.8993	.8998	.9004	.9009	.9015	.9020	.9025
8.0	.9031	.9036	.9042	.9047	.9053	.9058	.9063	.9069	.9074	.9079
8.1	.9085	.9090	.9096	.9101	.9106	.9112	.9117	.9122	.9128	.9133
8.2	.9138	.9143	.9149	.9154	.9159	.9165	.9170	.9175	.9180	.9186
8.3	.9191	.9196	.9201	.9206	.9212	.9217	.9222	.9227	.9232	.9238
8.4	.9243	.9249	.9253	.9258	.9263	.9269	.9274	.9279	.9284	.9289
8.5	.9294	.9299	.9304	.9309	.9315	.9320	.9325	.9330	.9335	.9340
8.6	.9345	.9350	.9355	.9360	.9365	.9370	.9375	.9380	.9385	.9390
8.7	.9395	.9400	.9405	.9410	.9415	.9420	.9425	.9430	.9435	.9440
8.8	.9445	.9450	.9455	.9460	.9465	.9469	.9474	.9479	.9484	.9489
8.9	.9494	.9499	.9504	.9509	.9513	.9518	.9523	.9528	.9533	.9538
9.0	.9542	.9547	.9552	.9557	.9562	.9566	.9571	.9576	.9581	.9586
9.1	.9590	.9595	.9600	.9605	.9609	.9614	.9619	.9624	.9628	.9633
9.2	.9638	.9643	.9647	.9652	.9657	.9661	.9666	.9671	.9675	.9680
9.3	.9685	.9689	.9694	.9699	.9703	.9708	.9713	.9717	.9722	.9727
9.4	.9731	.9736	.9741	.9745	.9750	.9754	.9759	.9763	.9768	.9773
9.5	.9777	.9782	.9786	.9791	.9795	.9800	.9805	.9809	.9814	.9818
9.6	.9823	.9827	.9832	.9836	.9841	.9845	.9850	.9854	.9859	.9863
9.7	.9868	.9872	.9877	.9881	.9886	.9890	.9894	.9899	.9903	.9908
9.8	.9912	.9917	.9921	.9926	.9930	.9934	.9939	.9943	.9948	.9952
9.9	.9956	.9961	.9965	.9969	.9974	.9978	.9983	.9987	.9991	.9996

Table 3 *Natural Logarithms (Base e)*

	0	1	2	3	4	5	6	7	8	9
1.0	0.0000	0.0100	0.0198	0.0296	0.0392	0.0488	0.0583	0.0677	0.0770	0.0862
1.1	0.0953	0.1044	0.1133	0.1222	0.1310	0.1398	0.1484	0.1570	0.1655	0.1740
1.2	0.1823	0.1906	0.1989	0.2070	0.2151	0.2231	0.2311	0.2390	0.2469	0.2546
1.3	0.2624	0.2700	0.2776	0.2852	0.2927	0.3001	0.3075	0.3148	0.3221	0.3293
1.4	0.3365	0.3436	0.3507	0.3577	0.3646	0.3716	0.3784	0.3853	0.3920	0.3988
1.5	0.4055	0.4121	0.4187	0.4253	0.4318	0.4383	0.4447	0.4511	0.4574	0.4637
1.6	0.4700	0.4762	0.4824	0.4886	0.4947	0.5008	0.5068	0.5128	0.5188	0.5247
1.7	0.5306	0.5365	0.5423	0.5481	0.5539	0.5596	0.5653	0.5710	0.5766	0.5822
1.8	0.5878	0.5933	0.5988	0.6043	0.6098	0.6152	0.6206	0.6259	0.6313	0.6366
1.9	0.6419	0.6471	0.6523	0.6575	0.6627	0.6678	0.6729	0.6780	0.6831	0.6881
2.0	0.6931	0.6981	0.7031	0.7080	0.7129	0.7178	0.7227	0.7275	0.7324	0.7372
2.1	0.7419	0.7467	0.7514	0.7561	0.7608	0.7655	0.7701	0.7747	0.7793	0.7839
2.2	0.7885	0.7930	0.7975	0.8020	0.8065	0.8109	0.8154	0.8198	0.8242	0.8286
2.3	0.8329	0.8373	0.8416	0.8459	0.8502	0.8544	0.8587	0.8629	0.8671	0.8713
2.4	0.8755	0.8796	0.8838	0.8879	0.8920	0.8961	0.9002	0.9042	0.9083	0.9123
2.5	0.9163	0.9203	0.9243	0.9282	0.9322	0.9361	0.9400	0.9439	0.9478	0.9517
2.6	0.9555	0.9594	0.9632	0.9670	0.9708	0.9746	0.9783	0.9821	0.9858	0.9895
2.7	0.9933	0.9969	1.0006	1.0043	1.0080	1.0116	1.0152	1.0188	1.0225	1.0260
2.8	1.0296	1.0332	1.0367	1.0403	1.0438	1.0473	1.0508	1.0543	1.0578	1.0613
2.9	1.0647	1.0682	1.0716	1.0750	1.0784	1.0818	1.0852	1.0886	1.0919	1.0953
3.0	1.0986	1.1019	1.1053	1.1086	1.1119	1.1151	1.1184	1.1217	1.1249	1.1282
3.1	1.1314	1.1346	1.1378	1.1410	1.1442	1.1474	1.1506	1.1537	1.1569	1.1600
3.2	1.1632	1.1663	1.1694	1.1725	1.1756	1.1787	1.1817	1.1848	1.1878	1.1909
3.3	1.1939	1.1969	1.2000	1.2030	1.2060	1.2090	1.2119	1.2149	1.2179	1.2208
3.4	1.2238	1.2267	1.2296	1.2326	1.2355	1.2384	1.2413	1.2442	1.2470	1.2499
3.5	1.2528	1.2556	1.2585	1.2613	1.2641	1.2669	1.2698	1.2726	1.2754	1.2782
3.6	1.2809	1.2837	1.2865	1.2892	1.2920	1.2947	1.2975	1.3002	1.3029	1.3056
3.7	1.3083	1.3110	1.3137	1.3164	1.3191	1.3218	1.3244	1.3271	1.3297	1.3324
3.8	1.3350	1.3376	1.3403	1.3429	1.3455	1.3481	1.3507	1.3533	1.3558	1.3584
3.9	1.3610	1.3635	1.3661	1.3686	1.3712	1.3737	1.3762	1.3788	1.3813	1.3838
4.0	1.3863	1.3888	1.3913	1.3938	1.3962	1.3987	1.4012	1.4036	1.4061	1.4085
4.1	1.4110	1.4134	1.4159	1.4183	1.4207	1.4231	1.4255	1.4279	1.4303	1.4327
4.2	1.4351	1.4375	1.4398	1.4422	1.4446	1.4469	1.4493	1.4516	1.4540	1.4563
4.3	1.4586	1.4609	1.4633	1.4656	1.4679	1.4702	1.4725	1.4748	1.4771	1.4793
4.4	1.4816	1.4839	1.4861	1.4884	1.4907	1.4929	1.4951	1.4974	1.4996	1.5019
4.5	1.5041	1.5063	1.5085	1.5107	1.5129	1.5151	1.5173	1.5195	1.5217	1.5239
4.6	1.5261	1.5282	1.5304	1.5326	1.5347	1.5369	1.5390	1.5412	1.5433	1.5454
4.7	1.5476	1.5497	1.5518	1.5539	1.5560	1.5581	1.5602	1.5623	1.5644	1.5665
4.8	1.5686	1.5707	1.5728	1.5748	1.5769	1.5790	1.5810	1.5831	1.5851	1.5872
4.9	1.5892	1.5913	1.5933	1.5953	1.5974	1.5994	1.6014	1.6034	1.6054	1.6074
5.0	1.6094	1.6114	1.6134	1.6154	1.6174	1.6194	1.6214	1.6233	1.6253	1.6273
5.1	1.6292	1.6312	1.6332	1.6351	1.6371	1.6390	1.6409	1.6429	1.6448	1.6467
5.2	1.6487	1.6506	1.6525	1.6544	1.6563	1.6582	1.6601	1.6620	1.6639	1.6658
5.3	1.6677	1.6696	1.6715	1.6734	1.6752	1.6771	1.6790	1.6808	1.6827	1.6845
5.4	1.6864	1.6882	1.6901	1.6919	1.6938	1.6956	1.6974	1.6993	1.7011	1.7029

Table 3 (Continued)

	0	1	2	3	4	5	6	7	8	9
5.5	1.7047	1.7066	1.7084	1.7102	1.7120	1.7138	1.7156	1.7174	1.7192	1.7210
5.6	1.7228	1.7246	1.7263	1.7281	1.7299	1.7317	1.7334	1.7352	1.7370	1.7387
5.7	1.7405	1.7422	1.7440	1.7457	1.7475	1.7492	1.7509	1.7527	1.7544	1.7561
5.8	1.7579	1.7596	1.7613	1.7630	1.7647	1.7664	1.7681	1.7699	1.7716	1.7733
5.9	1.7750	1.7766	1.7783	1.7800	1.7817	1.7834	1.7851	1.7868	1.7884	1.7901
6.0	1.7918	1.7934	1.7951	1.7967	1.7984	1.8001	1.8017	1.8034	1.8050	1.8066
6.1	1.8083	1.8099	1.8116	1.8132	1.8148	1.8165	1.8181	1.8197	1.8213	1.8229
6.2	1.8245	1.8262	1.8278	1.8294	1.8310	1.8326	1.8342	1.8358	1.8374	1.8390
6.3	1.8405	1.8421	1.8437	1.8453	1.8469	1.8485	1.8500	1.8516	1.8532	1.8547
6.4	1.8563	1.8579	1.8594	1.8610	1.8625	1.8641	1.8656	1.8672	1.8687	1.8703
6.5	1.8718	1.8733	1.8749	1.8764	1.8779	1.8795	1.8810	1.8825	1.8840	1.8856
6.6	1.8871	1.8886	1.8901	1.8916	1.8931	1.8946	1.8961	1.8976	1.8991	1.9006
6.7	1.9021	1.9036	1.9051	1.9066	1.9081	1.9095	1.9110	1.9125	1.9140	1.9155
6.8	1.9169	1.9184	1.9199	1.9213	1.9228	1.9242	1.9257	1.9272	1.9286	1.9301
6.9	1.9315	1.9330	1.9344	1.9359	1.9373	1.9387	1.9402	1.9416	1.9430	1.9445
7.0	1.9459	1.9473	1.9488	1.9502	1.9516	1.9530	1.9544	1.9559	1.9573	1.9587
7.1	1.9601	1.9615	1.9629	1.9643	1.9657	1.9671	1.9685	1.9699	1.9713	1.9727
7.2	1.9741	1.9755	1.9769	1.9782	1.9796	1.9810	1.9824	1.9838	1.9851	1.9865
7.3	1.9879	1.9892	1.9906	1.9920	1.9933	1.9947	1.9961	1.9974	1.9988	2.0001
7.4	2.0015	2.0028	2.0042	2.0055	2.0069	2.0082	2.0096	2.0109	2.0122	2.0136
7.5	2.0149	2.0162	2.0176	2.0189	2.0202	2.0215	2.0229	2.0242	2.0255	2.0268
7.6	2.0281	2.0295	2.0308	2.0321	2.0334	2.0347	2.0360	2.0373	2.0386	2.0399
7.7	2.0412	2.0425	2.0438	2.0451	2.0464	2.0477	2.0490	2.0503	2.0516	2.0528
7.8	2.0541	2.0554	2.0567	2.0580	2.0592	2.0605	2.0618	2.0631	2.0643	2.0656
7.9	2.0669	2.0681	2.0694	2.0707	2.0719	2.0732	2.0744	2.0757	2.0769	2.0782
8.0	2.0794	2.0807	2.0819	2.0832	2.0844	2.0857	2.0869	2.0882	2.0894	2.0906
8.1	2.0919	2.0931	2.0943	2.0956	2.0968	2.0980	2.0992	2.1005	2.1017	2.1029
8.2	2.1041	2.1054	2.1066	2.1078	2.1090	2.1102	2.1114	2.1126	2.1138	2.1150
8.3	2.1163	2.1175	2.1187	2.1199	2.1211	2.1223	2.1235	2.1247	2.1259	2.1270
8.4	2.1282	2.1294	2.1306	2.1318	2.1330	2.1342	2.1353	2.1365	2.1377	2.1389
8.5	2.1401	2.1412	2.1424	2.1436	2.1448	2.1459	2.1471	2.1483	2.1494	2.1506
8.6	2.1518	2.1529	2.1541	2.1552	2.1564	2.1576	2.1587	2.1599	2.1610	2.1622
8.7	2.1633	2.1645	2.1656	2.1668	2.1679	2.1691	2.1702	2.1713	2.1725	2.1736
8.8	2.1748	2.1759	2.1770	2.1782	2.1793	2.1804	2.1815	2.1827	2.1838	2.1849
8.9	2.1861	2.1872	2.1883	2.1894	2.1905	2.1917	2.1928	2.1939	2.1950	2.1961
9.0	2.1972	2.1983	2.1994	2.2006	2.2017	2.2028	2.2039	2.2050	2.2061	2.2072
9.1	2.2083	2.2094	2.2105	2.2116	2.2127	2.2138	2.2148	2.2159	2.2170	2.2181
9.2	2.2192	2.2203	2.2214	2.2225	2.2235	2.2246	2.2257	2.2268	2.2279	2.2289
9.3	2.2300	2.2311	2.2322	2.2332	2.2343	2.2354	2.2364	2.2375	2.2386	2.2396
9.4	2.2407	2.2418	2.2428	2.2439	2.2450	2.2460	2.2471	2.2481	2.2492	2.2502
9.5	2.2513	2.2523	2.2534	2.2544	2.2555	2.2565	2.2576	2.2586	2.2597	2.2607
9.6	2.2618	2.2628	2.2638	2.2649	2.2659	2.2670	2.2680	2.2690	2.2701	2.2711
9.7	2.2721	2.2732	2.2742	2.2752	2.2762	2.2773	2.2783	2.2793	2.2803	2.2814
9.8	2.2824	2.2834	2.2844	2.2854	2.2865	2.2875	2.2885	2.2895	2.2905	2.2915
9.9	2.2925	2.2935	2.2946	2.2956	2.2966	2.2976	2.2986	2.2996	2.3006	2.3016

$\ln 10 \approx 2.3026$

Table 4 Exponential Functions

x	e^x	e^{-x}	x	e^x	e^{-x}
0.00	1.0000	1.0000	1.5	4.4817	0.2231
0.01	1.0101	0.9901	1.6	4.9530	0.2019
0.02	1.0202	0.9802	1.7	5.4739	0.1827
0.03	1.0305	0.9705	1.8	6.0496	0.1653
0.04	1.0408	0.9608	1.9	6.6859	0.1496
0.05	1.0513	0.9512	2.0	7.3891	0.1353
0.06	1.0618	0.9418	2.1	8.1662	0.1225
0.07	1.0725	0.9324	2.2	9.0250	0.1108
0.08	1.0833	0.9231	2.3	9.9742	0.1003
0.09	1.0942	0.9139	2.4	11.023	0.0907
0.10	1.1052	0.9048	2.5	12.182	0.0821
0.11	1.1163	0.8958	2.6	13.464	0.0743
0.12	1.1275	0.8869	2.7	14.880	0.0672
0.13	1.1388	0.8781	2.8	16.445	0.0608
0.14	1.1503	0.8694	2.9	18.174	0.0550
0.15	1.1618	0.8607	3.0	20.086	0.0498
0.16	1.1735	0.8521	3.1	22.198	0.0450
0.17	1.1853	0.8437	3.2	24.533	0.0408
0.18	1.1972	0.8353	3.3	27.113	0.0369
0.19	1.2092	0.8270	3.4	29.964	0.0334
0.20	1.2214	0.8187	3.5	33.115	0.0302
0.21	1.2337	0.8106	3.6	36.598	0.0273
0.22	1.2461	0.8025	3.7	40.447	0.0247
0.23	1.2586	0.7945	3.8	44.701	0.0224
0.24	1.2712	0.7866	3.9	49.402	0.0202
0.25	1.2840	0.7788	4.0	54.598	0.0183
0.30	1.3499	0.7408	4.1	60.340	0.0166
0.35	1.4191	0.7047	4.2	66.686	0.0150
0.40	1.4918	0.6703	4.3	73.700	0.0136
0.45	1.5683	0.6376	4.4	81.451	0.0123
0.50	1.6487	0.6065	4.5	90.017	0.0111
0.55	1.7333	0.5769	4.6	99.484	0.0101
0.60	1.8221	0.5488	4.7	109.95	0.0091
0.65	1.9155	0.5220	4.8	121.51	0.0082
0.70	2.0138	0.4966	4.9	134.29	0.0074
0.75	2.1170	0.4724	5.0	148.41	0.0067
0.80	2.2255	0.4493	5.5	244.69	0.0041
0.85	2.3396	0.4274	6.0	403.43	0.0025
0.90	2.4596	0.4066	6.5	665.14	0.0015
0.95	2.5857	0.3867	7.0	1,096.6	0.0009
1.0	2.7183	0.3679	7.5	1,808.0	0.0006
1.1	3.0042	0.3329	8.0	2,981.0	0.0003
1.2	3.3201	0.3012	8.5	4,914.8	0.0002
1.3	3.6693	0.2725	9.0	8,103.1	0.0001
1.4	4.0552	0.2466	10.0	22,026	0.00005

Table 5 Trigonometric Functions of Real Numbers

Real Number x or θ radians	θ degrees	sin x or sin θ	csc x or csc θ	tan x or tan θ	cot x or cot θ	sec x or sec θ	cos x or cos θ
0.00	0°00'	0.0000	No value	0.0000	No value	1.000	1.000
.01	0°34'	.0100	100.0	.0100	100.0	1.000	1.000
.02	1°09'	.0200	50.00	.0200	49.99	1.000	0.9998
.03	1°43'	.0300	33.34	.0300	33.32	1.000	0.9996
.04	2°18'	.0400	25.01	.0400	24.99	1.001	0.9992
0.05	2°52'	0.0500	20.01	0.0500	19.98	1.001	0.9988
.06	3°26'	.0600	16.68	.0601	16.65	1.002	.9982
.07	4°01'	.0699	14.30	.0701	14.26	1.002	.9976
.08	4°35'	.0799	12.51	.0802	12.47	1.003	.9968
.09	5°09'	.0899	11.13	.0902	11.08	1.004	.9960
0.10	5°44'	0.0998	10.02	0.1003	9.967	1.005	0.9950
.11	6°18'	.1098	9.109	.1104	9.054	1.006	.9940
.12	6°53'	.1197	8.353	.1206	8.293	1.007	.9928
.13	7°27'	.1296	7.714	.1307	7.649	1.009	.9916
.14	8°01'	.1395	7.166	.1409	7.096	1.010	.9902
0.15	8°36'	0.1494	6.692	0.1511	6.617	1.011	0.9888
.16	9°10'	.1593	6.277	.1614	6.197	1.013	.9872
.17	9°44'	.1692	5.911	.1717	5.826	1.015	.9856
.18	10°19'	.1790	5.586	.1820	5.495	1.016	.9838
.19	10°53'	.1889	5.295	.1923	5.200	1.018	.9820
0.20	11°28'	0.1987	5.033	0.2027	4.933	1.020	0.9801
.21	12°02'	.2085	4.797	.2131	4.692	1.022	.9780
.22	12°36'	.2182	4.582	.2236	4.472	1.025	.9759
.23	13°11'	.2280	4.386	.2341	4.271	1.027	.9737
.24	13°45'	.2377	4.207	.2447	4.086	1.030	.9713
0.25	14°19'	0.2474	4.042	0.2553	3.916	1.032	0.9689
.26	14°54'	.2571	3.890	.2660	3.759	1.035	.9664
.27	15°28'	.2667	3.749	.2768	3.613	1.038	.9638
.28	16°03'	.2764	3.619	.2876	3.478	1.041	.9611
.29	16°37'	.2860	3.497	.2984	3.351	1.044	.9582
0.30	17°11'	0.2955	3.384	0.3093	3.233	1.047	0.9553
.31	17°46'	.3051	3.278	.3203	3.122	1.050	.9523
.32	18°20'	.3146	3.179	.3314	3.018	1.053	.9492
.33	18°54'	.3240	3.086	.3425	2.920	1.057	.9460
.34	19°29'	.3335	2.999	.3537	2.827	1.061	.9428
0.35	20°03'	0.3429	2.916	0.3650	2.740	1.065	0.9394
.36	20°38'	.3523	2.839	.3764	2.657	1.068	.9359
.37	21°12'	.3616	2.765	.3879	2.578	1.073	.9323
.38	21°46'	.3709	2.696	.3994	2.504	1.077	.9287
.39	22°21'	.3802	2.630	.4111	2.433	1.081	.9249
0.40	22°55'	0.3894	2.568	0.4228	2.365	1.086	0.9211
.41	23°29'	.3986	2.509	.4346	2.301	1.090	.9171
.42	24°04'	.4078	2.452	.4466	2.239	1.095	.9131
.43	24°38'	.4169	2.399	.4586	2.180	1.100	.9090
.44	25°13'	.4259	2.348	.4708	2.124	1.105	.9048
0.45	25°47'	0.4350	2.299	0.4831	2.070	1.111	0.9004
.46	26°21'	.4439	2.253	.4954	2.018	1.116	.8961
.47	26°56'	.4529	2.208	.5080	1.969	1.122	.8916
.48	27°30'	.4618	2.166	.5206	1.921	1.127	.8870
.49	28°04'	.4706	2.125	.5334	1.875	1.133	.8823
0.50	28°39'	0.4794	2.086	0.5463	1.830	1.139	.8776

Table 5 *(Continued)*

Real Number x or θ radians	θ degrees	sin x or sin θ	csc x or csc θ	tan x or tan θ	cot x or cot θ	sec x or sec θ	cos x or cos θ
0.50	28°39′	0.4794	2.086	0.5463	1.830	1.139	0.8776
.51	29°13′	.4882	2.048	.5594	1.788	1.146	.8727
.52	29°48′	.4969	2.013	.5726	1.747	1.152	.8678
.53	30°22′	.5055	1.978	.5859	1.707	1.159	.8628
.54	30°56′	.5141	1.945	.5994	1.668	1.166	.8577
0.55	31°31′	0.5227	1.913	0.6131	1.631	1.173	0.8525
.56	32°05′	.5312	1.883	.6269	1.595	1.180	.8473
.57	32°40′	.5396	1.853	.6410	1.560	1.188	.8419
.58	33°14′	.5480	1.825	.6552	1.526	1.196	.8365
.59	33°48′	.5564	1.797	.6696	1.494	1.203	.8309
0.60	34°23′	0.5646	1.771	0.6841	1.462	1.212	0.8253
.61	34°57′	.5729	1.746	.6989	1.431	1.220	.8196
.62	35°31′	.5810	1.721	.7139	1.401	1.229	.8139
.63	36°06′	.5891	1.697	.7291	1.372	1.238	.8080
.64	36°40′	.5972	1.674	.7445	1.343	1.247	.8021
0.65	37°15′	0.6052	1.652	0.7602	1.315	1.256	0.7961
.66	37°49′	.6131	1.631	.7761	1.288	1.266	.7900
.67	38°23′	.6210	1.610	.7923	1.262	1.276	.7838
.68	38°58′	.6288	1.590	.8087	1.237	1.286	.7776
.69	39°32′	.6365	1.571	.8253	1.212	1.297	.7712
0.70	40°06′	0.6442	1.552	0.8423	1.187	1.307	0.7648
.71	40°41′	.6518	1.534	.8595	1.163	1.319	.7584
.72	41°15′	.6594	1.517	.8771	1.140	1.330	.7518
.73	41°50′	.6669	1.500	.8949	1.117	1.342	.7452
.74	42°24′	.6743	1.483	.9131	1.095	1.354	.7385
0.75	42°58′	0.6816	1.467	0.9316	1.073	1.367	0.7317
.76	43°33′	.6889	1.452	.9505	1.052	1.380	.7248
.77	44°07′	.6961	1.436	.9697	1.031	1.393	.7179
.78	44°41′	.7033	1.422	.9893	1.011	1.407	.7109
.79	45°16′	.7104	1.408	1.009	.9908	1.421	.7038
0.80	45°50′	0.7174	1.394	1.030	0.9712	1.435	0.6967
.81	46°25′	.7243	1.381	1.050	.9520	1.450	.6895
.82	46°59′	.7311	1.368	1.072	.9331	1.466	.6822
.83	47°33′	.7379	1.355	1.093	.9146	1.482	.6749
.84	48°08′	.7446	1.343	1.116	.8964	1.498	.6675
0.85	48°42′	0.7513	1.331	1.138	0.8785	1.515	0.6600
.86	49°16′	.7578	1.320	1.162	.8609	1.533	.6524
.87	49°51′	.7643	1.308	1.185	.8437	1.551	.6448
.88	50°25′	.7707	1.297	1.210	.8267	1.569	.6372
.89	51°00′	.7771	1.287	1.235	.8100	1.589	.6294
0.90	51°34′	0.7833	1.277	1.260	0.7936	1.609	0.6216
.91	52°08′	.7895	1.267	1.286	.7774	1.629	.6137
.92	52°43′	.7956	1.257	1.313	.7615	1.651	.6058
.93	53°17′	.8016	1.247	1.341	.7458	1.673	.5978
.94	53°51′	.8076	1.238	1.369	.7303	1.696	.5898
0.95	54°26′	0.8134	1.229	1.398	0.7151	1.719	0.5817
.96	55°00′	.8192	1.221	1.428	.7001	1.744	.5735
.97	55°35′	.8249	1.212	1.459	.6853	1.769	.5653
.98	56°09′	.8305	1.204	1.491	.6707	1.795	.5570
.99	56°43′	.8360	1.196	1.524	.6563	1.823	.5487
1.00	57°18′	0.8415	1.188	1.557	0.6421	1.851	0.5403
1.01	57°52′	.8468	1.181	1.592	.6281	1.880	.5319
1.02	58°27′	.8521	1.174	1.628	.6142	1.911	.5234
1.03	59°01′	.8573	1.166	1.665	.6005	1.942	.5148
1.04	59°35′	.8624	1.160	1.704	.5870	1.975	.5062
1.05	60°10′	0.8674	1.153	1.743	0.5736	2.010	0.4976

Table 5 (Continued)

Real Number x or θ radians	θ degrees	sin x or sin θ	csc x or csc θ	tan x or tan θ	cot x or cot θ	sec x or sec θ	cos x or cos θ
1.05	60°10′	0.8674	1.153	1.743	0.5736	2.010	0.4976
1.06	60°44′	.8724	1.146	1.784	.5604	2.046	.4889
1.07	61°18′	.8772	1.140	1.827	.5473	2.083	.4801
1.08	61°53′	.8820	1.134	1.871	.5344	2.122	.4713
1.09	62°27′	.8866	1.128	1.917	.5216	2.162	.4625
1.10	63°02′	0.8912	1.122	1.965	0.5090	2.205	0.4536
1.11	63°36′	.8957	1.116	2.014	.4964	2.249	.4447
1.12	64°10′	.9001	1.111	2.066	.4840	2.295	.4357
1.13	64°45′	.9044	1.106	2.120	.4718	2.344	.4267
1.14	65°19′	.9086	1.101	2.176	.4596	2.395	.4176
1.15	65°53′	0.9128	1.096	2.234	0.4475	2.448	0.4085
1.16	66°28′	.9168	1.091	2.296	.4356	2.504	.3993
1.17	67°02′	.9208	1.086	2.360	.4237	2.563	.3902
1.18	67°37′	.9246	1.082	2.427	.4120	2.625	.3809
1.19	68°11′	.9284	1.077	2.498	.4003	2.691	.3717
1.20	68°45′	0.9320	1.073	2.572	0.3888	2.760	0.3624
1.21	69°20′	.9356	1.069	2.650	.3773	2.833	.3530
1.22	69°54′	.9391	1.065	2.733	.3659	2.910	.3436
1.23	70°28′	.9425	1.061	2.820	.3546	2.992	.3342
1.24	71°03′	.9458	1.057	2.912	.3434	3.079	.3248
1.25	71°37′	0.9490	1.054	3.010	0.3323	3.171	0.3153
1.26	72°12′	.9521	1.050	3.113	.3212	3.270	.3058
1.27	72°46′	.9551	1.047	3.224	.3102	3.375	.2963
1.28	73°20′	.9580	1.044	3.341	.2993	3.488	.2867
1.29	73°55′	.9608	1.041	3.467	.2884	3.609	.2771
1.30	74°29′	0.9636	1.038	3.602	0.2776	3.738	0.2675
1.31	75°03′	.9662	1.035	3.747	.2669	3.878	.2579
1.32	75°38′	.9687	1.032	3.903	.2562	4.029	.2482
1.33	76°12′	.9711	1.030	4.072	.2456	4.193	.2385
1.34	76°47′	.9735	1.027	4.256	.2350	4.372	.2288
1.35	77°21′	0.9757	1.025	4.455	0.2245	4.566	0.2190
1.36	77°55′	.9779	1.023	4.673	.2140	4.779	.2092
1.37	78°30′	.9799	1.021	4.913	.2035	5.014	.1994
1.38	79°04′	.9819	1.018	5.177	.1931	5.273	.1896
1.39	79°38′	.9837	1.017	5.471	.1828	5.561	.1798
1.40	80°13′	0.9854	1.015	5.798	0.1725	5.883	0.1700
1.41	80°47′	.9871	1.013	6.165	.1622	6.246	.1601
1.42	81°22′	.9887	1.011	6.581	.1519	6.657	.1502
1.43	81°56′	.9901	1.010	7.055	.1417	7.126	.1403
1.44	82°30′	.9915	1.009	7.602	.1315	7.667	.1304
1.45	83°05′	0.9927	1.007	8.238	0.1214	8.299	0.1205
1.46	83°39′	.9939	1.006	8.989	.1113	9.044	.1106
1.47	84°13′	.9949	1.005	9.887	.1011	9.938	.1006
1.48	84°48′	.9959	1.004	10.98	.0910	11.03	.0907
1.49	85°22′	.9967	1.003	12.35	.0810	12.39	.0807
1.50	85°57′	0.9975	1.003	14.10	0.0709	14.14	0.0707
1.51	86°31′	.9982	1.002	16.43	.0609	16.46	.0608
1.52	87°05′	.9987	1.001	19.67	.0508	19.69	.0508
1.53	87°40′	.9992	1.001	24.50	.0408	24.52	.0408
1.54	88°14′	.9995	1.000	32.46	.0308	32.48	.0308
1.55	88°49′	0.9998	1.000	48.08	0.0208	48.09	0.0208
1.56	89°23′	.9999	1.000	92.62	.0108	92.63	.0108
1.57	89°57′	1.000	1.000	1256	.0008	1256	.0008

594

Table 6 Trigonometric Functions of Angles

Angle	sin	cos	tan	cot	sec	csc	
0°00′	.0000	1.0000	.0000	—	1.000	—	90°00′
10	.0029	1.0000	.0029	343.8	1.000	343.8	50
20	.0058	1.0000	.0058	171.9	1.000	171.9	40
30	.0087	1.0000	.0087	114.6	1.000	114.6	30
40	.0116	.9999	.0116	85.94	1.000	85.95	20
50	.0145	.9999	.0145	68.75	1.000	68.76	10
1°00′	.0175	.9998	.0175	57.29	1.000	57.30	89°00′
10	.0204	.9998	.0204	49.10	1.000	49.11	50
20	.0233	.9997	.0233	42.96	1.000	42.98	40
30	.0262	.9997	.0262	38.19	1.000	38.20	30
40	.0291	.9996	.0291	34.37	1.000	34.38	20
50	.0320	.9995	.0320	31.24	1.001	31.26	10
2°00′	.0349	.9994	.0349	28.64	1.001	28.65	88°00′
10	.0378	.9993	.0378	26.43	1.001	26.45	50
20	.0407	.9992	.0407	24.54	1.001	24.56	40
30	.0436	.9990	.0437	22.90	1.001	22.93	30
40	.0465	.9989	.0466	21.47	1.001	21.49	20
50	.0494	.9988	.0495	20.21	1.001	20.23	10
3°00′	.0523	.9986	.0524	19.08	1.001	19.11	87°00′
10	.0552	.9985	.0553	18.07	1.002	18.10	50
20	.0581	.9983	.0582	17.17	1.002	17.20	40
30	.0610	.9981	.0612	16.35	1.002	16.38	30
40	.0640	.9980	.0641	15.60	1.002	15.64	20
50	.0669	.9978	.0670	14.92	1.002	14.96	10
4°00′	.0698	.9976	.0699	14.30	1.002	14.34	86°00′
10	.0727	.9974	.0729	13.73	1.003	13.76	50
20	.0756	.9971	.0758	13.20	1.003	13.23	40
30	.0785	.9969	.0787	12.71	1.003	12.75	30
40	.0814	.9967	.0816	12.25	1.003	12.29	20
50	.0843	.9964	.0846	11.83	1.004	11.87	10
5°00′	.0872	.9962	.0875	11.43	1.004	11.47	85°00′
10	.0901	.9959	.0904	11.06	1.004	11.10	50
20	.0929	.9957	.0934	10.71	1.004	10.76	40
30	.0958	.9954	.0963	10.39	1.005	10.43	30
40	.0987	.9951	.0992	10.08	1.005	10.13	20
50	.1016	.9948	.1022	9.788	1.005	9.839	10
6°00′	.1045	.9945	.1051	9.514	1.006	9.567	84°00′
10	.1074	.9942	.1080	9.255	1.006	9.309	50
20	.1103	.9939	.1110	9.010	1.006	9.065	40
30	.1132	.9936	.1139	8.777	1.006	8.834	30
40	.1161	.9932	.1169	8.556	1.007	8.614	20
50	.1190	.9929	.1198	8.345	1.007	8.405	10
7°00′	.1219	.9925	.1228	8.144	1.008	8.206	83°00′
10	.1248	.9922	.1257	7.953	1.008	8.016	50
20	.1276	.9918	.1287	7.770	1.008	7.834	40
30	.1305	.9914	.1317	7.596	1.009	7.661	30
40	.1334	.9911	.1346	7.429	1.009	7.496	20
50	.1363	.9907	.1376	7.269	1.009	7.337	10
8°00′	.1392	.9903	.1405	7.115	1.010	7.185	82°00′
10	.1421	.9899	.1435	6.968	1.010	7.040	50
20	.1449	.9894	.1465	6.827	1.011	6.900	40
30	.1478	.9890	.1495	6.691	1.011	6.765	30
40	.1507	.9886	.1524	6.561	1.012	6.636	20
50	.1536	.9881	.1554	6.435	1.012	6.512	10
9°00′	.1564	.9877	.1584	6.314	1.012	6.392	81°00′
	cos	sin	cot	tan	csc	sec	Angle

Table 6 (*Continued*)

Angle	sin	cos	tan	cot	sec	csc	
9°00′	.1564	.9877	.1584	6.314	1.012	6.392	81°00′
10	.1593	.9872	.1614	6.197	1.013	6.277	50
20	.1622	.9868	.1644	6.084	1.013	6.166	40
30	.1650	.9863	.1673	5.976	1.014	6.059	30
40	.1679	.9858	.1703	5.871	1.014	5.955	20
50	.1708	.9853	.1733	5.769	1.015	5.855	10
10°00′	.1736	.9848	.1763	5.671	1.015	5.759	80°00′
10	.1765	.9843	.1793	5.576	1.016	5.665	50
20	.1794	.9838	.1823	5.485	1.016	5.575	40
30	.1822	.9833	.1853	5.396	1.017	5.487	30
40	.1851	.9827	.1883	5.309	1.018	5.403	20
50	.1880	.9822	.1914	5.226	1.018	5.320	10
11°00′	.1908	.9816	.1944	5.145	1.019	5.241	79°00′
10	.1937	.9811	.1974	5.066	1.019	5.164	50
20	.1965	.9805	.2004	4.989	1.020	5.089	40
30	.1994	.9799	.2035	4.915	1.020	5.016	30
40	.2022	.9793	.2065	4.843	1.021	4.945	20
50	.2051	.9787	.2095	4.773	1.022	4.876	10
12°00′	.2079	.9781	.2126	4.705	1.022	4.810	78°00′
10	.2108	.9775	.2156	4.638	1.023	4.745	50
20	.2136	.9769	.2186	4.574	1.024	4.682	40
30	.2164	.9763	.2217	4.511	1.024	4.620	30
40	.2193	.9757	.2247	4.449	1.025	4.560	20
50	.2221	.9750	.2278	4.390	1.026	4.502	10
13°00′	.2250	.9744	.2309	4.331	1.026	4.445	77°00′
10	.2278	.9737	.2339	4.275	1.027	4.390	50
20	.2306	.9730	.2370	4.219	1.028	4.336	40
30	.2334	.9724	.2401	4.165	1.028	4.284	30
40	.2363	.9717	.2432	4.113	1.029	4.232	20
50	.2391	.9710	.2462	4.061	1.030	4.182	10
14°00′	.2419	.9703	.2493	4.011	1.031	4.134	76°00′
10	.2447	.9696	.2524	3.962	1.031	4.086	50
20	.2476	.9689	.2555	3.914	1.032	4.039	40
30	.2504	.9681	.2586	3.867	1.033	3.994	30
40	.2532	.9674	.2617	3.821	1.034	3.950	20
50	.2560	.9667	.2648	3.776	1.034	3.906	10
15°00′	.2588	.9659	.2679	3.732	1.035	3.864	75°00′
10	.2616	.9652	.2711	3.689	1.036	3.822	50
20	.2644	.9644	.2742	3.647	1.037	3.782	40
30	.2672	.9636	.2773	3.606	1.038	3.742	30
40	.2700	.9628	.2805	3.566	1.039	3.703	20
50	.2728	.9621	.2836	3.526	1.039	3.665	10
16°00′	.2756	.9613	.2867	3.487	1.040	3.628	74°00′
10	.2784	.9605	.2899	3.450	1.041	3.592	50
20	.2812	.9596	.2931	3.412	1.042	3.556	40
30	.2840	.9588	.2962	3.376	1.043	3.521	30
40	.2868	.9580	.2994	3.340	1.044	3.487	20
50	.2896	.9572	.3026	3.305	1.045	3.453	10
17°00′	.2924	.9563	.3057	3.271	1.046	3.420	73°00′
10	.2952	.9555	.3089	3.237	1.047	3.388	50
20	.2979	.9546	.3121	3.204	1.048	3.356	40
30	.3007	.9537	.3153	3.172	1.049	3.326	30
40	.3035	.9528	.3185	3.140	1.049	3.295	20
50	.3062	.9520	.3217	3.108	1.050	3.265	10
18°00′	.3090	.9511	.3249	3.078	1.051	3.236	72°00′
	cos	sin	cot	tan	csc	sec	Angle

Table 6 (Continued)

Angle	sin	cos	tan	cot	sec	csc	
18°00′	.3090	.9511	.3249	3.078	1.051	3.236	72°00′
10	.3118	.9502	.3281	3.047	1.052	3.207	50
20	.3145	.9492	.3314	3.018	1.053	3.179	40
30	.3173	.9483	.3346	2.989	1.054	3.152	30
40	.3201	.9474	.3378	2.960	1.056	3.124	20
50	.3228	.9465	.3411	2.932	1.057	3.098	10
19°00′	.3256	.9455	.3443	2.904	1.058	3.072	71°00′
10	.3283	.9446	.3476	2.877	1.059	3.046	50
20	.3311	.9436	.3508	2.850	1.060	3.021	40
30	.3338	.9426	.3541	2.824	1.061	2.996	30
40	.3365	.9417	.3574	2.798	1.062	2.971	20
50	.3393	.9407	.3607	2.773	1.063	2.947	10
20°00′	.3420	.9397	.3640	2.747	1.064	2.924	70°00′
10	.3448	.9387	.3673	2.723	1.065	2.901	50
20	.3475	.9377	.3706	2.699	1.066	2.878	40
30	.3502	.9367	.3739	2.675	1.068	2.855	30
40	.3529	.9356	.3772	2.651	1.069	2.833	20
50	.3557	.9346	.3805	2.628	1.070	2.812	10
21°00′	.3584	.9336	.3839	2.605	1.071	2.790	69°00′
10	.3611	.9325	.3872	2.583	1.072	2.769	50
20	.3638	.9315	.3906	2.560	1.074	2.749	40
30	.3665	.9304	.3939	2.539	1.075	2.729	30
40	.3692	.9293	.3973	2.517	1.076	2.709	20
50	.3719	.9283	.4006	2.496	1.077	2.689	10
22°00′	.3746	.9272	.4040	2.475	1.079	2.669	68°00′
10	.3773	.9261	.4074	2.455	1.080	2.650	50
20	.3800	.9250	.4108	2.434	1.081	2.632	40
30	.3827	.9239	.4142	2.414	1.082	2.613	30
40	.3854	.9228	.4176	2.394	1.084	2.595	20
50	.3881	.9216	.4210	2.375	1.085	2.577	10
23°00′	.3907	.9205	.4245	2.356	1.086	2.559	67°00′
10	.3934	.9194	.4279	2.337	1.088	2.542	50
20	.3961	.9182	.4314	2.318	1.089	2.525	40
30	.3987	.9171	.4348	2.300	1.090	2.508	30
40	.4014	.9159	.4383	2.282	1.092	2.491	20
50	.4041	.9147	.4417	2.264	1.093	2.475	10
24°00′	.4067	.9135	.4452	2.246	1.095	2.459	66°00′
10	.4094	.9124	.4487	2.229	1.096	2.443	50
20	.4120	.9112	.4522	2.211	1.097	2.427	40
30	.4147	.9100	.4557	2.194	1.099	2.411	30
40	.4173	.9088	.4592	2.177	1.100	2.396	20
50	.4200	.9075	.4628	2.161	1.102	2.381	10
25°00′	.4226	.9063	.4663	2.145	1.103	2.366	65°00′
10	.4253	.9051	.4699	2.128	1.105	2.352	50
20	.4279	.9038	.4734	2.112	1.106	2.337	40
30	.4305	.9026	.4770	2.097	1.108	2.323	30
40	.4331	.9013	.4806	2.081	1.109	2.309	20
50	.4358	.9001	.4841	2.066	1.111	2.295	10
26°00′	.4384	.8988	.4877	2.050	1.113	2.281	64°00′
10	.4410	.8975	.4913	2.035	1.114	2.268	50
20	.4436	.8962	.4950	2.020	1.116	2.254	40
30	.4462	.8949	.4986	2.006	1.117	2.241	30
40	.4488	.8936	.5022	1.991	1.119	2.228	20
50	.4514	.8923	.5059	1.977	1.121	2.215	10
27°00′	.4540	.8910	.5095	1.963	1.122	2.203	63°00′
	cos	sin	cot	tan	csc	sec	Angle

Table 6 *(Continued)*

Angle	sin	cos	tan	cot	sec	csc	
27°00′	.4540	.8910	.5095	1.963	1.122	2.203	63°00′
10	.4566	.8897	.5132	1.949	1.124	2.190	50
20	.4592	.8884	.5169	1.935	1.126	2.178	40
30	.4617	.8870	.5206	1.921	1.127	2.166	30
40	.4643	.8857	.5243	1.907	1.129	2.154	20
50	.4669	.8843	.5280	1.894	1.131	2.142	10
28°00′	.4695	.8829	.5317	1.881	1.133	2.130	62°00′
10	.4720	.8816	.5354	1.868	1.134	2.118	50
20	.4746	.8802	.5392	1.855	1.136	2.107	40
30	.4772	.8788	.5430	1.842	1.138	2.096	30
40	.4797	.8774	.5467	1.829	1.140	2.085	20
50	.4823	.8760	.5505	1.816	1.142	2.074	10
29°00′	.4848	.8746	.5543	1.804	1.143	2.063	61°00′
10	.4874	.8732	.5581	1.792	1.145	2.052	50
20	.4899	.8718	.5619	1.780	1.147	2.041	40
30	.4924	.8704	.5658	1.767	1.149	2.031	30
40	.4950	.8689	.5696	1.756	1.151	2.020	20
50	.4975	.8675	.5735	1.744	1.153	2.010	10
30°00′	.5000	.8660	.5774	1.732	1.155	2.000	60°00′
10	.5025	.8646	.5812	1.720	1.157	1.990	50
20	.5050	.8631	.5851	1.709	1.159	1.980	40
30	.5075	.8616	.5890	1.698	1.161	1.970	30
40	.5100	.8601	.5930	1.686	1.163	1.961	20
50	.5125	.8587	.5969	1.675	1.165	1.951	10
31°00′	.5150	.8572	.6009	1.664	1.167	1.942	59°00′
10	.5175	.8557	.6048	1.653	1.169	1.932	50
20	.5200	.8542	.6088	1.643	1.171	1.923	40
30	.5225	.8526	.6128	1.632	1.173	1.914	30
40	.5250	.8511	.6168	1.621	1.175	1.905	20
50	.5275	.8496	.6208	1.611	1.177	1.896	10
32°00′	.5299	.8480	.6249	1.600	1.179	1.887	58°00′
10	.5324	.8465	.6289	1.590	1.181	1.878	50
20	.5348	.8450	.6330	1.580	1.184	1.870	40
30	.5373	.8434	.6371	1.570	1.186	1.861	30
40	.5398	.8418	.6412	1.560	1.188	1.853	20
50	.5422	.8403	.6453	1.550	1.190	1.844	10
33°00′	.5446	.8387	.6494	1.540	1.192	1.836	57°00′
10	.5471	.8371	.6536	1.530	1.195	1.828	50
20	.5495	.8355	.6577	1.520	1.197	1.820	40
30	.5519	.8339	.6619	1.511	1.199	1.812	30
40	.5544	.8323	.6661	1.501	1.202	1.804	20
50	.5568	.8307	.6703	1.492	1.204	1.796	10
34°00′	.5592	.8290	.6745	1.483	1.206	1.788	56°00′
10	.5616	.8274	.6787	1.473	1.209	1.781	50
20	.5640	.8258	.6830	1.464	1.211	1.773	40
30	.5664	.8241	.6873	1.455	1.213	1.766	30
40	.5688	.8225	.6916	1.446	1.216	1.758	20
50	.5712	.8208	.6959	1.437	1.218	1.751	10
35°00′	.5736	.8192	.7002	1.428	1.221	1.743	55°00′
10	.5760	.8175	.7046	1.419	1.223	1.736	50
20	.5783	.8158	.7089	1.411	1.226	1.729	40
30	.5807	.8141	.7133	1.402	1.228	1.722	30
40	.5831	.8124	.7177	1.393	1.231	1.715	20
50	.5854	.8107	.7221	1.385	1.233	1.708	10
36°00′	.5878	.8090	.7265	1.376	1.236	1.701	54°00′
	cos	sin	cot	tan	csc	sec	Angle

Table 6 *(Continued)*

Angle	sin	cos	tan	cot	sec	csc	
36°00′	.5878	.8090	.7265	1.376	1.236	1.701	54°00′
10	.5901	.8073	.7310	1.368	1.239	1.695	50
20	.5925	.8056	.7355	1.360	1.241	1.688	40
30	.5948	.8039	.7400	1.351	1.244	1.681	30
40	.5972	.8021	.7445	1.343	1.247	1.675	20
50	.5995	.8004	.7490	1.335	1.249	1.668	10
37°00′	.6018	.7986	.7536	1.327	1.252	1.662	53°00′
10	.6041	.7969	.7581	1.319	1.255	1.655	50
20	.6065	.7951	.7627	1.311	1.258	1.649	40
30	.6088	.7934	.7673	1.303	1.260	1.643	30
40	.6111	.7916	.7720	1.295	1.263	1.636	20
50	.6134	.7898	.7766	1.288	1.266	1.630	10
38°00′	.6157	.7880	.7813	1.280	1.269·	1.624	52°00′
10	.6180	.7862	.7860	1.272	1.272	1.618	50
20	.6202	.7844	.7907	1.265	1.275	1.612	40
30	.6225	.7826	.7954	1.257	1.278	1.606	30
40	.6248	.7808	.8002	1.250	1.281	1.601	20
50	.6271	.7790	.8050	1.242	1.284	1.595	10
39°00′	.6293	.7771	.8098	1.235	1.287	1.589	51°00′
10	.6316	.7753	.8146	1.228	1.290	1.583	50
20	.6338	.7735	.8195	1.220	1.293	1.578	40
30	.6361	.7716	.8243	1.213	1.296	1.572	30
40	.6383	.7698	.8292	1.206	1.299	1.567	20
50	.6406	.7679	.8342	1.199	1.302	1.561	10
40°00′	.6428	.7660	.8391	1.192	1.305	1.556	50°00′
10	.6450	.7642	.8441	1.185	1.309	1.550	50
20	.6472	.7623	.8491	1.178	1.312	1.545	40
30	.6494	.7604	.8541	1.171	1.315	1.540	30
40	.6517	.7585	.8591	1.164	1.318	1.535	20
50	.6539	.7566	.8642	1.157	1.322	1.529	10
41°00′	.6561	.7547	.8693	1.150	1.325	1.524	49°00′
10	.6583	.7528	.8744	1.144	1.328	1.519	50
20	.6604	.7509	.8796	1.137	1.332	1.514	40
30	.6626	.7490	.8847	1.130	1.335	1.509	30
40	.6648	.7470	.8899	1.124	1.339	1.504	20
50	.6670	.7451	.8952	1.117	1.342	1.499	10
42°00′	.6691	.7431	.9004	1.111	1.346	1.494	48°00′
10	.6713	.7412	.9057	1.104	1.349	1.490	50
20	.6734	.7392	.9110	1.098	1.353	1.485	40
30	.6756	.7373	.9163	1.091	1.356	1.480	30
40	.6777	.7353	.9217	1.085	1.360	1.476	20
50	.6799	.7333	.9271	1.079	1.364	1.471	10
43°00′	.6820	.7314	.9325	1.072	1.367	1.466	47°00′
10	.6841	.7294	.9380	1.066	1.371	1.462	50
20	.6862	.7274	.9435	1.060	1.375	1.457	40
30	.6884	.7254	.9490	1.054	1.379	1.453	30
40	.6905	.7234	.9545	1.048	1.382	1.448	20
50	.6926	.7214	.9601	1.042	1.386	1.444	10
44°00′	.6947	.7193	.9657	1.036	1.390	1.440	46°00′
10	.6967	.7173	.9713	1.030	1.394	1.435	50
20	.6988	.7153	.9770	1.024	1.398	1.431	40
30	.7009	.7133	.9827	1.018	1.402	1.427	30
40	.7030	.7112	.9884	1.012	1.406	1.423	20
50	.7050	.7092	.9942	1.006	1.410	1.418	10
45°00′	.7071	.7071	1.000	1.000	1.414	1.414	45°00′
	cos	sin	cot	tan	csc	sec	Angle

Answers to Odd - Numbered Problems

Exercises 1.1

1. $0.8\overline{0}$ **3.** $0.4\overline{5}$ **5.** $6.1\overline{6}$ **7.** $1.\overline{428571}$ **9.** $\frac{2}{9}$ **11.** $\frac{321}{999} = \frac{107}{333}$ **13.** $\frac{3}{10}$ **15.** 6 (or $\frac{6}{1}$) **17.** $\frac{1,929}{900} = \frac{643}{300}$ **19.** Real number, rational number, integer **21.** None of these **23.** Real number, rational number, integer **25.** Real number, rational number, integer **27.** Real number, irrational number **29.** Real number, rational number **31.** Commutative property of addition **33.** Associative property of multiplication **35.** Multiplication identity property **37.** Commutative property of multiplication **39.** Distributive property **41.** Associative property of multiplication **43.** Commutative property of addition **45.** Commutative property of multiplication **47.** Multiplication inverse property **49.** Commutative property of addition **51.** -2 **53.** -14 **55.** -27 **57.** -126 **59.** -40 **61.** -3 **63.** 1 **65.** -52 **67.** -12 **69.** -15 **71.** $2x - 5y$ **73.** $-16a + 11b$ **75.** $b^3 - 3b^2 + b + 5$ **77.** $12a + 14$ **79.** $2a + 5b$ **81.** $3^5 = 243$ **83.** 1 **85.** $\frac{3}{4}$ **87.** $1/2^3 = 1/8$ **89.** $\frac{3}{4}$ **91.** x^7 **93.** $4p^2$ **95.** c^6 **97.** $1/(x + y)^3$ **99.** x^2 **101.** $25/y^2$ **103.** $-8/x^3$ **105.** $a^2x^2/4$ **107.** $-x^4/(4y^5z^8)$ **109.** xz^2/y^2 **111.** 2^{x+y} **113.** 2^{1-n} or $1/2^{n-1}$ **115.** 5^{2x^2} **117.** y^{7a} **119.** $(a - b)^{x+y}$ **121.** 1 **123.** $1/y$ **125.** x^{ap-1} **127.** x^a/y **129.** $(1 - x)^{a+2}$ **131.** $5 + \frac{4}{3}$;

$5 \boxed{+} 4 \boxed{=} \boxed{\div} 3 \boxed{=}$ **133.** $\frac{2}{15} \div 9$; $2 \boxed{\div} \boxed{(} 15 \boxed{\div} 9 \boxed{)} \boxed{=}$ **135.** Correct **137.** False, $\frac{1}{2}$ **139.** False, 1
141. \$470.73

Remember This (1.1)

1. 6 **2.** $81, 2^4, 2^6$ **3.** $-2, 0, 6$ **4.** -4 and 3 **5.** (a) $7x^3y^3$ (b) $10x^6y^6$ **6.** $x^3 - 2x^2 + 10x + 75$ **7.** True **8.** $a + (-b)$
9. Commutative property of multiplication **10.** Commutative and associative properties of addition

Exercises 1.2

1. $2x - 2y$ **3.** $-5x^4 + 5x^3 + 5x^2$ **5.** $-8x^2yz + 2xy^2z - 14xyz^2$ **7.** $p^4q^2 + p^2q^4$ **9.** $4n + 3x$ **11.** $2x + h$
13. $6x^3 + 8x^2 - 10x$ **15.** $a^2 + 7a + 12$ **17.** $6x^2 - 11x + 4$ **19.** $k^2 - 4k + 4$ **21.** $y^3 + y^2 - 21y + 4$ **23.** $x^3 - y^3$
25. $-6y^3 + 5y^2 + 3y - 2$ **27.** $x^2 - 2xy + y^2 - 2x + 2y + 1$ **29.** $x^4 - 2x^2y^2 + y^4$ **31.** $y(x + z)$ **33.** $3ax(3ax + 1)$
35. $(a - c)(b + d)$ **37.** $(x + 5)(3x + 2)$ **39.** $(6x + 1)(6x - 1)$ **41.** $(5p + 7q)(5p - 7q)$ **43.** $(6r^3 + k^2)(6r^3 - k^2)$
45. $(y + 3)(y^2 - 3y + 9)$ **47.** $(1 - x)(1 + x + x^2)$ **49.** $(3t - 5)(9t^2 + 15t + 25)$ **51.** $(a + 4)(a + 1)$ **53.** $(x + 4)(x - 3)$
55. $(7z - 6)(z - 1)$ **57.** $(x + 4y)(x + y)$ **59.** $(4y - 3)^2$ **61.** $(3a - 2b)^2$ **63.** $(x + y)(5 + a)$ **65.** $(n + 2)(3m + 1)$
67. $(x^2 + 1)(x^3 + 1)$ **69.** $x(1 + y)(1 - y)$ **71.** $3(k - 4)(k + 2)$ **73.** $(2 + x)(3 - x)$ **75.** $3x(x - 3)(x - 1)$
77. $4x^2(x + 6)(x - 6)$ **79.** $2x^2(x + 3)(x^2 - 3x + 9)$ **81.** $3(7x - 3)(-x - 3)$ **83.** $(x + 1)(x^2 - x + 1)(x - 1)(x^2 + x + 1)$
85. $(x^4 + y^4)(x^2 + y^2)(x + y)(x - y)$ **87.** $(x + 2 + y)(x + 2 - y)$ **89.** $(x + y + z)(x - y - z)$ **91.** 1, factorable
93. 985, not factorable **95.** $(3x + 5)(2x - 1)$ **97.** $(9b + 2)(b - 3)$ **99.** $(12x - 5y)(x - 2y)$ **101.** $x^{2n} + 7x^n + 10$
103. $z^{2a} + 6z^a + 9$ **105.** $x^{2a} - y^{2b}$ **107.** $a^{2bx} + 2 + a^{-2bx}$ **109.** $x^{3n} + y^{3n}$ **111.** $x^n(x + 1)$ **113.** $(y^n + 4)(y^n - 4)$ **121.** $\dfrac{x^2 + 2xh + h^2 + 1 - x^2 - 1}{h} = \dfrac{2xh + h^2}{h}$

115. $(1 - t^m)(1 + t^m + t^{2m})$ **117.** $x^m(x + 1)(x - 1)$ **119.** $x(x^n + 5)(x^n - 4)$

$= 2x + h$ **123.** $6 - h$ **125.** $2x + h + 2$ **127.** $(a + b)(a^2 - ab + b^2) = a^3 - a^2b + ab^2 + a^2b - ab^2 + b^3 = a^3 + b^3$
129. $(2k_1 + 1)(2k_2 + 1) = 4k_1k_2 + 2k_1 + 2k_2 + 1 = 2(2k_1k_2 + k_1 + k_2) + 1 = 2k + 1$

Remember This (1.2)

1. $-\frac{4}{3}$ **2.** $\frac{31}{35}$ **3.** 72 **4.** $1.\overline{54}$ **5.** -1 **6.** Undefined **7.** $-\frac{1}{27}$ **8.** $\frac{2}{3}$ **9.** x^{4b+3} **10.** $a(a + b)$, $(a + b)(a - b)$, $b(a - b)$

Exercises 1.3

1. $\frac{4}{11}$ **3.** $\frac{1}{3b+1}$ **5.** -1 **7.** $\frac{-1}{ax}$ **9.** $\frac{-1}{x+1}$ **11.** $\frac{a+1}{a-4}$ **13.** $\frac{7-y}{7+y}$ **15.** $\frac{3x(2x-5)}{2(x-4)}$ **17.** x^2+ax+a^2 **19.** $\frac{3-2x}{x^4}$

21. $\frac{y}{4x^2}$ **23.** $\frac{2}{a(a+3)}$ **25.** $\frac{(y-1)(y-2)}{(y+2)(y+1)}$ **27.** $\frac{(x-1)^2}{(x+1)^2}$ **29.** $\frac{1}{x}$ **31.** $\frac{n-1}{n-3}$ **33.** $\frac{(x+1)(x-1)^2}{7x}$ **35.** $\frac{1-x}{x(a+1)}$

37. 1 **39.** $\frac{3a}{xy}$ **41.** $\frac{x-1}{x+2}$ **43.** $\frac{3n-2}{n-6}$ **45.** $\frac{9a}{5}$ **47.** $\frac{7}{12x}$ **49.** $\frac{5k^2-3k+2}{k^3}$ **51.** $\frac{2x+47}{10x}$

53. $\frac{y-2}{6(3y+2)}$ **55.** $\frac{3a^2-3a+9}{a(a-3)}$ **57.** $\frac{6x+11}{(2x-3)(2x+3)}$ **59.** $\frac{2n^2-2n+1}{n(n-1)}$ **61.** $\frac{1}{x-y}$ **63.** $\frac{65}{72}$ **65.** $\frac{x+1}{x-1}$

67. $\frac{n^2+n+2}{n^2-n-2}$ **69.** $\frac{y+x}{x}$ **71.** $\frac{x^2+3x-4}{4}$ **73.** $\frac{1}{y-x}$ **75.** $\frac{-1}{2h}$ **77.** $x+a-1$ **79.** $\frac{1-x}{(x+1)^3}$ **81.** $\frac{x-y}{-4a^2}$ **83.** $-y$

85. $\frac{7y-4}{y^2-1}$ **87.** 0 **89.** $\frac{-1}{x(x+h)}$ **91.** $\frac{1}{(x+1)(x+h+1)}$ **93.** $\frac{x^2+xh-1}{x(x+h)}$ **95.** $\frac{3P}{12+2\pi}$ **97.** $\frac{n+1}{3n}$ **99.** $\frac{a^x-x-1}{x(a^x-1)}$

101. $\frac{x+2}{x^n+1}$ **103.** $\frac{-4}{2+x}$ **105.** 10 **107.** $\frac{2y^2+3y-4}{y^2}$ **109.** $\frac{8x+5}{(3x+1)^2}$ **111.** $y+x$ **113.** $\frac{xy}{y+x}$ **115.** 1 **117.** $\frac{R_1R_2}{R_2+R_1}$

119. $\frac{2s_g s_h}{s_h+s_g}$

Remember This (1.3)

1. (a) $>$ **(b)** $<$ **2.** a^2-b^2 **3.** x^2 **4.** y^8/x^6 **5.** $2x-4a$ **6.** $x^3-6x^2+12x-8$ **7.** $(x-2)(x^2+2x+4)$ **8.** $\frac{86}{9}$
9. $=$ **10.** $\sqrt{1000}$

Exercises 1.4

1. $\frac{9}{2}$ **3.** 16 **5.** 4 **7.** 9 **9.** $\frac{1}{5}$ **11.** $x^{1/3}$; $\sqrt[3]{x}$ **13.** $\frac{1}{x^{7/6}}$; $\frac{1}{\sqrt[6]{x^7}}$ **15.** $2xy^2$ **17.** $\frac{a}{2b^3}$ **19.** $\frac{125x^3}{343y^{24}}$ **21.** $\frac{4}{9}x^{5/2}y^{7/12}$; $\frac{4}{9}\sqrt{x^5}\sqrt[12]{y^7}$

23. $2^{7/12}$; $\sqrt[12]{2^7}$ **25.** $16^{2/3}$; $\sqrt[3]{16^2}$ **27.** $3^{3/4}$; $\sqrt[4]{3^3}$ **29.** $7^{1/(mn)}$; $\sqrt[mn]{7}$ **31.** $(2^2\cdot 3^3)^{1/6}$; $\sqrt[6]{2^2\cdot 3^3}$ **33.** $x^{1/2}y^{1/5}$; $\sqrt{x}\cdot\sqrt[5]{y}$

35. $k^{(1-y)/x}$; $\sqrt[x]{k^{1-y}}$ **37.** $x^{(p-q)/q}$; $\sqrt[q]{x^{p-q}}$ **39.** $x+x^{-1}-2=(x^2-2x+1)/x$ **41.** $x+2x^{1/2}y^{1/2}+y$; $x+2\sqrt{x}\sqrt{y}+y$

43. $\frac{2x-1}{2x^{3/2}}$; $\frac{2x-1}{2\sqrt{x^3}}$ **45.** $\frac{x+1}{x-1}$ **47.** $\frac{x-2}{2x^{3/2}}$; $\frac{x-2}{2\sqrt{x^3}}$ **49.** $\frac{x-2}{2x^{3/2}}$; $\frac{x-2}{2\sqrt{x^3}}$ **51.** $2x^3y^2\sqrt{3x}$ **53.** $xy\sqrt[4]{9x^2y^3}$ **55.** $\sqrt{5}$ **57.** $y\sqrt[3]{9y}$

59. $3\sqrt{x}/x$ **61.** $\sqrt[3]{xy^2}/y$ **63.** $-7\sqrt{2}$ **65.** $7\sqrt{2}$ **67.** $2\sqrt{6}/3$ **69.** $-29\sqrt[3]{3}/3$ **71.** $8\sqrt{2x}$ **73.** $(2y-6x)\sqrt{5x}$

75. $5x\sqrt{2xy}/2$ **77.** $\left(\frac{2}{y}-\frac{2}{x^2}+\frac{5x}{4}\right)\sqrt{2xy}$ **79.** $\left(\frac{1}{y}-1\right)\sqrt[4]{xy^3}$ **81.** $3\sqrt{5}$ **83.** -1 **85.** $3x\sqrt{10}$ **87.** $2xy\sqrt[3]{3xy}$

89. $\sqrt{6x}/x$ **91.** $\sqrt[4]{56x}/2$ **93.** $2\sqrt{3}/3$ **95.** $\frac{\sqrt{3}-3}{3}$ **97.** $3-2\sqrt{2}$ **99.** $\frac{x-\sqrt{xy}}{x-y}$ **101.** $\frac{\sqrt{x^2-1}}{x+1}$ **103.** $\frac{1}{\sqrt{x}+2}$

105. $\frac{1}{\sqrt{x+h}+\sqrt{x}}$ **107.** $\frac{2}{\sqrt{2x+2h+1}+\sqrt{2x+1}}$ **109.** $\frac{1+x^3}{x^5}$ **111.** $\frac{4(3x^4+2)}{x^5}$ **113.** $\frac{3(x-5)}{x^{2/3}}$ **115.** $\frac{4x^{3/2}+1}{2x^{1/2}}$

117. $\frac{x(x^2-5x-1)}{(x-5)^{1/2}}$ **119.** $1,200\sqrt{5}$ **121.** $2x+6$ **123.** $\frac{2\sqrt{x+1}-1}{2\sqrt{x+1}}=\frac{2x+2-\sqrt{x+1}}{2x+2}$ **125.** $\sqrt{a^2+b^2}$

127. $\frac{\pi(4R^2-2h^2)}{\sqrt{4R^2-h^2}}=\frac{\pi(4R^2-2h^2)\sqrt{4R^2-h^2}}{4R^2-h^2}$ **129.** $\frac{-1}{x^2\sqrt{x^2+1}}=\frac{-\sqrt{x^2+1}}{x^2(x^2+1)}$ **131.** 0 **133.** 1 **135.** $-b/a$

137. $|ab|=\sqrt{(ab)^2}=\sqrt{a^2}\cdot\sqrt{b^2}=|a|\cdot|b|$

Remember This (1.4)

1. Rational numbers: $\sqrt{9}$, $-\sqrt{9}$; real numbers: $\sqrt{9}$, $-\sqrt{9}$, $\sqrt{12}$ **2.** True **3.** 83 **4.** $\frac{-x^2-1}{x^2-1}$ **5.** $\frac{x}{a-x}$ **6.** $-2\sqrt{2}$
7. $10+5\sqrt{3}$ **8.** $a^2+2ab+b^2$ **9.** $a^2-b^2y^2$ **10.** a^4a^2

Exercises 1.5

1. $5i$ **3.** $8i$ **5.** $i\sqrt{22}$ **7.** $i\sqrt{3}$ **9.** $2 - 8i\sqrt{2}$ **11.** $2i$ **13.** $2i\sqrt{2}$ **15.** $(1/4) - (\sqrt{23}/4)i$ **17.** $-2 + i$
19. $(-1/5) - (\sqrt{14}/5)i$ **21.** $0 - 3i$ **23.** $0 + 0i$ **25.** $-3 - i$ **27.** $5 - 6i$ **29.** $-6 + 0i$ **31.** $-4 + 0i$ **33.** $10 - 10i$
35. $13 + 0i$ **37.** $-3 - 4i$ **39.** $\frac{80}{9} - 2i$ **41.** 0 **43.** 0 **45.** $-6 - 6i$ **47.** 0 **49.** $-27 - 27i$ **51.** $3 - 4i$ **53.** $-i$
55. -7 **57.** $\frac{1}{5} - \frac{2}{5}i$ **59.** $0 + \frac{2}{5}i$ **61.** $-\frac{4}{17} - \frac{1}{17}i$ **63.** $\frac{7}{10} + \frac{1}{10}i$ **65.** $(1/3) - (2\sqrt{2}/3)i$ **67.** $0 + i$ **69.** $\frac{1}{2} - \frac{1}{2}i$
71. $-i$ **73.** $-i$ **75.** 1 **77.** i **79. (a)** $\overline{(1 + i) + (2 - 3i)} = \overline{3 - 2i} = 3 + 2i; \overline{1 + i} + \overline{2 - 3i} = (1 - i) + (2 + 3i) = 3 + 2i$
(b) $\overline{(1 + i)(2 - 3i)} = \overline{5 - i} = 5 + i; \overline{1 + i} \cdot \overline{2 - 3i} = (1 - i)(2 + 3i) = 5 + i$
(c) $\overline{(1 + i)^2} = \overline{2i} = -2i; (\overline{1 + i})^2 = (1 - i)^2 = -2i$
81. $\overline{z + w} = \overline{(a + c) + (b + d)i} = (a + c) - (b + d)i = (a - bi) + (c - di) = \overline{z} + \overline{w}$
83. If z is a real number, $z = a + 0i$, so $\overline{z} = a - 0i$ and $\overline{z} = z$.
85. $\overline{z^2} = \overline{(a + bi)(a + bi)} = \overline{(a^2 - b^2) + 2abi} = (a^2 - b^2) - 2abi = (a - bi)^2 = (\overline{z})^2$

Remember This (1.5)

1. $2\sqrt{7}$ **2.** $(5x - 1)(x - 2)$ **3.** $x(5x - 11)$ **4.** $\dfrac{b^2 - 4ac}{4a^2}$ **5.** $-\sqrt{7}, \sqrt{1000}$ **6.** $-\frac{5}{3}$ **7.** $3t(3t - 1)$ **8.** All real numbers
9. No   **10.** True

Exercises 1.6

1. $\{3\}$ **3.** $\{2\}$ **5.** $\{-6\}$ **7.** Set of all real numbers   **9.** $\emptyset$ **11.** $\{5\}$ **13.** $\{-5\}$ **15.** $\{-1\}$ **17.** $\{\frac{11}{12}\}$ **19.** $\{-\frac{8}{55}\}$ **21.** $\{4\}$
23. $\{\frac{5}{3}\}$ **25.** Set of all real numbers except 2   **27.** $\emptyset$ **29.** $\{7\}$ **31.** $\{0,5\}$ **33.** $\{4,-2\}$ **35.** $\{5,\frac{1}{3}\}$ **37.** $\{1,\frac{2}{3}\}$ **39.** $\{\pm 2i\sqrt{2}\}$
41. $\{3,7\}$ **43.** $\{-1 \pm i\}$ **45.** $\{7 \pm i\sqrt{7}\}$ **47.** $\{1 \pm i\}$ **49.** $\{(2 \pm \sqrt{10})/3\}$ **51.** $\{-4,-1\}$ **53.** $\{(1 \pm \sqrt{17})/2\}$
55. $\{(1 \pm \sqrt{19})/3\}$ **57.** $\{(1 \pm i\sqrt{2})/3\}$ **59.** $\{-1,\frac{3}{4}\}$ **61.** $\{3,-2\}$ **63.** $\{\pm 2i\sqrt{5}/5\}$ **65.** $\{0,\frac{2}{3}\}$ **67.** $\{1 \pm \sqrt{5}\}$
69. $\{60,9\}$ **71.** $\{0,10\}$ **73.** $\{(3 \pm \sqrt{33})/6\}$ **75.** $\{\frac{3}{5},\frac{5}{3}\}$ **77.** $\{5,-3\}$ **79.** $\{3\}$ **81.** -15; conjugate complex numbers
83. 0; real, rational, equal   **85.** 136; real, irrational, unequal   **87.** 84; real, irrational, unequal   **89.** -3; conjugate complex numbers
91. $40°, 60°, 80°$ **93.** $\$1.05, \$0.05$ **95.** 271 gallons   **97.** 93.3 mi   **99.** 38 lb of 35 percent tin, 57 lb of 10 percent tin
101. 20 quarts   **103.** 24,000 miles   **105.** 84   **107.** 3.8 in., 7.8 in.   **109. (a)** 6 seconds   **(b)** 5 seconds, 1 second
(c) The projectile attains the given height going up and coming down.   **111.** 2.3 seconds   **113.** 11.1 in.

115. $x_1 + x_2 = \dfrac{-b + \sqrt{b^2 - 4ac}}{2a} + \dfrac{-b - \sqrt{b^2 - 4ac}}{2a} = -\dfrac{2b}{2a} = -\dfrac{b}{a}$

$x_1 \cdot x_2 = \dfrac{-b + \sqrt{b^2 - 4ac}}{2a} \cdot \dfrac{-b - \sqrt{b^2 - 4ac}}{2a} = \dfrac{b^2 - (b^2 - 4ac)}{4a^2} = \dfrac{4ac}{4a^2} = \dfrac{c}{a}$

Remember This (1.6)

1. $x^2 - 17$ **2.** $x^2 + 8x + 16$ **3.** $4 - 4\sqrt{x} + x$ **4.** -3 **5.** $x(3x + 2)(3x - 2)$ **6.** 7 **7.** The distance between -2 and 0 on
the number line is 2 units.   **8.** $0 + 9\sqrt{2}i$ **9.** $\frac{3}{13} + \frac{2}{13}i$ **10.** Real number, rational number, integer

Exercises 1.7

1. $\{-1\}$ **3.** $\emptyset$ **5.** $\{\frac{25}{3}\}$ **7.** $\{6\}$ **9.** $\{-9\}$ **11.** $\{4\}$ **13.** $\{-2\}$ **15.** $\{3\}$ **17.** $\{0\}$ **19.** $\{-3,1\}$ **21.** $\emptyset$ **23.** $\{0,4\}$ **25.** $\{\frac{1}{2}\}$
27. Set of all nonnegative real numbers   **29.** $\emptyset$ **31.** $\{1,-9\}$ **33.** $\{1\}$ **35.** $\{27 \pm 10\sqrt{2}\}$ **37.** $\{\pm 1, \pm \sqrt{5}\}$ **39.** $\{\pm \sqrt{2}/2\}$
41. $\{\pm \sqrt{2 \pm \sqrt{3}}\}$ **43.** $\{1,-2\}$ **45.** $\{\frac{7}{2}\}$ **47.** $\{-1\}$ **49.** $\{16,1\}$ **51.** $\{0,-2,-7\}$ **53.** $\{0,1,-1\}$ **55.** $\{0,\frac{1}{2},-\frac{1}{3}\}$
57. $\{2,-2,1,-1\}$ **59.** $\{1,-1,\frac{11}{3}\}$ **61.** $\{2,-1 \pm i\sqrt{3}\}$ **63.** $\{1,-1,(1 \pm i\sqrt{3})/2,(-1 \pm i\sqrt{3})/2\}$ **65.** $\{3,-1\}$ **67.** $\{34,2\}$
69. $\{\frac{3}{2},-\frac{3}{2}\}$ **71.** $[0,\infty)$ **73.** $(-\infty,5]$ **75.** $\{1,\frac{1}{3}\}$ **77.** $\{1\}$ **79.** $\{2,-\frac{4}{3}\}$ **81.** $y = \dfrac{4 - x}{2}$ **83.** $y = x^3 - 2$
85. $b = \dfrac{2A - hc}{h}$ **87.** $p = \dfrac{a}{1 + rt}$ **89.** $d^2 = \dfrac{kL}{R}$ **91.** $R = \dfrac{E - CS}{C}$ **93.** $Z_2 = \dfrac{ZZ_1}{Z_1 - Z}$ **95.** $a = \dfrac{fb}{b - f}$ **97.** $x = \dfrac{3}{y - 1}$
99. $y' = \dfrac{2x + y}{2y - x}$ **101.** $y' = \dfrac{-x^2 - 2xy}{x^2 + y^2}$ **103.** $\frac{21}{5}$ cm

Remember This (1.7)

1. >   **2.** Because $2 = 2$ is true   **3.** Negative   **4.** Nonnegative real numbers   **5.** Undefined   **6.** $\dfrac{x^2 - ax + a^2}{x - a}$ **7.** $\{4, -\frac{1}{2}\}$

8. $\frac{1}{2} + \frac{3}{2}i, \frac{1}{2} - \frac{3}{2}i$ **9.** Real, irrational, unequal   **10.** 11 seconds

Exercises 1.8

1. $(-\infty, -\frac{9}{2})$

3. $(-\infty, -6)$

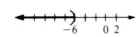

5. $(-\infty, \frac{12}{7})$

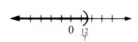

7. $(-8, \infty)$

9. Set of all real numbers

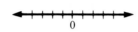

11. $(0, \infty)$

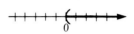

13. $\emptyset$

15. $\{x: 1 \le x < 4\}$

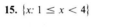

17. $(-\infty, 3]$

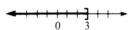

19. $(-3, 0); \{x: -3 < x < 0\}$

21. $\{x: x \ge 0\}$

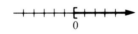

23. $(-\infty, -2) \cup (2, \infty); \{x: x < -2 \text{ or } x > 2\}$ **25.** $(2, 6)$ **27.** $[-5, 7]$ **29.** $(\frac{7}{4}, 4]$

31. $(-4, 4)$

33. $[2, 8]$

35. $(a - 1, a + 1)$

37. $(-\infty, -2) \cup (2, \infty)$

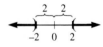

39. $(-\infty, 2] \cup [8, \infty)$

41. $(-\frac{13}{3}, 1)$ **43.** $[-6, 7]$ **45.** $(-\infty, -1) \cup (2, \infty)$ **47.** $\emptyset$ **49.** $(\frac{7}{4}, \frac{7}{2})$ **51.** $(-\infty, -1) \cup (2, \infty)$ **53.** $[0, 1]$
55. $(-3, 2)$ **57.** $(-\infty, -1] \cup [1, \infty)$ **59.** $(-\infty, -\sqrt{3}) \cup (\sqrt{3}, \infty)$ **61.** $(-\infty, 0) \cup (5, \infty)$ **63.** $[-2, 1]$
65. $(-\infty, -\frac{1}{2}] \cup [4, \infty)$ **67.** $(-\infty, \infty)$ **69.** $\{1\}$ **71.** $(-\infty, 2 - \sqrt{7}) \cup (2 + \sqrt{7}, \infty)$ **73.** $[(2 - \sqrt{10})/3, (2 + \sqrt{10})/3]$
75. $\emptyset$ **77.** $[0, 1] \cup [2, \infty)$ **79.** $(-\infty, 0)$ **81.** $(1, 2)$ **83.** $(-2, 0) \cup (2, \infty)$ **85.** $[-2, -1] \cup [1, 2]$
87. $[-3, -1] \cup [0, 1] \cup [3, \infty)$ **89.** $(-\infty, -1) \cup (1, \frac{5}{2})$ **91.** $(-2, 3]$ **93.** $(-\infty, -1] \cup (0, 1]$ **95.** $(-\infty, -2] \cup [2, 3)$
97. $(-\infty, 0) \cup (\frac{1}{2}, \infty)$ **99.** $(-\infty, -21] \cup (7, \infty)$ **101.** $(-\infty, 4) \cup [12, \infty)$ **103.** $(-3, 2)$
105. [24,500 mi, 25,500 mi]; [24,855 mi, 24,865 mi)   **107.** $(0°, 40°)$ **109.** $[71, 91]$ **111.** $(-7, -3) \cup (2, 6)$
113. (1 second, 5 seconds)   **115.** $(-\infty, -2] \cup [2, \infty)$ **117.** $\{k: -6 < k < 6\}$ **119.** $a < b$ implies $b - a$ is positive. Then $b - a = (b + c) - (a + c)$, so $(b + c) - (a + c)$ is positive, which implies $a + c < b + c$.   **121.** If $a < b$ and $b < c$, then $b - a$ and $c - b$ are both positive. Therefore, the sum $(b - a) + (c - b)$, which simplifies to $c - a$, is positive, which implies $a < c$.
123. $(\sqrt{b} - \sqrt{a})^2 \ge 0$ implies $b - 2\sqrt{ab} + a \ge 0$. Then $-2\sqrt{ab} \ge -(a + b)$, so $\sqrt{ab} \le (a + b)/2$.

Remember This (1.8)

1. $(-\infty,\infty)$ **2.** $\{0,2,\frac{3}{2}\}$ **3.** $\{1\}$ **4.** $\dfrac{-(3x-1)(x+3)}{9}$ **5.** $\dfrac{z^2}{x^{1/2}y}$ **6.** $y=\pm\sqrt{1-x^2}$ **7.** $x=ab/(a+b)$ **8.** Solution, since

$(5-i)^2+26=50-10i$ and $10(5-i)=50-10i$ **9.** $[-2,2]$ **10.** Height: 5 cm; base: 12 cm

Chapter 1 Review Exercises

1. $(1-x)^{2b}$ **3.** $\dfrac{2x}{3y^3}$ **5.** $\dfrac{x^2+1}{4x}$ **7.** $\dfrac{72y^2z^3}{x^4}$ **9.** $16+12x-2x^2-2x^3$ **11.** $-\frac{19}{8}$ **13.** $2R\sqrt{3}$ **15.** $\dfrac{s-1}{s+1}$

17. $\dfrac{-2}{(x+1)(x+h+1)}$ **19.** $\dfrac{x-y}{x+y}$ **21.** $x+a+1$ **23.** $\dfrac{2x^{2/3}+1}{2x^{1/3}}$ **25.** $\dfrac{3x-3}{2x^{1/2}}$ **27.** $\dfrac{2x^2}{\sqrt{1-x^2}}=\dfrac{2x^2\sqrt{1-x^2}}{1-x^2}$

29. $\dfrac{-4}{u^2\sqrt{u^2+4}}=\dfrac{-4\sqrt{u^2+4}}{u^2(u^2+4)}$ **31.** $\{2\}$ **33.** $[0,\infty)$ **35.** $\{\frac{25}{2}\}$ **37.** $(-5,5)$ **39.** $\{-7\}$ **41.** $(-\infty,1]\cup[2,\infty)$

43. $\{200,-200\}$ **45.** $\{-5\pm3i\}$ **47.** $\{6,-1\}$ **49.** $\{0,-2\}$ **51.** $(-\infty,-3)\cup(0,3)$ **53.** $\{\frac{3}{2},-4\}$ **55.** $\{1,-1,5\}$

57. $(-\infty,1)$ **59.** $(-3,1)\cup[5,\infty)$ **61.** $t^3(1+x)(1-x)$ **63.** $(x+1)(x^2+1)$ **65.** $\dfrac{5(2-3x^3)}{x^6}$ **67.** False, $\sqrt{2}$ **69.** $\frac{467}{99}$

71. Real, rational   **73.** **(a)** $(-\infty,0]\cup(6,\infty)$, $\{x: x\le0$ or $x>6\}$ **(b)** $(-4,-1)$, $\{x: -4<x<-1\}$ **75.** 10   **77.** $6-8i$

79. $y=\dfrac{x-1}{x+2}$ **81.** $y'=-(y/x)^{4/3}$ **83.** $\dfrac{\sqrt{x^2-4}}{x+2}$ **85.** $\dfrac{-1}{\sqrt{x}\sqrt{x+h}(\sqrt{x}+\sqrt{x+h})}$ **87.** $\dfrac{4x^2-2x+2x-1-2x^2+2x}{(2x+1)^2}$

$=\dfrac{2x^2+2x-1}{(2x+1)^2}$ **89.** $\dfrac{xy(1+x-2y)}{(x-2y)^3}$ **91.** $\dfrac{R_1R_2R_3}{R_2R_3+R_1R_3+R_1R_2}$ **93.** $(-\infty,-3]\cup[1,\infty)$ **95.** 9 ft   **97.** \$85,106

99. 2 seconds; [0 seconds,1.5 seconds]   **101.** (d)   **103.** (a)   **105.** (c)   **107.** (d)   **109.** (a)

Chapter 1 Test

1. Real number, rational number, integer   **2.** x/y^a **3.** $2x(x+2)(x-2)$ **4.** $2x+h$ **5.** $\dfrac{-x^2+4x-1}{x^2-1}$ **6.** $\dfrac{a}{b-a}$ **7.** $\frac{29}{3}$

8. $\dfrac{x^{2/3}+1}{x^{1/3}}$ **9.** $\dfrac{3+\sqrt{5}}{2}$ **10.** $\frac{1}{2}+\frac{1}{2}i$ **11.** $\{-\frac{23}{24}\}$ **12.** $\{0,-1,-\frac{2}{3}\}$ **13.** $(-5,4)$ **14.** $(0,\frac{2}{3})$ **15.** -1 **16.** $C_1=\dfrac{CC_2}{C_2-C}$

17. $(-\infty,-1]\cup[0,\infty)$ **18.** 120 mi/hour   **19.** 4.8 ft   **20.** 0.6 second

Exercises 2.1

1. Not a function   **3.** Function   **5.** Function   **7.** Not a function   **9.** Function   **11.** Not a function   **13.** $D: (0,\infty)$; $R: (0,\infty)$

15. $D: (-\infty,\infty)$; $R: [3,\infty)$ **17.** $D: (-\infty,\infty)$; $R: [0,\infty)$ **19.** $D: (-\infty,\infty)$; $R: (-\infty,1]$ **21.** $D:$ Set of all real numbers except -1;

$R:$ Set of all real numbers except 0   **23.** $D: (-\infty,\infty)$; $R: \{0\}$ **25.** $D: [1,\infty)$; $R: [0,\infty)$ **27.** $D: [-2,2]$; $R: [-2,0]$ **29.** $D: (-\infty,4]$;

$R: (0,4]$ **31.** $(-2,-5)$, $(1,4)$ **33.** $(1,1)$, $(-1,1)$ **35.** $(0,7)$, $(\frac{7}{3},0)$, $(-5,22)$, $(\frac{2}{3},5)$ **37.** 0   **39.** 2   **41.** Function   **43.** Function

45. Not a function   **47.** $D: \{-2,-1,1,2\}$; $R: \{1,4\}$ **49.** 11, 5, -3 **51.** 19, 4, 7   **53.** $\frac{2}{5}$, -2, undefined   **55.** **(a)** 3   **(b)** 1

(c) -1 **57.** **(a)** 2   **(b)** -10 **(c)** 10   **(d)** -4 **(e)** 3   **(f)** $\frac{1}{2}$ **(g)** 4   **(h)** 4   **(i)** 2   **(j)** x^2+1 **59.** **(a)** 3   **(b)** -1 **(c)** No

61. **(a)** -2 **(b)** -1 **(c)** Undefined   **63.** **(a)** 3   **(b)** 0   **(c)** 3   **65.** **(a)** $2(x+h)^2-1$ **(b)** $4xh+2h^2$ **(c)** $4x+2h$

67. $2x+h$ **69.** $-2x-h$ **71.** 2   **73.** 0   **75.** $-1/[x(x+h)]$ **77.** $A=s^2$; $(0,\infty)$ **79.** $L=A/5$; $[25,\infty)$

81. $s=P/4$; $(0,\infty)$ **83.** $d=40t$; $[0,\infty)$ **85.** $e=28n$; $[0,\infty)$ **87.** $e=0.06p$; $[0,\infty)$ **89.** $c=5x+400$;

$\{x: x\ge0$, where x is an integer$\}$ **91.** $a=10,000-50n$; $[0,200]$ **93.** $t=310+0.19(i-2,000)$; $[2,000,4,000]$

95. $c=\begin{cases}1.75 & \text{if } 0\le n\le 12 \\ 1.75+0.0382(n-12) & \text{if } 12<n\le 48;\end{cases}$ $[0,48]$ **97.** $A=(12-2x)x$; $(0,6)$ **99.** $A=(100-2x)x$; $(0,50)$

101. $A=\left(\dfrac{100}{w}+10\right)(w+6)$; $(0,10)$

Remember This (2.1)

1. 5   **2.** 164   **3.** 13 ft   **4.** $4\sqrt{3}$ **5.** $2\sqrt{10}$ **6.** $-(x-5)$ **7.** -8 **8.** 3   **9.** $[-4,\infty)$ **10.** $[0,\infty)$

Exercises 2.2

1.

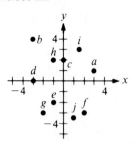

3. (a) 2 **(b)** 3 **(c)** 4 **(d)** 5 **(e)** 13 **(f)** $\sqrt{17}$ **(g)** $\sqrt{40}$ or $2\sqrt{10}$ **(h)** $\sqrt{41}$ **(i)** $\sqrt{2}$ **(j)** $\sqrt{17}$ **(k)** 10 **(l)** 13 **5.** $(4,-4)$; 30 **7.** $d_1 = \sqrt{40}$, $d_2 = \sqrt{90}$, $d_3 = \sqrt{130}$; since $(d_3)^2 = (d_1)^2 + (d_2)^2$, the Pythagorean theorem ensures a right triangle. **9.** $\sqrt{13}$

11.

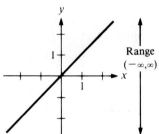

13.

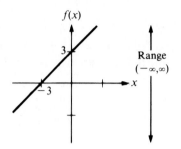

15.

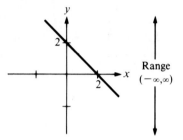

17.

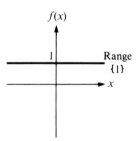

19.

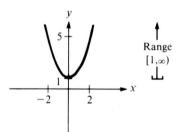

21.

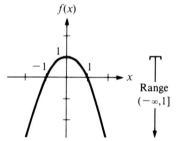

23.

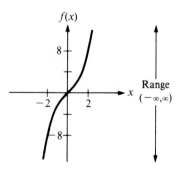

25.

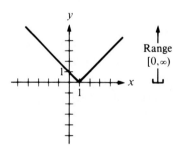

27.

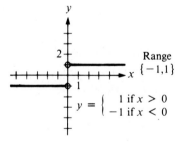

29.

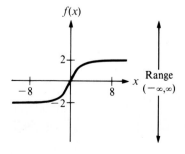

31.

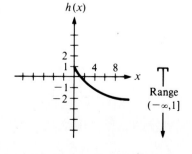

33.

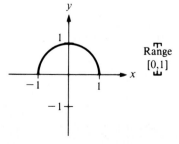

35. (a)

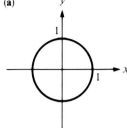

(b)

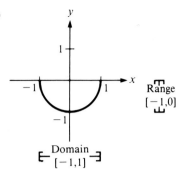

Range
[−1,0]

Domain
[−1,1]

37. (a) 0   **(b)** 1   **(c)** −2   **(d)** −4
(e)

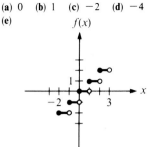

39.

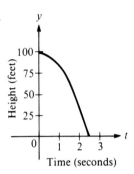

41.

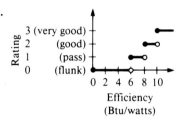

43.

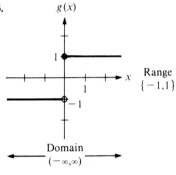

Range
{−1,1}

Domain
$(-\infty,\infty)$

45.

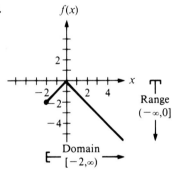

Domain
[−2,∞)

Range
$(-\infty,0]$

47.

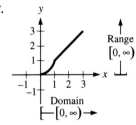

Range
[0,∞)

Domain
[0,∞)

49.

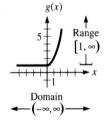

Range
[1,∞)

Domain
$(-\infty,\infty)$

51. Domain: $(-\infty,\infty)$; range: $(0,\infty)$ **53. (a)** $(-\infty,\infty)$ **(b)** $(-\infty,4]$ **(c)** $f(0) = -5; f(5) = 0$ **(d)** {3}   **(e)** {1,5}   **(f)** {x: 1 < x < 5}
(g) {x: x < 1 or x > 5}   **(h)** {x: 0 < x < 6}   **55.** Function   **57.** Not a function   **59.** Not a function   **61.** Function   **63.** Function

Remember This (2.2)

1. Identity   **2.** $-x^2 + 2$ **3.** Yes   **4. (a)** 28   **(b)** 7   **5. (a)** $a^2 + 4a + 4$ **(b)** $a^2 + 2$ **6.** Yes   **7.** $-\frac{1}{2}$
8. All real numbers except 4   **9.** {−2,−1,0}   **10.** $A = b^2/4$

Exercises 2.3

1. Even   **3.** Odd   **5.** Neither   **7.** Even   **9. (a)**

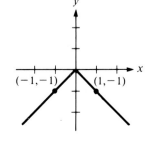

(−1,−1)   (1,−1)

(b)

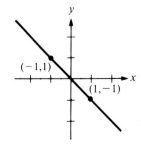

(−1,1)

(1,−1)

11. (a)

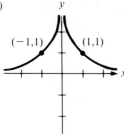

(b)

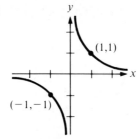

13. Odd   **15.** Even   **17.** Neither
19. Even   **21.** Odd

23. Even

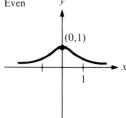

25.

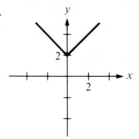

27.

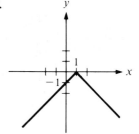

29.

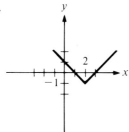

31.

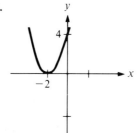

33.

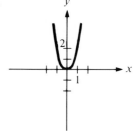

35.

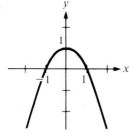

37.

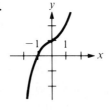

39.

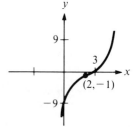

41.

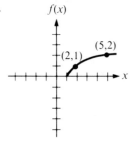

43.

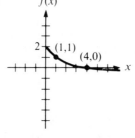

45.

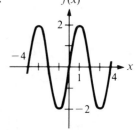

47.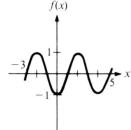

49.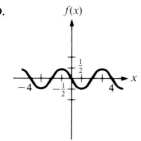

Remember This (2.3)

1. $\{2,3\}$ **2.** $\dfrac{x^2 + x + 1}{x(x + 1)}$ **3.** $-3x^3 + 18x^2 + 2x - 12$ **4.** $-9x + 5$ **5.** $\sqrt{x}$ **6. (a)** $[0,\infty)$ **(b)** $(0,\infty)$ **7.** $\{0,-3,-5\}$

8. $3x^2 + 6xh + 3h^2 - x - h + 5$ **9.** 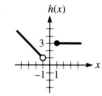   **10.** 12π

Exercises 2.4

1. $(f + g)(x) = 3x - 1$, $(f - g)(x) = x + 1$, $(f \cdot g)(x) = 2x^2 - 2x$, $(f/g)(x) = 2x/(x - 1)$, $(f \circ g)(x) = 2x - 2$, $(g \circ f)(x) = 2x - 1$; $D_{f/g}$ is the set of all real numbers except 1, otherwise the domain is $(-\infty,\infty)$.   **3.** $(f + g)(x) = 4x^2 - x - 2$, $(f - g)(x) = 4x^2 + x - 8$, $(f \cdot g)(x) = -4x^3 + 12x^2 + 5x - 15$, $(f/g)(x) = (4x^2 - 5)/(-x + 3)$, $(f \circ g)(x) = 4x^2 - 24x + 31$, $(g \circ f)(x) = -4x^2 + 8$; $D_{f/g}$ is the set of all real numbers except 3, otherwise the domain is $(-\infty,\infty)$.   **5.** $(f + g)(x) = x^2 + 1$, $(f - g)(x) = x^2 - 1$, $(f \cdot g)(x) = x^2$, $(f/g)(x) = x^2$, $(f \circ g)(x) = 1$, $(g \circ f)(x) = 1$; Domain in all cases is $(-\infty,\infty)$.   **7.** $(f + g)(x) = (x^2 + x + 1)/(x^2 + x)$, $(f - g)(x) = (1 + x - x^2)/(x^2 + x)$, $(f \cdot g)(x) = 1/(1 + x)$, $(f/g)(x) = (1 + x)/x^2$, $(f \circ g)(x) = (1 + x)/x$, $(g \circ f)(x) = 1/(x + 1)$; Domain in all cases is the set of all real numbers except 0 and -1.   **9.** $(f + g)(x) = x^2 - 1 + \sqrt{x + 1}$, $(f - g)(x) = x^2 - 1 - \sqrt{x + 1}$, $(f \cdot g)(x) = (x^2 - 1)\sqrt{x + 1}$, $(f/g)(x) = (x - 1)\sqrt{x + 1}$, $(f \circ g)(x) = x$, $(g \circ f)(x) = |x|$; $D_{f+g} = D_{f-g} = D_{f \cdot g} = D_{f \circ g} = [-1,\infty)$, $D_{f/g} = (-1,\infty)$, $D_{g \circ f} = (-\infty,\infty)$ **11.** $f + g = \{(2,3), (3,5)\}$, $f - g = \{(2,5), (3,5)\}$, $f \cdot g = \{(2,-4), (3,0)\}$, $f/g = \{(2,-4)\}$, $f \circ g = \{(3,2), (4,3), (5,4)\}$, $g \circ f = \{(0,-1), (1,0), (2,1), (3,2)\}$; $D_{f+g} = D_{f-g} = D_{f \cdot g} = \{2,3\}$, $D_{f/g} = \{2\}$, $D_{f \circ g} = \{3,4,5\}$, $D_{g \circ f} = \{0,1,2,3\}$ **13. (a)** -3 **(b)** 20 **(c)** Undefined   **(d)** 4 **15.** -1 **17.** -2 **19.** 0 **21.** -1 **23.** $(f \circ g)(x) = x$, $(g \circ f)(x) = x$

25. $(g \circ f)(x) = (5x - 4)^3$ **27.** $(f \circ g)(t) = 36\pi t^2$ **29.** $g(x) = 4x - 1$; $f(x) = x^3$ **31.** $g(x) = \dfrac{x - 1}{x + 1}$; $f(x) = x^{1/2}$

33. $g(x) = 2x + 1$; $f(x) = \sqrt[3]{x}$ **35.** $g(x) = 3 - x$; $f(x) = 2x^4$ **37.** $g(x) = 1 - x$; $f(x) = x^3 + 6x^2$ **39.** $g(x) = 2x^2 + 2$

41. -12 **43.** $(f \circ g \circ h)(x) = 2(x - 2)^2$ **45.** $(f \circ g \circ h)(x) = \sqrt{1/(1 - x)}$

47. $(f + g)(-x) = f(-x) + g(-x) = -f(x) - g(x) = -[f(x) + g(x)] = -(f + g)(x)$ **49.** Let f be an even function and g be an odd function. Then $(f \cdot g)(-x) = f(-x) \cdot g(-x) = f(x) \cdot [-g(x)] = -[f(x) \cdot g(x)] = -(f \cdot g)(x)$, so the product is an odd function.

51. (a) $y = 1{,}760m$ **(b)** $f = 3y$ **(c)** $f = 3(1{,}760m) = 5{,}280m$ **53.** $(f \circ g)(t) = \frac{1}{6}\pi t^3$; $V = (f \circ g)(t) = \frac{1}{6}\pi t^3$ gives the volume of the balloon t seconds after the start of inflation.

Remember This (2.4)

1. $y = -9x + 4$ **2.** $y = x^3 + 5$ **3.** x **4.** No; $\sqrt{x^2} = |x|$ since $\sqrt{x^2} = x$ if $x \ge 0$, and $\sqrt{x^2} = -x$ if $x < 0$.   **5.** No   **6.** No

7. 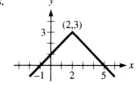   **8.**   **9.** $[0,\infty)$ **10. (a)** No   **(b)** Yes

Exercises 2.5

1. $f^{-1} = \{(7,1),(8,2),(9,3)\}$; $D_{f^{-1}} = \{7,8,9\}$; $R_{f^{-1}} = \{1,2,3\}$; $D_f = \{1,2,3\}$; $R_f = \{7,8,9\}$ **3.** $f^{-1} = \{(10,-2),(0,0),(-1,5)\}$;
$D_{f^{-1}} = \{10,0,-1\}$; $R_{f^{-1}} = \{-2,0,5\}$; $D_f = \{-2,0,5\}$; $R_f = \{10,0,-1\}$ **5.** $f^{-1} = \{(4,4),(5,5),(6,6)\}$; $D_{f^{-1}} = R_{f^{-1}} = D_f = R_f = \{4,5,6\}$
7. No   **9.** Yes   **11.** No   **13.** No   **15.** Yes   **17.** Yes

19.

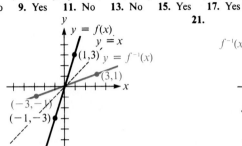

21.

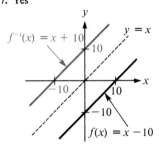

23.

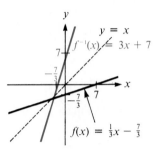

25. $y = x^2$ is not a one-to-one function, so no inverse function exists.
27. $f(x) = 2$ is not a one-to-one function, so no inverse function exists.

29.

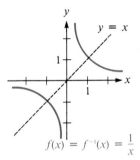

31.

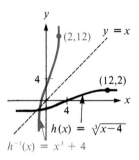

33.

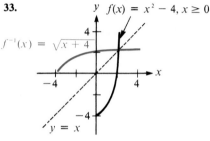

35.

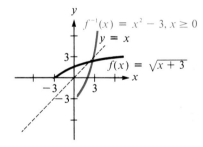

37. $f(x) = \sqrt{1 - x^2}$ is not a one-to-one function, so no inverse function exists.   **39.** $\frac{2}{7}$ **41.** $f^{-1}(x) = x^2 + 1$, $x \geq 0$;
$D_f = R_{f^{-1}} = [1,\infty)$, $R_f = D_{f^{-1}} = [0,\infty)$ **43.** $f[g(x)] = 3(\frac{1}{3}x + \frac{2}{3}) - 2 = x$; $g[f(x)] = \frac{1}{3}(3x - 2) + \frac{2}{3} = x$

45. $f[g(x)] = \dfrac{1}{[(1 - 4x)/x] + 4} = \dfrac{x}{1 - 4x + 4x} = x$; $g[f(x)] = \dfrac{1 - 4[1/(x + 4)]}{1/(x + 4)} = \dfrac{(x + 4) - 4}{1} = x$

47. $f[g(x)] = \sqrt{(x^2 - 1) + 1} = \sqrt{x^2} = x$ (since $x \geq 0$); $g[f(x)] = (\sqrt{x + 1})^2 - 1 = (x + 1) - 1 = x$

49. $f[f(x)] = \dfrac{2[(2x + 1)/(3x - 2)] + 1}{3[(2x + 1)/(3x - 2)] - 2} = \dfrac{2(2x + 1) + (3x - 2)}{3(2x + 1) - 2(3x - 2)} = \dfrac{7x}{7} = x$

51. Not inverses since $f[g(x)] = 3(\frac{1}{3}x + 1) - 1 = x + 2$ **53.** $y = \sqrt{x}$, $x > 0$; This is the formula for the side length (y) of a square
in terms of the area (x).

Remember This (2.5)

1. $24xy/d^2$ **2.** $k = 441$ **3.** $y = k/x$ **4.** $T = 0.06p$ **5.** $\ell = 30/w$ **6.** Domain: Set of all real numbers except 3; range: $\{2,-2\}$

7. Odd   **8.** 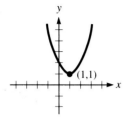   **9.** 37   **10.** $(g \circ f)(x) = \dfrac{8x + 3}{x}$

Exercises 2.6

1. $y = \frac{7}{3}x; \frac{70}{3}$ **3.** $y = \frac{3}{4}x^2; 12$ **5.** $y = 72/x; 3$ **7.** $y = 24/x^3; \frac{8}{9}$ **9.** 60   **11.** 49   **13.** 4 in.   **15.** 2,000,000 tons
17. 3 in.   **19.** 600 rpm   **21.** $500\pi/3$ cubic units   **23.** 181 lb   **25.** Exposure time is multiplied by 4.   **27.** $\frac{5}{3}$ atm
29. Force is multiplied by 24.

Remember This (2.6)

1. 5   **2.** 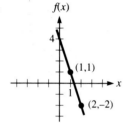   **3.** Yes   **4.** $f^{-1}(x) = 6x + 6$ **5.** $g(x) = 5x - 2; f(x) = 3x^5$

6. $(g - f)(x) = \dfrac{x^3 - 1}{x}$; Set of all real numbers except 0   **7.** $3\sqrt{5}$ **8.**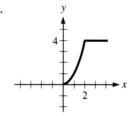

9. 0   **10.** $A = \dfrac{C^2}{4\pi}$

Chapter 2 Review Exercises

1. $(0,4), (\frac{5}{2},-1), (-1,6), (2,0), (3,-2)$ **3.** Yes   **5.** Yes; no   **7.** $-\frac{1}{2}$ **9.** $y = \frac{5}{12}x; \frac{25}{3}$ **11.** Set of all real numbers except
0 and -1 **13.** $A = d^2/2$ **15.** Yes   **17.** No   **19.** Yes   **21.** Odd function
23. 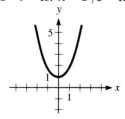   **25.** 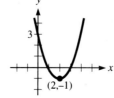   **27.** $\frac{9}{4}$ **29.** $A = \pi d^2/4$

31.

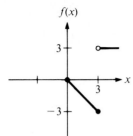

33. $\{y: y = 3 \text{ or } -3 \leq y \leq 0\}$

35.

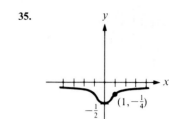

37.

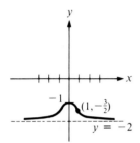

39.

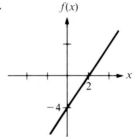

41.

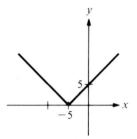

43. Even function   **45. (a)** $g^{-1}(x) = -2x + 5$ **(b)** 7   **47.** $[-\pi/2, \pi/2]$ **49.** 0   **51.** (0,1]   **53.**

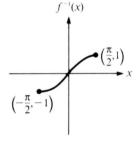

55. $6x + 3h$ **57.** -1 **59.** $A = (\sqrt{3}/4)x^2$ **61.** $(f \cdot g)(x) = x$; set of all real numbers except 0
63. $e = \begin{cases} 500 & \text{if } 0 \leq a \leq 2{,}000 \\ 500 + 0.09(a - 2{,}000) & \text{if } a > 2{,}000 \end{cases}$
65. $[-c,c]$ **67.** Even function   **69.** $\{-c,0,c\}$ **71.**

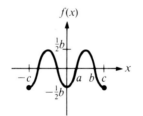

73. (a) $D: (-\infty,\infty); R: [-1,1]$ **(b)** $D: [-3,3]; R: [-3,0]$ **75.** $f[g(x)] = \frac{1}{2}(2x + 6) - 3 = x; g[f(x)] = 2(\frac{1}{2}x - 3) + 6 = x$
77. (b)   **79. (d)**   **81. (b)**   **83. (a)**   **85. (a)**

Chapter 2 Test

1. Set of all real numbers except 9   **2.** $(-\infty,6]$ **3.** 38   **4.** $2x + h - 5$ **5.** $A = x(400 - x)$ **6.** 10π **7.** Yes   **8.** Yes
9. Domain: $[-5,5]$; range: $[0,5]$
10.

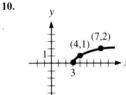

11.

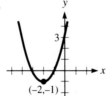

12.

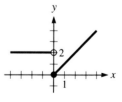

13. Even   **14.** $\{(0,2), (1,-2)\}$ **15.** $(f \circ g)(x) = 6x^2 + 3$ **16.** $g(x) = 25 - x^2; f(x) = \sqrt{x}$ **17.** $f^{-1}(x) = \sqrt[3]{x + 5}$
18. $f[g(x)] = 7(\frac{1}{7}x + \frac{2}{7}) - 2 = x; g[f(x)] = \frac{1}{7}(7x - 2) + \frac{2}{7} = x$ **19.** 6.4 in.   **20.** Exposure time is divided by 9.

Exercises 3.1

1. Yes; 3   **3.** Yes; 1   **5.** No   **7.** No   **9.** Yes; 0   **11.** 1   **13.** $\frac{4}{7}$ **15.** $-\frac{11}{2}$ **17.** 0   **19.** Undefined   **21.** $y = 5x - 17$
23. $y = \frac{1}{2}x + 1$ **25.** $y = -\frac{1}{2}x$ **27.** $y = -x + 5$ **29.** $y = -x - 6$ **31.** 1; (0,7)   **33.** 5; (0,0)   **35.** 0; (0,-2)
37. $-\frac{2}{3}$; $(0,-\frac{2}{3})$ **39.** 6; (0,7)

41.

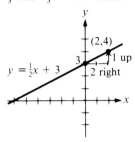

43.

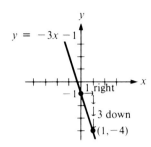

45.

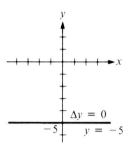

47. $f(x) = \frac{2}{3}x - 2$ **49.** $f(x) = -\frac{4}{3}x - \frac{13}{3}$ **51.** $f(x) = 1$ **53.** -1 **55.** $-\frac{11}{3}$ **57.** (a) $y = x + 1$ (b) $y = -x + 3$
59. (a) $y = -2x + 6$ (b) $y = \frac{1}{2}x - \frac{3}{2}$ **61.** (a) $y = -\frac{1}{7}x$ (b) $y = 7x$ **63.** (a) $y = \frac{5}{3}x - \frac{14}{3}$ (b) $y = -\frac{3}{5}x - \frac{12}{5}$
65. (a) $y = -\frac{1}{4}x + \frac{13}{4}$ (b) $y = 4x - 18$ **67.** (a) $y = \frac{3}{2}x - \frac{13}{12}$ (b) $y = -\frac{2}{3}x$ **69.** (a) -1 (b) $\frac{1}{2}$ (c) 0   (d) Undefined
(e) 3   **71.** (a) $y = 0.6x + 40$ (b) \$55   (c) \$40   (d) \$0.60   **73.** (a) $v = -32t - 220$ (b) -348 ft/second   (c) -220 ft/second
(d) $+$ ball rising, $-$ ball falling   **75.** $m_{AB} = 4$, $m_{BC} = -\frac{1}{4}$; Since $m_{AB} \cdot m_{BC} = -1$, the segments are perpendicular, ensuring a right
triangle.   **77.** 5   **79.** $m_1 = -A_1/B_1$, $m_2 = -A_2/B_2$; Since the lines are parallel, $-A_1/B_1 = -A_2/B_2$, so $A_1B_2 = A_2B_1$.
81. m; The quotient is the slope of the graph of f.

Remember This (3.1)

1.

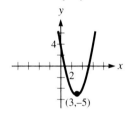

2.

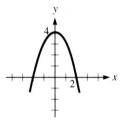

3. $\{3 \pm \sqrt{11}\}$ **4.** $\{-3,1\}$ **5.** $(-\infty,-3) \cup (1,\infty)$ **6.** $\emptyset$ **7.** 180   **8.** $\frac{1}{4}$ **9.** 0   **10.** $[-1,\infty)$

Exercises 3.2

1.

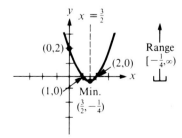

3.

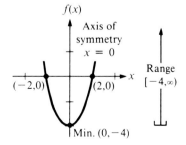

5.

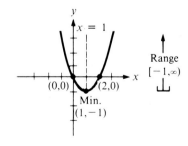

7.

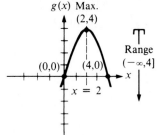

9.

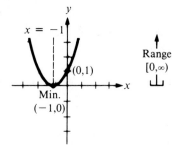

11.

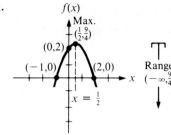

13.

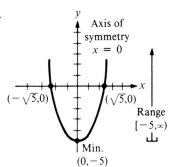

15.

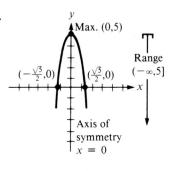

17.

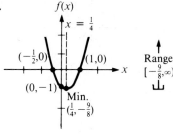

19.

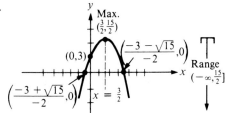

21. $(-2, -3)$; $x = -2$ **23.** $(0, -4)$; $x = 0$ **25.** $(1,1)$; $x = 1$ **27.** $(\frac{3}{2}, -\frac{23}{4})$; $x = \frac{3}{2}$ **29.** $(\frac{1}{2}, \frac{9}{4})$; $x = \frac{1}{2}$

31. Set of all real numbers   **33.** 144 ft; 6 seconds   **35.** 360 watts   **37.** 10, 10   **39.** 10, 10   **41.** 200 yd by 100 yd   **43.** 3 in.

45. \$15   **47.** $(-\infty, -1) \cup (5, \infty)$ **49.** $(-1, 5)$ **51.** $(-\infty, -1] \cup [2, \infty)$ **53.** $(-\infty, -2] \cup [2, \infty)$ **55.** $(0,3)$

57. $[-2, \frac{3}{2}]$ **59.** $(-1, \frac{5}{3})$ **61.** Set of all real numbers except -1 **63.** $\emptyset$ **65.** $[-1, 1]$

Remember This (3.2)

1. $25x - 4$ **2.** $-\frac{35}{8}$ **3.** 0   **4.** $2x^3 - 3x^2 - 2x - 5$ **5.** 5   **6.** Using dividend = (divisor)(quotient) + remainder,

$266 = 9(29) + 5$ checks.   **7.** $-\frac{3}{8}$ **8.** $y = -\frac{3}{8}x - \frac{9}{4}$ **9.** $\frac{5}{2}$; $(0, -4)$ **10.** $y = -\frac{1}{3}x - 6$

Exercises 3.3

1. $x^2 + 7x - 2 = (x + 5)(x + 2) - 12$ **3.** $6x^3 - 3x^2 + 14x - 7 = (2x - 1)(3x^2 + 7) + 0$

5. $3x^4 + x - 2 = (x^2 - 1)(3x^2 + 3) + x + 1$ **7.** $\dfrac{x^2 - 5}{x + 1} = x - 1 + \dfrac{-4}{x + 1}$ **9.** $\dfrac{3x^4 - 5x^2 + 7}{x^2 + 2x + 1} = 3x^2 - 6x + 4 + \dfrac{-2x + 3}{x^2 + 2x + 1}$

11. $\dfrac{x^3 + 1}{x(x - 1)} = x + 1 + \dfrac{x + 1}{x(x - 1)}$ **13.** $x^3 - 5x^2 + 2x - 3 = (x - 1)(x^2 - 4x - 2) - 5$

15. $2x^3 + 9x^2 - x + 14 = (x + 5)(2x^2 - x + 4) - 6$ **17.** $7 + 6x - 2x^2 - x^3 = (x + 3)(-x^2 + x + 3) - 2$

19. $2x^3 + x - 5 = (x + 1)(2x^2 - 2x + 3) - 8$ **21.** $x^4 - 16 = (x - 2)(x^3 + 2x^2 + 4x + 8) + 0$

23. 0, 0   **25.** $-9, 87$ **27.** 0, 0   **29.** $-\frac{176}{27}, -\frac{10}{3}$

Remember This (3.3)

1. Yes   **2.** $(r_1, 0)$, $(r_2, 0)$ **3.** 0   **4.** 4   **5.** $-6 + 4i$ **6.** $-\sqrt{2}, \sqrt[3]{-2}, -2 + \sqrt{3}$ **7.** 14   **8.** $-\frac{1}{3}$ **9.** $x = -6$ **10.** $(\frac{3}{2}, \frac{9}{4})$

Exercises 3.4

1. Degree 2; -3 is a zero of multiplicity 2.   **3.** Degree 3; -4, $\sqrt{2}$, and $-\sqrt{2}$ are zeros of multiplicity 1.
5. $P(x) = (x - 1)(x - 2)(x - 3)$ **7.** $P(x) = x(x + 4)(x - i)(x + i)$ **9.** $P(x) = (x - 1)^2(x - 5)^2$

11.

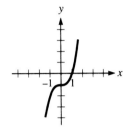

13.

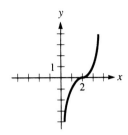

15.

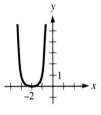

17.

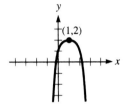

19.

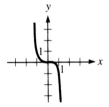

21.

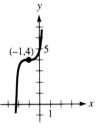

23.

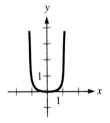

25.

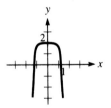

27.

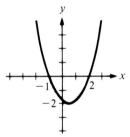

29.

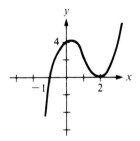

31.

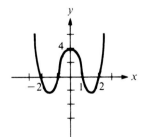

33.

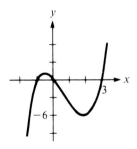

35.

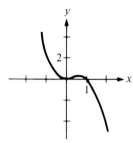

37. $P(x) = (x - 2)(x - \sqrt{3})(x + \sqrt{3})$
39. $P(x) = x(x - (4 - \sqrt{3}))(x - (4 + \sqrt{3}))(x - (2 + 3i))(x - (2 - 3i))$
41. $P(x) = (x - 3)(x + 3)(x - 2i)(x + 2i)$ **43.** $2 - \sqrt{7}$ **45.** $-i\sqrt{2}$
47. $2 - 2i$ **49.** $-\frac{1}{2}$ and -1 (multiplicity 2)   **51.** $\pm\sqrt{2}$ **53.** $\pm i\sqrt{2}, -\sqrt{2}$
55. $1, -1, -i$

57.

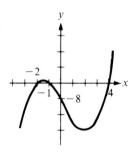

59. True   **61.** False   **63.** Since $x - b$ is a factor of $P(x)$, we know $P(x)$
$= (x - b)Q(x)$. Then $P(b) = (b - b)Q(b) = 0$, so b is a zero of $y = P(x)$.

Remember This (3.4)

1. $\frac{2}{3}, -\frac{1}{2}, -5, 0$ **2.** $p^{n-1}q$ **3. (a)** $5x^n$ **(b)** $-5x^n$ **4.** 0   **5.** $\{-3 \pm \sqrt{5}\}$ **6.**

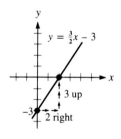

$y = \frac{3}{2}x - 3$

3 up

-3 2 right

7. $(-\infty, 3 - \sqrt{3}) \cup (3 + \sqrt{3}, \infty)$ **8.** $[0, \infty)$ **9.** Let x be one number, then $16 - x$ is the other. The product formula is given by
$P = x(16 - x) = 16x - x^2$, and P is a maximum when $x = -16/[2(-1)] = 8$. The numbers are 8 and 8.   **10.** 615

Exercises 3.5

1. $\pm 20, \pm 10, \pm 5, \pm 4, \pm 2, \pm 1$ **3.** $\pm 6, \pm 3, \pm 2, \pm\frac{3}{2}, \pm 1, \pm\frac{3}{4}, \pm\frac{1}{2}, \pm\frac{1}{4}$ **5.** Positive: 3; negative: 0
7. Positive: 1; negative: 3   **9.** $4, -1, -2$ **11.** $-\frac{1}{3}, \pm i\sqrt{5}$ **13.** $\frac{1}{2}$ (multiplicity 3)   **15.** $-1, 2, (1 \pm i\sqrt{79})/8$
17. $-2, -\frac{1}{2}, (-1 \pm i\sqrt{35})/2$ **19. (a)** $3, -1 \pm i\sqrt{3}$ **(b)** One; 3   **(c)** One; 3   **(d)** Three; $3 + 0i, -1 + \sqrt{3}i, -1 - \sqrt{3}i$
21. (a) $-2, \frac{1}{3}, \pm\sqrt{7}$ **(b)** Two; $-2, \frac{1}{3}$ **(c)** Four; $-2, \frac{1}{3}, \sqrt{7}, -\sqrt{7}$ **(d)** Four; $-2 + 0i, \frac{1}{3} + 0i, \sqrt{7} + 0i, -\sqrt{7} + 0i$
23. $P(0) = -8, P(-1) = 5$; Since $P(0)$ and $P(-1)$ have opposite signs, the location theorem guarantees at least one real zero between 0
and -1.   **25.** $P(1) = -4, P(2) = 2$; Since $P(1)$ and $P(2)$ have opposite signs, the location theorem guarantees at least one real zero
between 1 and 2. Similarly, since $P(-1) = -16$ and $P(-2) = 50$, there is at least one real zero between -1 and -2.   **27.** 0.7
29. 0.3   **31.** -1.3 **33.** $105 = x(x + 2)(2x + 1)$, so $2x^3 + 5x^2 + 2x - 105 = 0$. Then 3 is a root while the reduced equation
$2x^2 + 11x + 35 = 0$ has no real-number solutions. Unique dimensions: 7 in. by 5 in. by 3 in.   **35.** 3 in. or $6 - 3\sqrt{2}$ in.

Remember This (3.5)

1. No   **2.** $\dfrac{x^2 - 8}{x + 2} = x - 2 + \dfrac{-4}{x + 2}$ **3.** All real numbers except 0

4. (a)

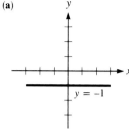

(b)

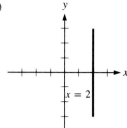

5. $\{-10\}$ **6.** $\emptyset$ **7.** $f(x) = \frac{3}{2}x + \frac{7}{2}$ **8.** $(\frac{2}{3},0), (-\frac{1}{2},0), (0,-2)$ **9.** $4x^2 + 5x - 4$ **10.** 0 is a zero of multiplicity 1, -3 is a zero of multiplicity 4.

Exercises 3.6

1. $x = \frac{3}{2}$ **3.** $x = 6, x = -1$

5.

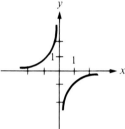

Vertical asymptote: $x = 0$
Horizontal asymptote: $y = 0$

7.

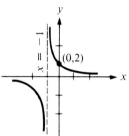

Vertical asymptote: $x = -1$
Horizontal asymptote: $y = 0$

9.

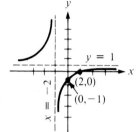

Vertical asymptote: $x = -2$
Horizontal asymptote: $y = 1$

11.

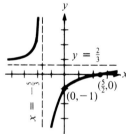

Vertical asymptote: $x = -\frac{5}{3}$
Horizontal asymptote: $y = \frac{2}{3}$

13.

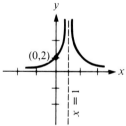

Vertical asymptote: $x = 1$
Horizontal asymptote: $y = 0$

15.

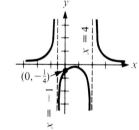

Vertical asymptotes: $x = -1$,
$x = 4$
Horizontal asymptote: $y = 0$

17.

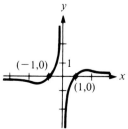

Vertical asymptote: $x = 0$
Horizontal asymptote: $y = 0$

19.

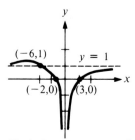

Vertical asymptote: $x = 0$
Horizontal asymptote: $y = 1$

21.

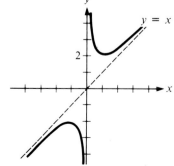

Vertical asymptote: $x = 0$
Horizontal asymptote: none

23.

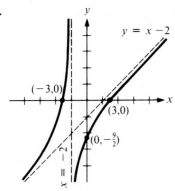

Vertical asymptote: $x = -2$
Horizontal asymptote: none

25.

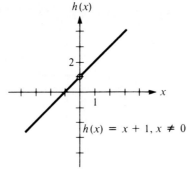

$h(x) = x + 1, x \neq 0$

27.

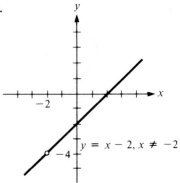

$y = x - 2, x \neq -2$

29.

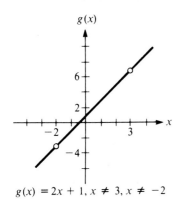

$g(x) = 2x + 1, x \neq 3, x \neq -2$

31.

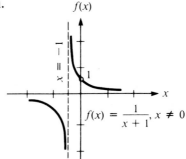

$f(x) = \dfrac{1}{x + 1}, x \neq 0$

Vertical asymptote: $x = -1$
Horizontal asymptote: $y = 0$

33.

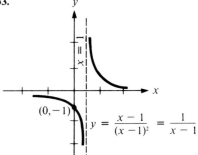

$y = \dfrac{x - 1}{(x - 1)^2} = \dfrac{1}{x - 1}$

Vertical asymptote: $x = 1$
Horizontal asymptote: $y = 0$

Remember This (3.6)

1. 1.72   **2. (a)** $\frac{1}{64}$ **(b)** 16   **(c)** 1   **3. (a)** 2^{x-1} **(b)** b^{2x} **4.** $(\frac{1}{3})^x = (3^{-1})^x = 3^{-1x} = 3^{-x}$ **5.** The graph of $y = f(x + 2)$ is the
graph of $y = f(x)$ shifted 2 units to the left.   **6.** $\pm 2i\sqrt{3}$ **7.** 0   **8.** $2, 3, \frac{2}{3}$ **9.** $y = 3x + 23$ **10.**

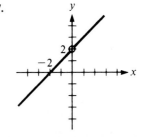

Chapter 3 Review Exercises

1. Yes   **3.** No   **5.** No   **7.** $0, \frac{3}{2}$ **9.** $2, i, -i$ **11.** $(5 \pm \sqrt{17})/4$

13.

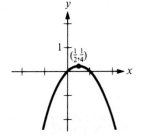

15.

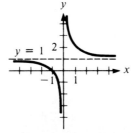

17.

19.

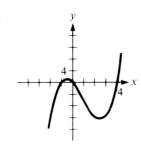

21. $(-\infty,-3] \cup [2,\infty)$ **23.** Set of all real numbers except $-\frac{1}{2}$ **25.** $(-5,-1)$;
$x = -5$ **27.** $y = \frac{2}{3}$; $(\frac{13}{2},\frac{2}{3})$ **29.** $y = \frac{2}{3}x - 2$ **31. (a)** -16 **(b)** $\frac{7}{5}$
33. $P(x) = x(x - 5)(x + 2)(x - 3)$ **35.** 16   **37.** $x - 3$ **39. (a)** 4   **(b)** 0
41. $x^2 - x - 6 = 0$ **43.** 1   **45.** 2.3   **47.** 108 ft (front) by 81 ft
49. (a) $v = -32t + 320$ **(b)** 32 ft/second   **(c)** 320 ft/second   **(d)** 10 seconds;
Projectile reaches its highest point when $t = 10$ seconds.   **51. (b)**   **53. (a)**   **55. (b)**
57. (c)   **59. (a)**

Chapter 3 Test

1.

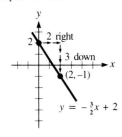

2. $y = 2x + 9$ **3.** $f(x) = \frac{3}{5}x - 3$ **4.** $(-1,0)$, $(\frac{3}{2},0)$ **5.** $(-\infty,9]$
6. $[1 - \sqrt{7}, 1 + \sqrt{7}]$ **7.** 256 ft   **8.** $-3x + 32$ **9.** $2x^3 - 6x^2 + 17x - 51$
10. (a) -13 **(b)** -13 **11.** 0 is a zero of multiplicity 2; -1 is a zero of multiplicity 3.

12.

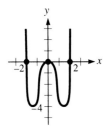

13. $\pm\sqrt{6}$ **14.** ± 2, ± 1, $\pm\frac{2}{3}$, $\pm\frac{1}{3}$ **15.** 1   **16.** $-\frac{3}{2}$, $\pm i\sqrt{3}$ **17.** $x = 0$, $x = 6$
18. $(2,0)$, $(-2,0)$ **19.** $y = x - 2$ **20.**

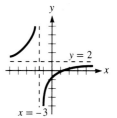

Exercises 4.1

1. 16; $\frac{1}{4}$; 8   **3.** $\frac{16}{81}$; $\frac{27}{8}$; 1   **5.** $\{6\}$ **7.** $\{\frac{1}{2}\}$ **9.** $\{-3\}$ **11.** $\{-2\}$ **13.** $\{4\}$ **15.** $\{-\frac{11}{4}\}$ **17.** $\{\frac{3}{5}\}$ **19.** $\{\frac{4}{9}\}$
21.

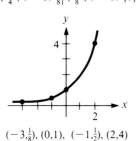

$(-3,\frac{1}{8})$, $(0,1)$, $(-1,\frac{1}{2})$, $(2,4)$

23.

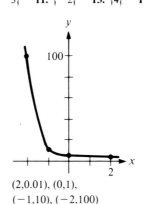

$(2,0.01)$, $(0,1)$,
$(-1,10)$, $(-2,100)$

25.

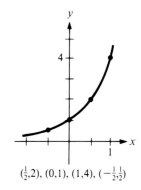

$(\frac{1}{2},2)$, $(0,1)$, $(1,4)$, $(-\frac{1}{2},\frac{1}{2})$

27.

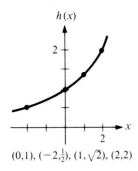

$(0,1), (-2,\frac{1}{2}), (1,\sqrt{2}), (2,2)$

29.

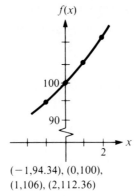

$(-1,94.34), (0,100),$
$(1,106), (2,112.36)$

31.

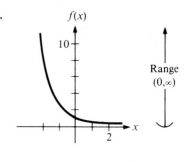

33.

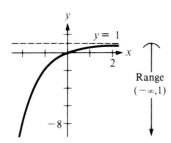

35.

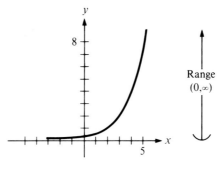

37.

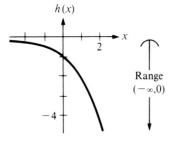

39.

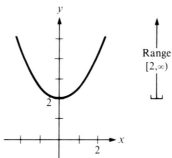

41. 4   **43.** $\frac{1}{3}$ **45.** 16   **47.** $\frac{1}{8}$ **49.** $\frac{1}{8}$ **51.** If $f(x) = b^x$, then $f(x_1 + x_2) = b^{x_1 + x_2} = b^{x_1} \cdot b^{x_2} = f(x_1)f(x_2)$.
53. (a) $y = 100(1.05)^t$ (b) \$115.76   **55.** (a) $f(t) = 10,000(0.8)^t$ (b) \$5,120   **57.** (a) $y = 16(\frac{1}{2})^{2t}$ (b) 4 g

Remember This (4.1)

1. $\{729\}$ **2.** $\{-\frac{1}{2}\}$ **3.** 3.8×10^{-4} **4.** 0.00000421   **5.** $(-2,\infty)$ **6.**

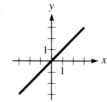

7. $f^{-1} = \{(1,0), (3,1), (9,2)\}$
8. $f^{-1}(x) = x^3$ **9.** No   **10.** Domain: $(0,\infty)$; range: $(-\infty,\infty)$

Exercises 4.2

1. $\log_3 9 = 2$ **3.** $\log_{1/2}(\frac{1}{4}) = 2$ **5.** $\log_4(\frac{1}{16}) = -2$ **7.** $\log_{25} 5 = \frac{1}{2}$ **9.** $\log_7 1 = 0$ **11.** $5^1 = 5$ **13.** $(\frac{1}{3})^2 = \frac{1}{9}$

15. $2^{-2} = \frac{1}{4}$ **17.** $49^{1/2} = 7$ **19.** $100^0 = 1$ **21.** 2   **23.** 1   **25.** -1 **27.** $\frac{1}{2}$ **29.** 0   **31.** 3   **33.** $\frac{1}{5}$ **35.** 5   **37.** 3

39. $b > 0, b \neq 1$ **41.** $3; -1; 0$ **43.** $4; 0; -7$

45.

47.

49.

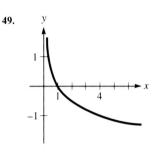

Note: $y = -\log_4 x = \log_{\frac{1}{4}} x$
(see Exercise 47)

51.

53.

55.

57. $(-\infty, -3) \cup (3, \infty)$ **59.** Set of all real numbers except 1   **61.** 7   **63.** 3.6   **65.** 1   **67.** 7.9×10^{-3} **69. (a)** 0   **(b)** 10
(c) 20   **(d)** 30; If the power ratio is multiplied by 10, add 10 dB to loudness.   **71.** 25 dB

Remember This (4.2)

1. 3^{u+v} **2.** 5^x **3.** b^{15} **4.** $\sqrt[3]{x^2}$ **5.** $L^{1/2}$ **6.** $\{\frac{6}{5}\}$ **7.** $\{16\}$ **8.** $(-\infty, 0)$ **9.** $f[f^{-1}(x)] = x; f^{-1}[f(x)] = x$ **10.** \$4,915.20

Exercises 4.3

1. $\log_{10} 7 + \log_{10} 5$ **3.** $\log_6 3 - \log_6 5$ **5.** $16 \log_5 11$ **7.** $\frac{1}{2} \log_2 3$ **9.** $\frac{1}{5} \log_{10} 16$ **11.** $2 + 3 \log_4 3$ **13.** $\frac{1}{2}(\log_b x + \log_b y)$

15. $\frac{5}{3} \log_b x$ **17.** $\frac{1}{4}(\log_b x + 2 \log_b y - 3 \log_b z)$ **19.** $\log_2 12$ **21.** $\log_4 4$ or 1   **23.** $\log_7 9$ **25.** $\log_{10} \frac{3}{8}$ **27.** $\log_b \sqrt[3]{xy^2}$

29. $\log_b(x - 1)$ **31.** $\log_b \sqrt{\dfrac{xz^3}{y^5}}$ **33.** 5   **35.** $k + \log_{10} m$ **37.** 3   **39.** $\dfrac{n}{x} x^n = nx^{n-1}$ **41. (a)** 0.7781   **(b)** 0.6020

(c) 0.2386   **(d)** -0.1761 **(e)** -0.3010 **(f)** 0.6990   **43.** $\log_b(1/a) = \log_b a^{-1} = -\log_b a$ **45.** Let $x = b^m$ and $y = b^n$,

then $m = \log_b x$ and $n = \log_b y$.   **(a)** $\dfrac{x}{y} = \dfrac{b^m}{b^n} = b^{m-n}$; Thus, $\log_b \dfrac{x}{y} = m - n = \log_b x - \log_b y$.   **(b)** $(x)^k = (b^m)^k = b^{km}$;

Thus, $\log_b(x)^k = km = k \log_b x$.

Remember This (4.3)

1. $\log_a s = r$ **2.** $10^{-3} = x$ **3.** $\{5, -1\}$ **4.** $x = \dfrac{-2m - n}{m - 2n}$ **5.** False   **6.** $(0, \infty)$ **7.** 0   **8.** $\frac{1}{27}$ **9.** $-\frac{1}{3}$ **10.** 3.2×10^{-8}

Exercises 4.4

1. $\{3.322\}$ **3.** $\{1.044\}$ **5.** $\{-3.170\}$ **7.** $\{3.376\}$ **9.** $\{-0.7304\}$ **11.** $\{-1.631\}$ **13.** $\{0.0159\}$ **15.** $\{4.351\}$ **17.** $\{0.0830\}$
19. $\{18.58\}$ **21.** $\{100\}$ **23.** $\{10\}$ **25.** $\{\frac{9}{10}\}$ **27.** $\{5\}$ **29.** $\{2\sqrt{2}\}$ **31.** $\{\frac{1}{2}\}$ **33.** $\{2\}$ **35.** $\{\pm 3\}$ **37.** $\{2\}$ **39.** $\{4\}$ **41.** $\{5\}$
43. $\{3\}$ **45.** $\{5\}$ **47.** $\emptyset$ **49.** $(0,\infty)$ **51.** 3.170   **53.** 1.465   **55.** -4.248 **57.** -3.190 **59.** $\frac{1}{3}$ **61.** 2   **63.** 18.9 percent

Remember This (4.4)

1. True   **2.** x **3.** The graphs of f and g are reflections of each other about the line $y = x$.   **4.**

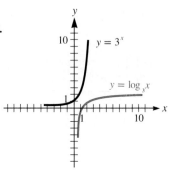

5. 2   **6.** 1   **7.** $\log_b \sqrt[3]{xy^2}$ **8.** $\log x - 4 \log y$ **9.** $\frac{1}{2}(x + y + 1)$ **10.** $15,004.15

Exercises 4.5

1. 1.1735   **3.** $\{3.9318\}$ **5.** $\{0.00552\}$ **7.** $\{5\}$ **9.** 2   **11.** $\{0.4343\}$ **13.** $\{9\}$ **15.** $\{0.3466\}$ **17.** $\{7.259\}$ **19.** $-0.6931/k$
21. (a) $1,194.05   **(b)** $6,333.85   **(c)** $3,166.07   **23.** 4.1 years   **25. (a)** $1,197.22   **(b)** $57,466.23   **(c)** $245,325.30
(d) $216,224.53   **27.** 9.9 percent   **29. (a)** 1,078   **(b)** 24.6 hours   **31.** 111,430   **33.** 7.6 g   **35. (a)** -0.000124 **(b)** 78 percent
37. 7:10 A.M.   **39.** $-\ln x$ **41.** $\frac{1}{2} + \ln 3$ **43.** $\ln x + \ln(x - 1) - \ln(x + 2)$ **45.** $\ln x^2$ **47.** $\ln(x^2 + 3x)$ **49.** $\ln \sqrt{x/a}$
51. $\{e^2 - 3\}$ **53.** $\{\frac{8}{9}\}$ **55.** $\{3\}$ **57.** $\{(1 + \sqrt{4e^2 + 1})/2\}$ **59.** 2.303   **61.** $t = \dfrac{1}{k} \ln\left(\dfrac{T - t_\alpha}{D_0}\right)$

Remember This (4.5)

1. $\{4\}$ **2.** $\{-3\}$ **3.** $\{-1.631\}$ **4.** $\{0.8959\}$ **5.** $\{64\}$ **6.** $\{0.6065\}$ **7.** $\{10\}$ **8.** $\{1.356\}$ **9.** $t = \dfrac{\log y}{\log a}$ **10.** $t = \ln y$

Chapter 4 Review Exercises

1. $\{7\}$ **3.** $\{32\}$ **5.** $\{3.43 \times 10^7\}$ **7.** $\{0.00384\}$ **9.** $\{3\}$ **11.** $\ln(x^2 \cdot \sqrt[3]{y})$ **13.** $\frac{1}{2}[3 \log_b x - \log_b y]$ **15.** $(\frac{7}{3}, \infty)$ **17.** $2^3 = 8$

19.

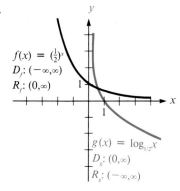

21. $(3, \frac{1}{64})$, $(-1, 4)$, $(0, 1)$, $(\frac{3}{2}, \frac{1}{8})$, $(-2, 16)$ **23.** $3,664.21
25. $f^{-1} = \{(-1,3), (-2,4)\}$; $D_f = R_{f^{-1}} = \{3,4\}$; $D_{f^{-1}} = R_f = \{-1,-2\}$ **27.** M^2
29. $\frac{1}{3}(x + y)$ **31.** 0   **33.** Domain: $(-\infty,\infty)$; range: $(-2,\infty)$

35.

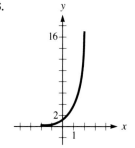

37.

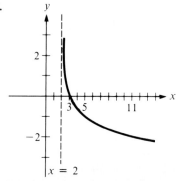

39. $\{-\frac{3}{2}\}$ **41.** 100   **43.** $\frac{13}{2}$ **45.** -0.098 **47.** $y = r + ce^{-kt}; \dfrac{y-r}{c} = e^{-kt}; -kt = \ln\dfrac{y-r}{c}; t = \dfrac{-1}{k}\ln\dfrac{y-r}{c} = \dfrac{1}{k}\ln\left(\dfrac{y-r}{c}\right)^{-1}$

$= \dfrac{1}{k}\ln\dfrac{c}{y-r}$ **49.** (pH of X) = (pH of Y) $- 1$ **51.** (d)   **53.** (c)   **55.** (c)   **57.** (a)   **59.** (c)

Chapter 4 Test

1. 9   **2.** $\{-\frac{3}{2}\}$ **3.** $(-3,\infty)$ **4.** \$1,344.56   **5. (a)** $\log_8 4 = \frac{2}{3}$ **(b)** $10^{-2} = 0.01$ **6.** $(-2,\infty)$

7.

8. 3.2   **9.** $\log_b 5 + \frac{1}{2}\log_b x$ **10.** $x - 2y$ **11.** $\log_b 2$ **12.** $\frac{1}{2}$ **13.** $\{1.893\}$

14. $\{4\}$ **15.** 2.322   **16.** $\{0.0247\}$ **17.** $\{6\}$ **18.** $\{-0.00047\}$ **19.** \$4,508.64

20. 12.2 years

Exercises 5.1

1. 5   **3.** 8 yd   **5.** 9.6 meters   **7.** 5   **9.** $\dfrac{11}{2}, \dfrac{11}{4\pi}$ **11.** $\dfrac{25\pi}{2}$ in.^{2} **13.** 8 ft^2 **15.** $\dfrac{\pi}{6}$ **17.** $\dfrac{\pi}{3}$ **19.** $\dfrac{2\pi}{3}$ **21.** $\dfrac{5\pi}{6}$ **23.** $\dfrac{7\pi}{6}$

25. $\dfrac{4\pi}{3}$ **27.** $\dfrac{5\pi}{3}$ **29.** $\dfrac{11\pi}{6}$ **31.** $\dfrac{10\pi}{9}$ **33.** $\dfrac{\pi}{9}$ **35.** $\dfrac{61\pi}{36}$ **37.** $60°$ **39.** $40°$ **41.** $120°$ **43.** $140°$ **45.** $432°$ **47.** $70°$

49. $115°$ **51.** $229°$ **53.** $332°$ **55.** 6 rad/second   **57. (a)** $(200\pi/3)$ rad/minute   **(b)** 400π in./minute   **59.** 300π in./second

61. π in.

Remember This (5.1)

1. $\sqrt{2}/2$ **2.** $2\sqrt{3}/3$ **3.** $\pi/3$ **4.** Q$_1$, Q$_4$ **5.** Q$_3$, Q$_4$ **6.** b **7.** $-b$ **8.** Domain: $(-\infty,\infty)$; range: $[-3,\infty)$ **9.** $[-1,1]$

10. $x^2 + y^2 = 1$

Exercises 5.2

1. 1   **3.** 0   **5.** 0   **7.** 0   **9.** Undefined   **11.** 1   **13.** -1 **15.** $\frac{1}{2}$ **17.** 1   **19.** $\frac{1}{2}$ **21.** $\dfrac{\sqrt{3}}{3}$ **23.** $\dfrac{\sqrt{3}}{2}$ **25.** $-\dfrac{2\sqrt{3}}{3}$

27. $\dfrac{\sqrt{3}}{3}$ **29.** $-\dfrac{\sqrt{3}}{2}$ **31.** $-\dfrac{\sqrt{3}}{2}$ **33.** $\sqrt{3}$ **35.** $-\dfrac{\sqrt{2}}{2}$ **37.** 2   **39.** $\frac{1}{2}$ **41.** $\frac{1}{2}$ **43.** -2 **45.** $-\dfrac{\sqrt{2}}{2}$ **47.** $-\sqrt{3}$

49. $-\dfrac{\sqrt{3}}{2}$ **51.** -0.4176 (C: -0.4161)   **53.** 1.162 (C: 1.158)   **55.** -0.7643 (C: -0.7664)   **57.** 1.560 (C: 1.566)

59. 0.2085 (C: 0.2116)   **61.** 0.6749 (C: 0.6702)   **63.** -1.041 (C: -1.042)   **65.** 0.9521 (C: 0.9511)   **67.** 0.2190 (C: 0.2225)

69. -1.081 (C: -1.082)

Remember This (5.2)

1. $\sqrt{29}$ **2.** $2\sqrt{29}/29$ **3.** $\sqrt{29}$ **4.** 15   **5.** Q$_4$ **6.** $90°$ **7.** $\cot s = 1/\tan s$ **8.** True   **9.** $-\frac{1}{2}$ **10.** $[-1,1]$

Exercises 5.3

	sin θ	csc θ	cos θ	sec θ	tan θ	cot θ
1.	$-4/5$	$-5/4$	$3/5$	$5/3$	$-4/3$	$-3/4$
3.	$5/13$	$13/5$	$-12/13$	$-13/12$	$-5/12$	$-12/5$
5.	$-4/5$	$-5/4$	$3/5$	$5/3$	$-4/3$	$-3/4$
7.	$\sqrt{2}/2$	$\sqrt{2}$	$\sqrt{2}/2$	$\sqrt{2}$	1	1
9.	$2\sqrt{5}/5$	$\sqrt{5}/2$	$-\sqrt{5}/5$	$-\sqrt{5}$	-2	$-1/2$

11. Q_4 **13.** Q_3 **15.** Q_2

	sin θ	csc θ	cos θ	sec θ	tan θ	cot θ
17.	$-3/5$	$-5/3$	$-4/5$	$-5/4$	$3/4$	$4/3$
19.	$-\sqrt{3}/2$	$-2\sqrt{3}/3$	$1/2$	2	$-\sqrt{3}$	$-\sqrt{3}/3$
21.	$-3/4$	$-4/3$	$-\sqrt{7}/4$	$-4\sqrt{7}/7$	$3\sqrt{7}/7$	$\sqrt{7}/3$
23.	$1/3$	3	$2\sqrt{2}/3$	$3\sqrt{2}/4$	$\sqrt{2}/4$	$2\sqrt{2}$

25. $\sin A = \cos B = \frac{5}{13}$, $\cos A = \sin B = \frac{12}{13}$, $\tan A = \cot B = \frac{5}{12}$, $\cot A = \tan B = \frac{12}{5}$, $\sec A = \csc B = \frac{13}{12}$, $\csc A = \sec B = \frac{13}{5}$ **27.** $\sin \theta = \cos \beta = \frac{4}{5}$, $\cos \theta = \sin \beta = \frac{3}{5}$, $\tan \theta = \cot \beta = \frac{4}{3}$, $\cot \theta = \tan \beta = \frac{3}{4}$, $\sec \theta = \csc \beta = \frac{5}{3}$, $\csc \theta = \sec \beta = \frac{5}{4}$ **29.** $\frac{x}{y}$ **31.** $\frac{40}{9}$ **33.** $\frac{y}{x}$ **35.** sec X or csc Z **37.** $\csc \theta = 2$, $\cos \theta = \frac{\sqrt{3}}{2}$, $\sec \theta = \frac{2\sqrt{3}}{3}$, $\tan \theta = \frac{\sqrt{3}}{3}$, $\cot \theta = \sqrt{3}$ **39.** $\tan \theta = \frac{5}{3}$, $\sin \theta = \frac{5\sqrt{34}}{34}$, $\csc \theta = \frac{\sqrt{34}}{5}$, $\cos \theta = \frac{3\sqrt{34}}{34}$, $\sec \theta = \frac{\sqrt{34}}{3}$ **41.** $\cot \theta = \frac{1}{2}$, $\sin \theta = \frac{2\sqrt{5}}{5}$, $\csc \theta = \frac{\sqrt{5}}{2}$, $\cos \theta = \frac{\sqrt{5}}{5}$, $\sec \theta = \sqrt{5}$ **43.** 20   **45.** 36   **47.** $6\sqrt{2}$ **49.** $\cos A = \frac{12}{13}$ **51.** 1

Remember This (5.3)

1. $70°$ **2.** 60   **3.** -1 **4.** $\sqrt{2}/2$ **5.** $-\sqrt{3}/3$ **6.** $45°$ **7.** $\sqrt{2}$ units   **8.** $\sqrt{3}$ units   **9.** Yes   **10.** 9 ft²

Exercises 5.4

1. No   **3.** No   **5.** Yes   **7.** $2\sqrt{3}$, 4 (hypotenuse)   **9.** cos 73°   **11.** cot 8°30′   **13.** sec 21.9°   **15.** 0.9925   **17.** 0.1228   **19.** 5.396   **21.** 1.287   **23.** 0.2130   **25.** 45°00′ or 45.0°   **27.** 38°20′ or 38.3°   **29.** 55°50′ or 55.8°   **31.** 52°30′ or 52.6°   **33.** 3°40′ or 3.6°   **35.** -1 **37.** Undefined   **39.** 0   **41.** -1 **43.** 1   **45.** 25°   **47.** 60°   **49.** 83°   **51.** 25°30′   **53.** $-\frac{1}{2}$ **55.** $2\sqrt{3}/3$ **57.** -1 **59.** $\frac{1}{2}$ **61.** $-\sqrt{2}/2$ **63.** $-\frac{1}{2}$ **65.** $2\sqrt{3}/3$ **67.** $-\sqrt{3}$ **69.** $-\sqrt{2}/2$ **71.** $-\sqrt{2}/2$ **73.** $-\sqrt{3}/2$ **75.** $\sqrt{3}$ **77.** 1   **79.** $2\sqrt{3}/3$ **81.** 1   **83.** -0.5299 **85.** 3.487   **87.** 1.923   **89.** -0.8557 **91.** -0.5354 **93.** 0.0494   **95.** -1.360 **97.** -0.9983 **99.** -0.6111 **101.** 0.3346   **103.** 0.1564   **105.** 1.150   **107.** 0.4848   **109.** -0.6787 **111.** 1.881   **113.** 69 ft

Remember This (5.4)

1. -1 **2.** -4 **3.** $-\pi/4$ **4.** 4π **5.** 3   **6.** $\{x: 0 \le x < 2\pi\}$ **7.** 1   **8.** True   **9.** The graph of $y = -f(x)$ is the graph of $y = f(x)$ reflected about the x-axis.   **10.** The graph of $y = f(x - \pi/2)$ is the graph of f shifted $\pi/2$ units to the right.

Exercises 5.5

1.

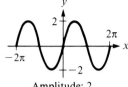

Amplitude: 2
Period: 2π

3.

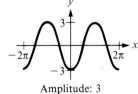

Amplitude: 3
Period: 2π

5.

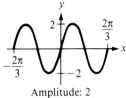

Amplitude: 2
Period: $2\pi/3$

7.

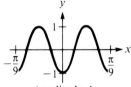

Amplitude: 1
Period: $\pi/9$

9.

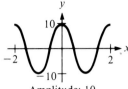

Amplitude: 10
Period: 2

11.

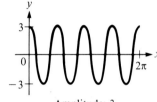

Amplitude: 3
Period: $\pi/2$

13.

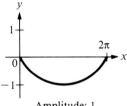

Amplitude: 1
Period: 4π

15.

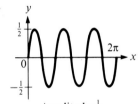

Amplitude: $\frac{1}{2}$
Period: $2\pi/3$

17.

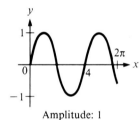

Amplitude: 1
Period: 4

19.

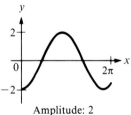

Amplitude: 2
Period: 6

21.

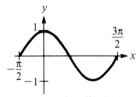

Amplitude: 1
Period: 2π
Phase shift: $-\pi/2$

23.

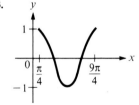

Amplitude: 1
Period: 2π
Phase shift: $\pi/4$

25.

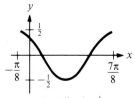

Amplitude: $\frac{1}{2}$
Period: π
Phase shift: $-\pi/8$

27.

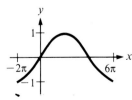

Amplitude: 1
Period: 8π
Phase shift: -2π

29.

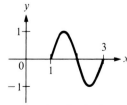

Amplitude: 1
Period: 2
Phase shift: 1

31. $y = 3 \sin 2x$ **33.** $y = -4 \sin 3x$ **35.** $y = 1.5 \cos \frac{1}{2}x$ **37.** $y = -2 \cos 10x$ **39.** $y = 10 \sin \pi x$ **41.** (d)   **43.** (d)
45. (c)   **47.** (d)   **49.** (d)   **51.** (d)   **53.** (b)   **55.** (d)   **57.** (d)   **59.** (a)

Remember This (5.5)

1. 1.7   **2.** -1 **3.** $(-\infty,\infty)$ **4.** $\csc x = 1/\sin x$ **5.** $\sec 2\pi x = 1/\cos 2\pi x$ **6.** True   **7.** -1 **8.** 1/0 is undefined because no number times 0 equals 1.   **9.** $\{(0,3),(1,0)\}$ **10.** The graph of $y = f(x) + 3$ is the graph of $y = f(x)$ shifted 3 units up.

Exercises 5.6

x	0	$\dfrac{\pi}{6}$	$\dfrac{\pi}{3}$	$\dfrac{\pi}{2}$	$\dfrac{2\pi}{3}$	$\dfrac{5\pi}{6}$	π	$\dfrac{7\pi}{6}$	$\dfrac{4\pi}{3}$	$\dfrac{3\pi}{2}$	$\dfrac{5\pi}{3}$	$\dfrac{11\pi}{6}$	2π
1. $\cot x$	und.	1.7	0.6	0	-0.6	-1.7	und.	1.7	0.6	0	-0.6	-1.7	und.
3. $\csc x$	und.	2	1.2	1	1.2	2	und.	-2	-1.2	-1	-1.2	-2	und.

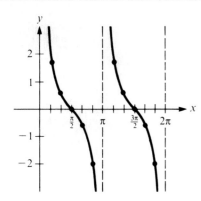
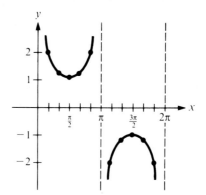

5. *D:* set of all real numbers except $x = \dfrac{\pi}{2} + k\pi$ (k any integer); *R:* $(-\infty,-1]\cup[1,\infty)$; period: 2π **7.** Inc., dec., dec., inc.
9. Always increasing   **11.** Inc., inc., dec., dec.   **13.** The other trigonometric functions do not attain a maximum value.

15.

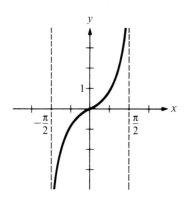

17.

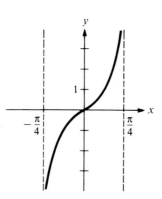

19.

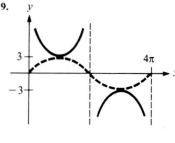

21.

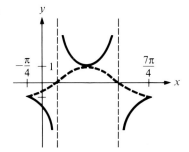

23.

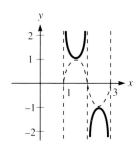

25.

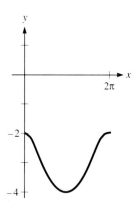

27.

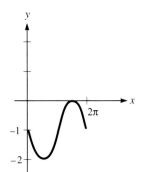

29.

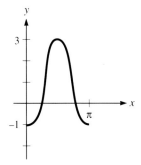

31.

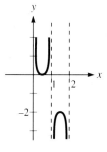

33.

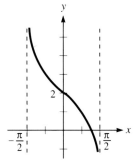

35.

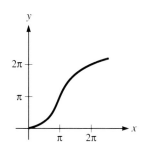

37.

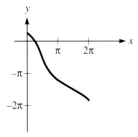

39.

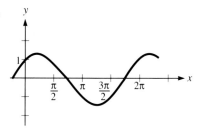

41.

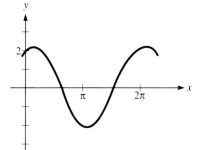

43.

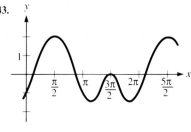

Remember This (5.6)

1. $f^{-1} = \{(1,0),(0,\pi/2)\}$ **2.** Domain: $[-1,1]$; range: $[-\pi/2,\pi/2]$ **3.** $f^{-1}(x) = (x + 2)/3$ **4.** $f[f^{-1}(x)] = x, f^{-1}[f(x)] = x$
5. If f and g are inverse functions, then their graphs are reflections of each other about the line $y = x$.   **6.** $(-\infty,\infty)$
7. $\sin 0 = \sin \pi = \sin(-\pi) = 0$ **8.** $(-\pi/2,\pi/2)$ **9.**

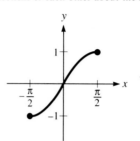

10.

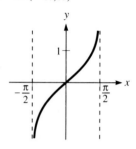

Exercises 5.7

1. $\dfrac{\pi}{3}$ **3.** $-\dfrac{\pi}{3}$ **5.** $\dfrac{\pi}{4}$ **7.** $-\dfrac{\pi}{2}$ **9.** $\dfrac{\pi}{4}$ **11.** $-\dfrac{\pi}{3}$ **13.** 0.32   **15.** 2.15   **17.** -1.05 **19.** 0.64   **21.** 1   **23.** $\frac{1}{2}$ **25.** 0

27. 0   **29.** -0.6552 (C: -0.6616)   **31.** $\frac{3}{5}$ **33.** $\frac{12}{5}$ **35.** $\dfrac{1}{\sqrt{x^2 + 1}}$ **37.** $\dfrac{\sqrt{1 - x^2}}{x}$ **39.** $\dfrac{x - 2}{\sqrt{9 - (x - 2)^2}}$ **41.** $\arcsin x$

43. $\arcsec\left(\dfrac{x - 2}{3}\right)$ **45.** $-\dfrac{\sqrt{x^2 + 4}}{4x}$ **47.** $-\dfrac{x + 2}{\sqrt{4(x + 2)^2 - 1}}$ **49.** $\theta = \arcsin\left(\dfrac{1}{m}\right)$ **51.** $\theta = \arcsin\left(\dfrac{Tg}{2v_0}\right)$ **53.** $[-\pi/2,\pi/2]$

55. $[0,\pi]$ **57.** $(-\infty,\infty)$ **59.** 0   **61.** -1 **63.** $\pi/2$ **65.** -1 **67.** No value

69.

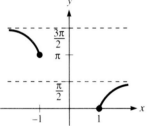

71.

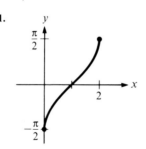

73. (a) $\theta = \arccos(a/25)$ (b) $\theta = \arcsin(b/25)$
75. $\pi/4$

Remember This (5.7)

1. $24°30'$ **2.** 146   **3.** 343.8   **4.** $65°30'$ **5.** Amplitude: 2; period: $\frac{1}{5}$ **6.** 6   **7.** $-5\sqrt{61}/61$ **8.** $60°$ **9.** $-\sqrt{2}/2$
10. $(-\infty,-1] \cup [1,\infty)$

Exercises 5.8

1. $b = 87$ ft, $c = 100$ ft, $B = 60°$ **3.** $a = 13$ ft, $b = 7.5$ ft, $B = 30°$ **5.** $a = 8.1$ ft, $b = 24$ ft, $A = 19°$
7. $a = 20$ ft, $b = 14$ ft, $B = 35°$ **9.** $b = 85.1$ ft, $c = 121$ ft, $B = 44.5°$ **11.** $a = 379$ ft, $b = 497$ ft, $A = 37°20'$
13. $b = 975$ ft, $c = 975$ ft, $A = 1°50'$ **15.** $b = 14$ ft, $A = 24°$, $B = 66°$ **17.** $c = 1.4$ ft, $A = 45°$, $B = 45°$
19. $a = 23.1$ ft, $A = 62°30'$, $B = 27°30'$ **21.** 25 ft   **23.** 52 yd   **25.** 89 ft   **27.** 400 ft   **29.** $10°$ **31.** 22 ft   **33.** 14.1 in.
35. $60°$ **37.** 530 ft   **39.** 9.8 ft   **41.** Amplitude: 3 cm; period: $\frac{1}{4}$ second; frequency: 4 cycles per second
43. Amplitude: 1.5 cm; period: $\pi/3$ seconds; frequency $3/\pi$ cycles per second   **45.** Amplitude: 3 cm; period: 2π seconds; frequency: $1/2\pi$
cycles per second   **47.** $x = 1.7 \cos 2t$ **49.** (a) $\pi/6$ seconds   (b) $x = 3 \cos 12t$

Remember This (5.8)

1. $\sec x = 1/\cos x$ **2.** $\sin(-x) = -\sin x$ **3.** $(1 + s)(1 - s)$ **4.** $\dfrac{s^2 + c^2}{c^2}$ **5.** $\dfrac{s^2 - c^2}{s^2 + c^2}$ **6.** $[-1,1]$ **7.** $[-1,1]$ **8.** $\frac{15}{17}$ **9.** 0
10. $\pi/4$

Chapter 5 Review Exercises

1. $\sqrt{3}/2$ **3.** 0   **5.** $-\sqrt{3}/3$ **7.** $-\pi/4$ **9.** 0   **11.** 0.0175   **13.** -3.11 (C: -3.08)   **15.** 1.35   **17.** No value   **19.** 1.37

21.
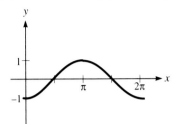

23.
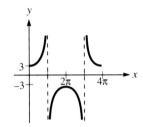

25.
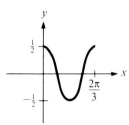

Amplitude: $\frac{1}{2}$
Period: $2\pi/3$
Phase shift: 0

27.
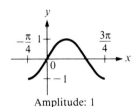

Amplitude: 1
Period: π
Phase shift: $-\pi/4$

29.

31.
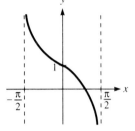

33.
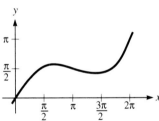

35.
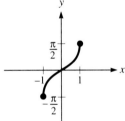

37. Domain: set of all real numbers except $x = (\pi/2) + k\pi$ (k any integer); range: $(-\infty,\infty)$ **39.** Domain: set of all real numbers except $x = k\pi$ (k any integer); range: $(-\infty,-1] \cup [1,\infty)$ **41.** Q_2 **43.** (a) $7\pi/6$ (b) $72°$ **45.** $\sqrt{2}/2$ **47.** $[-1,1]$ **49.** 2

51. $\dfrac{9}{2} \arcsin\left(\dfrac{x}{3}\right) - \dfrac{x\sqrt{9-x^2}}{4}$ **53.** $x = 2\cos 12t$ **55.** 78 ft   **57.** 49 mi   **59.** 56.3°   **61.** (d)   **63.** (c)   **65.** (d)   **67.** (a)

69. (c)

Chapter 5 Test

1. $[-\pi/2,\pi/2]$ **2.** Set of all real numbers except $x = (\pi/2) + k\pi$ (k any integer)   **3.** $-\sqrt{3}$ **4.** -1.23

5.
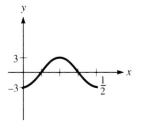

6.
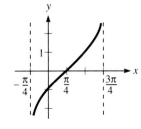

7.
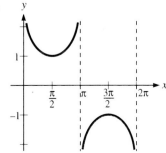

8. Amplitude: 1; period: $2\pi/3$; phase shift: $\pi/6$ **9.** $-\sqrt{15}/4$ **10. (a)** $3\pi/2$ **(b)** $315°$ **11.** $25\pi/3$ cm   **12.** $-\pi/3$
13. $\sqrt{1-x^2}/x$ **14.** $\cos 75°$ **15.** $\frac{3}{4}$ **16.** 6.6 m   **17.** x/r or $x/\sqrt{x^2+y^2}$ **18.** $[-\pi/2,\pi/2]$ **19.** Q_2
20. One example: $y = \sin 6x$

Exercises 6.1

41. $\cos \theta$ **43.** $3 \sec \theta$ **45.** $\tan \theta$ **47.** $\cos \theta$ **49.** $27 \tan^3 \theta$ **51.** $\frac{13}{12}$ **53.** $\frac{12}{5}$ **55.** $-\sqrt{1-a^2}$ **57.** $1/a$ **59.** $\csc x = \frac{17}{8}$,
$\cos x = \frac{15}{17}$, $\sec x = \frac{17}{15}$, $\tan x = \frac{8}{15}$, $\cot x = \frac{15}{8}$ **61.** $\cos x = -3/4$, $\sin x = \sqrt{7}/4$, $\csc x = 4\sqrt{7}/7$, $\tan x = -\sqrt{7}/3$,
$\cot x = -3\sqrt{7}/7$ **63.** $\pm\sqrt{1-\sin^2 x}$ **65.** $\pm\dfrac{1}{\sqrt{1-\sin^2 x}}$ **67.** $\pm\sqrt{1-\cos^2 x}$ **69.** $\dfrac{1}{\cos x}$

Remember This (6.1)

1. False   **2.** $\sin(\pi/2) = 1$, $\cos(\pi/2) = 0$ **3.** $\sqrt{3}$ **4.** $\sin(x + h)$ **5.** $\cos \theta = \sin(90° - \theta)$ **6.** $1/\sqrt{10}$ or $\sqrt{10}/10$
7. $(\sqrt{3} - 1)/2$ **8.** $-\sqrt{3}/2$ **9.** 1   **10.** Yes

Exercises 6.2

27. $(\sqrt{2} + \sqrt{6})/4$ **29.** $(\sqrt{2} - \sqrt{6})/4$ **31.** $(\sqrt{6} - \sqrt{2})/4$ **33.** $2 - \sqrt{3}$ **35. (a)** $\sqrt{3}/2$ **(b)** $1/2$ **(c)** -1 **(d)** $-\sqrt{3}/2$
37. $\frac{36}{325}$ **39.** $\frac{253}{325}$ **41.** $-\frac{36}{323}$ **43.** $\frac{117}{125}$ **45.** $-\frac{63}{65}$ **55.** $\frac{1}{3}$
57. $\dfrac{\cos(x + h) - \cos x}{h} = \dfrac{\cos x \cos h - \sin x \sin h - \cos x}{h}$
$$= \dfrac{\cos x(\cos h - 1) - \sin x \sin h}{h}$$
$$= \cos x\left(\dfrac{\cos h - 1}{h}\right) - \sin x\left(\dfrac{\sin h}{h}\right)$$

Remember This (6.2)

1. $-\frac{12}{5}$ **2.** $-\frac{24}{25}$ **3.** $\sqrt{3}/2$ **4.** $\sin^2 x = 1 - \cos^2 x$ **5.** $\sin \theta$ is positive, $\cos \theta$ and $\tan \theta$ are negative.
6. $\dfrac{\sqrt{2}/2}{1 + (\sqrt{2}/2)} = \dfrac{\sqrt{2}}{2 + \sqrt{2}} = \dfrac{\sqrt{2}}{2 + \sqrt{2}} \cdot \dfrac{2 - \sqrt{2}}{2 - \sqrt{2}} = \dfrac{2\sqrt{2} - 2}{4 - 2} = \sqrt{2} - 1$ **7.** If $x = 30°$, $\tan 2(30°) = \tan 60° = \sqrt{3}$ and
$2 \tan 30° = 2\sqrt{3}/3$, so $\tan 2(30°) \neq 2 \tan 30°$.   **8.** $\cos x\left(\dfrac{\cos x}{\sin x}\right) + \sin x = \dfrac{\cos^2 x + \sin^2 x}{\sin x} = \dfrac{1}{\sin x} = \csc x$ **9.** $2 \sec \theta$
10. $2 \cos^2 x - 1 = 2(1 - \sin^2 x) - 1 = 2 - 2 \sin^2 x - 1 = 1 - 2 \sin^2 x$

Exercises 6.3

1. $\sin 2x = \frac{24}{25}$, $\cos 2x = -\frac{7}{25}$, $\tan 2x = -\frac{24}{7}$ **3.** $\sin 2x = -3\sqrt{7}/8$, $\cos 2x = -1/8$, $\tan 2x = 3\sqrt{7}$
5. $\sin 2x = \frac{4}{5}$, $\cos 2x = \frac{3}{5}$, $\tan 2x = \frac{4}{3}$ **17.** $\sin(x/2) = \sqrt{17}/17$, $\cos(x/2) = -4\sqrt{17}/17$, $\tan(x/2) = -1/4$
19. $\sin(x/2) = \sqrt{8 + 2\sqrt{15}}/4$, $\cos(x/2) = -\sqrt{8 - 2\sqrt{15}}/4$, $\tan(x/2) = -4 - \sqrt{15}$ **21.** $\sqrt{2 + \sqrt{3}}/2$ **23.** $\sqrt{2 + \sqrt{2}}/2$

25. $2 - \sqrt{3}$ **27.** $\sin \dfrac{\pi}{12} = \sin \dfrac{\pi/6}{2} = \sqrt{\dfrac{1 - \cos(\pi/6)}{2}} = \sqrt{\dfrac{1 - (\sqrt{3}/2)}{2}} = \sqrt{\dfrac{2 - \sqrt{3}}{4}} = \dfrac{\sqrt{2 - \sqrt{3}}}{2}$
29. $\tan \dfrac{7\pi}{12} = \tan \dfrac{7\pi/6}{2} = \dfrac{1 - \cos(7\pi/6)}{\sin(7\pi/6)} = \dfrac{1 - (-\sqrt{3}/2)}{-1/2} = -2 - \sqrt{3}$ **31.** $\dfrac{b^2 - a^2}{c^2}$ **33.** $\dfrac{2ab}{b^2 - a^2}$ **35.** $-2a\sqrt{1 - a^2}$
37. $\dfrac{-2a\sqrt{1 - a^2}}{2a^2 - 1}$ **39.** $-\sqrt{\dfrac{1 + a}{2}}$ **41.** $\cos 12t$ **43.** $\tan 2x$ **45.** $\cos 5\theta$ **47.** $\frac{1}{2} \sin 4x$ **49.** $4 \cos 18t$

75. $\sin 3x = 3 \sin x - 4 \sin^3 x$ **77.** $V^2 \dfrac{\sin A \sin B}{16} = V^2 \dfrac{\sin A \cos A}{16} = \dfrac{1}{32} V^2(2 \sin A \cos A) = \dfrac{1}{32} V^2 \sin 2A$

Remember This (6.3)

1. $\sqrt{3}/2$ **2.** $-\pi/3$ **3.** Amplitude: 2; period: π; phase shift: $\pi/6$ **4.** $2\cos x \sin y$ **5.** A^2 **6.** $\tan 2x$ **7.** $-\sqrt{1 - b^2}$
8. $(4\sqrt{3} - 3)/10$

9. $\sqrt{2} \sin\left(x + \dfrac{\pi}{4}\right) = \sqrt{2}\left(\sin x \cos \dfrac{\pi}{4} + \cos x \sin \dfrac{\pi}{4}\right)$

$= \sqrt{2}\left(\dfrac{\sqrt{2}}{2} \sin x + \dfrac{\sqrt{2}}{2} \cos x\right)$

$= \sin x + \cos x$

10. $\dfrac{1 + \cot^2 x}{\cot^2 x} = \dfrac{1}{\cot^2 x} + \dfrac{\cot^2 x}{\cot^2 x} = \tan^2 x + 1 = \sec^2 x$

Exercises 6.4

1. $\frac{1}{2}\cos x - \frac{1}{2}\cos 5x$ **3.** $2 \sin 5x - 2 \sin 3x$ **5.** $\frac{1}{2}\sin 3\theta - \frac{1}{2}\sin 2\theta$ **7.** $\frac{1}{2}\cos 2A + \frac{1}{2}\cos 2B$ **9.** $\frac{1}{4}$

11. $(2 + \sqrt{3})/4$ **13.** $2 \sin 4x \cos x$ **15.** $2 \sin 2x \sin x$ **17.** $2 \cos t \cos \frac{1}{2}t$ **19.** $2 \cos \theta$ **21.** $\sqrt{6}/2$ **23.** $\sqrt{2}/2$

33. $2\sqrt{2} \sin\left(x + \dfrac{\pi}{4}\right)$ **35.** $2 \sin\left(x - \dfrac{\pi}{3}\right)$ **37.** $-\sin\left(2t - \dfrac{\pi}{3}\right)$ **39.** $5 \sin(x + 0.64)$

41.

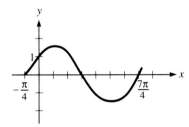

Amplitude: $\sqrt{2}$
Period: 2π
Phase shift: $-\dfrac{\pi}{4}$

43.

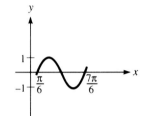

Amplitude: 1
Period: π
Phase shift: $\dfrac{\pi}{6}$

Remember This (6.4)

1. $\pi/6$ **2.** $60°$ **3.** $\pi/4$ **4.** 2π **5.** Yes   **6.** No   **7.** Identity   **8.** $\{1, -\frac{1}{2}\}$ **9.** -1

10. $\dfrac{1 - \cos 2x}{1 + \cos 2x} = \dfrac{1 - (1 - 2\sin^2 x)}{1 + (2\cos^2 x - 1)} = \dfrac{2\sin^2 x}{2\cos^2 x} = \tan^2 x$

Exercises 6.5

1. $\left\{\dfrac{\pi}{3}, \dfrac{2\pi}{3}\right\}$ **3.** $\left\{\dfrac{2\pi}{3}, \dfrac{4\pi}{3}\right\}$ **5.** $\left\{\dfrac{3\pi}{4}, \dfrac{7\pi}{4}\right\}$ **7.** $\{0.12, 3.02\}$ **9.** $\{1.88, 5.02\}$ **11.** $\left\{\dfrac{\pi}{4}, \dfrac{7\pi}{4}\right\}$ **13.** $\emptyset$ **15.** $\{0.32, 3.46\}$

17. $\left\{x: x = \dfrac{\pi}{6} + k2\pi \text{ or } x = \dfrac{5\pi}{6} + k2\pi, k \text{ any integer}\right\}$ **19.** $\{x: x = 3.60 + k2\pi \text{ or } x = 5.82 + k2\pi, k \text{ any integer}\}$

21. $\{60°, 300°\}$ **23.** $\{60°, 240°\}$ **25.** $\{240°, 300°\}$ **27.** $\{210°, 330°\}$ **29.** $\{45°, 225°\}$
31. $\{\theta: \theta = 45° + k \cdot 360° \text{ or } \theta = 135° + k \cdot 360°, k \text{ any integer}\}$
33. $\{\theta: \theta = 199°30' + k \cdot 360° \text{ or } \theta = 340°30' + k \cdot 360°, k \text{ any integer}\}$ **35.** $\{\theta: \theta = 113°30' + k \cdot 360° \text{ or } \theta = 246°30' + k \cdot 360°,$
$k \text{ any integer}\}$ **37.** $\emptyset$ **39.** $\{\theta: \theta = 180° + k \cdot 360°, k \text{ any integer}\}$

41. $\left\{\dfrac{\pi}{4}, \dfrac{3\pi}{4}, \dfrac{5\pi}{4}, \dfrac{7\pi}{4}\right\}$ **43.** $\left\{0, \dfrac{\pi}{2}, \dfrac{3\pi}{2}\right\}$ **45.** $\left\{\dfrac{\pi}{6}, \dfrac{\pi}{2}, \dfrac{5\pi}{6}, \dfrac{3\pi}{2}\right\}$ **47.** $\{1.25, 1.71, 4.39, 4.85\}$ **49.** $\left\{\dfrac{\pi}{6}, \dfrac{\pi}{2}, \dfrac{5\pi}{6}, \dfrac{7\pi}{6}, \dfrac{3\pi}{2}, \dfrac{11\pi}{6}\right\}$

51. $\left\{\dfrac{\pi}{12}, \dfrac{5\pi}{12}, \dfrac{13\pi}{12}, \dfrac{17\pi}{12}\right\}$ **53.** $\left\{\dfrac{3\pi}{4}\right\}$ **55.** $\left\{0, \dfrac{\pi}{4}, \dfrac{\pi}{2}, \pi, \dfrac{5\pi}{4}, \dfrac{3\pi}{2}\right\}$ **57.** $\{0.90, 5.38\}$ **59.** $\emptyset$ **61.** $\left\{\dfrac{\pi}{4}, \dfrac{5\pi}{4}\right\}$ **63.** $\{0, \pi\}$ **65.** $\left\{\dfrac{\pi}{4}, \dfrac{5\pi}{4}\right\}$

67. $\left\{\dfrac{\pi}{6}, \dfrac{5\pi}{6}\right\}$ **69.** $\left\{\dfrac{\pi}{2}, \dfrac{7\pi}{6}, \dfrac{11\pi}{6}\right\}$ **71.** $\left\{\dfrac{7\pi}{6}, \dfrac{3\pi}{2}, \dfrac{11\pi}{6}\right\}$ **73.** $\{1.84, 2.20, 4.98, 5.34\}$ **75.** $\left\{0, \dfrac{2\pi}{3}, \pi, \dfrac{4\pi}{3}\right\}$ **77.** $\left\{\dfrac{7\pi}{12}, \dfrac{11\pi}{12}, \dfrac{19\pi}{12}, \dfrac{23\pi}{12}\right\}$

79. $\left\{0,\dfrac{2\pi}{3},\dfrac{4\pi}{3}\right\}$ **81.** $\left\{\dfrac{\pi}{8},\dfrac{5\pi}{8},\dfrac{9\pi}{8},\dfrac{13\pi}{8}\right\}$ **83.** $\left\{0,\dfrac{\pi}{2},\pi,\dfrac{3\pi}{2}\right\}$ **85.** $\left\{0,\dfrac{\pi}{4},\dfrac{\pi}{2},\dfrac{3\pi}{4},\pi,\dfrac{5\pi}{4},\dfrac{3\pi}{2},\dfrac{7\pi}{4}\right\}$ **87.** $43°$ or $47°$

Remember This (6.5)

1. 2   **2.** False

3. $\tan(90° - \theta) = \dfrac{\sin(90° - \theta)}{\cos(90° - \theta)} = \dfrac{\sin 90° \cos \theta - \cos 90° \sin \theta}{\cos 90° \cos \theta + \sin 90° \sin \theta} = \dfrac{\cos \theta}{\sin \theta} = \cot \theta$

4. $\dfrac{\csc^2 x}{\csc^2 x - 1} = \dfrac{\csc^2 x}{\cot^2 x} = \dfrac{1/\sin^2 x}{\cos^2 x/\sin^2 x} = \dfrac{1}{\cos^2 x} = \sec^2 x$ **5.** There is no solution for θ if $\sin \theta > 1$ because the range of the sine function is $[-1,1]$.   **6.** $\{26.0°,154.0°\}$ **7.** $h = b \sin A$ **8.** (c)   **9.** $3\overline{0}$ **10.** $A = 65°$, $a = 51$ ft, $c = 57$ ft

Chapter 6 Review Exercises

21. $-\dfrac{3}{5}$ **23.** $\dfrac{5}{4}$ **25.** $\dfrac{7}{25}$ **27.** $-\dfrac{24}{7}$ **29.** $\sqrt{10}/10$ **31.** Conditional equation   **33.** $\dfrac{7}{8}$ **35.** $a/(1 + 2a^2)$ **37.** $5 \sec \theta$
39. $\sin x$ **41.** $\csc x$ **43.** $\dfrac{1}{2} \sin 4\theta + \dfrac{1}{2} \sin 2\theta$ **45.** $\dfrac{117}{125}$
47. $\sin(x_1 - x_2) = \sin[x_1 + (-x_2)]$
$\qquad\qquad\quad = \sin x_1 \cos(-x_2) + \cos x_1 \sin(-x_2)$
$\qquad\qquad\quad = \sin x_1 \cos x_2 + \cos x_1(-\sin x_2)$
$\qquad\qquad\quad = \sin x_1 \cos x_2 - \cos x_1 \sin x_2$
49. $\sin 4x = 8 \cos^3 x \sin x - 4 \sin x \cos x$ **51.** $\{7\pi/6,11\pi/6\}$ **53.** $\{\pi/6,5\pi/6,\pi/2\}$ **55.** $\{0,\pi/3,\pi,5\pi/3\}$ **57.** $\{30°,150°,210°,330°\}$
59. $\left\{x: x = \dfrac{\pi}{4} + k \cdot \dfrac{\pi}{2}, k \text{ any integer}\right\}$ **61.** (b)   **63.** (b)   **65.** (c)   **67.** (d)   **69.** (b)

Chapter 6 Test

1. $\dfrac{1 + \cot^2 x}{\cot^2 x} = \dfrac{\csc^2 x}{\cot^2 x} = \dfrac{1/\sin^2 x}{\cos^2 x/\sin^2 x} = \dfrac{1}{\cos^2 x} = \sec^2 x$ **2.** $\cos^2 x$ **3.** $-\dfrac{12}{5}$ **4.** $4 \cos \theta$

5. $\cos(x - \pi) = \cos x \cos \pi + \sin x \sin \pi = (-1)\cos x + (0)\sin x = -\cos x$ **6.** $\dfrac{\tan x - 1}{1 + \tan x}$ **7.** $\dfrac{\sqrt{6} - \sqrt{2}}{4}$ **8.** $\dfrac{4}{5}$

9. $\dfrac{1 - \tan^2 x}{1 + \tan^2 x} = \dfrac{1 - (\sin^2 x/\cos^2 x)}{1 + (\sin^2 x/\cos^2 x)} = \dfrac{\cos^2 x - \sin^2 x}{\cos^2 x + \sin^2 x} = \dfrac{\cos^2 x - \sin^2 x}{1} = \cos 2x$ **10.** $\dfrac{1}{2} \sin 6x$

11. $4 \sin^2 4\theta = 4\left(\dfrac{1 - \cos 2(4\theta)}{2}\right) = 2 - 2 \cos 8\theta$ **12.** $-5\sqrt{26}/26$ **13.** $(2 - \sqrt{3})/4$ **14.** $2 \sin \dfrac{5}{2}x \cos \dfrac{1}{2}x$

15. $\dfrac{1}{2} \cos 4\theta + \dfrac{1}{2} \cos 2\theta$ **16.** $2\sqrt{2} \sin\left(x - \dfrac{\pi}{4}\right)$ **17.** $\left\{x: x = \dfrac{7\pi}{6} + k \cdot 2\pi \text{ or } x = \dfrac{11\pi}{6} + k \cdot 2\pi, k \text{ any integer}\right\}$
18. $\{\theta: \theta = 53.1° + k \cdot 360° \text{ or } \theta = 233.1° + k \cdot 360°, k \text{ any integer}\}$ **19.** $\{\pi/12,7\pi/12,3\pi/4,5\pi/4,17\pi/12,23\pi/12\}$
20. $\{0,2\pi/3,4\pi/3\}$

Exercises 7.1

1. $b = 35$ ft, $c = 48$ ft, $C = 105°$ **3.** $b = 140$ ft, $c = 10\overline{0}$ ft, $A = 20°$ **5.** $b = 34.6$ ft, $c = 23.5$ ft, $C = 41°20'$
7. $c = 110$ ft, $B = 26°$, $C = 109°$ **9.** $c = 57$ ft, $A = 42°$, $C = 108°$ or $c = 12$ ft, $A = 138°$, $C = 12°$
11. $a = 26$ ft, $A = 8°$, $B = 22°$ **13.** $\dfrac{3}{4}$ **15.** 6   **17.** $\dfrac{7}{2}$ **19.** $10\sqrt{2}$ **21.** $b = 10$, $c = 10\sqrt{3}$ **23.** Two triangles
25. One triangle   **27.** No triangle   **29.** 433 ft   **31.** 95 mi   **33.** $\theta = 15°$, $\alpha = 57°$, $y = 23$, $\beta = 12°$, $x = 110$
(*Note:* $\theta = 81°$, $\alpha = 123°$, $y = 86$, $\beta = 49°$, $x = 16$ also satisfy the data but do not correspond to the figure and the condition that α is an acute angle.)   **35.** Since $\sin C = \sin 90° = 1$, $\sin A/\sin 90° = a/c$ becomes $\sin A = a/c$ and $\sin B/\sin 90° = b/c$ becomes $\sin B = b/c$.
37. If $A = B$, then $\sin A = \sin B$, so $a/\sin A = b/\sin B$ implies $a = b$.

Remember This (7.1)

1. ABC is a right triangle with $B = 90°$.   **2.** (a) Obtuse angle   (b) Acute angle   **3.** $36°$ **4.** $43°$ **5.** $\sqrt{13}$ **6.** 40 square units
7. $84\sqrt{2}$ **8.** $128°$ **9.** (d)   **10.** Ambiguous case

Exercises 7.2

1. $c = 14$ ft, $A = 49°$, $B = 71°$ **3.** $b = 27.4$ ft, $A = 162°30'$, $C = 7°10'$ **5.** $c = 23.1$ ft, $A = 55°50'$, $B = 28°30'$
7. $A = 128°$, $B = 20°$, $C = 32°$ **9.** $A = 37°40'$, $B = 93°30'$, $C = 48°50'$ **11.** 31 m   **13.** 1,300 ft
15. $r^2 = s^2 + t^2 - 2st \cos R$ **17.** $\cos T = \dfrac{r^2 + s^2 - t^2}{2rs}$ **19.** $\frac{1}{5}$ **21.** $\sqrt{37}$ **23.** $a = 11$ ft, $c = 43$ ft, $B = 78°$
25. $c = 78$ ft, $A = 26°$, $B = 34°$ **27.** $b = 243$ ft, $c = 152$ ft, $A = 33°00'$ **29.** $A = 104°$, $B = 29°$, $C = 47°$
31. $a = 165$ ft, $A = 164°00'$, $C = 8°50'$ or $a = 17.4$ ft, $A = 1°40'$, $C = 171°10'$ **33.** $A = 46°30'$, $B = 58°00'$, $C = 75°30'$
35. 99 mi   **37.** 34 m   **39.** 7.0 ft²   **41.** 79 m²   **43.** 2.97 km²   **45.** 9.8 ft²   **47.** 126 m²   **49.** 2.02 km²
51. $44\sqrt{3}$ square units   **53.** $9\sqrt{3}$ square units   **55.** Since $\cos C = \cos 90° = 1$, $c^2 = a^2 + b^2 - 2ab \cos 90°$ becomes $c^2 = a^2 + b^2$
57. 195 mm²

Remember This (7.2)

1. Case 1: Measures for three sides are known. Case 2: Measures for two sides and the angle between these two sides are known.
2. Case 1: Measures for two angles and one side are known. Case 2: Measures for two sides and the angle opposite one of them are known.
3. 53°   **4.** 45 ft   **5.** 108°   **6.** $\sqrt{2}$ **7.** 96   **8.** 16°   **9.** $\overline{20}$ **10.** 81

Exercises 7.3

1. (a) 13 lb (b) 23°   **3.** (a) 18 lb (b) 34°   **5.** (a) 16 mi/hour (b) 18°   **7.** (a) 510 mi (b) 81°   **9.** (a) 13° (b) 310 mi/hour
11. v, $\overline{50}$ lb; h, 87 lb   **13.** v, 8.2 lb; h, 16 lb   **15.** $R = 7.1$ lb, $\theta = 47°$ **17.** $R = 22$ lb, $\theta = 149°$ **19.** $R = 24$ lb, $\theta = 196°$
21. (a) 340 mi/hour (b) 91 mi/hour   **23.** 34 lb   **25.** 26 lb   **27.** (a) $4\overline{0}$ lb (b) 12°   **29.** (a) 321 lb (b) 5°50'   **31.** 23 lb
33. 30°, 56°, 86°

Remember This (7.3)

1. Law of sines   **2.** $\cos B = \dfrac{a^2 + c^2 - b^2}{2ac}$ **3.** 12 m²   **4.** 85 ft²   **5.** Distributive property   **6.** $6 \cos 150° = -3\sqrt{3}$,
$6 \sin 150° = 3$ **7.** $(a - c)x + (b - d)y$ **8.** $\sqrt{a_1^2 + a_2^2}$ **9.** 270°   **10.** 191°

Exercises 7.4

1. $\sqrt{2}$; 135°   **3.** 2; 210°   **5.** 1; 90°   **7.** 13; 293°   **9.** 1; 53°   **11.** $\langle -1, -2 \rangle$ **13.** $\langle -15, 3 \rangle$ **15.** $\langle -30, 17 \rangle$ **17.** $\langle 5, -1 \rangle$
19. $\langle 9, -4 \rangle$ **21.** $\langle -33, 11 \rangle$

23.

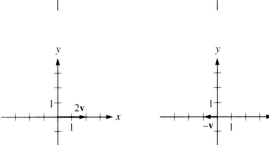
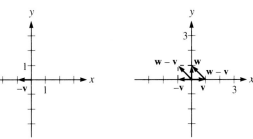

25.

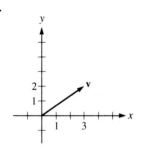
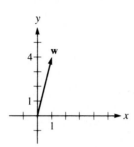
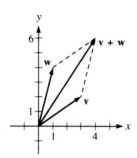

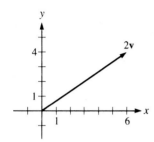
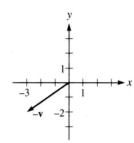
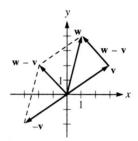

27. (a) $4\mathbf{i} + 3\mathbf{j}$ (b) $-2\mathbf{i} - \mathbf{j}$ (c) $13\mathbf{i} + 8\mathbf{j}$ **29.** (a) $-4\mathbf{i} - \mathbf{j}$ (b) $6\mathbf{i} - 5\mathbf{j}$ (c) $-27\mathbf{i} + 10\mathbf{j}$ **31.** (a) $\mathbf{i} + 5\mathbf{j}$ (b) $\mathbf{i} - 5\mathbf{j}$

(c) $-2\mathbf{i} + 25\mathbf{j}$ **33.** $4\mathbf{i} + 7\mathbf{j}$ **35.** $-5\mathbf{j}$ **37.** $-6\mathbf{i}$ **39.** $-\frac{25}{2}\mathbf{i} + \frac{25}{2}\sqrt{3}\mathbf{j}$ **41.** $h, 2; v, -8$ **43.** $h, -5\sqrt{2}/2; v, 5\sqrt{2}/2$

45. $h, 4\sqrt{3}; v, -4$ **47.** $3\sqrt{2}; 315°$ **49.** $7; 90°$

51. $\mathbf{u} + \mathbf{0} = \langle u_1, u_2 \rangle + \langle 0, 0 \rangle = \langle u_1 + 0, u_2 + 0 \rangle$
$= \langle u_1, u_2 \rangle = \mathbf{u}$

53. $\mathbf{u} + \mathbf{v} = \langle u_1, u_2 \rangle + \langle v_1, v_2 \rangle = \langle u_1 + v_1, u_2 + v_2 \rangle$
$= \langle v_1 + u_1, v_2 + u_2 \rangle = \mathbf{v} + \mathbf{u}$

55. $c(d\mathbf{u}) = c(d\langle u_1, u_2 \rangle) = c\langle du_1, du_2 \rangle = \langle cdu_1, cdu_2 \rangle$
$= cd\langle u_1, u_2 \rangle = (cd)\mathbf{u}$

57. $(\mathbf{u} + \mathbf{v}) + \mathbf{w} = (\langle u_1, u_2 \rangle + \langle v_1, v_2 \rangle) + \langle w_1, w_2 \rangle$
$= \langle u_1 + v_1, u_2 + v_2 \rangle + \langle w_1, w_2 \rangle$
$= \langle (u_1 + v_1) + w_1, (u_2 + v_2) + w_2 \rangle$
$= \langle u_1 + (v_1 + w_1), u_2 + (v_2 + w_2) \rangle$
$= \langle u_1, u_2 \rangle + \langle v_1 + w_1, v_2 + w_2 \rangle$
$= \mathbf{u} + (\mathbf{v} + \mathbf{w})$

59. $-\mathbf{u} = \langle -u_1, -u_2 \rangle = \langle (-1)u_1, (-1)u_2 \rangle = (-1)\langle u_1, u_2 \rangle = (-1)\mathbf{u}$

61. $\|-2\mathbf{v}\| = \|\langle -2v_1, -2v_2 \rangle\| = \sqrt{(-2v_1)^2 + (-2v_2)^2}$
$= \sqrt{4v_1^2 + 4v_2^2} = \sqrt{4}\sqrt{v_1^2 + v_2^2} = 2\|\mathbf{v}\|$

63. $\|\mathbf{i}\| = \|\langle 1, 0 \rangle\| = \sqrt{1^2 + 0^2} = 1$, so $\mathbf{i}$ is a unit vector. **65.** 2 **67.** 0 **69.** $4.808\mathbf{i} + 5.175\mathbf{j}; 7.1$ lb; $47°$ **71.** $-19.10\mathbf{i} + 11.28\mathbf{j}$;
22 lb; $149°$ **73.** $-22.89\mathbf{i} - 6.434\mathbf{j}; 24$ lb; $196°$ **75.** (a) 5.0 lb (b) $37°$ **77.** (a) 390 mi/hour due west (b) 69 mi/hour due north

Remember This (7.4)

1. Law of cosines **2.** Two **3.** $106°$ **4.** 22 ft **5.** 37 lb **6.** 3 and -3 are each 3 units from zero on the number line,
so $|-3| = |3|$. **7.** $3 - 4i$ **8.** $a + bi = c + di$ if and only if $a = c$ and $b = d$. **9.** $\sin(\theta_1 + \theta_2) = \sin \theta_1 \cos \theta_2 + \cos \theta_1 \sin \theta_2$
10. $\cos(\theta_1 + \theta_2) = \cos \theta_1 \cos \theta_2 - \sin \theta_1 \sin \theta_2$

Exercises 7.5

1.
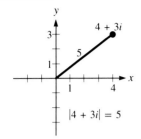

3.
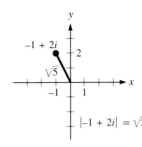

5.
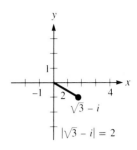

7.
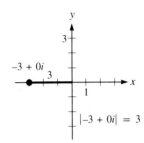

9.
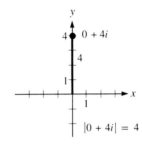

11. $3(\cos 0° + i \sin 0°)$ **13.** $2(\cos 270° + i \sin 270°)$ **15.** $\sqrt{2}(\cos 315° + i \sin 315°)$ **17.** $5(\cos 36°50' + i \sin 36°50')$
19. $\sqrt{13}(\cos 213°40' + i \sin 213°40')$ **21.** $2(\cos 120° + i \sin 120°)$ **23.** $2(\cos 315° + i \sin 315°)$ **25.** $0 + 3i$ **27.** $-4 + 0i$
29. $-\dfrac{\sqrt{3}}{2} + \dfrac{3}{2}i$ **31.** $-\sqrt{2} - \sqrt{2}i$ **33.** $0.6157 + 0.7880i$ **35.** (a) $8(\cos 63° + i \sin 63°)$ (b) $\frac{1}{2}(\cos 41° + i \sin 41°)$
(c) $2[\cos(-41°) + i \sin(-41°)]$ **37.** (a) $\cos 270° + i \sin 270°$ (b) $\cos(-90°) + i \sin(-90°)$ (c) $\cos 90° + i \sin 90°$
39. (a) $36(\cos 336° + i \sin 336°)$ (b) $\frac{1}{4}[\cos(-74°) + i \sin(-74°)]$ (c) $4(\cos 74° + i \sin 74°)$

41. (a) $2\sqrt{2}(\cos 345° + i \sin 345°)$ (b) $\sqrt{2}(\cos 255° + i \sin 255°)$ (c) $\left(\dfrac{\sqrt{2}}{2}\right)[\cos(-255°) + i \sin(-255°)]$
43. (a) $\cos(-65°) + i \sin(-65°)$ (b) $\cos 25° + i \sin 25°$ (c) $\cos(-25°) + i \sin(-25°)$
45. $|z| = |a + bi| = \sqrt{a^2 + b^2}$ and $|\bar{z}| = |a - bi| = \sqrt{a^2 + (-b)^2} = \sqrt{a^2 + b^2}$, so $|\bar{z}| = |z|$.
47. $-z = -1 \cdot z = (\cos \pi + i \sin \pi) \cdot r(\cos \theta + i \sin \theta)$
$= r[\cos(\theta + \pi) + i \sin(\theta + \pi)]$
49. $z^2 = (r \cos \theta + i \sin \theta) \cdot (r \cos \theta + i \sin \theta)$
$= r \cdot r[\cos(\theta + \theta) + i \sin(\theta + \theta)]$
$= r^2(\cos 2\theta + i \sin 2\theta)$
51. $\cos 2\theta + i \sin 2\theta$

Remember This (7.5)

1. $\cos R = \dfrac{p^2 + q^2 - r^2}{2pq}$ **2.** $5\sqrt{2}; 135°$ **3.** $\langle -16, 7 \rangle$ **4.** $9\sqrt{2}$ square units   **5.** 4 **6.** $64^{1/3} = \sqrt[3]{64} = 4$ **7.** $-\sqrt{2}/2$
8. $4(\cos 270° + i \sin 270°)$ **9.** $\frac{25}{2}\sqrt{3} + \frac{25}{2}i$ **10.** $\{i, -i\}$

Exercises 7.6

1. $4\sqrt{3} + 4i$ **3.** $512\sqrt{2} - 512\sqrt{2}i$ **5.** $16 + 0i$ **7.** $16 + 16\sqrt{3}i$ **9.** $1 + 0i$ **11.** $\dfrac{1}{4} - \dfrac{\sqrt{3}}{4}i$ **13.** $\dfrac{\sqrt{2}}{16} - \dfrac{\sqrt{2}}{16}i$

15. $5(\cos 30° + i \sin 30°) = \dfrac{5\sqrt{3}}{2} + \dfrac{5}{2}i$, $5(\cos 210° + i \sin 210°) = -\dfrac{5\sqrt{3}}{2} - \dfrac{5}{2}i$ **17.** $2(\cos 20° + i \sin 20°) \approx 1.8794 + 0.6840i$,
$2(\cos 110° + i \sin 110°) \approx -0.6840 + 1.8794i$, $2(\cos 200° + i \sin 200°) \approx -1.8794 - 0.6840i$, $2(\cos 290° + i \sin 290°)$
$\approx 0.6840 - 1.8794i$ **19.** $\cos 0° + i \sin 0° = 1 + 0i$, $\cos 120° + i \sin 120° = -\dfrac{1}{2} + \dfrac{\sqrt{3}}{2}i$, $\cos 240° + i \sin 240° = -\dfrac{1}{2} - \dfrac{\sqrt{3}}{2}i$

21. $\cos 135° + i \sin 135° = -\dfrac{\sqrt{2}}{2} + \dfrac{\sqrt{2}}{2}i$, $\cos 315° + i \sin 315° = \dfrac{\sqrt{2}}{2} - \dfrac{\sqrt{2}}{2}i$ **23.** $2(\cos 67.5° + i \sin 67.5°) \approx 0.7654 + 1.8478i$,
$2(\cos 157.5° + i \sin 157.5°) \approx -1.8478 + 0.7654i$, $2(\cos 247.5° + i \sin 247.5°) \approx -0.7654 - 1.8478i$, $2(\cos 337.5° + i \sin 337.5°)$
$\approx 1.8478 - 0.7654i$ **25.** $\sqrt[5]{2}(\cos 12° + i \sin 12°) \approx 1.1236 + 0.2388i$, $\sqrt[5]{2}(\cos 84° + i \sin 84°) \approx 0.1201 + 1.1424i$,
$\sqrt[5]{2}(\cos 156° + i \sin 156°) \approx -1.0494 + 0.4672i$, $\sqrt[5]{2}(\cos 228° + i \sin 228°) \approx -0.7686 - 0.8536i$, $\sqrt[5]{2}(\cos 300° + i \sin 300°)$
$\approx 0.5743 - 0.9948i$ **27.** $1 + \sqrt{3}i, -2 + 0i, 1 - \sqrt{3}i$ **29.** $1 + 0i, 0 + i, -1 + 0i, 0 - i$ **31.** $\cos 36° + i \sin 36°$,
$\cos 108° + i \sin 108°, -1 + 0i, \cos 252° + i \sin 252°, \cos 324° + i \sin 324°$
33.

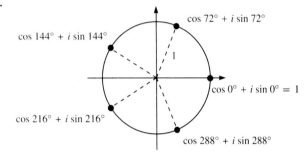

Remember This (7.6)

1. $16\sqrt{3}$ square units   **2.** $\sin Q = \dfrac{q \sin P}{p}$ **3.** $4; 300°$ **4.** $\sqrt{2}\mathbf{i} - \sqrt{2}\mathbf{j}$ **5.** 93 lb   **6.** $405°, 765°, -315°$ **7.** $5\pi/6$
8. $\sin 120° \approx 0.87; \sin 300° \approx -0.87$ **9.** $135°, 315°$
10. $\sin(180° - \theta) = \sin 180° \cos \theta - \cos 180° \sin \theta = 0 \cdot \cos \theta - (-1)\sin \theta = \sin \theta$

Exercises 7.7

1.

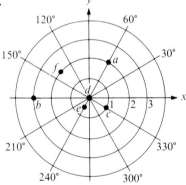

3. (a) $(1, 330°)$ **(b)** $(2, 11\pi/6)$ **(c)** $(1, \pi/3)$ **(d)** $(3, 225°)$ **(e)** $(1, 5\pi/4)$ **(f)** $(2, 180°)$

5.

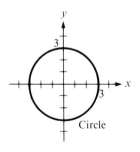

Circle

7.

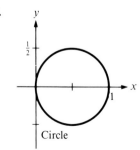

Circle

9.

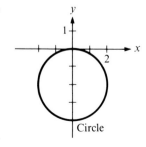

Circle

11.

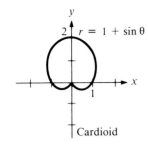

Cardioid

13.

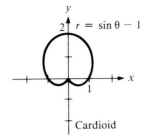

Cardioid

15.

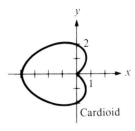

Cardioid

17.

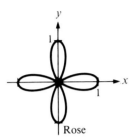

Rose

19.

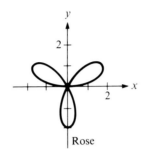

Rose

21.

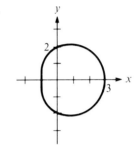

23.

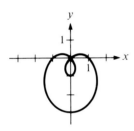

25.

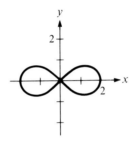

27.

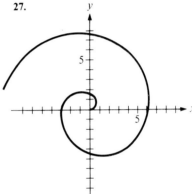

29. $(-1,0)$ **31.** $(3,0)$ **33.** $(-1,-1)$ **35.** $(1,180°)$ **37.** $(\sqrt{2},315°)$ **39.** $(2,150°)$ **41.** $r = 2 \sec \theta$ **43.** $r = 3$
45. $r = \cos \theta$ **47.** $r = \tan \theta \sec \theta$ **49.** $r^2 = 4 \sec 2\theta$ **51.** $x^2 + y^2 = 9$ **53.** $x = 2$ **55.** $x^2 + y^2 = -3y$ **57.** $y = 2$
59. $3y + 2x = 1$

Remember This (7.7)

1. 25   **2.** 25   **3.** 210°   **4.** 210°   **5.** $2\sqrt{2}(\cos 315° + i \sin 315°)$ **6.** $(2\sqrt{2},315°)$ **7.** $1 - \sqrt{3}i$ **8.** $(1,-\sqrt{3})$ **9.** $-\sqrt{3}$
10. $-8 + 0i$

Chapter 7 Review Exercises

1. 109°   **3.** 48°50'   **5.** $\sqrt{37} \approx 6.1$ **7.** $8(\cos 90° + i \sin 90°); 0 + 8i$ **9.** $\cos 450° + i \sin 450°; 0 + i$

11.

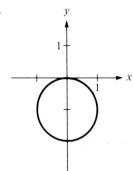

13.

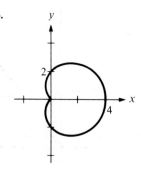

15. $(\sqrt{2}, 225°)$ **17.** 21 lb   **19.** 79 m²   **21.** $2x - y = 3$ **23.** $3(\cos 60° + i \sin 60°)$, $3(\cos 180° + i \sin 180°)$,
$3(\cos 300° + i \sin 300°)$ **25.** 2; 300°   **27.** 5; 270°   **29.** $\langle 28, 7 \rangle$ **31.** $-i - 17j$ **33.** LF: 193 ft; CF: 276 ft; RF: 266 ft
35. 30°, 56°, 94°   **37.** 116°   **39.** By the law of cosines $r^2 = p^2 + q^2 - 2pq \cos R$. If $r^2 = p^2 + q^2$, then $2pq \cos R = 0$,
so $\cos R = 0$ and $R = 90°$. Thus, PQR is a right triangle.   **41.** (a)   **43.** (b)   **45.** (b)   **47.** (c)   **49.** (c)

Chapter 7 Test

1. True   **2.** $24\sqrt{2}$ **3.** 68.8° or 111.2°   **4.** 101.3°   **5.** $C = 49°$, $A = 55°$, $a = 38$ ft   **6.** 62.4 cm   **7.** 8.5 cm²   **8.** 6; 330°
9. $\langle 23, -17 \rangle$ **10.** $-2i + 2\sqrt{3}j$ **11.** 34 lb   **12.** 39 lb   **13.** $17(\cos 152° + i \sin 152°)$ **14.** $-18 + 0i$ **15.** $32i$
16. $\dfrac{5}{2} + \dfrac{5\sqrt{3}}{2}i$, $-5 + 0i$, $\dfrac{5}{2} - \dfrac{5\sqrt{3}}{2}i$ **17.** $(3, 45°)$, $(-3, -135°)$ **18.**

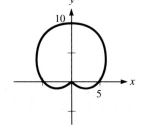

19. $(-3, 3\sqrt{3})$

20. $r = 9 \cos \theta$

Exercises 8.1

1. $x - 3y + 11 = 0$ **3.** $x^2 + y^2 + 4x + 6y - 3 = 0$ **5.** $y^2 + 4y + 8x - 12 = 0$ **7.** $3x^2 + 4y^2 - 12 = 0$
9. $4x^2 - 12y^2 + 27 = 0$ **11.** $(2, 1)$ **13.** $(0, -\frac{7}{2})$ **15.** $(0, \frac{3}{2})$ **17.** $y = -\frac{3}{2}x + 10$ **19.** $y = -x - 9$ **21.** $y = \frac{5}{7}x + \frac{6}{7}$
23. $y = \frac{7}{3}x - \frac{16}{3}$ **25.** $\overline{OB} = \sqrt{a^2 + b^2}$; $\overline{AC} = \sqrt{a^2 + b^2}$; thus, $\overline{OB} = \overline{AC}$.   **27.** $m_{OB} = b/a$ and $m_{AC} = -b/a$. If the diagonals are
perpendicular, $(b/a)(-b/a) = -1$, so $a^2 = b^2$, which implies $a = b$. Thus, the rectangle is a square.

29. Midpoint of OB is $\left(\dfrac{a + b}{2}, \dfrac{c}{2}\right)$; midpoint of AC is
$\left(\dfrac{a + b}{2}, \dfrac{c}{2}\right)$; thus, OB and AC bisect each other.

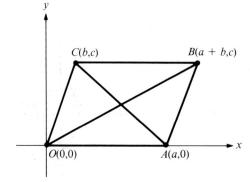

31. If C and D are the midpoints of OB and AB, then
$C = \left(\dfrac{b}{2},\dfrac{c}{2}\right)$ and $D = \left(\dfrac{a+b}{2},\dfrac{c}{2}\right)$. Since OA and CD both
have slope 0, CD is parallel to OA. Also, $\overline{OA} = a$ and
$\overline{CD} = \left|\dfrac{a+b}{2} - \dfrac{b}{2}\right| = \dfrac{a}{2}$, so $\overline{CD}$ is one-half of $\overline{OA}$.

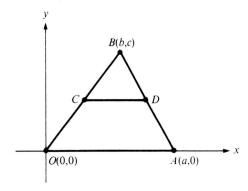

Remember This (8.1)

1. 12   **2.** $4\sqrt{3}$ **3.** $\sqrt{34}$ **4.** 36   **5.** $(y-4)^2$ **6.** $d = 2r, C = 2\pi r, A = \pi r^2$ **7.** No
8.   **9.**   **10.**

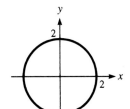

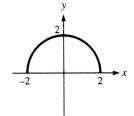

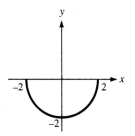

Exercises 8.2

1. $(x+3)^2 + (y-4)^2 = 4$ **3.** $x^2 + y^2 = 1$ **5.** $(x+4)^2 + (y+1)^2 = 9$ **7.** $(x-3)^2 + (y+3)^2 = 18$
9. $(x+1)^2 + (y+1)^2 = 58$ **11.** (2,5); 4   **13.** (3,0); $2\sqrt{5}$ **15.** (0,0); 3   **17.** (5,3); 7   **19.** $(-4,1)$; $3\sqrt{2}$
21. $(-\frac{7}{2},-\frac{3}{2})$; $\dfrac{\sqrt{42}}{2}$ **23.** $(-1,2)$; $\frac{7}{2}$

Remember This (8.2)

1. $3\sqrt{3}$ **2.** $\sqrt{3}/2$ **3.** 16   **4.** $2a$ **5.** $9(y^2 - 6y)$ **6.** $9(x+2)^2$ **7.** $4x^2$ **8.** $(2,-2)$ **9.** The midpoint of AC and the
midpoint of BD are both $\left(\dfrac{b+c}{2},\dfrac{d}{2}\right)$. Thus, AC and BD bisect each other.   **10.** $3x^2 + 4y^2 - 48 = 0$

Exercises 8.3

1.   **3.**   **5.**

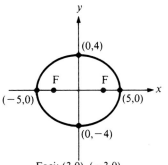

Foci: (3,0), $(-3,0)$

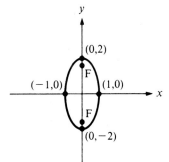

Foci: $(0,\sqrt{3})$, $(0,-\sqrt{3})$

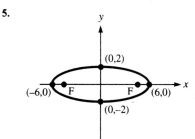

Foci: $(4\sqrt{2},0)$, $(-4\sqrt{2},0)$

7.

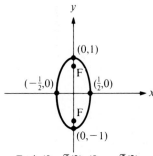

Foci: $(0, \sqrt{3}/2), (0, -\sqrt{3}/2)$

9.

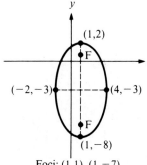

Foci: $(1,1), (1,-7)$
Center: $(1,-3)$

11.

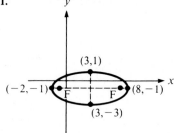

Foci: $(3 - \sqrt{21}, -1), (3 + \sqrt{21}, -1)$
Center: $(3,-1)$

13.

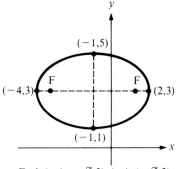

Foci: $(-1 - \sqrt{5}, 3), (-1 + \sqrt{5}, 3)$
Center: $(-1,3)$

15. $\dfrac{x^2}{16} + \dfrac{y^2}{25} = 1$ **17.** $\dfrac{x^2}{36} + \dfrac{y^2}{20} = 1$ **19.** $\dfrac{x^2}{9} + \dfrac{y^2}{4} = 1$

21. $\dfrac{(x-2)^2}{16} + \dfrac{(y-2)^2}{25} = 1$ **23.** $\dfrac{(x-2)^2}{25} + \dfrac{(y-3)^2}{16} = 1$

25. (a) 186,000,000 mi   **(b)** 3,000,000 mi

Remember This (8.3)

1. 51   **2.** 3   **3.** $-9(y^2 - 6y)$ **4.** $y = 0$. As $|x|$ gets larger, the graph of f approaches the line $y = b$.

5.

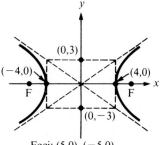

6. $y = -\frac{2}{3}x + \frac{5}{3}$ **7.** $(-6,3)$ **8.** $x^2 + (y+5)^2 = 7$

9. $(x - \frac{1}{2})^2 + (y - \frac{3}{2})^2 = \frac{3}{2}$ **10.** $3x^2 - y^2 - 3 = 0$

Exercises 8.4

1.

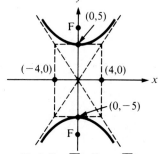

Foci: $(5,0), (-5,0)$
Asymptotes: $y = \pm\frac{3}{4}x$

3.

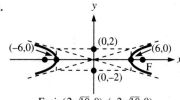

Foci: $(0, \sqrt{41}), (0, -\sqrt{41})$
Asymptotes: $y = \pm\frac{5}{4}x$

5.

Foci: $(2\sqrt{10}, 0), (-2\sqrt{10}, 0)$
Asymptotes: $y = \pm\frac{1}{3}x$

7.

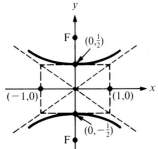

Foci: $(0, \sqrt{5}/2), (0, -\sqrt{5}/2)$
Asymptotes: $y = \pm\frac{1}{2}x$

9.

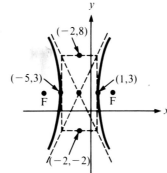

Foci: $(-2 + \sqrt{34}, 3), (-2 - \sqrt{34}, 3)$
Center: $(-2, 3)$
Asymptotes: $y - 3 = \pm\frac{5}{3}(x + 2)$

11.

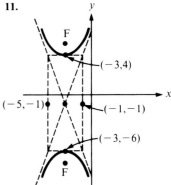

Foci: $(-3, -1 + \sqrt{29}), (-3, -1 - \sqrt{29})$
Center: $(-3, -1)$
Asymptotes: $y + 1 = \pm\frac{5}{2}(x + 3)$

13.

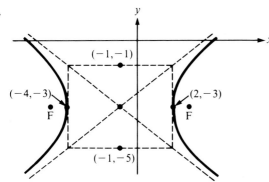

Foci: $(-1 + \sqrt{13}, -3), (-1 - \sqrt{13}, -3)$
Center: $(-1, -3)$
Asymptotes: $y + 3 = \pm\frac{2}{3}(x + 1)$

15. $\dfrac{y^2}{16} - \dfrac{x^2}{9} = 1$ **17.** $\dfrac{x^2}{9} - \dfrac{y^2}{7} = 1$ **19.** $\dfrac{x^2}{9} - \dfrac{y^2}{25} = 1$

21. $\dfrac{(x + 3)^2}{9} - \dfrac{(y + 4)^2}{16} = 1$ **23.** $\dfrac{(y - 1)^2}{4} - \dfrac{(x - 3)^2}{5} = 1$

Remember This (8.4)

1. Each graph is the reflection of the other about the x-axis.

2.

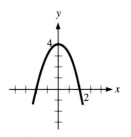

3. (a)

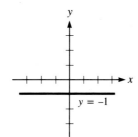

(b)

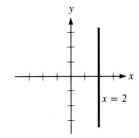

4. $(2, 4)$ **5.** $x = 3$ **6.** 13   **7.** $(x - 2)^2 + (y - 4)^2 = 18$ **8.** $\dfrac{(x + 2)^2}{64} + \dfrac{(y - 2)^2}{9} = 1$ **9.** $\frac{8}{3}$ **10.** $y^2 - 16x = 0$

Exercises 8.5

1.

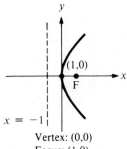

Vertex: (0,0)
Focus: (1,0)
Directrix: $x = -1$

3.

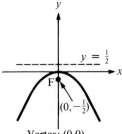

Vertex: (0,0)
Focus: $(0, -\frac{1}{2})$
Directrix: $y = \frac{1}{2}$

5.

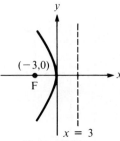

Vertex: (0,0)
Focus: $(-3,0)$
Directrix: $x = 3$

7.

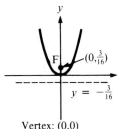

Vertex: (0,0)
Focus: $(0, \frac{3}{16})$
Directrix: $y = -\frac{3}{16}$

9.

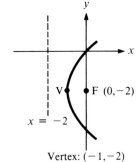

Vertex: $(-1, -2)$
Focus: $(0, -2)$
Directrix: $x = -2$

11.

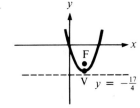

Vertex: $(2, -4)$
Focus: $(2, -\frac{15}{4})$
Directrix: $y = -\frac{17}{4}$

13.

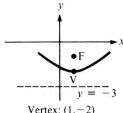

Vertex: $(1, -2)$
Focus: $(1, -1)$
Directrix: $y = -3$

15. $x^2 = 12y$ **17.** $y^2 = 2x$ **19.** $x^2 = 16y$ **21.** $(y - 1)^2 = 12(x + 1)$
23. $(x - 3)^2 = 12y$ **25.** (a) $x^2 = 12y$ if the vertex is at (0,0).   (b) (0,3)   **27.** 21 ft

Remember This (8.5)

1. (b)   **2.** (d)   **3.** (a)   **4.** (c)   **5.** x^2 and y^2 are never negative, so their sum can never be -1.   **6.** $(x + 2)^2 + (y - 3)^2 = 49$
7. $\dfrac{(x - 2)^2}{18} + \dfrac{(y + 2)^2}{8} = 1$ **8.** $\dfrac{(x + 3)^2}{9} - \dfrac{(y - 2)^2}{25} = 1$ **9.** $y - 2 = \pm\frac{1}{3}(x - 1)$ **10.** $x - y + 1 = 0$; line

Exercises 8.6

1. Parabola   **3.** Ellipse   **5.** Hyperbola   **7.** Parabola   **9.** Circle   **11.** Two intersecting straight lines: $y + 2 = \pm x$ **13.** Circle
15. Parabola   **17.** Point: $(-1,1)$ **19.** No graph

Remember This (8.6)

1.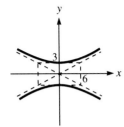
2.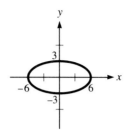
3.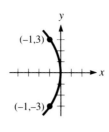
4.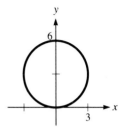

5. $(0, 3\sqrt{5}), (0, -3\sqrt{5})$ **6.** $(3\sqrt{3}, 0), (-3\sqrt{3}, 0)$ **7.** $(-\frac{1}{2}, 0)$ **8.** $(2, -2)$ **9.** $x = \frac{1}{2}$ **10.** $(x - 5)^2 + (y + 6)^2 = 49$; circle

Chapter 8 Review Exercises

1. $(x + 1)^2 = -16(y - 1)$ **3.** $(-2, 3); 5$ **5.** $(0, 3), (0, -3)$ **7.** $y = \pm\frac{4}{5}x$ **9.** $(0, -\frac{7}{4})$ **11.** $(x - 1)^2 + (y - 2)^2 = 9$

13.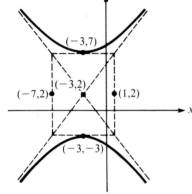

15. $(-1, -1), (7, -1)$ **17.** (c)   **19.** (d)

Chapter 8 Test

1. $(x + 4)^2 + (y - 3)^2 = 25$ **2.** $y = 2x - 8$ **3.** Position square $OABC$ with vertices $O(0,0)$, $A(a,0)$, $B(a,a)$, and $C(0,a)$. Then $m_{OB} = 1$ and $m_{AC} = -1$. Because $m_{OB} \cdot m_{AC} = -1$, the diagonals are perpendicular to each other.   **4.** $(x - 1)^2 + (y + 2)^2 = 5$

5. $(-5, 1); 6$ **6.** $4\sqrt{2}$ **7.** $(0, \sqrt{5}), (0, -\sqrt{5})$ **8.** $\dfrac{(x + 1)^2}{36} + \dfrac{(y - 1)^2}{4} = 1$ **9.** $4\sqrt{3}$ ft

10.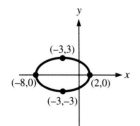

11.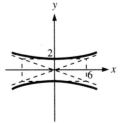

12. $(\sqrt{5}/2,0), (-\sqrt{5}/2,0)$ **13.** $y = \pm\frac{1}{2}x$ **14.** $\dfrac{(y-4)^2}{16} - \dfrac{x^2}{20} = 1$ **15.** $(3,-2)$ **16.** $y = \frac{5}{2}$ **17.** $(x-2)^2 = 12(y-1)$

18.

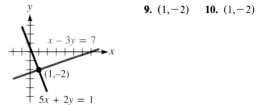

y

x

(2,–1)

19. Hyperbola   **20.** Point: $(2,-1)$

Exercises 9.1

1. $(-2,-4)$ **3.** $(-\frac{3}{2},-\frac{1}{2})$ **5.** $(0,4)$ **7.** $(-6,-6)$ **9.** Dependent   **11.** $(4,21)$ **13.** $(-1,1)$
15. No solution, inconsistent   **17.** $(5,1)$ **19.** $(\frac{67}{41},-\frac{155}{41})$ **21.** 46, 24   **23.** $140°, 40°$ **25.** 200 g
27. $v_0 = 100$ ft/second; $a = -32$ ft/second^2 **29.** $F_1 = 12$ lb; $F_2 = 6$ lb   **31.** 3 mi/hour; 9 mi/hour
33. \$8,000 at 6 percent, \$2,000 at 11 percent   **35.** More than 1,000   **37.** $p = (d-b)/(a-c)$; positive: a, d; negative: b, c

Remember This (9.1)

1. -3 **2.** -9 **3.** 3   **4.** No   **5.** -1 for x, -2 for y, 1 for z **6.** 2   **7.** -1 **8.** $y - 3z$ **9.** $-10z$ **10.** $x - 4y + 5z$

Exercises 9.2

1. $(\frac{5}{2},\frac{9}{2})$ **3.** $a = \frac{1}{4}, b = \frac{3}{4}, c = 1$ **5.** $(-27,59)$ **7.** No unique solution (dependent)   **9.** $x = 7, y = -3, z = -5$
11. $x = 3, y = -1, z = -1$ **13.** $x_1 = 2, x_2 = -2, x_3 = -2$ **15.** $A = \frac{1}{2}, B = -1, C = \frac{1}{2}$ **17.** No unique solution
(inconsistent)   **19.** $a = 2, b = -1, c = -1, d = 0$ **21.** $a = -1, b = 5, c = -2$ **23.** $a = -1, b = 0, c = 7, d = -5$
25. $r_1 = 11$ m, $r_2 = 3$ m, $r_3 = 5$ m

Remember This (9.2)

1. 62   **2.** -50 **3.** -2 **4.** Undefined   **5.** -1 **6.** 1   **7.** $\dfrac{qa - pc}{ad - bc}$ **8.**   **9.** $(1,-2)$ **10.** $(1,-2)$

y

$x - 3y = 7$

x

(1,–2)

$5x + 2y = 1$

Exercises 9.3

1. -10 **3.** 2   **5.** 26   **7.** -3 **9.** 0   **11.** -52 **13.** 17   **15.** 254   **17.** 97   **19.** 0   **21.** $(4,21)$ **23.** $(-1,1)$
25. No solution, inconsistent   **27.** $(5,1)$ **29.** $(\frac{67}{41},-\frac{155}{41})$ **31.** $x = 1, y = 2, z = 0$ **33.** $x = -3, y = 1, z = 4$ **35.** $x = \frac{1}{2},$
$y = -\frac{3}{2}, z = 1$ **37.** $x = 2, y = 2, z = -2$ **39.** No unique solution   **41.** $I_1 = -\frac{8}{73}$ amp; $I_2 = \frac{55}{73}$ amp; $I_3 = -\frac{47}{73}$ amp
43. $I_1 = \frac{73}{101}$ amp; $I_2 = -\frac{35}{101}$ amp; $I_3 = -\frac{38}{101}$ amp

Remember This (9.3)

1. Commutative property of multiplication   **2.** $a + (-b)$ **3.** $-a$ **4.** 1   **5.** $x = \dfrac{1}{4}(a - b)$ **6.** Yes   **7.** $(1,-2)$

8. $x = -3, y = \frac{1}{2}, z = -4$ **9.** $\begin{bmatrix} -1 & 1 & 0 & 14 \\ 0 & 1 & -1 & 11 \\ 1 & 0 & 1 & -9 \end{bmatrix}$ **10.** $x = -6, y = 8, z = -3$

Exercises 9.4

1. 2×4 (read "2 by 4")   **3.** 2×1 (read "2 by 1")   **5.** $a = 10, b = -3, c = 5, d = 4$ **7.** $a = 2, b = 0, c = -3$

9. $\begin{bmatrix} -1 & -3 \\ 4 & 10 \end{bmatrix}$ **11.** $\begin{bmatrix} 0 & 2 & -1 \\ 4 & -3 & 1 \end{bmatrix}$ **13.** $\begin{bmatrix} 10 & 2 \\ -1 & -5 \end{bmatrix}$ **15.** Both AB and BA are 2×2.   **17.** AB is 2×2, BA is 3×3.

19. Both AB and BA are undefined.   **21.** $\begin{bmatrix} 11 & 0 \\ -2 & 3 \end{bmatrix}$ **23.** $\begin{bmatrix} 1 & 3 \\ -1 & 0 \end{bmatrix}$ **25.** $\begin{bmatrix} -10 \\ 7 \\ 0 \end{bmatrix}$ **27.** $\begin{bmatrix} 1 & -4 & -3 \\ -2 & -4 & -9 \end{bmatrix}$ **29.** Undefined

31. B **33.** $\begin{bmatrix} 0 & 4 \\ -6 & -10 \end{bmatrix}$ **35.** I **37.** $\begin{bmatrix} -3 & -2 & -9 \\ -2 & -6 & -5 \end{bmatrix}$ **39.** $\begin{array}{l} x + y = 0 \\ 5x - 2y = 3 \end{array}$ **41.** $\begin{bmatrix} 3 & 1 \\ -7 & -2 \end{bmatrix}$ **43.** $\frac{1}{13}\begin{bmatrix} 2 & 3 \\ 1 & -5 \end{bmatrix}$

45. A^{-1} does not exist.   **47.** $\begin{bmatrix} 1 & 0 & -1 \\ 1 & 1 & -1 \\ 1 & 1 & 0 \end{bmatrix}$ **49.** A^{-1} does not exist.   **51.** $(4,21)$ **53.** $(-27,59)$ **55.** $\left(\frac{67}{41}, -\frac{155}{41}\right)$

57. $x = 2, y = 2, z = -2$ **59.** (a) $(29,17)$ (b) $(-25,-15)$

Remember This (9.4)

1. $\dfrac{8x - 1}{(x - 2)(x + 1)}$ **2.** $\dfrac{-2x + 3}{(x + 3)^2}$ **3.** $\dfrac{4x + 3}{x(x^2 + 3)}$ **4.** $x(2B + C)$ **5.** $x^2(x - 3)^2$ **6.** $\dfrac{x^2 + 6}{x^2 + 3} = 1 + \dfrac{-3x + 6}{x^2 + 3x}$

7. $\dfrac{x^5 - x^3 + x - 1}{x^4 + x^2} = x + \dfrac{-2x^3 + x - 1}{x^4 + x^2}$ **8.** $A = 2, B = -1$ **9.** $A = 2, B = -2, C = 4$ **10.** $-\frac{1}{4}$

Exercises 9.5

1. $A = -\frac{1}{4}, B = \frac{1}{4}$ **3.** $A = 3, B = -3, C = 3$ **5.** $A = 2, B = -4$ **7.** $A = 1, B = -2, C = -5, D = 5$

9. $A = -2, B = 1, C = 2, D = 1$ **11.** $\dfrac{1/2}{1 + x} + \dfrac{1/2}{1 - x}$ **13.** $\dfrac{1}{x + 1} + \dfrac{-1}{(x + 1)^2}$ **15.** $x + \dfrac{2}{x + 2} + \dfrac{2}{x - 2}$

17. $\dfrac{1}{x} + \dfrac{-3}{x - 2} + \dfrac{2}{x + 3}$ **19.** $\dfrac{1}{x} + \dfrac{-x}{x^2 + 1}$ **21.** $\dfrac{1}{x + 1} + \dfrac{1}{x - 1} + \dfrac{-2x}{x^2 + 1}$ **23.** $\dfrac{x + 4}{x^2 + 1} + \dfrac{-x}{x^2 + 2}$

Remember This (9.5)

1. $\{\pm 2\sqrt{2}\}$ **2.** $\{5, -1\}$ **3.** $\{(-1 \pm \sqrt{21})/2\}$ **4.** $\{\pm 2, \pm \sqrt{2}/2\}$
5.   **6.**

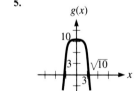

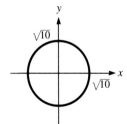

7. 2×2 (read "2 by 2")   **8.** -1 **9.** $\begin{bmatrix} 7 & -9 \\ -4 & 5 \end{bmatrix}$ **10.** $\begin{bmatrix} 5 & 9 \\ 4 & 7 \end{bmatrix}$

Exercises 9.6

1. $(0,0), (1,1)$ **3.** $(3,1), (-3,1)$ **5.** $(2,0), (-2,0)$ **7.** $(0,10), (6,-8)$ **9.** $(2,2)\ (2,-2), (-2,2), (-2,-2)$

11. $(2\sqrt{2},4), (2\sqrt{2},-4), (-2\sqrt{2},4), (-2\sqrt{2},-4)$ **13.** $(5,3), (3,5)$ **15.** $(1,2), (-1,-2), (2\sqrt{2}, \sqrt{2}/2), (-2\sqrt{2}, -\sqrt{2}/2)$

17. $(1,1), (1,-1)$ **19.** $(0,-2), (5,3)$ **21.** $(1,4), (2,1), (-\frac{2}{3},9)$ **23.** $\left(\sqrt{\dfrac{-1 + \sqrt{21}}{2}}, \dfrac{-1 + \sqrt{21}}{2} \right),$

$$\left(-\sqrt{\frac{-1+\sqrt{21}}{2}},\frac{-1+\sqrt{21}}{2}\right)$$ **25.** $(2,0),(-2,0)$ **27.** $(\frac{1}{7},-\frac{1}{3})$ **29.** $(\frac{1}{2},-\frac{1}{3},-\frac{1}{2})$

31. $(1,1),(1,-1),(-1,1),(-1,-1)$ **33.** $(2.4,1.7),(2.4,-1.7),(-2.4,1.7),(-2.4,-1.7)$ **35.** $4\sqrt{6}$ ft by $3\sqrt{6}$ ft
37. $b=9$ m, $h=8$ m **39.** A: 4 hours, B: 12 hours, C: 6 hours

Remember This (9.6)

1. True **2.**

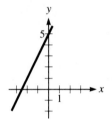

3. $(0,5)$ **4.** $(680,600)$ **5.** $(1,7)$ **6.** 3×1 (read "3 by 1")

7. $\begin{bmatrix} 20 \\ 10 \\ 70 \end{bmatrix}$ **8.** 0 **9.** Because $|A|=0$, A^{-1} does not exist. **10.** $\dfrac{-2}{x}+\dfrac{9}{x+2}$

Exercises 9.7

1.

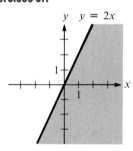

3.

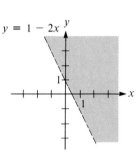

5.

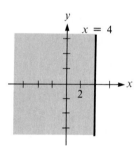

7.

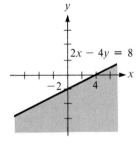

9.

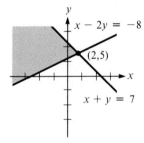

11.

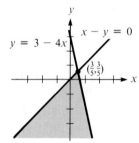

13.

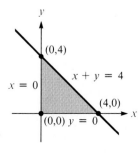

15.

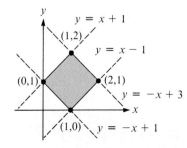

17. Max: 8; Min: -12 **19.** (2,1)   **21.** 38,000   **23.** $C = 70$ at $x = 20$, $y = 30$ **25.** $0 \le 90x + 50y + 30z \le 20,000$; $x, y, z \ge 0$
27. $0 \le Bx + Cy \le A$; $x, y \ge 0$ **29.** 80 type A, 400 type B, maximum profit: \$37,200   **31.** Food A: none; food B: $\frac{9}{2}$ ounces

Remember This (9.7)

1. True   **2.** False   **3.** False   **4.** False   **5.** True   **6.** True   **7.** False   **8.** $(3\sqrt{3}, \sqrt{37})$, $(3\sqrt{3}, -\sqrt{37})$, $(-3\sqrt{3}, \sqrt{37})$,
$(-3\sqrt{3}, -\sqrt{37})$ **9.** $\dfrac{3}{x + 3} + \dfrac{-9}{(x + 3)^2}$ **10.** $a = -4$, $b = 5$, $c = -3$

Chapter 9 Review Exercises

1. $(-1, -5)$ **3.** $(6, 0)$ **5.** $(13, -3)$ **7.** $x = 5$, $y = -1$, $z = -1$ **9.** $x = 15$, $y = 13$, $z = -5$ **11.** $(1, 2)$ **13.** $(\frac{1}{2}, \frac{1}{3})$
15. $(3 + \sqrt{5}, 3 - \sqrt{5})$, $(3 - \sqrt{5}, 3 + \sqrt{5})$ **17.** $(4, 2)$ **19.** $x = 1$, $y = -1$, $z = 0$ **21.** $x = 0$, $y = 0$, $z = 0$ **23.** -11
25. 0   **27.** $-\begin{vmatrix} c & d \\ a & b \end{vmatrix} = -(cb - da) = ad - bc = \begin{vmatrix} a & b \\ c & d \end{vmatrix}$ **29.** $\begin{bmatrix} -7 & -5 \\ -3 & -7 \end{bmatrix}$ **31.** Undefined   **33.** $\begin{bmatrix} 12 & 15 \\ 9 & 12 \end{bmatrix}$ **35.** $\begin{bmatrix} 13 & 17 \\ 8 & 25 \end{bmatrix}$

37.

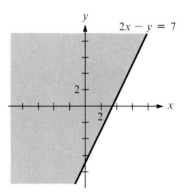

39. $P = 13,750$ at $x = 50$, $y = 375$ **41.** $a = 2$, $b = -3$, $c = 4$ **43.** For $r \ge \sqrt{5}$
the curves intersect at the four points given by $(\pm \sqrt{(r^2 + 5)/2}, \pm \sqrt{(r^2 - 5)/2})$.
45. 8   **47.** False   **49.** $A = -4$, $B = 7$ **51.** $1 + \dfrac{-3/2}{x + 3} + \dfrac{3/2}{x - 3}$

Chapter 9 Test

1. 61   **2.** $\begin{bmatrix} -7 & 4 \\ -5 & 3 \end{bmatrix}$ **3.** $\begin{bmatrix} -11 & 3 \\ 30 & -8 \end{bmatrix}$ **4.** $(1, 6)$, $(-2, 9)$ **5.** $\dfrac{1/2}{x + 2} + \dfrac{7/2}{x - 2}$ **6.** $x = -5$ **7.** $x = -5$, $y = 2$, $z = -10$
8. \$40,000   **9.** $a = 3$, $b = -5$, $c = 1$ **10.** 90

Exercises 10.1

1. 1, 3, 5, 7; 99   **3.** 1, 4, 9, 16; 49   **5.** $1, -\frac{1}{2}, \frac{1}{3}, -\frac{1}{4}; -\frac{1}{10}$ **7.** $\frac{1}{2}, \frac{1}{4}, \frac{1}{8}, \frac{1}{16}; \frac{1}{32}$ **9.** 1, 3, 6, 10; 36   **11.** AP; $d = 5$; 27, 32
13. Neither   **15.** GP; $r = 1.02$; $(1.02)^5$, $(1.02)^6$ **17.** GP; $r = 0.1$; 0.00003, 0.000003   **19.** Neither   **21.** $a_n = 4n - 1$; 159
23. $a_n = -5n + 14$; -11 **25.** $a_n = 0.3n + 0.8$; 8.3   **27.** $a_n = \frac{4}{3}n - 1$; 39   **29.** $a_n = \frac{3}{8}n + \frac{29}{8}$; $\frac{53}{8}$
31. $a_n = 2(3)^{n-1}$; 1,458   **33.** $a_n = (-\frac{1}{2})^{n-1}$; $-\frac{1}{32}$ **35.** $a_n = \sqrt{2}(\sqrt{2})^{n-1}$; 16   **37.** $a_n = 6(\frac{2}{3})^{n-1}$; $\frac{128}{243}$
39. $a_n = (-\frac{1}{3})^{n-1}$; $-\frac{1}{243}$ **41.** $4, \frac{9}{2}, 5, \frac{11}{2}$ **43.** 6   **45.** 7   **47.** $\frac{5}{9}$ or $-\frac{5}{9}$ **49.** $a_1 = \frac{15}{2}$, $d = \frac{3}{2}$

Remember This (10.1)

1. $a_n = n + 1$ **2.** $a_n = 10n - 8$ **3.** $a_n = n^2$ **4.** $a_n = 2^n$ **5.** 17.2   **6.** $\frac{63}{32}$ **7.** $\frac{3,367}{1,024}$ **8.** 586.6601   **9.** 655.398   **10.** $a_n r$

Exercises 10.2

1. 210   **3.** 2,186   **5.** About 5.42   **7.** $\frac{3,367}{1,024}$ **9.** 31.9   **11.** $1 + 2 + 3 + 4 + \cdots + 9 + 10 = 55$
13. $-1 + 1 - 1 + 1 - 1 + 1 = 0$ **15.** $1 + 3 + 5 + 7 + \cdots + 21 + 23 = 144$ **17.** $4 + \frac{9}{2} + 5 + \frac{11}{2} = 19$

19. $1 + \frac{1}{3} + \frac{1}{9} + \frac{1}{27} + \frac{1}{81} = \frac{121}{81}$ **21.** $\sum_{i=3}^{10} i$ **23.** $\sum_{i=1}^{4} \frac{1}{2^i}$ **25.** $\sum_{i=1}^{5} \frac{i}{i+1}$ **27.** $\sum_{i=1}^{6} x^i$ **29.** $\sum_{i=1}^{n} i^3$ **31.** $n(n+1)$

33. $1,600$ ft   **35.** 254 **37.** $\frac{3,389}{81}$ ft   **39.** $\$8,318.13$

Remember This (10.2)

1. $-\frac{2}{3}$ **2.** 0.01 **3.** $\frac{9}{10}$ **4.** $\frac{643}{300}$ **5.** $0.\overline{63}$ **6.** $\frac{5}{11}$ **7.** $(-1,1)$ **8.** $-\frac{1}{64}$ **9.** $243\sqrt{3}$ **10.** 2.25

Exercises 10.3

1. 2 **3.** $\frac{1}{4}$ **5.** $\frac{10}{3}$ **7.** $\frac{9}{10}$ **9.** $\frac{50}{9}$ **11.** $\frac{2}{9}$ **13.** $\frac{7}{99}$ **15.** $\frac{107}{333}$ **17.** 6 **19.** $\frac{643}{300}$ **21.** 45 ft
23. $-\frac{3}{2} < x < -\frac{1}{2}$; $(x+1)/(-2x-1)$

Remember This (10.3)

1. $2(k+1)^2$ **2.** $a^k(x-a)$ **3.** $2(k+1)(k+2)$ **4.** $(1+r)^{k+1}$ **5.** $\dfrac{2k^2 + 7k + 6}{6}$ **6.** Neither   **7.** $a_n = 4n - 2$

8. $S_n = 2n^2$ **9.** $1 + 4 + 7 + 10 + 13 = 35$ **10.** 394.91472

Exercises 10.4

1. *Step 1:* $1 \overset{?}{=} \dfrac{1(1+1)}{2}$; $1 = 1$

   Step 2: $1 + 2 + 3 + \cdots + k + (k+1)$

   $\underbrace{\qquad\qquad\qquad}_{\text{induction assumption}}$

   $= \quad \dfrac{k(k+1)}{2} \quad + (k+1)$

   $= (k+1)\left(\dfrac{k}{2} + 1\right)$

   $= \dfrac{(k+1)(k+2)}{2}$

   $= \dfrac{(k+1)[(k+1)+1]}{2}$

3. *Step 1:* $2(1) \overset{?}{=} 1(1+1)$; $2 = 2$

   Step 2: $2 + 4 + \cdots + 2k + 2(k+1)$

   $\underbrace{\qquad\qquad\qquad}_{\text{induction assumption}}$

   $= \quad k(k+1) \quad + 2(k+1)$

   $= (k+1)(k+2)$

   $= (k+1)[(k+1)+1]$

5. *Step 1:* $4(1) \overset{?}{=} 2(1)(1+1)$; $4 = 4$

   Step 2: $4 + 8 + \cdots + 4k + 4(k+1)$

   $\underbrace{\qquad\qquad\qquad}_{\text{induction assumption}}$

   $= \quad 2k(k+1) \quad + 4(k+1)$

   $= 2(k+1)(k+2)$

   $= 2(k+1)[(k+1)+1]$

7. *Step 1:* $2^1 \overset{?}{=} 2(2^1 - 1)$; $2 = 2$

   Step 2: $2 + 2^2 + \cdots + 2^k + 2^{k+1}$

   $\underbrace{\qquad\qquad\qquad}_{\text{induction assumption}}$

   $= \quad 2(2^k - 1) \quad + 2^{k+1}$

   $= 2^{k+1} - 2 + 2^{k+1}$

   $= 2(2^{k+1}) - 2$

   $= 2(2^{k+1} - 1)$

9. *Step 1:* $1^3 \overset{?}{=} \dfrac{1^2(1+1)^2}{4}$; $1 = 1$

       Step 2: $1^3 + 2^3 + \cdots + k^3 + (k+1)^3$

                        $\underbrace{\qquad\qquad\qquad}_{\text{induction assumption}}$

$$= \frac{k^2(k+1)^2}{4} + (k+1)^3$$

$$= (k+1)^2 \left[\frac{k^2}{4} + (k+1) \right]$$

$$= (k+1)^2 \left[\frac{k^2 + 4k + 4}{4} \right]$$

$$= \frac{(k+1)^2(k+2)^2}{4}$$

$$= \frac{(k+1)^2[(k+1)+1]^2}{4}$$

11. *Step 1:*   If $n = 1$, $2^1 > 1$, which is true.

      Step 2: $2^k > k$ (induction assumption)

               $2 \cdot 2^k > 2 \cdot k$

               $2^{k+1} > 2k$

               $2^{k+1} > k + 1$ (since $2k \ge k + 1$, for $k \ge 1$)

13. *Step 1:*   If $n = 5$, $5^2 < 2^5$, which is true.

      Step 2: $k^2 < 2^k$ (induction assumption)

$$k^2 + (2k+1) < 2^k + (2k+1)$$

$$(k+1)^2 < 2^k + 2k + k \text{ (since } k > 1)$$

$$= 2^k + 3k$$

$$< 2^k + k^2 \text{ (since } k > 3)$$

$$< 2^k + 2^k \text{ (induction assumption)}$$

$$= 2(2^k) = 2^{k+1}$$

15. *Step 1:*   If $n = 2$, $\log(x_1 \cdot x_2) = \log x_1 + \log x_2$, which is true.

      Step 2: $\log(x_1 x_2 \cdots x_n x_{n+1})$

$$= \log[(x_1 x_2 \cdots x_n)x_{n+1}]$$

$$= \underbrace{\log(x_1 x_2 \cdots x_n)}_{\text{induction assumption}} + \log x_{n+1}$$

$$= (\log x_1 + \log x_2 + \cdots + \log x_n) + \log x_{n+1}$$

$$= \log x_1 + \log x_2 + \cdots + \log x_n + \log x_{n+1}$$

17. *Step 1:* If $n = 1$, $0 < a < 1$, which is given.

      Step 2: $0 < a^k < 1$ (induction assumption)

$$a \cdot 0 < a \cdot a^k < a \cdot 1$$

$$0 < a^{k+1} < a$$

$$0 < a^{k+1} < 1 \text{ (since } 0 < a < 1)$$

19. *Step 1:* True for $n = 1$, since $x - a$ is divisible by $x - a$.

      Step 2: $x^{k+1} - a^{k+1}$

$$= x^{k+1} - xa^k + xa^k - a^{k+1}$$

$$= x(x^k - a^k) + a^k(x - a)$$

Then $a^k(x - a)$ is obviously divisible by $x - a$, while $x(x^k - a^k)$ is divisible by $x - a$ since $x - a$ divides $x^k - a^k$ by the induction assumption. Thus, $x^{k+1} - a^{k+1}$ is divisible by $x - a$.

Remember This (10.4)

1. $25x^2 - 20x + 4$ **2.** $a^3 + 3a^2b + 3ab^2 + b^3$ **3.** Four   **4.** 0   **5.** $-720x^2y^3$ **6.** 11   **7.** $\sum_{i=1}^{4} \frac{(-1)^{i+1}}{3^{i-1}}$ **8.** 1,370   **9.** $\frac{1,031}{625}$

10. $\frac{5}{2}$

Exercises 10.5

1. $(x + y)^6 = x^6 + 6x^5y + 15x^4y^2 + 20x^3y^3 + 15x^2y^4 + 6xy^5 + y^6$
3. $(x - y)^5 = x^5 - 5x^4y + 10x^3y^2 - 10x^2y^3 + 5xy^4 - y^5$ **5.** $(x + h)^4 = x^4 + 4x^3h + 6x^2h^2 + 4xh^3 + h^4$
7. $(x - 1)^7 = x^7 - 7x^6 + 21x^5 - 35x^4 + 35x^3 - 21x^2 + 7x - 1$ **9.** $(2x + y)^3 = 8x^3 + 12x^2y + 6xy^2 + y^3$
11. $(3c - 4d)^4 = (3c)^4 + 4(3c)^3(-4d) + 6(3c)^2(-4d)^2 + 4(3c)(-4d)^3 + (-4d)^4$
$= 81c^4 - 432c^3d + 864c^2d^2 - 768cd^3 + 256d^4$
13. $(x + y)^{15} = x^{15} + 15x^{14}y + 105x^{13}y^2 + 455x^{12}y^3 + \cdots$
15. $(x - 3y)^{12} = x^{12} + 12x^{11}(-3y) + 66x^{10}(-3y)^2 + 220x^9(-3y)^3 + \cdots$
$= x^{12} - 36x^{11}y + 594x^{10}y^2 - 5,940x^9y^3 + \cdots$
17. 1, 3, 3, 1; The binomial coefficients 1, 3, 3, 1 match the entries in row 3 of Pascal's triangle.

19. $\binom{6}{0}, \binom{6}{1}, \binom{6}{2}, \binom{6}{3}, \binom{6}{4}, \binom{6}{5}, \binom{6}{6}$ **21.** 10   **23.** 126   **25.** 1   **27.** 190   **29.** 43,758

31. $18x^{17}y$ **33.** $-702x^2y^{11}$ **35.** $126x^8$ **37.** $80x$ **39.** $4,320x^6y^9$ **41.** 330   **43.** -960
45. $(1.2)^4 = 1 + 4(0.2) + 6(0.2)^2 + 4(0.2)^3 + (0.2)^4 = 2.0736$

47. $\binom{n}{n-1} = \frac{n!}{(n-1)![n-(n-1)]!} = \frac{n!}{(n-1)!} = \frac{n(n-1)!}{(n-1)!} = n$

Remember This (10.5)

1. 720   **2.** 35   **3.** $12! \approx 4.79 \times 10^8$ **4.** $n(n-1)$ **5.** Yes   **6.** $\pm 4\sqrt{2}$ **7.** 9   **8.** $\frac{n(n+1)}{2}$ **9.** $\frac{58}{33}$
10. *Step 1:* If $n = 1$, $6(1) - 3 = 3(1)^2$, which is true.
Step 2: $3 + 9 + \cdots + (6k - 3) + [6(k + 1) - 3]$
$\underbrace{}_{\text{induction assumption}}$
$= \underbrace{3k^2} + [6(k + 1) - 3]$
$= 3k^2 + 6k + 3$
$= 3(k + 1)^2$

Exercises 10.6

1. 720   **3.** 126   **5.** 120   **7.** 6,720   **9.** 10   **11.** (a) $5!/0! = 120$ (b) $5!/(0!5!) = 1$ (c) $5!/(5!0!) = 1$
13. 8   **15.** 12

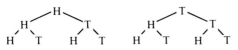

HHH HHT HTH HTT   THH THT   TTH TTT

17. 34,650   **19.** 720   **21.** 56   **23.** 495   **25.** 18,564   **27.** 15   **29.** 45   **31.** 10   **33.** 80   **35.** 113,400   **37.** 36   **39.** 120

41. 5   **43.** $_nC_{n-r} = \frac{n!}{[n-(n-r)]!(n-r)!} = \frac{n!}{r!(n-r)!} = _nC_r$ **45.** (a)

Remember This (10.6)

1. $0.4, \frac{2}{5}$ **2.** $\frac{9}{13}$ **3.** $\frac{7}{12}$ **4.** $\frac{1}{120}$ **5.** 210   **6.** $625x^4 - 500x^3y + 150x^2y^2 - 20xy^3 + y^4$

7. $x^n + nx^{n-1}h + \frac{n(n-1)}{2}x^{n-2}h^2 + \frac{n(n-1)(n-2)}{6}x^{n-3}h^3$

8. *Step 1:* If $n = 0$, $\log a^0 = 0 \log a$, which is true since $a^0 = 1$ and $\log 1 = 0$, which gives $0 = 0$.

$\quad$ *Step 2:* $\log a^{k+1} = \log(a^k \cdot a)$

$$\begin{aligned}
&= \log a^k + \log a \\
&= k \log a + \log a \quad \text{(induction assumption)} \\
&= (k + 1) \log a
\end{aligned}$$

9. $\frac{653}{16}$ ft   **10.** 56 ft

Exercises 10.7

1. $\frac{5}{12}$ **3.** (a) $\frac{1}{52}$ (b) $\frac{1}{4}$ (c) $\frac{3}{13}$ (d) $\frac{25}{26}$ (e) $\frac{2}{13}$ (f) $\frac{11}{13}$ (g) $\frac{4}{13}$ (h) $\frac{9}{13}$ **5.** $\frac{17}{19}$ **7.** $\frac{1}{9}$ **9.** (a) $\frac{4}{9}$ (b) $\frac{1}{3}$ (c) $\frac{4}{9}$ **11.** $\frac{1}{12}$

13. $\frac{1}{120}$ **15.** $\frac{3}{8}$ **17.** $\frac{21}{44}$ **19.** $\frac{1}{2}$

Remember This (10.7)

1. 11,880   **2.** 495   **3.** 495   **4.** 30   **5.** $\frac{5}{8}$ **6.** 60,300   **7.** 1,024   **8.** (a) 24   (b) 3,360   **9.** $-4,320x^3$

10. *Step 1:* If $n = 1$, $1^3 + 2(1) = 3$, which is divisible by 3.

$\quad$ *Step 2:* $(k + 1)^3 + 2(k + 1)$

$$\begin{aligned}
&= (k^3 + 3k^2 + 3k + 1) + (2k + 2) \\
&= (k^3 + 2k) + (3k^2 + 3k + 3)
\end{aligned}$$

$\quad$ Then $3k^2 + 3k + 3$ is obviously divisible by 3, while $k^3 + 2k$ is divisible by 3 by the induction assumption. Thus, $(k + 1)^3 + 2(k + 1)$ is divisible by 3.

Chapter 10 Review Exercises

1. $-1, 2, -4, 8; 32$ **3.** 2,485   **5.** $\displaystyle\sum_{i=1}^{5} \frac{1}{i^2}$

7. $\quad$ *Step 1:* If $n = 1$, $a = \frac{1}{2}[2a + (1 - 1)d]$, which is true.

$\quad\quad$ *Step 2:* $\quad a + (a + d) + \cdots + [a + (k - 1)d] + (a + kd)$

$$\begin{aligned}
&\overbrace{= \quad \frac{k}{2}[2a + (k - 1)d]}^{\text{induction assumption}} \quad + (a + kd) \\
&= ka + \frac{k^2 d}{2} - \frac{kd}{2} + a + kd \\
&= \frac{k^2 d + kd + 2ka + 2a}{2} \\
&= \frac{kd(k + 1) + 2a(k + 1)}{2} \\
&= \frac{k + 1}{2}(2a + kd)
\end{aligned}$$

9. (a) $c = 2b - a$ (b) $c = b^2/a$ **11.** $\frac{1}{2}$ **13.** 55   **15.** 5,040; 720; 144   **17.** $d = \dfrac{a_n - a_1}{n - 1}$ **19.** Yes   **21.** 15   **23.** $\frac{2}{3}$

25. 210   **27.** $n = 10, r = 3$ **29.** $35x^2$ **31.** 0.01   **33.** $1 - 15x + 105x^2 - 455x^3$ **35.** (c)   **37.** (a)   **39.** (c)   **41.** (a)

43. (c)

Chapter 10 Test

1. Geometric progression   **2.** 281   **3.** $a_n = 27(-\frac{2}{3})^{n-1}$ **4.** 2,235   **5.** 733.5929   **6.** $\frac{191}{8}$ ft   **7.** $\frac{28}{5}$ **8.** $\frac{214}{99}$

9. $x^4 - 12x^3y + 54x^2y^2 - 108xy^3 + 81y^4$ **10.** $y^{11} + 22y^{10} + 220y^9 + 1320y^8$ **11.** $-21x^4$ **12.** 9,000,000   **13.** 360

14. 56   **15.** $\frac{1}{18}$ **16.** $\frac{7}{8}$ **17.** $\frac{8}{13}$

18. *Step 1:* If $n = 1$, $10(1) = 5(1)(1 + 1)$, which is true.

    Step 2: $10 + 20 + \cdots + 10k + 10(k + 1)$

$$\underbrace{}_{\text{induction assumption}}$$

$$= \quad 5k(k + 1) \quad + 10(k + 1)$$
$$= 5(k + 1)(k + 2)$$
$$= 5(k + 1)[(k + 1) + 1]$$

Exercises A.1

1. 7.4   **3.** -2 **5.** 1.23×10^7 **7.** 45   **9.** 0.000000908   **11.** 4.2×10^1 **13.** 3.4251×10^4 **15.** 5.9×10^{12} mi
17. 1.26×10^6 **19.** 5.3×10^{-23} g   **21.** 92,000   **23.** 42.1   **25.** 580,000,000 mi   **27.** 6,280,000,000,000,000,000
29. 0.00000000000000000000000003 g   **31.** 5.58×10^7 mi   **33.** 3.6×10^{-25}

Exercises A.2

1. Five   **3.** One   **5.** Three   **7.** Three   **9.** Two   **11.** Three   **13.** Two   **15.** Three   **17.** Four   **19.** Two   **21. (a)** 12.35
(b) 12   **23. (a)** 59,370   **(b)** 59,000   **25. (a)** 29,700   **(b)** $30,\overline{0}00$ **27. (a)** 84.10   **(b)** 84   **29. (a)** 0.06855   **(b)** 0.069
31. (a) 0.002001   **(b)** 0.0020   **33. (a)** 423   **(b)** 0.004   **35. (a)** Same accuracy   **(b)** 40.2   **37. (a)** 6.430   **(b)** Same precision
39. (a) Same accuracy   **(b)** 0.0002   **41. (a)** Same accuracy   **(b)** 48.2   **43. (a)** $30\overline{0}$ **(b)** $3\overline{0}0$ **45.** 14.1   **47.** -7.4 **49.** 1.9
51. 1.2   **53.** 32,800   **55.** 200,000   **57.** 200,000   **59.** 11.6   **61.** 90   **63.** 9.75

Exercises A.3

1. 0.3692   **3.** -0.6308 **5.** 3.3692   **7.** 1.9542   **9.** -0.9547 **11.** 6.59   **13.** 0.659   **15.** 36.0   **17.** 0.0387   **19.** 0.304
21. 1,130   **23.** 63.5   **25.** 335   **27.** 9.85   **29.** 761,000   **31.** 6.34   **33.** 1.72   **35.** 55.8   **37.** 6.69   **39.** 3.85×10^{-6}
41. 1.9459   **43.** 3.7377   **45.** 8.7796   **47.** -0.5798 **49.** -7.7288 **51.** 1.2092   **53.** 9.9742   **55.** 0.7866   **57.** 0.0247

Exercises A.4

1. $y = x^2$ is a power function with $m = 2$ and $a = 1$. The graph is a straight line with slope 2. The intercept at $x = 1$ is 1.

3. $y = 4\sqrt{x}$ is a power function with $m = \frac{1}{2}$ and $a = 4$. The graph is a straight line with slope $\frac{1}{2}$. The intercept at $x = 1$ is 4.

5. $xy = 1$, so $y = 1/x = x^{-1}$. The function is a power function with $m = -1$ and $a = 1$. The graph is a straight line with slope -1. The intercept at $x = 1$ is 1.   **7.** $y = 7.0x^{-0.35}$ **9.** $y = 3.0x^{0.65}$ **11.** The function $y = 2^{-x}$ is an exponential function and its graph on semilogarithmic paper is a straight line with negative slope.   **13.** The function $y = 0.5(3^x)$ is an exponential function and its graph on semilogarithmic paper is a straight line with positive slope.

Index

A

Absolute value
    of complex number, 400
    in equations, 74
    in inequalities, 82
    of real numbers, 6, 74
Absolute value function, 112
    graph of, 112
Accuracy, of number, 572
ac method of factoring, 24
Acute angle, 249
Addition
    of algebraic expressions, 10
    of complex numbers, 53
    of fractions, 30
    of functions, 130
    of matrices, 485
    of radicals, 43
    of real numbers, 6
Algebraic expression, 8
Algebraic fraction, 28
Algebraic Operating System (AOS), 8
Ambiguous case (SSA), 371
AM (amplitude modulation), 292
Amplitude
    of periodic function, 295
    of sine and cosine, 288
Angle
    coterminal, 280
    of depression, 313
    of elevation, 313
    measure of, 249
    reference, 281
    standard position of, 267
Angular velocity, 252
Annuity, 530
Approximate numbers, 570
    rules for computations, 572–75
Archimedes, 15, 474
Arc length, 250
Area, of sector of circle, 252
Area, of triangle, 378–81
Argument, of complex number, 401
Aristarchus, 248
Associative property
    of addition, 4
    of multiplication, 4
Asymptotes
    horizontal, 202, 204
    of hyperbola, 440–42
    oblique, 205
    vertical, 201, 204
Augmented matrix, 468
Axes, 108

Axis
    conjugate, 440
    of ellipse, 434
    of symmetry, 167, 445
    transverse, 440

B

Bar
    above approximate number, 571
    above repeating decimal, 3
Binomial, 18
Binomial coefficient, 546
Binomial theorem, 544, 547
    general term of, 548
Bounds of real zeros, 194
Breakeven point, 460

C

Calculator computation, 8–9
Carbon 14 dating, 243
Cardioid, 415
Cartesian coordinate system, 108
Change of base of logarithm, 232
Characteristic, 576
Circle, definition of, 431
Circumference, of earth, 68
Closed interval, 81
Closure property
    of addition, 4
    of multiplication, 4
Coefficient, 9
Coefficient matrix, 468
Cofactor, 479
Cofunctions, 276
Combination, 554
Common logarithm, 223
Commutative property
    of addition, 4
    of multiplication, 4
Complementary angles, 276
Completing the square, 61
Complex fractions, 33
Complex numbers
    conjugate of, 51
    definition of, 49
    geometric form of, 400
    imaginary part of, 49
    real part of, 49
    summary of operational rules, 53
    trigonometric form of, 401
Components of vector, 386, 392
Composite function, 131
    domain of, 132
Compound interest, 217, 235–39

Conic sections, 430
    characteristics of different, 431, 451
Conjugate axis, 440
Conjugate of complex number, 51
Conjugate-pair theorems, 189
Conjugates, properties of, 51
Constant
    absolute, 7
    arbitrary, 7
    definition of, 7
Constant function, 100
    graph of, 111
Constant of variation, 143
Continued fraction, 70
Continuity, 185
Convex set, 510
Coordinates, 109
Corner point theorem, 510
Cosecant function
    definition of, 257, 268, 271
    graph of, 297
    inverse, 305
Cosine function
    definition of, 257, 268, 271
    graph of, 290
    inverse, 305
Cotangent function
    definition of, 257, 268, 271
    graph of, 297
    inverse, 305
Coterminal angles, 280
Cramer's rule, 477, 481
Critical number, 84

D

Damped sine wave, 302
Decibels (dB), 225
Decimals, repeating, 3, 535
Degree
    of polynomial, 18
    of polynomial function, 153
    to radian measure, 251
Degree (1°), 249
De Moivre's theorem, 406
Denominator, rationalizing, 44–45
Dependent
    system of equations, 459
    variable, 99
Descartes, Rene, 108
    rule of signs, 193
Determinant, 476
Digits, significant, 571
Dimension, 485
Diophantus, 69

Direction angle of a vector, 393
Directrix, 445
Direct variation, 143
Discriminant, 23, 64
Distance formula, 110
Distributive property, 4
Division
   of algebraic expressions, 16
   of complex numbers, 53, 402
   of fractions, 30
   of functions, 130
   long, 175
   of polynomials, 175
   of radicals, 43
   of real numbers, 6
   synthetic, 176
   by zero, 2
Domain, 98, 102
Dot product, 398
Double-angle formulas, 338

E

e, definition of, 236
Elementary row operations, 468
Element of matrix, 468, 485
Ellipse
   applications of, 437
   axes of, 434
   definition of, 433
Empty set, 57
Equal monthly payment formula, 15
Equation
   conditional, 55
   definition of, 55
   equivalent, 56
   exponential, 216, 231
   fractional, 57
   linear, 56
   logarithmic, 232
   quadratic, 58
   with quadratic form, 73
   radical, 70
   rules to obtain equivalent, 56
   solution set of, 55
   trigonometric, 351
Equilibrium point, market, 463
Equilibrium price, 463
Equivalent matrices, 468
Eratosthenes, 68
Evaluation, of algebraic expressions, 8
Even function, 122
Exact number, 575
Exponential functions
   definition of, 211
   graphed (on semilog paper), 582–83
   graphs of, 215
Exponents
   laws of, 11
   negative, 12
   positive integer, 7
   rational, 39
   zero, 11
Extraneous solutions, 71

F

Factor, 7
Factorability, test for, 22
Factorial, 544
Factoring, 18–25
Factor theorem, 181
Feasible solution, 510
Fibonacci, Leonardo, 523
FM (frequency modulation), 292
Focal chord, 438, 444, 450
Foci
   of ellipse, 433
   of hyperbola, 439
Focus, of parabola, 445
FOIL, multiplication method, 17
Fourier, Joseph, 292
Fractions, fundamental principle, 28
Functional notation, 102–4
Function hexagon, 330
Functions
   composite, 131
   composite, domain of, 132
   composition of, 131
   constant, 100, 111
   difference quotient of, 103
   even, 122
   exponential, 211
   graph of, 110
   graphical test for, 115
   inverse, 136
   linear, 154
   logarithmic, 222
   odd, 123
   one-to-one, 136
   operations with, 130
   ordered pair, definition of, 102
   ordered pair, domain of, 102
   ordered pair, range of, 102
   periodic, 287
   polynomial, 153
   quadratic, 165
   rational, 200
   rule definition, 98
   rule definition, domain of, 98
   rule definition, machine analogy, 99
   rule definition, range of, 98
   trigonometric, 257, 268, 271
   value of, 103
   zero of, 181
Fundamental counting principle, 551
Fundamental identities, 325–26
Fundamental principle
   of analytic geometry, 116, 424
   of fractions, 28
Fundamental theorem of algebra, 187
Fundamental theorem of arithmetic, 15

G

Garfield, James, 27
Gauss, Carl Friedrich, 190, 526
Gaussian elimination, 466
Gauss-Jordan elimination, 471
GCF (greatest common factor), 19
General term, of binomial formula, 548

Golden ratio, 47, 70
Golden rectangle, 47
Graphs
   definition of, 110
   on logarithmic paper, 581–82
   on semilogarithmic paper, 582–83
Greater than, 5
Greater than or equal to, 5
Greatest integer function, 117
Grouping, removing symbols of, 10

H

Half-angle formulas, 341
Half-plane, 507
Harmonic mean, 36
Harmonic motion, 315
Heron's area formula, 380
Hooke's law, 143
Horizontal
   asymptote, 202, 204
   line test, 137
   shift, 125
Hyperbola, definition of, 439

I

i, definition of, 49
Identity, definition of, 55
Identity matrix, 491
Identity property
   of addition, 4
   of multiplication, 4
If and only if, 6
Imaginary axis, 400
Imaginary number, 49
Imaginary part of complex number, 49
Impedance triangle, 318
Inconsistent system of equations, 459
Independent events, 560
Independent variable, 99
Index, 38
Inequalities
   properties of, 79
   quadratic, 83, 172
Initial ray, 249
Integers, 2
Intermediate value theorem, 197
Interval notation, 81
Inverse function
   definition of, 136, 140
   symbol for, 136
   tests for, 136, 137
Inverse of a matrix, 492
Inverse property
   of addition, 4
   of multiplication, 4
Inverse trigonometric functions, 303–9
Inverse variation, 144
Irrational numbers, 3
Irreducible polynomial, 23

J

j, definition of, 55
Joint variation, 145

L

Law of cosines, 375
Law of sines, 366
Law of tangents, 374
LCD (least common denominator), 31
Less than, 5
Less than or equal to, 5
Line
    general form, 161
    point-slope form, 157
    slope of, 154
    slope-intercept form, 158
Linear combination, 396
Linear function
    definition of, 154
    graph of, 111, 154
Linear programming, 506–12
Linear system, 457
Linear velocity, 252
Lines
    parallel, 160
    perpendicular, 160
Location theorem, 198
Logarithmic function
    definition of, 221
    graph of, 222
Logarithms
    change of base, 232
    characteristic of, 576
    common (base 10), 223
    computations using, 578–79
    definition of, 220
    mantissa of, 576
    natural (base e), 238
    properties of, 227
    solving equations with, 230–32
Long division, 175
Loyd, Sam, 474

M

Magnitude of a vector, 392
Major axis of ellipse, 434
Mantissa, of logarithm, 576
Market equilibrium point, 463
Mathematical induction, 537
Matrix, 467
    addition, 485
    augmented, 468
    coefficient, 468
    dimension of, 485
    element of, 468
    equality, 485
    inverse, 492
    multiplication, 488
    principal diagonal of, 491
    square, 476
    subtraction, 485
Maximum-minimum problems, 170–71
Midpoint formula, 427
Minor, 479
Minor axis of ellipse, 434
Minute (angular measure), 249
Mollweide's check formulas, 374
Monomial, 18

Multiple-angle formulas, 338
Multiplication
    of algebraic expressions, 16
    of complex numbers, 53, 402
    of fractions, 29
    of functions, 130
    of matrices, 488
    of radicals, 43
    of real numbers, 6
    rules for signs, 6
Musical sounds, 292
Mutually exclusive events, 560

N

Natural logarithm, 238
Negative, of a number, 5
Negative exponent, 12
Newton's law
    of cooling, 241
    of gravitation, 146
Nonlinear systems, 502–4
nth root, 38
nth root formula, 409
nth roots of unity, 409
Number of zeros theorem, 187
Number(s)
    complex, 49
    exact, 575
    imaginary, 49
    integers, 2
    irrational, 3
    negative of a, 5
    precision of a, 571
    rational, 2
    real, 4
    real number line, 5

O

Objective function, 510
Obtuse angle, 249
Odd function, 122
One-to-one function, 136
Open interval, 81
Operations, order of, 7
Ordered pair, 101
Origin, 108
Oscilloscope, 292

P

Parabola, 114, 166
    applications of, 170–71, 448–49
    definition of, 445
Parallel lines, 160
Partial fraction, 496
    distinct linear factors, 499
    distinct quadratic factors, 499
    repeated factors, 500
Pascal's triangle, 543
Period
    of any periodic function, 287
    of sine and cosine, 288
Periodic function, 287
Permutations, 552
    distinguishable, 554

Perpendicular lines, 160
pH (hydrogen potential), 224–25
Phase angle, 318
Phase shift, 291
pi (π), 3
Point-of-division formula, 429
Point-slope equation, 157
Polar axis, 411
Polar coordinates, 411
Pole, 411
Polynomial, 18
Polynomial function
    definition of, 153
    degree of, 153
    graphs of, 183–87
    rational zeros of, 192
Population growth, 240
Position vector, 392
Power, 7
Power function
    definition of, 581
    graphs of, on log paper, 582
Power reduction formulas, 340
Precision of number, 571
Principal diagonal, 491
Principal of powers, 70
Principal root, 38
Probability, 558
Product-sum formulas, 346–47
Progression
    arithmetic, 521
    geometric, 522
Properties
    of inequalities, 79
    of logarithms, 227
    of matrices, 487
    of radicals, 42
    of real numbers, 4
    of scalar multiplication, 395
    of vector addition, 395
Proportion, 65
Pythagorean identities, 326
Pythagorean theorem, 27

Q

Quadrant, 108
Quadrantal angle, 279
Quadratic equation
    definition of, 58
    nature of solutions, 64
    solution by factoring, 59
    solution by formula, 63
Quadratic formula, 63
Quadratic function
    definition of, 165
    graphs of, 114, 166
    maximum or minimum of, 167, 170
Quadratic inequalities, 83, 172

R

Radian
    definition of, 250
    to degree measure, 251
Radical equation, 70

Radicals, properties of, 42
Radical sign, 38
Radicand, 38
Radioactive decay, 239
Radio waves, 292
Range, 98, 102
Ratio, 65
Rational exponent, 39
Rational functions, 200
Rationalizing denominators, 44–45
Rationalizing numerators, 45
Rational numbers, 2
Rational zero theorem, 192
Ray, 249
Real axis, 400
Real number exponent property, 213
Real number line, 5
Real numbers, 4
Real part of complex number, 49
Reduced row-echelon matrix, 472
Reduction formula, 348
Reference angle, 281
Reference number, 262
Reflecting, 126
Reflection, law of, 318
Reflection property
    in ellipse, 437
    in parabola, 448
Refraction, law of, 285
Remainder theorem, 179
Removing symbols of grouping, 10
Repeating decimals, 3, 535
Resolving a vector, 388
Resultant, 384
Root
    of an equation, 55
    principal, 38
Rose, 415
Rounding off, 571

S

Sample space, 558
Scalar multiplication, 393, 485
Scientific notation, 568
Secant function
    definition of, 257, 268, 271
    graph of, 297
    inverse, 305
Sense of inequality, 78
Sequence
    arithmetic, 521
    definition of, 520
    Fibonacci, 523
    finite, 520
    geometric, 522
    infinite, 520
Series
    arithmetic, 525
    definition of, 525
    finite, 526
    geometric, 526
    infinite geometric, 533
    sigma notation, 529
    sum, of arithmetic, 527

sum, of geometric, 528
sum, of infinite geometric, 534
Set, 2
Set-builder notation, 80
Shrinking, 126
Sigma notation, 529
Significant digits, 571
Similar terms, 10
Similar triangles, 268
Simple harmonic motion, 315
Simplified radical, 42
Sine function
    definition of, 257, 268, 271
    graph of, 287
    inverse, 304
Slope, 154
    of parallel lines, 160
    of perpendicular lines, 160
Slope-intercept equation, 158
Snell's law, 285
Square matrix, 476
Square root property, 60
Standard form
    of circle, 115, 431
    of ellipse, 434, 436
    of hyperbola, 440–42
    of parabola, 445–47
Standard position of angle, 267
Stretching, 126
Subset, 4
Substitution property, 8
Subtraction
    of algebraic expressions, 10
    of complex numbers, 53
    of fractions, 30
    of functions, 130
    of matrices, 485
    of radicals, 43
    of real numbers, 6
Sum and difference formulas, 331
Supply and demand, 463
Symmetry
    axis of, 167, 445
    origin, 122
    y-axis, 121
Synthetic division, 176
System of equations
    addition-elimination method, 461
    dependent, 459
    determinant method, 475–82
    determinant method, three variables,
        479–82
    determinant method, two variables,
        475–78
    inconsistent, 459
    solution of, 458
    substitution method, 459

T

Tangent function
    definition of, 257, 268, 271
    graph of, 296
    inverse, 305

Terminal ray, 249
Terms, 9
Test point, 508
Transitive property, 79
Transverse axis, 440
Tree diagram, 550
Triangular form, 466
Trichotomy property, 5
Trigonometric functions
    as line segments, 274
    of general angles, 268
    inverse, 303–9
    of real numbers, 257
    in right triangles, 271
    signs of, 262, 269
Trigonometric identities
    definition of, 259, 324
    guidelines for proving, 327
    summary of, 361
Trinomial, 18
Turning points, 186

U

Union, 82
Unit vector, 396
Upper and lower bounds theorem, 194
Upper triangular matrices, 484

V

Value of function, 103
Variable
    definition of, 7
    dependent, 99
    independent, 99
Variation
    constant of, 143
    direct, 143
    inverse, 144
    joint, 145
Vector
    addition, 384, 388, 393
    components of, 386, 392
    definition of, 383, 392
    equality, 392
    resolving, 386
Vertex, of parabola, 167, 445
Vertical asymptote, 201, 204
Vertical line test, 115
Vertical shift, 125
Vertices
    of ellipse, 434
    of hyperbola, 440

Z

Zero
    division by, 2
    exponent, 11
    of a function, 181
    multiple, 182
    real bounds of, 194
Zero product principle, 59
Zero vector, 393